Challenges in the Guidance, Navigation and Control of Autonomous and Transport Vehicles

Challenges in the Guidance, Navigation and Control of Autonomous and Transport Vehicles

Editors

Nadjim Horri
William Holderbaum
Fabrizio Giulietti

MDPI

Basel • Beijing • Wuhan • Barcelona • Belgrade • Novi Sad • Cluj • Manchester

Editors

Nadjim Horri
University of Leicester
Leicester
United Kingdom

William Holderbaum
University of Salford
Manchester
United Kingdom

Fabrizio Giulietti
Bologna University
Bologna
Italy

Editorial Office
MDPI AG
Grosspeteranlage 5
4052 Basel, Switzerland

This is a reprint of articles from the Special Issue published online in the open access journal *Applied Sciences* (ISSN 2076-3417) (available at: https://www.mdpi.com/journal/applsci/special_issues/Autonomous_Transport_Vehicles).

For citation purposes, cite each article independently as indicated on the article page online and as indicated below:

Lastname, A.A.; Lastname, B.B. Article Title. *Journal Name* **Year**, *Volume Number*, Page Range.

ISBN 978-3-7258-2183-9 (Hbk)
ISBN 978-3-7258-2184-6 (PDF)
doi.org/10.3390/books978-3-7258-2184-6

Contents

About the Editors

Nadjim Horri

Nadjim Horri is a lecturer (assistant professor) in aerospace control at the University of Leicester. He specialises in the navigation and control of unmanned aerial vehicles, manned aircraft and spacecraft. He has delivered funded research work on the control of satellites, unmanned aircraft and autonomous vehicles including automotive applications. His current research interests include state and fault estimation, fault tolerant control of unmanned aerial vehicles using adaptive or robust control approaches, optimal control under flight envelope or actuator constraints including model predictive control, UAV path planning and following, wind disturbance rejection for aircraft, and the optimal and collaborative control of formations of unmanned aircrafts or spacecrafts.

William Holderbaum

William Holderbaum is a professor in control systems at the University of Salford. He has led high-profile UK funded research projects including aspects of sensor fusion, and his research interests include energy forecasting and the application of mathematical modelling and control techniques to different fields from healthcare and housing to autonomous and connected vehicles. His research experience includes the optimal control of different types of vehicles from unmanned aircraft control to the optimal control of autonomous underwater vehicles in Lie groups and the stochastic model for the predictive control of high-power vehicles such as electrified cranes. He is also interested in the use of artificial intelligence and machine learning for applications such as vehicle fault diagnosis.

Fabrizio Giulietti

Fabrizio Giulietti is a professor in flight mechanics at the University of Bologna. His research interests include the modelling, simulation and control system design of atmospheric and space vehicles, the preliminary sizing and performance analysis of electrically powered unmanned aerial systems, and the attitude control of space vehicles. He is the coordinator of the Flight Mechanics Laboratory. He has co-authored more than 70 scientific papers, and he has been involved in several national and international collaborative projects related to civil applications of remotely piloted aircraft. Since 2018, he is the director of second cycle degree of aerospace engineering.

applied sciences

MDPI

Editorial

Challenges in the Guidance, Navigation and Control of Autonomous and Transport Vehicles

Nadjim Horri [1,*], William Holderbaum [2] and Fabrizio Giulietti [3]

1 School of Engineering, University of Leicester, Leicester LE1 7RH, UK
2 School of Science, Engineering and the Environment, University of Salford, Salford M5 4WT, UK; w.holderbaum@salford.ac.uk
3 Department of Industrial Engineering, Bologna University, 40126 Bologna, Italy; fabrizio.giulietti@unibo.it
* Correspondence: nmh22@leicester.ac.uk

Citation: Horri, N.; Holderbaum, W.; Giulietti, F. Challenges in the Guidance, Navigation and Control of Autonomous and Transport Vehicles. *Appl. Sci.* **2024**, *14*, 6635. https://doi.org/10.3390/app14156635

Received: 16 July 2024
Accepted: 21 July 2024
Published: 30 July 2024

1. Introduction

In recent years, autonomous and transportation vehicles have been evolving towards more electric or hybrid electric systems, with new challenges arising in their guidance, navigation and control (GNC) systems. Some of these challenges are focused on allowing vehicles to operate more autonomously and efficiently but also more safely in challenging or increasingly congested city environments. Some of the challenges also involve anticipated requirements for collaborative GNC to allow new capabilities such as platooning or to benefit from the new mission capabilities that can be made possible through the coordination of different vehicle types. Some answers to these challenges have been developed in recent years, such as simultaneous localisation and mapping (SLAM) approaches [1]. Increasing levels of autonomy are also driving a need for advances in sensor fusion, trajectory optimisation and control, while new actuation and control technologies are giving rise to new control problems. These include torque vectoring [2], for which energy-optimised control allocation approaches have been developed but with little consideration of lateral motion control, for example. Another control challenge is the development of regenerative braking solutions for electric or hybrid vehicles in driving modes such as eco-approach [3] and autonomous lane management. Other challenges are associated with the new environments where some vehicles are increasingly required to evolve. These challenges must be addressed while meeting requirements regarding safety, efficiency and emissions for different types of autonomous or transportation vehicles, ranging from mobile robots to ground vehicles and aircraft. In the field of electric aviation, policies are shaping some of the priorities in the fields of GNC and vehicle autonomy. The Federal Aviation Authority (FAA) was recently (May 2024) authorised by the United States (US) House of Representatives to issue approvals for beyond visual line of sight (BVLOS) electric vertical take-off and landing (eVTOL) and drone operations, which will imply advances in their flight safety and autonomy. In Europe, the European Union has also adopted legislation on drones and eVTOLs, including air taxis, as part of a wider strategy towards reaching net-zero emissions by 2050. In the United Kingdom (UK), routine BVLOS drone operations by 2027 and fully autonomous eVTOL flights by 2030 are part of the UK Future of Flight Action Plan [4]. Many of the proposed innovative drone and eVTOL designs use hybrid aircraft configurations with new fault tolerance, aerodynamics and control challenges, combining the ability to cruise like a plane with VTOL capability. In car manufacturing, similar electrification trends are being observed, and the UK has, for example, mandated that zero-emission vehicles will represent 80% of all car manufacturers' sales by 2030 and 100% by 2035. The majority of road vehicles will also offer autonomy level 2+ (as defined by the Society of Automobile Engineers (SAE) International) or higher from 2030 [5], which will include new features such as lane-keeping assistance, safe distance maintenance, and automatic lane change manoeuvres. The proportion of vehicles with level 3 to level 5 autonomy will also significantly increase compared to current levels. A direct implication is the need for higher-performance integrated sensor fusion using radar, Lidar and visual navigation technologies and higher-level computation performance, which also implies efficient GNC algorithm implementation. These trends are global; a fast-growing electric vehicle and eVTOL industry is also emerging in China, for example. Advances in vehicle autonomy also require the development of sense-and-avoid technologies and algorithms, the ability to autonomously handle constraints and recover from faults or sensor signal interruption, and an efficient combination of manned and automated control systems. Optimal and robust navigation and control systems are required to meet the performance challenges and energy requirements of the modern day. Increased levels of autonomy will also require advanced multimode navigation and control system architectures, including hierarchical mode-switching control strategies to handle challenges such as autonomous driving at junctions and roundabouts [6] and sensor fusion-based navigation algorithms in global navigation satellite system (GNSS) degraded environments [7]. Computationally efficient and optimal machine learning-based methods are also increasingly employed for autonomous vehicle path planning and following in urbanised or challenging environments or to deal with challenges linked to the vehicles themselves, as in the case of differential-drive ground vehicles [8]. When multiple vehicles are involved, some of the future GNC challenges will involve collaborative navigation and control problems, with accuracy, synchronicity and connectivity requirements, involving groups of similar (e.g., platooning with communication delays, as in [9]) or heterogenous vehicles to offer real-world solutions that other configurations cannot offer [10].

The Special Issue that led to this book captures the diversity of the research, hot topics and latest advances in GNC systems for a good range of transport and autonomous vehicles. It brings together some of the theoretical developments, as well as simulation-based and experiment-based studies in the field. This editorial aims to encourage the reader to explore these publications and to appreciate the commonalities between the main emerging GNC research priorities for a diverse range of transport and autonomous vehicles.

2. Growing Research Areas in Vehicle Navigation and Control

From this Special Issue on the GNC of autonomous and transport vehicles and from global policy trends, it is possible to identify areas for which research and development interest is growing in vehicle navigation and control, driven by the needs of the industry and by policy decisions in the field of future transport.

The articles collected in this book cover air, ground and sea vehicles over the time horizon from 2022 to 2024. The current trends identified in these articles indicate that navigation challenges in cities or challenging environments represent the leading research direction, followed by trajectory optimisation and control challenges. The navigation and trajectory optimisation challenges are mostly focused on sensor fusion, collaborative localisation or navigation in challenging or urbanised environments. Control challenges are often associated with new actuation and control system technologies.

Sensor fusion and sensing for specific environments are the focus of a number of contributions. Contribution 1 introduces a machine learning algorithm that consists of a new version of the light gradient boosting machine (GBM) regression algorithm. This algorithm displays improvements as compared to a random forest algorithm, through faster training and more accurate prediction of the vehicle's position during a global navigation satellite system (GNSS) outage, by using the inertial navigation system (INS) in a GNSS/INS implementation. In Contribution 4, multiple positioning systems including GNSS, simultaneous localisation and mapping (SLAM) and INS are used, and error residuals are used for fault detection and correction in an autonomous vehicle. In Contribution 13, issues with poor GNSS signal quality are also handled for application to forestry vehicles. To deal with a similar problem, Contribution 12 uses computer vision based on a YOLOv4 image detection architecture and an nVidia Jetson AX Xavier device with the ability to handle image quality issues and detect landmarks in the Arctic region. Another computer vision-based approach is proposed in Contribution 21, using Kalman filtering with dynamically adjusted measurement noise to allow for precise navigation using multiple tags in environments where GNSS signals are not available. Apart from environment perception issues using sensors, the importance of scene understanding is also emphasised in Contribution 16 for autonomous robot navigation in unstructured environments. Multimodal data fusion is another challenge and is the focus of Contribution 17, wherein a generative adversarial network (GAN) model framework to fuse different types of images and data is proposed for autonomous vehicles.

Connected autonomous vehicles are at the centre of several contributions to this Special Issue. In Contribution 3, the focus is on the secure platooning of multiple vehicle convoys against denial-of-service jamming attacks that exploit the fact that such convoys rely on wireless communications. To overcome this cybersecurity challenge, a behaviour-based approach is used where sensor inputs are fed to a behaviour manager that is used to arbitrate between possible actions. The behaviour manager leverages layered costmaps with Vector Field Histogram (VFH) path planning. In Contribution 11, the focus is on energy efficiency. In Contribution 7, the focus is on the specific problem of cooperative control at roundabouts through the optimisation of merge-in flows under vehicle safety constraints. The cost function being optimised accounts for the waiting time and the number of vehicles at the roundabout, with improved queue lengths and safety. Collaborative networks of unmanned aerial vehicles (UAVs) and unmanned ground vehicles (UGVs) are another emerging hot topic. Contribution 5 tackles the cooperative localisation problem in a UAV–UGV vehicle network in three dimensions. In Contribution 6, a UAV–UGV multicast cluster-based routing approach is used to increase the communication throughput and reduce delays in landmine detection and message dissemination between different vehicles. Cluster maintenance algorithms are used to incorporate motion (speed, direction) information and update the record (group) of vehicles in the cluster.

Control challenges are also presented in this book to address emerging challenges with advances in power, actuation and control technologies. Contribution 8 is focused on the use of nonlinear programming to optimise hybrid electric vehicle (HEV) power distributions. Contribution 15 focuses on torque vectoring strategies for hybrid vehicles with in-wheel rear motors and presents a comparison of different optimal and sliding-mode control strategies. In Contribution 10, a fuzzy logic rule-based approach is used to determine the torque of the powertrain system and manage regenerative braking and eco-approach, using a conventional pure pursuit lateral controller and realistic powertrain models. In Contribution 19, a new actuator that uses shape-memory alloys is used for idle air control, with PID control to manage actuation stroke delays.

Path planning and trajectory optimisation issues are also covered in this book. In Contribution 14, an improved hybrid A* graph-based global path planning algorithm is used for a nonholonomic mobile robot using a robot operating system (ROS) implementation and model predictive control (MPC) for local path planning to avoid obstacles, with reduced path lengths and search times as compared to A* + DWA (dynamic window approach). A new path planning algorithm is also proposed in Contribution 18, using an improved bidirectional search Gaussian (BSG) A* algorithm for differential-drive mobile robots, with reduced path lengths and search times as compared to D* lite (dynamic A*), the genetic algorithm (GA) and ant colony optimisation (ACO). Machine learning is another research trend in this field, particularly for increasingly autonomous vehicles. Reinforcement learning is used in Contribution 20 for the dynamic trajectory optimisation of a self-driving vehicle.

Other areas that are less strongly present in this Special Issue but where research will be essential include the handling of connectivity constraints in vehicle trajectory optimisation and control problems, particularly the ones requiring communication between vehicles, as in the case of unmanned aerial vehicle (UAV) and unmanned ground vehicle (UGV) collaboration, but more generally in the case of vehicles navigating within sensed environments and obtaining information from a ground sensor network using advances from the Internet of Things (IoT). The human–machine aspects of interfacing human operator input with automation remain important, and Contribution 2 of this Special Issue illustrates an example of the effect of onboard training in simulated maritime environments on the navigation performance of autonomous surface vessels. Contribution 9 focuses on scenario designs for advanced driver assistance systems (ADASs) and autonomous driving systems (ADSs) applied to longitudinal motion using MPC-based driver models. To construct the models, human drivers' characteristics were derived using a large driving database. Future systems are expected to allow both humans and machines to learn from each other, even though increasingly autonomous systems are also expected to operate with reduced human operator input.

3. Conclusions

This book presents a collection of articles spanning the diversity of GNC with applications to different types of autonomous or transport vehicles. Many of the foreseen challenges lie in the development of robust and fault-tolerant navigation systems allowing the vehicles to operate in challenging environments, including GNSS-denied or degraded environments, urbanised environments, and environments that are difficult to access due to severe weather or other conditions. To address these challenges, sensor fusion methods and systems will be needed to advance emerging capabilities, such as the ability to sense and avoid obstacles. Other challenges are linked to system complexity and connectivity issues when multiple vehicles are required to operate in the same environment and collaborate to perform missions. Control challenges are emerging from the new power systems and control technologies that are being developed, given the increased reliance of autonomous and transport vehicles on electric power to meet emissions and energy efficiency requirements. Trajectory optimisation is also becoming increasingly challenging, with an increased reliance on artificial intelligence and machine learning approaches, even though there is still a need to further develop the robustness and performance of mainstream optimisation techniques for new vehicle types and their applications.

Acknowledgments: The authors of this book would like to thank all contributors to the Special Issue https://www.mdpi.com/journal/applsci/special_issues/Autonomous_Transport_Vehicles, (URL accessed on 15 July 2024) without whom this book would not have been possible.

Conflicts of Interest: The authors declare no conflicts of interest.

List of Contributions:

1. Li, B.; Chen, G.; Si, Y.; Zhou, X.; Li, P.; Li, P.; Fadiji, T. GNSS/INS Integration Based on Machine Learning LightGBM Model for Vehicle Navigation. *Appl. Sci.* **2022**, *12*, 5565. https://doi.org/10.3390/app12115565.
2. Hwang, H.; Hwang, T.; Youn, I. Effect of Onboard Training for Improvement of Navigation Skill under the Simulated Navigation Environment for Maritime Autonomous Surface Ship Operation Training. *Appl. Sci.* **2022**, *12*, 9300. https://doi.org/10.3390/app12189300.
3. Cheung, C.; Rawashdeh, S.; Mohammadi, A. Jam Mitigation for Autonomous Convoys via Behavior-Based Robotics. *Appl. Sci.* **2022**, *12*, 9863. https://doi.org/10.3390/app12199863.
4. Chang, W.; Juang, R.; Lin, Y. Research on Data Fusion of Positioning System with a Fault Detection Mechanism for Autonomous Vehicles. *Appl. Sci.* **2022**, *12*, 11339. https://doi.org/10.3390/app122211339.
5. Oliveros, J.; Ashrafiuon, H. Application and Assessment of Cooperative Localization in Three-Dimensional Vehicle Networks. *Appl. Sci.* **2022**, *12*, 11805. https://doi.org/10.3390/app122211805.
6. Farooq, W.; Islam, S.; Khan, M.; Rehman, S.; Gulzari, U.; Boudjadar, J. UGAVs-MDVR: A Cluster-Based Multicast Routing Protocol for Unmanned Ground and Aerial Vehicles Communication in VANET. *Appl. Sci.* **2022**, *12*, 11995. https://doi.org/10.3390/app122311995.
7. Wang, C.; Wang, Y.; Peeta, S. Cooperative Roundabout Control Strategy for Connected and Autonomous Vehicles. *Appl. Sci.* **2022**, *12*, 12678. https://doi.org/10.3390/app122412678.
8. Lee, J.; Lee, H. A New HEV Power Distribution Algorithm Using Nonlinear Programming. *Appl. Sci.* **2022**, *12*, 12724. https://doi.org/10.3390/app122412724.

9. Cho, K.; Park, C.; Lee, H. A Study on Longitudinal Motion Scenario Design for Verification of Advanced Driver Assistance Systems and Autonomous Driving Systems. *Appl. Sci.* **2023**, *13*, 716. https://doi.org/10.3390/app13020716.
10. Bautista-Montesano, R.; Galluzzi, R.; Mo, Z.; Fu, Y.; Bustamante-Bello, R.; Di, X. Longitudinal Control Strategy for Connected Electric Vehicle with Regenerative Braking in Eco-Approach and Departure. *Appl. Sci.* **2023**, *13*, 5089. https://doi.org/10.3390/app13085089.
11. Katsanikakis, A.; Bechlioulis, C. Design and Implementation of an Energy-Efficient Vehicle Platoon Control Algorithm Using Prescribed Performance and Extremum Seeking Control. *Appl. Sci.* **2023**, *13*, 5650. https://doi.org/10.3390/app13095650.
12. Chuprov, S.; Belyaev, P.; Gataullin, R.; Reznik, L.; Neverov, E.; Viksnin, I. Robust Autonomous Vehicle Computer-Vision-Based Localization in Challenging Environmental Conditions. *Appl. Sci.* **2023**, *13*, 5735. https://doi.org/10.3390/app13095735.
13. Rybansky, M.; KratochvÃl, V.; Dohnal, F.; Gerold, R.; Kristalova, D.; Stodola, P.; Nohel, J. GNSS Signal Quality in Forest Stands for Off-Road Vehicle Navigation. *Appl. Sci.* **2023**, *13*, 6142. https://doi.org/10.3390/app13106142.
14. Chi, Z.; Yu, Z.; Wei, Q.; He, Q.; Li, G.; Ding, S. High-Efficiency Navigation of Nonholonomic Mobile Robots Based on Improved Hybrid A* Algorithm. *Appl. Sci.* **2023**, *13*, 6141. https://doi.org/10.3390/app13106141.
15. de Carvalho Pinheiro, H.; Carello, M.; Punta, E. Torque Vectoring Control Strategies Comparison for Hybrid Vehicles with Two Rear Electric Motors. *Appl. Sci.* **2023**, *13*, 8109. https://doi.org/10.3390/app13148109.
16. Wijayathunga, L.; Rassau, A.; Chai, D. Challenges and Solutions for Autonomous Ground Robot Scene Understanding and Navigation in Unstructured Outdoor Environments: A Review. *Appl. Sci.* **2023**, *13*, 9877. https://doi.org/10.3390/app13179877.
17. Ignatious, H.; El-Sayed, H.; Khan, M.; Kulkarni, P. A Generic Framework for Enhancing Autonomous Driving Accuracy through Multimodal Data Fusion. *Appl. Sci.* **2023**, *13*, 10749. https://doi.org/10.3390/app131910749.
18. Yao, M.; Deng, H.; Feng, X.; Li, P.; Li, Y.; Liu, H. Global Path Planning for Differential Drive Mobile Robots Based on Improved BSGA* Algorithm. *Appl. Sci.* **2023**, *13*, 11290. https://doi.org/10.3390/app132011290.
19. Turabimana, P.; Sohn, J.; Choi, S. Design and Control of a Shape Memory Alloy-Based Idle Air Control Actuator for a Mid-Size Passenger Vehicle Application. *Appl. Sci.* **2024**, *14*, 4784. https://doi.org/10.3390/app14114784.
20. Kim, Y.; Ahn, W.; Jang, S.; Lim, M.; Pae, D. A Reinforcement Learning Approach to Dynamic Trajectory Optimization with Consideration of Imbalanced Sub-Goals in Self-Driving Vehicles. *Appl. Sci.* **2024**, *14*, 5213. https://doi.org/10.3390/app14125213.
21. Liu, Z.; Qin, W.; Li, Z.; Zhou, G. Multi-Tag Fusion Localization Method Based on Geometric Constraints. *Appl. Sci.* **2024**, *14*, 5480. https://doi.org/10.3390/app14135480.

References

1. Gamini Dissanayake, M.W.M.; Newman, P.; Clark, S.; Durrant-Whyte, H.F.; Csorba, M. A solution to the simultaneous localization and map building (SLAM) problem. *IEEE Trans. Robot. Autom.* **2001**, *17*, 229–241. [CrossRef]
2. Carello, M.; Bonansea, P.; D'Auria, M. Driveline Optimization for a Hybrid Electric City Vehicle to Minimize Fuel Consumption. In Proceedings of the SAE 2014 World Congress & Exhibition, Detroit, MI, USA, 8–10 April 2014. [CrossRef]
3. Wu, G.; Hao, P.; Wang, Z.; Jiang, Y.; Boriboonsomsin, K.; Barth, M.; McConnell, M.; Qiang, S.; Stark, J. Eco-Approach and Departure along Signalized Corridors Considering Powertrain Characteristics. *SAE Int. J. Sustain. Transp. Energy Environ. Policy* **2021**, *2*, 25–40. [CrossRef]
4. Department for Transport, UK Future of Flight Action Plan, March 2024. Available online: https://www.gov.uk/government/publications/future-of-flight-action-plan (accessed on 15 July 2024).
5. Hodgson, J. *A Scalable Approach to ADAS and Autonomous Driving*; ABI Research White Paper; 2023. Available online: https://www.amd.com/en/solutions/automotive/adas/adas-and-autonomous-driving-white-paper.html (accessed on 15 July 2024).
6. Horri, N.; Haas, O.; Wang, S.; Foo, M.; Fernandez, M.S. Mode Switching Control Using Lane Keeping Assist and Waypoints Tracking for Autonomous Driving in a City Environment. *Transp. Res. Rec.* **2022**, *2676*, 712–727. [CrossRef]
7. Hernandez, G.E.V.; Petrunin, I.; Shin, H.S.; Gilmour, J. Robust Multi-Sensor Navigation in GNSS Degraded Environments. In Proceedings of the AIAA Scitech Forum, National Harbor, DA, USA, 23–27 January 2023. [CrossRef]
8. Stefek, A.; Van Pham, T.; Krivanek, V.; Pham, K.-L. Energy Comparison of Controllers Used for a Differential Drive Wheeled Mobile Robot. *IEEE Access* **2020**, *8*, 170915–170927. [CrossRef]

9. Wu, L.; Zhang, L.; Zhou, Q. Event-Based Control and Scheduling of a Platoon of Vehicles in VANETs. *IEEE Access* **2021**, *9*, 166223. [CrossRef]
10. Ding, Y.; Xin, B.; Chen, J. A Review of Recent Advances in Coordination Between Unmanned Aerial and Ground Vehicles. *J. Unmanned Syst.* **2021**, *9*, 97–117. [CrossRef]

applied sciences

MDPI

Article

GNSS/INS Integration Based on Machine Learning LightGBM Model for Vehicle Navigation

Bangxin Li [1,2], Guangwu Chen [1,2,*], Yongbo Si [1,2], Xin Zhou [1,2], Pengpeng Li [1,2], Peng Li [1,2] and Tobi Fadiji [3]

1 School of Automation and Electrical Engineering, Lanzhou Jiaotong University, Lanzhou 730070, China; 12201800@stu.lzjtu.edu.cn (B.L.); syb8838182@mail.lzjtu.cn (Y.S.); 0219718@stu.ljztu.edu.cn (X.Z.); lipengpeng1002@126.com (P.L.); pengleework@163.com (P.L.)

2 Key Laboratory of Plateau Traffic Information Engineering and Control of Gansu Province, Lanzhou Jiaotong University, Lanzhou 730070, China

3 Centre for Applied Data Science, University of Johannesburg, Johannesburg 2092, South Africa; tobifadiji@gmail.com

* Correspondence: cgwyjh1976@126.com

Abstract: To solve the problem of data accuracy degradation of vehicle GNSS/INS integrated navigation systems when the GNSS signal is unavailable or there is a GNSS outage, this paper improves the existing GNSS/INS integration methodology for land vehicle navigation based on the AI method. First, a GNSS/INS integration methodology for land vehicle navigation based on position update architecture (PUA) using LightGBM regression for predicting the position of a vehicle during a GNSS outage is presented. It uses LightGBM to model the relationship between INS data and vehicle position changes. On-board INS and GNSS data are collected when the GNSS signal is available and are used to train the PUA-LightGBM model; in the event of a GNSS outage, INS data are used as the input to the PUA-LightGBM to predict the change in vehicle position. Second, a vehicle navigation data acquisition system was designed for model validation. This included a self-developed GNSS/INS integrated navigation system and a Novatel pwrpak7-e1 GNSS/INS integrated navigation system for data acquisition on six road segments. Finally, the collected data were used for machine learning training of the PUA-LightGBM model and the existing PUA-RandomForest model. As a result, the PUA-LightGBM predicts the vehicle position with less error in the event of a GNSS outage and takes less time to train. It was also demonstrated that by allowing the model to be dynamically trained or updated while the vehicle is moving the PUA-LightGBM could adapt perfectly to the predictions of vehicle position changes in different complex road segments.

Keywords: LightGBM; integrated navigation algorithm; global navigation satellite system; inertial navigation system; GNSS outage

Citation: Li, B.; Chen, G.; Si, Y.; Zhou, X.; Li, P.; Li, P.; Fadiji, T. GNSS/INS Integration Based on Machine Learning LightGBM Model for Vehicle Navigation. *Appl. Sci.* **2022**, *12*, 5565. https://doi.org/10.3390/app12115565

Academic Editors: Fabrizio Giulietti, Nadjim Horri and William Holderbaum

Received: 7 May 2022
Accepted: 29 May 2022
Published: 30 May 2022

1. Introduction

Navigation and positioning are among the most widely used technologies in intelligent transportation systems, advanced vehicle control, and vehicle safety systems [1]. The main idea behind their algorithms is to use satellite data and vehicle dynamics data to calculate the current specific position of the vehicle. An algorithm's robustness determines the accuracy of the information output from the vehicle navigation system and its ability to adapt to the environment. GNSS/INS integrated navigation is a low-cost, highly accurate, and versatile integrated navigation solution [2]. Global navigation satellite systems (GNSS) provide real-time information on the position of a vehicle, primarily via satellites. The inertial navigation system first gathers information about angular and linear motion relative to inertial space. It then uses inertial navigation differential equations to calculate vehicle speed and position changes. The GNSS and INS are complementary in many ways [3]: (1) The raw vehicle dynamics data measured by the INS are often noisy. The vehicle position information obtained through independent INS solving can drift over time, requiring external assistance to calibrate the INS data. GNSS systems, on the other hand, can provide long-term stable data when the signal is available and can be used to limit INS errors. (2) There could be GNSS outages in practical operation due to signal blockage or interference, such as the presence of buildings, tunnels, forest cover, etc. In this case, the vehicle cannot obtain the data provided by the GNSS. Hence, continuity data provided by the INS can compensate for the lack of position data in the event of a GNSS outage. (3) The data provided by the GNSS for the vehicle are updated less frequently, while the INS provides accurate short-term data at a very high rate, which can be used to interpolate GNSS trajectories. (4) The INS provides the complete vehicle state (e.g., position, speed, attitude, etc.), whereas a single GNSS receiver cannot provide angular information about the vehicle. These are also the reasons for creating vehicle integrated navigation algorithms that integrate GNSS and INS. However, the accuracy of the integrated navigation algorithm could be inevitably reduced by a long-time outage of GNSS. As a result, the solution to this problem has emerged as a hot topic in integrated navigation algorithm research.

The approaches to vehicle integrated navigation algorithms can be classified into two categories. One category is the vehicle integrated navigation algorithms based on Bayesian filtering techniques. Kalman filtering (KF) is one of the most common implementations. Liu et al. applied a dual-filter smoother to an extended Kalman filter (EKF)-based INS/GPS integrated navigation algorithm for vehicles. The authors found that the algorithm provided accurate navigation parameters in the presence of GPS signal blockage, thereby improving the overall data fusion [4]. Yang et al. proposed a MEMS-INS/GNSS integrated navigation algorithm for vehicles combined with Allan variance analysis under non-complete constraints. The resulting algorithm is applicable to the loose/tight integrated navigation method for vehicles with specific fault detection and troubleshooting capability [5]. Sasani et al. proposed the attitude and heading reference system (AHRS) algorithm for an INS/GPS loose integrated navigation system, which was found to suppress the divergence of the INS solution during a GPS outage [6]. Wang et al. proposed an SINS/OD integrated navigation algorithm for vehicles based on the state transformation extended Kalman filter (ST-EKF). The authors discovered that it achieved higher positioning and heading angle accuracy than the conventional EKF algorithm [7]. Cui et al. used an improved cubature Kalman filter (CKF) algorithm based on the sigma point update framework for GNSS/INS integrated navigation and found that it improved the reliability of the navigation system in a GNSS-challenged environment [8]. Bijjahalli and Sabatini proposed a novel low-cost high-performance vehicle integrated navigation algorithm. It employed EKF to fuse data from GNSS, INS, visual odometry, and vehicle dynamics models (as virtual sensors) [9]. The other category is artificial intelligence (AI) technology-based integrated navigation algorithms for vehicles, which aim to find linear and non-linear relationships between sensor data during vehicle operation, thus completing the process of data fusion. Chiang et al. used multilayer feedforward neural networks and backpropagation learning algorithms to improve the INS/DGPS integrated navigation method for vehicles. The resulting algorithm eliminates the effects of neural network parameters and random noise on positioning accuracy [10]. Aggarwal et al. proposed a D-S neural network for GPS/INS data fusion by combining D-S evidence theory with a neural network model. The authors found that the model improved the positioning accuracy of the vehicle integrated navigation system under GPS outage conditions [11]. Noureldin et al. implemented an autoregressive process in a stochastic modeling approach based on microelectromechanical systems inertial sensor (MEMS-INS) errors to reduce INS errors. The authors used AI to improve the accuracy of information fusion by enhancing the conventional KF module, resulting in improved performance of a low-cost MEMS-INS/GPS integrated navigation system for vehicles [12]. Chiang and Huang proposed an INS/GPS integrated navigation method that combines artificial neural networks and uses historical information. This method prevents the continuous growth of INS errors when GPS is unavailable for long periods [13]. Yue et al. proposed a two-layer, highly robust algorithm for vehicle integrated navigation systems that fuses GNSS and INS data via support vector machine regression (SVR) and adaptive

Kalman filtering (AKF). This resulted in an algorithm that improves vehicle integrated navigation system positioning accuracy when GNSS signals are weak [14]. Liu et al. proposed a GPS/INS neural network (GI-NN) deep learning network structure as an INS aid when GPS is unavailable; this algorithm uses historical data from the IMU and exploits both spatial and temporal features in the IMU to achieve better localization results [15]. Using a T2FHNN (type-2 fuzzy Hammerstein neural network), Khankalantary et al. proposed a data fusion scheme for a SINS/GNSS integrated navigation system with better navigation accuracy and stability [16].

In summary, some researchers have made significant progress in improving vehicle positioning accuracy through improved integrated navigation algorithms. However, little information is available on vehicle integrated navigation algorithms based on efficient AI methods. The machine learning methods used in the existing literature for vehicle position prediction models almost did not consider the efficiency optimization issue, such as the cost of time and computational resources required for model training [12,15]. In addition, most previous AI models for predicting a vehicle's position in the event of a GNSS outage cannot be updated online after the initial machine learning training. This results in these models only having good position prediction accuracy when the vehicle is driving on certain types of road segments covered in the training set. Moreover, the models cannot adapt to complex and dynamic changes in road segments. Overall, these models lack robustness. The AI method based on an improved position update architecture (PUA) for GNSS/INS integrated navigation can predict the position of a vehicle in the case of a GNSS outage. As a result, the objective of this study is to propose a new LightGBM machine learning algorithm based on an improved PUA for GNSS/INS integrated navigation that predicts vehicle position in the event of a GNSS outage. Additionally, the study proposes to solve the problem of the inefficient model training used in AI-based integrated navigation algorithms by improving model training efficiency. Furthermore, the PUA-LightGBM model is being proposed for the first time.

The structure of the paper is as follows: The LightGBM framework and the PUA-LightGBM model are described in Section 2. Section 3 describes the methodology and process of the experiments and the workflow of the PUA-LightGBM model in detail. The results and discussions of the experiments are presented in Section 4. Section 5 provides a conclusion.

2. Introduction to LightGBM Machine Learning Model Based on PUA

2.1. LightGBM Machine Learning Model

Gradient boosting decision tree (GBDT) is a common machine learning model used in artificial intelligence that employs several basic learners that are iteratively trained to produce the optimal model. LightGBM is a distributed and efficient framework for implementing the gradient boosted decision tree (GBDT) algorithm [17], which fits a new decision tree by approximating the negative gradient of the loss function as the residual of the current decision tree.

To improve the efficiency of machine learning model training while losing as little accuracy as possible in model prediction, the following seven optimization methods exist for LightGBM: (1) histogram-based decision tree algorithm, (2) gradient-based one-side sampling (GOSS), (3) exclusive feature bundling (EFB), (4) leaf-wise with max depth limitation, (5) categorical feature support, (6) parallel learning support, and (7) increase cache hit chance. Methods (1)–(4) are optimization algorithms, and (5)–(7) are optimizations for engineering applications.

GOSS and EFB, two main techniques used by LightGBM, improve efficiency by reducing the number of samples and features, respectively. The GOSS algorithm retains all instances with large gradients and samples with small gradients at random. EFB bundles mutually exclusive features and can losslessly reduce feature dimensionality.

2.2. Position Update Architecture and PUA-LightGBM Model

The position update architecture is a data fusion architecture for GNSS/INS integrated navigation that uses the AI method, which defines the AI-based integrated navigation model's inputs and outputs (Figure 1A). The architecture trains the AI model based on INS data and GNSS data, with the inputs to the model being the velocity (V_{INS}) and azimuth (θ_{INS}) given by the INS and the outputs being the differences in longitude (δE) and latitude (δN) between the two epochs. Based on the linear and non-linear relationships between the motion data and vehicle position coordinates, PUA provides a training model for predicting vehicle position changes. Hence, it is reasonable to use PUA to determine the inputs and outputs of AI. PUA makes it possible to apply AI methods to find relationships between vehicle dynamics data and position changes [18]. Initially, El-Sheimy et al. implemented an AI-based integrated navigation algorithm using ANN and PUA; in addition, AI-based integrated navigation algorithms using PUA have been proposed successively: DSNN [11], RFR [19], DS-SVM [20], etc.

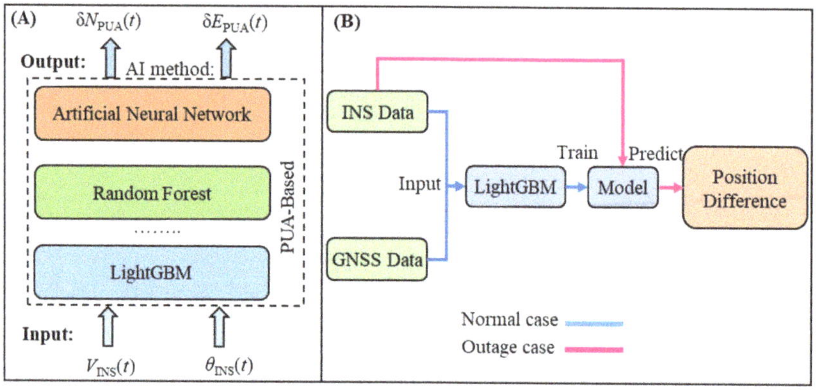

Figure 1. PUA-based LightGBM machine learning model. (**A**) Position update architecture (PUA), t—time; $V_{INS}(t)$—Velocity provided by INS at t; $\theta_{INS}(t)$—Heading angle provided by INS at t; $\delta N_{PUA}(t)$—Latitude difference between epoch t and epoch $t-1$; $\delta E_{PUA}(t)$—Longitude difference between epoch t and epoch $t-1$; (**B**) Scheme of the LightGBM machine learning model based on the PUA.

This study proposes to use the PUA-based LightGBM machine learning model to predict changes in vehicle position during GNSS outage conditions. During vehicle operation, the data from the INS are combined with data from the GNSS to construct a training dataset and train a position change prediction model. When there is a GNSS outage, the INS data are used in the model to predict the change in vehicle position in real-time, as shown in Figure 1B. It is important to note that this paper uses vehicle three-axis acceleration as a feature input based on PUA. One of the main advantages of LightGBM is its high training efficiency [17]. From a practical deployment point of view, LightGBM models for vehicle position prediction in the event of a GNSS outage could collect more data. In this way, the model can update or expand the dataset while the vehicle moves, allowing it to be retrained at a lower cost and with greater adaptability.

3. Experimental Analysis and Method

3.1. Data Acquisition

A vehicle navigation data acquisition system was designed to acquire vehicle dynamics and position data before and after the GNSS outage (Figure 2A). The system consisted of a self-developed GNSS/INS integrated navigation system (Kalman-filter-based, loosely coupled integration is used), a Novatel pwrpak7-e1 GNSS/INS integrated navigation system, and a computer. The self-developed GNSS/INS integrated navigation system included a No. 2 satellite antenna, MEMS inertial measurement unit, data acquisition board, data fusion board, and power supply unit. The data acquisition board was equipped with an integrated UB4B0M GNSS board (accuracy: better than 1.5 m for single-point positioning). The IMU200A inertial measurement unit (IMU) consisted of a gyroscope (range, ±300°/s; zero bias, 25.2°/hr; angular random walk, 144°/$\sqrt{hr}$) and an accelerometer (range, ±17 g; zero bias, 0.2 mg; velocity random walk, 12 m/s/$\sqrt{hr}$), which were fixed to a platform in the vehicle. Foam was used as a buffer material between the IMU and the platform to minimize external interference during IMU operation. Its working process was as follows: While the vehicle was moving, the No. 2 antenna installed on the vehicle's roof searched for and tracked satellite signals. When the number of satellites searched reached four or more, the GNSS board embedded in the data acquisition board demodulated the received signals. Then, it calculated the GNSS data, which included the vehicle's position coordinates. Simultaneously, the vehicle's three-axis angular motion data and three-axis acceleration data output from the IMU were transmitted in real time to the data acquisition board, which intercepted and converted the GNSS data and INS raw data (IMU output) to the preset data format to complete the data acquisition process. Then, the pre-processed data were transferred to the data fusion board, where they were solved and fused to obtain the final GNSS and INS data. Subsequently, these data were transferred to the computer via the serial port of the data fusion board.

The Novatel pwrpak7-e1 GNSS/INS integrated navigation system consisted of No. 2 and No. 3 satellite antennas, a Novatel pwrpak7-e1 navigation unit with an integrated satellite receiver, and an Epson G320 MEMS-IMU (gyroscope: range, ±150°/s; zero bias, 3.5°/hr; angle random walk, 0.1°/$\sqrt{hr}$. Accelerometer: range, ±5 g; zero bias, 0.1 mg; velocity random walk, 0.5 m/s/$\sqrt{hr}$). It was possible to capture a vehicle's trajectory, i.e., its location, with greater accuracy. The procedure was as follows: The No. 1 and No. 3 satellite antennas searched for and track the satellite signals. When the satellite positioning conditions were met, the Novatel pwrpak7-e1 performed the initial alignment to establish the initial INS benchmark, then completed the process of collecting, calculating, and fusing the GNSS and INS data and finally transmitted the data to the computer via the communication interface.

For the experiments, an open closed-loop urban road trajectory (perimeter x road width: 8168.0 × 20.6 m) was first selected (Figure 2B) to ensure that GNSS outages would not occur automatically. While the vehicle was moving and the GNSS signal was available, the position, speed, heading angle, pitch angle, and three-axis acceleration data were collected simultaneously at a frequency of 50 Hz by two integrated navigation systems installed on the vehicle. The data collected by the self-developed integrated navigation system were then used to build a training set for the LightGBM model, which was used to predict vehicle position in the event of a GNSS outage. Furthermore, to closely match all realistic driving scenarios encountered by a typical land vehicle, the data acquisition board of the self-developed integrated navigation system was manually disconnected from the satellite antenna in several scenarios. These scenarios included uphill straight, uphill bend, downhill straight, downhill bend, and two flat straight segments, which simulated six GNSS outage situations encountered by the vehicle, with each outage lasting 20 s. When simulating an outage in the self-developed integrated navigation system, only the INS system can properly output data, as the input of the LightGBM model for vehicle position prediction is under GNSS outage.

Because the outage would not occur in regular operation when a Novatel pwrpak7-e1 integrated navigation system was employed, it provided accurate vehicle position data throughout, which were used as a baseline reference against the predicted output of the LightGBM machine learning model.

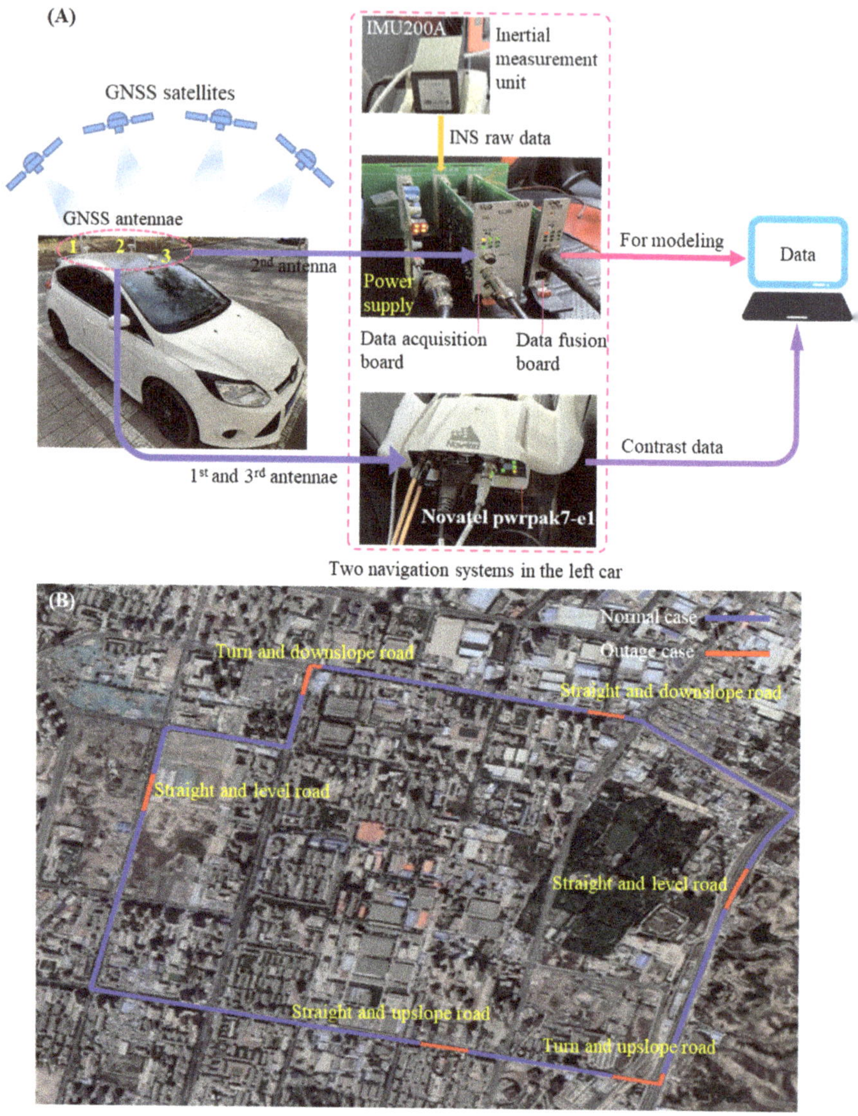

Figure 2. Experimental device and method. (**A**) Two integrated navigation systems, 1—1st antennae; 2—2nd antennae; 3—3rd antennae; (**B**) closed-loop experimental road trajectory.

3.2. Model Training

First, data pre-processing was performed. A dataset was constructed using the output data of the self-developed integrated navigation system; when the GNSS signal was available, incomplete strips of data were removed. Then, 80% of the dataset was divided into a training set, and the remaining 20% was used as a test set. The latitude and longitude data were compared to obtain the differences in longitude and latitude between the two epochs. It is worth noting that LightGBM is a high-performance framework for implementing GBDT, which is a tree model. Because the numerical scaling of the tree model does not affect the position of the split point or its structure, data normalization is not required. The training environment used was Windows 10 with an Intel i7-6700HQ processor, and the programming language was Python version 3.7 (Python Software Foundation, Fredericksburg, VA, USA).

The LightGBM model training process was then carried out, as shown in Figure 3 [21]. The inputs and outputs of the PUA-LightGBM model for vehicle position prediction in the event of a GNSS outage were determined according to Section 2.2. This means that the inputs were the velocity (V_{INS}) and azimuth (θ_{INS}) given by the INS and the three-axis acceleration of the vehicle, and the outputs were the differences in longitude (δE) and latitude (δN) between the two epochs. The longitude difference prediction and latitude difference prediction models were trained separately to implement a multiple-output regression. The parameter settings used in the model training and the descriptions of the relevant parameters are given in Table 1.

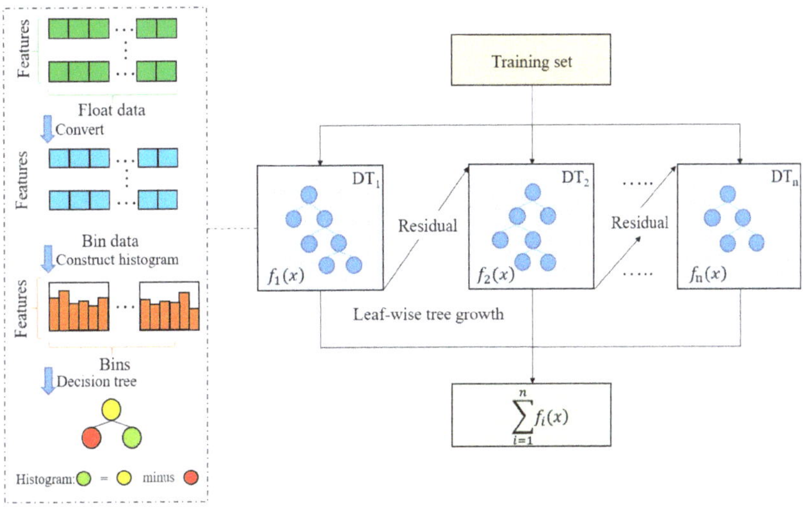

Figure 3. Training process of PUA-LightGBM model and histogram-based decision tree algorithm. DT—Regression Decision Tree; $f_i(x)$—Output of DT_i; $\sum_{i=1}^{n} f_i(x)$ —final output of the PUA-LightGBM model.

Table 1. The parameters of the LightGBM model.

Parameter Name	Value	Parameter Implication
learning_rate	0.05	Learning rate
num_leaves	63	The number of leaves in a tree
num_iterations	300	Number of basic learners
bagging_fraction	0.8	The fraction of data to be considered for each iteration.
feature_fraction	0.9	The fraction of features to be considered in each iteration.

The output data were first used to initialize a regression decision tree (DT) based on the loss function. Following that, *n* regression decision trees were created. A negative gradient, or pseudo-residual, was calculated using the output value of the initialized DT with the set sample output value. This residual was then used as the output value of the set sample to train the first DT. The best split point needed to be found when training the DT, and LightGBM discretized the continuous floating-point feature values and generated a histogram (bins). As the data were traversed, the cumulative statistics of each discrete value in the histogram were counted. Feature selection was accomplished by traversing the histogram based on the discrete values to determine the best split point. The algorithm cost *O(#data* × *#feature)* for histogram building and *O(#bin* × *#feature)* for split point finding. Since *#bin* is usually much smaller than *#data*, the histogram building dominated the computational complexity [17]. DT growth employs a leaf-wise technique, which locates and divides the leaf node with the maximum splitting gain relative to the present leaf node. In addition, during DT construction, the histogram of a leaf node can be obtained from the difference between its parent node's histogram and its sibling node's histogram. When DT reaches a depth limit or a limit on the number of leaf nodes, the residuals are fitted by assigning leaf node weights to fit this time. After training, the LightGBM model for vehicle position prediction in the event of a GNSS outage can be updated, and the process can be repeated until n regression decision trees are constructed. Following this process, the final LightGBM model for vehicle location prediction in the event of a GNSS outage was generated. In Figure 3, $f_i(x)$ is the output of DT_i. The final output of the LightGBM model for vehicle position prediction in the event of a GNSS outage is $\sum_{i=1}^{n} f_i(x)$, which is the prediction of the differences in longitude (δE) or latitude (δN).

3.3. Model Predictions

The model is ready for use once it has been trained. When there is a GNSS outage, the IMU can continue to work regardless of the external environment. The velocity (V_{INS}) and azimuth (θ_{INS}) provided by INS as well as the vehicle's three-axis acceleration were employed in the model to obtain the model's prediction of the differences in longitude (δE) and latitude (δN) between the two epochs. The longitude and latitude information provided before the GNSS outage were added to the prediction model's longitude difference and latitude difference outputs, respectively, to obtain the vehicle's latitude and longitude location information.

In this study, the performance of the GNSS outage position prediction model was evaluated by two metrics: the root mean square error (*RMSE*) and the coefficient of determination (R^2), which are both commonly used in regression tasks. The equations are shown in Equations (1) and (2).

$$RMSE = \sqrt{\frac{1}{n}\sum_{i=1}^{n}(y_i - \hat{y}_i)^2} \tag{1}$$

$$R^2 = 1 - \frac{\sum_{i=1}^{n}(y_i - \hat{y}_i)^2}{\sum_{i=1}^{n}(y_i - \overline{y}_i)^2} \tag{2}$$

where n is the total number of data points, y_i is the actual value, $\hat{y}_i$ is the predicted value, $\overline{y}_i$ is the mean of the observed data, *RMSE* is the root mean square error, and R^2 is the coefficient of determination. The smaller the *RMSE* value, the higher the model prediction accuracy; the closer the R^2 value is to 1, the better the model fits the data.

In practical application, the model is used to predict the position and trajectory of a vehicle when there is a GNSS outage, and it is compared with the actual vehicle trajectory. Then, it gives the average error of the predicted trajectory compared to the real trajectory. Model training time is critical in practice because the data acquisition system can capture new data that are generated while the vehicle moves with a strong GNSS signal. A subset of this new data can be used to augment the dataset. The shorter the training time for each model, the less time it takes to retrain the LightGBM model for vehicle position prediction during a GNSS outage. This means new models can be deployed faster and adapt to more driving conditions. As a result, the training time of the model was also used as the final evaluation metric.

4. Results and Discussion

4.1. Performance of the PUA-LightGBM Model

Table 2 presents the performance comparison of the PUA-LightGBM model to the PUA-RandomForest model. Both the PUA-LightGBM model and the PUA-RandomForest model are position prediction models for vehicles in the event of a GNSS outage. PUA-RandomForest is an INS/GPS integrated method for position prediction based on PUA using a random forest regression model. PUA-RandomForest reduces position errors compared to PUA-ANN, which has been proven to work better than the Kalman filter [19]. Because LightGBM only supports a single output, training the PUA-LightGBM model requires sequential training of the PUA-LightGBM (δE) for longitude difference prediction and PUA-LightGBM (δN) for latitude difference prediction, resulting in a total training time of 3.076 s. The performance metrics (root mean square error and coefficient of determination) of the PUA-LightGBM model for longitude and latitude change prediction are evidently similar to those of the PUA-RandomForest model. However, the training time of the PUA-LightGBM model for longitude and latitude variation prediction is only 3.7% of the training time of the PUA-RandomForest model. This demonstrates the ability of the PUA-LightGBM model to have a short training time while maintaining accuracy. This is attributed to the existence of algorithms and strategies for efficiency optimization in LightGBM, such as the histogram-based decision tree algorithm, leaf-wise leaf growth strategy with max depth limitation, gradient-based one-side sampling (GOSS) algorithm, exclusive feature bundling (EFB), etc. This is in line with the findings of Chen et al. [22] and Ju et al. [23]. The former used LightGBM to efficiently predict protein interactions, while the latter combined convolutional neural networks and LightGBM to improve the accuracy and efficiency of ultra-short-term wind power prediction models.

Table 2. Performance comparison of the PUA-LightGBM model to the PUA-RandomForest model.

Model Name	*RMSE*	R^2	Training Time (s)
PUA-LightGBM (δE)	1.927×10^{-7}	0.983	1.563
PUA-LightGBM (δN)	1.334×10^{-7}	0.986	1.513
PUA-RandomForest	1.417×10^{-7}	0.989	83.082

Note: PUA-LightGBM (δE)—The PUA-LightGBM model for predicting the difference in longitude between two epochs; PUA-LightGBM (δN)—The PUA-LightGBM model for predicting the difference in latitude between two epochs; *RMSE*—The root mean square error; R^2—The coefficient of determination.

The existing algorithms for vehicle position prediction can give more accurate and reliable navigation solutions than traditional machine learning algorithms. However, these algorithms have disadvantages such as long design time and high computational burdens [15], complex and insufficiently streamlined structures [12], and complex modelling processes [16]. PUA-LightGBM model has a high-efficiency working process while maintaining a high level of accuracy. An efficient training process is critical in application scenarios of vehicle position prediction during a GNSS outage. Firstly, the shorter training time allows the model dataset to be expanded, and an updated model is retrained while the vehicle is moving. Meanwhile, the GNSS outage vehicle position prediction model can be updated at a low cost, enabling the prediction of position changes in the presence of more dynamic changes in the vehicle. However, this may necessitate an analysis of data characteristics and the removal of similar training data to avoid wasting resources due to data redundancy. Secondly, the models can be applied to various land vehicles and do not require data to be collected and modeled in advance for the type of vehicle or driving characteristics. Data are simply collected while the vehicle is moving, and the model is quickly trained when the dataset reaches a certain volume. This allows for the prediction of vehicle position changes in the event of a GNSS outage.

A major requirement for safe travel is fast and accurate navigation assistance [1]. From the transport provider's point of view, using the PUA-LightGBM model provides accurate vehicle positioning at a low hardware cost. Even if the vehicle encounters GNSS outages (e.g., obstruction of GNSS signals by tall buildings, forests, and tunnels), the vehicle's position information is still provided. Furthermore, based on the high adaptability of PUA-LightGBM, positioning products using PUA-LightGBM are easily scalable.

To the best of our knowledge, the LightGBM algorithm has not been used to predict vehicle position in the event of a GNSS failure. Furthermore, updating the GNSS outage vehicle position prediction model while the vehicle is moving has not been proposed yet, which may be useful for future research.

4.2. The Field Test Analysis

Figure 4 shows the vehicle trajectories outputted by PUA-LightGBM (in yellow) and Novatel pwrpak7-e1 (in red) on the six road segments of the field test. The trajectory output from PUA-LightGBM is the model-predicted trajectory, and the Novatel pwrpak7-e1 integrated navigation system provides the trajectory as the vehicle reference trajectory. As can be seen from the trajectory figure, the PUA-LightGBM model predicted the position changes of vehicles on straight roads (Figure 4B,C,E,F) better than those of vehicles on curved road (Figure 4A,D). This is in line with the analysis by Chiang and Huang, which found that the results predicted by the position prediction model during U-turns or sharp turns (where there are significant dynamic changes in the vehicle's movement) produce significant position errors [13]. The vehicle motion is more stable when travelling on a straight road, and the PUA-LightGBM model is better at predicting position changes. However, the two turning segments chosen for the experiments were both sharp turning segments, where the vehicle experienced significant dynamic changes in motion and vibration of the vehicle body. The INS data at this point did not accurately reflect the vehicle's turning state, and the model's training set did not contain enough vehicle turning situations, causing the PUA-LightGBM model to be less effective in predicting the position change of the vehicle when driving around a turn.

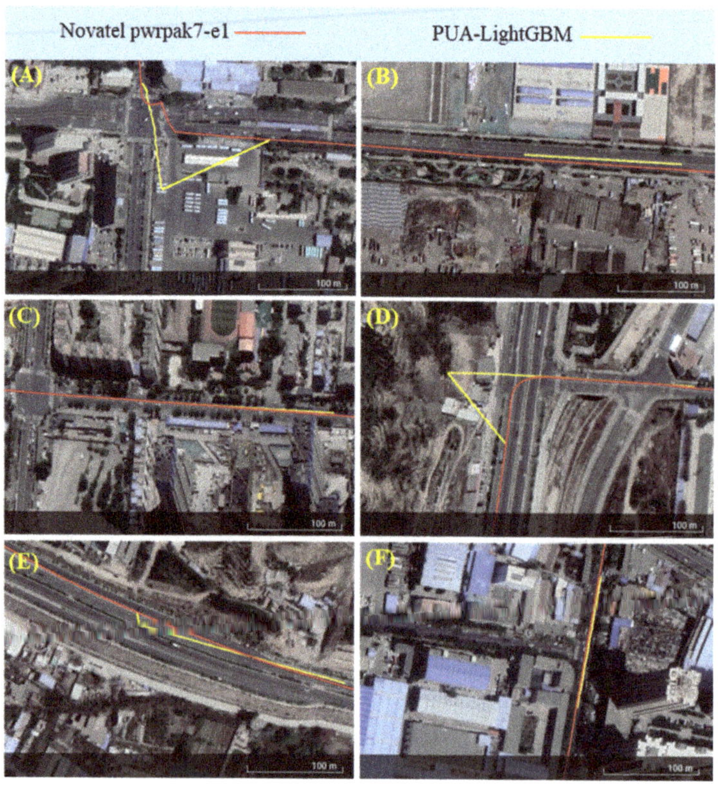

Figure 4. The vehicle trajectories outputted by PUA-LightGBM (in yellow) and Novatel pwrpak7-e1 (in red) on six road segments of the field test. (**A**) Turn and downslope road, (**B**) straight and level road, (**C**) straight and upslope road, (**D**) turn and upslope road, (**E**) straight and level road, (**F**) straight and downslope road.

Table 3 shows the average absolute error between the predicted and true positions of the PUA-LightGBM and PUA-RandomForest models for the GNSS outage scenario. PUA-LightGBM outperformed PUA-RandomForest on road segments a, b, and f in terms of position prediction. On road segments c, d, and e, the position prediction accuracy of PUA-LightGBM was slightly lower than that of PUA-RandomForest. Overall, the prediction error of PUA-LightGBM was less different from PUA-RandomForest. The PUA-LightGBM model, on the other hand, had a much shorter training time than the PUA-RandomForest model (Table 2). This suggests that PUA-LightGBM is more efficient in training, costs less, and has the potential for dynamic updating. The model does not need to be trained in advance and can be dynamically trained or updated while the vehicle is moving. This provides some flexibility to the PUA-LightGBM model and allows it to predict vehicle position changes on a variety of complex road segments via dynamic updating of the model.

Table 3. Average absolute positional error for the PUA-LightGBM model and the PUA-RandomForest model.

Outage Road Segment	Road Length (m)	Average Absolute Positional Error (m)		Percentage Improvement (%)
		PUA-RandomForest	PUA-LightGBM	
a	191.18	61.85	59.71	+3.5
b	222.58	22.11	21.90	+0.95
c	215.24	17.43	17.55	−0.69
d	246.46	32.84	33.52	−2.1
e	231.65	23.33	24.48	−4.9
f	146.88	22.03	21.92	+0.50

5. Conclusions

This paper proposes a prediction model of vehicle position change in the event of a GNSS outage based on position update architecture using LightGBM. The model solves the problem of data accuracy degradation of the GNSS/INS integrated navigation system for a vehicle when the satellite signal quality is weak or there is a GNSS outage and improves the existing AI-based GNSS/INS integrated navigation algorithm. When the vehicle is capable of receiving available GNSS signals, on-board INS and GNSS data are collected, and the PUA-LightGBM model is trained. In the event of a GNSS outage, INS data are used as the input to PUA-LightGBM to obtain the changes in longitude and latitude between the two epochs output by the model, thus predicting the vehicle's position at this time. Experiments on six road segments were conducted using a self-designed vehicle navigation data acquisition system. The results show that the error in predicting vehicle position in the event of a GNSS outage for PUA-LightGBM is less than that of PUA-RandomForest. However, the time required to train the PUA-LightGBM model is only 3.076 s, which is significantly less than the time required to train the PUA-RandomForest model. The authors believe that two factors should be considered in future research to enhance the predictive model's effectiveness in the event of a GNSS outage. One is reducing the noise in INS data to obtain better position predictions. The other is the higher training efficiency of PUA-LightGBM, which can save time and enable the model to be dynamically trained or updated during vehicle operation. Thus, the PUA-LightGBM can adapt to the prediction of vehicle position change on various complex road segments. The limitation of this study is that the validation of the PUA-LightGBM model was not real-time. It might be better to develop a real vehicle navigation system for embedding the PUA-LightGBM model in the future.

Author Contributions: B.L. and G.C. designed and performed the experiments and wrote the manuscript. Y.S., X.Z. and P.L. (Pengpeng Li) supervised the work. P.L. (Peng Li) and T.F. reviewed and edited the manuscript. All authors have read and agreed to the published version of the manuscript.

Funding: This research was supported in part by the National Natural Science Foundation of China (62141303), The Science and Technology Guidance Plan of Gansu Province (2020-61-14) and the Gansu Provincial Science and Technology Major Project of China (21ZD4WA018).

Data Availability Statement: All datasets generated for this study are included in the article.

Conflicts of Interest: The authors declare no conflict of interest.

References

1. Vörös, F.; Gartner, G.; Peterson, M.P.; Kovács, B. What Does the Ideal Built-In Car Navigation System Look Like?—An Investigation in the Central European Region. *Appl. Sci.* **2022**, *12*, 3716. [CrossRef]
2. Onyekpe, U.; Palade, V.; Kanarachos, S. Learning to localise automated vehicles in challenging environments using Inertial Navigation Systems (INS). *Appl. Sci.* **2021**, *11*, 1270. [CrossRef]
3. Angrisano, A. GNSS/INS Integration Methods. Ph.D. Thesis, Universita'degli Studi di Napoli Parthenope, Naple, Italy, 2010; p. 21.
4. Liu, H.; Nassar, S.; El-Sheimy, N. Two-filter smoothing for accurate INS/GPS land-vehicle navigation in urban centers. *IEEE Trans. Veh. Technol.* **2010**, *59*, 4256–4267. [CrossRef]
5. Yang, L.; Li, Y.; Wu, Y.; Rizos, C. An enhanced MEMS-INS/GNSS integrated system with fault detection and exclusion capability for land vehicle navigation in urban areas. *GPS Solut.* **2014**, *18*, 593–603. [CrossRef]
6. Sasani, S.; Asgari, J.; Amiri-Simkooei, A. Improving MEMS-IMU/GPS integrated systems for land vehicle navigation applications. *GPS Solut.* **2016**, *20*, 89–100. [CrossRef]
7. Wang, M.; Wu, W.; He, X.; Li, Y.; Pan, X. Consistent ST-EKF for long distance land vehicle navigation based on SINS/OD integration. *IEEE Trans. Veh. Technol.* **2019**, *68*, 10525–10534. [CrossRef]
8. Cui, B.; Wei, X.; Chen, X.; Li, J.; Li, L. On sigma-point update of cubature Kalman filter for GNSS/INS under GNSS-challenged environment. *IEEE Trans. Veh. Technol.* **2019**, *68*, 8671–8682. [CrossRef]
9. Bijjahalli, S.; Sabatini, R. A high-integrity and low-cost navigation system for autonomous vehicles. *IEEE Trans. Intell. Transp. Syst.* **2019**, *22*, 356–369. [CrossRef]
10. Chiang, K.-W.; Noureldin, A.; El-Sheimy, N. Multisensor integration using neuron computing for land-vehicle navigation. *GPS Solut.* **2003**, *6*, 209–218. [CrossRef]
11. Aggarwal, P.; Bhatt, D.; Devabhaktuni, V.; Bhattacharya, P. Dempster Shafer neural network algorithm for land vehicle navigation application. *Inf. Sci.* **2013**, *253*, 26–33. [CrossRef]
12. Noureldin, A.; Karamat, T.B.; Eberts, M.D.; El-Shafie, A. Performance enhancement of MEMS-based INS/GPS integration for low-cost navigation applications. *IEEE Trans. Veh. Technol.* **2008**, *58*, 1077–1096. [CrossRef]
13. Chiang, K.-W.; Huang, Y.-W. An intelligent navigator for seamless INS/GPS integrated land vehicle navigation applications. *Appl. Soft Comput.* **2008**, *8*, 722–733. [CrossRef]
14. Yue, S.; Cong, L.; Qin, H.; Li, B.; Yao, J. A robust fusion methodology for MEMS-based land vehicle navigation in GNSS-challenged environments. *IEEE Access* **2020**, *8*, 44087–44099. [CrossRef]
15. Liu, Y.; Luo, Q.; Zhou, Y. Deep Learning-Enabled Fusion to Bridge GPS Outages for INS/GPS Integrated Navigation. *IEEE Sens. J.* **2022**, *22*, 8974–8985. [CrossRef]
16. Khankalantary, S.; Rafatnia, S.; Mohammadkhani, H. An adaptive constrained type-2 fuzzy Hammerstein neural network data fusion scheme for low-cost SINS/GNSS navigation system. *Appl. Soft Comput.* **2020**, *86*, 105917. [CrossRef]
17. Ke, G.; Meng, Q.; Finley, T.; Wang, T.; Chen, W.; Ma, W.; Ye, Q.; Liu, T.-Y. Lightgbm: A highly efficient gradient boosting decision tree. In Proceedings of the Advances in Neural Information Processing Systems 30 (NIPS 2017), Long Beach, CA, USA, 4–9 December 2017; Volume 30.
18. El-Sheimy, N.; Chiang, K.-W.; Noureldin, A. The utilization of artificial neural networks for multisensor system integration in navigation and positioning instruments. *IEEE Trans. Instrum. Meas.* **2006**, *55*, 1606–1615. [CrossRef]
19. Adusumilli, S.; Bhatt, D.; Wang, H.; Bhattacharya, P.; Devabhaktuni, V. A low-cost INS/GPS integration methodology based on random forest regression. *Expert Syst. Appl.* **2013**, *40*, 4653–4659. [CrossRef]
20. Bhatt, D.; Aggarwal, P.; Devabhaktuni, V.; Bhattacharya, P. A novel hybrid fusion algorithm to bridge the period of GPS outages using low-cost INS. *Expert Syst. Appl.* **2014**, *41*, 2166–2173. [CrossRef]
21. Liang, W.; Luo, S.; Zhao, G.; Wu, H. Predicting hard rock pillar stability using GBDT, XGBoost, and LightGBM algorithms. *Mathematics* **2020**, *8*, 765. [CrossRef]
22. Chen, C.; Zhang, Q.; Ma, Q.; Yu, B. LightGBM-PPI: Predicting protein-protein interactions through LightGBM with multi-information fusion. *Chemom. Intell. Lab. Syst.* **2019**, *191*, 54–64. [CrossRef]
23. Ju, Y.; Sun, G.; Chen, Q.; Zhang, M.; Zhu, H.; Rehman, M.U. A model combining convolutional neural network and LightGBM algorithm for ultra-short-term wind power forecasting. *IEEE Access* **2019**, *7*, 28309–28318. [CrossRef]

applied sciences

MDPI

Article

Research on Data Fusion of Positioning System with a Fault Detection Mechanism for Autonomous Vehicles

Wei-Hsuan Chang [1,2,*], Rong-Terng Juang [3] and You-Sian Lin [2]

1 Ph.D. Program of Electrical and Communication Engineering, Feng-Chia University, Taichung 40724, Taiwan
2 Autmotive Research and Testing Center, Changhua 50544, Taiwan
3 Department of Electronic Engineering, Feng-Chia University, Taichung 40724, Taiwan
* Correspondence: P0965854@o365.fcu.edu.tw or brant.chang@artc.org.tw

Abstract: A positioning system provides useful positioning information for applications. The ability to maintain high-quality positioning information in real time is crucial. Autonomous vehicles require positioning information to plan control strategies. Inadequate positioning accuracy affects control and safety. This article explores the effect of an error detection and correction method on coordinate information fusion. This method can be used to efficiently correct positioning errors in the presence of faults. The fault detection mechanism is not limited to positioning systems with certain types of sensor applications. and it has certain flexibility in various applications. This positioning system detection method uses more than two positioning systems. Primarily, one positioning system is used to monitor the other. When the error of the positioning system differs excessively from the system setting value, the positioning fusion algorithm will make appropriate adjustments to output the corrected coordinate value. Therefore, this method integrates different positioning technologies, including a global position system (GPS), simultaneous localization and mapping (SLAM), inertial measurement units (IMUs), and mileage meters. Autonomous vehicles already have the required sensors deployed, and our method can be implemented at no additional cost. We integrated two positioning systems to support the positioning applications and designed a fault detection method to detect errors in the positioning system. This allowed the system to have some reference information by which to correct the coordinate information. According to the experimental findings, positioning information errors were corrected at a rate of 89.8%. The positioning system with the added detection function was able to effectively filter the data without too many errors. Additionally, the results showed that this method can effectively correct positioning errors. The results indicate that adding an error detection mechanism can effectively improve the accuracy of coordinate data fusion. Positioning systems with an error detection function can thus improve applications' performance.

Keywords: positioning system; SALM; GPS; IMU

Citation: Chang, W.-H.; Juang, R.-T.; Lin, Y.-S. Research on Data Fusion of Positioning System with a Fault Detection Mechanism for Autonomous Vehicles. *Appl. Sci.* **2022**, *12*, 11339. https://doi.org/10.3390/app122211339

Academic Editors: Nadjim Horri, William Holderbaum and Fabrizio Giulietti

Received: 3 October 2022
Accepted: 4 November 2022
Published: 8 November 2022

1. Introduction

Vehicles are important tools in everyday life. An increasing number of vehicle applications are being developed, such as autonomous vehicles. Many manufacturers have joined the research and development of autonomous vehicles. These vehicles are based on the existing vehicle architecture for innovation. The vehicle requires a control center to plan the driving strategy. The computer used for this strategy requires some position information to allow it to understand where the vehicle is. The positioning system can provide the vehicle with location information. There are various methods by which to provide reliable location information for autonomous vehicles. Every application system has the possibility of the occurrence of a system fault. The positioning system also has the problem of potential failures. Vehicle applications should consider this situation and design functions to detect faults. On-board diagnostics (OBD) is one such approach [1], but there are few discussions of fault detection methods for positioning systems in the study of the application of positioning systems.

In this paper, the influence of the fault detection mechanism on the data fusion of the positioning system is discussed. This method is developed based on the sensor configured by the WinBus [2]. The WinBus is an autonomous vehicle that integrates many application systems. This fault detection mechanism is discussed using positioning methods developed by on-board sensors. Different sensing systems have different coordinate values. The coordinate system must be unified for this mechanism to work. The coordinate system uses 3D coordinates (X, Y, and Z). This coordinate system is unified in the WinBus application. This coordinate system allows this method to have the same benchmark that can be used for cross-validation. In addition to the sensors provided by the vehicle, there are also many studies using the "Internet of vehicles" system to complete the positioning method. The experimental field did not have road equipment linked to the Internet of vehicles. Therefore, this paper does not address the issues related to the Internet of vehicles; instead, it focuses more on the positioning system application of the vehicle itself.

The fault detection mechanism proposed in this paper is an extension based on the abovementioned method to improve the reliability of the positioning system. This paper consists of five sections. Section 2 is the literature review. Section 3 describes the positioning system's experimental platform, i.e., the WinBus. Section 4 presents the positioning fault detection methods. The method used includes a procedure and algorithm that are described in this section. Section 5 presents the verification results. The final section reports the conclusions and development plan.

2. Literature Review

Each positioning system uses different sensors to implement the navigation functions. The vehicles deploy many sensors that provide the autonomous vehicle strategy center with information required to plan control strategies. For example, real-time kinematic (RTK) positioning, inertial measurement units (IMU), 2D lidar, 3D lidar, radars, and cameras are the sensors of the applications. These sensors can also provide useful information for different positioning methods. RTK provides coordinate information to end applications. There are various studies regarding the development of technology based on this system [3–8]. Some methods integrate many sensors, while others use more RTK systems to reduce data loss. Researchers try to improve the usefulness and accuracy of RTK information, which is the goal. These methods require many application systems to support them to correct coordinate information. Each application system has a damage probability. In this paper, we propose a system fault-finding mechanism, and the coordinate accuracy is verified based on this mechanism. This mechanism references some related application systems' information for implementation. The reference system is part of WinBus, so there is no additional cost of the sensor in this method.

Simultaneous localization and mapping (SLAM) is another currently applied technology in autonomous vehicle positioning. Some studies have discussed this technology. This technology is based on the 3D lidar for implementation [9–15]. There are many types of research that discuss the 3D lidar SLAM. The 3D lidar provides scan information of the surrounding environment to support the SLAM calculation. SLAM uses some deep learning methods for implementation, such as convolutional neural networks (CNN). This method provides some position information for the applications. There is autonomous vehicle research based on this method for implementation. This method supports autonomous vehicles in a scenario where there is no GPS signal support. This is one of the advantages of the autonomous vehicle application described in this study. Systems based on 3D lidar applications require many hardware computing resources. SLAM alone cannot achieve autonomous vehicles according to the previous application experience and literature. Similarly, when the system responsible for SLAM fails, how the decision-making control system of the self-driving car is informed is relatively less discussed. The fault detection mechanism proposed in this paper is a possible solution for autonomous vehicle applications.

Communication is an indispensable element in autonomous vehicle applications. The communication interface includes wired communication and wireless communication. Control area networks (CAN) and the eEhernet are the wired communication interfaces [16–18]. The CAN bus is used for message transmission between the power chain, chassis, and body. The International Standards Organization also defines the corresponding communication specifications, but the information on the positioning system applications is only partially defined. The CAN bus bandwidth cannot transmit a lot of information from the sensor. The autonomous vehicle requires the communication interface to be the Ethernet. In recent years, there has been much research on positioning systems in the Internet of vehicles (IoV). Communication interfaces used in the Internet of vehicles, such as dedicated short-range communication (DSRC), cellular-vehicle-to-everything (C-V2X), and fifth-generation mobile networks (5G) have positioning-related information [19–21]. Most of the positioning-related information follows the specifications of SAE J2735 [22]. A part of the communication interface applied by IoV belongs to the wireless communication interface. The IoV method is based on RSU and vehicle coordinates for implementation. There are also studies implementing positioning systems using the SLAM and IoV techniques discussed earlier. The deployment of hardware facilities is the key to the positioning system implemented using these technologies. This research used the wired network inside the vehicle to collect and transmit information. Limited by the problem of hardware deployment in the experimental field, discussion of the positioning system discussion of the Internet of vehicles is temporarily excluded from this research.

Some accurate and reliable location data are necessary for autonomous vehicles' requirements. It can be seen from the previous discussion that positioning information alone cannot meet these requirements. Researchers have proposed many information fusion methods to solve the accuracy problem [23–25]. The information fusion method can solve some of the system's disadvantages. This method integrates multiple sensor systems that can provide some positioning information in different scenarios. To address these shortcomings, researchers have used different approaches to solve them. Information fusion combines some of the benefits of other systems, making the positioning system more precise and reliable. GPS and IMU integrate the extended Kalman filter (EKF) as one of the methods. In addition, the coordinate data fusion method has the function of repairing inaccurate information coordinates. The additional error detection mechanism of the positioning system correcting the coordinate information is a method by which to solve the error of the positioning system. There are still some differences between these methods.

We propose a method that can be used to improve the stability and reliability of positioning systems; it supports the positioning system with an error detection mechanism. Considering the sensing systems and methods configured on the vehicle, this mechanism is based on the RTK, SLAM, IMU, and chassis systems for implementation of the function.

3. Positioning System Experimental Platform

3.1. WinBus

The WinBus is the experimental platform integrated by the Automotive Research and Testing Center (ARTC) and related manufacturers. The sensors deployed in the WinBus are shown in Figure 1.

Two 3D lidars are deployed on the front-end and back-end roof. Lidar is a kind of optical device. The device is based on a laser beam that is used to detect the obstacle. Lidar measures the time gap between the laser beam transmission and reception to better understand the obstacle's distance. It predicts the shape of the target using the reflected signals of multiple beams. It can also use the reflectivity of the received beam to judge the material of the target surface. The front, back, and side effective ranges are each 60 m, and the side range is 20 m. The position, relative speed, and ID index constitute the obstacle data. The technique used here captures 3D lidar point cloud data. Two radars are mounted, one at the front and one at the back of the WinBus. Similar to the 3D lidar, the radar has the ability to identify obstacles. This information helps the vehicle locate the obstruction. This sensor can deliver information regarding obstacles, including their location and relative speed. Both the 3D lidar and the radar can simultaneously find obstacles in the path of the vehicle. These two sensors can aid in the development of a control rule that will boost safety. The cameras were mounted all over the body. The front camera shows information about lanes and traffic signals. The GPS and IMU deliver data on the dynamics and location of the vehicle. For radio signal transmission, the GPS uses 24 to 32 Earth-orbiting satellites. The GPS receiver can obtain the position, time, and speed with the assistance of the signal. An appropriate received signal quality is necessary for GPS accuracy. There are many factors affecting the positioning signal, including the environment, the ionosphere, and electromagnetic interference. The GNSS network is the foundation of real-time dynamic technology [26], which assesses the positioning inaccuracy of the region serviced by the reference station. The IMU determines the attitude of the item by monitoring the gyroscope and G-sensor. The information provides the positioning system with an angular velocity and acceleration reference.

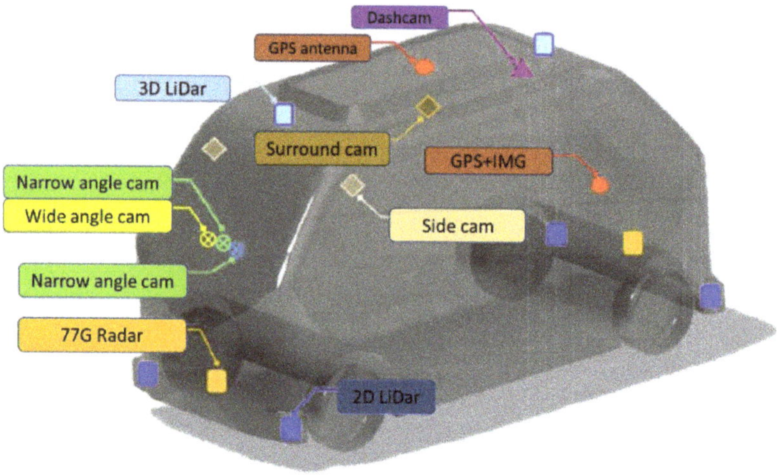

Figure 1. Sensors deployed in the WinBus.

3.1.1. LiDAR

Lidar is an optical sensing technology that illuminates a beam of light. Lidar measures the time interval between the emitted and received beam signals to count and calculate the distance to the target. It predicts the shape of the target using the reflected signals of multiple beams. It can also use the reflectivity of the received beam to determine the target surface material. The WinBus deploys two 3D lidars on the vehicle roof. 3D lidar sensors are also known as 3D laser scanners. This sensor captures their surroundings almost continuously. Therefore, 3D lidar sensors are ideal for collision protection for autonomous vehicles or for the detection of objects. In the previously discussed literature, 3D lidars were used as sensors for SLAM. The lidar mounted on the WinBus is used to record the scene point cloud. The system creates a recorded point cloud by overlaying the map. The completed map is the point cloud map data of the comparison ends. During the comparison, the lidar scans the scene environment in real-time. The SLAM algorithm compares the current point cloud with the point cloud image. This method finds feature locations that match the current point cloud, and then calculates the vehicle coordinates.

3.1.2. Global Positioning System (GPS)

GPS is the only fully operational global navigation satellite system in the world. This system has about 24 to 32 Earth-orbiting satellites. GPS satellites are usually located at an altitude of 20,000 km from the Earth and operate at a speed of 14,000 km per hour. GPS signals cannot penetrate walls. Signals in buildings are also blocked. It is less effective when the user uses it in an area with tall buildings or dense trees. Therefore, GPS signals are rarely used directly in practical applications.

The real-time dynamic technology uses a GNSS network composed of multiple satellite positioning reference stations to evaluate the positioning error of the area covered by the reference stations. Then, it matches the observation data of the nearest physical base station to produce a virtual base station as the RTK master station. The autonomous vehicle does not receive actual observations from a physical base station. The error-corrected virtual observation data are the information used by the applications. The WinBus integrates RTK to provide location information for applications. RTK is one of the positioning systems for the WinBus.

3.1.3. Simultaneous Localization and Mapping (SLAM)

SLAM is another positioning system used by the WinBus. Mapping and matching are the two parts of the SLAM method. The quality of the mapping determines whether this method is accurate because the positioning data source is from the feature of the map. The intricacy of the environment and the number of dynamic objects are two factors that influence the SLAM's outcome. The clarity of the mapping will be affected if the characteristics of the environment are ambiguous or the base map environment changes too frequently. This kind of application has some risks for autonomous vehicle applications. In the drawing part, for scenes where one want to position oneself, first, create a map. The lidar mounted on the car will record the scene point cloud at a slow speed. The recorded point cloud is created by overlaying the map, and the completed map is the base map at the end of the comparison. Then, the scene environment is instantly scanned during the comparison. SLAM compares the current point cloud with the map point cloud. Additionally, it finds feature locations that match the current point cloud. The output coordinates are unified three-dimensional coordinates for the WinBus application. The WinBus uses SLAM to support the positioning system. Increasing the positioning system's reliability is the goal.

3.1.4. Inertial Measurement Unit (IMU)

The IMU uses three-direction gyroscopes and three-direction accelerometers to measure the three-axis angular velocity and acceleration changes. It uses these data to calculate the pose of the object. The IMU is usually mounted on the object's center of gravity to ensure correct calculations. Some IMUs integrate other sensors to extend its functions, such as magnetometers or barometers, and to calculate bearing and altitude. The WinBus integrates the IMU to increase the reliability of the positioning system.

3.2. Positioning System of the WinBus

The positioning system of the WinBus fuses different information from different sensor systems. Figure 2 shows the positioning system function block. GPS, SLAM, and reference chassis data are all combined into one positioning system. The information for the applications is the coordinates, both absolute and relative. The two elements that make up the absolute positioning system are GPS and SLAM. Autonomous cars may locate their position using these two technologies' absolute latitude and longitude information. The relative coordinates are from the wheel speedometer, steering angle, and IMU. The positioning system predicts coordinates and guarantees stability through dynamic models.

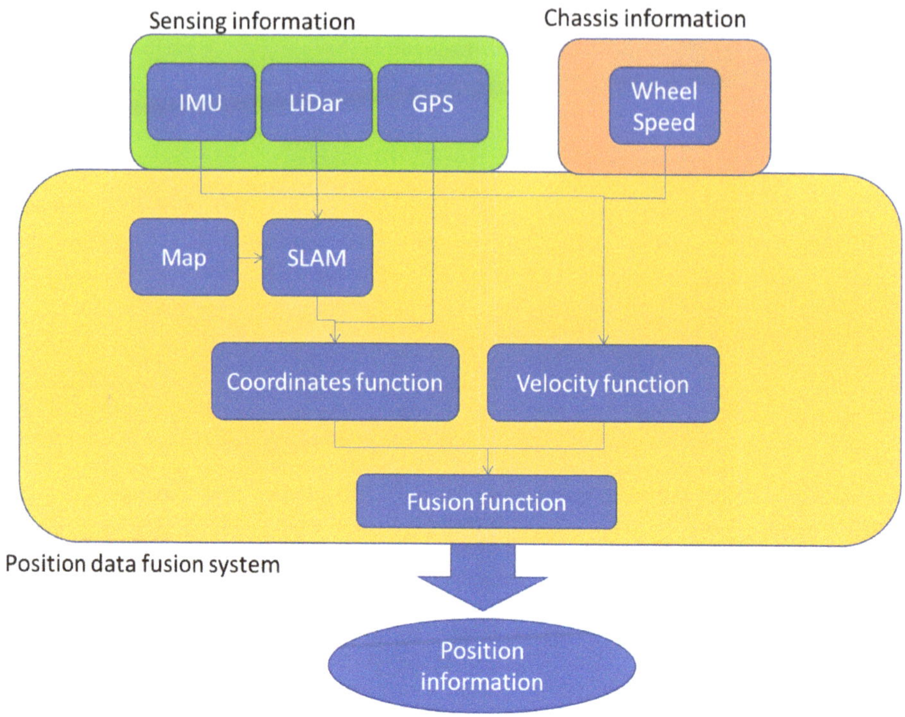

Figure 2. Positioning system.

4. Positioning Fault Detection Methods

How to increase positioning accuracy is a problem that researchers and developers wish to solve. This positioning system combines RTK and SLAM to aid the autonomous vehicle's strategy-building process. The positioning fusion algorithm references the RTK and SLAM information to provide the location data. For precision, these two systems work together as a team. However, every application system has the possibility of failure. In this research, we propose a strategy by which to determine the fault detection mechanism and discuss the relationship between coordinate data fusion and faults. Here, we take advantage of the interdependence of the data to achieve this. Finding faults requires some mechanisms and a relatively stable device to support them. Taking speed information as an example, there are devices in the vehicle that provide speed information, such as the ABS (antilock braking system), IMU, and GPS. The ABS is a more stable provider of speed signals than the IMU and GPS. Based on this concept, the speed of the fault detection method references the ABS speed information. The detection method integrates the IMU and ABS to help the system to find the faults. A prediction algorithm with heading and velocity information can provide reference information to the positioning fusion algorithm. This information is a reference message that helps the fusion algorithm determine the correctness of the positioning data from the RTK. The SLM fault detection method references the RTK information to detect the fault.

Figure 3 shows the functional blocks of the positioning system and the fault detection method. The time t_n is the current time, and t_{n+1} is the following time. The predicted coordinates reference the previous credible coordinates for the fault detection method of RTK and SLAM and are used as reference sources. The prediction data reference the vehicle speed and heading. Coordinate data, which must be reliable information derived from a prediction method, are referred to by the positioning system fault detection algorithm. The speed and heading information are from the ABS and IMU. The application of these two devices on a vehicle results in a relatively stable device. This method is suitable for use as a reference for estimating messages in positioning system fault detection.

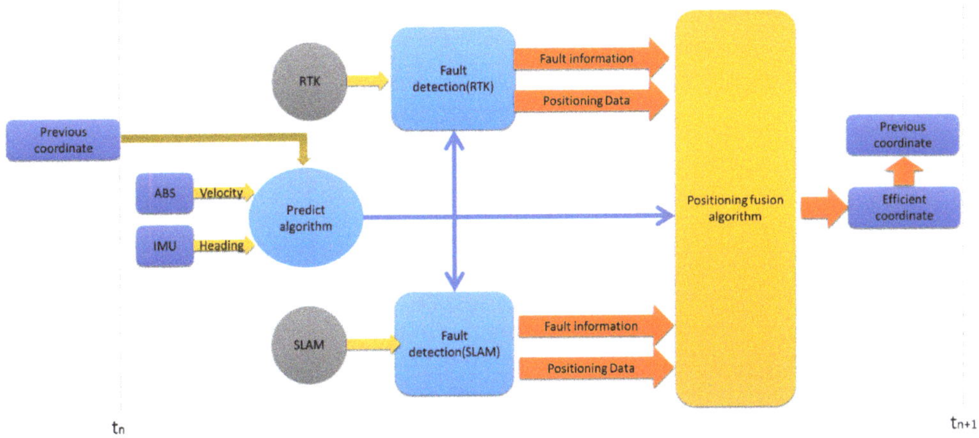

Figure 3. Functional blocks of the positioning system.

The fault detection function block is shown in Figure 4. This method of obtaining reference data must be chosen following the positioning system's features, for example, whether the fault detection algorithm used in the RTK uses the input coordinate data from the RTK or whether the method used in SLAM and the input coordinate data will be replaced with SLAM coordinate data. The reference coordinates for SLAM and RTK are both from the prediction coordinates. The positioning error condition is the error level condition. This condition defines the allowable error range of the positioning system. The data output are fault information and positioning data that the positioning fusion algorithm will reference.

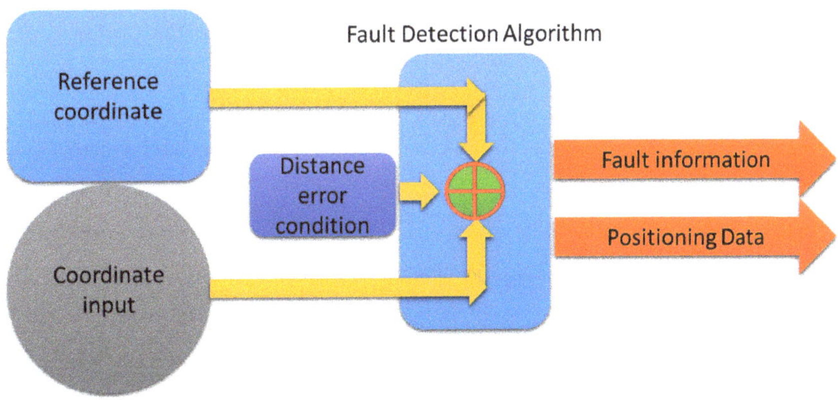

Figure 4. Fault detection method function block.

The coordinate information allows the system to have a reference position. The positioning system of the WinBus fuses the RTK and SLM information to generate location information for the autonomous application. These data are used by the vehicle to determine its location. The original fusion algorithm refers to Equation (1), where P_f is the position data that can be latitude or longitude information. This position information is calculated from the fusion algorithm. *Prtk* is the position information from RTK. *Pslam* is the position information from SLAM. *Prtk* and *Pslam* can be the latitude or longitude information. *Wrtk* is the weight for the RTK. *Wslam* is the weight for SLAM. The weight parameter can adjusted by the application. RTK has a large weight in this system. Each application has the primary trust information. The weight parameter can allow the strategy process to switch the primary position system when a system is damaged.

$$P_f = \frac{Prtk * Wrtk + Pslam * Wslam}{Wrtk + Wslam} \tag{1}$$

We designed a prediction method to find reference information for autonomous vehicle applications. Reference information must be stable and reliable. This method integrates some information from the chassis and other sensors, such as the positioning system, the ABS sensor, and the IMU. We took into account that the RTK is subject to weather. From the perspective of system stability, the ABS and IMU on the vehicle are more stable than the GPS-based positioning system. Referring to these two pieces of information in the fault detection mechanism is more reliable. The positioning system provides previous valid position data. The ABS sensor provides the final speed information. The IMU provides the heading information to the algorithm. The latitude and longitude prediction algorithm equation is represented by Equation (2). The parameter PD_{x+1} is the estimated latitude data, and P_x is the current latitude data. The parameter PD_{y+1} is the estimated longitude data, and P_y is the current longitude data. The parameter V is the vehicle speed. The parameter θ is the heading of the vehicle. Equation (3) defines PD_{x+1} and PD_{y+1} as belonging to P_{f+1} . The distance calculation algorithm is shown in Equation (4) The parameter D_f is the distance between the prediction position and the current position.

$$\begin{cases} PD_{x+1} = P_x + \int_0^t V * \cos\theta dt \\ PD_{y+1} = P_y + \int_0^t V * \sin\theta dt \end{cases} \tag{2}$$

$$PD_{f+1} \in \{ PD_{x+1}, PD_{y+1} \} \tag{3}$$

$$D_f\left(P_f, PD_{f+1}\right) = \text{R}* \cos^{-1}\left[\cos P_y * \cos\ PD_{y+1} * \cos\ (\ P_x - PD_{x+1}) + \sin P_y * \sin\ PD_{y+1}\ \right] \tag{4}$$

The fault result is shown in Equation (5), which is another indicator function. The parameter D_r is the fault detection result. D_e can be a level that triggers the fault condition. If D_e is less than D_f, the result is 1. The fault detection algorithm generates an error message to the self-driving application stating that an error has occurred in the positioning system. Equation (5) fits different applications that can be rewritten. If the fault detection mechanism is used in SLAM, the parameter D_r can be rewritten as $D_{r(SLAM)}$. If the fault detection mechanism is used in RTK, the parameter D_r can be rewritten as $D_{r(RTK)}$.

$$D_r\left(D_{e,}, D_f\right) = \begin{cases} 1\ ,\ if\ D_e > D_f \\ 0\ ,\ otherwise \end{cases} \tag{5}$$

Equation (1) can be rewritten as Equation (6) based on Equation (5). This equation integrates the positioning fusion algorithm and fault detection algorithm for the autonomous application. It allows the strategy process to have a positional reference for planning the control rule to control the vehicle.

$$P_f = \frac{\left(PD_{f+1} * Wpd\right) + (Prtk * Wrtk)^{\ D_{r(RTK)}\ (D_{e,}\ ,D_f)} + (Pslam * Wslam)^{\ D_{r(SLAM)}\ (D_{e,}\ ,D_f)}}{Wpd + (Wrtk)^{\ D_{r(RTK)}\ (D_{e,}\ ,D_f)} + (Wslam)^{\ D_{r(SLAM)}\ (D_{e,}\ ,D_f)}} \tag{6}$$

5. Results

The experimental field was in the Chang Hua Coastal Industrial Park. The route is shown in Figure 5. The route is 7.5 km long. The positioning system was deployed in the WinBus. The fault detection algorithm was integrated with the system in the WinBus. This experiment obtained three results from three positioning methods: the RTK, SLAM, and prediction. The results were used to compare the different positioning methods for the coordinate results' effect.

Figure 5. The experimental route.

The result compares the SLAM and RTK information shown in Figure 6. The figure was generated from the latitude and longitude data. The red line presents the SLAM result, and the yellow line shows the RTK. This finding shows that these two positioning systems indicate the two trajectories in the figure. By comparing the actual route with the results, we can see that the SLAM system contained some errors. We discuss the failure detection mechanism below, as well as how to use this mechanism to increase the accuracy of the positioning system.

Figure 6. Comparison of SLAM and RTK results.

Figure 7 compares the RTK, SLAM, fusion, and prediction results. The fusion result is the blue line from Equation (1), shown in Figure 7. The prediction result is the green line, shown in Figure 7. This result is from Equations (2) and (3). There were some errors detected without a fault detection method that were compared the RTK result. The prediction result showed that the track was similar to the RTK. Another comparison between the RTK and prediction data showed that the information was similar to the actual trajectory. This also proved that the prediction results and RTK are reliable information for autonomous applications in this field.

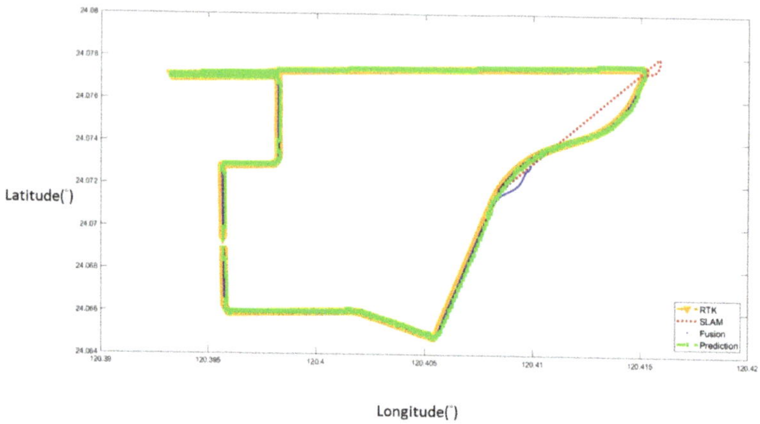

Figure 7. Comparison of SLAM, RTK, fusion, and prediction results.

Figure 8 includes all of the above results for comparison. The purple line is the fusion with the fault detection mechanism. Here, we discuss the impact of the fault detection mechanism of the positioning system. From the previous discussion, it can be seen that this prediction method is feasible. Finally, we compared the actual path, the predicted path, and the results of the fusion algorithm based on the fault detection mechanism. The results showed that the fault detection mechanism can effectively correct the coordinate results of the positioning system.

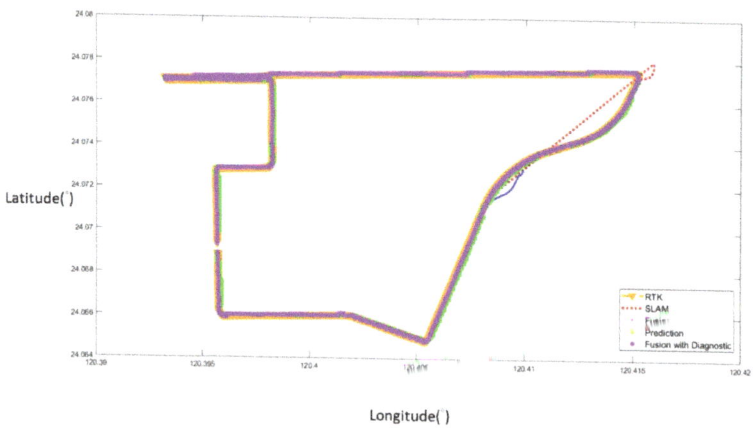

Figure 8. Comparison of the RTK, SLAM, fusion prediction, and fusion with the diagnostic mechanism results.

Figure 9 (longitude) and Figure 10 (latitude) compare each result. In the figures, lines of different colors represent RTK, SLAM, fusion, prediction, and fusion with the diagnostic mechanism results. The result showed that the SLAM and fusion without any fault detection mechanisms had some problems.

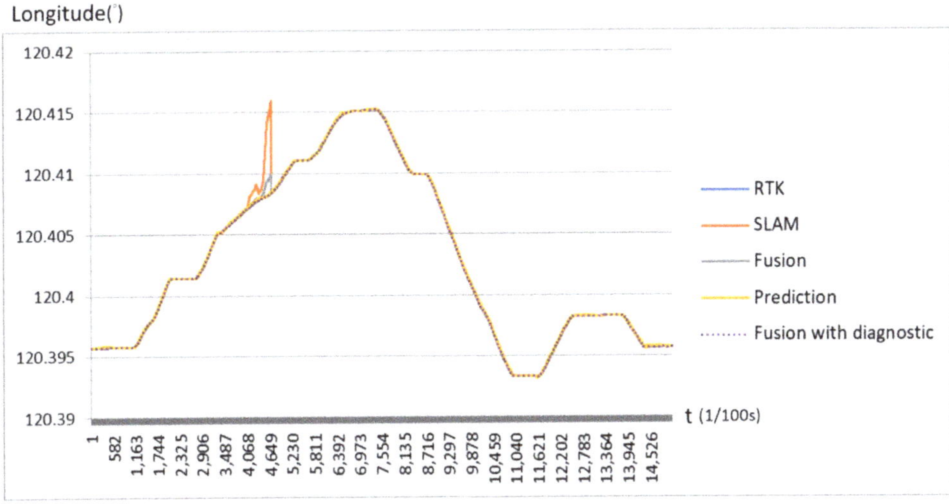

Figure 9. The results from each positioning method (longitude).

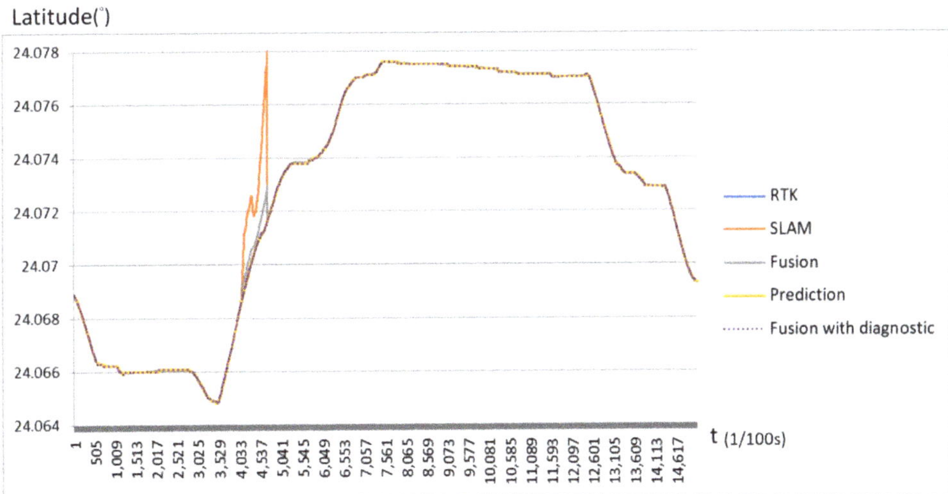

Figure 10. The results from each positioning method (latitude).

We also performed a more detailed comparison regarding the location of anomalous coordinates. Figures 11 and 12 show the parts with abnormal coordinates in Figures 9 and 10, respectively. Figures 11 and 12 show the results of the longitude and latitude coordinates of each method's coordinates, respectively. These diagrams show the coordinates of the anomalous parts for further discussion. The results showed that the coordinates results of SLAM cannot satisfy the requirements for autonomous vehicle applications. SLAM has the greatest number of mistakes after comparing all the methods. Due to the deviation of the coordinates, the system's operation may be faulty.

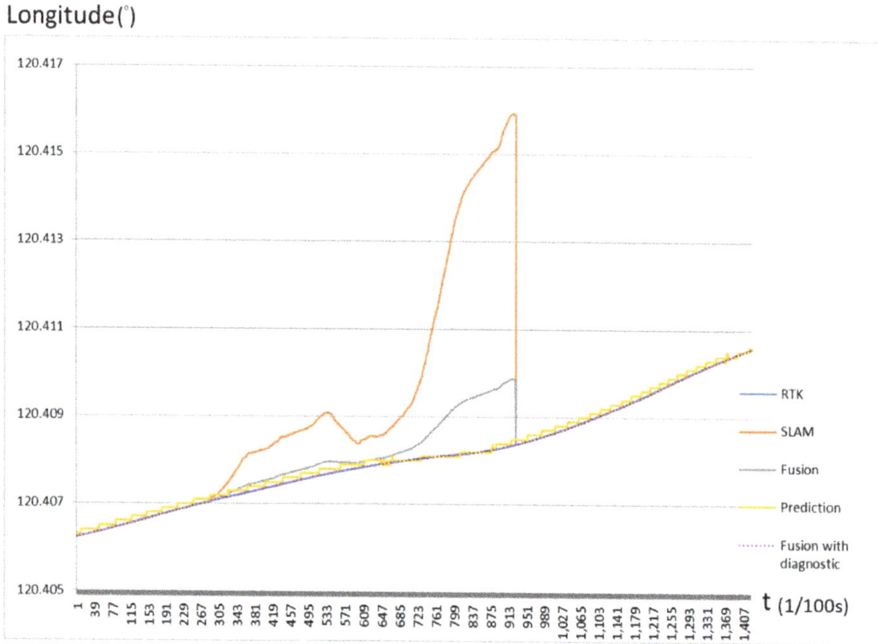

Figure 11. The result from each positioning method (longitude).

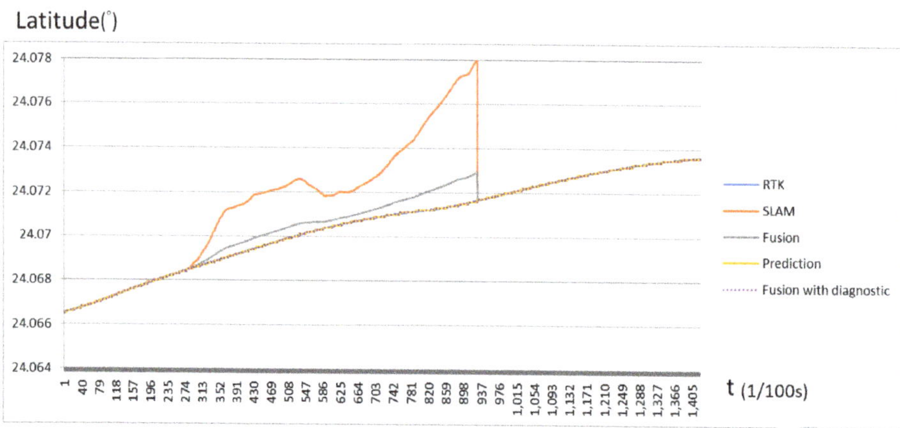

Figure 12. The result from each positioning method (latitude).

The [illegible] fusion method was based on Equation (1). The grey line shows the result obtained by the original fusion algorithm. This result is compared with the SLAM and RTK results, shown in Figures 11 and 12. The original fusion result was without reference fault information. This result showed that the fusion algorithm was able to correct the biases in the coordinate information, but it was not sufficient. The original fusion algorithm only considered the weights occupied by the various positioning systems. The erroneous value was still included in the evolutionary algorithm's calculations when one of the positioning systems failed. Due to the influence of SLAM, the trajectory of the coordinates was biased.

The correct path in this situation was the most similar to that of the RTK. Thus, the result of using the RTK here is the benchmark. The coordinate prediction results (yellow line) were similar to the RTK results, proving that the prediction mechanism was efficient. The fusion results and the diagnosis mechanism (light blue line) were combined with reference to the results of RTK, SLAM, and data fusion. Equation (7) describes the positioning system results. This result is close to the coordinates of the correct route.

The positioning error was calculated from Equation (4). The RTK position coordinates were compared with those obtained using SLAM, the original fusion algorithm, prediction, and fusion with a diagnostic mechanic. The results are shown in Figure 13. Table 1 shows a comparison between the values of each method. In order to compare trends and reduce errors caused by noise, 100 records were averaged from the peak in the comparison of the results. The average distance by which SLAM deviated in this test was 91.8 m.

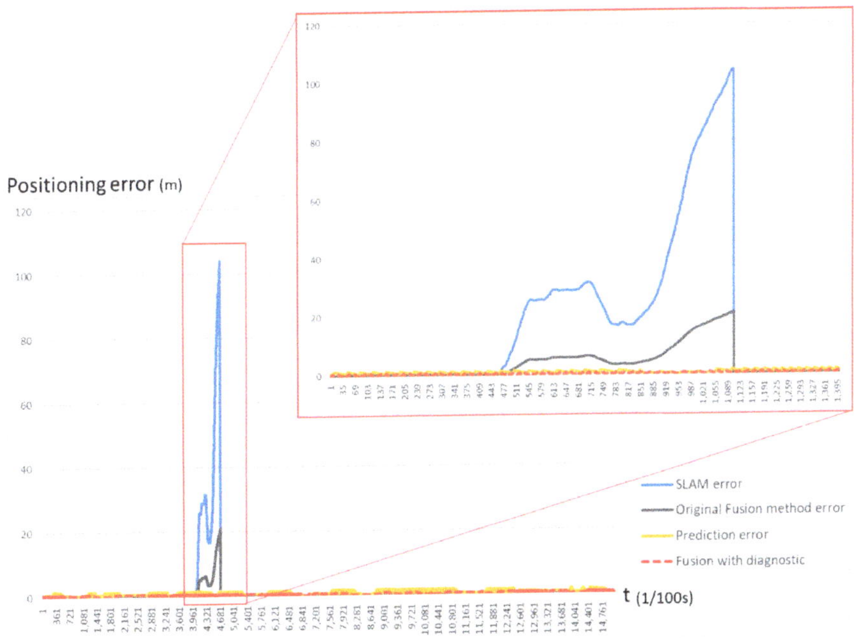

Figure 13. Comparison of the results based on the RTK results.

Table 1. The comparison between each method based on the max positioning error of 91.8m.

Error Condition	Positioning Error (m)	Correction Rate (%)
Limited two-meter error	89.80	97.82
RTK—original fusion algorithm	18.30	80.06
RTK—prediction	0.64	99.30
RTK—fusion with diagnostic	0.25	99.73
RTK—fusion with diagnostic (Considering the limited two-meter error)	2.25	89.89

The original algorithm was able to correct the error value correctly. However, the magnitude of the correction was limited. The maximum error was 18.3 m after correction. The original algorithm corrected the error at a rate of 80%. The prediction coordinate was from Equation (2). The prediction of the coordinates was based on the previously valid coordinates, with reference to the speed and heading of the vehicle. The error between the RTK coordinates and the predicted value was about 0.64 m. As can be seen from Figure 13, this result is very similar to the coordinate value of RTK. It can be used as reference data for the fault detection mechanism. Equation (6) presents the fusion algorithm with the diagnostic mechanism. The coordinate result of this algorithm differed from the RTK coordinate value by 0.25 m. The results showed that the error was further corrected.

Finally, the positioning errors of the algorithms were considered. A 2 m lateral error was set in the experiment by considering the lane width. Comparing the values, the difference between the RTK and the diagnostic mechanism was 2.25 m. This method corrected 89.8% of the errors. This result proves that our proposed method is useful.

6. Conclusions

Autonomous vehicles use relatively accurate positioning information to formulate control strategies and improve vehicle safety. Positioning systems have the problem of positioning errors. There are many methods implemented using only a single positioning method. There are certain risks in autonomous vehicles that rely on a single positioning system. Therefore, scholars have discussed a variety of options. Coordinate information fusion is one possible solutions. There are still some problems to be solved in positioning systems, such as fault problems.

In this paper, we have proposed a fault detection method to support coordinate fusion and compared the differences between the errors of various positioning methods and the proposed method. In the application, the concept of data fusion was used for error correction so that the autonomous vehicle had relatively accurate coordinate values as a reference. The coordinate information obtained by the original fusion algorithm still had an error problem. This was due to the fact that the original algorithm only focused on the weight allocation of the coordinate information. It did not consider the impact of system failures on information fusion. The fault detection mechanism referenced a prediction coordinate. SLAM was found to have more errors in all the positioning systems after comparing this coordinate. After observing these phenomena, we changed part of the algorithm. The adjusted coordinate fusion algorithm corrected 89.8% of errors. The experiments demonstrated that the method was able to detect errors when the positioning system had failed. The fault detection and correction mechanism was used to correct the coordinate. The results showed that this mechanism was able to support the positioning system in certain conditions and was able to correct the coordinate data.

We used the WinBus as the experimental platform. The scope of the WinBus experiment was mainly limited to Chang Hua Coastal Industrial Park. The use of sensors configured in the vehicle and the application equipment configured in the experiment field were limited. Not every vehicle had the same sensors. In this method, there will be gaps in the parameters of the vehicle dynamics, and there will be differences in the threshold values of the positioning system's fault detection. Artificial intelligence (AI) may be used to support the proposed fault detection and correction methods in the future. There are AI tools available that can be used to determine the most suitable setting value for each car.

Author Contributions: Conceptualization, W.-H.C.; methodology, W.-H.C.; software, W.-H.C. and Y.-S.L.; validation, W.-H.C. and Y.-S.L.; formal analysis, W.-H.C. and R.-T.J.; investigation, Y.-S.L.; resources, W.-H.C. and Y.-S.L.; data curation, Y.-S.L.; writing—original draft preparation, W.-H.C.; writing—review and editing, W.-H.C. and R.-T.J.; visualization, W.-H.C.; supervision, R.-T.J.; project administration, W.-H.C.; funding acquisition, W.-H.C. and Y.-S.L. All authors have read and agreed to the published version of the manuscript.

Funding: This work cooperated with the Automotive Research and Testing Center under grant number 111-EC-17-A-25-1588 in the Intelligent Vehicle Technology and Autonomous Driving System Development Project (1/4) of the Ministry of Economic Affairs, Taiwan ROC.

Conflicts of Interest: The authors declare no conflict of interest.

References

1. Cao, D.; Zhao, B.; Han, H.; Gan, W. Design on Remote Diagnosis System of Refrigerated Containers. In Proceedings of the 2008 International Symposium on Knowledge Acquisition and Modeling, Wuhan, China, 20–21 December 2008; pp. 375–379.
2. Available online: https://www.artc.org.tw/tw/research/detail/39 (accessed on 6 June 2022).
3. Bakuła, M.; Pelc-Mieczkowska, R.; Walawski, M. Reliable and redundant RTK positioning for applications in hard observational conditions. *Artif. Satell.* **2021**, *47*, 23–33. [CrossRef]
4. Speth, T.; Kamann, A.; Brandmeier, T.; Jumar, U. Precise relative ego-positioning by stand-alone RTK-GPS. In Proceedings of the 13th Workshop Positioning Navigation Communications (WPNC), Bremen, Germany, 19–20 October 2016; pp. 1–6.
5. Langley, R.B. RTK GPS. *GPS World* **1998**, *9*, 70–76.
6. Milanés, V.; Naranjo, J.E.; González, C.; Alonso, J.; de Pedro, T. Autonomous vehicle based in cooperative GPS and inertial systems. *Robotica* **2008**, *26*, 627–633. [CrossRef]
7. Jo, K.; Chu, K.; Sunwoo, M. Interacting multiple model filter-based sensor fusion of GPS with in-vehicle sensors for real-time vehicle positioning. *IEEE Trans. Intell. Transp. Syst.* **2012**, *13*, 329–343.
8. Song, Y.; Li, Q.; Li, B.; Xu, J.; Zheng, N. Real-Time Dynamic Display on Spatial Data for Car Navigation Service. In Proceedings of the International Conference on Intelligent Transportation System Telecommunications Proceedings, Toronto, ON, Canada, 17–20 September 2006; pp. 94–97.
9. Bavle, H.; De La Puente, P.; How, J.P.; Campoy, P. VPS-SLAM: Visual planar semantic SLAM for aerial robotic systems. *IEEE Access* **2020**, *8*, 60704–60718. [CrossRef]
10. Tateno, K.; Tombari, F.; Laina, I.; Navab, N. CNN-SLAM: Real-time dense monocular SLAM with learned depth prediction. *IEEE CVPR* **2017**, 6243–6252.
11. Available online: https://www.artc.org.tw/tw/research/detail/32 (accessed on 11 September 2022).
12. Wang, L.; Zhang, Y.; Wang, J. Map-based localization method for autonomous vehicles using 3D-LIDAR. In Proceedings of the 20th IFAC World Congress, Toulouse, France, 14 July 2017; Volume 50, pp. 276–281.
13. Kohlbrecher, S.; Stryk, O.; Meyer, J.; Klingauf, U. A Flexible and Scalable SLAM System with Full 3D Motion Estimation. In Proceedings of the International Conference on Safety Security and Rescue Robotics (SSRR), Kyoto, Japan, 1–5 November 2011; pp. 155–160.
14. Pengpeng, S.S.; Zhao, X.; Xu, Z.; Wang, R.; Min, H. A 3D LiDAR data-based dedicated road boundary detection algorithm for autonomous vehicles. *IEEE Access* **2019**, *7*, 29623–29638.
15. Brock, O.; Trinkle, F.; Ramos, J. *Model-Based Vehicle Tracking for Autonomous Driving in Urban Environments*; MIT Press: Cambridge, MA, USA, 2009; Volume 1, pp. 175–182.
16. *ISO 11898-1:2016*; Road Vehicles—Controller Area Network (CAN)—Part 1: Data Link Layer and Physical Signaling. ISO: Palm Bay, FL, USA, 2015.
17. *ISO 11898-1:2016*; Road Vehicles—Controller Area Network (CAN)—Part 2: High-Speed Medium Access Unit. ISO: Palm Bay, FL, USA, 2016.
18. *ISO 11898-1:2006*; Road vehicles—Controller Area Network (CAN)—Part 3: Low-Speed, Fault-Tolerant, Medium-Dependent Interface. ISO: Palm Bay, FL, USA, 2016.
19. Miletiev, R.; Iontchev, E.; Yordanov, R. Multisensor precision positioning based on inertial and GNSS systems via 5G network. In Proceedings of the 45th International Spring Seminar on Electronics Technology (ISSE), Vienna, Austria, 11–15 May 2022; pp. 11–15.
20. Mostafavi, S.S.; Sorrentino, S.; Guldogan, M.B.; Fodor, G. Vehicular Positioning Using 5G Millimeter Wave and Sensor Fusion in Highway Scenarios. In Proceedings of the IEEE International Conference on Communications (ICC), Dublin, Ireland, 7–11 June 2020; pp. 7–11.
21. Eckermann, F.; Gorczak, P.; Wietfeld, C. Cooperative Validation of CAM Position Information Using C-V2X. In Proceedings of the IEEE 92nd Vehicular Technology Conference (VTC2020-Fall), Antwerp, Belgium, 25–28 May 2020; p. 15.
22. Hedges, C.; Perry, F. *Overview and Use of SAE j2735 Message Sets for Commercial Vehicles*; SAE Technical Paper: Troy, MI, USA, 2008; No. 2008-01-2650.
23. El-Shafie, A.; Hussain, A.; Eldin, A.E.N. ANFIS-Based Model for Real-time INS/GPS Data Fusion for Vehicular Navigation System. In Proceedings of the International Conference on Computer Technology and Development, Kota Kinabalu, Malaysia, 13–15 November 2009.
24. Raveena, C.; Sravya, R.; Kumar, R.; Chavan, A. Sensor Fusion Module Using IMU and GPS Sensors For Autonomous Car. In Proceedings of the IEEE INOCON, Virtual Conference, 6–9 July 2020.
25. Min, H.; Wu, X.; Cheng, C.; Zhao, X. Kinematic and dynamic vehicle model-assisted global positioning method for autonomous vehicles with low-cost GPS/camera/in-vehicle sensors. *Sensors* **2019**, *19*, 5430. [CrossRef] [PubMed]
26. Hashim, H.A.; Eltoukhy, A.E.E. Landmark and IMU data fusion: Systematic convergence geometric nonlinear observer for SLAM and velocity bias. *IEEE Trans. Intell. Transp. Syst.* **2020**, *23*, 3292–3301. [CrossRef]

applied sciences

MDPI

Article

Robust Autonomous Vehicle Computer-Vision-Based Localization in Challenging Environmental Conditions

Sergei Chuprov [1], Pavel Belyaev [2], Ruslan Gataullin [2], Leon Reznik [1,*], Evgenii Neverov [2] and Ilia Viksnin [2]

[1] Department of Computer Science and Global Cybersecurity Institute, Rochester Institute of Technology, Rochester, NY 14623, USA; sc1723@rit.edu

[2] Department of Computer Science, Saint Petersburg Electrotechnical University "LETI", St. Petersburg 197022, Russia; piubeliaev@etu.ru (P.B.); rusfiner@mail.ru (R.G.); wixnin@mail.ru (I.V.)

* Correspondence: leon.reznik@rit.edu

Abstract: In this paper, we present a novel autonomous vehicle (AV) localization design and its implementation, which we recommend to employ in challenging navigation conditions with a poor quality of the satellite navigation system signals and computer vision images. In the case when the GPS signal becomes unstable, other auxiliary navigation systems, such as computer-vision-based positioning, are employed for more accurate localization and mapping. However, the quality of data obtained from AV's sensors might be deteriorated by the extreme environmental conditions too, which infinitely leads to the decrease in navigation performance. To verify our computer-vision-based localization system design, we considered the Arctic region use case, which poses additional challenges for the AV's navigation and might employ artificial visual landmarks for improving the localization quality, which we used for the computer vision training. We further enhanced our data by applying affine transformations to increase its diversity. We selected YOLOv4 image detection architecture for our system design, as it demonstrated the highest performance in our experiments. For the computational platform, we employed a Nvidia Jetson AGX Xavier device, as it is well known and widely used in robotic and AV computer vision, as well as deep learning applications. Our empirical study showed that the proposed computer vision system that was further trained on the dataset enhanced by affine transformations became robust regarding image quality degradation caused by extreme environmental conditions. It was effectively able to detect and recognize images of artificial visual landmarks captured in the extreme Arctic region's conditions. The developed system can be integrated into vehicle navigation facilities to improve their effectiveness and efficiency and to prevent possible navigation performance deterioration.

Keywords: computer vision; robust vehicle localization; data quality; machine learning

Citation: Chuprov, S.; Belyaev, P.; Gataullin, R.; Reznik, L.; Neverov, E.; Viksnin, I. Robust Autonomous Vehicle Computer-Vision-Based Localization in Challenging Environmental Conditions. *Appl. Sci.* **2023**, *13*, 5735. https://doi.org/10.3390/app13095735

Academic Editors: Nadjim Horri, William Holderbaum and Fabrizio Giulietti

Received: 30 March 2023
Revised: 25 April 2023
Accepted: 28 April 2023
Published: 6 May 2023

1. Introduction

Navigation is one of the most important autonomous vehicle (AV) functions that performs positioning in the environment and decides on further mobility actions. The failure of accurate localization decreases AV effectiveness and efficiency in other operations, and this might result in safety accidents. Contemporary AVs primarily rely on global navigation satellite systems (GNSSs) to perform navigation and localization, such as GPS or GLONASS, which provide a global coverage and a high accuracy position. For example, contemporary GPS-based methods, which are the most commonly used GNSS systems, are capable of achieving a subcentimetric accuracy (e.g., RTK DGPS [1]). This level of accuracy, which could be further complemented by the measurements from other AV's on-board sensors, is sufficient for AVs to safely and efficiently navigate their surroundings. However, in some territories, e.g., dense urban or low populated areas with frequent extreme environmental conditions such as the Arctic region, the GPS signal can become unstable or even absent, which exacerbates navigational operations. In addition to communication obstructions and environmental conditions, other factors can lead to GPS failures, such as electronic interference that may be caused by a variety of sources, e.g., other electronic devices or power lines; signal jamming that can be intentionally activated by various parties or malicious attackers; and software and hardware malfunctions in GPS-related components that can cause failures. To mitigate these factors, multi-sensor perception may be employed for localization and mapping improvement. The data from multi-modality sensors, such as an on-board camera and an inertial measurement units (IMUs), might be fused in order to determine the AV's position in relation to the surrounding objects. However, these sensor devices might also fail or be maliciously attacked, which leads to data quality (DQ) degradation and jeopardizes the AV's security and safety [2,3]. Obviously, the localization accuracy in this case depends on the quality of data acquired from the sensors. The data from multiple sensors might be fused to enhance its accuracy and quality; however, the sensors need to be properly selected in this case [4]. This operation might employ various sensor selection techniques, such as genetic-algorithms-based ones [5].

In this paper, we consider computer-vision-based AV localization as a complimentary system to perform navigational operations. Since computer vision utilizes intelligent machine learning (ML) techniques, its operation involves initial procedures of the ML model training and validating. In most cases, such procedures commonly employ only high quality training data [6], which allows ML-based systems to achieve reliable performance on high quality testing images. However, our previous investigations demonstrated that using real life images with the degraded quality images dramatically affects the performance of ML-based computer vision classification [7,8]. To prevent ML performance degradation, multiple techniques might be leveraged, including the increase of training set diversity and the re-training of ML-based systems on samples of varying DQ [9].

The AV's navigation operation largely depends on how the sensor system is designed. One of the well-known sensors used for AV navigation is LiDAR. For example, Krishnamoorthi et al. [10] considered a navigation and localization system consisting of LiDARs and on-board cameras. The system performed path segmentation from point cloud data, which could be further used in computer vision and AV localization. AVs, as well as unmanned aerial vehicles, can rely on their own sensors to analyze the environment. The navigation and localization system allows AVs to make decisions about further actions without the operator. Arash et al. [11] discussed ways to improve the navigation system for the Internet of Drones using ML and deep learning algorithms. Their survey [12] refers to 37 articles devoted to assuring the system's reliability and fault tolerance. Data from sensors can be processed by computer vision algorithms to solve navigation and monitoring problems, as presented in [13]. The practical applications of autonomous robots may be diverse. Such robots are especially useful when the manual operator's control is associated with significant risks to their health; while our research studies the employment of AVs in Arctic region extreme conditions, other authors, for example, consider AV operation in underground mines [14] and in radioactive conditions [15].

The problem of designing a reliable navigation system is related to the task of ensuring the safety of the vehicle. Amiri et al. [12] deliberated on AV navigation security issues related to the system's reliability and fault tolerance. The task of controlling an autonomous vehicle was considered by Liu et al. [16]. The technology was focused on the evolution from vehicle state estimation and trajectory tracking control in AVs at the microscopic level to collaborative control in connected and automated vehicles at the macroscopic level. The task of developing a navigation system for electric vehicles is more specific. Various scenarios for driving electric vehicles were considered by Chen et al. [17]. The results presented in our work also contribute to improving the safety of using AVs in various applications, e.g., autonomous transport navigation in extreme weather conditions.

The issue of sensor fusion is well covered in the literature. For example, Xia et al. [18] discussed AV kinematic and dynamic synthesis for sideslip angle estimation based on a consensus Kalman filter. Gao et al. [19] studied slip angle estimation by considering a signal measurement characteristic.

In our work, we investigated methods of improving navigation performance in extreme environmental and poor GPS signal communication conditions. We designed and developed a complimentary AV computer-vision-based localization system to make it robust in harsh environmental conditions that cause the degradation of images obtained from the on-board camera. This computer vision recognition system should integrate with the conventional GPS or even replace it when it becomes unavailable. In particular, we investigated and verified our design on an Arctic region use case, as, nowadays, many authorities seek to automate their transportation operations in this region, and AVs are deemed to become a prominent solution in this field [20]. An example of the resource excavation facilities cluster in the Arctic region is demonstrated in Figure 1. Highly limited natural visual landmarks and changing weather conditions in this region make navigation a real challenge for the AVs. In our particular use case, we considered specialized artificial visual tags allocated over the AV operation area. These tags contain GPS coordinates encoded via AprilTag [21], which is needed to be detected and properly recognized by AVs in order to perform localization in case of GPS unavailability. To implement our design, we employed Nvidia Jetson AGX Xavier as the computing device for the developed ML-based system verification. To enhance our ML-based model robustness toward DQ degradation, we generated images of the artificial visual tags affected by the various lightning and harsh weather conditions, such as blizzards and heavy rain. In addition, we employed several types of affine transformations to these images to diversify our training set. We utilized the produced set of images to re-train our ML-based image classification model. Then, we evaluated the performance of the developed ML-based model to detect and recognize the artificial visual tags from the varied distances.

Figure 1. An example of the fossil fuels excavation facilities cluster located in the Arctic region. As one can see, the landscape is covered with snow and ice and might be characterized as mostly flat and uniform, with very few distinctive features. The extreme weather conditions and low levels of precipitation in the region result in a lack of any vegetation forms that could serve as visual markers. The absence of large geological formations such as mountains or canyons also contributes to the lack of visual landmarks.

This paper's major contribution is our computer-vision-based localization system design that might be employed by AVs in order to improve their localization and mapping accuracy in case of GPS signal unavailability. The advantage of the proposed system is its robustness regarding GPS signal instability and image quality degradation caused by extreme environmental conditions. Our verification use case demonstrated that the developed solution is able to effectively detect and recognize images of artificial visual landmarks in the extreme conditions of Arctic region.

2. Background

In our study, we considered the Arctic region as a great use case that poses both challenges of input DQ degradation and GPS signal instability. This region experiences a long winter, lasting up to 10 months, which creates severe problems for AV operation. The air temperature can drop to −60 degrees Celsius, and the strong cold winds can reach up to 30 m/s. The increased frequency of snowfall and blizzards, as well as the small possibility of manual search and rescue in the case of emergency situations, exacerbate the challenges. Such climatic conditions significantly affect AV technical component operation. The excess resource consumption, fast battery discharge, and failures of electronic devices that are not adapted for low temperatures are some of the issues faced. The remote and inaccessible nature can limit the GPS signal's availability, thus making navigation challenging for AVs. The extreme weather conditions, low visibility, and harsh terrain can further compound the challenge of GPS signal loss [22], thus potentially leaving AVs stranded and unable to navigate safely.

To address the challenges of AV navigation in such arduous areas as the Arctic region, AVs may use alternative navigation and localization systems in addition to GPS, e.g., IMU that relies on inertial sensors to track changes in position and movement direction or a computer-vision-based system, which employs on-board cameras to identify and track the position based on visual landmarks. These systems can complement GPS-based localization, thereby providing redundancy and ensuring that AVs can continue to navigate safely and accurately, even in the event of GPS signal loss. Some developments have been introduced to realize the AV's computer-vision-based localization system. Marinho et al. [23] presented an approach to improve the mobile robot navigation via extracting the information from the topological map. The approach was designed mainly for outdoor areas; however, it requires a sufficient number of visual reference points to navigate the robot effectively.

Data-driven ML methods are also widely used to approach AV navigation [24,25]. Here, learning approaches relying on expert reference [26] or based on reinforcement learning [27] may be employed. These learning approaches make it possible to improve AV navigation performance when the navigation experience is increasing over the operation time. This might happen in the conditions of a dynamic environment, which is typical for operating in the real world. However, ML-based navigation approaches may poorly generalize and easily "forget" previous knowledge that is displaced by recent experience. This inevitably leads to AV overall navigation performance degradation. The analysis of existing AV computer-vision-based navigation and localization approaches demonstrates that these methods usually require a substantial number of visual landmarks in order to maintain the level of navigation performance that is acceptable by the user and application requirements. Most of the presented approaches are primarily acceptable for indoor navigation, as its environment can often be characterized by a large number of informative reference points. Outdoor navigation is much more challenging, as its environment usually provides insufficient visual information for accurate positioning and localization, especially in specific areas such as the Arctic region. In these conditions, computer-vision-based navigation and localization methods might not be effective. To approach this challenge, additional artificial visual landmarks may be introduced to the AV's operating environment for more accurate localization. The AprilTag is one of the known examples of such artificial landmarks [21]. These tags may carry various information that might be used by AVs to perform localization and mapping on the terrain; for example, they might provide the current GPS coordinates where these tags are allocated. However, artificial visual landmarks are still required to be properly recognized by the AV's computer-vision-based system in extreme environmental conditions. Petrov [28] proposed a system for navigating AVs in a winter terrain by extracting the vector of the visual navigation marker. This approach might be effectively employed when the visual conditions are more or less stable, without significant environmental condition degradation. Ali and Shahhood [29] developed a system for navigation in a 3D space, including open terrain areas. The disadvantage of the solution is that it also does not consider the AV's operating conditions, which makes it unsuitable for complex environments.

In this paper, we focus on rather complex, dynamic, and weather-dependent environmental conditions that affect the quality of the obtained visual information in various ways. Specifically, we developed a computer-vision-based AV localization system and enhanced its robustness regarding the on-board camera image quality degradation caused by harsh environmental conditions. One of the significant challenges we addressed is the low informative features visibility caused by the unstable environment. The extreme weather can cause camera lenses to freeze or fog, which can further degrade the image quality and affect the computer-vision-based localization performance. Additionally, the reflective properties of snow and ice can create glare and distortions in the images captured by the system, thereby further degrading the DQ. We present a solution capable of adapting to those complex weather conditions by employing an ML model for a computer-vision-based localization system. As the visual information extractor, we employed a convolutional neural network (CNN), as it demonstrated the highest performance in our empirical investigation. To enhance the robustness of our ML-based model, we produced images affected by various complex environmental conditions and employed them to re-train the model. To further assure the ML model performance on the low quality data, we expanded the training dataset by modifying it using affine transformations. Below, we describe the architecture and major components of the developed AV computer-vision-based localization system and how it might be integrated in the overall navigation process.

3. Integration of the Computer-Vision-Based Localization into the AV's Navigation Stack

The generic AV navigation process can be described as follows: if GPS is available, it is employed as the primary AV navigation system to perform route planning and proceeding to the target destination. By default, computer-vision-based localization is considered as the auxiliary navigation system (ANS). Depending on the AV's composition, it might also employ other ANSs, such as IMU or radar. ANSs might be used in combination with GPS or with each other (which is a typical practice) or individually if one or more navigation components fail [30]. ANSs are routinely used to correct the GPS navigation error and for precise positioning and obstacles detection. Based on the available data from the navigation and localization system components, the AV makes decisions on further driving actions. Figure 2 schematically represents the routine process of decision making depending on the navigation and localization components' availability and their performance. For the sake of demonstration, an IMU was used here as another ANS that the AV was equipped with. Initially, the availability of the GPS signal was verified. With a successful response from the GPS, the system state switches to the IMU availability verification. The main problem with relying solely on the IMU for the navigation and localization operations is the complex environmental conditions in which the AV has to operate. When primarily relying on the IMU, the navigation error is rapidly accumulated [31], which ultimately leads to the AV's disorientation. In this case, such an error can be corrected by the GPS coordinates; however, the GPS system might be unavailable for a considerable time. In this case, localization and further navigation might be performed based on computer vision in combination with other available ANSs.

The problem of time synchronization for a IMU/GNSS is also challenging for reliable navigation, and it is addressed in the following way: the AV's sensors operate in series, not in parallel. In the sequential operation of sensors, the data is transmitted from one system to another, thus receiving additional parameters for work. The information first passes through the IMU, then through the GNSS. The use of AprilTags makes it possible to verify inaccuracies in the IMU and correct their impact on the navigation performance.

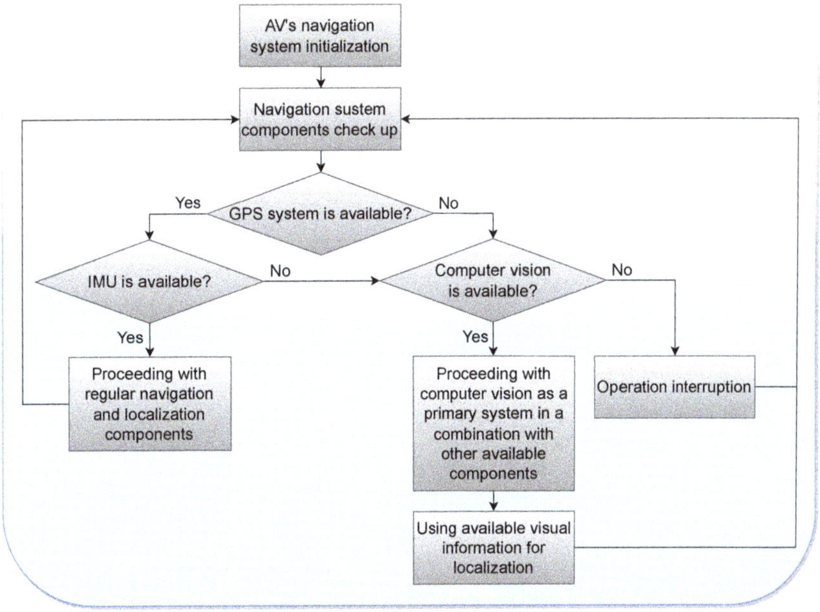

Figure 2. Schematic representation of how the proposed computer-vision-based localization might be integrated into the AV navigation system process—our design example. The process might be represented as an iterative algorithm that incorporates verification of the navigation and localization facilities the AV is equipped with, and it employs them if available.

The computer-vision-based localization system should be capable of detecting and retrieving the information from the artificial visual landmark in the current location. The architecture of the proposed computer-vision-based localization incorporates the following basic modules:

- A detection module;
- An approach module;
- A retrieval module.

The architecture of these modules and the connections between them are represented in Figure 3. The detection module operates based on the implemented and trained ML model, the approach module is implemented based on the Nvidia Isaac "Followme" module [32], and the retrieval module utilizes the AprilTag recognition function [21]. Each module performs unique functions within the overall AV navigation system. The detection module is responsible for detecting the artificial visual landmark on which the AprilTag with the visually encoded GPS coordinates is printed. The approach module allows for pointing and holding the detected landmark in the field of the AV's on-board camera view while the AV is traveling closer to properly recognize the tag. The retrieval module is responsible for retrieving the information on the current GPS coordinates from the visual tag, and it performs the AV's localization adjustment according to these coordinates. In the case of two or more landmarks appearing in the on-board camera's field of view simultaneously, the one that is detected with a higher performance is selected for proceeding toward it. If multiple visual landmarks have a similar detection performance, then their bounding boxes are compared, and the decision is made based on the higher intersection over union (IoU) value. If the IoU value is also similar, the visual landmark is selected randomly. In our empirical study, we assumed that the visual landmarks were allocated with the distance of 20 m between each other, which makes multiple detection almost impossible. The detection module is the initial component that informs the other two modules if there are any artificial visual landmarks detected. The other two modules operate in parallel in order to reduce the time for the information retrieval. Such a combination of actions makes it possible to minimize the time required to perform localization and mapping. Instead of approaching within a certain distance to the tag to properly recognize it, the retrieval process is constantly performed throughout the AV's approach.

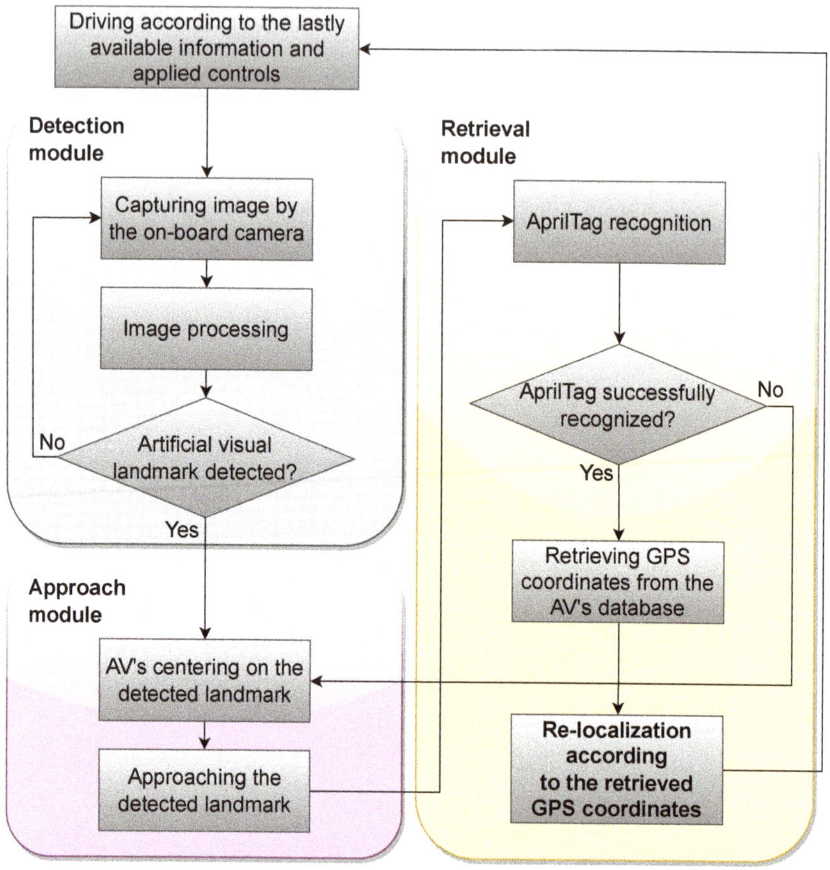

Figure 3. Schematic representation of the proposed computer-vision-based system architecture and its operation—our design example. The proposed architecture is composed of three modules: detection, approach, and retrieval. The detection module is responsible for analyzing the visual data obtained from the on-board camera and detecting artificial landmarks. The approach module is responsible for keeping the visual landmark in the AV's on-board camera field of view by adjusting the movement while approaching. The retrieval module is responsible for recognizing the AprilTag and retrieving the information on the GPS coordinates.

The detection module utilized the YOLOv4 ML object detection model [33] that was pre-trained on the MS COCO dataset [34]. This is a reliable version of a well-known architecture that outperforms other popular ML models in terms of both performance and efficiency [33].

The approach module operation was based on the software provided by the Nvidia Isaac "Followme" software development kit [32]. This software allows for AVs to perform the recognition of the detected object and correct the movement actions while approaching it. The original programming logic for the "Followme" component implies that the AV should be close enough to the AprilTag to recognize it. When considering extreme environmental conditions, this might be impractical, as it will deteriorate the AV's operation effectiveness and efficiency. To minimize the approach time in order to improve the overall navigation and localization system efficiency, we introduced our modifications into the "Followme" source code. Instead of initially detecting the AprilTag and using it as a reference point for the approach, we switched to the visual artificial landmark as the initial object of interest for detection. Harsh weather conditions significantly deteriorate the AprilTag's visibility and detectability, especially from a large distances. Artificial visual landmarks themselves are rather large and more visible in such conditions, which makes them good candidates to be utilized as reference points. Hence, the AV can detect these artificial visual landmarks from a longer distance and spend less resources for this detection, which makes navigation more efficient.

Once the visual landmark is detected, it is approached until the information is retrieved successfully from the AprilTag, which is printed on this landmark. The successful information retrieval means that the current location is extracted from the AprilTag. The AV stores the ID of the AprilTag, which is related to the GPS coordinates represented in the format of *DD.DDDDD, DD.DDDDD*, where *DD* stands for the longitude and latitude values. An example of the encoded data is 40.17312, 42.84121. When the AprilTag is recognized, the corresponding coordinates are retrieved from the AV's database. After retrieving the GPS coordinates, the current coordinates and path of the AV is updated, and all passed landmarks are recorded. After re-localization according to the retrieved coordinates, the system responsible for the AV's routing updates the route to the target destination by considering the current location, and the AV proceeds to the target destination according to the process displayed in Figure 3.

4. Employed Data and Transformations

Despite the numerous publicly available image datasets of various sizes, categories, and designs for different knowledge domains, those datasets are usually composed of high quality data. For our considered specific Arctic region use case, to produce the ML model that is robust regarding possible DQ degradation, we also needed images affected by the extreme environmental conditions. We found the available public datasets not suitable for our specific use case, so we developed the visual artificial landmark model with the printed AprilTag by employing the Nvidia Isaac Sim application [32]. Alongside this application, we also utilized the Unity3D software package [35] to generate images of the visual tags in various extreme environment conditions that were captured from various distances and angles. The Nvidia Isaac Sim made it possible to configure numerous parameters, such as object location, illumination, overlapping with other objects, etc. With these settings, it was possible to produce a substantial collection of the unique data samples to train ML models. An example of the designed visual artificial landmark is demonstrated in Figure 4. Considering the features of the Arctic region terrain, we produced a substantial set of images capturing the artificial visual landmark in various complex environmental conditions, such as snow drifts, blizzards, insufficient illumination, etc. In Figure 4, we represent a few instances of the produced images. The constructed artificial visual landmark consisted of two parts: the holder and the panel for the AprilTag. The height of the holder was 1 m, the size of the panel was 40 by 40 cm, and the size of the printed AprilTag was 30 by 30 cm.

From the initial $\approx$4000 original good quality images, we generated around 20,000 samples of various qualities that were captured from various angles and distances. To further increase the image diversity, we also employed affine transformation techniques. Random affine transformations are geometrical vector modifications that are introduced to existing images. Such transformation methods as rotation, scale modification, shift, and reflection might be applied, which make is possible to produce a visually distinct image without discarding the essential details and structure of the image. Affine transformations are widely used in computer vision applications to expand the diversity of the samples in order to train more generalizable models. This approach is especially useful in training deep neural networks that require a large amount of diverse data to achieve acceptable performance. In our work, we employed such affine transformations as rotation, scale modification, shift, and reflection alongside the changing environment to produce images that would be infeasible to obtain in real conditions. By employing the affine transformations, we expanded our data by 3126 additional samples. Then, we manually labeled the produced data samples to employ them for further training procedures. The produced data was split into 70% training, 15% validation, and 15% test batches.

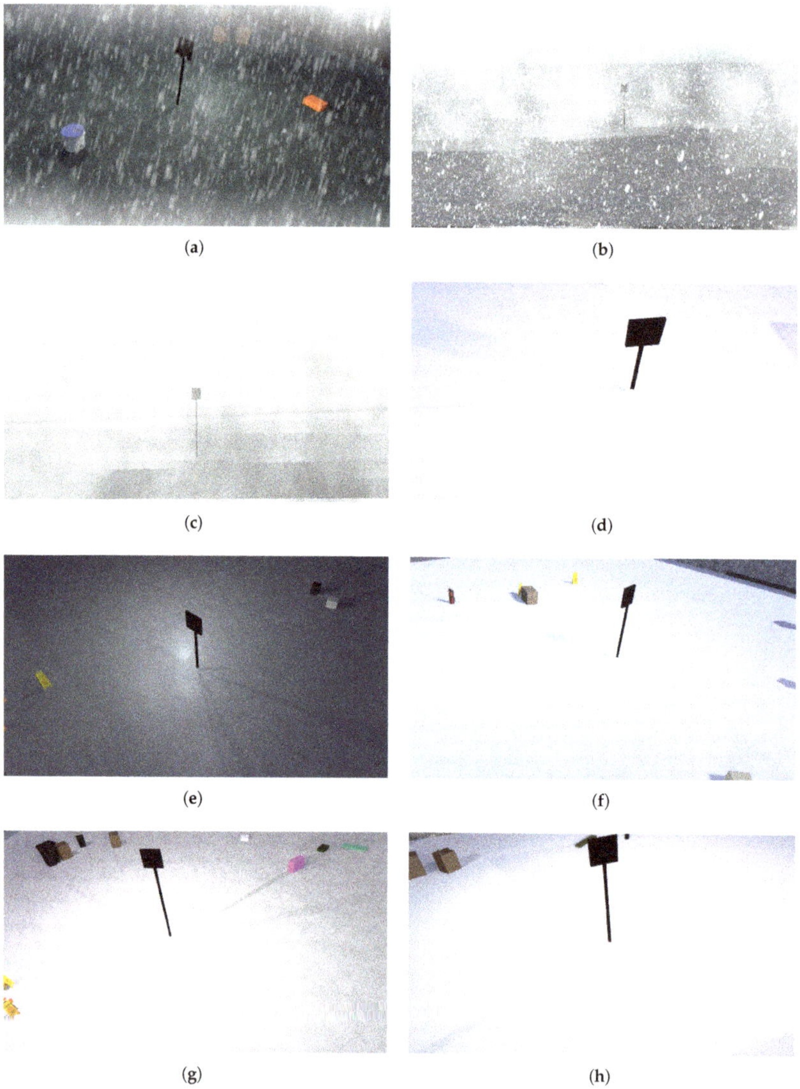

Figure 4. Artificial visual landmark image samples generated in Unity3D software package. The figure demonstrates multiple image examples of varying quality under diverse environmental conditions: (**a**) heavy rain; (**b**) heavy snowfall; (**c**) blizzard; (**d–h**) captured from varied distances, angles, and in various illumination conditions.

5. Empirical Study

The produced labels were represented by the class ID and point coordinates of the visual landmark in the $[x, y, x + w, y + h]$ format, where w and h are the width and height of the bounding box, respectively, x is the initial coordinate of the bounding box on the horizontal axis, and y is the initial coordinate of the bounding box on the vertical axis. As the employed ML model was initially pre-trained on the MS COCO dataset [34], we only re-trained the fully connected layer, which used the Softmax activation function. As a computing device for training, we employed the Nvidia Jetson AGX Xavier [36]. This device is known for its low power consumption of only 10W, which allows it to be employed in autonomous mobile robotic devices. The device makes it possible to utilize CUDA cores for real-time calculations. The employed device possesses the following technical characteristics:

- AI performance: 32 TOPS (trillion operations per second);
- GPU: Nvidia Volta architecture with 512 Nvidia CUDA cores and 64 tensor cores;
- CPU: 8-core 64-bit Nvidia Carmel processor with Armv8.2 architecture, 8 MB L2, and 4 MB L3;
- Deep learning accelerator: 2 NVDLA Engines;
- Computer vision accelerator: 2 PVA accelerators.

For the ML model performance evaluation metric, we employed Average Precision (AP), which incorporates precision and recall and is usually employed for evaluating ML performance [37]. AP can be characterized as the area under the precision–recall curve, which shows how the accuracy and completeness of the algorithm change at various classification probability thresholds.

The IoU is a metric that is usually employed to evaluate the object detection algorithm's performance. Typical threshold values indicate correct detection when the IoU is more than 50%. The IoU value is calculated as the relationship between the intersection areas of the predicted bounding box and the actual ground truth bounding box. Since we are evaluating accuracy for the detection and classification algorithm, we also employed the mean AP (mAP) metric, which can be calculated by Equation (1):

$$mAP = \sum_{q=1}^{Q} AP(q), \tag{1}$$

where Q is the number of classes; q is the class; and $AP(q)$ is the AP value for the q's class.

To select an ML model architecture for our use case, we performed an ablation study where we tested how various ML model architectures performed on a testing set of the produced data samples. We tested ML architectures based on a CNN (YOLOv4); Haar cascades; and based on ML with a histogram of directional gradients (HOG) used as a descriptor for a support vector machine (SVM). We pre-trained these models on the same cohort of the produced data and then tested them over the prepared testing set. We employed ML accuracy as the evaluation metric for this ablation study. The ML performance results are presented in Table 1. According to the results, the YOLOv4 object detector that is based on CNN architecture demonstrated the highest accuracy in the studied conditions. Based on these results, we selected the YOLOv4 as an image processor for the designed AV computer-vision-based localization system. Furthermore, the choice of the YOLOv4 was justified by the fact that, at the time of the research, this architecture was recommended for achieving robust real-time computation performance using the employed Nvidia Jetson AGX Xavier computational device.

Table 1. Results of the ablation study for evaluating the performance of three distinct ML architectures over the data samples produced with Unity3D. The testing data includes image samples affected by the extreme weather conditions and captured from various angles.

	YOLOv4, %	**Haar Cascades, %**	**HOG + SVM, %**
Accuracy	78.75	42.39	49.17

Initially, we re-trained our model for 2000 epochs on the images without affine transformations and tested it on the produced images of the artificial visual landmark captured from the varied distances (from 5 to 20 m) and affected by various environmental conditions. The first column in Table 2 represents the performance results achieved by this model. As one can see, the model poorly detected the target objects in the images and provided unacceptable performance. We then continued to further re-train the model, but, in this case, it was only trained on the images produced with affine transformations. After further training, we test the model again over the similar images employed in the former case. These testing results are represented in the second column of Table 2. The model was able to achieve substantial performance improvement with the 83.8% for the mAP and 78.73% for the average IoU.

Table 2. Results comparing the effect of re-training the ML model on the data modified with affine transformations. The affine transformations include such geometrical vector modifications as rotation, scale modification, shift, and reflection.

	Model without Affine Transformations, %	**Model with Affine Transformations, %**
mAP	25.8	83.8
Average IoU	46.31	78.73

To evaluate the performance of the produced model in various weather conditions, we formed four distinct testing data cohorts based on the distance to the artificial visual landmarks in the images. In our experiments, we employed images captured from the distance of 10, 20, 30, and 40 m. Figure 5 represents the performance demonstrated by the model on these data cohorts. For the sake of comparison, we also presented the results for the model trained on the data without affine transformations. From the results, one can see the employment of various augmentation techniques over the training data, wherein the quality was significantly deteriorated by extreme environmental conditions, which made it possible to substantially enhance the ML performance in the case of testing sample quality degradation. However, as the distance between the visual landmark and the camera increased, the tag recognition performance dropped. A distance of 30 m from the tag contributed to a 60% ML performance decline. Furthermore, when starting from a distance of 40 m away from the tag, the model completely stopped recognizing the tag.

In addition to extreme weather conditions, we also considered changing illumination as another quality degradation factor. In the Arctic region, daylight is highly limited in most times of the year, which poses an additional challenge for AV computer-vision-based localization systems. To evaluate how our model was robust regarding the changes in illumination, we formed a number of testing data cohorts. They were generally divided into two categories: daylight, which incorporates images captured in the lightning conditions of 155 LUX, and nighttime, with images captured in the extremely limited lightning conditions of 5 LUX. Each of these categories included three sub-categories of images captured from the varied distances of 0.5, 2, and 5 m. Table 3 demonstrates the results of the ML model performance over the employed images cohorts.

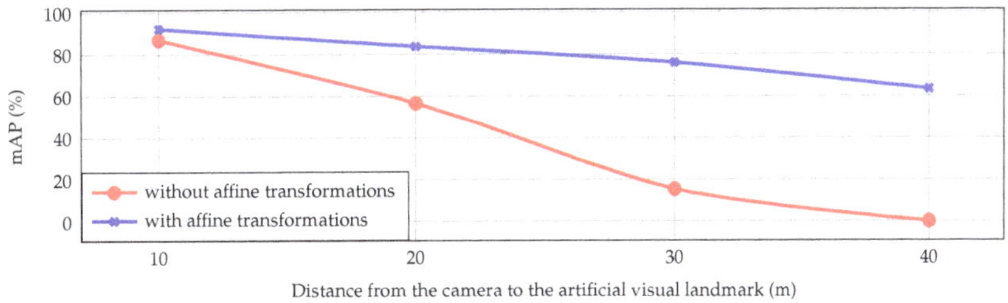

Figure 5. The tag recognition performance by the ML model trained with and without affine transformations tested on images affected by various extreme weather conditions and captured from different distances. The results are presented for the model trained only on the produced images without re-training on data modified by affine transformations (orange plot) and for the models further re-trained on the data modified by the affine transformations (blue plot).

As one can see, in the case of the 0.5 m distance, the ML model demonstrated high performance in both daylight and nighttime conditions, with an accuracy of 97% and 94%, respectively. However, increasing the distance from the camera to 2 m affected the performance significantly, and even critically, in the nighttime conditions, while, in case of the daytime, the performance was still over 80%. For the nighttime images, it dropped below 50%, which substantially exacerbated the AV's computer-vision-based localization. Further increasing the distance to 5 m completely prevented the ML model from selecting the correct operation in the nighttime. However, for the images captured in the daytime, even though the demonstrated performance was impractical, it was still above the 50% boundary.

Table 3. Comparison of the performance demonstrated by the ML model in recognizing AprilTags captured under different illumination conditions and from various distances. The distance varied from 0.5 to 5 m, and the illumination was 155 LUX for the daytime and 5 LUX for the nighttime.

Distance, m	Accuracy (Daylight), %	Accuracy (Nighttime), %
0.5	97	94
2	83	43
5	54	0

6. Limitations and Potential Future Directions

Both our empirical study and the developed solution have several limitations. First, as the use case scenario for our solution, we considered the AV's operation in a structured environment with natural or artificial visual landmarks allocated in advance over the operation area. An AV has to possess prior knowledge on the environment, which might be realized with an off-line map and relationships between the known landmarks and GPS coordinates stored in the AV's memory. Second, instead of producing and employing real data for training the ML model, we utilized the well-known Unity3D software package to generate artificial samples designed to represent affects by extreme weather conditions. This allowed us to significantly extend the size and the diversity of the dataset in order to train the model to be more robust in varying conditions. In addition, in our research, we concentrated only on improving the robustness of the computer-vision-based localization system against the low quality data affected by the harsh weather conditions. We did not consider additional influences that these conditions might have had on the AV's navigation and localization quality, such as failures of the other AV systems and elements responsible for driving control.

Our approach improved the value and contributions developed in our work. As we showed in our experiments, pre-trained ML models further trained on augmented low quality data become more robust regarding DQ degradation. Since Unity provides a reliable simulation environment for designing and testing AI-based solutions for industrial robotic systems before their actual implementation and deployment, the artificial samples we produced can be considered as robust and close-to-real. Similar training operations might be performed with real images in the case of implementing our solution in practice. In our development, we concentrated only on enhancing the robustness of computer-vision-based localization against the DQ degradation caused by extreme weather conditions, which we effectively demonstrated in our results. Despite our value for the detrimental effects of other processes and factors that may influence the navigation, localization, and driving operations, they are out of our paper's scope.

As the potential future research directions, we can propose research on the optimization of the landmarks' design, their spatial distribution, and the sensor fusion. The optimal size and visual characteristics of the artificial visual landmarks play important roles in their early detection and optimal approaching by the AVs. Comprehensive research resulting in recommendations on the shape, appearance, and other visual characteristics for improving artificial visual landmark detection in extreme environmental conditions by the computer-vision-based systems would help to enhance the navigation and localization quality and performance. Another direction is the landmarks' optimal allocations over the AV's operation area. This is a complex problem that requires the consideration of numerous factors, such as the AV's technical characteristics, landscape features, dynamic environment conditions, operation area dimensions, starting and target destination locations, etc. One more direction is sensor fusion optimization aimed at selecting appropriate sensors and integrating the data acquired from them in order to improve the overall DQ. Progress in this direction would make it possible to effectively produce data whose quality satisfies the user and application requirements. This would allow for enhancing the quality of intelligent navigation.

7. Conclusions

AV navigation is a challenging problem in the conditions of GNSS signal instability or absence, which might often be the case in certain regions as diverse as dense architecture urban areas or low populated areas with extreme environmental conditions. To address this problem, data from other AV sensor systems are commonly employed to maintain the required localization and mapping accuracy. However, the quality of measurements produced by those sensors might be affected by various factors; for example, the quality of images from the on-board camera might be degraded by harsh weather conditions. This leads to the AV's navigation performance deterioration, decreases the AV's effectiveness and efficiency, and might jeopardize the safety of AVs and surrounding objects. In this paper, we proposed an enhanced design of a computer-vision-based localization system that can be integrated into AV navigation facilities to improve their localization and mapping accuracy in extreme environmental conditions where GPS signals are unstable or absent. Based on our empirical study results, the following conclusions can be formulated. A CNN-based ML model architecture (in particular, YOLOv4), demonstrated performance that was higher than the two other investigated architectures and can be employed as an image processor for AV localization applications. However, as our research showed, even the model trained on a comprehensive set of images representing affects by various extreme environmental conditions, was insufficient to achieve performance acceptable for effective and efficient AV operation. Employing data augmentation techniques such as affine transformations and further training the model on this data significantly improved the model robustness regarding testing image quality degradation. Indeed, the further trained model was able to demonstrate a performance higher than 80% for the degraded images captured from the distance of up to 20 m. In the changing illumination conditions, the model was able to achieve high performance in the distance of up to 2 m in daylight. Unfortunately, in the nighttime, when the distance increased above 0.5 m, the model was hardly able to properly recognize the tag.

Author Contributions: Conceptualization, L.R. and I.V.; methodology, S.C.; software, P.B.; validation, P.B.; formal analysis, R.G.; investigation, P.B.; resources, E.N.; data curation, P.B. and E.N.; writing—original draft preparation, S.C.; writing—review and editing, R.G. and L.R.; visualization, S.C.; project administration, L.R., R.G., E.N. and I.V.; funding acquisition, L.R. and I.V. All authors have read and agreed to the published version of the manuscript.

Funding: The work of Sergei Chuprov and Leon Reznik was supported by the United States Military Academy (USMA), Grant Number W911NF-20-1-0337. The work of Pavel Belyaev, Ruslan Gataullin, Evgenii Neverov, and Ilia Viksnin was supported by the Ministry of Science and Higher Education of the Russian Federation, known as "Goszadanie" №075-01024-21-02 of 29.09.2021 (project FSEE-2021-0014).

Institutional Review Board Statement: Not applicable.

Informed Consent Statement: Not applicable.

Data Availability Statement: The data presented in this study is available on request from the authors: I.V. and P.B. The data is not publicly available due to it is a commercial company's intellectual property.

Conflicts of Interest: The authors declare no conflict of interest.

References

1. Stateczny, A.; Specht, C.; Specht, M.; Brčić, D.; Jugović, A.; Widźgowski, S.; Wiśniewska, M.; Lewicka, O. Study on the positioning accuracy of GNSS/INS systems supported by DGPS and RTK receivers for hydrographic surveys. *Energies* **2021**, *14*, 7413. [CrossRef]
2. Chuprov, S.; Viksnin, I.; Kim, I.; Melnikov, T.; Reznik, L.; Khokhlov, I. Improving Knowledge Based Detection of Soft Attacks Against Autonomous Vehicles with Reputation, Trust and Data Quality Service Models. In Proceedings of the 2021 IEEE International Conference on Smart Data Services (SMDS), Chicago, IL, USA, 5–10 September 2021; pp. 115–120.
3. Chuprov, S.; Viksnin, I.; Kim, I.; Reznikand, L.; Khokhlov, I. Reputation and trust models with data quality metrics for improving autonomous vehicles traffic security and safety. In Proceedings of the 2020 IEEE Systems Security Symposium (SSS), Crystal City, VA, USA, 1 July 2020–1 August 2020; pp. 1–8.
4. Khokhlov, I.; Chuprov, S.; Reznik, L. Integrating Security with Accuracy Evaluation in Sensors Fusion. In Proceedings of the 2022 IEEE Sensors, Dallas, TX, USA, 30 October 2022–2 November 2022; pp. 1–4.
5. Chuprov, S.; Reznik, L.; Khokhlov, I.; Manghi, K. Multi-Modal Sensor Selection with Genetic Algorithms. In Proceedings of the 2022 IEEE Sensors, Dallas, TX, USA, 30 October 2022–2 November 2022; pp. 1–4.
6. Reznik, L. *Intelligent Security Systems: How Artificial Intelligence, Machine Learning and Data Science Work for and against Computer Security*; IEEE Press-John Wiley & Sons: Hoboken, NJ, USA, 2022.
7. Chuprov, S.; Reznik, L.; Obeid, A.; Shetty, S. How Degrading Network Conditions Influence Machine Learning End Systems Performance? In Proceedings of the IEEE INFOCOM 2022-IEEE Conference on Computer Communications Workshops (INFOCOM WKSHPS), New York, NY, USA, 2–5 May 2022; pp. 1–6.
8. Chuprov, S.; Satam, A.N.; Reznik, L. Are ML Image Classifiers Robust to Medical Image Quality Degradation? In Proceedings of the 2022 IEEE Western New York Image and Signal Processing Workshop (WNYISPW), Rochester, NY, USA, 4 November 2022; pp. 1–4.
9. Chuprov, S.; Khokhlov, I.; Reznik, L.; Shetty, S. Influence of Transfer Learning on Machine Learning Systems Robustness to Data Quality Degradation. In Proceedings of the 2022 International Joint Conference on Neural Networks (IJCNN), Padua, Italy, 18–23 July 2022; pp. 1–8.
10. Rajathi, K.; Gomathi, N.; Mahdal, M.; Guras, R. Path Segmentation from Point Cloud Data for Autonomous Navigation. *Appl. Sci.* **2023**, *13*, 3977. [CrossRef]
11. Heidari, A.; Jafari Navimipour, N.; Unal, M.; Zhang, G. Machine Learning Applications in Internet-of-Drones: Systematic Review, Recent Deployments, and Open Issues. *ACM Comput. Surv.* **2023**, *55*, 1–45. [CrossRef]
12. Amiri, Z.; Heidari, A.; Navimipour, N.; Unal, M. Resilient and dependability management in distributed environments: a systematic and comprehensive literature review. *Clust. Comput.* **2022**, *26*, 1565–1600. [CrossRef]
13. Kim, I.; Matos-Carvalho, J.P.; Viksnin, I.; Simas, T.; Correia, S.D. Particle Swarm Optimization Embedded in UAV as a Method of Territory-Monitoring Efficiency Improvement. *Symmetry* **2022**, *14*, 1080. [CrossRef]
14. Kim, H.; Choi, Y. Development of Autonomous Driving Patrol Robot for Improving Underground Mine Safety. *Appl. Sci.* **2023**, *13*, 3717. [CrossRef]
15. Mascarich, F.; Kulkarni, M.; De Petris, P.; Wilson, T.; Alexis, K. Autonomous mapping and spectroscopic analysis of distributed radiation fields using aerial robots. *Auton. Robots* **2022**, *47*, 139–160. [CrossRef]
16. Liu, W.; Hua, M.; Deng, Z.; Huang, Y.; Hu, C.; Song, S.; Gao, L.; Liu, C.; Xiong, L.; Xia, X. A systematic survey of control techniques and applications: From autonomous vehicles to connected and automated vehicles. *arXiv* **2023**, arXiv:2303.05665.

17. Chen, G.; Hua, M.; Liu, W.; Wang, J.; Song, S.; Liu, C. Planning and Tracking Control of Full Drive-by-Wire Electric Vehicles in Unstructured Scenario. *arXiv* **2023**, arXiv:2301.02753.
18. Xia, X.; Hashemi, E.; Xiong, L.; Khajepour, A. Autonomous Vehicle Kinematics and Dynamics Synthesis for Sideslip Angle Estimation Based on Consensus Kalman Filter. *IEEE Trans. Control Syst. Technol.* **2022**, *31*, 179–192. [CrossRef]
19. Gao, L.; Xiong, L.; Xia, X.; Lu, Y.; Yu, Z.; Khajepour, A. Improved vehicle localization using on-board sensors and vehicle lateral velocity. *IEEE Sens. J.* **2022**, *22*, 6818–6831. [CrossRef]
20. Ryghaug, M.; Haugland, B.T.; Søraa, R.A.; Skjølsvold, T.M. Testing Emergent Technologies in the Arctic: How Attention to Place Contributes to Visions of Autonomous Vehicles. *Sci. Technol. Stud.* **2022**, *35*, 4–21. [CrossRef]
21. Olson, E. AprilTag: A robust and flexible visual fiducial system. In Proceedings of the 2011 IEEE International Conference on Robotics and Automation, Shanghai, China, 9–13 May 2011; pp. 3400–3407.
22. Yastrebova, A.; Höyhtyä, M.; Boumard, S.; Lohan, E.S.; Ometov, A. Positioning in the Arctic region: State-of-the-art and future perspectives. *IEEE Access* **2021**, *9*, 53964–53978. [CrossRef]
23. Marinho, L.B.; Rebouças Filho, P.P.; Almeida, J.S.; Souza, J.W.M.; Souza Junior, A.H.; de Albuquerque, V.H.C. A novel mobile robot localization approach based on classification with rejection option using computer vision. *Comput. Electr. Eng.* **2018**, *68*, 26–43. [CrossRef]
24. Zeng, J.; Ju, R.; Qin, L.; Hu, Y.; Yin, Q.; Hu, C. Navigation in Unknown Dynamic Environments Based on Deep Reinforcement Learning. *Sensors* **2019**, *19*, 3837. [CrossRef] [PubMed]
25. Wang, Y.; He, H.; Sun, C. Learning to Navigate Through Complex Dynamic Environment With Modular Deep Reinforcement Learning. *IEEE Trans. Games* **2018**, *10*, 400–412. [CrossRef]
26. Kahn, G.; Villaflor, A.; Ding, B.; Abbeel, P.; Levine, S. Self-supervised deep reinforcement learning with generalized computation graphs for robot navigation. In Proceedings of the 2018 IEEE International Conference on Robotics and Automation (ICRA), Brisbane, Australia, 21–25 May 2018; pp. 5129–5136.
27. Han, S.H.; Choi, H.J.; Benz, P.; Loaiciga, J. Sensor-Based Mobile Robot Navigation via Deep Reinforcement Learning. In Proceedings of the 2018 IEEE International Conference on Big Data and Smart Computing (BigComp), Shanghai, China, 15–17 January 2018; pp. 147–154.
28. Dmitriy, P. Navigation System of an Autonomous Mobile Robot for Work in the Far North. Master's Thesis, Tomsk Polytechnic University, Tomsk, Russia, 2020. Available online: http://earchive.tpu.ru/handle/11683/61872 (accessed on 29 March 2023).
29. Ali, D.A.; Shahhood, F. Modular and hierarchical approach to agent training to explore a three-dimensional environment. In Proceedings of the Student Science Spring, Moscow, Russia, 1–30 April 2021; pp. 251–252.
30. Chuprov, S.; Marinenkov, E.; Viksnin, I.; Reznik, L.; Khokhlov, I. Image processing in autonomous vehicle model positioning and movement control. In Proceedings of the 2020 IEEE 6th World Forum on Internet of Things (WF-IoT), New Orleans, LA, USA, 2–16 June 2020; pp. 1–6.
31. Liu, X.; Zhou, Q.; Chen, X.; Fan, L.; Cheng, C.T. Bias-error accumulation analysis for inertial navigation methods. *IEEE Signal Process. Lett.* **2021**, *29*, 299–303. [CrossRef]
32. Monteiro, F.F.; Vieira, A.L.B.; Teixeira, J.M.X.N.; Teichrieb, V. Simulating real robots in virtual environments using NVIDIA's Isaac SDK. In Proceedings of the Anais Estendidos do XXI Simpósio de Realidade Virtual e Aumentada, Rio de Janeiro, Brazil, 28–31 October 2019; pp. 47–48.
33. Bochkovskiy, A.; Wang, C.Y.; Liao, H.Y.M. Yolov4: Optimal speed and accuracy of object detection. *arXiv* **2020**, arXiv:2004.10934.
34. Lin, T.Y.; Maire, M.; Belongie, S.; Hays, J.; Perona, P.; Ramanan, D.; Dollár, P.; Zitnick, C.L. Microsoft COCO: Common Objects in Context. In Proceedings of the Computer Vision—ECCV 2014, Zurich, Switzerland, 6–12 September 2014.
35. Ogre Development Team. UNITY COMPUTER VISION—Free Tools and Content for Generating Synthetic Data. Available online: https://unity.com/products/computer-vision (accessed on 25 March 2023).
36. NVIDIA Jetson AGX Xavier. AI Platform for Autonomous Machines. Available online: https://www.nvidia.com/en-us/autonomous-machines/jetson-agx-xavier/ (accessed on 20 April 2023).
37. Padilla, R.; Netto, S.L.; da Silva, E.A.B. A Survey on Performance Metrics for Object-Detection Algorithms. In Proceedings of the 2020 International Conference on Systems, Signals and Image Processing (IWSSIP), Niteroi, Brazil, 1–3 July 2020; pp. 237–242.

applied sciences

MDPI

Article

GNSS Signal Quality in Forest Stands for Off-Road Vehicle Navigation

Marian Rybansky [1,*], Vlastimil Kratochvíl [1], Filip Dohnal [1], Robin Gerold [1], Dana Kristalova [1], Petr Stodola [2] and Jan Nohel [2]

1 Department of Military Geography and Meteorology, Faculty of Military Technology, University of Defence, Kounicova 65, 662 10 Brno, Czech Republic; vlastimil.kratochvil@unob.cz (V.K.); filip.dohnal@unob.cz (F.D.); robin.gerold@unob.cz (R.G.); dana.kristalova@unob.cz (D.K.)

2 Department of Intelligence Support, Faculty of Military Leadership, University of Defence, Kounicova 65, 662 10 Brno, Czech Republic; petr.stodola@unob.cz (P.S.); jan.nohel@unob.cz (J.N.)

* Correspondence: marian.rybansky@unob.cz

Abstract: One of the basic possibilities of orientation in forest stands is the use of global navigation satellite systems (GNSS). Today, these systems are used for pedestrian orientation and also for off-road vehicle navigation. This article presents the results of research aimed at measuring the quality of GNSS signal in different types of coniferous and deciduous vegetation for the purpose of optimizing the navigation of off-road vehicles. To determine the structure (density) of the forest stand, tachymetry was chosen as the reference method. The Trimble Geo 7X cm edition device with Tornado for 7X antenna devices using real time VRS (virtual reference station) method was used to measure GNSS signal quality. This article presents the results of recorded numbers of GNSS satellites (GPS, GLONASS, Galileo and BeiDou) during the driving of a terrain vehicle in two different forest locations. Significant presented results include the deviations of vehicle positions determined by GNSS from tachymetrically precisely measured and marked routes along which the vehicle was moving. The authors of the article focused on the accuracy of determining the position of the vehicle using GNNS, as the most commonly used device for off-road vehicle navigation. The measurement results confirmed the assumption that the accuracy of positioning was better in deciduous forest than in coniferous (spruce) or mixed vegetation. This research was purposefully focused on the possibilities of navigation of military vehicles, but the achieved results can also be applied to the navigation of forestry, rescue and other types of off-road vehicles.

Keywords: GNSS signal quality; navigation in forest; cross-country mobility; terrain vehicles

Citation: Rybansky, M.; Kratochvíl, V.; Dohnal, F.; Gerold, R.; Kristalova, D.; Stodola, P.; Nohel, J. GNSS Signal Quality in Forest Stands for Off-Road Vehicle Navigation. *Appl. Sci.* **2023**, *13*, 6142. https://doi.org/10.3390/app13106142

Academic Editors: Nadjim Horri, William Holderbaum, Fabrizio Giulietti and Yutaka Ishibashi

Received: 24 March 2023
Revised: 9 May 2023
Accepted: 11 May 2023
Published: 17 May 2023

1. Introduction

Rescue and military operations usually include both movement over existing roads and also through natural terrain. In cases when drivers are not able to use some segments of roads (damaged or destroyed objects, traffic jam, etc.), they have to be provided with complete cross-country mobility (CCM) analyses considering all main geographical factors to solve transportation problems [1,2]. The cross-country movement of off-road vehicles is affected by various terrain features such as slope, terrain roughness, soil type, vegetation, seasonal changes to the ground surface, rivers, etc. [3–13]. One of the most difficult environments in which to maneuver is forested terrain. Two-thirds of European countries are at least 30% forested. In Sweden, Finland, Estonia, Latvia, Bosnia and Herzegovina, and Slovenia, vegetation covers over 50% of the country—see Figure 1 and [14,15]—severely restricting the off-road movement of terrain vehicles. Types of vegetation in a specific area can also give an indication of climatic conditions, hydrologic transformation, drainage systems, water supply, and soil type. The ability of a vehicle to cross a forested area depends on whether or not it can either maneuver between tree stems or override individual trees. The planning and success of moving off-road vehicles through a forest stand depends not only on vegetation factors such as stem spacing, tree diameter, and root structure (by tree overpassing), and vehicle parameters (width, length, turning radius, weight, and traction force), but also on the quality of navigation.

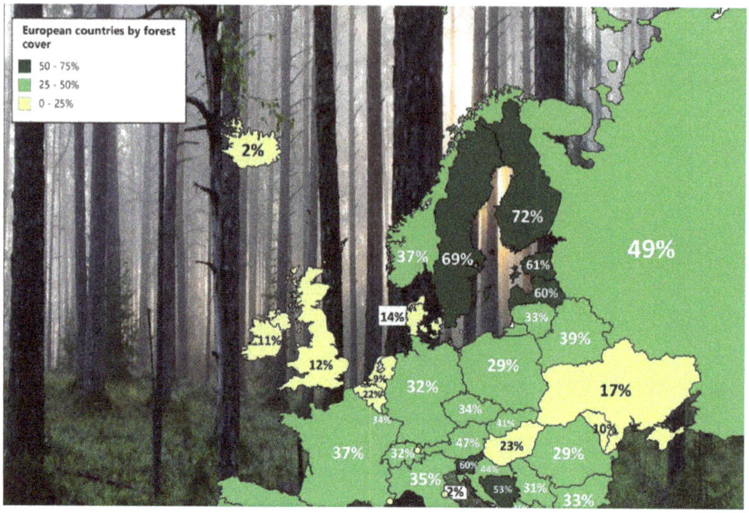

Figure 1. Forest cover in Europe.

There are several possible ways to navigate off-road vehicles. One option is to use global satellite navigation systems (GNSS). A satellite navigation system is a system that uses satellites to provide autonomous geospatial positioning. It allows small electronic receivers to determine their location (longitude, latitude, and altitude/elevation) with high precision, using time signals transmitted along a line of sight by radio from satellites. Current fully-operational GNSS applications include the United States' Global Positioning System (GPS), Russia's Global Navigation Satellite System (GLONASS), China's BeiDou Navigation Satellite System (BDS), and the European Union's Galileo. Japan's Quasi-Zenith Satellite System (QZSS) is a (U.S.) GPS satellite-based augmentation system to enhance the accuracy of GPS, with satellite navigation independent of GPS, scheduled for 2023. The Indian Regional Navigation Satellite System (IRNSS) plans to expand to a global version in the long term. Outside of GNSS, multisensory navigation systems are increasingly being used which routinely include the sensing of non-GNSS (acceleration, radio frequency signals, gravity, magnetic fields, imaging, barometric pressure, sound waves, etc.). However, each sensor type has its own set of limitations. The aim of this presented study was to verify the quality of the GNSS signal under forest vegetation, with a focus on coniferous and deciduous vegetation of varying densities in the temperate vegetation zone of Central Europe.

The prerequisite for GNSS navigation is that data of the structure of the forest will be available, including the coordinates of the position of individual trees. We are currently seeing a growing number of studies dealing with the use of LIDAR for mapping the structure of forest vegetation, e.g., Hyyppä and Inkinen [16], Persson et al. [17], and Peuhkurinen et al. [18]. Navigation of an off-road vehicle using GNSS methods in a forest depends both on the accuracy of mapping the positions of individual trees and on the quality of the navigation itself. Both the determination of tree positions and the quality of the GNSS signal depend on many factors, such as the density of the trees, their branching, and meteorological conditions, etc. There are several studies that focus on positioning under tree crowns using various GNSS systems. Zheng et al. [19] measured GPS performance by two aspects—position accuracy and data update frequency—using image processing for measuring canopy densities. Piedallu and Gégout [20] tested different

GPS survey components under different vegetation density covers. Valbuena et al. [21] described the accuracy of measurement using differential GPS and GLONASS receivers under mountainous Pinus sylvestris canopy. Galán et al. [22] analyzed the influence of forest stand variables on the accuracy of measurements performed by global positioning systems (GPS) receivers, using genetic algorithms. Bastos and Hasegawa [23] observed interactions between signal interruption probability (SIP), canopy opening index, observation period, number of available satellites, and positional dilution of precision (PDOP). Ucar et al. [24] studied the seasonal impact of oak–hickory forest on the positioning level using recreation-grade GPS receivers. Brach and Zasada [25] raised the antenna of a GNSS receiver during measurements, in order to reduce the multipath effect in the highest part of various forest conditions. Bakula et al. [26] presented the rapid static and real-time kinematic (RTK) technology for GPS/GLONASS surveying under forest environments. Luengo et al. [27] presented the first-ever dual-frequency multi-constellation Global Navigation Satellite Systems Reflectometry (GNSS-R) polarimetric measurements over boreal forests and lakes from the stratosphere. According to the study by Næsset and Gjevestad [28], positioning accuracy in a coniferous forest stand ranged from 0.27 to 3.48 m for observation periods ranging from 120 min to 15 min, respectively, depending on the time of measurement and number of tries. Camps et al. [29] presented an analysis of the vegetation impact on GPS L1 C/A (coarse acquisition code) signals in terms of attenuation and depolarization. Lamprecht et al. [30] developed a post-processing strategy to improve the positional accuracy of GNSS-measured sample plot centers and to develop a method to automatically match trees within a terrestrial sample plot to aerially detected trees. Jensen et al. [31] examined contemporaneous ALOS2 PALSAR-2 L-band imaging radar, CYGNSS GNSS reflections, and ground measurements to assess associated advantages and challenges to mapping inundation dynamics, particularly in regions under dense tropical forest canopies. Wright et al. in [32] presented GPS (Global Positioning System)-based methods to measure L-band GPS signal-to-noise ratios (SNRs) through different forest canopy conditions using hemispherical sky-oriented photos (HSOPs) along with GPS receivers.

Terrestrial mobile laser scanning (MLS) with GNSS navigation support has been developing dynamically in recent years. Nevalainen et al. [33] proposed a two-phase on-board process, where tree stem registration produces a sparse point cloud (PC) which is then used for simultaneous location and mapping (SLAM). A field test was carried out using a harvester with a laser scanner and a global navigation satellite system (GNSS) performing forest thinning over a 520 m strip route. Čerňava et al. [34] proposed an approach to the processing of MLS data of forest scanned from different views with two mobile laser scanners under heavy canopy. Qian et al. in [35] analyzed how the highly precise attitude and velocity information from GNSS/INS could be integrated with SLAM to improve its robustness and accuracy for high-precision positioning in forests. Kaartinen et al. [36] tested the accuracy of various instruments utilizing GNSS in motion under forest canopies of varying densities using several different combinations of GNSS and inertial measurement unit (IMU).

GNSS devices are mainly used for the navigation of off-road vehicles in forest cover. The accuracy of navigation depends on the quality of GNSS receivers, on the number and location of satellites, and also on the type of vegetation, the properties of which can change due to meteorological conditions. Verification of this accuracy is relatively time-consuming, requiring appropriate selection of representative stands, and measurements in different seasons and different meteorological conditions. The mentioned factors can significantly influence the results of research in this area.

The aim of this study was to verify the reliability of GNSS navigation of off-road vehicles in a temperate vegetation zone, focusing on a typical monocultural coniferous and deciduous forest.

2. Material and Methods

Two sites were selected to verify the quality of the GNSS signal for off-road vehicle navigation. The first measurement took place in locality 1 (Figure 2) in 2020. The second measurement took place in locality 2 (Figure 2) in 2021. Both testing areas are located in the eastern part of the Czech Republic, about 40 km northeast of Brno.

Figure 2. Study areas: 1—deciduous forest, 2—coniferous forest, both in the military training area Brezina (www.mapy.cz).

2.1. *Characteristics of Localities No. 1 and No. 2*

The forest in locality 1 (WGS 84 coordinates: B = 16°58′2,29″, L = 49°19′55,68″; 642,930 E, 5,466,240 N, UTM 33N; H = 456 m) consists of the predominant winter oak (Quercus petraea), with mixed white birch (*Betula pendula*) and aspen (*Populus tremula*), which are either mixed or form more independent vegetation parts, surrounded by oak—see Figure 3. The growth is probably around 50 years old. The spatial structure of the stand is diverse, the dispersion of individuals is from individual to cluster. The location is on the top plateau, slightly exposed to the south, almost flat, and the terrain surface is slightly bent in places (possibly anthropically). Geologically, the locality is located on the Drahansky kulm, the soil-forming rock is consolidated sediments, the character is shale and debris. The measurement in this locality took place on 25 August 2020 between 9:00 and 16:00 (UTC+2). The temperature fluctuated between 18.2 °C in the morning and 21.9 °C in the afternoon at the end of the measurement; the medium humidity during GNSS measurement was 60%.

Figure 3. Study area 1—deciduous forest.

The forest in locality 2 (WGS 84 coordinates: B = 17°00′6,57″ L = 49°22′25,92″; 643,315 E, 5,470,945 N, UTM 33N; H = 392 m) is situated on the small plateau northwest of the village of Podivice in the central part of the military training area (MTA) Brezina. Vegetation consists mostly of conifers, predominantly with Scots pine (*Pinus silvestris*) and Norway spruce (Picea abies), mixed winter oak (*Quercus petraea*), and places with aspen (*Populus tremula*), and white birch (*Betula pendula*) or beech (*Fagus sylvatica*), and larch (*Larix decidua*)—see Figure 4. The measurement in this locality took place on 26 August and 9 September 2021 between 9:00 and 16:00 (UTC+2). The temperature fluctuated between 16.2 °C in the morning and 18.9 °C in the afternoon at the end of the measurement; the medium humidity was 90%.

Figure 4. Study area 2—mostly coniferous forest.

2.2. Measurement Methodology

When designing the methodology for determining the accuracy of navigation of a moving vehicle, the authors based the above-mentioned literature on article [36], in which the authors determined the accuracy of navigation from data measured by GNSS, IMU and laser scanner (territory of Finland) and from publications focused on the combination of different navigation methods—see [37–40].

Since determining the navigation accuracy of a moving vehicle also depends on the regional characteristics of forest units, the authors chose for their own experiment two representative localities of the temperate climate zone (coniferous and deciduous forest) with typical tree species for this area. Another important factor is the quality of the GNSS signal, resulting from the quantity and distribution of satellites during the vehicle's route, therefore, the analysis of the quantity and quality of the GNSS signal was part of the proposed methodology.

The methodology for determining the accuracy of off-road vehicle navigation using GNSS systems consisted of the following steps, see also Figure 5:

- Demarcation of forest stand for testing with a focus on finding a relatively homogeneous part of the forest in which there were representative tree types of stand;
- Surveying the coordinates of the reference ground control points around the forest and in the middle of the forest (places not covered by vegetation) using the GNSS, Trimble Geo XR with Zephyr model 2 Geo XR antenna, the Trimble Geo 7X cm version with antenna, Tornado pro 7X devices with 2–3 cm estimated position error, and using the real time VRS (virtual reference station) method. Selected parameters of receiver Geo 7X were: firmware GNSS: 4.50.8.; software GNSS: Terra Sync version 5.86.; number of channels: 142 (max. 220); GNSS types: GPS L1/L2 (L1E, L2C), GLONASS G1/G2, Galileo E1, and BeiDou B1; SBAS types: WAAS and EGNOS;
- For field measurements, the GNSS receiver was set as: measurement mode: 1; external GNSS: 2; SBAS: 3; autonomous mode; logging interval: 1 s; elevation mask: 5°; PDOP mask: 99; signal to noise ratio (SNR): 12; and height of the external antenna location: 1 m;
- The number of reference ground control points depended on the density of forest stands;
- Measuring tree positions using tachymetry. The measurement was performed from reference ground control points, from which there was good visibility to the surrounding trees. First, the distance was measured on a pole with a surveying prism touching the tree from the side. Then the pole was moved in front of the center of the tree and the azimuth was measured. This avoided measuring eccentricities, i.e., diameter at breast high (DBH) measurement. During the measurement, the surveying prism was placed at a height of 1.3 m above the ground surface;
- Measurement of vehicle positions while driving between pre-marked trees. The GNSS Tornado pro 7X antenna of the Geo 7X receiver and the device itself were attached to the front of the vehicle body at a height of 1.0 m. The vehicle was moving between reference trees for which position coordinates had previously been measured on a route previously targeted by the tachymetric method. Before the actual testing, the route along which the vehicle was moving was also marked with a spray. This route was also measured using tachymetry in order to determine the deviations between the reference route and the routes located using the GNSS method. Deviations of vehicle movement from the reference marked path depended on the terrain configuration and vegetation density. The speed of the moving vehicle was 3–5 km·h^{-1}, the distances between the trees were 3 m to 5 m, the height of the trees was approximately 15 m. For most of the vehicle's path, the sky was obscured by tree branches. The receiving of satellite signals was characterized by frequent short-term outages;
- Evaluation of signal-to-noise ratio (SNR) as a function of the influence of vegetation density on the quality of the received GNSS signal;

- Evaluation of the position accuracy of a moving vehicle, determined by a GNNS receiver and compared with the coordinates of a pre-targeted vehicle route.

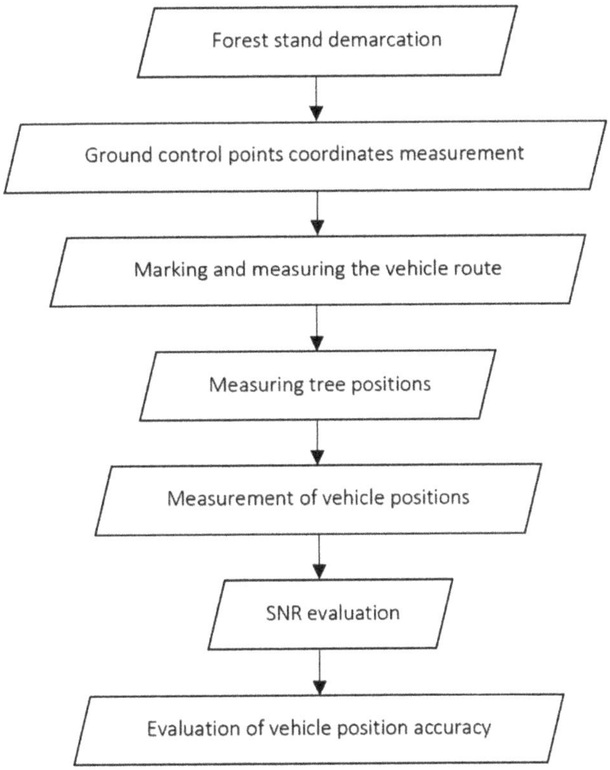

Figure 5. Methodology of determining the accuracy of off-road vehicle navigation using GNSS.

3. Results

The results of the presented research included targeting and displaying tree positions without measuring eccentricity (the distance was measured on a reflecting prism attached to the side of the tree and the azimuth to the center of the tree). Furthermore, the tracks along which the vehicle moved were targeted and marked by tachymetry and its positions were measured (Figures 6 and 7). Additionally presented analyses include the quality of the GNSS signal during vehicle travel (Figures 8 and 9), determination of the number of satellites during the course (Figures 10 and 11), and determination of the estimation of receiver positioning accuracy (σ_{HZ}) and position deviations (d_{HZ}) detected as the difference between the position recorded by the GNSS receiver and the position determined terrestrially (Figures 12–16 and Tables 1 and 2).

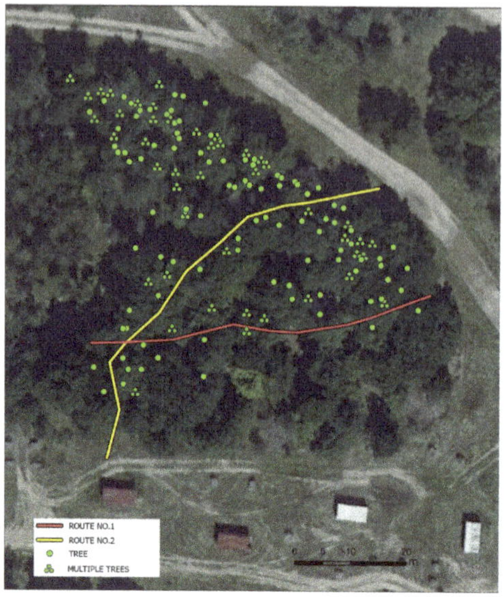

Figure 6. Tree positions in area 1—deciduous forest (Ortophoto ČÚZK).

Figure 7. Tree positions in area 2—mostly coniferous forest (Ortophoto ČÚZK).

```
FON:·Begin·Line·Feature:····Time:·25.08.2020·8:03:06,000····Version:·4¶
ATT:·Feature····Description:·2·sekundy····ID:·0003····Flags:··0·hex¶
STR:·Time·of·Week:·25.08.2020·8:08:05,000····Vozidlo····ctyrkolka¶
SGM:·Time:·25.08.2020·8:03:06,000····Sigma·North:·2,177····Sigma·East:·1,691····Sigma·Up:·4,861¶
POS:·____U:····25.08.2020·8:03:06,000····642993,175····5466237,117····457,937¶
OBS:·Time:·25.08.2020·8:03:07,000·····Clock·Bias:·3 119,015·Meters·····No.·of·SVs:·24¶
OBS:·Time:·25.08.2020·8:03:08,000·····Clock·Bias:·3 174,421·Meters·····No.·of·SVs:·24¶
SGM:·Time:·25.08.2020·8:03:08,000····Sigma·North:·2,178····Sigma·East:·1,686····Sigma·Up:·4,819¶
POS:·____U:····25.08.2020·8:03:08,000····642993,142····5466237,117····457,930¶
OBS:·Time:·25.08.2020·8:03:09,000·····Clock·Bias:·3 229,755·Meters·····No.·of·SVs:·24¶
OBS:·Time:·25.08.2020·8:03:10,000·····Clock·Bias:·3 285,083·Meters·····No.·of·SVs:·24¶
SGM:·Time:·25.08.2020·8:03:10,000····Sigma·North:·2,178····Sigma·East:·1,707····Sigma·Up:·5,245¶
POS:·____U:····25.08.2020·8:03:10,000····642993,037····5466236,981····458,026¶
OBS:·Time:·25.08.2020·8:03:11,000·····Clock·Bias:·3 340,273·Meters·····No.·of·SVs:·24¶
OBS:·Time:·25.08.2020·8:03:12,000·····Clock·Bias:·3 395,497·Meters·····No.·of·SVs:·24¶
SGM:·Time:·25.08.2020·8:03:12,000····Sigma·North:·3,133····Sigma·East:·2,393····Sigma·Up:·9,477¶
POS:·____U:····25.08.2020·8:03:12,000····642991,159····5466235,623····458,030¶
OBS:·Time:·25.08.2020·8:03:13,000·····Clock·Bias:·3 450,690·Meters·····No.·of·SVs:·24¶
OBS:·Time:·25.08.2020·8:03:14,000·····Clock·Bias:·3 506,049·Meters·····No.·of·SVs:·24¶
SGM:·Time:·25.08.2020·8:03:14,000····Sigma·North:·3,160····Sigma·East:·2,169····Sigma·Up:·9,726¶
```

Figure 8. Part of the content of the *.SSF file converted to text format.

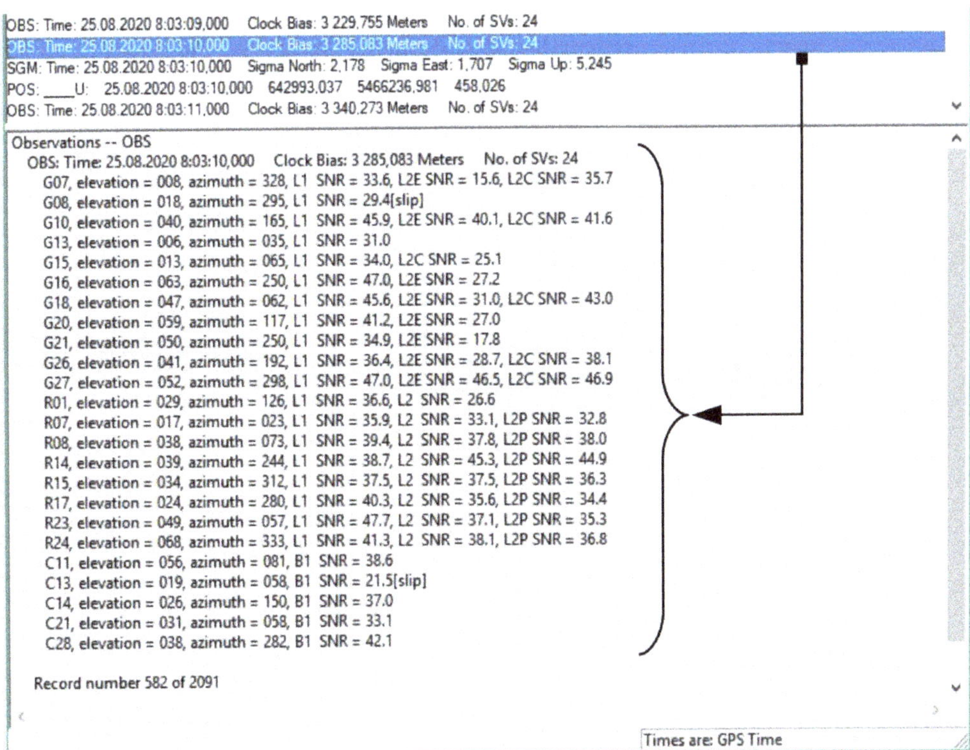

Figure 9. Example of a specific GNSS measurement record.

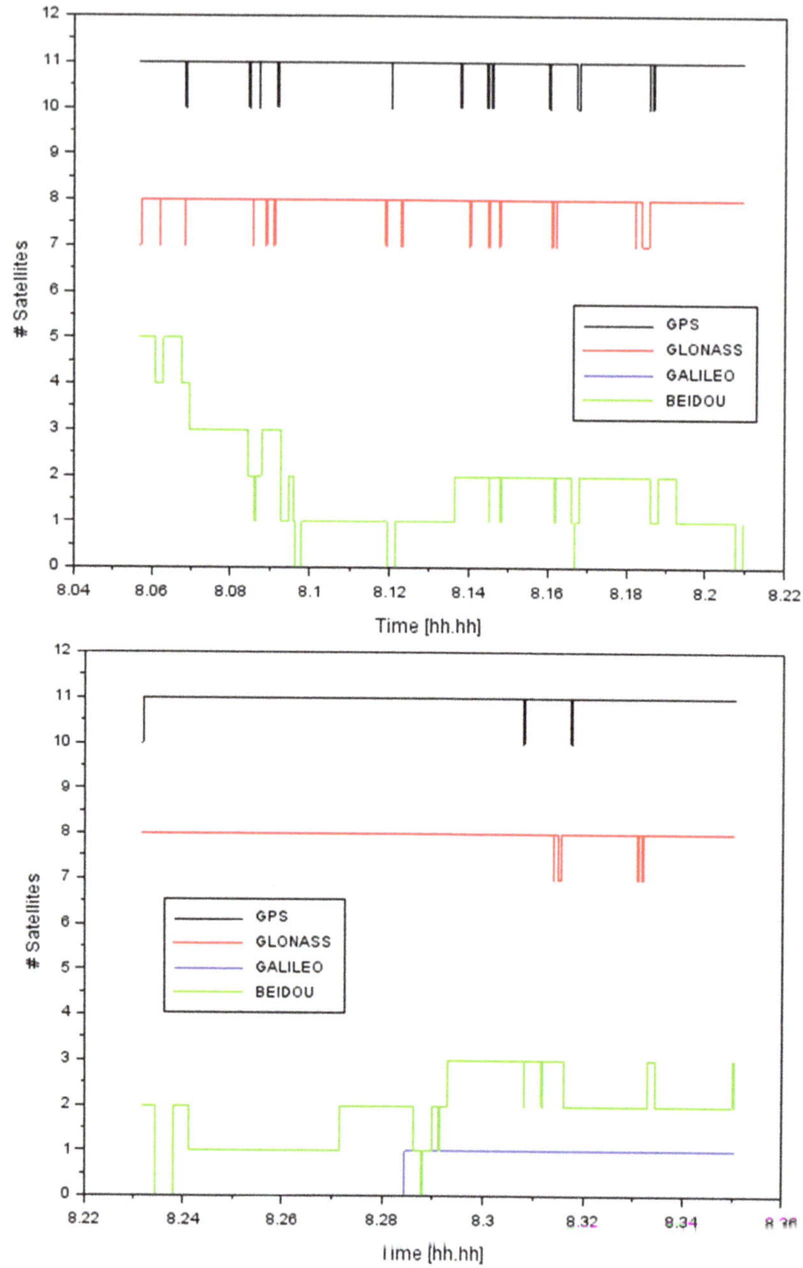

Figure 10. *Cont.*

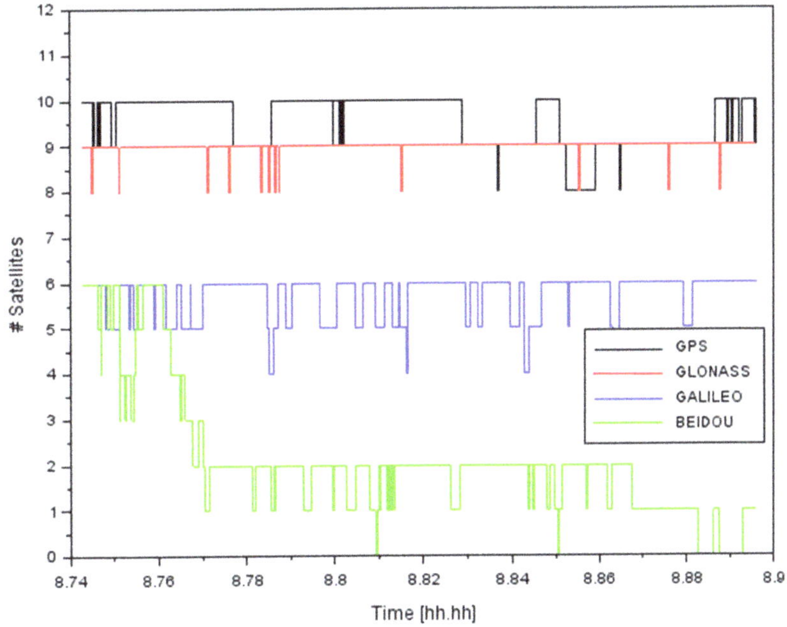

Figure 10. Number of individual GNSS satellites during measurement in area 1.

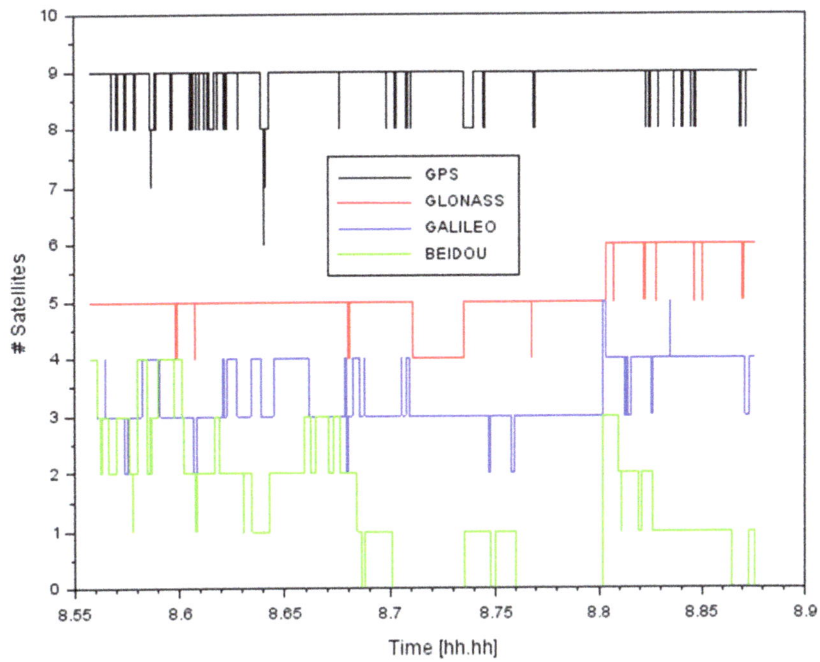

Figure 11. Number of individual GNSS satellites during measurement in area 2.

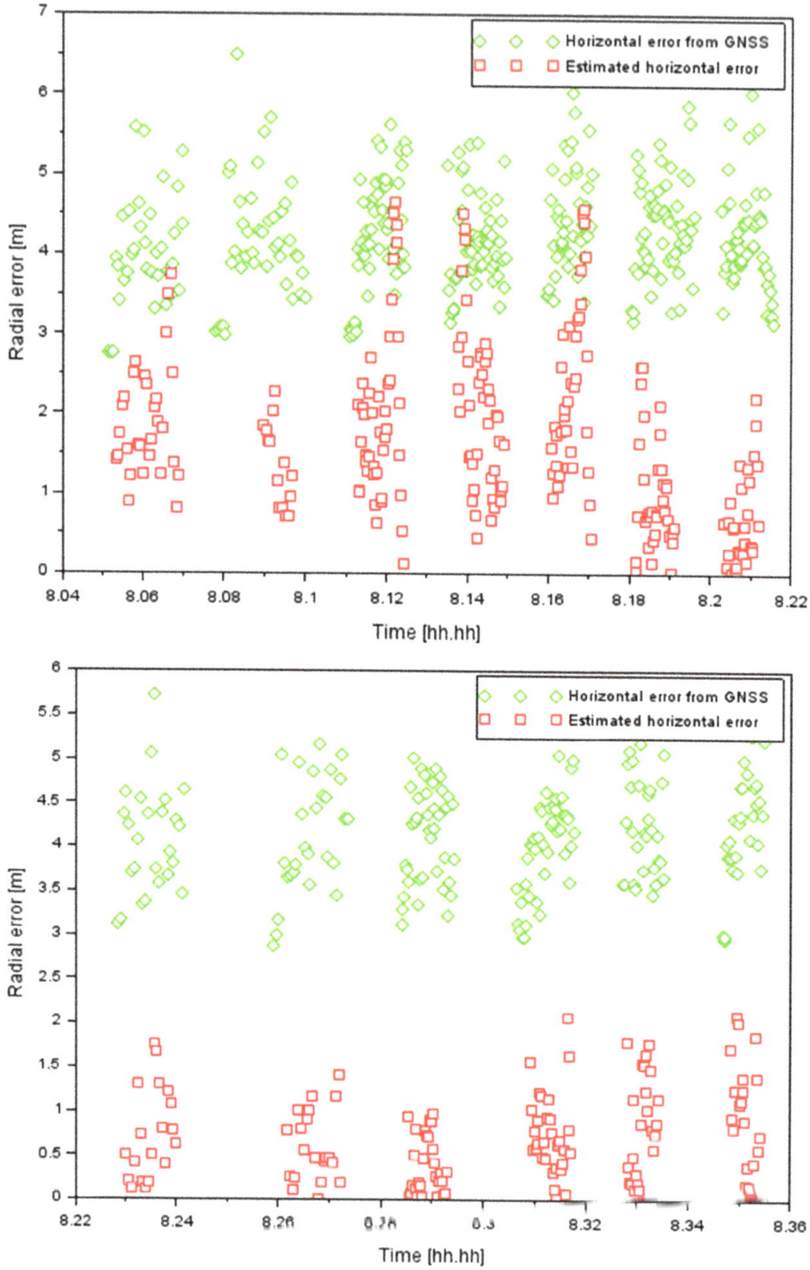

Figure 12. Comparison of d_{Hz} horizontal deviations (red squares) with d_{Hz} horizontal errors (green color) derived from GNSS file data—area 1.

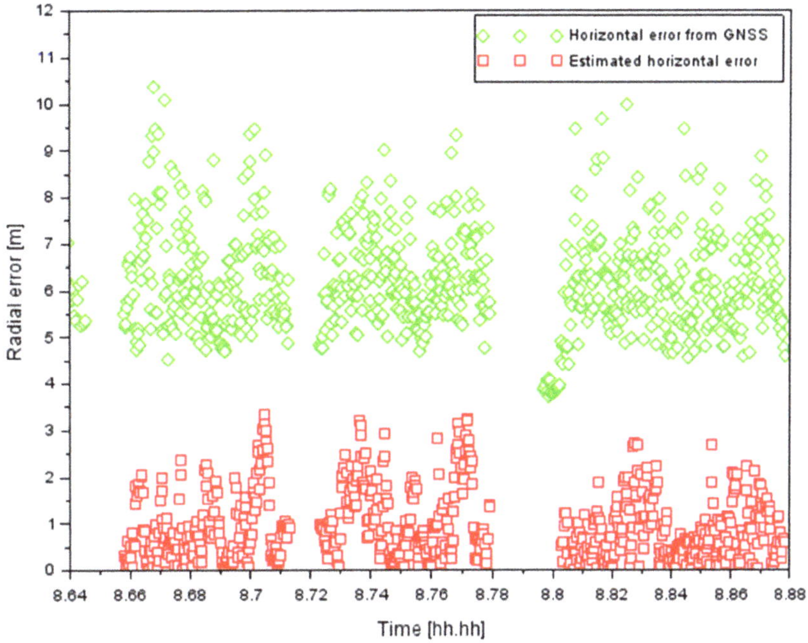

Figure 13. Comparison of d_{Hz} horizontal deviations (red squares) with d_{Hz} horizontal errors (green color) derived from GNSS file data—area 2.

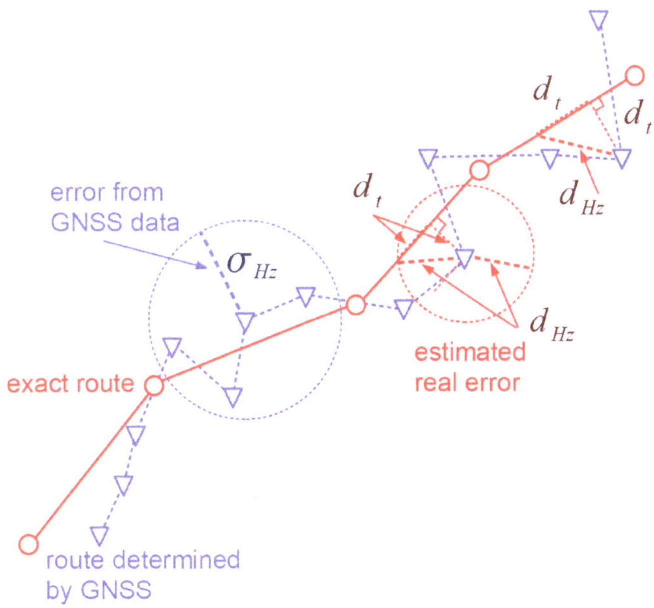

Figure 14. The principle of estimating the accuracy of the vehicle path from GNSS measurements.

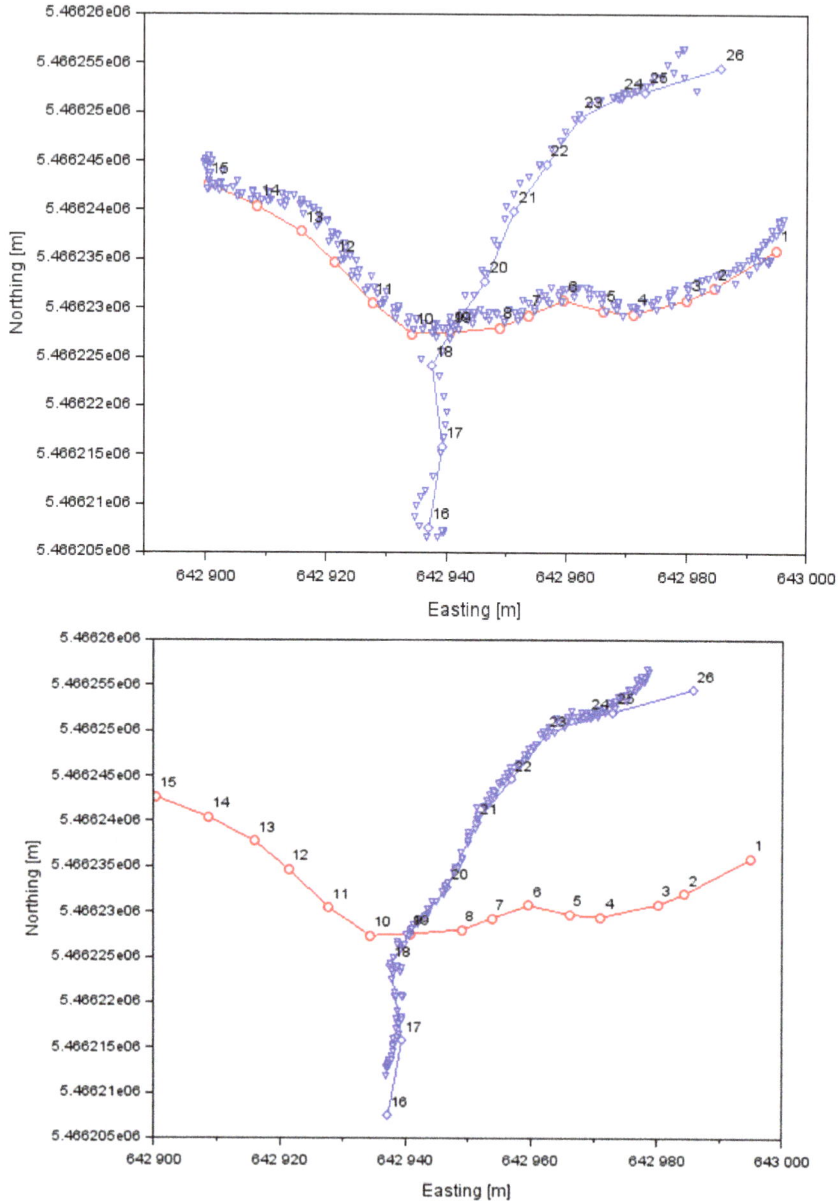

Figure 15. Projected routes (broken line) and GNSS focused route (blue triangles).

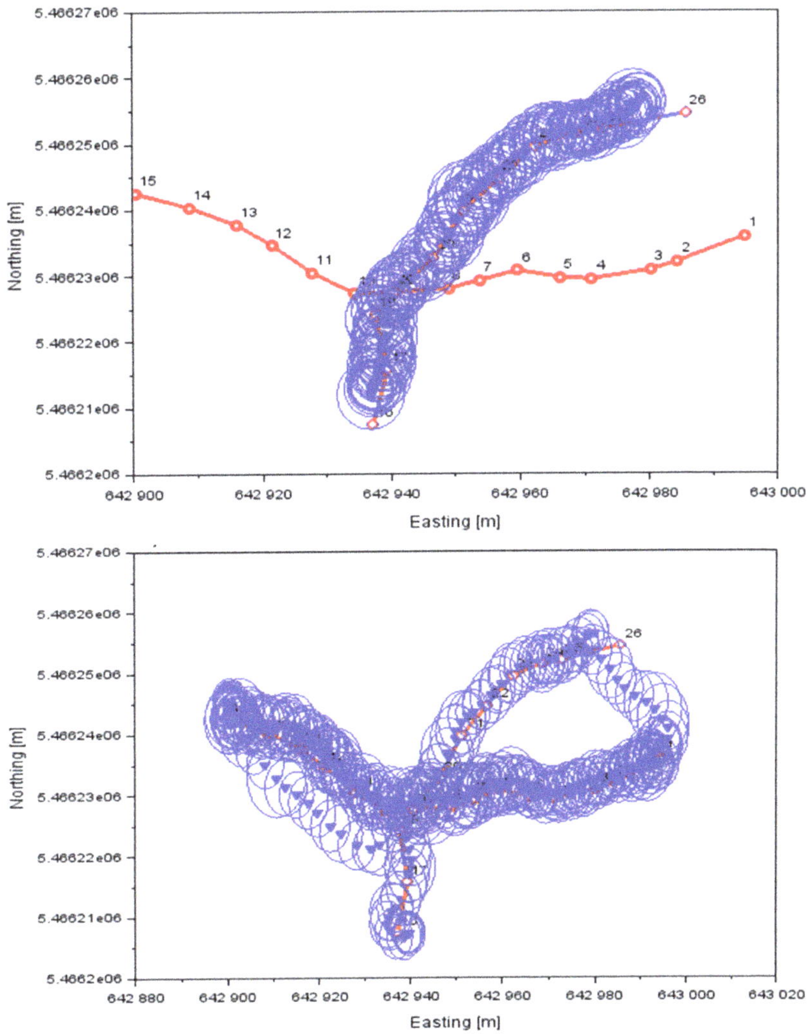

Figure 16. Horizontal errors σ_{Hz} obtained from GNSS measurements (blue circles).

Table 1. Characteristics of the accuracy of off-road vehicle navigation using GNSS means.

Area 1		**Area 1**		**Area 2**	
$\bar{\sigma}_{Hz}$ **[m]**	$\bar{t}_{Hz}$ **[m]**	$\bar{\sigma}_{Hz}$ **[m]**	$\bar{t}_{Hz}$ **[m]**	$\bar{\sigma}_{Hz}$ **[m]**	$\bar{t}_{Hz}$ **[m]**
4.3	2.0	4.2	0.9	6.3	1.2

Table 2. Characteristics of the accuracy of off-road vehicle navigation using GNSS means–expressed in percentiles.

Area 1			**Area 1**			**Area 2**		
Percentile	σ_{Hz} **[m]**	t_{Hz} **[m]**	**Percentile**	σ_{Hz} **[m]**	t_{Hz} **[m]**	**Percentile**	σ_{Hz} **[m]**	t_{Hz} **[m]**
63%	4.4	1.9	63%	4.4	0.8	63%	6.4	1.0
95%	5.5	4.0	95%	5.1	1.8	95%	8.4	2.4

3.1. Tachymetric Determination of Tree Positions

Figure 6 shows the positions of the deciduous trees, consisting of the predominant winter oak (*Quercus petraea*), with mixed white birch (*Betula pendula*) and aspen (*Populus tremula*), that were targeted by tachymetry. Although the accuracy of the tachymetric method for determining the positions of regular objects is in the order of millimeters, the positions of trees were measured with an accuracy of centimeters because some tree trunks were tilted; and from some tachymeter locations, the more distant trees could not be seen well because they were obscured by branches of the vegetation. Nevertheless, these determined positions of trees can be considered sufficient with respect to the GNNS method of determining the position of a moving vehicle, which oscillated in decimeters to meters depending on the position of the satellites, the structure of the vegetation, and meteorological conditions.

Figure 7 shows the positions of the trees, consisting mostly of conifers, predominantly with Scots pine (*Pinus silvestris*) and Norway spruce (*Picea abies*), mixed winter oak (*Quercus petraea*), places with aspen (*Populus tremula*), white birch (*Betula pendula*) or beech (Fagus sylvatica), and larch (*Larix decidua*). The tachymetric determination of the positions of coniferous trees was more accurate than that of deciduous trees, primarily because spruces and pines form a sparser stand with upright trunks. Predominantly coniferous stands allowed better conditions for targeting the positions of trees, as there was usually no undergrowth that hindered the targets for the determined trees.

3.2. GNSS Signal Quality Evaluation

The evaluation of the quality of the GNSS signal was performed using the GNSS Trimble Geo XR with Zephyr model 2 Geo XR antenna, and the Trimble Geo 7X cm version with antenna Tornado for 7X devices—see Section 2.2 Measurement Methodology for more details. The files recorded by the receiver were used for the evaluation (analysis) of satellite signals. The recorded binary files of type *.SSF from area 1 were transferred to text format by the SSF Editor application, which is part of the PathFinder Office program. A sample of part of the content of the records taken during the measurement in area 1 is shown in Figure 8.

The following information was obtained and evaluated from the text files:

- Number of satellites of individual GNSS types;
- GNSS receiver antenna coordinates in 1 s intervals;
- Accuracy estimates (standard deviations) of coordinates in the direction "north-south"—sigma north (σ_N), "east-west"—sigma east (σ_E), and height—sigma up;
- Average values of PDOP and SNR of satellite signals during measurement;
- GPS connection time to all records.

A specific example of one measurement record (no. 582) related to GPS time (25.8. 2020, 8:03:10,000) is shown in Figure 9.

The number of individual GNSS satellites from which data was received during vehicle movement measurement in area 1 is shown in Figure 10.

Figure 10 shows that the largest number of GNSS satellites in area 1 belonged to GPS (9–11), followed by GLONASS (7–9). In the case of Galileo or BeiDou, the number of satellites were rather marginal. Figure 11 shows that the largest number of GNSS satellites in area 2 belonged to GPS (6–9), followed by GLONASS (4–6), Galileo (2–4), and BeiDou (0–6). It can be assumed that currently the number of correresponding satellites will be much more favorable. It is worth noting that all positioning was performed in autonomous mode.

3.3. Accuracy of Vehicle Positioning Using the GNSS Method

Figure 12 (area 1) and Figure 13 (area 2) show six (respectively, four) groups of points (clusters) that represent individual vehicle trips along the same marked route section. The green rhombuses represent the estimation of receiver positioning accuracy (σ_{Hz})

derived from data contained in *.SSF files from area 1, $\sigma_{Hz} = \sqrt{\sigma_E^2 + \sigma_N^2}$, where σ_E is the estimate of accuracy in the "east-west" direction and σ_N is the estimate in the "north-south" direction—see also detail in Figure 14. The red squares show the position deviations, d_{Hz}, detected as the difference between the position recorded by the GNSS receiver and the position determined terrestrially. The terrestrially determined positions (coordinates) were assumed as reference data due to the assumed accuracy of the GNSS measurement results. These route reference points (along which the vehicle drove) were measured terrestrially using tachymetry and signaled on the terrain surface (spray on the ground). Vehicle movement time on the route was displayed in hundredths of seconds. Note that the positional (horizontal) deviation (σ_{Hz}) was calculated according to the formula $d_{Hz} = d_t\sqrt{2}$, where d_t is the distance of the GNSS determined position from the reference route (Figure 14). It was clear that the positional accuracy σ_{Hz} was very pessimistic, $\sigma_{Hz} > d_{Hz}$ was almost exclusively valid. The size of the actual positional deviations from the route was approximately 2× to 4× smaller. However, a general statement could not be made from the described fact. The statistical set was from a small scale of space and time.

An overview of the reliability of the data of the horizontal error d_{Hz} with respect to the actual course of the vehicle route can be made using Figures 15 and 16. In Figure 15, the projected (planned, terrestrially focused) route of the vehicle is shown by line segments. Blue triangles show the position of the vehicle determined by GNSS measurement. It should be noted that the GNSS position shows the movement of the vehicle including its approach to the projected route.

In Figure 16, circles show the horizontal errors, σ_{Hz}, which are contained in *.SSF files (see Figure 8). It follows from the figures that the σ_{Hz} data, obtained from GNSS measurement files, always included the projected route of the vehicle directed terrestrially (using tachymetry).

One of the prerequisites for GNSS signal quality analysis is the signal-to-noise ratio (SNR) evaluation as the important factor of influence of signal quality on off-road vehicle navigation in a wooded area. SNR is a measure comparing the level of a desired signal to the level of background noise. SNR is defined as the ratio of signal power to noise power (SNR = P_{signal}/P_{noise}), often expressed in decibels. A ratio higher than 1:1 (greater than 0 dB) indicates more signal than noise.

The following summary characteristics were chosen for the individual locations (1 and 2):

1. Square mean of radial (horizontal) deviations:

$$\bar{\sigma}_{Hz} = \sqrt{\frac{1}{n}\sum \sigma_{Hz}^2} \quad (1)$$

$$\bar{t}_{Hz} = \sqrt{\frac{1}{m}\sum t_{Hz}^2} \quad (2)$$

where n, (m) represent the number of values obtained from GNSS measurement records, respectively, from terrestrial measurements.

The calculated values of radial (horizontal) deviations are shown in Table 1. In area 1, two groups of accuracy characteristics are presented, which correspond to two measurement sections—see Figures 6 and 7 (Figures 12 and 13).

2. Percentiles of values σ_{Hz} and t_{Hz}. Two percentile levels were chosen: 63%, which theoretically corresponded to the value $\bar{\sigma}_{Hz}$, and 95%, theoretically corresponding to the value $\bar{t}_{Hz}$.

The calculated percentile values are shown in Table 2.

4. Discussion

The results of vehicle location analyses within deciduous forest (oak stand) and mostly coniferous forest (spruce and pine stand) confirmed that the accuracy of the measurement depends mainly on stand density, season (deciduous forest), vegetation age and meteorological conditions, which can significantly reduce the quality of the received GNSS signal, especially during precipitation. For this reason, it is necessary to take into account the results obtained, which were not affected by the same humidity in both tested localities (60% and 90%).

During the experiment, the number of satellites (GPS and GLONASS) varied between 15 and 20 and was never less than 11 (see Figures 10 and 11). The horizontal DOP (dilution of precision) was in the interval of 0.8–4. Only two position records out of more than 400 records had DOP = 17. The dependence of the accuracy of determining the horizontal position (σ_{HZ} or d_{HZ}) on HDOP parameters or total number of satellites was not statistically proven (significant).

The average values of the actual deviations of vehicle positions determined using GNSS from two terrestrially measured routes reached values of 2.0 m and 1.9 m in locality 1 (deciduous forest), and 1.2 m in locality 2 (mainly coniferous forest). These values are sufficient for the navigation of an off-road vehicle, provided that we evaluate the possibilities of movement using measured distances and directions (azimuths) from the start to the destination. If we are to consider the possibility of moving a vehicle in a dense forest, we have to count on greater errors in determining the position and the fact that the possibility of moving the vehicle will decrease with its increasing dimensions and with turning radius—see [41].

The achieved positioning was sufficient for locating a moving vehicle, however, ultimately, the type of vegetation and terrain topography (such as deep valleys) will have a significant effect on the ability of the vehicle to maneuver between trees, but this factor was not the subject of research.

Another factor that can significantly affect the movement of the vehicle is the time of day. At night, orientation in the forest is much more difficult.

The canopy differences may be best described by the hemispheric photographs taken with the camera pointed vertically at the forest stand—see Figures 17 and 18.

Figure 17. *Cont.*

Figure 17. Hemispheric photographs—deciduous forest.

Figure 18. Hemispheric photographs—coniferous forest.

The achieved results of determining the accuracy of vehicle navigation corresponded to the character of both locations where the tests were carried out. These results can be used to assess the possibilities of navigation in similar stands that are typical for the Central European region, that is, for forests in a temperate climate zone. It can be assumed that in more northerly latitudes, drier arid regions, and inland, where vegetation is sparser, the accuracy of vehicle navigation determination will also be better. On the other hand, in denser vegetation in a monsoon forest region and in the equatorial climate zone, where the vegetation is denser and air humidity is greater, the accuracy of vehicle navigation will be worse.

In the future, the authors of this article would like to focus on research to determine the accuracy of vehicle navigation in different types of vegetation with different parameters of tree density, the nature of their crowns (density, height), and in different seasons.

5. Conclusions

The aim of this study was to investigate the accuracy of off-road vehicle navigation using GNSS receivers in typical forest stands of the temperate climate zone of Central Europe. For this purpose, two localities were chosen in which reference elements (trees and vehicle routes) were precisely located by tachymetry, which were then combined to determine the accuracy of the moving vehicle. Positioning accuracy (σ_{HZ}) and position deviations (d_{HZ}) detected as the difference between the position recorded by the GNSS receiver and the position determined terrestrially, were determined.

The introduced research studies prove that GNSS navigation devices can be successfully used for approximate localization of moving vehicles, especially when it comes to movement in a sparse forest. Bigger problems will arise if we seek to accurately determine vehicle–tree collisions while maneuvering the vehicle between trees. In this case, it will be necessary to install terrestrial laser or radar sensors of tree positions and calculate the vehicle's route in real time. However, this type of navigation has limited possibilities given the cover of trees and the fact that trees cannot be mapped beyond the horizon. Therefore, support using remote sensing means will also be necessary, or support using autonomous navigation, based on measuring acceleration, or determining the position using an odometer and gyroscope.

Additionally, an unscented Kalman filter (UKF)-based multi-sensor optimal data fusion methodology for INS/GNSS/CNS (inertial navigation system/global navigation satellite system/celestial navigation system) integration based on a nonlinear system model can significantly improve navigation options in terrain covered with vegetation and can provide a reliable positioning solution in case of GNSS (GPS) blockage—see [37–39]. In addition, the methods based on the fundamental improvement of the fusion mechanism using matrix weighted fusion can significantly improve the quality of determining navigation in vegetation in the future—see [42,43].

The GNSS measurement issue under poor environments can also be addressed by identification of abnormal measurements. Some methods have been proposed to optimize off-road vehicle navigation deficiencies caused by these abnormal measurements using, e.g., a novel robust fault-tolerant federated filter based on a strapdown inertial navigation system (SINS) [37,44], global navigation satellite system (GNSS), celestial navigation system (CNS), and doppler velocity log (DVL) [40].

Author Contributions: Organization, M.R.; conceptualization and methodology, M.R. and V.K.; measurement, V.K., F.D., M.R., J.N., P.S. and D.K.; validation, M.R., R.G., V.K. and F.D.; writing the paper M.R., V.K.; original draft preparation, M.R. All authors have read and agreed to the published version of the manuscript.

Funding: This research received no external funding.

Institutional Review Board Statement: Not applicable.

Informed Consent Statement: Not applicable.

Data Availability Statement: Map materials were prepared using mapy.cz application, orthophoto maps using https://ags.cuzk.cz/geoprohlizec/, accessed on 10 May 2023, other data were measured and analyzed by the authors of this article.

Acknowledgments: This paper is a particular result of the defense research project DZRO VAROPS managed by the University of Defence in Brno, the NATO—STO Support Project CZE-AVT-2019 and the specific research project 2021–2023 at Department K-210 managed by the University of Defence, Brno.

Conflicts of Interest: The authors declare no conflict of interest. The funders had no role in the design of the study; in the collection, analyses, or interpretation of data; in the writing of the manuscript, or in the decision to publish the results.

References

1. Rada, J.; Rybansky, M.; Dohnal, F. The Impact of the Accuracy of Terrain Surface Data on the Navigation of Off-Road Vehicles. *ISPRS Int. J. Geo-Inf.* **2021**, *10*, 106. [CrossRef]
2. Rada, J.; Rybansky, M.; Dohnal, F. Influence of Quality of Remote Sensing Data on Vegetation Passability by Terrain Vehicles. *ISPRS Int. J. Geo-Inf.* **2020**, *9*, 684. [CrossRef]
3. Rybansky, M.; Hofmann, A.; Hubacek, M.; Kovarik, V.; Talhofer, V. Modelling of cross-country transport in raster format. *Environ. Earth Sci.* **2015**, *74*, 7049–7058. [CrossRef]
4. Rybansky, M. Determination the ability of military vehicles to override vegetation. *J. Terramechan.* **2020**, *91*, 129–138. [CrossRef]
5. Pokonieczny, K.; Rybansky, M. Method of Developing the Maps of Passability for Unmanned Ground Vehicles. In Proceedings of the IOP Conference Series: Earth and Environmental Science, 9th IGRSM International Conference and Exhibition on Geospatial & Remote Sensing (IGRSM 2018), Kuala Lumpur, Malaysia, 24–25 April 2018; pp. 1–9.
6. Stodola, P.; Drozd, J.; Nohel, J.; Michenka, K. Model of Observation Posts Deployment in Tactical Decision Support System. In *Artificial Intelligence and Lecture Notes in Bioinformatics*; Lecture Notes in Computer Science (Including Subseries Lecture Notes); Springer: Berlin/Heidelberg, Germany, 2020; Volume 11995 LNCS, pp. 231–243. [CrossRef]
7. Nohel, J.; Stodola, P.; Flasar, Z. Combat UGV Support of Company Task Force Operations. In *Artificial Intelligence and Lecture Notes in Bioinformatics*; Lecture Notes in Computer Science (Including Subseries Lecture Notes); Mazal, J., Fagiolini, A., Vasik, P., Turi, M., Eds.; Springer: Cham, Switzerland; Science and Business Media Deutschland GmbH: Baden-Wuerttemberg, Germany, 2021; Volume 12619 LNCS, pp. 29–42, ISBN 978-3-030-70739-2. [CrossRef]
8. Dawid, W.; Pokonieczny, K. Analysis of the Possibilities of Using Different Resolution Digital Elevation Models in the Study of Microrelief on the Example of Terrain Passability. *Remote Sens.* **2020**, *12*, 4146. [CrossRef]
9. Dawid, W.; Pokonieczny, K. Methodology of Using Terrain Passability Maps for Planning the Movement of Troops and Navigation of Unmanned Ground Vehicles. *Sensors* **2021**, *21*, 4682. [CrossRef]
10. Drozd, J.; Stodola, P.; Křišťálová, D.; Kozůbek, J. Experiments with the UAS Reconnaissance Model in the Real Environment. In *Modelling and Simulation for Autonomous Systems*; Springer International Publishing: Cham, Switzerland, 2018; pp. 340–349, ISBN 978 3 319 76071 1.
11. Heštera, H.; Pahernik, M. Physical-geographic factors of terrain trafficability of military vehicles according to Western World methodologies. *Hrvat. Geogr. Glas. Geogr. Bull.* **2018**, *80*, 5–31. [CrossRef]
12. Hoskova-Mayerova, S. Geospatial Data Reliability, Their Use in Crisis Situations. In Proceedings of the 21st International Scientific Conference on The Knowledge-Based Organization (KBO), Sibiu, Romania, 15–17 June 2023; pp. 694–698.
13. Hoskova-Mayerova, S.; Talhofer, V.; Otrisal, P.; Rybansky, M. Influence of Weights of Geographical Factors on the Results of Multicriteria Analysis in Solving Spatial Analyses. *ISPRS Int. J. Geo-Inf.* **2020**, *9*, 489. [CrossRef]
14. Raši, R. *State of Europe's Forests 2020*; Ministerial Conference on the Protection of Forests in Europe—FOREST EUROPE: Zvolen, Slovakia, 2020; Available online: https://foresteurope.org/wp-content/uploads/2016/08/SoEF_2020.pdf (accessed on 15 October 2022).
15. European Union. *Forestry in the EU and the World*; Publications Office of the European Union: Luxembourg, 2011.
16. Hyyppä, J.; Inkinen, M. Detecting and estimating attributes for single trees using laser scanner. *Photogramm. J. Finl.* **1999**, *16*, 27–42.
17. Persson, A.; Holmgren, J.; Söderman, U. Detecting and measuring individual trees using an airborne laser scanner. *Photogramm. Eng. Remote Sens.* **2002**, *68*, 925–932.
18. Peuhkurinen, J.; Maltamo, M.; Malinen, J.; Pitkänen, J.; Packalén, P. Preharvest measurement of marked stands using airborne laser scanning. *For. Sci.* **2007**, *53*, 653–661.
19. Zheng, J.; Wang, Y.; Nihan, N.L. *Quantitative Evaluation of GPS Performance under Forest Conopies*; Department of Civil and Environmental Engineering, University of Washington: Seattle, WA, USA, 2002.
20. Piedallu, C.; Gégout, J.C. Effect of forest environment and survey protocol on GPS accuracy. *Photogramm. Eng. Remote Sens.* **2005**, *71*, 1071–1078. [CrossRef]
21. Valbuena, R.; Mauro, F.; Rodriguez-Solano, R.; Manzanera, J.A. Accuracy and precision of GPS receivers under forest canopies in a mountainous environment. *Span. J. Agric. Res.* **2010**, *8*, 1047–1057. [CrossRef]

22. Galán, C.O.; Rodríguez-Pérez, J.R.; Torres, J.M.; García Nieto, P.J. Analysis of the influence of forest environments on the accuracy of GPS measurements by using genetic algorithms. *Math. Comput. Model.* **2011**, *54*, 1829–1834. [CrossRef]
23. Bastos, A.S.; Hasegawa, H. Behavior of GPS signal interruption probability under tree canopies in different forest conditions. *Eur. J. Remote Sens.* **2013**, *46*, 613–622. [CrossRef]
24. Ucar, Z.; Bettinger, P.; Weaver, S.; Merry, K.L.; Faw, K. Dynamic accuracy of recreation-grade GPS receivers in oak-hickory forests. *Forestry* **2014**, *87*, 504–511. [CrossRef]
25. Brach, M.; Zásada, M. The effect of Mounting Height on GNSS Receiver Positioning Accuracy in Forest Conditions. *Croat. J. For. Eng.* **2014**, *35*, 245–253.
26. Bakula, M.; Przestrzelski, P.; Kazmierczak, R. Reliable Technology of Centimeter GPS/GLONASS Surveying in Forest Environments. *IEEE Trans. Geosci. Remote Sens.* **2015**, *53*, 1029–1038. [CrossRef]
27. Luengo, H.C.; Amèzaga, A.; Vidal, D.; Olivé, R.; Munoz, J.F.; Camps, A. First Polarimetric GNSS-R Measurements from a Stratospheric Flight over Boreal Forests. *Remote Sens.* **2015**, *7*, 13120–13138. [CrossRef]
28. Næsset, E.; Gjevestad, J.G. Performance of GPS precise point positioning under conifer forest canopies. *Photogramm. Eng. Remote Sens.* **2008**, *74*, 661–668. [CrossRef]
29. Camps, A.; Alonso-Arroyo, A.; Park, H.; Onrubia, R.; Pascual, D.; Querol, J. L-Band Vegetation Optical Depth Estimation Using Transmitted GNSS Signals: Application to GNSS-Reflectometry and Positioning. *Remote Sens.* **2020**, *12*, 2352. [CrossRef]
30. Lamprecht, L.; Hill, A.; Stoffels, J.; Udelhoven, T. A Machine Learning Method for Co-Registration and Individual Tree Matching of Forest Inventory and Airborne Laser Scanning Data. *Remote Sens.* **2017**, *9*, 505. [CrossRef]
31. Jensen, K.; McDonald, K.; Podest, E.; Rodriguez-Alvarez, N.; Horna, V.; Steiner, N. Assessing L-Band GNSS-Reflectometry and Imaging Radar for Detecting Sub-Canopy Inundation Dynamics in a TropicalWetlands Complex. *Remote Sens.* **2018**, *10*, 1431. [CrossRef]
32. Wright, W.; Wilkinson, B.; Cropper, W., Jr. Development of a GPS Forest Signal Absorption Coefficient Index. *Forests* **2018**, *9*, 226. [CrossRef]
33. Nevalainen, P.; Li, Q.; Melkas, T.; Riekki, K.; Westerlund, T.; Heikkonen, J. Navigation and Mapping in Forest Environment Using Sparse Point Clouds. *Remote Sens.* **2020**, *12*, 4088. [CrossRef]
34. Čerňava, J.; Mokroš, J.; Tuček, J.; Antal, M.; Slatkovská, Z. Processing Chain for Estimation of Tree Diameter from GNSS-IMU-Based Mobile Laser Scanning Data. *Remote Sens.* **2019**, *11*, 615. [CrossRef]
35. Qian, C.; Liu, H.; Tang, J.; Chen, Y.; Kaartinen, H.; Kukko, A.; Zhu, L.; Liang, X.; Chen, L.; Hyyppä, J. An Integrated GNSS/INS/LiDAR-SLAM Positioning Method for Highly Accurate Forest Stem Mapping. *Remote Sens.* **2017**, *9*, 3. [CrossRef]
36. Kaartinen, H.; Hyyppä, J.; Vastaranta, M.; Kukko, A.; Jaakkola, A.; Yu, X.; Pyörälä, J.; Liang, X.; Liu, J.; Wang, Y.; et al. Accuracy of Kinematic Positioning Using Global Satellite Navigation Systems under Forest Canopies. *Forests* **2015**, *6*, 3218–3236. [CrossRef]
37. Gao, B.; Hu, G.; Gao, S.; Zhong, Y.; Gu, C. Multi-sensor Optimal Data Fusion for INS/GNSS/CNS Integration Based on Unscented Kalman Filter. *Int. J. Control. Autom. Syst.* **2018**, *16*, 129–140. [CrossRef]
38. Dai, H.-F.; Bian, H.-W.; Wang, R.-Y.; Ma, H. An INS/GNSS integrated navigation in GNSS denied environment using recurrent neural network. *Defence Technol.* **2020**, *16*, 334–340. [CrossRef]
39. Gao, G.; Gao, B.; Gao, S.; Hu, G.; Zhong, Y. A Hypothesis Test-Constrained Robust Kalman Filter for INS/GNSS Integration with Abnormal Measurement. *IEEE Trans. Veh. Technol.* **2023**, *72*, 1662–1673. [CrossRef]
40. Xiong, H.; Bian, R.; Li, Y.; Du, Z.; Mai, Z. Fault-Tolerant GNSS/SINS/DVL/CNS Integrated Navigation and Positioning Mechanism Based on Adaptive Information Sharing Factors. *IEEE Syst. J.* **2020**, *14*, 3744–3754. [CrossRef]
41. Rybansky, M. Determination of Forest Structure from Remote Sensing Data for Modeling the Navigation of Rescue Vehicles. *Appl. Sci.* **2022**, *12*, 3939. [CrossRef]
42. Hu, G.; Gao, S.; Zhong, Y.; Gao, B.; Subic, A. Matrix weighted multisensor data fusion for INS/GNSS/CNS integration. *Proc. Inst. Mech. Eng. Part G J. Aerosp. Eng.* **2016**, *230*, 1011–1026. [CrossRef]
43. Zhang, P.; Qi, W. Distributed Weighting Fusion and Covariance Intersection Fusion Kalman Smoother for Systems with Colored Measurement Noises. In Proceedings of the International Conference on Intelligent Control and Computer Application (ICCA), Zhengzhou, China, 16–17 January 2016. [CrossRef]
44. Gao, G.; Zhong, Y.; Gao, S.; Gao, B. Double-Channel Sequential Probability Ratio Test for Failure Detection in Multisensor Integrated Systems. *IEEE Trans. Instrum. Meas.* **2021**, *70*, 3514814. [CrossRef]

applied sciences

MDPI

Review

Challenges and Solutions for Autonomous Ground Robot Scene Understanding and Navigation in Unstructured Outdoor Environments: A Review

Liyana Wijayathunga, Alexander Rassau * and Douglas Chai

School of Engineering, Edith Cowan University, Perth, WA 6027, Australia;
lwijayat@our.ecu.edu.au (L.W.); d.chai@ecu.edu.au (D.C.)
* Correspondence: a.rassau@ecu.edu.au

Abstract: The capabilities of autonomous mobile robotic systems have been steadily improving due to recent advancements in computer science, engineering, and related disciplines such as cognitive science. In controlled environments, robots have achieved relatively high levels of autonomy. In more unstructured environments, however, the development of fully autonomous mobile robots remains challenging due to the complexity of understanding these environments. Many autonomous mobile robots use classical, learning-based or hybrid approaches for navigation. More recent learning-based methods may replace the complete navigation pipeline or selected stages of the classical approach. For effective deployment, autonomous robots must understand their external environments at a sophisticated level according to their intended applications. Therefore, in addition to robot perception, scene analysis and higher-level scene understanding (e.g., traversable/non-traversable, rough or smooth terrain, etc.) are required for autonomous robot navigation in unstructured outdoor environments. This paper provides a comprehensive review and critical analysis of these methods in the context of their applications to the problems of robot perception and scene understanding in unstructured environments and the related problems of localisation, environment mapping and path planning. State-of-the-art sensor fusion methods and multimodal scene understanding approaches are also discussed and evaluated within this context. The paper concludes with an in-depth discussion regarding the current state of the autonomous ground robot navigation challenge in unstructured outdoor environments and the most promising future research directions to overcome these challenges.

Keywords: unstructured environments; mobile robots; robot navigation; perception; scene understanding; path planning

Citation: Wijayathunga, L.; Rassau, A.; Chai, D. Challenges and Solutions for Autonomous Ground Robot Scene Understanding and Navigation in Unstructured Outdoor Environments: A Review. *Appl. Sci.* **2023**, *13*, 9877. https://doi.org/10.3390/app13179877

Academic Editors: Fabrizio Giulietti, Nadjim Horri and William Holderbaum

Received: 13 April 2023
Revised: 28 August 2023
Accepted: 29 August 2023
Published: 31 August 2023

1. Introduction

The use of autonomous mobile robotic systems is rapidly expanding in many scientific and commercial fields, and the capabilities of these robots have been growing due to continuous research and industrial efforts. The range of different research application areas in autonomous mobile robotics is wide, including areas such as surveillance, space exploration, defence, petrochemical, industrial automation, disaster management, construction, marine, personal assistance, extreme environments, sports entertainment, agriculture, transportation logistics, and many other industrial and non-industrial applications [1]. Many different types of robot platforms have been and continue to be developed for these applications, and autonomous mobile robotics is a global and continuously evolving scientific field. Robots that move in contact with ground surfaces are commonly referred to as mobile ground robots, and these robots may be deployed in different working environments, including indoor or outdoor, structured or unstructured, and in proximity to static or dynamic actors. Each of these environments creates various challenges for robot applications. The overall system configuration of a robot mostly depends on the relevant operating environment;

for example, a robot designed for a static environment may not effectively adapt to dynamic situations that arise within the environment [2]. The development of ground robotic systems is particularly challenging when the intended application area of autonomous vehicles/robots falls into the unstructured off-road category, making this a very active area of research. The difficulties mainly arise due to the weak scene understanding of robots in these unstructured environment scenarios. Mobile ground robots have recently shown increasingly sophisticated scene-understanding capabilities in indoor environments (e.g., Amazon Robotics warehouse robots [3]). However, in outdoor robot navigation, the environments are generally much less structured and more unpredictable (e.g., lighting, shadows, occlusions and texture) than in indoor robot navigation application scenarios. Due to these particular challenges and the more limited level of progress achieved to date to address them, this work seeks to provide a thorough review of existing research works in this field in order to provide directions for advancing robot navigation in unstructured outdoor environments. The agriculture industry is one example of a relevant application area to deploy scene understanding based on off-road autonomous robotic systems. In recent times, the agriculture industry has shown a growing adoption of robotics technologies, but in general, the involvement of novel technologies depends on economic sustainability. Ground mobile robots and manipulators are already used in precision farming to pick fruits, harvest vegetables, and remove undesirable vegetation, but their application areas are generally fairly narrow due to limited scene understanding capability. In the past few years, self-driving vehicles have gradually grown in technological capability and market size within the automobile industry [3,4]. Despite these developments, autonomous driving is still a very challenging problem, and in complex scenarios, the performance level remains below that of an average human operator. This is particularly true when the intended application area of autonomous vehicles/robots includes off-road areas. This low performance mainly arises due to the weak external environment understanding of robots. In general, application areas such as disaster management, environment exploration, defence, mining and transportation are associated with complex and unstructured environments. Therefore, these fields are also driving further research on mobile robot scene understanding as it relates to the broader topic of autonomous robot navigation. The following challenges for the classical autonomous robot navigation pipeline remain for current robotic systems:

- Perception;
- Localisation and mapping; and
- Scene understanding.

In robot perception, robots sense environments using different sensors and extract actionable information via the sensor data. Perception plays an important role in the realisation of autonomous robots. For robots to perform effectively in unstructured outdoor environments in real time, it is essential that they possess accurate, reliable, and robust perception capabilities. To achieve these characteristics, in general, autonomous robots in complex scenarios are equipped with several sensor modalities (that can be exteroceptive or proprioceptive) [5,6]. Different modalities such as sound, pressure, temperature, light, and contact have been used in robot environmental perception applications [7]. Sensor fusion (combining different sensor modalities) has been applied in many recent autonomous mobile robot/self-driving applications [8]. Multimodal sensor fusion brings the complementary properties of different sensors together to achieve better environment perception across a range of conditions [9]. Many recent deep learning-based sensor fusion methods have shown higher robustness in perception than conventional mono-sensor methodologies [6,10]. Camera, Light Detection and Ranging (LiDAR), radar, ultrasonic, Global Navigation Satellite System (GNSS), inertial measurement unit (IMU) and odometry sensors are used in many mobile ground robot perception applications. The most frequently used robot vision-based sensor fusion methods combine camera images with LiDAR point clouds. Research into sensor fusion techniques remains an important component of achieving better sensing capabilities for unstructured outdoor environments.

In the second stage of the autonomous robot navigation pipeline, the information retrieved from perception sensors is used for robot localisation and to map their external environments. Autonomous mobile robots require accurate and reliable environmental mappings and localisation at a sophisticated level based on the application context. The Simultaneous Localisation and Mapping (SLAM) approach is a commonly used technique in autonomous mobile robot systems to represent the robot positions and the map of their external environments when both the robot pose and environmental map are previously unknown. Many SLAM systems use LiDAR sensors and vision-based sensors such as Red-Green-Blue (RGB)/RGB-Depth (RGB-D) cameras that can retrieve visual or proximity information, and the fusion of these sensors has achieved better robustness than using either camera or LiDAR sensors alone [11]. LiDAR sensors use highly accurate Time-of-Flight laser range finding for the detection of objects. An example of a lightweight and low-cost semantic environment classification approach for indoor robot navigation is introduced in [12]. Semantic classification becomes challenging with the variation of the lighting, shadows, occlusions, texture and the number of unseen semantic classes. Dynamic agents present another problem, [13] has proposed a LiDAR-based concept to segment large areas with dynamic agents. It can use LiDAR range view representations and further categorise objects in the images into semantic classes. Several research works have incorporated the fusion of camera and LiDAR data for improving scene understanding of outdoor environments [14].

A fundamental aim of robotic vision is to interpret the semantic information present in a scene to provide scene understanding. Scene understanding goes beyond object detection and requires analysis and elaboration of the data retrieved by the sensors [15]. This concept is used in many practical applications, such as self-driving vehicles, surveillance, transportation, mobile robot navigation, and activity recognition. Understanding scenes using images or videos is a complex problem, however, and requires more steps than just recording and extracting features [16]. Scene understanding can be aided by taking advantage of multiple sensor modalities, which is usually termed multimodal scene understanding [17]. One of the requirements for autonomous robots operating in unstructured environments is the capacity to understand the surrounding environment. Scene understanding comprises subtasks such as depth estimation, scene categorisation, object detection, object tracking, and event categorisation [18]. These scene-understanding sub-tasks can describe different aspects of a scene acquired by perception sensors. In scene understanding, a representation is given to a scene by carrying out some of the above tasks jointly to obtain a holistic understanding of the retrieved scene. To generate this overall representation, the information observed from the above-mentioned scene understanding subtasks must be combined meaningfully. Some of the early methods used to obtain scene understanding included using a block world assumption [19] or bottom-up, top-down inference [20]. Many of these early works depended on heuristics rather than learning-based methods and, thus, were not suitable for generalisation to unstructured real-world scenarios [21]. The most used recent methods try to acquire information from deep learning-based (i.e., convolutional neural networks (CNNs), graph neural networks, vision transformers) approaches. Scene understanding in unstructured outdoor environments using multimodal scene understanding concepts is a challenging task. Many modern scene understanding methods use feature-based high-level representations of environments. In unstructured environments, however, the detection of useful object features is challenging. Therefore, it is difficult to reliably interpret visual information from unstructured or dynamic environments [22]. However, to attain real-world effectiveness, robots should understand their operating environments up to a level that is accurate enough to execute real-time and goal-oriented decisions.

One of the key capabilities required for autonomous robot operation is autonomous path planning for robot navigation. Many ground robot navigation applications in unstructured environments require traversal of uneven terrains. Some challenges in robot unstructured environment navigation, including traversal of previously unseen unstructured grounds, can be addressed by adequately designing robot structure, path planning and control [23,24]. Path planning is generally separated into global and local planning. For global planning, previous knowledge of the operating environment is necessary, and this planning method is also identified as an off-line mode for robot path planning. Robot local path planning, also known as online robot path planning, allows for real-time decisions to be made by the robot in response to the perception of the local environment. Autonomous robot local path planning for optimal terrain traversal in unstructured outdoor environments is an important challenge to solve for robots operating in off-road conditions. This is due to the limitations of standard path planning algorithms, which are incapable of performing the desired tasks in dynamic or unstructured environments where the system lacks prior knowledge and/or already existing maps [25]. Artificial potential fields, simulated annealing, fuzzy logic, artificial neural networks, and dynamic window [26] approaches are some of the algorithms that have been used in robot local path planning [21]. An optimal local path planning approach should enable robots to adaptably deal with their environments, such as assisting in avoiding dynamic obstacles or identifying traversable routes through varying terrain conditions. In unstructured environments, classical path planning concepts such as Rapidly exploring Random Tree (RRT) variants, Batch Informed Tree (BIT), D* algorithm variants, artificial potential field methods, A* algorithm variants and learning-based path planning methods have been used [27]. Over the past years, deep learning has been used to enhance the performance of sensor fusion [28], multimodal scene understanding [29], and robot local path planning [30] techniques.

For autonomous navigation, a robot must have the ability to understand both its pose and enough about the external environment to determine an optimal and traversable path to safely reach a goal position without human assistance in the robot control loop. However, fully autonomous navigation in unstructured environments has not yet been achieved despite significant advancements in computing and engineering technology. Autonomous vehicle navigation on urban roads has been of great interest due to its emerging commercial applications around the world. As a result, autonomous perception technologies are developing continuously due to the competitiveness of this industry [31]. Compared to the level of development of methods for autonomous navigation in urban settings [32], however, autonomous navigation in off-road scenarios has not been studied to the same extent, meaning there are significant opportunities for new research in this area. A large number of autonomous navigation techniques have been explored, and these can be broadly divided into two subsections according to the approach used to execute control commands following the processing of the input sensory data.

As shown in Figure 1, autonomous navigation methods can be categorised as the classical modular pipeline or end-to-end learning-based approaches. The modular pipeline architecture includes intermediate representations that humans can interpret and that can provide information related to the failure modes of the approach. Furthermore, the system development can be parallelised among several expert teams because of its modular nature. Perception, localisation and mapping, path planning and robot control modules may consist of classical, learning-based or hybrid methods [8,33]. Due to algorithmic and modelling limitations, however, the modular approach may not be optimal for general autonomous navigation applications. In addition, machine learning-based modules must be separately trained and validated using auxiliary loss functions, which can create suboptimality across the system as a whole. In contrast, end-to-end learning-based systems learn policies from observations of the outcomes that result from the actions of autonomous systems [18,34]. In these systems, Deep Reinforcement Learning (DRL) is used to refine the control algorithms used to determine autonomous actions. However, commonly employed methods like imitation learning can suffer from overfitting and cause problems with the poor generalisation of the system behaviour when deploying in different environment scenarios [11]. In addition, when errors occur in an end-to-end system, it can be complex to investigate because the origin of these errors is hidden in the holistic neural network-based architectures [8,35]. It is clear, however, that future development of autonomous robot navigation will rely on further advancements in robot environmental perception and machine learning.

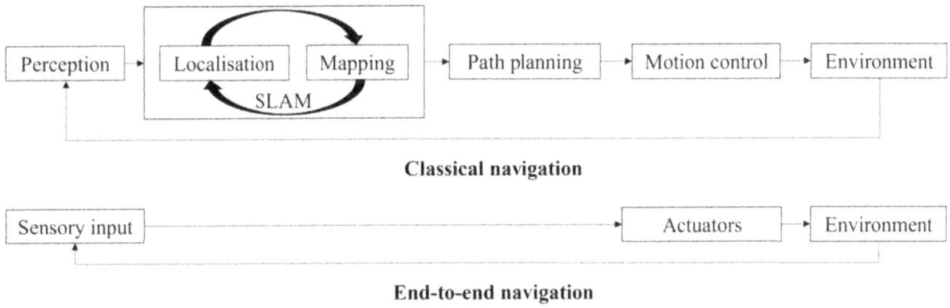

Figure 1. Classical and end-to-end autonomous navigation approaches.

This paper provides a comprehensive review of the current state of research and the most promising advancements relevant to the field of ground robot navigation in unstructured environments, including perception and sensor fusion, scene understanding, and path planning. The most promising state-of-the-art methods used in classical outdoor robot navigation pipelines are discussed and evaluated, along with emerging learning-based approaches; promising future research directions are then highlighted. Readers of this paper will gain insights and knowledge regarding current state-of-the-art capabilities, as well as trends and emerging opportunities in the autonomous robot outdoor scene understanding and navigation research field. The remainder of this paper is structured in the following manner. A discussion of robot vision and the types of active ranging sensors that are used for robot environment perception is presented in Section 2. This section also discusses deep learning-based camera and LiDAR sensor fusion methods for depth estimation, object detection, and semantic and instance segmentation. Section 3 describes modern mobile ground robot scene understanding techniques and scene representations that are utilised in robot navigation. Robot path planning algorithms at the global and local levels are discussed in Section 4. In Section 5, the details of robot vision and ranging sensors, fusion methods, scene understanding concepts, and local navigation approaches are summarised. Section 6 provides an overview of the research challenges related to the topic. Additionally, it offers some potential future research directions. Finally, the conclusion of the review can be found in Section 7.

2. Robot Environment Perception for Navigation

Robot external and internal environment sensing by extraction of raw sensor data and their interpretation is the basic principle of robot perception. In the modular or end-to-end robot navigation approach, sensors play a critical role in capturing the environment or internal robot attributes for robot perception. Therefore, to achieve better perception, a wide range of different sensor modalities have been investigated by researchers. A sensor modality represents a sensor that inputs a particular form of energy and processes the signal using similar methods. Modalities include raw input types for sensors like sound, pressure, light (infrared, visible), or magnetic fields. Robot perception sensor modalities commonly include cameras (infrared, RGB or depth), LiDAR, radar, sonar, GNSS, IMU, and odometry sensors.

2.1. Vision and Ranging Sensors

Camera sensors are usually incorporated in vision applications by researchers to retrieve environmental information for the use of mobile robots. However, LiDAR sensors have shown more reliability in low-light environmental conditions than cameras and produce highly accurate depth measurements. LiDAR sensors come with 2D or 3D mapping capability [36], and these sensors can generate high-fidelity point clouds of outdoor environments. However, these unstructured point clouds tend to become increasingly sparse as the sensor range is increased.

- Vision-based Sensor Types

Vision is crucial for a robot to navigate in unknown environments. Robots are equipped with vision-based sensors like cameras to understand environmental information. Cameras are generally cheaper than LiDAR sensors and are a type of passive sensor (although some depth cameras use active sensing). The monocular configuration is highly used in standard RGB cameras, for example, the GoPro Hero camera series. They are compact passive sensors that consume low power, approximately 5 watts, depending on resolution and mode settings. Monocular SLAM has been explored in the research literature due to its simple hardware design [32,37]. However, the algorithms that are used for it are very complex because depth measurements cannot be directly retrieved from static monocular images. Monocular cameras also suffer from pose estimation problems [38]. The pose of the camera is obtained by referencing previous poses. Hence, errors in pose estimation propagate through the process, and this phenomenon is called scale drift [28,29,39].

Stereo cameras are inspired by human eyes and use two lenses and separate passive image sensors to obtain two perspectives of the same scenes. They have been used in indoor and outdoor SLAM applications. These camera types use the disparity of the two images to calculate the depth information. Stereo camera vision does not suffer from scale drift. Available popular stereo cameras are Bumblebee 2, Bumblebee XB3, Surveyor stereo vision system, Capella, Minoru 3D Webcam, Ensenso N10, ZED 2, and PCI nDepth vision system. Stereo camera power consumption is generally between 2 to 15 watts. The maximum range of stereo cameras varies from 5 to 40 m at different depth resolution values. The accuracy of these sensors varies from around a few millimetres to 5 cm at the maximum range [40]. The cost of these sensors varies from one hundred to several thousand Australian dollars. RGB-D camera sensors consist of monocular or stereo cameras coupled to infrared transmitters and receivers. Kinect camera from Microsoft is a relatively inexpensive RGB-D sensor that provides colour images and depth information of image pixels. The Kinect sensor is mainly used for indoor robot applications due to the saturation of infrared receivers in outdoor scenarios from sunlight [41]. The Kinect sensor has three major versions: Kinect 1, Kinect 2, and Azure Kinect (the latest version). Kinect 1 uses structured light for depth measurement, and the other models use Time-Of-Flight (TOF) as the depth measuring principle. The newest Kinect model, Azure Kinect, also generates substantial noise in outdoor bright light conditions with a practical measuring range below 1.5 m [42]. In general, RGB-D sensors utilise three depth measurement principles: structured light, TOF, and active infrared stereo techniques. The structured light RGB-D sensors underperform compared to TOF techniques in measuring the range of distant objects. The structured light technique is vulnerable to multi-device interference. The TOF methods suffer from multi-path interference and motion artefacts. The active infrared stereo principle has drawbacks due to common stereo matching problems such as occluded pixels and flying pixels near contour edges [43]. Active infrared stereo cameras are identified as an extension of passive stereo cameras. They offer more reliable performance in indoor and some outdoor scenarios. However, they require high computation capability. Table 1 shows the main depth measurement methods that are used in RGB-D sensors.

Event cameras are asynchronous sensors that use different visual information acquisition principles than the standard frame-based image sensors. This camera type samples light based on the changes in the scene dynamics (asynchronously measuring brightness per pixel). These cameras currently cost several thousands of dollars. Event cameras have advantages like very high temporal resolution, high dynamic range, low latency (in microseconds), and lower power consumption than standard cameras. However, this camera type is not suitable for the detection of static objects. The main burden is the requirement of new methods (algorithms and hardware, e.g., neuromorphic-based approaches or spiking neural networks) to process event camera outputs to acquire data and information because traditional image processing methods are not directly applicable [44].

Table 1. Depth sensor modalities.

Techniques	Typical Sensors	Advantages	Disadvantages
Structured light	Kinect 1, Xtion PROLive, RealSense SR300 and F200	High accuracy and precision in indoor environments	Limited range, not suitable for outdoor environment due to noise from ambient light, interference from reflections and other light sources
TOF	Kinect 2	Good for indoor outdoor applications, long range, robust to illumination changes	Lower image resolution than structured light cameras, high power consumption, cost varies with resolution, rain fog can affect sensor performance
Active infrared stereo	RealSense R200, RealSense D435, D435i	Compact, lightweight, dense depth images	Stereo matching requires high processing power, struggle at high occlusions and featureless environments, relatively low range especially outdoors

Omnidirectional cameras are utilised in robotic applications where more information is needed about surrounding environments. These cameras provide a wider-angle view than conventional cameras, which typically have a limited field of view. Due to the lens configuration of omnidirectional cameras, the obtained images have distortions. Therefore, these cameras require different mathematical models, like the unified projection model, to correct image distortions. A summary of these different camera types and their advantages and disadvantages is shown in Table 2.

Table 2. Camera configurations.

Configurations	Advantages	Disadvantages
Monocular	Compactness, low hardware requirements	No direct depth measurements
Stereo	Depth measurements, low occlusions	Fails in featureless environments, CPU intensive, accuracy/range depends on camera quality
RGB-D	Colour and depth information per pixel	Limited range, reflection problems on transparent, shiny, or very matte and absorbing objects
Event	High temporal resolution, suitable for changing light intensities, low latency [45]	No direct depth information, costly, not suitable for static scenes, requires non-traditional algorithms
Omni-directional	Wide angle view (alternative to rotating cameras)	Lower resolution, needs special methods to compensate for image distortions

- Active Ranging Sensors

Active sensors emit energy to the environment and measure the return signal from the environment. There are several types of active-ranging sensors, such as reflectivity, ultrasonic, laser rangefinder (e.g., LiDAR), optical triangulation (1D), and structured light (2D). The commonly used LiDAR sensor is an active-ranging sensor that shows improved performance compared to ultrasonic TOF sensors for perception in autonomous navigation systems [46]. LiDAR imaging is one of the most studied topics in the optoelectronics field. The TOF measurement principle is used by LiDAR sensors for obtaining depth measurements in different environments. Rotating LiDAR imagers were the first type to successfully achieve acceptable performance using a rotating-mirror mechanical configuration with multiple stacked laser detectors [47]. New trends in LiDAR technology are in the development of low-cost, compact, solid-state commercial LiDAR sensors [48,49]. The three most widely used LiDAR techniques utilise pulsed, amplitude-modulated, and frequency-modulated laser beams. The most common commercially available method is to use a pulsed laser beam, which directly measures the time taken by the pulsed signal to return to the sensor. In such sensors, time measurements require resolutions in picoseconds (high-speed photon detectors). Therefore, the cost of pulsed LiDAR sensors is comparably higher than the other two methods for equivalent range and resolution [50]. The range of a pulsed LiDAR is limited by the signal-to-noise ratio (SNR) of the sensor. Pulsed LiDAR sensors are suitable in indoor and outdoor environments due to their instantaneous higher peak pulse contrast to ambient irradiance noise. However, Amplitude Modulated Continuous Wave (AMCW) LiDAR sensors with the same average signal power have lower continuous wave peaks and, hence, are vulnerable to solar irradiance. AMCW LiDAR sensors are popular in indoor robot applications due to low SNR. In general, there is no significant depth resolution difference between pulsed LiDAR and AMCW LiDAR, but pulsed LiDAR may outperform AMCW LiDAR accuracy at the same optical power levels because of higher SNR. Increasing the modulation frequency of AMCW LiDAR sensors improves depth resolution but reduces ambiguity distance. Thus, pulsed sensors have higher ranges compared to AMCW sensors. AMCW sensors also have slow frame rates relative to pulsed sensors and usually underperform in dynamic environments. Frequency Modulated Continuous Wave (FMCW) sensors have higher depth resolution in comparison to pulsed and AMCW LiDAR sensors. These sensors can measure the depth and velocity of targets simultaneously, a highly advantageous feature for the autonomous vehicle industry. These sensors can also avoid interference from other LiDAR sensors because of the frequency modulation. FMCW sensors, however, require accurate frequency sweeps to generate emitter signals, which is a challenging task. FMCW sensor technology has been in continuous development but has not yet established itself at the commercial level. A summary of these three LiDAR technologies is provided in Table 3.

Table 3. LiDAR sensor types.

Configurations	Advantages	Disadvantages
Pulsed	High frame rate	Low depth resolution, higher inference from other LiDAR sensors
AMCW	Not limited by low SNRs; however, not effective at very low SNRs	Low accuracy than FMCW, lower depth resolution than FMCW
FMCW	Velocity and range detection in a single shot, higher accuracy than AMCW, higher depth resolution	Currently at the research and development stage

The concept of robotic perception pertains to various robotics applications that utilise sensory information and modern deep learning methods. These applications include identifying objects, creating representations of the environment, comprehending scenes, detecting humans/pedestrians, recognising activities, categorising locations based on meaning, scene reconstruction, and others. Towards the goal of fully autonomous robot navigation, robust and accurate environmental perception is a necessity. Passive RGB cameras are relatively low cost and can capture rich environment details. However, their perception abilities are vulnerable to environmental illumination changes and occlusions. Therefore, sensor fusion methods are important to gain robust perception.

2.2. LiDAR and Camera Data Fusion

Achieving a reliable real-time understanding of external 3D environments is very important for safe robot navigation. Visual perception using cameras is commonly employed in many mobile robotic systems. Camera images can be efficiently and often effectively processed by CNN-based deep learning architectures. However, relying on one sensor can lead to robustness challenges, particularly in applications like self-driving vehicles, autonomous field robots, etc. Therefore, different sensor modalities are often combined to achieve better reliability and robustness for autonomous systems. The fusion of LiDAR and camera data is one of the most investigated sensor fusion areas in the multimodal perception literature [51]. Camera and LiDAR fusion has been applied in various engineering fields and has shown better robustness than camera-only robot navigation approaches [52]. This fusion strategy is more effective and popular than other sensor fusion approaches such as radar-camera, LiDAR-radar, and LiDAR-thermal camera. Still, the technical challenges and cost of sensors and processing power requirements have constrained the application of these methods in more general human activities. Recent deep-learning algorithms have significantly improved the performance of camera-LiDAR fusion methods [53], with monocular and stereo cameras mainly used with LiDAR sensors to fuse images and point cloud data [54].

Deep learning-based LiDAR and camera sensor fusion methods have been applied in depth completion, object detection, and semantic and instance segmentation. Image and point cloud scene representations include volumetric, 2D multi-view projection, graph-based and point-based. In general, most of the early methods fuse LiDAR and camera image semantics using 2D CNNs. Many 2D-based networks project LiDAR points onto respective image planes to process feature maps through 2D convolutions [55–58]. Several works have used point cloud processing techniques such as PointNet [59] to extract features or 3D convolutions [60] to detect objects in volumetric representations [61]. Some other LiDAR and image fusion methods use 2D LiDAR representations for feature fusion and then cluster and segment 3D LiDAR points to generate 3D region proposals [62]. Voxel-based representations and multi-view camera-LiDAR fusion approaches are utilised to generate 3D proposals in object detection. State-of-the-art camera-LiDAR semantic segmentation methods employ feature fusion methods to obtain 2D and 3D voxel-based segmentation results. Multi-view approaches map RGB camera images onto the LiDAR Bird's-Eye-View (BEV) plane to align respective features from the RGB image plane to the BEV plane [63–66], and several other methods propose to combine LiDAR BEV features with RGB image features directly [51,54]. These direct mapping methods use trained CNNs to align image features with LiDAR BEV features from different viewpoints.

Computer vision has been a rapidly growing field in the past decade, and the development of machine learning methods has only accelerated this. Recently, deep learning strategies have influenced the rapid advancement of various computer vision algorithms. Computer vision includes subtopics like object detection, depth estimation, semantic segmentation, instance segmentation, scene reconstruction, motion estimation, object tracking, scene understanding and end-to-end learning [41]. Computer vision methods have been applied to a great extent in emerging autonomous navigation applications. However, these vision techniques may be less effective in previously unseen or complex environments and highly rely on the trained domain. Therefore, continuous improvements are being made towards the development of fully autonomous systems. Several state-of-the-art benchmarking datasets have been utilised to compare the performance of different autonomous driving vision methods. KITTI [67], Waymo [68], A2D2 [69], nuScenes [70], and Cityscapes [71] are some examples of these state-of-the-art autonomous driving datasets.

- Dense Depth Prediction

Dense depth completion is a technique that estimates dense depth images from sparse depth measurements. Achieving depth perception is of significant importance in many different engineering industries and research applications such as autonomous robotics, self-driving vehicles, augmented reality, and 3D map construction. LiDAR sensors, monocular cameras, stereo cameras, and RGB-D cameras have been the most utilised in dense depth estimation applications, but these sensors have specific limitations in the estimation process. LiDAR sensors with high accuracies are costly to use in large-scale applications. Three main challenges have been identified in LiDAR depth completion [72]. The first relates to the fact that, in general, even expensive LiDAR sensors produce sparse measurements for distant objects. In addition, LiDAR points are irregularly spaced compared to monocular RGB images. Therefore, it is non-trivial to increase the depth prediction accuracy using the corresponding colour image. Secondly, there are difficulties in combining multiple sensor modalities. The third challenge is that depth estimation using deep learning-based approaches has limitations with regard to the availability of pixel-level ground truth depth labels for training networks. Another possible approach to depth estimation is by using stereo cameras. Stereo cameras, however, require accurate calibration, demand high computational requirements, and fail in featureless or uniformly patterned environments. RGB-D cameras are capable of depth sensing but have a limited measuring range and poor performance in outdoor environments. A technique called depth inpainting can be used as a depth completion method for structured light sensors like the Microsoft Kinect 1, and these sensors produce relatively dense depth measurements but are generally only usable in indoor environments. Dense depth estimation techniques generally up-sample sparse and irregular LiDAR depth measurements to dense and regular depth predictions. Depth completion methods, however, still have a variety of problems that need to be overcome. These challenges are primarily sensor-dependent, and solutions should overcome respective difficulties at the algorithm development stage.

Many state-of-the-art dense depth prediction networks combine relatively dense depth measurements or sparse depth maps with RGB images to assist the prediction process. In general, retrieving dense depth detail from relatively dense depth measurements is easier than from sparse depth maps. In relatively dense depth images, a higher percentage of pixels (typically over 80%) have observed depth values. Therefore, in similar scenarios, predicting dense depth is relatively less complex. However, in autonomous navigation applications, 3D LiDAR sensors account for only approximately 4% of pixels when depth measurements are mapped onto the camera image space, which creates challenges in generating reliable dense depth images [72].

- Dense Depth from Monocular Camera and LiDAR Fusion

Depth estimation based solely on monocular images is not reliable or robust. Therefore, to address these monocular camera limitations, the LiDAR-monocular camera fusion-based depth estimation process has been proposed by researchers. Using monocular RGB images and sparse LiDAR depth maps, a residual network learning-based autoencoder decoder network was introduced by [73] to estimate dense maps. However, this method needs a ground truth depth image when retrieving sparse depth images during the network training process. In practice, obtaining such ground truth images is not simple or easily scalable [72]. To mitigate the requirement of a ground truth depth image, ref. [72] presented a self-supervised model-based network that only requires a monocular RGB image sequence and LiDAR sparse depth images in the network training step. This network consists of a deep regression model to identify a one-to-one transformation from a sparse LiDAR depth map to a dense map. This method achieved state of the art performance on the KITTI dataset and considers the pixel-level depth estimation as a deep regression problem in machine learning. LiDAR sparse depth maps use per-pixel depth, and pixels without measured depth are set to zero. The proposed network follows an encoder-decoder architecture. The encoder has a sequence of convolutions, and the decoder has a set of transposed convolutions to up sample feature spatial resolutions. Convolved sparse depth data and colour images are concatenated into a single tensor and input to residual blocks of ResNet-34 [74]. The self-supervised training framework requires only colour/intensity images from monocular cameras and sparse depth images from LiDAR sensors. In the network training step, a separate RGB supervision signal is used (a nearby frame). LiDAR sparse depth can be used as a supervision signal. However, this framework requires a static environment to be able to warp the second RGB frame to the first one.

With this implementation, the root-mean-square of the depth prediction error reduces as a power function with the increment of the resolution of the LiDAR sensor (i.e., the number of scan lines). One of the limitations of this approach is that the observed environment needs to be stationary. If not, the network will not generate accurate results. Large moving objects and surfaces that have specular reflectance can cause the process of training the network to fail. These reasons reduce the applicability of this method in dynamic situations that are often present in outdoor environments. In addition, this network training process may become stuck in the local minimums of the photometric loss function due to improper network weight initialisation. This effect may result in output depth images that are not close enough to the ground truth because of the erroneous training process.

In [75], a real-time sparse-to-dense map is constructed by using a Convolutional Spatial Propagation Network (CSPN). The propagation process preserves LiDAR sparse depth input values in the final depth map. This network aims to extract the affinity matrix for respective images. The introduced method learns the affinity matrix by using a deep convolution neural network. The training process of the network model is achieved by incorporating a stochastic gradient descent optimiser. This network implementation showed memory paging cost as a dominant factor when larger images were fed into the PyTorch-based network. The CSPN network has shown good performance in real-time and thus is well-suited for applications such as robotics and autonomous driving. CSPN++ [76] is an improved version of the CSPN network with adaptively learning convolutional kernel sizes and numbers of iterations for propagation. The network training experiments were carried out using four NVIDIA Tesla P40 Graphic Processing Units (GPUs) on the KITTI dataset. This research shows that hyper-parameter learning from weighted assembling can lead to significant accuracy improvements, and weighted selection could reduce the computational resource with the same or better accuracy compared to the CSPN network.

- Dense Depth from Stereo Camera and LiDAR Fusion

Estimating depth using stereo cameras provides more reliable results compared to monocular cameras. LiDAR sensors can produce depth measurements with improved accuracy over increased ranges and varying lighting conditions. The fusion of LiDAR and stereo camera sensors produces more accurate 3D mappings of environments than LiDAR-monocular camera depth completion. However, stereo cameras commonly have shorter detection ranges, and depth estimation becomes challenging in textureless environments and high occlusion scenarios. One of the significant works in LiDAR-stereo camera fusion is presented in [77]. This paper presents the first unsupervised LiDAR-stereo fusion network. The network does not require dense ground truth maps, and training is done in an end-to-end manner that shows a broad generalisation capability in various real-world scenarios. The sparsity of LiDAR depth measurements can vary in real-world applications. One advantage of the network proposed in this work is that it handles a high range of sparsity up to the point where the LiDAR sensor has no depth measurements. A feedback loop has been incorporated into the network to connect outputs with inputs to compensate for noisy LiDAR depth measurements and misalignments between the LiDAR and stereo sensors. This network is currently regarded as one of the state-of-the-art methods for LiDAR-stereo camera fusion.

- Multimodal Object Detection

Reliable object detection is a vital part of the autonomous navigation of robots/vehicles. Object detection in autonomous navigation is described as identifying and locating various objects in an environment scene in the form of bounding boxes, including dynamic objects and static objects. Object detection may become difficult due to sensor accuracy, lighting conditions, occlusions, shadows, reflections, etc. One major challenge in object detection is occlusion, which consists of different types. The main occlusion types are self-occlusion, inter-object occlusion, and object-to-background occlusion [78]. Early image-based object detection algorithms commonly included two steps. The first stage was dividing the image into multiple smaller sections. Then, these sections were conveyed into an image classifier to identify whether the image section contained an object or not. If any object was detected in an image section, the respective portion of the original image was marked with the relevant object label. The sliding window approach is one way of achieving the above-mentioned first step [79].

A different set of algorithms uses a technique in contrast to the sliding window approach by grouping similar pixels of an image to form a region. These regions are then fed to a classifier to identify semantic classes (with grouping done using image segmentation methods). Further improved image segmentation can be achieved by using the selective search algorithm [80]. The selective search algorithm emphasises a hierarchical grouping-based segmentation algorithm. In this method, initially detected image regions are merged in a stepwise manner by selecting the most similar segments until the whole image represents a single region. These regions resulting from each step are added to the region proposals and fed to a classifier. The classifier performance depends on the used region proposal method. This object detection approach does not produce real-time performance suitable for autonomous navigation applications. However, advances such as SSPnet [81], Fast Regional-based Convolutional Neural Networks (R-CNNs) [82], and Faster R-CNN [83] were introduced to address this issue. The Faster R-CNN network generates a feature map by utilising the CNN layer output, and the region proposal generation is achieved by sliding a window (comprising three different aspect ratios and sizes) over the feature map. Each sliding window is mapped to a vector and fed to two parallel classification and regression networks. The classification network calculates the probability of region proposals containing objects, and the regression network indicates the coordinates of each of the proposals.

Object detection research has been mainly employed in the autonomous vehicle industry (for vehicle and pedestrian detection [84]) and mobile robotics. In contrast to camera-only object detection, sensor fusion has been implemented in different real-world applications to obtain more accurate and robust detection results. As previously discussed, LiDAR and camera sensor fusion are some of the most used and highest-performing sensor fusion methods. LiDAR and camera sensor fusion object detection approaches consist of two main techniques. These are the sequential and one-step models [54]. Sequential models use 2D proposal-based methods or direct 3D proposals to detect objects. In the sequential approach, 2D/3D regions are proposed in the first stage, and then 3D bounding box regression is done in the second stage. The 2D/3D region proposal stage incorporates fused image-point cloud regions that may contain objects. In the bounding box regression stage, feature extraction from region proposals and bounding box prediction is done. One-step models generate region proposals and achieve bounding box regression in parallel in a single step. The 2D proposal-based sequential approach uses 2D image semantics to generate a 2D proposal and point cloud processing methods to detect dynamic objects. This approach utilises already developed image processing models to identify 2D object proposals and then project these proposals to LiDAR 3D point cloud space for object detection.

Two approaches are mainly used to manipulate image-based 2D object proposals and irregular 3D LiDAR data. In the first method, image-based 2D bounding boxes are projected to the LiDAR 3D point cloud to implement 3D point cloud object detection algorithms. The second approach utilises point cloud projections on the 2D images and applies 2D semantic segmentation techniques to achieve point-wise semantic labels of the points within the semantic regions [54]. LiDAR-camera 2D sequential object detection methods include result, feature, and multi-level fusion strategies. These 2D proposal-based result-level fusion methods incorporate image object detection algorithms to retrieve 2D region proposals. These retrieved 2D object bounding boxes are then mapped onto 3D LiDAR point clouds. The enclosing points in frustum proposals are transferred into a point cloud-based 3D object detection algorithm [57]. The overall performance of this object detection approach depends on the modular behaviour of the 2D detection architecture.

Sequential fusion may lose complementary data in LiDAR point clouds due to initial 2D image object proposal detection. Two-dimensional proposal-based feature-level fusion uses a mapping from 3D LiDAR point clouds onto the respective 2D images and employs image-based techniques for image feature extraction and object detection. One of these approaches appends per-point 2D semantic details as additional channels of LiDAR 3D point clouds and uses an existing LiDAR-based object detection method [85]. However, this approach is not optimal for identifying objects in a three-dimensional world because 3D details in the point clouds may be lost due to the projection.

Multi-level fusion combines 2D result-level fusion with feature-level fusion. This approach uses already available 2D object detectors and generates 2D bounding boxes. Then, points within these bounding boxes are detected. Subsequently, image and point cloud features are combined within the bounding boxes to estimate 3D objects. LiDAR and camera object detection using 3D proposal-based sequential models avoid 2D to 3D proposal transformations and directly generate 3D proposals from 2D or 3D data. This technique consists of two approaches, namely multi-view and voxel-based. MV3D [66] is a multi-view object detection network that uses LiDAR and camera data to predict the full 3D envelope of objects in the 3D environment. A deep fusion scheme was proposed to fuse features from multiple views in respective regions. The detection network comprises two networks: the 3D proposal network and the region-based fusion network. As the inputs, the LiDAR BEV, LiDAR front view and the RGB camera image are fed to the network. The LiDAR BEV is fed to the 3D proposal network to retrieve 3D box proposals. These proposals are used to extract features from the LiDAR front-view and camera RGB image inputs. Then, using these extracted features, the deep fusion network predicts object size, orientation, and location in the 3D space. The network was built on the 16-layer VGGnet [86], and the KITTI dataset was used for the training process.

One of the drawbacks of the multi-level fusion method is the loss of small objects in the detection stage due to feature map down-sampling. Combining image and point cloud feature maps by RoI (Regions of Interest)-pooling decreases the fine-grained spatial details. MVX-Net [65] introduces a method to fuse point cloud and image data voxel-wise or point-wise. Two-dimensional CNNs are used for the image feature extraction process, and a VoxelNet [87] based network detects 3D objects in the voxel representation. The input 3D LiDAR point cloud is mapped to the 2D image for image feature extraction in the point-wise fusion method, and then voxelisation and processing are done using VoxelNet. In voxel-wise fusion, the point cloud is firstly voxelised and then projected onto the image-based 2D feature representation to extract features. This sequential approach achieved state-of-the-art performance for 3D object detection at the time of its publication. Object detection utilising one-stage models performs object proposal retrieval and bounding box prediction in a single process. These detection models are suitable for real-time autonomous robot decision-making scenarios. State-of-the-art single-stage object detection methods, such as [88], simultaneously process depth images and RGB images to fuse points with image features, and then the generated feature map is used for bounding box prediction. The introduced method [88] utilises, two CNN-based networks to parallelly process point cloud and RGB front-view images. One CNN identifies LiDAR features, and the other CNN extracts RGB image features. Then, these RGB image features are mapped into the LiDAR range view. Finally, mapped RGB and LiDAR image features are concatenated and fed into LaserNet [89] for object detection and semantic segmentation. This network has been trained in an end-to-end manner. The network training was done for 300K iterations with a batch size of 128, distributed over 32 GPUs. The image fusion, object detection, and semantic segmentation process took 38 milliseconds on an Nvidia Titan Xp GPU.

- Multimodal Semantic Segmentation

Scene semantic segmentation assigns a semantic category label to each pixel in a scene image, and it can be regarded as a refinement of object detection [90]. A scene can incorporate obstacles, free space, and living creatures (not necessarily limited to these categories). The complete semantic segmentation of images applies these semantic categories to all pixels across an image. Many recent computer vision methods rely on CNN architectures. These networks favour dense depth image data over sparse data. In general, LiDAR sensors produce irregular, sparse depth measurements. Ref. [73] introduced a technique to utilise LiDAR sparse data and RGB images to achieve depth completion and semantic segmentation in a 2D view. This network can work with sparse depth measurement densities as low as 0.8%, and at the time of its publication, this method showed state-of-the-art performance on the KITTI benchmark dataset. The base network of this prior work was adopted from NASNet [91], which has an encoder and decoder architecture. Using LiDAR and RGB image feature fusion, ref. [92] proposed a novel method to achieve 2D semantic segmentation in 2D images. This method introduced a self-supervised network that suits different object categories, geometric locations, and environmental contexts. This self-supervised network uses two sensor modality-specific encoder streams, which concatenate to a single intermediate encoder and then connect into a decoder to fuse the complementary features. The segmentation part is achieved with a network termed AdapNet++ [92] that consists of an encoder-decoder architecture. All these network models were implemented using the deep learning TensorFlow library. Another high-performing deep learning-based LiDAR-Camera 2D semantic segmentation method was presented in [93]. In this method, the generated 3D LiDAR point data is mapped to the 2D image and up-sampled to retrieve a 2D image set that consists of spatial information. Then, fully convolutional networks are used to segment the image using three approaches: signal level, feature, and cross-fusion. In the cross-fusion method, the network was designed to learn directly from the input data.

The techniques discussed up to now have been 2D semantic segmentation methods. In contrast to 2D methods, 3D semantic segmentation approaches provide a realistic 3D inference of environments. An early approach for a 3D scene semantic segmentation network is presented in [94] termed 3DMV. This method requires relatively dense depth scans along with RGB images, and it was developed to map indoor scenarios. Voxelised 3D geometries are built by using LiDAR depth scans. Two-dimensional feature maps are extracted from the RGB images using CNNs, and these image feature maps are mapped in a voxel-wise manner with the 3D grid geometry. This fused 3D geometry is then fed into 3D CNNs to obtain a per-voxel semantic prediction. The overall performance of the approach depends on the voxel resolution, and real-time processing is challenging for higher voxel resolutions. Therefore, this dense volumetric grid becomes impractical for high resolutions. The system was implemented using PyTorch and utilised 2D and 3D convolution layers already provided by the application programming interface. Semantic segmentation of point clouds is challenging for structureless and featureless regions [95]. A point-based 3D semantic segmentation framework has been introduced by [95]. This approach effectively optimises geometric construction and pixel-level features of outdoor scenes. The network projects features of detected RGB images into LiDAR space and learns 2D surface texture and 3D geometric attributes. These multi-viewpoint features are extracted by implementing a semantic segmentation network and then fused point-wise in the point cloud. Then, this point data is passed to a PointNet++ [96] based network to identify per-point semantic label predictions. A similar approach was followed by Multi-view Point-Net [97] to fuse RGB semantics with 3D geometry to obtain per LiDAR point semantic labels.

Instead of localised or point cloud representations, ref. [98] have used a high dimensional lattice representation for LiDAR and camera data processing. This representation reduces memory usage and computational cost by utilising bilateral convolutional layers. Then, these layers employ convolutions for unoccupied sections in the generated lattice representation. Firstly, the identified features of point clouds are mapped to a high-dimensional lattice, and then convolutions are used. Following this, CNNs are applied to detect image features from multi-view images, and these features are projected to the lattice representation to combine with three-dimensional lattice features. The generated lattice feature map was assessed by using 3D CNNs to obtain point-wise labels. A spatial-aware and hierarchical learning strategy has been incorporated to learn 3D features. The introduced network was capable of training in an end-to-end manner.

- Multimodal Instance Segmentation

Instance segmentation identifies individual instances within a semantic category. It is considered a more advanced semantic segmentation method. This method not only provides per-pixel semantic categories but also distinguishes object instances, which is more advantageous for robot scene understanding. However, instance segmentation in autonomous navigation introduces more challenges than semantic segmentation. Instance segmentation approaches based on fused LiDAR-camera sensor data show proposal-based and proposal-free architectures. A voxel-wise 3D instance segmentation approach was introduced by [99] that consists of two-stage 3D CNN networks. A feature map was extracted from the low-resolution voxel grid by implementing 3D CNNs. Another feature map was obtained from the RGB multi-view images using 2D CNNs and projected onto the associated voxels in the volumetric grid to append with respective 3D geometry features. Then, object classes and 3D bounding boxes are predicted by feeding these fused features to a 3D CNN architecture. In the second phase, another 3D CNN estimates the per-pixel object instances using already identified features, object classes and bounding boxes.

These voxel-based segmentation methods are constrained by the voxel resolution and require increased computation capabilities with higher grid resolutions. The application of instance segmentation in LiDAR-camera fusion for real-time systems is challenging. Some research studies had limitations, such as the system developed in [100], which does not support dynamic environments. A proposal-free deep learning framework that jointly realises 3D semantic and instance segmentation is presented in [101]. This method performs 3D instance segmentation in the BEV of point clouds. However, this approach is less effective in identifying vertically oriented objects because of the BEV segmentation approach. This method first extracts a 2D semantic and instance map from a 2D BEV representation of the observed point cloud. Then, using the mean shift algorithm [102], and semantic features of the 2D BEV, instance segmentation is achieved by propagating 3D features onto the point cloud. It should be noted that the instance segmentation approaches discussed have been developed to segment 3D point clouds from static indoor environments, and these methods have not shown any segmentation capabilities in dynamic environments.

Overall, while significant progress has been made in the perception capabilities of autonomous robots, particularly in regard to object detection and scene segmentation, many of the existing approaches have only been tested in relatively structured indoor environments, and substantial additional work may be required to adapt these techniques to be used in unstructured outdoor environments. Nonetheless, some very useful research directions have been identified that have the potential to significantly advance this field.

3. Robot Scene Understanding for Navigation Planning

For effective navigation at local and global scales, robots build maps to represent their external environments and to assist with making safe and accurate decisions at local levels. SLAM refers to the real-time combination of robot localisation and external environment mapping. SLAM is highly focused on mapping techniques. Early mobile ground robot map-building approaches comprised feature-based and photometric error-based methods. Robots could efficiently localise their positions in sparse or dense maps by implementing these mapping techniques. However, autonomous robots require more than basic 2D or 3D maps for effective task planning. Therefore, semantic SLAM techniques have been developed and investigated. Topological maps represent the environment as abstract graphs in contrast to classical metric maps. Classical metric maps are used in robotics to visualise the environment as a two-dimensional grid of cells, and each of these cells represents a specific location or area. The occupancy of a cell is typically represented by a binary value. However, these maps have limitations, such as their assumption of a flat and static environment and their inability to illustrate sensor data uncertainty.

To overcome these limitations, researchers have developed probabilistic mapping techniques, such as occupancy grid mapping, which use probabilities to represent occupancy and can handle more complex and dynamic environments. The metric maps include details such as visited locations by a robot, proximity to features and general arrangement of external environments. One of the challenges in classical 3D geometric mapping is the accumulation of position errors over time at the global map level. Topological maps are not directly affected by these error-driven shifts in the geometrical maps, which means that topological mapping is more adaptable to different environmental contexts than geometric mapping. Classical and topological mapping methods have their strengths and weaknesses. Hybrid methods combine metric and topological mapping techniques to leverage advantages of each method. Many hybrid representations have been utilised by indoor robots to generate environmental mappings. Hybrid mapping representations from recent research works generally consist of two or more layers in a hierarchy (e.g., places, categories, ontology). The research work presented in [103] uses depth measurements, semantic segmentation and scene flow to map a 3D environment. This concept can detect the dynamic behaviour of objects. Neural implicit scalable encoding for SLAM [104] introduces a SLAM framework that utilises a hierarchical feature representation. It uses a hierarchical feature grid with encoder-decoder networks to predict occupancies. The geometric maps of indoor environments are developed by using extracted coarse, mid, and fine-level feature grids. The middle and fine-level feature grids represent observed scene geometry. The feature grid in the coarse level detects indoor geometries such as the floor and walls. The coarse level occupancy is used in the scene reconstruction of unobserved environment regions. In the reconstruction process, coarse-level occupancy is optimised in the mid-level feature grid and refined at the fine level.

Kimera [105] presents a dynamic scene graph that consults a five-layer hierarchical representation. The method combines a metric semantic mesh with a topological spatial representation. It is capable of segmenting indoor structures and is robust in overcrowded environments. The five layers of the graph include metric-semantic mesh, objects and agents, places and structures, rooms, and buildings. To effectively implement these hybrid hierarchical maps in real-world applications, the representations need to be updated as the environment changes and should accurately map the interconnections between environment attributes. These hierarchical mapping techniques were implemented in indoor environments where semantic information extraction is less complex, however, and have not yet been applied to unstructured outdoor environments. Kimera-multi [106] is a state-of-the-art metric semantic SLAM system for multi-robot systems. The method is an extension of Kimera [105] and was tested in computer-simulated environments based on SLAM datasets and outdoor datasets collected by ground robots. This concept creates a 3D mesh that labels outdoor environments using semantic labels to gain high-level scene understanding. Kimera-multi outperforms the accuracy of Kimera [105] visual-inertial odometry SLAM. Overall, Robot semantic understanding requires the classification of both objects and places. To extract these predefined object models, conditional random field models, CNN-based scene and place classification algorithms, and scene graphs (e.g., robot scene graphs [107], dynamic scene graphs [105]) are utilised.

SLAM techniques can produce sparse or dense geometric or semantic maps of outdoor environments. However, these maps may not have adequate detail for safe local robot navigation where terrains consist of abrupt deviations and vegetation. Therefore, deep learning and image-based scene understanding are necessary directions to identify terrain properties and suitable environment regions for robot navigation in challenging outdoor unstructured environments. Recent robot navigation trends have emerged to segment camera egocentric images using computer vision approaches such as visual attention mechanisms [108]. Ref. [109] presents a camera-only approach for segmenting egocentric images to assist robot navigation. The segmentation procedure consists of soft labelling of the images according to three levels of driveability. For every pixel, soft labels are mapped by assessing original semantic classes. The SegNet [110] based deep convolutional encoder-decoder architecture was used for the pixel classification. Another paper [111] introduces an RGB-based method to classify terrains into navigability levels by extracting multiscale features. This work achieved state-of-the-art performance on the RUGD [112] and RELLIS-3D [113] datasets at the time of publication. A multi-head self-attention network architecture was used to classify the terrain into smooth, rough, bumpy, forbidden, obstacle, and background regions. This network requires further improvements to identify drastic terrain geometry changes (i.e., slopes) to gain higher accuracy in navigability segmentation.

In general, 2D scene understanding can lose essential details from the mapping process of 2D data to the 3D real world. Therefore, 3D scene understanding-based methods are used with the most recent robot navigation applications. Ref. [114] presented a 3D semantic segmentation method using a CNN encoder-decoder architecture. This approach inputs RGB images and depth images for the fusion of feature maps, with the FuseNet [115] architecture taken as the base for the CNN. Visual dynamic object-aware SLAM [116] combines SLAM with dynamic modelling of scenes, enabling robots to operate effectively in dynamic environments. This work introduces instance segmentation of dynamic objects with the camera trajectory. The model can identify and track stationary structures and dynamic objects and integrate them into the traditional SLAM framework. This framework does not need prior knowledge of objects and their shapes or models. The method uses monocular depth estimation to achieve instance-level segmentation and optical flow. Then, static and dynamic object features are tracked to trace the trajectories. The local map is updated in each consecutive frame. Environmental scene understanding using only monocular cameras can be challenging in mapping applications. As a result, semantic segmentation using LiDAR, radar, and stereo cameras has been investigated. The development of deep learning techniques has increased the usage of point cloud data in robot 3D scene understanding applications. VoxNet [61] is a 3D CNN architecture that was designed to obtain voxel grid representations of the environment using point cloud data. Although this network is capable of detecting objects, object detection alone is not sufficient for scene understanding. Ref. [117] used a LeNet [118] based architecture for semantic segmentation of point clouds. This approach could segment outdoor scenes into seven object categories by voxelisation of input point cloud data. Ref. [119] proposed a method for semantic segmentation of indoor scenes using only depth images. This network jointly predicts the 3D occupancy grid and semantic categories. However, some of the objects were missing during the inference due to a lack of depth details for those objects. In general, point cloud data is irregular; therefore, only representing point cloud data using uniform voxel grids might not accurately interpret real-world information. To address this adverse effect, ref. [120] proposed the OctNet architecture to represent point cloud data in a hierarchical partitioning of space using voxel representations. One of the problems of using voxel grid representations is the requirement of high computational power for varying sparsity of irregular point cloud data. Multi-view representations of 3D point clouds can be employed to reduce the adverse effects of volumetric grid representations. Ref. [121] attempted to generate 2D views of point cloud data, which were then utilised to analyse large urban point cloud data in 3D scenes using deep CNNs. Then, "tangent convolutions" were applied to the point cloud data to obtain an image by projecting local geometry to a 2D plane tangent to a point.

In general, 2D projective methods are more efficient and scalable than 3D volumetric representations. However, there may be a degree of information loss in the final 3D structure due to inaccuracies in those projections. Instead of converting point cloud data into uniform volumetric representations, processing these point clouds as unstructured data is also a feasible approach [122]. Ref. [59] proposed PointNet, one of the significant point cloud semantic segmentation works. This work shows a degree of permutation invariance to the input point cloud data. It can classify objects by allocating them to categories (e.g., TV, table) and carry out semantic segmentation in scenes. To try to improve on these point-level processing techniques, several works have been carried out that use deep learning methods [123–127]. Computer memory requirements for 3D voxel representations tend to vary according to the level of voxel resolution. In the presence of sensor noise, the estimation of the occupancy of voxels might be challenging. Instead of using 3D volumetric representations, graph and tree-based representations have been proposed to recover 3D structures from point cloud data. Ref. [128] introduces a 3D graph neural network to semantically segment the environment using 2D RGB image and point cloud data. This method uses both CNNs and graph neural networks to predict semantic classes, with the CNN used to extract features from 2D RGB images. The Superpoints graph method was introduced by [129] to learn scene contextual information and geometry. Superpoints are interest points in a point cloud that are detected and described using a neural network. They are densely sampled and designed to be efficient and robust for use in SLAM applications. First, the input point cloud is separated into geometric elements (named Superpoints), and then these Superpoints are combined considering their mutual features. Finally, graph convolutions are implemented to generate contextual information and semantic labels. This method was shown to be appropriate for the segmentation of large point clouds. Another significant work that processes orderless point cloud data using a permutation invariant deep learning architecture is So-Net [130]. This network extracts features from point clouds hierarchically. The receptive field of the CNN is controlled systematically by performing a point-to-node k-nearest neighbour search to retrieve the point cloud spatial information. This network conserves the topology attributes of the point representations.

Overall, a number of approaches show promise for improving the scene understanding capabilities of autonomous robots, although questions remain regarding how effectively they can be applied in outdoor unstructured environments, particularly with dense vegetation or other complex terrain features. Improvements in both sensor capabilities and algorithms are likely to be necessary to achieve significant progress in this area, but some promising research directions have been highlighted.

4. Mobile Robot Local Path Planning

Approaches for path planning of robots in local environments can be broadly classified into either classical methods or learning-based methods that attempt to modify the path-planning based on factors such as environmental conditions and prior experience [131,132]. The classical methods follow a modular architecture with environmental perception, planning of paths relative to generated global maps, and trajectory following. Many classical techniques are applicable for robot navigation in static environments. However, their suitability can substantially diminish in unstructured or dynamic environments [133]. In general, these methods can be used effectively in indoor mobile robot applications but may not be suitable for outdoor off-road navigation conditions (e.g., terrain with grass might be traversable despite appearing blocked, and ground with mud or sand may not be suitable for pathing, despite appearing clear), hence the requirement to investigate machine learning-based navigation approaches. Several global and local motion planning methods for ground robot navigation are compared in Tables 4 and 5.

The A* and Dijkstra algorithms have been well-researched in the past decades, and these methods have shown their potential by extensively being applied with the Robot Operating System (ROS) in many real-world robotics applications. Combining these two path-planning methods with heuristic searching is effective in relatively low-complexity 2D environments. However, these methods require heavy computational capabilities in high-dimensional environments or may struggle in unstructured and dynamic working environments.

Random sampling based path planning algorithms generally consist of BITs and RRTs. Regionally Accelerated Batch Informed Trees (RABIT) are more widely used and perform better in high dimensional and dynamic environments [134] in comparison to graph search-based path planning algorithms. Bionic-based intelligent robot path planning methods simulate the behaviours of insects to generate evolving paths. These evolutionary methods include the ant colony, particle swarm optimisation, and genetic and artificial bee colony algorithms. Many other optimised versions of these algorithms have been proposed to improve calculation efficiency and avoid local minimum problems. A welding robot system developed in [135] had a combination of genetic and particle swarm optimisation algorithms to solve for the shortest route while avoiding obstacles. The artificial potential field method, ant colony optimisation, and geometry optimisation path methods were combined in another research study [136] to search the globally optimal path in 2D scenarios using computer simulations. This method showed fast solutions and reduced the risk of trapping robots in local minimum points. A multi-objective hierarchical particle swarm optimisation method has been proposed by [137] to plan global optimal path trajectories in cluttered environments. This method utilises three layers to generate the robot navigation trajectories. The triangular decomposition method [138] is applied in the first layer, the Dijkstra algorithm is used in the second, and a modified particle swarm optimisation algorithm is applied in the last layer.

Table 4. Global path planning algorithms.

Algorithms	Advantages	Disadvantages
Dijkstra	The calculation strategy is not complex and gives the shortest path	The increment of traversal nodes complicates the calculations
A*	In static environments, the algorithm search efficiency is high	Not appropriate for dynamic environments
D*	Good for dynamic environment path planning and more efficient than A*	Planning longer paths via D* creates challenges
RRT	Fast convergence, high search capability	Algorithm efficiency is low in unstructured environments
Genetic	Appropriate for complex environments, good for finding optimal paths	Low algorithm convergence speed, low search ability in local paths
Ant colony	Appropriate for complex environments, can be combined with other heuristic-based path planners	Slow convergence rate, easily trapped in local minima
Particle swarm optimisation	High convergence rate, good robustness	Frequently, solutions converge into local optimal solutions

Table 5. Local path planning algorithms.

Algorithms	Advantages	Disadvantages
Artificial potential field	Can be implemented for 3D path planning, and can solve the local minimum problem	Cannot guarantee the optimal solution
Simulated annealing	Flexible and easy implementation, can deal with noisy data and non-linear models	Can produce unstable results, the trade-off between accuracy and speed
Fuzzy logic	Strong robustness, decrease the dependencies between environmental data	Needs accurate prior knowledge, poor learning capabilities
Neural network	Strong robustness, and learning ability from experiences	Low path planning efficiency
Dynamic window	Good self-adaptation to environments	Not appropriate for unstructured complex environments

In general, the local path planning strategies that have been discussed use the available sensor data of robots regarding their surroundings to map, understand, and generate local paths while avoiding obstacles. These local path planning methods are effective in mobile robotic applications because the data captured by sensors varies in real-time in response to the dynamics of environments. In comparison to global path planning strategies, local path planning is more critical for practical robot operation and usually serves as a bridge between global planning and the direct control of robots. However, local path planners have one notable drawback in that they often lead robots to local minimum points. Many classical local path planning algorithms can generate optimal paths in the local environments while avoiding local minimum problems. These methods include techniques such as the Fuzzy Logic algorithm, artificial potential field method, and simulated annealing algorithm. In general, these methods do not evaluate the relative velocities between robots and dynamic environment objects, however, which can lead to difficulties. In many worst-case scenarios, even the velocity profiles of these obstacles can be hard to acquire for the robots. Visual-inertial odometry methods show success in outdoor environment navigation scenarios but are not suitable for navigating in off-road conditions without pre-built maps or GPS assistance.

Machine learning methods have been applied for mobile robot navigation to learn semantic information [139–141] and statistical patterns [142,143] of environments. Several other research works [144–147] have used machine learning to achieve robustness in path following. In recent years, many Reinforcement Learning (RL) and imitation learning methods and approaches based on self-supervised learning [148–151] have been applied in mobile robot navigation policy design and for training supervision. Many classical modular and deep learning-based approaches have been utilised in the navigation modules of outdoor mobile robots [152]. RL uses strategies to learn optimal robot decisions from experiences. The interconnection between environments and robots is modelled as a Markov Decision Process (MDP). Robots receive rewards as feedback signals for training while traversing different environments. The basic RL process is illustrated in Figure 2. RL methods can be separated into model-free and model-based learning scenarios. In model-free RL, the robots are not required to evaluate the MDP model rewards or policies directly and can obtain these directly through what the robot experiences. Model-free RL approaches have several subcategories such as value-based, policy-based, and Actor-Critic (improved versions of the policy-based algorithms) [153,154].

In value-based methods, optimal policies are obtained by iteratively updating the value functions. Policy gradient-based methods directly approximate a policy network and update the policy parameters to obtain an optimal policy that maximises the reward value. Deep Q network (DQN) and Double DQN are the two main value-based DRL methods. A DQN-based end-to-end navigation method has been introduced in [155]. In this work, a feature-extracting network used an edge segmentation method to improve the efficiency of the network training process. The simulated models were transferred to the real world without significant performance loss. Discrete robot actions were implemented based on a grid map. The robot exploration framework is divided into decision, planning, and mapping modules. The learning-based decision module has shown good performance, efficiency, and adaptability in novel environments. A graph-based technique has been implemented for the mapping module. The applied path planning module includes the A* algorithm as the global planner along with a timed elastic band [156] local planner.

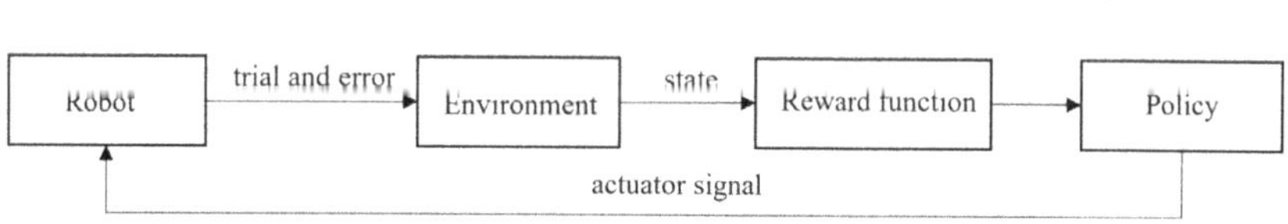

Figure 2. Robot RL approach.

Value-based DRL methods produce discrete actions, meaning they are not appropriate for the continuous robot action space. The policy-based DRL methods provide continuous motion commands for robots. Combining the policy gradient with the value function creates the Actor-Critic RL method. In general, Actor-Critic algorithms are well-suited for the continuous motion space of autonomous mobile robots. The Actor approximates the policy used to generate actions in the environment. The Critic architecture is accountable for using a reward function to evaluate robot actions iteratively and guides the Actor in successive iterations. The Critic uses deep learning value functions such as DQN and Double DQN to evaluate each iteration step. The Actor-Critic approach includes learning methods such as Depth Deterministic Policy Gradient algorithm (DDPG), Trust Region Policy Optimisation (TRPO), Proximal Policy Optimisation (PPO), Asynchronous Advantage Actor-Critic (A3C), and Soft Actor-Critic (SAC) [21]. An extended Actor-Critic algorithm was proposed in [157]. The visual navigation module of the deep neural network consisted of depth map prediction and semantic segmentation auxiliary tasks. The proposed learning network requires an image of the target and an observed image as inputs. This network architecture was proposed to obtain a visual navigation policy for indoor environments. A summary of DRL motion planning methods is included in Table 6.

Table 6. DRL motion planning methods.

Algorithms	Advantages	Disadvantages
DQN	Updates are done offline, are not complex, and are reliable	Only discrete motions
DDPG	High sample efficiency, less data correlation and faster convergence compared to DQN	The poor generalisation of novel environments
TRPO	Ensure stable convergence	Too many assumptions, may create large errors
PPO	Simplified solution process, good performance and easier to implement compared to TRPO	Low sampling efficiency
A3C	Asynchronous parallel network training, fast convergence, suitable for multi-robot systems	Require large training data, difficult to migrate model to real world
SAC	Better robustness and sample efficiency compared to the above methods	Bulky model size

Model-based robot RL methods generate models of external environments using supervised training and implement value functions to learn actions that maximise the return. Model-based RL has faster convergence and high sample efficiency. Ref. [158] presents a biped robot that learns on a rotating platform by combining model-based and model-free machine-learning methods. This paper has addressed the overfitting problem of model-based robots. The research has simulated the robot in a 2D scenario and has shown a reduction in learning time compared to model-free RL algorithms. Imitation learning is another related approach that uses demonstrations by experts operating robots for specific tasks to obtain policy functions. Mobile robots learn mappings between observations of these demonstrations and appropriate robot actions. Path trajectories can be rapidly generated by manipulating learned policies from expert demonstrations. However, the capacity of the robot to practically record these demonstration results from real-world experiments can be challenging. Ref. [159] has introduced a method to acquire a mapping from actions to states using ego-centric videos collected by a human demonstrator using a mobile phone camera. The policy learning step was executed within a simulation platform; then, the developed navigation policy was tested on a Clearpath Jackal wheeled robot in an indoor environment. The robot and the trained videos had view-point mismatches, but the model was robust to those changes. The system was successfully able to map camera sensor inputs to actuator commands using the developed imitation learning policy.

Inverse RL is another approach capable of learning reward functions from expert demonstrations. However, inverse RL is more appropriate for exploring novel environments because imitation learning tries to directly follow a demonstrator rather than improve beyond that knowledge level. The creation of an inverse RL network by employing a vision-based imitation learning method was presented by [160]. This method approximates a value function from one middle layer of a policy trained by imitation learning. The related experiments were carried out on a real-world setup and using the ROS Gazebo simulation platform. This proposed method has shown the capability of generalising the system for unseen environments by producing usable cost maps. Traditional geometric-based SLAM procedures lack the capacity to capture dynamic objects in external environments and are thus only suitable for planning robot actions in static environments. RL-based techniques, however, can learn policies by considering dynamic obstacles in outdoor environments, although the sample efficiency is lower compared to imitation learning and model-based learning. The implementation of learning-based maps and traditional path planners is one way to improve sample efficiency. A mobile robot affordance map generation using the RL policy-based method was introduced by [161]. This research used metric cost maps (attaching semantics and geometry) for robot navigation. The A* classical path planner was implemented for path generation. This learning-based SLAM approach was implemented using simulation software. Semantic, dynamic, and behavioural attributes of unseen environments have been learnt by the model using the simulation scheme. Local path planning remains a highly challenging task, particularly in environments with dynamic actors or challenging terrain features such as unpredictably varying topography or varying ground conditions. Modern machine learning approaches, coupled with advanced SLAM techniques, show promise for enabling effective navigation planning, but methods that can effectively handle all of these challenges do not yet exist, and considerable further work will be required to enable fully autonomous navigation in all conditions.

5. Summary of the Current State-of-the-Art Techniques

Autonomous navigation can be broadly categorised into the classical modular pipeline methods and end-to-end learning approaches. The modular pipeline approach has limitations due to the requirement for high levels of human intervention in designing the modules, the loss of information through each module, and the overall lack of robustness when conditions vary beyond anticipated limits. The end-to-end learning approach suffers from problems such as over-fitting and poor generalisation. Both modular and end-to-end robot navigation approaches rely on sensors to capture information about the environment or internal robot attributes. Researchers have investigated different sensor modalities to improve robot perception, including raw input types such as sound, pressure, light, and magnetic fields, as well as common robot perception sensor modalities such as cameras, LiDAR, radar, sonar, GNSS, IMU, and odometry sensors. It is crucial to have a reliable real-time understanding of external 3D environments to ensure safe robot navigation. While cameras are commonly used in mobile robotics, relying on a single sensor is not always robust. Different sensor modalities are often combined in robotic applications to improve perception reliability and robustness. Camera and LiDAR fusion has been a popular approach in the robotics community. This fusion method has shown better performance than other vision-based dual sensor fusion approaches, and recent advances in deep learning algorithms have further improved its performance. Monocular and stereo cameras are commonly used with LiDAR sensors to fuse images and point cloud data. However, the technical challenges and cost of sensors and processing power requirements still limit the widespread application of these methods for regular use. Computer vision has experienced rapid growth in the past decade, with deep learning methods accelerating its development. Object detection, depth estimation, semantic segmentation, instance segmentation, scene reconstruction, motion estimation, object tracking, scene understanding, and end-to-end learning are among the subtopics of computer vision. These methods have been widely applied in autonomous navigation, but their accuracy and reliability can be limited, prompting ongoing efforts to improve their quality. Benchmarking datasets, such as KITTI, Waymo, A2D2, nuScenes, Cityscapes, RELLIS-3D, RUGD, Freiburg [29], and WildDash [162], have been used to compare the performance of different vision methods in autonomous urban driving and off-road driving applications.

Semantic SLAM techniques provide geometric and semantic details of external environments. However, existing semantic SLAM techniques may not be adequate for the safe navigation of robots in outdoor unstructured environments due to challenges in feature detection, uneven terrain conditions (which can lead to robot localisation errors, detection errors, etc.) and effects of wind on vegetation, which can lead to sensor noise and ambiguous results. Therefore, terrain traversability analysis and improved scene understanding using deep learning methods have been developed that are showing promise in assisting robot navigation in outdoor unstructured environments.

Many classical robot path-planning approaches are used for navigation in local environments. These classical approaches rely on a modular architecture to perceive the environment, plan paths relative to generated maps, and follow trajectories. However, these classical methods are often challenged in dynamic or deformable environments and are not well-suited for off-road navigation conditions where terrains may be unpredictable. Learning-based methods have been developed to improve robot path-planning abilities under different environmental conditions and have shown promise in addressing these challenges. In particular, learning-based navigation methods have been shown to be better suited for off-road navigation conditions, where terrains can be highly variable and unpredictable.

6. Research Challenges and Future Directions in Unstructured Outdoor Environment Navigation

Despite the significant progress that has been made in improving the abilities of robots to perceive their environment, localise themselves within it, and plan paths to navigate through it; significant challenges remain with the application of these methods to unstructured outdoor environments. Many limitations exist with respect to perception capability, particularly when sensor and computation costs are a factor, and effective methods for robust scene understanding within such environments have not yet been demonstrated, substantially hindering the ability to deploy fully autonomous robots into such environments. This section of the paper explores the current state of these significant research challenges and provides suggestions for the most promising directions for future research efforts to overcome them.

6.1. Research Challenges

SLAM methods have been extensively used for robot navigation applications in indoor environments. Urban autonomous navigation systems also implement SLAM algorithms successfully for sophisticated motion planning. Only a few applications have demonstrated the use of SLAM for dirt roads and off-road uneven terrain navigation. Many visual SLAM techniques are unstable in off-road environments with trees, bushes, uneven terrain and cluttered objects. Application of classical SLAM approaches in off-road environments with vegetation is challenging because of the irregular and dynamically varying structure of the natural environment. LiDAR-based methods may detect traversable regions as geometric obstacles. Therefore, existing SLAM research on robot navigation in off-road unstructured environments is limited. Several feature-based SLAM algorithms fail due to reasons such as the absence of specific features or the loss of detected features caused by the dynamic movement of vegetation due to wind. In general, mobile robots in off-road environments in the presence of vegetation use GNSS with IMUs to estimate the real-time position of robots. In relatively open tree canopies, GNSS sensors can establish a consistent communication link with satellites. The GNSS positioning accuracy in sparse tree canopy environments can be decimeter level [163]. However, under dense tree canopies, GNSS may lose location data due to potential signal interruptions.

Visual SLAM (VSLAM) uses features or pixel intensity in images to generate information about environments. Many visual SLAM methods employ feature-based (e.g., points, lines, corners, planes) algorithms for detecting objects or developing a higher-level understanding of environments. In general, monocular, stereo and RGB-D cameras are used in VSLAM. RGB-D camera SLAM is restricted to indoor applications because of its sensitivity to ambient light conditions. SLAM approaches that use event cameras are still not well established. Many point-feature-based SLAM methods are not robust in low-textured scenes such as plane surfaces and are vulnerable to illumination changes. Line features are less sensitive to changes in lighting conditions than point features. Plane features are often found in the artificial environment but are not regularly present in natural environments.

Using pixel intensities instead of features in SLAM is more robust to changes in lighting conditions and results in more scene information. Modern semantic VSLAM methods are mostly restricted to specific environments such as indoor or urban driving. Many VSLAM approaches focus on feature-based techniques to generate maps and localise robots. In an off-road environment with vegetation, feature extraction is challenging. The number of distinguishable features in these environments might be low due to the unstructured nature of natural environments. The presence of occlusions, shadows and illumination variation in off-road unstructured environments can obstruct useful feature detection in VSLAM. Therefore, obtaining a higher level of semantic understanding is more complex than understanding urban driving scenarios. In order to improve VSLAM performance in vegetated, outdoor, and unstructured environments, one can train deep learning models to understand features using a large dataset. However, there are difficulties in adapting these networks to unseen environments [164].

LiDAR SLAM is more robust than VSLAM in varying illumination conditions and under shadows. Point cloud registration and feature-based methods are the most common LiDAR SLAM approaches [165]. The LiDAR SLAM point cloud registration approach is more robust in vegetated natural environment mapping than the feature-based method. An adverse effect of LiDAR SLAM is the loss of semantic details in the environment. Researchers have introduced camera and LiDAR sensor fusion-based SLAM concepts to compensate for these issues. There are two separate LiDAR-camera fused SLAM directions, loosely coupled and tightly coupled. In loosely coupled SLAM systems, the modalities use two independent approaches for extracting features and high-level information to combine the two information streams. Computations in loosely coupled systems are less complex and generate faster results than in tightly coupled systems. However, these systems may produce low accuracies due to complementary information losses in independent sensor data processing steps. Tightly-coupled models fuse the two forms of sensor data into one framework to generate feature detection or higher-level semantic understanding. Therefore, this approach is more robust in challenging environments. However, these methods are more complex and need high computer memory and GPU requirements.

In unstructured outdoor environments, robots require perception, scene understanding and identification of terrain regions suitable for robot navigation. If robot working environments are dynamic, achieving these tasks will be more challenging. Robot scene understanding requires a robot to be able to perceive, analyse, and interpret its working environment. Scene understanding relies on subtopics such as object recognition, semantic segmentation, instance segmentation and scene representations of images or videos. Robot scene understanding is a continuous process, and robots need to be able to learn and adapt to new environments and situations over time. Implementation of machine learning algorithms in scene understanding has increased the possibilities of generating more sophisticated learning-based architectures. However, in general, these methods have high computational requirements. Object recognition in unstructured environments is challenging due to the higher frequency of object occlusions, similar features in different objects, indistinct object boundaries and varying lighting conditions. Semantic segmentation of an unstructured environment is another essential part of understanding the environment, and it is more complex than segmentation in urban environment scenarios due to the variation of different semantic classes. It is challenging to extract fine-grained semantic details from unstructured outdoor environments. These environments have wide semantic diversity, and training robots using large, annotated datasets is challenging due to the limited availability and difficulty of creating such datasets.

Researchers use instance segmentation algorithms to acquire a semantically rich understanding of environments, relationships between objects, and contexts. However, limited research has been conducted on robot instance segmentation of urban road environments, and these methods require high computational requirements and are not robust in off-road scenarios. Robots should learn how to robustly interpret environments over time using scene-understanding information to overcome these limitations. Point-based, graph-based, tree-based, multi-view projections, image-based, and volumetric representations have been developed to achieve scene understanding beyond basic perception. These techniques can comprehend environments beyond what can be inferred from a single image. Further investigation of the above-mentioned scene representations is necessary to learn and evolve scene interpretations over time. Novel concepts need to be developed and investigated to extract information from scenes at the object and semantic levels.

Robot domain identification and adaptation are valuable for real-world applications to improve performance and achieve more sophisticated navigation capability. In general, robots require the adaptability of learning-based navigation models. Therefore, networks that are trained on one domain (e.g., simulated environment [166], urban city environment) need to identify and adapt to changes in the application domain (e.g., real world, rural off-road environment). These domain gaps can be expressed as one challenge for robot navigation in unstructured environments. The domain gap refers to the significant deviations in environment attributes of the known or trained domains relative to the actual operating domains of the robot. These differences in application domain can occur due to variations in weather, texture, previously unseen terrain conditions, environment context changes (e.g., static to dynamic) or different factors in the robot operating environments. Deep learning architectures such as reinforcement learning can generate robust performance in one target domain but may subsequently fail due to an unforeseen drastic change in that environment.

Robots can be designed to be more adaptable if they are trained on large and diverse real-world data. However, the collection of real-world data to achieve such diversity can be challenging. Thus, there is a requirement for the development of more efficient data collection and data augmentation techniques. One option to address this is to train robots in computer simulators using data collected synthetically instead of training on real-world data. This method has benefits such as a reduction of the number of real-world experiments and a more economical and safer deployment in comparison to real-world systems. However, the trained models might have low adaptation to the real world due to overfitting to the simulation environment conditions. If the physics and the environments of simulators are significantly different from the target domain, the learned policies and loss functions in the simulator domains might not be directly applicable to the real world, thus leading to low performance. Novel techniques are needed to ensure that learning-based models, policies, and loss functions are transferable to the application domain to bridge this gap. Real-time performance is a critical aspect of robots working effectively in unstructured environments. The models trained in simulators might be unnecessarily complex for application in the real world. Therefore, techniques of domain adaptation must be transferable, economically sustainable, and computationally efficient in making decisions. In safety-critical applications, domain adaptation must be performed while ensuring safety, which requires careful validation and testing to ensure that the adapted model behaves safely in the real world.

6.2. Future Research Directions

Multimodal sensor fusion for robot vision provides more robust and reliable environment perception and scene understanding results than mono-sensor robot vision strategies in unstructured environments. Cameras and LiDARs are usually employed to retrieve RGB and depth data of external environments, respectively. However, high-resolution LiDAR sensors are costly, and their depth data sparsity increases when the observation range is increased. In contrast to mechanical LiDARs, solid-state LiDARS can provide highly accurate and higher-resolution depth data. Their costs are also likely to be significantly lower than mechanical spinning LiDARs, but these sensors are still at the research stage. The development and commercialisation of solid-state LiDARs will likely be pivotal for future advancements in robotic unstructured environment perception.

In robot navigation, domain change can involve training a model on simulated data and then fine-tuning it on real-world data to improve its performance in the real world. Robot domain adaptability can be enhanced by using techniques such as transfer learning, adversarial training, or domain adversarial neural networks. Researchers have proposed techniques such as data augmentation, feature alignment, or domain adaptation loss functions to address the domain gap. However, domain adaptation remains an active area of research in robotics, and there is still much to be done to improve the performance of machine-learning models in real-world settings. End-to-end learning architectures for robot navigation are appropriate for understanding robot tasks in the local motion planning context. For long-range navigation and in different environment domains, learning-based methods in combination with the classical hierarchical navigation pipeline show promise to produce more robust and safe results.

Robot environment perception, analysis and scene understanding are critical for robots to generate useful maps and make intelligent decisions. Recent research works such as Kimera [105] and Kimera-multi [106] are able to generate high-level geometric and hierarchical semantic environment understanding for robots. Robot systems that consist of geometric and hierarchical scene representations combined with incremental scene understanding will be advantageous in future robot outdoor unstructured navigation applications. However, the degree of scalability and adaptability of these approaches in unstructured environments with vegetation remains an open research question. In vegetated unstructured environments, obtaining feature-based scene understanding is challenging. Many SLAM methods provide rich geometric data but may incorrectly identify grass-covered terrains as not being traversable and muddy areas as traversable. Therefore, the combination of LiDAR-based geometric/semantic maps with the assistance of reinforcement learning of ego-centric images for robot local path planning could be a promising direction for more robust outdoor unstructured environment navigation. This approach will combine robot SLAM with image-based scene understanding to generate higher levels of spatial understanding of local environments for robots to be able to generate more accurate and safe movements. Overall, our comprehensive review of the literature has indicated that further research into multimodal sensor fusion techniques and deep learning-based scene understanding and task planning methods provides the most promise towards achieving the goal of fully autonomous navigation in outdoor unstructured environments.

7. Conclusions

This paper has provided a comprehensive review of the current state-of-the-art in autonomous mobile ground robot navigation, identified research gaps and challenges, and suggested promising future research directions for improved autonomous navigation in outdoor unstructured terrains. A broad review of robot sensing, camera-LiDAR sensor fusion, robot scene understanding, and local path planning techniques has been provided to deliver a comprehensive discussion of their essential methodologies and current capabilities and limitations. The use of deep learning, multimodal sensor fusion, incremental scene understanding concepts, scene representations that preserve input data topology and spatial geometry, and learning-based hierarchical path planning concepts are identified as promising research domains to investigate in order to realise fully autonomous navigation in unstructured outdoor environments. Our review has indicated that applying these techniques to outdoor unstructured terrain robot navigation research can likely improve robot domain adaptability, scene understanding and conscious decision-making abilities.

Author Contributions: Conceptualisation, A.R., D.C. and L.W.; writing—original draft, L.W.; writing—review and editing, A.R. and D.C.; supervision, A.R. and D.C. All authors have read and agreed to the published version of the manuscript.

Funding: This research received no external funding.

Institutional Review Board Statement: Not applicable.

Informed Consent Statement: Not applicable.

Data Availability Statement: Not applicable.

Conflicts of Interest: The authors declare no conflict of interest.

References

1. Rubio, F.; Valero, F.; Llopis-Albert, C. A review of mobile robots: Concepts, methods, theoretical framework, and applications. *Int. J. Adv. Robot. Syst.* **2019**, *16*, 1729881419839596. [CrossRef]
2. Abdulsaheb, J.A.; Kadhim, D.J. Classical and heuristic approaches for mobile robot path planning: A survey. *Robotics* **2023**, *12*, 93. [CrossRef]
3. Bogue, R. Warehouse robot market boosted by Covid pandemic and technological innovations. *Ind. Robot. Int. J. Robot. Res. Appl.* **2022**, *49*, 181–186. [CrossRef]
4. Quarles, N.; Kockelman, K.M.; Lee, J. America's fleet evolution in an automated future. *Res. Transp. Econ.* **2021**, *90*, 101107. [CrossRef]
5. Zhang, S.; Yao, J.; Wang, R.; Liu, Z.; Ma, C.; Wang, Y.; Zhao, Y. Design of intelligent fire-fighting robot based on multi-sensor fusion and experimental study on fire scene patrol. *Robot. Auton. Syst.* **2022**, *154*, 104122. [CrossRef]
6. Li, Q.; Kroemer, O.; Su, Z.; Veiga, F.F.; Kaboli, M.; Ritter, H.J. A review of tactile information: Perception and action through touch. *IEEE Trans. Robot.* **2020**, *36*, 1619–1634. [CrossRef]
7. Alatise, M.B.; Hancke, G.P. A review on challenges of autonomous mobile robot and sensor fusion methods. *IEEE Access* **2020**, *8*, 39830–39846. [CrossRef]
8. Feng, D.; Haase-Schütz, C.; Rosenbaum, L.; Hertlein, H.; Glaeser, C.; Timm, F.; Wiesbeck, W.; Dietmayer, K. Deep multi-modal object detection and semantic segmentation for autonomous driving: Datasets, methods, and challenges. *IEEE Trans. Intell. Transp. Syst.* **2020**, *22*, 1341–1360. [CrossRef]
9. Hu, C.; Yang, C.; Li, K.; Zhang, J. A forest point cloud real-time reconstruction method with single-line LiDAR based on visual-IMU fusion. *Appl. Sci.* **2022**, *12*, 4442. [CrossRef]
10. Jin, X.B.; Su, T.L.; Kong, J.L.; Bai, Y.T.; Miao, B.B.; Dou, C. State-of-the-art mobile intelligence: Enabling robots to move like humans by estimating mobility with artificial intelligence. *Appl. Sci.* **2018**, *8*, 379. [CrossRef]
11. Huang, Z.; Lv, C.; Xing, Y.; Wu, J. Multi-modal sensor fusion-based deep neural network for end-to-end autonomous driving with scene understanding. *IEEE Sens. J.* **2021**, *21*, 11781–11790. [CrossRef]
12. Alenzi, Z.; Alenzi, E.; Alqasir, M.; Alruwaili, M.; Alhmiedat, T.; Alia, O.M. A semantic classification approach for indoor robot navigation. *Electronics* **2022**, *11*, 2063. [CrossRef]
13. Zhao, H.; Liu, Y.; Zhu, X.; Zhao, Y.; Zha, H. Scene understanding in a large dynamic environment through a laser-based sensing. In Proceedings of the 2010 IEEE International Conference on Robotics and Automation, Anchorage, AK, USA, 3–7 May 2010; pp. 127–133.
14. Zhang, F.s.; Ge, D.y.; Song, J.; Xiang, W.j. Outdoor scene understanding of mobile robot via multi-sensor information fusion. *J. Ind. Inf. Integr.* **2022**, *30*, 100392. [CrossRef]
15. Moon, J.; Lee, B.H. PDDL planning with natural language-based scene understanding for UAV-UGV cooperation. *Appl. Sci.* **2019**, *9*, 3789. [CrossRef]
16. Yang, M.; Rosenhahn, B.; Murino, V. *Multimodal Scene Understanding: Algorithms, Applications and Deep Learning*; Academic Press: London, UK, 2019; pp. 1–7.
17. Zhang, Y.; Sidibé, D.; Morel, O.; Mériaudeau, F. Deep multimodal fusion for semantic image segmentation: A survey. *Image Vis. Comput.* **2021**, *105*, 104042. [CrossRef]
18. Janai, J.; Güney, F.; Behl, A.; Geiger, A. Computer vision for autonomous vehicles: Problems, datasets and state of the art. *Found. Trends® Comput. Graph. Vis.* **2020**, *12*, 1–308. [CrossRef]
19. Gupta, A.; Efros, A.A.; Hebert, M. Blocks world revisited: Image understanding using qualitative geometry and mechanics. In Proceedings of the 11th European Conference on Computer Vision, Heraklion, Crete, Greece, 5–11 September 2010; pp. 482–496.
20. Wang, W.; Shen, J.; Cheng, M.M.; Shao, L. An iterative and cooperative top-down and bottom-up inference network for salient object detection. In Proceedings of the IEEE/CVF Conference on Computer Vision and Pattern Recognition (CVPR), Long Beach, CA, USA, 15–20 June 2019; pp. 5968–5977.

21. Sun, H.; Zhang, W.; Yu, R.; Zhang, Y. Motion planning for mobile robots—Focusing on deep reinforcement learning: A systematic review. *IEEE Access* **2021**, *9*, 69061–69081. [CrossRef]
22. Kocić, J.; Jovičić, N.; Drndarević, V. Sensors and sensor fusion in autonomous vehicles. In Proceedings of the 26th Telecommunications Forum (TELFOR), Serbia, Belgrade, 20–21 November 2018; pp. 420–425.
23. Chen, Z.; Li, J.; Wang, S.; Wang, J.; Ma, L. Flexible gait transition for six wheel-legged robot with unstructured terrains. *Robot. Auton. Syst.* **2022**, *150*, 103989. [CrossRef]
24. Chen, Z.; Li, J.; Wang, J.; Wang, S.; Zhao, J.; Li, J. Towards hybrid gait obstacle avoidance for a six wheel-legged robot with payload transportation. *J. Intell. Robot. Syst.* **2021**, *102*, 60. [CrossRef]
25. Muñoz-Bañón, M.Á.; Candelas, F.A.; Torres, F. Targetless camera-LiDAR calibration in unstructured environments. *IEEE Access* **2020**, *8*, 143692–143705. [CrossRef]
26. Li, A.; Cao, J.; Li, S.; Huang, Z.; Wang, J.; Liu, G. Map construction and path planning method for a mobile robot based on multi-sensor information fusion. *Appl. Sci.* **2022**, *12*, 2913. [CrossRef]
27. Santos, L.C.; Santos, F.N.; Pires, E.S.; Valente, A.; Costa, P.; Magalhães, S. Path planning for ground robots in agriculture: A short review. In Proceedings of the IEEE International Conference on Autonomous Robot Systems and Competitions (ICARSC), Ponta Delgada, Portugal, 15–16 April 2020; pp. 61–66.
28. Fayyad, J.; Jaradat, M.A.; Gruyer, D.; Najjaran, H. Deep learning sensor fusion for autonomous vehicle perception and localization: A review. *Sensors* **2020**, *20*, 4220. [CrossRef]
29. Valada, A.; Oliveira, G.L.; Brox, T.; Burgard, W. Deep multispectral semantic scene understanding of forested environments using multimodal fusion. In Proceedings of the International Symposium on Experimental Robotics, Tokyo, Japan, 3–6 October 2016; pp. 465–477.
30. Lei, X.; Zhang, Z.; Dong, P. Dynamic path planning of unknown environment based on deep reinforcement learning. *J. Robot.* **2018**, *2018*, 5781591. [CrossRef]
31. Crespo, J.; Castillo, J.C.; Mozos, O.M.; Barber, R. Semantic information for robot navigation: A survey. *Appl. Sci.* **2020**, *10*, 497. [CrossRef]
32. Galvao, L.G.; Abbod, M.; Kalganova, T.; Palade, V.; Huda, M.N. Pedestrian and vehicle detection in autonomous vehicle perception systems—A review. *Sensors* **2021**, *21*, 7267. [CrossRef]
33. Hewawasam, H.; Ibrahim, M.Y.; Appuhamillage, G.K. Past, present and future of path-planning algorithms for mobile robot navigation in dynamic environments. *IEEE Open J. Ind. Electron. Soc.* **2022**, *3*, 353–365. [CrossRef]
34. Martini, M.; Cerrato, S.; Salvetti, F.; Angarano, S.; Chiaberge, M. Position-agnostic autonomous navigation in vineyards with deep reinforcement learning. In Proceedings of the IEEE 18th International Conference on Automation Science and Engineering (CASE), Mexico City, Mexico, 20–24 August 2022; pp. 477–484.
35. Hamza, A. Deep Reinforcement Learning for Mapless Mobile Robot Navigation. Master's Thesis, Luleå University of Technology, Luleå, Sweden, 2022.
36. Carrasco, P.; Cuesta, F.; Caballero, R.; Perez-Grau, F.J.; Viguria, A. Multi-sensor fusion for aerial robots in industrial GNSS-denied environments. *Appl. Sci.* **2021**, *11*, 3921. [CrossRef]
37. Li, R.; Wang, S.; Gu, D. DeepSLAM: A robust monocular SLAM system with unsupervised deep learning. *IEEE Trans. Ind. Electron.* **2020**, *68*, 3577–3587. [CrossRef]
38. Aguiar, A.; Santos, F.; Sousa, A.J.; Santos, L. Fast-fusion: An improved accuracy omnidirectional visual odometry system with sensor fusion and GPU optimization for embedded low cost hardware. *Appl. Sci.* **2019**, *9*, 5516. [CrossRef]
39. Li, Y.; Brasch, N.; Wang, Y.; Navab, N.; Tombari, F. Structure-SLAM: Low-drift monocular SLAM in indoor environments. *IEEE Robot. Autom. Lett.* **2020**, *5*, 6583–6590. [CrossRef]
40. Zaffar, M.; Ehsan, S.; Stolkin, R.; Maier, K.M. Sensors, SLAM and long-term autonomy: A review. In Proceedings of the NASA/ESA Conference on Adaptive Hardware and Systems (AHS), Edinburgh, UK, 6–9 August 2018; pp. 285–290.
41. Sabattini, L.; Levratti, A.; Venturi, F.; Amplo, E.; Fantuzzi, C.; Secchi, C. Experimental comparison of 3D vision sensors for mobile robot localization for industrial application: Stereo-camera and RGB-D sensor. In Proceedings of the 12th International Conference on Control Automation Robotics & Vision (ICARCV), Guangzhou, China, 5–7 December 2012; pp. 823–828.
42. Tölgyessy, M.; Dekan, M.; Chovanec, L.; Hubinský, P. Evaluation of the Azure Kinect and its comparison to Kinect V1 and Kinect V2. *Sensors* **2021**, *21*, 413. [CrossRef] [PubMed]
43. Evangelidis, G.D.; Hansard, M.; Horaud, R. Fusion of range and stereo data for high-resolution scene-modeling. *IEEE Trans. Pattern Anal. Mach. Intell.* **2015**, *37*, 2178–2192. [CrossRef] [PubMed]
44. Glover, A.; Bartolozzi, C. Robust visual tracking with a freely-moving event camera. In Proceedings of the IEEE/RSJ International Conference on Intelligent Robots and Systems (IROS), Vancouver, BC, Canada, 24–28 September 2017; pp. 3769–3776.
45. Gallego, G.; Delbrück, T.; Orchard, G.; Bartolozzi, C.; Taba, B.; Censi, A.; Leutenegger, S.; Davison, A.J.; Conradt, J.; Daniilidis, K.; et al. Event-based vision: A survey. *IEEE Trans. Pattern Anal. Mach. Intell.* **2020**, *44*, 154–180. [CrossRef] [PubMed]
46. Yuan, W.; Li, J.; Bhatta, M.; Shi, Y.; Baenziger, P.S.; Ge, Y. Wheat height estimation using LiDAR in comparison to ultrasonic sensor and UAS. *Sensors* **2018**, *18*, 1–20. [CrossRef]
47. Moosmann, F.; Stiller, C. Velodyne SLAM. In Proceedings of the IEEE Intelligent Vehicles Symposium, Baden-Baden, Germany, 5–9 June 2011; pp. 393–398.

48. Li, K.; Li, M.; Hanebeck, U.D. Towards high-performance solid-state-lidar-inertial odometry and mapping. *IEEE Robot. Autom. Lett.* **2021**, *6*, 5167–5174. [CrossRef]
49. Poulton, C.V.; Yaacobi, A.; Cole, D.B.; Byrd, M.J.; Raval, M.; Vermeulen, D.; Watts, M.R. Coherent solid-state LIDAR with silicon photonic optical phased arrays. *Opt. Lett.* **2017**, *42*, 4091–4094. [CrossRef]
50. Behroozpour, B.; Sandborn, P.A.; Wu, M.C.; Boser, B.E. LiDAR system architectures and circuits. *IEEE Commun. Mag.* **2017**, *55*, 135–142. [CrossRef]
51. Xu, X.; Zhang, L.; Yang, J.; Cao, C.; Wang, W.; Ran, Y.; Tan, Z.; Luo, M. A review of multi-sensor fusion SLAM systems based on 3D LIDAR. *Remote Sens.* **2022**, *14*, 2835. [CrossRef]
52. Li, Y.; Yu, A.W.; Meng, T.; Caine, B.; Ngiam, J.; Peng, D.; Shen, J.; Lu, Y.; Zhou, D.; Le, Q.V.; et al. Deepfusion: LiDAR-camera deep fusion for multi-modal 3D object detection. In Proceedings of the IEEE/CVF Conference on Computer Vision and Pattern Recognition (CVPR), New Orleans, LA, USA, 18–24 June 2022; pp. 17182–17191.
53. Zheng, W.; Xie, H.; Chen, Y.; Roh, J.; Shin, H. PIFNet: 3D object detection using joint image and point cloud features for autonomous driving. *Appl. Sci.* **2022**, *12*, 3686. [CrossRef]
54. Cui, Y.; Chen, R.; Chu, W.; Chen, L.; Tian, D.; Li, Y.; Cao, D. Deep learning for image and point cloud fusion in autonomous driving: A review. *IEEE Trans. Intell. Transp. Syst.* **2021**, *23*, 722–739. [CrossRef]
55. Du, X.; Ang, M.H.; Karaman, S.; Rus, D. A general pipeline for 3D detection of vehicles. In Proceedings of the 2018 IEEE International Conference on Robotics and Automation (ICRA), Brisbane, Australia, 21–25 May 2018; pp. 3194–3200.
56. Yang, Z.; Sun, Y.; Liu, S.; Shen, X.; Jia, J. Ipod: Intensive point-based object detector for point cloud. *arXiv* **2018**, arXiv:1812.05276. [CrossRef]
57. Qi, C.R.; Liu, W.; Wu, C.; Su, H.; Guibas, L.J. Frustum pointnets for 3D object detection from RGB-D data. In Proceedings of the IEEE/CVF Conference on Computer Vision and Pattern Recognition, Salt Lake City, UT, USA, 18–23 June 2018; pp. 918–927.
58. Shin, K.; Kwon, Y.P.; Tomizuka, M. Roarnet: A robust 3D object detection based on region approximation refinement. In Proceedings of the IEEE Intelligent Vehicles Symposium (IV), Paris, France, 9–12 June 2019; pp. 2510–2515.
59. Qi, C.R.; Su, H.; Mo, K.; Guibas, L.J. PointNet: Deep learning on point sets for 3D classification and segmentation. In Proceedings of the IEEE Conference on Computer Vision and Pattern Recognition (CVPR), Honolulu, HI, USA, 21–26 July 2017; pp. 652–660.
60. Tran, D.; Bourdev, L.; Fergus, R.; Torresani, L.; Paluri, M. Learning spatiotemporal features with 3D convolutional networks. In Proceedings of the IEEE International Conference on Computer Vision (ICCV), Santiago, Chile, 7–13 December 2015; pp. 4489–4497.
61. Maturana, D.; Scherer, S. Voxnet: A 3D convolutional neural network for real-time object recognition. In Proceedings of the IEEE/RSJ International Conference on Intelligent Robots and Systems (IROS), Hamburg, Germany, 28 September–2 October 2015; pp. 922–928.
62. Xu, D.; Anguelov, D.; Jain, A. Pointfusion: Deep sensor fusion for 3D bounding box estimation. In Proceedings of the IEEE/CVF Conference on Computer Vision and Pattern Recognition, Salt Lake City, UT, USA, 18–23 June 2018; pp. 244–253.
63. Ku, J.; Mozifian, M.; Lee, J.; Harakeh, A.; Waslander, S.L. Joint 3D proposal generation and object detection from view aggregation. In Proceedings of the IEEE/RSJ International Conference on Intelligent Robots and Systems (IROS), Madrid, Spain, 1–5 October 2018; pp. 1–8.
64. Liang, M.; Yang, B.; Wang, S.; Urtasun, R. Deep continuous fusion for multi-sensor 3D object detection. In Proceedings of the 15th European Conference on Computer Vision (ECCV), Munich, Germany, 8–14 September 2018; pp. 641–656.
65. Sindagi, V.A.; Zhou, Y.; Tuzel, O. MVX-Net: Multimodal voxelnet for 3D object detection. In Proceedings of the International Conference on Robotics and Automation (ICRA), Montreal, QC, Canada, 20–24 May 2019; pp. 7276–7282.
66. Chen, X.; Ma, H.; Wan, J.; Li, B.; Xia, T. Multi-view 3D object detection network for autonomous driving. In Proceedings of the IEEE Conference on Computer Vision and Pattern Recognition (CVPR), Honolulu, HI, USA, 21–26 July 2017; pp. 1907–1915.
67. Geiger, A.; Lenz, P.; Stiller, C.; Urtasun, R. Vision meets robotics: The KITTI dataset. *Int. J. Robot. Res.* **2013**, *32*, 1231–1237. [CrossRef]
68. Sun, P.; Kretzschmar, H.; Dotiwalla, X.; Chouard, A.; Patnaik, V.; Tsui, P.; Guo, J.; Zhou, Y.; Chai, Y.; Caine, B.; et al. Scalability in perception for autonomous driving: Waymo open dataset. In Proceedings of the IEEE/CVF Conference on Computer Vision and Pattern Recognition (CVPR), Seattle, WA, USA, 13–19 June 2020; pp. 2446–2454.
69. Geyer, J.; Kassahun, Y.; Mahmudi, M.; Ricou, X.; Durgesh, R.; Chung, A.S.; Hauswald, L.; Pham, V.H.; Mühlegg, M.; Dorn, S.; et al. A2d2: Audi autonomous driving dataset. arXiv 2020, arXiv:2004.06320 [CrossRef]
70. Caesar, H.; Bankiti, V.; Lang, A.H.; Vora, S.; Liong, V.E.; Xu, Q.; Krishnan, A.; Pan, Y.; Baldan, G.; Beijbom, O. nuscenes: A multimodal dataset for autonomous driving. In Proceedings of the IEEE/CVF Conference on Computer Vision and Pattern Recognition (CVPR), Seattle, WA, USA, 13–19 June 2020; pp. 11621–11631.
71. Cordts, M.; Omran, M.; Ramos, S.; Rehfeld, T.; Enzweiler, M.; Benenson, R.; Franke, U.; Roth, S.; Schiele, B. The cityscapes dataset for semantic urban scene understanding. In Proceedings of the IEEE Conference on Computer Vision and Pattern Recognition (CVPR), Las Vegas, NV, USA, 27–30 June 2016; pp. 3213–3223.
72. Ma, F.; Cavalheiro, G.V.; Karaman, S. Self-supervised sparse-to-dense: Self-supervised depth completion from lidar and monocular camera. In Proceedings of the International Conference on Robotics and Automation (ICRA), Montreal, QC, Canada, 20–24 May 2019; pp. 3288–3295.

73. Ma, F.; Karaman, S. Sparse-to-dense: Depth prediction from sparse depth samples and a single image. In Proceedings of the IEEE international conference on robotics and automation (ICRA), Brisbane, Australia, 21–25 May 2018; pp. 4796–4803.
74. He, K.; Zhang, X.; Ren, S.; Sun, J. Deep residual learning for image recognition. In Proceedings of the IEEE Conference on Computer Vision and Pattern Recognition (CVPR), Las Vegas, NV, USA, 27–30 June 2016; pp. 770–778.
75. Cheng, X.; Wang, P.; Yang, R. Depth estimation via affinity learned with convolutional spatial propagation network. In Proceedings of the European Conference on Computer Vision (ECCV), Munich, Germany, 8–14 September 2018; pp. 103–119.
76. Cheng, X.; Wang, P.; Guan, C.; Yang, R. CSPN++: Learning context and resource aware convolutional spatial propagation networks for depth completion. In Proceedings of the 34th AAAI Conference on Artificial Intelligence, New York, NY, USA, 7–12 February 2020; pp. 10615–10622.
77. Cheng, X.; Zhong, Y.; Dai, Y.; Ji, P.; Li, H. Noise-aware unsupervised deep LiDAR-stereo fusion. In Proceedings of the IEEE/CVF Conference on Computer Vision and Pattern Recognition (CVPR), Long Beach, CA, USA, 15–20 June 2019; pp. 6339–6348.
78. Jalal, A.S.; Singh, V. The state-of-the-art in visual object tracking. *Informatica* **2012**, *36*, 1–22.
79. Tang, P.; Wang, X.; Wang, A.; Yan, Y.; Liu, W.; Huang, J.; Yuille, A. Weakly supervised region proposal network and object detection. In Proceedings of the 15th European Conference on Computer Vision (ECCV), Munich, Germany, 8–14 September 2018; pp. 352–368.
80. Uijlings, J.R.; Van De Sande, K.E.; Gevers, T.; Smeulders, A.W. Selective search for object recognition. *Int. J. Comput. Vis.* **2013**, *104*, 154–171. [CrossRef]
81. Hong, M.; Li, S.; Yang, Y.; Zhu, F.; Zhao, Q.; Lu, L. SSPNet: Scale selection pyramid network for tiny person detection from UAV images. *IEEE Geosci. Remote Sens. Lett.* **2021**, *19*, 1–5. [CrossRef]
82. Girshick, R. Fast R-CNN. In Proceedings of the IEEE International Conference on Computer Vision (ICCV), Santiago, Chile, 7–13 December 2015; pp. 1440–1448.
83. Ren, S.; He, K.; Girshick, R.; Sun, J. Faster R-CNN: Towards real-time object detection with region proposal networks. *IEEE Trans. Pattern Anal. Mach. Intell.* **2015**, *39*, 1–9. [CrossRef]
84. Kim, J.; Cho, J. Exploring a multimodal mixture-of-YOLOs framework for advanced real-time object detection. *Appl. Sci.* **2020**, *10*, 612. [CrossRef]
85. Gupta, S.; Girshick, R.; Arbeláez, P.; Malik, J. Learning rich features from RGB-D images for object detection and segmentation. In Proceedings of the 13th European Conference on Computer Vision (ECCV), Zurich, Switzerland, 6–12 September 2014; pp. 345–360.
86. Simonyan, K.; Zisserman, A. Very deep convolutional networks for large-scale image recognition. In Proceedings of the International Conference on Learning Representations, San Diego, CA, USA, 7–9 May 2015.
87. Zhou, Y.; Tuzel, O. Voxelnet: End-to-end learning for point cloud based 3D object detection. In Proceedings of the IEEE/CVF Conference on Computer Vision and Pattern Recognition (CVPR), Salt Lake City, UT, USA, 18–23 June 2018; pp. 4490–4499.
88. Meyer, G.P.; Charland, J.; Hegde, D.; Laddha, A.; Vallespi-Gonzalez, C. Sensor fusion for joint 3D object detection and semantic segmentation. In Proceedings of the IEEE/CVF Conference on Computer Vision and Pattern Recognition Workshops (CVPRW), Long Beach, CA, USA, 16–17 June 2019; pp. 1–8.
89. Meyer, G.P.; Laddha, A.; Kee, E.; Vallespi-Gonzalez, C.; Wellington, C.K. Lasernet: An efficient probabilistic 3D object detector for autonomous driving. In Proceedings of the IEEE/CVF Conference on Computer Vision and Pattern Recognition (CVPR), Long Beach, CA, USA, 15–20 June 2019; pp. 12677–12686.
90. Guo, Z.; Huang, Y.; Hu, X.; Wei, H.; Zhao, B. A survey on deep learning based approaches for scene understanding in autonomous driving. *Electronics* **2021**, *10*, 471. [CrossRef]
91. Zoph, B.; Vasudevan, V.; Shlens, J.; Le, Q.V. Learning transferable architectures for scalable image recognition. In Proceedings of the IEEE/CVF Conference on Computer Vision and Pattern Recognition (CVPR), Salt Lake City, UT, USA, 18–23 June 2018; pp. 8697–8710.
92. Valada, A.; Mohan, R.; Burgard, W. Self-supervised model adaptation for multimodal semantic segmentation. *Int. J. Comput. Vis.* **2020**, *128*, 1239–1285. [CrossRef]
93. Caltagirone, L.; Bellone, M.; Svensson, L.; Wahde, M. LIDAR–camera fusion for road detection using fully convolutional neural networks. *Robot. Auton. Syst.* **2019**, *111*, 125–131. [CrossRef]
94. Dai, A.; Nießner, M. 3DMV: Joint 3D-multi-view prediction for 3D semantic scene segmentation. In Proceedings of the European Conference on Computer Vision (ECCV), Munich, Germany, 8–14 September 2018; pp. 452–468.
95. Chiang, H.Y.; Lin, Y.L.; Liu, Y.C.; Hsu, W.H. A unified point-based framework for 3D segmentation. In Proceedings of the International Conference on 3D Vision (3DV), Québec, QC, Canada, 16–19 September 2019; pp. 155–163.
96. Qi, C.R.; Yi, L.; Su, H.; Guibas, L.J. Pointnet++: Deep hierarchical feature learning on point sets in a metric space. *Adv. Neural Inf. Process. Syst.* **2017**, *30*, 1–10.
97. Jaritz, M.; Gu, J.; Su, H. Multi-view pointnet for 3D scene understanding. In Proceedings of the IEEE/CVF International Conference on Computer Vision Workshop (ICCVW), Seoul, Republic of Korea, 27–28 October 2019; pp. 1–9.
98. Su, H.; Jampani, V.; Sun, D.; Maji, S.; Kalogerakis, E.; Yang, M.H.; Kautz, J. Splatnet: Sparse lattice networks for point cloud processing. In Proceedings of the IEEE/CVF Conference on Computer Vision and Pattern Recognition, Salt Lake City, UT, USA, 18–23 June 2018; pp. 2530–2539.

99. Hou, J.; Dai, A.; Nießner, M. 3D-SIS: 3D semantic instance segmentation of RGB-D scans. In Proceedings of the IEEE/CVF Conference on Computer Vision and Pattern Recognition (CVPR), Long Beach, CA, USA, 15–20 June 2019; pp. 4421–4430.
100. Narita, G.; Seno, T.; Ishikawa, T.; Kaji, Y. Panopticfusion: Online volumetric semantic mapping at the level of stuff and things. In Proceedings of the IEEE/RSJ International Conference on Intelligent Robots and Systems (IROS), Macau, China, 4–8 November 2019; pp. 4205–4212.
101. Elich, C.; Engelmann, F.; Kontogianni, T.; Leibe, B. 3D bird's-eye-view instance segmentation. In Proceedings of the 41st DAGM German Conference on Pattern Recognition, Dortmund, Germany, 10–13 September 2019; pp. 48–61.
102. Comaniciu, D.; Meer, P. Mean shift: A robust approach toward feature space analysis. *IEEE Trans. Pattern Anal. Mach. Intell.* **2002**, *24*, 603–619. [CrossRef]
103. Kochanov, D.; Ošep, A.; Stückler, J.; Leibe, B. Scene flow propagation for semantic mapping and object discovery in dynamic street scenes. In Proceedings of the IEEE/RSJ International Conference on Intelligent Robots and Systems (IROS), Daejeon, Republic of Korea, 9–14 October 2016; pp. 1785–1792.
104. Yue, Y.; Zhao, C.; Li, R.; Yang, C.; Zhang, J.; Wen, M.; Wang, Y.; Wang, D. A hierarchical framework for collaborative probabilistic semantic mapping. In Proceedings of the IEEE International Conference on Robotics and Automation (ICRA), Paris, France, 31 May–31 August 2020; pp. 9659–9665.
105. Rosinol, A.; Violette, A.; Abate, M.; Hughes, N.; Chang, Y.; Shi, J.; Gupta, A.; Carlone, L. Kimera: From SLAM to spatial perception with 3D dynamic scene graphs. *Int. J. Robot. Res.* **2021**, *40*, 1510–1546. [CrossRef]
106. Tian, Y.; Chang, Y.; Arias, F.H.; Nieto-Granda, C.; How, J.P.; Carlone, L. Kimera-multi: Robust, distributed, dense metric-semantic slam for multi-robot systems. *IEEE Trans. Robot.* **2022**, *38*, 2022–2038. [CrossRef]
107. Kim, U.H.; Park, J.M.; Song, T.J.; Kim, J.H. 3-D scene graph: A sparse and semantic representation of physical environments for intelligent agents. *IEEE Trans. Cybern.* **2019**, *50*, 4921–4933. [CrossRef]
108. Liu, H.; Yao, M.; Xiao, X.; Cui, H. A hybrid attention semantic segmentation network for unstructured terrain on Mars. *Acta Astronaut.* **2023**, *204*, 492–499. [CrossRef]
109. Humblot-Renaux, G.; Marchegiani, L.; Moeslund, T.B.; Gade, R. Navigation-oriented scene understanding for robotic autonomy: Learning to segment driveability in egocentric images. *IEEE Robot. Autom. Lett.* **2022**, *7*, 2913–2920. [CrossRef]
110. Badrinarayanan, V.; Kendall, A.; Cipolla, R. Segnet: A deep convolutional encoder-decoder architecture for image segmentation. *IEEE Trans. Pattern Anal. Mach. Intell.* **2017**, *39*, 2481–2495. [CrossRef] [PubMed]
111. Guan, T.; Kothandaraman, D.; Chandra, R.; Sathyamoorthy, A.J.; Weerakoon, K.; Manocha, D. GA-Nav: Efficient terrain segmentation for robot navigation in unstructured outdoor environments. *IEEE Robot. Autom. Lett.* **2022**, *7*, 8138–8145. [CrossRef]
112. Wigness, M.; Eum, S.; Rogers, J.G.; Han, D.; Kwon, H. A RUGD dataset for autonomous navigation and visual perception in unstructured outdoor environments. In Proceedings of the IEEE/RSJ International Conference on Intelligent Robots and Systems (IROS), Macau, China, 4–8 November 2019; pp. 5000–5007.
113. Jiang, P.; Osteen, P.; Wigness, M.; Saripalli, S. RELLIS-3D dataset: Data, benchmarks and analysis. In Proceedings of the IEEE international conference on robotics and automation (ICRA), Xi'an, China, 31 May–4 June 2021; pp. 1110–1116.
114. Ma, L.; Stückler, J.; Kerl, C.; Cremers, D. Multi-view deep learning for consistent semantic mapping with RGB-D cameras. In Proceedings of the IEEE/RSJ International Conference on Intelligent Robots and Systems (IROS), Vancouver, BC, Canada, 24–28 September 2017; pp. 598–605.
115. Hazirbas, C.; Ma, L.; Domokos, C.; Cremers, D. Fusenet: Incorporating depth into semantic segmentation via fusion-based CNN architecture. In Proceedings of the 13th Asian Conference on Computer Vision, Taipei, Taiwan, 20–24 November 2016; pp. 213–228.
116. Henein, M.; Zhang, J.; Mahony, R.E.; Ila, V. Dynamic SLAM: The need for speed. In Proceedings of the 2020 IEEE International Conference on Robotics and Automation (ICRA), Paris, France, 31 May–31 August 2020; pp. 2123–2129.
117. Huang, J.; You, S. Point cloud labeling using 3D convolutional neural network. In Proceedings of the 23rd International Conference on Pattern Recognition (ICPR), Cancun, Mexico, 4–8 December 2016; pp. 2670–2675.
118. LeCun, Y.; Bottou, L.; Bengio, Y.; Haffner, P. Gradient-based learning applied to document recognition. *Proc. IEEE* **1998**, *86*, 2278–2324. [CrossRef]
119. Song, S.; Yu, F.; Zeng, A.; Chang, A.X.; Savva, M.; Funkhouser, T. Semantic scene completion from a single depth image. In Proceedings of the IEEE Conference on Computer Vision and Pattern Recognition (CVPR), Honolulu, HI, USA, 21–26 July 2017; pp. 1746–1754.
120. Riegler, G.; Osman Ulusoy, A.; Geiger, A. OctNet: Learning deep 3D representations at high resolutions. In Proceedings of the IEEE Conference on Computer Vision and Pattern Recognition (CVPR), Honolulu, HI, USA, 21–26 July 2017; pp. 3577–3586.
121. Tatarchenko, M.; Park, J.; Koltun, V.; Zhou, Q.Y. Tangent convolutions for dense prediction in 3D. In Proceedings of the IEEE/CVF Conference on Computer Vision and Pattern Recognition, Salt Lake City, UT, USA, 18–23 June 2018; pp. 3887–3896.
122. Wang, F.; Yang, Y.; Wu, Z.; Zhou, J.; Zhang, W. Real-time semantic segmentation of point clouds based on an attention mechanism and a sparse tensor. *Appl. Sci.* **2023**, *13*, 3256. [CrossRef]
123. Wu, W.; Qi, Z.; Fuxin, L. Pointconv: Deep convolutional networks on 3D point clouds. In Proceedings of the IEEE/CVF Conference on Computer Vision and Pattern Recognition (CVPR), Long Beach, CA, USA, 15–20 June 2019; pp. 9621–9630.
124. Hua, B.S.; Tran, M.K.; Yeung, S.K. Pointwise convolutional neural networks. In Proceedings of the IEEE/CVF Conference on Computer Vision and Pattern Recognition (CVPR), Salt Lake City, UT, USA, 18–23 June 2018; pp. 984–993.

125. Zamorski, M.; Zięba, M.; Klukowski, P.; Nowak, R.; Kurach, K.; Stokowiec, W.; Trzciński, T. Adversarial autoencoders for compact representations of 3D point clouds. *Comput. Vis. Image Underst.* **2020**, *193*, 102921. [CrossRef]
126. Ye, X.; Li, J.; Huang, H.; Du, L.; Zhang, X. 3D recurrent neural networks with context fusion for point cloud semantic segmentation. In Proceedings of the 15th European Conference on Computer Vision (ECCV), Munich, Germany, 8–14 September 2018; pp. 403–417.
127. Wang, Y.; Sun, Y.; Liu, Z.; Sarma, S.E.; Bronstein, M.M.; Solomon, J.M. Dynamic graph CNN for learning on point clouds. *ACM Trans. Graph.* **2019**, *38*, 1–12. [CrossRef]
128. Qi, X.; Liao, R.; Jia, J.; Fidler, S.; Urtasun, R. 3D graph neural networks for RGB-D semantic segmentation. In Proceedings of the IEEE International Conference on Computer Vision (ICCV), Venice, Italy, 22–29 October 2017; pp. 5199–5208.
129. Landrieu, L.; Simonovsky, M. Large-scale point cloud semantic segmentation with superpoint graphs. In Proceedings of the IEEE/CVF Conference on Computer Vision and Pattern Recognition, Salt Lake City, UT, USA, 18–23 June 2018; pp. 4558–4567.
130. Li, J.; Chen, B.M.; Lee, G.H. SO-Net: Self-organizing network for point cloud analysis. In Proceedings of the IEEE/CVF Conference on Computer Vision and Pattern Recognition, Salt Lake City, UT, USA, 18–23 June 2018; pp. 9397–9406.
131. Thrun, S. Probabilistic robotics. *Commun. ACM* **2002**, *45*, 52–57. [CrossRef]
132. Siegwart, R.; Nourbakhsh, I.R.; Scaramuzza, D. *Introduction to Autonomous Mobile Robots*, 2nd ed.; MIT Press: Cambridge, MA, USA, 2011.
133. Sivakumar, A.N.; Modi, S.; Gasparino, M.V.; Ellis, C.; Velasquez, A.E.B.; Chowdhary, G.; Gupta, S. Learned visual navigation for under-canopy agricultural robots. In Proceedings of the 17th Robotics: Science and Systems, Virtual . 12–16 July 2021.
134. Atas, F.; Cielniak, G.; Grimstad, L. Benchmark of sampling-based optimizing planners for outdoor robot navigation. In *International Conference on Intelligent Autonomous Systems*; Springer: Cham, Switzerland, 2022; pp. 231–243.
135. Wang, X.; Shi, Y.; Ding, D.; Gu, X. Double global optimum genetic algorithm–particle swarm optimization-based welding robot path planning. *Eng. Optim.* **2016**, *48*, 299–316. [CrossRef]
136. Zhu, S.; Zhu, W.; Zhang, X.; Cao, T. Path planning of lunar robot based on dynamic adaptive ant colony algorithm and obstacle avoidance. *Int. J. Adv. Robot. Syst.* **2020**, *17*, 1729881419898979. [CrossRef]
137. Mac, T.T.; Copot, C.; Tran, D.T.; De Keyser, R. A hierarchical global path planning approach for mobile robots based on multi-objective particle swarm optimization. *Appl. Soft Comput.* **2017**, *59*, 68–76. [CrossRef]
138. Ghita, N.; Kloetzer, M. Trajectory planning for a car-like robot by environment abstraction. *Robot. Auton. Syst.* **2012**, *60*, 609–619. [CrossRef]
139. Zhu, Y.; Mottaghi, R.; Kolve, E.; Lim, J.J.; Gupta, A.; Fei-Fei, L.; Farhadi, A. Target-driven visual navigation in indoor scenes using deep reinforcement learning. In Proceedings of the IEEE international conference on robotics and automation (ICRA), Singapore, 29 May–3 June 2017; pp. 3357–3364.
140. Wijmans, E.; Kadian, A.; Morcos, A.S.; Lee, S.; Essa, I.; Parikh, D.; Savva, M.; Batra, D. DD-PPO: Learning near-perfect pointgoal navigators from 2.5 billion frames. In Proceedings of the International Conference on Learning Representations, New Orleans, LA, USA, 6–9 May 2019.
141. Gupta, S.; Davidson, J.; Levine, S.; Sukthankar, R.; Malik, J. Cognitive mapping and planning for visual navigation. *Int. J. Comput. Vis.* **2020**, *128*, 1311–1330. [CrossRef]
142. Datta, S.; Maksymets, O.; Hoffman, J.; Lee, S.; Batra, D.; Parikh, D. Integrating egocentric localization for more realistic point-goal navigation agents. In Proceedings of the 4th Conference on Robot Learning (CoRL), Virtual, 16–18 November 2020; pp. 313–328.
143. Kumar, A.; Gupta, S.; Fouhey, D.; Levine, S.; Malik, J. Visual memory for robust path following. *Adv. Neural Inf. Process. Syst.* **2018**, *31*, 1–10.
144. Pan, Y.; Cheng, C.A.; Saigol, K.; Lee, K.; Yan, X.; Theodorou, E.A.; Boots, B. Imitation learning for agile autonomous driving. *Int. J. Robot. Res.* **2020**, *39*, 286–302. [CrossRef]
145. Sadeghi, F.; Levine, S. CAD2RL: Real single-image flight without a single real image. In Proceedings of the Robotics: Science and Systems, Cambridge, MA, USA, 12–16 July 2017.
146. Ross, S.; Melik-Barkhudarov, N.; Shankar, K.S.; Wendel, A.; Dey, D.; Bagnell, J.A.; Hebert, M. Learning monocular reactive UAV control in cluttered natural environments. In Proceedings of the IEEE International Conference on Robotics and Automation (ICRA), Karlsruhe, Germany, 6–10 May 2013; pp. 1765–1772.
147. Gandhi, D.; Pinto, L.; Gupta, A. Learning to fly by crashing. In Proceedings of the IEEE/RSJ International Conference on Intelligent Robots and Systems (IROS), Vancouver, BC, Canada, 24–28 September 2017; pp. 3948–3955
148. Gasparino, M.V.; Sivakumar, A.N.; Liu, Y.; Velasquez, A.E.; Higuti, V.A.; Rogers, J.; Tran, H.; Chowdhary, G. Wayfast: Navigation with predictive traversability in the field. *IEEE Robot. Autom. Lett.* **2022**, *7*, 10651–10658. [CrossRef]
149. Sathyamoorthy, A.J.; Weerakoon, K.; Guan, T.; Liang, J.; Manocha, D. TerraPN: Unstructured terrain navigation using online self-supervised learning. In Proceedings of the IEEE/RSJ International Conference on Intelligent Robots and Systems (IROS), Kyoto, Japan, 23–27 October 2022; pp. 7197–7204.
150. Hirose, N.; Shah, D.; Sridhar, A.; Levine, S. ExAug: Robot-conditioned navigation policies via geometric experience augmentation. In Proceedings of the 2023 IEEE International Conference on Robotics and Automation (ICRA), London, UK, 29 May–2 June 2023; pp. 4077–4084.
151. Kahn, G.; Abbeel, P.; Levine, S. Badgr: An autonomous self-supervised learning-based navigation system. *IEEE Robot. Autom. Lett.* **2021**, *6*, 1312–1319. [CrossRef]

152. Gyagenda, N.; Hatilima, J.V.; Roth, H.; Zhmud, V. A review of GNSS-independent UAV navigation techniques. *Robot. Auton. Syst.* **2022**, *152*, 104069. [CrossRef]
153. Zhu, K.; Zhang, T. Deep reinforcement learning based mobile robot navigation: A review. *Tsinghua Sci. Technol.* **2021**, *26*, 674–691. [CrossRef]
154. Mousavi, S.S.; Schukat, M.; Howley, E. Deep reinforcement learning: An overview. In Proceedings of the SAI Intelligent Systems Conference (IntelliSys): London, UK, 6–7 September 2018; Springer: Cham, Switzerland, 2018; Volume 2, pp. 426–440.
155. Li, H.; Zhang, Q.; Zhao, D. Deep reinforcement learning-based automatic exploration for navigation in unknown environment. *IEEE Trans. Neural Networks Learn. Syst.* **2019**, *31*, 2064–2076. [CrossRef]
156. Wu, J.; Ma, X.; Peng, T.; Wang, H. An improved timed elastic band (TEB) algorithm of autonomous ground vehicle (AGV) in complex environment. *Sensors* **2021**, *21*, 8312. [CrossRef]
157. Kulhánek, J.; Derner, E.; De Bruin, T.; Babuška, R. Vision-based navigation using deep reinforcement learning. In Proceedings of the European Conference on Mobile Robots (ECMR), Prague, Czech Republic, 4–6 September 2019; pp. 1–8.
158. Xi, A.; Mudiyanselage, T.W.; Tao, D.; Chen, C. Balance control of a biped robot on a rotating platform based on efficient reinforcement learning. *IEEE/CAA J. Autom. Sin.* **2019**, *6*, 938–951. [CrossRef]
159. Karnan, H.; Warnell, G.; Xiao, X.; Stone, P. Voila: Visual-observation-only imitation learning for autonomous navigation. In Proceedings of the 2022 International Conference On Robotics And Automation (ICRA), Philadelphia, PA, USA, 23–27 May 2022; pp. 2497–2503.
160. Lee, K.; Vlahov, B.; Gibson, J.; Rehg, J.M.; Theodorou, E.A. Approximate inverse reinforcement learning from vision-based imitation learning. In Proceedings of the IEEE International Conference on Robotics and Automation (ICRA), Xi'an, China, 31 May–4 June 2021; pp. 10793–10799.
161. Qi, W.; Mullapudi, R.T.; Gupta, S.; Ramanan, D. Learning to move with affordance maps. In Proceedings of the International Conference on Learning Representations, Addis Ababa, Ethiopia, 26–30 April 2020.
162. Zendel, O.; Honauer, K.; Murschitz, M.; Steininger, D.; Dominguez, G.F. Wilddash-creating hazard-aware benchmarks. In Proceedings of the 15th European Conference on Computer Vision (ECCV), Munich, Germany, 8–14 September 2018; pp. 402–416.
163. Tang, J.; Chen, Y.; Kukko, A.; Kaartinen, H.; Jaakkola, A.; Khoramshahi, E.; Hakala, T.; Hyyppä, J.; Holopainen, M.; Hyyppä, H. SLAM-aided stem mapping for forest inventory with small-footprint mobile LiDAR. *Forests* **2015**, *6*, 4588–4606. [CrossRef]
164. Chen, W.; Shang, G.; Ji, A.; Zhou, C.; Wang, X.; Xu, C.; Li, Z.; Hu, K. An overview on visual SLAM: From tradition to semantic. *Remote Sens.* **2022**, *14*, 3010. [CrossRef]
165. Chghaf, M.; Rodriguez, S.; Ouardi, A.E. Camera, LiDAR and multi-modal SLAM systems for autonomous ground vehicles: A survey. *J. Intell. Robot. Syst.* **2022**, *105*, 2. [CrossRef]
166. Xue, H.; Hein, B.; Bakr, M.; Schildbach, G.; Abel, B.; Rueckert, E. Using deep reinforcement learning with automatic curriculum learning for mapless navigation in intralogistics. *Appl. Sci.* **2022**, *12*, 3153. [CrossRef]

applied sciences

MDPI

Article

A Generic Framework for Enhancing Autonomous Driving Accuracy through Multimodal Data Fusion

Henry Alexander Ignatious [1], Hesham El-Sayed [1,2,*], Manzoor Ahmed Khan [1,2] and Parag Kulkarni [1]

[1] College of Information Technology, United Arab Emirates University, Al-Ain 15551, United Arab Emirates; 201990006@uaeu.ac.ae (H.A.I.); man_khilji@hotmail.com (M.A.K.); parag@uaeu.ac.ae (P.K.)

[2] Emirates Center for Mobility Research, United Arab Emirates University, Al-Ain 15551, United Arab Emirates

* Correspondence: helsayed@uaeu.ac.ae

Abstract: Higher-level autonomous driving necessitates the best possible execution of important moves under all conditions. Most of the accidents in recent years caused by the AVs launched by leading automobile manufacturers are due to inadequate decision-making, which is a result of their poor perceivance of environmental information. In today's technology-bound scenarios, versatile sensors are used by AVs to collect environmental information. Due to various technical and natural calamities, the environmental information acquired by the sensors may not be complete and clear, due to which the AVs may misinterpret the information in a different context, leading to inadequate decision-making, which may then lead to fatal accidents. To overcome this drawback, effective preprocessing of raw sensory data is a mandatory task. Pre-processing the sensory data involves two vital tasks, namely data cleaning and data fusion. Since the raw sensory data are complex and exhibit multimodal characteristics, more emphasis is given to data preprocessing. Since more innovative models have been proposed for data cleaning, this study focused on data fusion. In particular, this study proposed a generic data fusion engine, which classifies different formats of sensory data and fuses them accordingly to improve accuracy. This study proposed a generic framework to fuse the text, image, and audio data. In the first stage of this research, an innovative hybrid model was proposed to fuse multispectral image and video data. Simple and efficient models to extract the salient image features were also proposed. The hybrid image fusion model that was proposed did not yield satisfactory outcomes when combining 3D point cloud data, and its performance declined when evaluating large datasets. To address this issue, the study expanded by introducing an advanced generative adversarial network (GAN) to transform the hybrid image fusion model into a machine learning model capable of handling substantial datasets. Additionally, customized kernel functions were suggested to fuse 3D point cloud data effectively. The performance of the proposed models was assessed using standard metrics and datasets, comparing them with existing popular models. The results revealed that the proposed image fusion model outperformed the other models.

Keywords: autonomous vehicles (AVs); data fusion; situation awareness; data-preprocessing; machine learning (ML)

Citation: Ignatious, H.A.; El-Sayed, H.; Khan, M.A.; Kulkarni, P. A Generic Framework for Enhancing Autonomous Driving Accuracy through Multimodal Data Fusion. *Appl. Sci.* **2023**, *13*, 10749. https://doi.org/10.3390/app131910749

Academic Editors: Nadjim Horri, William Holderbaum and Fabrizio Giulietti

Received: 10 May 2023
Revised: 20 June 2023
Accepted: 22 June 2023
Published: 27 September 2023

1. Introduction

Autonomous cars have the ability to cut traffic accidents significantly. This hypothesis is based on the fact that autonomous cars reduce the influence of a human component on the likelihood of a traffic collision. Autonomous vehicles are being tested in real-life traffic circumstances all around the world [1]. According to [1], if AVs become accessible in 2025, the market in the United States of America might reach 8 million in ten years. The widespread deployment of autonomous vehicles is intended to minimize collisions, relieve traffic congestion, increase fuel efficiency, reduce parking demands, and provide mobility to individuals who are unable to drive [2]. The same survey predicted that by 2040 about 33 million autonomous vehicles will be on the road. This just means that people will have to become acclimated to self-driving automobiles. This is supported by facts and data, yet further testing is needed to ensure everyone's safety.

In order to avoid accidents caused by the AVs, the decisions taken by the AVs to overcome roadside events must be instant and accurate [3]. For accurate decision-making, the environmental data perceived by the AVs must be clear and accurate. Perception systems must be exact in order to provide a precise comprehension of the world. They must be able to work in adverse conditions and even when particular sensors are broken or defective. In order to collect environmental data, as well as data relating to autonomous vehicle characteristics, sensor systems must be functional. However, data received from a variety of devices, including sensors, thermal cameras, radars, and so on, have heterogeneous multimodal features, making it difficult to achieve the correct perception. In order to enhance the situation awareness of the AV, there has to be more emphasis on the data preprocessing tasks of the AVs. Data preprocessing involves data cleaning and multimodal fusion. The upcoming paragraphs highlight the relationship between situation awareness and data preprocessing.

According to [4], situation awareness (SA) is defined as "the perception of the elements in the environment within a volume of time and space, the comprehension of their meaning, and the projection of their status in the near future". SA bridges the gap between what is known about the environment and what is occurring in it, and what could happen in the future. To establish the situation awareness of their perceived environment, AVs collect the environmental data using external and internal sensors, namely LiDAR, radar, and ultrasonic, followed by stereo and thermal cameras. The vehicular information collected can exist in different formats, such as textual, image, video, and audio, and also exhibit many discrepancies. The discrepancies exhibited by the vehicular data might be in the form of irrelevant data, missing data, outliers, and duplicate data. In addition, images may contain missing data in the form of blurred images and audio data that contains noisy irrelevant data. Hence, preprocessing the sensory data is essential to improve the accuracy. Since data cleaning has been well studied, this proposal focuses on the second task of data preprocessing, namely data fusion. The sensory data exhibit multimodal characteristics, hence an effective data fusion model is mandatory to convert the heterogeneous data into a unique format and also to enhance their accuracy for further effective processing.

Most of the studies fail to propose a generic multimodal fusion methodology to handle the diversity existing among different datasets. The relevant research literature also does not clearly explain key operations such as feature selection and dimensionality reduction of multimodal data, the mechanisms for 2D to 3D multimodal data transformation and storage, and the methodology for converting multimodal data to a single unique data format. Further very few contributions have been carried out to fuse multispectral environment data collected from sensors and satellites. Versatile fusion models with advanced image processing and machine learning techniques are required to fuse the multispectral high-resolution data. The accuracy level reached by most of the referred decision-making frameworks and models is around 85%. This statement proves that more emphasis must be given to preprocessing, especially in data fusion tasks to improve the data accuracy, which has an impact on enhancing the situation awareness of AVs to improve their accuracy of decision-making. The studies related to data fusion are elaborated on in the related work section.

In light of the mentioned limitations, this study suggests a hybrid fusion model that combines traditional data fusion methods with modern technology in order to merge high-spectral and high-resolution images. The study introduces simple and versatile models for extracting significant image features such as color, edge, height, and width. However, the proposed hybrid image fusion model had its own drawbacks. It encountered challenges in managing large volumes of image data and particularly struggled with fusing 3D point cloud data. To address these issues, this study has expanded the research by introducing a GAN model along with customized kernel functions. This transformation converts the proposed hybrid image fusion model into a machine learning model, capable of effectively handling vast amounts of image data, including both 2D images and point cloud data. The proposed GAN model successfully accomplishes tasks such as separating images into spatial and temporal data, extracting features from the separated data, and merging them using customized kernel functions to enhance the accuracy of the image data. To evaluate the proposed models, the nuScenes and GeoTiles datasets were employed.

The rest of the paper is organized as follows: Section 2 covers the basics of sensor technology, gives an overview, and discusses the importance of data fusion to improve the accuracy of the data. Section 3 discusses the existing literature and its pitfalls that motivated us to pursue this proposal. Section 4 elaborates on the development of the proposed fusion and feature extraction models. Evaluation of the proposed models along with the outcomes is discussed in Section 5. The overall summary, outcomes, and future directions of this paper are enclosed in Section 6.

2. Background and Motivation

2.1. Overview about Sensors

Figure 1 depicts the overall architecture of the autonomous driving (AD) system, and Figure 2 illustrates the functionality of the AD system.

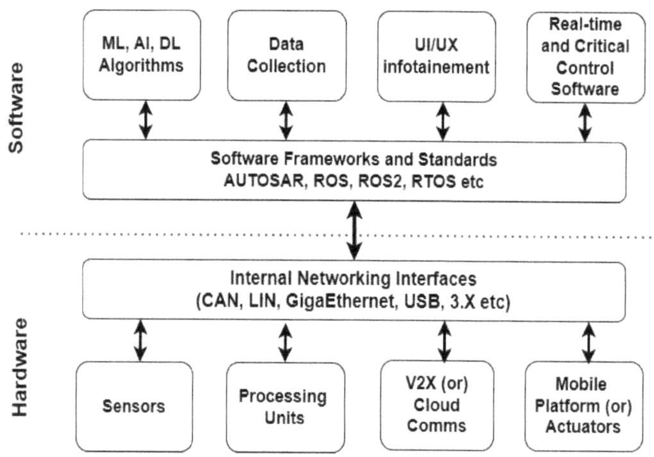

Figure 1. Architecture of an autonomous driving (AD) system.

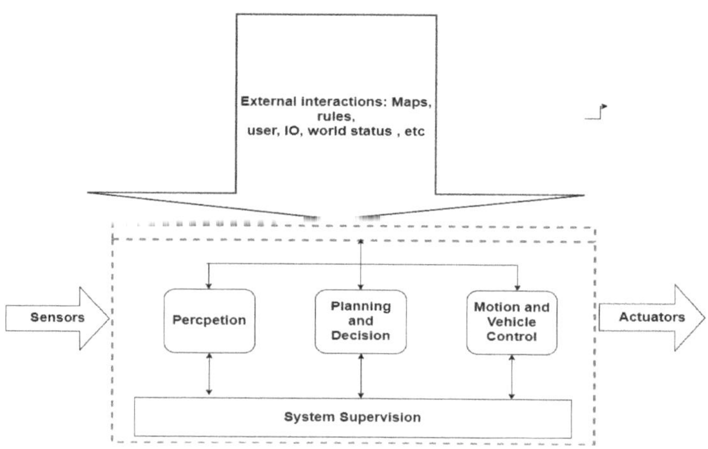

Figure 2. A functional perspective that describes four main functional blocks.

Further, this section discusses the basics of sensor technology, how AVs perceive their environment data using the sensors, various types of sensors, and a brief introduction to data fusion. While AV systems differ slightly from one another, they are usually complex systems with several sub-components. The layout of an AV model may be divided into two stages, according to several perspectives: the hardware and software tools in the initial step, and the functional perspective required for the AD system's operating units in the second stage. From a technical perspective, the two main tiers are hardware and software, with individual components representing various facets of the entire AV platform in each layer. Communication infrastructure is provided by a number of AD sub-components that operate between the hardware and software levels. Clarification, agility, automobile controller, system tracking, organization, and policy-making are the main operational components of AVs. These operational units' duties center on information flow from perceived data to vehicle control, data collection, and data analysis.

As discussed in the introduction the AVs' surrounding environment data are collected using advanced sensors such as LiDAR and radar also using thermal and stereo cameras. This section highlights the functioning and characteristics of sensors and cameras. Sensors convert events or changes in the environment into mathematical computations that may be processed later. Sensors are classified into two major categories based on their functionality. The first category denotes internal state sensors, also known as proprioceptive sensors, which store and detect the states of a mobile system, such as force, angular rate, wheel load, battery voltage, and other variables. Inertial measurement units (IMUs), encoders, inertial sensors (gyroscopes and magnetometers), and position sensors (global navigation satellite system (GNSS) receivers) are examples of proprioceptive sensors. Complete localization refers to the vehicle's location with respect to a global reference frame, whereas relative localization addresses the vehicle's coordinates that align with the surrounding landmarks (world). The external sensors detect information about the external world that is relevant to the AVs. For precise and accurate object detection and to conduct dependable and safe operations, individual and relative sensor orientation is critical. In general, getting enough data from a single reputable source in AD is challenging. The internal health state of the components, as well as the exterior surrounding information of the AVs, are critical for successful decision-making to ensure safety.

2.1.1. Sensor Characteristics

It is important to first establish the general characteristics of these sensors before exploring the specifics of the many sensors utilized in AVs. The selection of sensors in fused and other techniques is critically influenced by the following technical characteristics [5]:

1. Accuracy: the difference between the actual and measured values, recorded by the sensor. Inadequate data and improper environmental information acquired by the sensors will affect the accuracy of evaluations.
2. Resolution: The negligible dissimilarity between two measured values is far less accurate than the sensor's actual accuracy.
3. Sensitivity: the nominal value that is recognized and calculated. The slope of the output response curve or, more generally, the minimal input of a parameter that will result in a discernible output change is used to characterize a sensor's sensitivity.
4. Dynamic range: the least and highest values attained from the sensors, obtainable with precision.
5. Perspective: The term "field of view" is frequently used to describe this (FoV).
6. Active and passive: an inert sensor trusts ambient conditions to distribute statistics, whereas a dynamic sensor releases energy to observe an AV's surroundings.
7. Time scale: the pace of the sensor's frame rate and measuring range rate over time.
8. Output interface: the sensor's output, which might take the form of a cognate energy, automated signal, direct data flow, or data broadcast movement.

LiDar: LiDAR, or light detection and ranging, was first developed and launched in the 1960s and has since become widely employed in the alignment of aeronautical and aerospace topography. Laser scanner manufacturers produced and sold the first commercial LiDARs with 2000 to 25,000 pulses per second (PPS) for topographic mapping applications in the mid-1990s. LiDAR technology has advanced at an astonishing rate in recent years, and it is now one of the most important sensing technologies for autonomous driving. LiDAR is a distant sensing technique that scatters ray from target objects using infrared/laser beam impulses.

RADAR: RADAR stands for radio detection and ranging technology, and is a device that uses radio waves to detect things within a certain range. The surface of an object scatters waves back to the RADAR antenna when waves traveling across it (FoV) collide with it during transmission. The RADAR then picks up the backscattered signal (echo) from the object

Ultrasonic Sensors: In industrial surroundings, ultrasonic sensors are extensively used for various identification errands. They are able to find particles that are hard, fluid, grainy, or crushed. For industrial applications, the sensors generate sonic waves in the 40 kHz to 70 kHz range using sound transducers. Humans may safely hear frequencies in this range since they are above their hearing threshold. A car's parking system may produce sound pressure levels of above 100 dB to assure clear reception, which is identical to the acoustic signal pressure generated by a jet engine. Most ultrasonic sensors detect the time of flight (ToF) of sonic waves between transmission and reception [6].

Cameras: One of the most extensively utilized devices for monitoring the environment is a camera. A camera detects lights emitted by the surroundings on a photosensitive surface (image plane) through a camera lens to produce clear pictures of the surroundings (placed in front of the sensor). With the right software, affordable cameras can snap high-resolution pictures related to their surrounding environment as well as recognize mobile and immovable impediments in their range of view. These characteristics enable the vehicle's vision system to recognize various objects in off-road vehicles in addition to road signs, traffic signals, road lane markings, and obstacles in on-road vehicles. One of our articles [7] provides a comprehensive analysis of various sensors, outlining their distinct characteristics. Figure 3 illustrates the overview of sensors fixed in AVs.

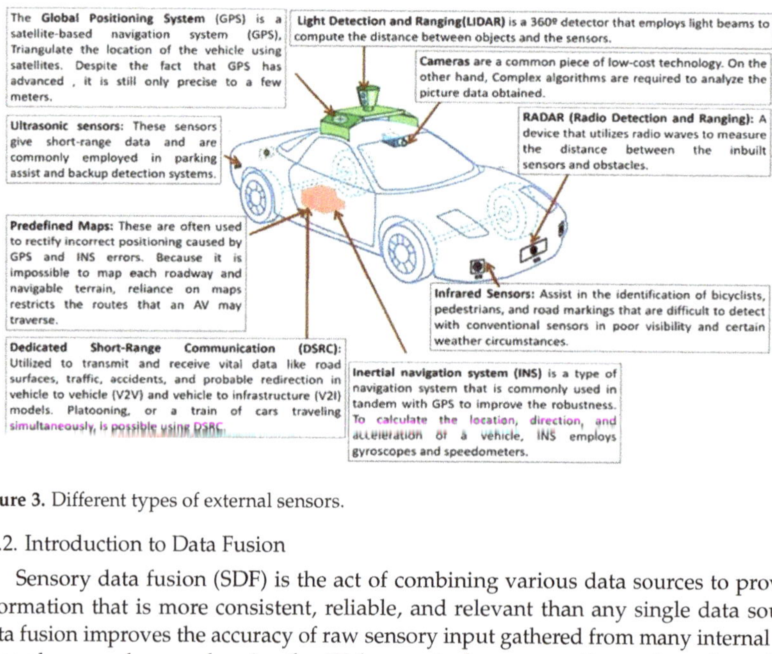

Figure 3. Different types of external sensors.

2.1.2. Introduction to Data Fusion

Sensory data fusion (SDF) is the act of combining various data sources to provide information that is more consistent, reliable, and relevant than any single data source. Data fusion improves the accuracy of raw sensory input gathered from many internal and external sensors, hence enhancing the AVs' conceptual awareness. Sensor data often exhibit multimodal characteristics. Multimodal data are difficult to process, and hence they must be fused and converted into a unique data format that simplifies further processing. Due to the complexity prevailing in data fusion, advanced mathematical models are used for transformations and calculations.

Figure 4 illustrates the process of data fusion. There are different modules, performing different activities. The first phase is the data acquisition stage, where different formats of heterogeneous data are collected from various sensors. The second phase is the data preprocessing stage, where various discrepancies existing in the data are identified and treated further using robust data-cleaning models. This task is followed by the feature extraction process, where the mandatory features are extracted from the data to minimize the fusion process. The next stage is data fusion. There are two types of fusion: (i) early fusion, where data are fused with partial preprocessed data, and (ii) high-level fusion, where strongly preprocessed data are used for fusion. The next phase is the situation refinement task, where the context of the perceived environment data is created for the AVs using the enhanced fused data. The information related to sensors, data objects, and conceptual awareness is represented and stored using advanced data structures, and appropriate models are used to process the data. The main advantage of data fusion is that it combines data from a variety of sensors and sources to create something more intelligent, decisive, reasonable, and precise. The data from each sensor may not make much sense on their own. Computing the (N) independent observations offers a statistical benefit of fusion; the data should be integrated as efficiently as feasible.

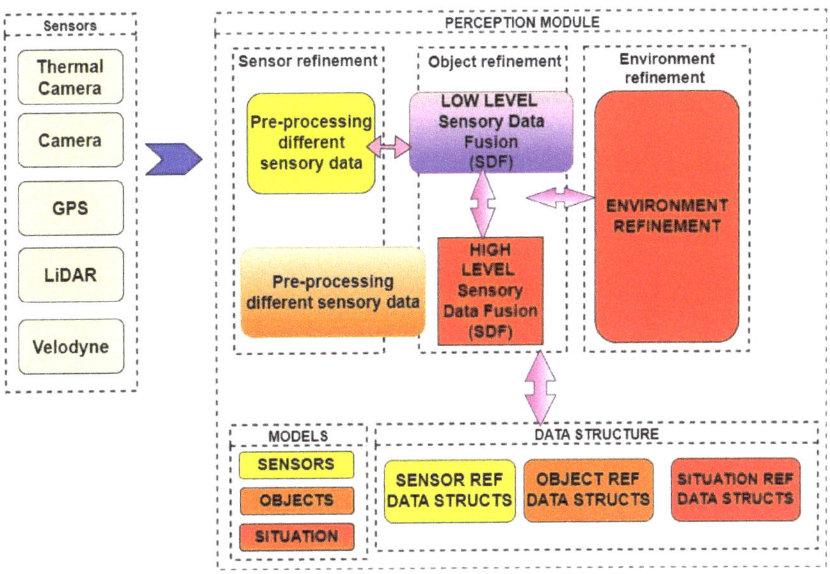

Figure 4. Overall architecture of data fusion.

Making extremely low-power sensors that do not require battery replacements during their lifetime is a crucial requirement in IoT, and this has increased demand for energy-efficient sensors. High-precision sensors are well-recognized for using a lot of energy. A collection of low-accuracy sensors with low power consumption can be used to solve this challenge. Data fusion enables the generation of exceptionally precise data. Another important advantage of data fusion is that it assists in the concealing of critical data or semantics that are responsible for the fused results.

Based on the mathematical approaches used, data fusion strategies can be divided into three categories:

- Artificial intelligence (AI)-based approaches such as classical machine learning, fuzzy logic, artificial neural networks (ANN), and genetic assessment.
- Probability-based methods such as Bayesian analysis, statistics, and recursive operators.
- Evidence-based data fusion strategies based on theory

Among fusing different data formats, fusing image and audio data is more difficult. Normally, image data fusion is performed at three levels: (i) pixel level fusion, which provides extensive image detail information that cannot be obtained at any other level; (ii) feature level, which is the intermediate level and not only stores but also compresses information; and (iii) decision-level fusion, which is the highest level and most sophisticated, with a reduced demand for picture registration. Recursive approaches, multiplicative algorithms, principal component analysis, high PostFilter, the Brovey transform image fusion technique, the color rotated technique, and discrete wavelet transformation are some of the most often used image data fusion techniques. The key image features are color, corners, the SIFT, SURF, blobs, and edges. Table 1 illustrates the operation of the popular image fusion models along with their pros and cons. The next section elaborates on the existing literature.

Table 1. Description of various popular image fusion models.

Method	Advantages	Disadvantages
Multiplicative [8]	• Stretches the histogram of all MS bands • Decreases the standard deviation • Used to fuse small targets	• Difficult to interpret and restore the original image color
Brovey simple method [9]	• Simple and fast • Restoration of high-resolution multispectral image • Useful for visual interpretation	• Ignores high-quality synthesis of spectral information • More spectral distortion
Subtractive method [8]	• Fast, user-friendly, and radiometrically accurate technique for fusing PAN and MS data.	• The output fused image has a lighter tone than the input MS image. • Output image has less clarity
Wavelength [10]	• Improves spatial and preserves spectral characteristics	• Color not smoothly dispersed in spatial features • Edges of small objects are lost
Intensity hue saturation (HIS) method [11]	• Enhanced fusion used for desperate datasets • Converts color images into its RGB of HIS spectrum	• Limited bands • Difficult to fuse multispectral bands
Principe component analysis [11]	• Redundancy of data is decreased • Large amount of input information is reduced without information loss	• Low interpretability of principal components • Tradeoff between information loss and dimensionality reduction
High pass filter [12]	• Smooth texture of output images	• Less sharpness • Low clarity of objects
Ehler method [13]	• Provides all basic spatial information • High resolution output	• More computational time • Decreases with variation in the resulting image

3. Related Work

An extensive survey is presented to explore different image, video, and audio data fusion techniques proposed by eminent scholars to identify their major contributions and existing gaps. The details gathered will be a stepping stone to progress this research further to reduce the existing gaps.

3.1. Hybrid Image Fusion Models

B. Shahian Jahromi et al. [14] have proposed a novel hybrid multi-sensor fusion pipeline configuration for autonomous cars that handles environment perception tasks such as road segmentation, obstacle identification, and tracking. A suggested encoder–decoder-based fully convolutional neural network (FCNx) and a standard extended Kalman filter (EKF) nonlinear state estimator approach are used in this fusion framework. It also employs an optimal camera, LiDAR, and radar sensor configurations for each fusion approach. The purpose of this hybrid architecture is to create a fusion system that is cost-effective, lightweight, adaptable, and resilient (in the event of a sensor failure). It employs the FCNx algorithm, which improves road identification accuracy above benchmark models while preserving real-time efficiency in an embedded computer for autonomous vehicles. D. Jia et al. [15] have presented a hybrid spatiotemporal fusion (STF) technique based on a deep learning model called the hybrid deep-learning-based spatiotemporal fusion model (HDLSFM). With a minimum amount of input, the method develops a hybrid framework for the reliable fusion of morphological and physiological data that explains the physical material at the surface of the earth. To handle radiation discrepancies across various types of satellite pictures, the suggested method combines a regressive deep-learning-based related radiometric normalization, a deep-learning-based super-resolution, and a linear-based fusion. Using Fit-FC as a benchmark, the HDLSFM's propensity to predict phenological and land-cover change has been demonstrated. Meanwhile, HDLSFM is immune to changes in radiation across different types of satellite images as well as the time interval between the forecast and base dates, assuring its usefulness in the synthesis of fused time-series data.

Y. Wang et al. [16] have proposed a hybrid fusion strategy that takes into consideration the geographical and semantic properties of sensor inputs concerning occurrences. To achieve this, the authors have used Cmage, an image-based representation for both physical and social sensor data that describes the situation of certain visual notions (e.g., "crowdedness" and "people marching"). The authors have proposed a fusion model that describes sparse sensor information using a Gaussian process based on the acquired Cmage representation, which combines multimodal event signals with a Bayesian method and integrates spatial links between the sensor and social data. A. V. Malawade et al. [17] have proposed a selective sensor fusion framework, namely HydraFusion, which learns to recognize the present driving environment and then combines the appropriate mix of sensors to enhance robustness without sacrificing efficiency. HydraFusion is the first method to suggest dynamically shifting between early fusion, late fusion, and combinations in between, so modifying both how and when fusion is used. On the industry-standard Nvidia Drive PX2 AV hardware platform, the authors show that HydraFusion outperforms early and late fusion techniques by 13.66 percent and 14.54 percent, respectively, without increasing computing complexity or energy consumption. Both static- and deep-learning-based context identification algorithms are proposed and evaluated by the authors.

Y. Zhao et al. [18] has proposed a hybrid spatial-temporal-spectral image fusion model (HSTSFM) for simultaneously generating synthetic satellite data with high spatial, temporal, and spectral resolution (STSR), which blends the high spatial resolution from the panchromatic image of the Landsat-8 Operational Land Imager (OLI), the high temporal resolution from the multispectral image of the Moderate Resolution Imaging Spectroradiometer (MODIS), and the high spectral resolution from the hyper-spectral image of Hyperion to produce high spatial–spectral image fusion, high spatial–temporal image fusion, and high temporal–spectral image fusion, which are the three fusion modules included in the proposed HSTSFM. To show the performance of the proposed technique, a set of test data containing both phenological and land cover type changes in Beijing suburbs, China, are used. B. Latreche et al. [19] have suggested an effective hybrid image fusion approach based on the integer lifting wavelet transform (ILWT) and the discrete cosine transformer (DCT) that are suited for video streaming networks (VSNs). There are two phases in the proposed fusion algorithm. To begin, the ILWT approximation coefficients (low frequencies) are fused by using the variance as an activity level measure in the DCT domain. Second, the high-frequency detail coefficients are fused using the best-weighted average based on the correlation between coefficients in the ILWT domain. The suggested solution addresses information loss, computational complexity, time and energy consumption, and memory space due to integer operations in the ILWT domain. Extensive tests have been carried out to show that the suggested method outperforms other picture fusion algorithms in the literature, both intuitively and numerically.

X. Zhang et al. [20] have proposed a multi-focus image fusion benchmark (MFIFB), that includes a test set of 105 picture pairings, a code library of 30 MFIF algorithms, and 20 evaluation measures. MFIFB is the first MFIF benchmark, providing a forum for the community to assess MFIF algorithms thoroughly. To understand the performance of these algorithms further, extensive tests have been carried out utilizing the suggested MFIFB. Effective MFIF algorithms are found by examining the experimental findings. More significantly, some remarks on the current state of the MFIF field are provided, which might aid in a better understanding of this topic. D. Kaimaris and A. Kandylas [21] have suggested an innovative mechanism to obtain multispectral image data using UAVs and fuse them to improve the accuracy of the data. The photos from Parrot's tiny multispectral (MS) camera Sequoia+ are examined at two ancient sites: a Byzantine wall (ground application) in Thessaloniki, Greece, and a mosaic floor (aerial application) at the archaeological site of Dion, Greece. The camera obtains RGB and MS pictures at the same time, which prevents image fusion, as is the case with the conventional use of panchromatic (PAN) and MS images in satellite passive systems. Using the image fusion methods of satellite PAN and MS pictures, this research shows that effective digital processing of the images (RGB and MS) of tiny MS cameras may result in a fused image with a high spatial resolution that maintains a considerable proportion of the original MS image's spectral information. The great spectrum fidelity of the fused pictures allows for high-precision digital measurements in ancient sites, such as precise digital item separation, area measurements, and recovery of information not apparent with standard RGB sensors using MS and RGB data from tiny MS sensors. Reference [22] have proposed a versatile hybrid fusion model to fuse infrared and visible image fusion models. They have used the combined concepts of visibility enhancement and multiscale decomposition to fuse the images. Initially, the authors proposed an effective preprocessing model followed by a decomposition model to decompose the information to the layers of their customized CNN model. Further, they have integrated the concepts of a visual saliency illumination map (VSIM) to retain the contrast information and enhance the fusion process. Reference [12] have proposed a hybrid image fusion model to fuse medical images that exhibit multimodal characteristics. They have used the dual combination of nonsubsampled contourlet transform (NSCT) and discrete wavelet transform (DTCWT) approaches to fuse the images. The authors have used an advanced CNN model to create weight maps to monitor the pixel movement of the images. Further, the authors have included an advanced comparison-based method to convert the fusion mode to the appropriate coefficients required for the CNN model.

Since the study has extended its work by proposing a versatile GAN model to fuse all types of advanced image data, the research has explored the contributions of researchers in image fusion using GAN models. Reference [23] have proposed a novel hybrid image fusion model using GAN techniques called as PAN-GAN. The author's model is used to fuse panchromatic images. The PAN-GAN model uses a separate adversarial mechanism that establishes a bond with the discriminators to preserve the spectral and spatial information of the fused images. Similarly reference [24] have proposed an innovative fusion model, namely GAN-FM, which uses the GAN principle to fuse infrared and visible images. The authors have designed a full-scale skip-connected generator along with discriminators based on Markovian principles for extracting features at different scales and to establish a link with the generators to retain the contrast of the fused images. In yet another interesting study, reference [25] have proposed an innovative hybrid image fusion model, namely THFuse, which uses GAN approaches to fuse infrared and visible images. The authors have used advanced fusion strategies such as transformer and hybrid feature extraction concepts to process both global and local image information. Reference [26] have proposed a versatile image fusion model called mask deep fusion network for visible and infrared image fusion (MDFN). The authors have proposed a novel mechanism to compute the weight score for every pixel to estimate the contributions of the two input source images. This operation transfers valuable information from source to fused images, helping them to retain their contrast. Reference [27] have suggested a hybrid image fusion model, namely the pair feature difference guided network (FDGNet), to fuse multimodal medical images. The authors have proposed a weight-guided mechanism to extract the features from complex medical images efficiently. Further, the authors have introduced a factor, namely hybrid loss, composed of weight fidelity loss and feature difference loss to train the network effectively.

Since this study focuses on the feature extraction process, detailed literature related to some of the proposed image feature extraction methods is analyzed. This proposed study plans to extract four important image features, namely color, edge, height, and width. In their recent publication, Li et al. [28] introduced a cutting-edge generative adversarial network named MSAt-GAN. This novel model incorporates multiscale feature extraction and deep attention techniques to merge infrared and visible images seamlessly. By utilizing three distinct fields for feature extraction, the model enhances the accuracy of data fusion. Moreover, the deep attention mechanism facilitates the extraction of multi-level features through spatial and channel attention, thus enabling effective data fusion. Reference [29] in their work, have introduced a versatile fusion model called multi-exposure image fusion on generative adversarial networks (MEF-GAN) with the aim of effectively fusing image data. The proposed model consists of two components: a generator and a discriminator network, which are trained concurrently to form an adversarial network. The generator is responsible for producing synthesized fused images that resemble the source image, while the discriminator is trained to differentiate between the source image and the fake fused images generated by the generator. This adversarial relationship helps to preserve data integrity and prevents information loss in the fused image, ultimately leading to a fused image probability distribution that closely approximates reality.

In their recent study, reference [30] proposed a robust fusion model called correlation driven feature decomposition fusion (CDDFuse) as an effective solution. The authors employed Restormer blocks to extract cross-modality image features and seamlessly integrate them with an advanced convolutional neural network (CNN) model. Additionally, Lite Transformer (LT) blocks were incorporated to extract low-level features. To establish the correlation between low-frequency and high-frequency features, the authors introduced a correlation-based loss factor. By leveraging the proposed LT model and invertible neural networks (INN), the authors successfully fused the low- and high-frequency features, resulting in the generation of the fused image. Reference [31] in their recent publication, introduced a multi-focus image fusion model that combines the principles of Transformers and an advanced CNN model to fuse multimodal image data effectively. By incorporating both local information from the CNN model and global information from the transformers, the accuracy of fusion is significantly improved. Furthermore, the authors proposed a feedback mechanism that maximizes the utilization of features, thereby enhancing the performance of the networks in feature extraction.

3.2. Feature Extraction Models (Image Data)

P. Tiede et al. [32] have proposed a novel universal image feature extraction approach called variational image domain analysis, which is used for a wide range of very long baseline interferometry (VLBI) image reconstructions. Variational image domain analysis, unlike earlier methods, may be used for any image reconstruction, independent of its structure. The authors' approach gives clear ideas on how to extract salient image features such as color and edge. Y. Liu, H et al. [33] have customized a CNN model to extract deep features of images related to food. The CNN model, when paired with nondestructive detection techniques and a computer vision system, has great potential for identifying and analyzing complex food matrices. CNN-based features outperform handmade or machine-learning-based features. N. Liang et al. [34] have proposed a multi-view structural feature extraction approach to provide a thorough characterization of spectral–spatial structures of various objects, which consists mostly of the stages below. First, the original image's spectral number is reduced using the minimum noise fraction (MNF) approach, and then the local structural feature is extracted from the dimension-reduced data using a relative total variation. The nonlocal structural characteristics from intra-view and inter-view are then produced using a superpixel segmentation approach that takes into account the intra- and inter-similarities of superpixels. The final picture characteristics for classification are formed by combining the local and nonlocal structural features. S. Barburiceanu et al. [35] have presented a texture feature extraction approach with increased discriminating power for volumetric pictures. The technique is used to classify textured volumetric data. The authors employ feature vectors obtained from local binary patterns (LBP) and the gray-level co-occurrence matrix-based approach to combine two complementing types of information. R. Ahmed Bhuiyan et al. [36] have provided a feature extraction methodology for human activity recognition that is both efficient and low in dimension. The enveloped power spectrum (EPS) is employed in this feature extraction approach to recover impulse components of the signal utilizing frequency domain analysis, which is more robust and noise intolerant. For human activity recognition, linear discriminant analysis (LDA) is utilized as a dimensionality reduction approach to extract the smallest amount of discriminant characteristics from the envelope spectrum, human activity recognition (HAR). A multi-class support vector machine (MCSVM) is used to recognize human activities using the derived characteristics. To extract robust features, Bo do et al. [37] have used a stacked convolutional denoising autoencoder (SCDAE), which reduces susceptibility to partially damaged data, or inputs data that are partially missing. Trial-and-error experiments were used to optimize SCDAE parameters such as network depth, number of convolution layers, number of convolution kernels, and convolution kernel size.

The analysis identified some gaps in both the explored image fusion models and feature extraction models. Regarding image fusion, most of the referred image models do not produce better accuracy. A generic fusion model to fuse all formats of data is missing. Most referred studies use the minimum dataset to evaluate their models. Complicated operations such as image transformation (2D to 3D) and other image functionalities are not transparent in many studies. Moreover, there are minimum contributions related to multispectral image fusion. Most of the referred models are complex and require advanced algorithms and techniques. There is a need for developing computationally efficient fusion algorithms that can operate in real-time or near real-time scenarios without sacrificing the quality of the fused images. Exploring techniques such as model compression, hardware acceleration, and parallel processing can help bridge this gap.

Implementing and fine-tuning these models can be challenging, requiring significant computational resources and expertise. Image fusion is a subjective task, and the quality of the fused image can vary depending on individual preferences and application requirements. Most referred models involve multiple parameters and design choices, making it difficult to determine an optimal fusion result that satisfies everyone. There is still room for exploring more efficient and effective deep-learning architectures specifically designed for hybrid image fusion. Research should focus on developing novel network architectures, attention mechanisms, and loss functions that can capture complementary information from multiple input images and improve the fusion quality. Most models often lack interpretability and explainability. It is challenging to understand the decision-making process and the contribution of different input images in the fusion result. Further research is needed to develop techniques that can provide insights into the fusion process, visualize the information fusion at different stages, and offer explanations for the final fusion outcome.

Regarding the referred feature extraction models, most of the referred models are complicated and require high computational costs. Most models depend on CPU utilization time and memory. Further, the feature detection models depend on the experiences of the designer. Having the specified gaps as the point of motivation, this study proposes a generic data fusion engine to fuse all formats of data and also proposes innovative strategies to extract the salient features of image and audio data. Many feature extraction models are trained and optimized for specific datasets or domains. However, there is a need for models that can generalize well across different domains, such as medical imaging, natural images, satellite imagery, and more. Developing domain-agnostic feature extraction models that can capture and represent diverse types of data effectively remains a challenge. With the increasing demand for real-time and large-scale applications, there is a need for feature extraction models that are efficient and scalable. Developing lightweight architectures and techniques for efficient feature extraction, model compression, and hardware acceleration is an ongoing research direction to enable faster and more resource-efficient feature extraction.

To address the identified gaps, this study introduces effective models for feature extraction that can extract image features from all types of image data. Additionally, a hybrid image fusion model is proposed to fuse 2D and 3D multispectral image data. The study presents advanced projection and image transformation formulas to enhance the efficiency of the image fusion process. However, the performance of the proposed hybrid image fusion model is found to be unsatisfactory when applied to 3D point cloud data and when dealing with large image datasets. To overcome these limitations, the study expands its scope by proposing an innovative image fusion model that incorporates advanced concepts from the GAN (Generative Adversarial Network) model. This new model performs various tasks, including advanced feature extraction to capture both spatial and spectral information, as well as generator and discriminator modules to facilitate fusion tasks while preserving image quality. Customized kernel functions are introduced for the CNN (Convolutional Neural Network) layers to execute the specified tasks.

4. Proposed Study

4.1. Proposed Framework

Figure 5 illustrates the overall functioning of the proposed framework. At the initial stage, raw sensory data are collected from different sensors. The collected data are preprocessed in the next stage to remove the discrepancies such as irrelevant data, missing data, duplicate data, outliers, and noisy data. In the next stage, the data are classified based on two attributes, namely the file extension (or) the data type. In the subsequent stage, mandatory features are extracted using the proposed feature extraction model. Using the extracted features, the data are fused in the following stage. In the final stage, the fused data are used by the AVs to frame intelligent decision rules for effective decision-making. For textual data, a model for data preprocessing has already been proposed, which is discussed in the article [38]. Hence this paper focuses more on image fusion.

4.2. Proposed Feature Extraction Models

For image data, the study plans to extract four mandatory features, namely color, edge, height, and width, from the image data for effective fusion. Two innovative models have been proposed for edge detection and color identification.

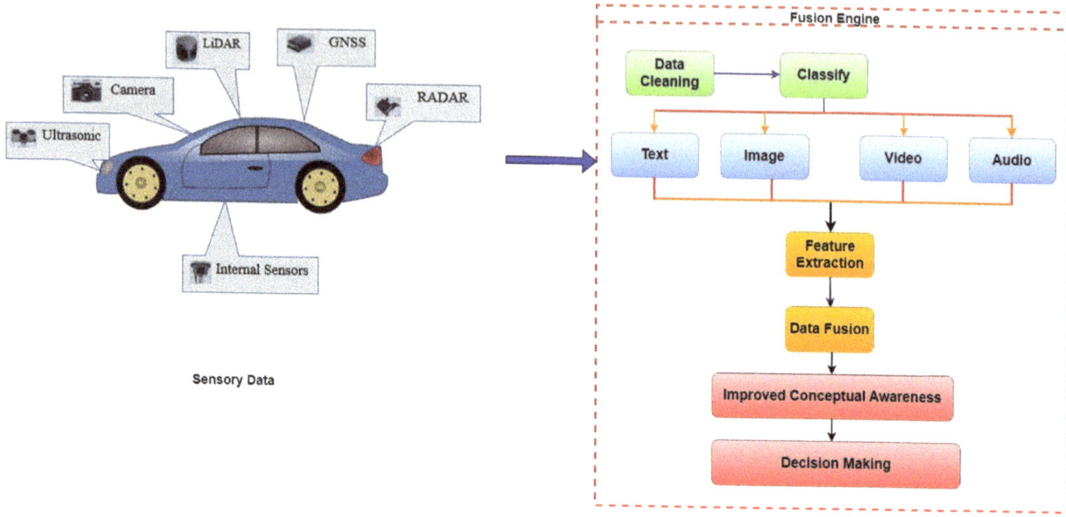

Figure 5. Proposed Framework.

4.2.1. Proposed Edge Detection Method

The study proposes an enhanced Sobel edge detection method. Sobel's approach uses the below-listed matrices to detect the horizontal and vertical edges of an image.

$$T_x = \begin{bmatrix} -1 & 0 & 1 \\ -2 & 0 & 2 \\ -1 & 0 & 1 \end{bmatrix} T_y = \begin{bmatrix} -1 & -2 & -1 \\ 0 & 0 & 0 \\ 1 & 2 & 1 \end{bmatrix}$$

However Sobel's method has a major drawback in that, aside from horizontal and vertical directions, they never traverse the pixels that represent the edges from other directions [39]. Hence, for images with good texture, the accuracy of edge detection using Sobel's method is not good. Hence, this research introduces new templates for edge detection that cover all dimensions of image pixels that are represented as two-dimensional matrices. It is a 5 × 5 matrix represented as four templates, namely T_x, T_y, $T_{i,j}$, and $T_{i-1,j-1}$, denoting vertical, horizontal, and diagonal wise horizontal and vertical representations of pixel weights, respectively. The weight of each template position is ascertained by its distance from the center as well as its directions. The weight of the equidistant points is the same. The following are the derived templates:

$$T_x = \begin{bmatrix} 2 & 3 & 0 & -3 & 2 \\ 3 & 4 & 0 & -4 & -3 \\ 6 & 6 & 0 & -6 & 6 \\ 3 & 4 & 0 & -4 & -3 \\ 2 & 3 & 0 & -3 & 2 \end{bmatrix} T_y = \begin{bmatrix} 2 & 3 & 6 & 3 & 2 \\ 3 & 4 & 6 & 4 & 3 \\ 0 & 0 & 0 & 0 & 0 \\ 3 & -4 & 0 & -4 & -3 \\ -2 & -3 & -6 & -3 & -2 \end{bmatrix}$$

$$T_{i,j} = \begin{bmatrix} 0 & -2 & -3 & -2 & -6 \\ 2 & 0 & -4 & -6 & -2 \\ 3 & 4 & 0 & -4 & -3 \\ 2 & 6 & 4 & 0 & -2 \\ 6 & 2 & 3 & 2 & 0 \end{bmatrix} T_{(i-1,j-1)} = \begin{bmatrix} -6 & -2 & -3 & -2 & 0 \\ 2 & -6 & -4 & 0 & 2 \\ -3 & -4 & 0 & 4 & 3 \\ -2 & 0 & 4 & 6 & 2 \\ 0 & 2 & 3 & 2 & 6 \end{bmatrix}$$

In the proposed approach, the template that produces the largest value for the input image is selected for further processing. The pixel value that produces the maximum threshold value is considered an edge. Generally, there are three ways of threshold estimation: (i) overall threshold, (ii) local threshold, and (iii) dynamic threshold estimation techniques. This study follows the overall threshold estimation method. In the proposed approach, gene-mutation-based techniques are used to detect an edge, and hence an overall threshold estimation approach is followed in this research. In the first stage, the input image is either in 2D $f(x,y)$ (or) 3D $f(x,y,z)$ and is represented in the gray range, say $0,1,\ldots L-1$. If the input images are in a 2D format then the study uses the same transformation mechanism followed in the proposed hybrid image fusion model to convert 2D to 3D images, else the study follows the following steps to detect the presence of an edge in the input image. Then, before initiating the edge detection, the entire image is divided into two categories of thresholds, namely $t : C_0 = (0,1\ldots t)$ and $C_1 = (t+1, t+2\ldots L-1)$. For a gray image of (N) pixels, the probability, P_i, of each gray level in a gray image is calculated using Equation (1)

$$P_i = \frac{n_i}{N}, i = 0,1,2\ldots L-1 \tag{1}$$

Now the square error between the two classes is calculated using the below equation.

$$\gamma^2 = \omega_0 \times \omega_1 \times ((\mu_0 - \mu 1)^2) \tag{2}$$

where ($\omega_0 = \sum_0^{t-1} P_i$, $\omega_1 = \sum_{t+1}^{L-1} 1 - \omega_0$,$\mu_0 = \sum_0^{t-1} P_i/\omega_0$ and $\mu_1 = \sum_{t+1}^{L-1} P_i/\omega_0$). ω_0 denotes the pixels whose gray value is less than the threshold value, ω_1 indicates the pixels whose gray value is greater than the threshold value, μ_0 represents the mean gray value of pixels whose image gray value is less than the threshold value, and μ_1 denotes the average gray values of the pixels whose image gray values are greater than the threshold value. The image gray values must be substituted in the four templates to estimate the best template that maximizes Equation (2). Any pixel value that has a value greater than or equal to the maximum threshold value is considered an edge.

Since it is a complicated task, this study proposes an innovative genetic algorithm, which minimizes the time and the steps involved in finding the best template and which uses the image pixel values to maximize Equation (2). The pixel value that represents a template that maximizes Equation (2) is considered an edge. However, all the templates do not maximize the threshold function; rather, a suitable template that maximizes the threshold function has to be selected. The proposed genetic algorithm selects the best template based on the crossover function represented in Equation (3). This research proposes an innovative cross-mutation function, which helps the genetic algorithm to switch efficiently between different templates. Any template that has a value greater than the crossover function will be the next template used by the image pixels to identify the presence of an edge.

$$CF = 0.9, \frac{f_{max} - f}{f_{max} - \widehat{f}} \tag{3}$$

where f_{max} denotes the maximum fitness function, f represents the mean fitness function, and $\widehat{f}$ depicts the fitness function of an individual pixel.

Figure 6 illustrates the flow of the proposed genetic-algorithm-based edge detection model, and Algorithm 1 elaborates the overall functioning of the genetic algorithm. Equation (3) displays the proposed crossover function, whose value determines whether the genetic algorithm must switch between different templates to determine the maximum fitness function.

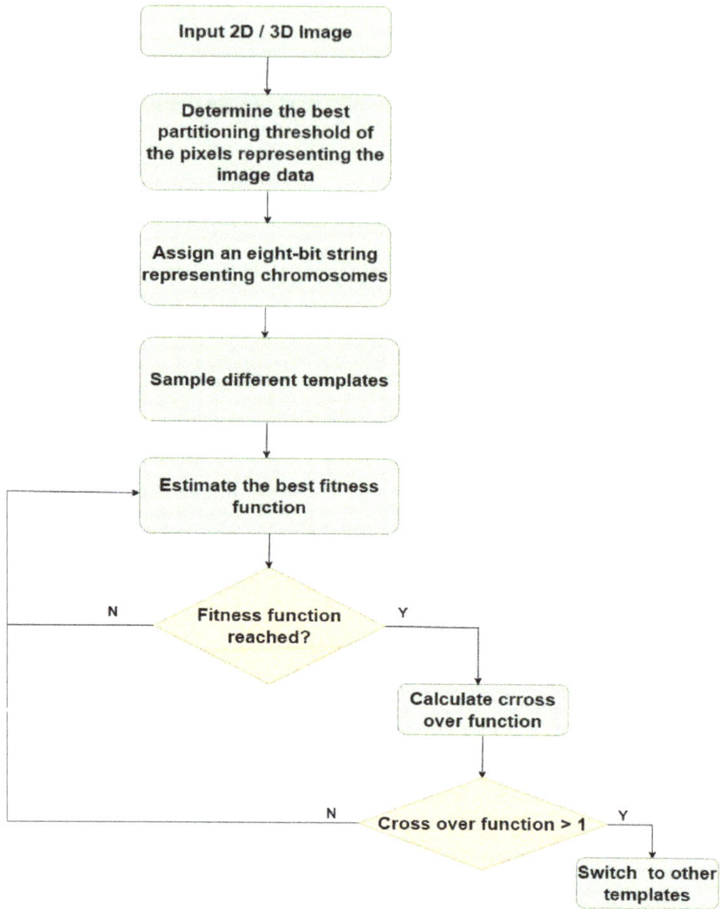

Figure 6. Flow of proposed edge detection model.

4.2.2. Proposed Color Detection Method

Figure 7 illustrates the mechanism behind the proposed color detection method. The pixel colored in green signifies the pivot pixel, while the pixels colored in red denote its neighboring pixels. The pivot pixel whose color is to be examined is taken into consideration. Its corresponding neighboring pixel values are compared with the gradient value, which is computed using the equation specified in Algorithm 2. If the pivot element's neighboring pixel value is greater than or equal to the gradient, then its original pixel value is replaced with the value of one, else a zero value is substituted with its original pixel value. Now, the binary bit patterns obtained from the sub-matrix give the actual color value of the pivot pixel since most of the RGB values of the color are between the numeric range of 0 and 255. The logic adapted in the proposed edge detection model is used to find the height and the width of the image. The first and last edge present in the column of the image matrix determines the height of the image while the first and last edge present in either of the row values of the image matrix represents the width of the image [40,41].

4.2.3. Proposed Hybrid Fusion Model

Algorithm 2 explains how the color value of a pixel is examined in image data. Figure 8 illustrates the functionality of the proposed hybrid image fusion model. In the first stage, the sentinel 2D images are acquired from different sensors. Then the image is normalized using the weighted median filter method to convert 2D pixel intensity to 3D pixel intensity. After normalization, the mandatory features, namely color, edge, height, and width, are extracted from the images using the proposed edge and color detection models. In the third step, appropriate proposed projection and transformations are applied to convert 2D pixel and wavelength intensities to corresponding 3D information. The data obtained are successfully fused with a sample 3D image. This mechanism converts the acquired 2D image to a 3D image. The obtained multispectral 3D image lacks in its clarity related to the RGB colors. In order to improve the RGB color intensities, the obtained 3D image information from the suggested fusion model is integrated with the Brovey fusion method, which increases the RGB color intensity of the fused image to an appreciable extent [42,43]. As discussed in the above paragraph, the suggested hybrid fusion model involves key matrix and vector operations such as projection, transformation, and transposition.

Algorithm 1: Proposed genetic-algorithm-based edge detection approach.

```
Input
2D (or) 3D image
Assign Variables
A: No of the wavelength of image
G_Val: Array to store Gray values of the image
Chr_Val: Array representing genetic chromes
BFit: Best Fitness function
CF: Cross Over Function
Step1: If (input image is 2D use template to convert to 3D)
Else
For i = 1 to Sizeof(A)
Step 2: G_Val[i] = Convert pixel values to corresponding Gray values using
  proposed templates
End For
Step 3: Assign an 8-bit string to Chr_Val
Step 4: Randomly generate 30 sets of templates using values of G_Val
Step 5: Substitute in the Equation (2)
Step 6: Estimate the best fitness function (BFit)
For i = 1 to Sizeof(A)
Step 7: If (G_Val[i] > BFit)
Edge Identified
End For
Else
Step 8: Compute (CF) for different templates using Equation (3)
Step 9: If (CF > 1)
Step 10: Switch to the template which has the highest CF value
Step 11: Repeat Step 2
Step 12: Else
Step 13: Switch to the next template
```

115	114	13
14	77	111
12	134	193
For the original red pixel set neighbor to 1 if the neighbor value is greater (or) equal to pixel value, zero otherwise		

1		
0	X	
0	1	
CLLBP operator Binary = 1001 Decimal = 9		

Figure 7. Proposed color estimation technique.

Algorithm 2: Proposed color detection algorithm.

```
Get the color image
Step 1: Extract the RGB components
Step 2: For each component initialize the extreme counter to zero
Step 3: For each pixel in each component calculate the threshold value as shown in
  Figure 7
Step 4: Calculate the gradient using the below equation
T= (P(i,j+1) + P(i+1,j) + P(i,j-1) + P(i-1,j) + P(i+1,j+1) + P(i+1,j-1) +
P(i-1,j+1) + P(i-1, j-1)-8 * P(i,,j))/9
Step 5: If (value of the neighboring pixel) > T
Step 6: Assign 1 to the matrix
Else
Step 7: Assign 0.
The patterns obtained give the intensity value of the specified color.
```

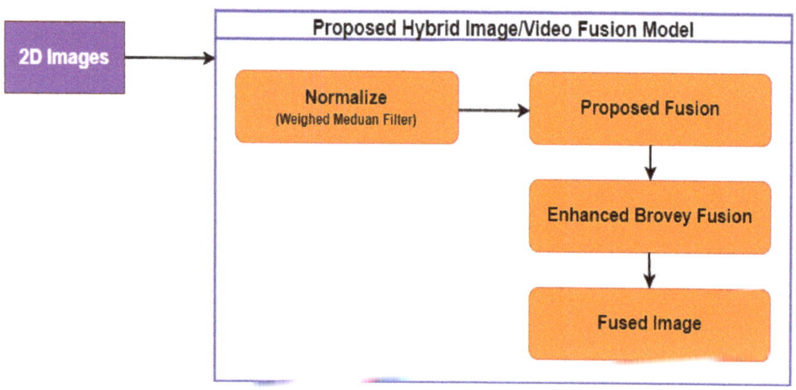

Figure 8. Proposed hybrid fusion framework.

Table 2 describes the mandatory parameters used in the proposed image fusion model. Figure 9 illustrates the flow of the proposed fusion model. Initially, the 2D sentinel model acquired is normalized using the weighted mean filter method. This process converts the 2D pixel to 3D pixel information. Later, to convert the 2D image to a 3D image, the study uses advanced vector projections and matrix transformations. Finally, to acquire the original 3D image the research uses the QR() decomposition method to obtain the inverse matrix information of the fused 3D image.

Table 2. Notations used in proposed image fusion model.

Parameters	Meaning
A	No. of the wavelength of 2D image
B	No. of the wavelength of 3D image
N_A	No. of pixels in 2D image
N_B	No. of pixels in 3D image
N_F	No. of pixels in fused 3D image

Before processing the fusion process initially, the intensity of each wavelength is subtracted from the values of both the 2D sentinel and 3D images. The mean intensity wavelength for both the 2D and the 3D images is calculated using the weighted mean filter method. In terms of mathematics, the weighted mean filter (WMF) is equivalent to global optimization. It can successfully filter images without causing significant edge blur. Within a local window, it is an operator that replaces the current pixel with the weighted median of nearby pixels. In processing pixel (p) in the image (I), only pixels inside the local window, R(p), of radius (r) centered at (p) are considered. WMF connects each pixel, $q\varepsilon R(p)$, with a weight, ϖpq, in the appropriate feature map (f), i.e.,

$$\varpi pq = g(f(p), f(q)) \tag{4}$$

where $f(p)$ and $f(q)$ are features of pixels p and q, and g is a typical influence function between p and q.

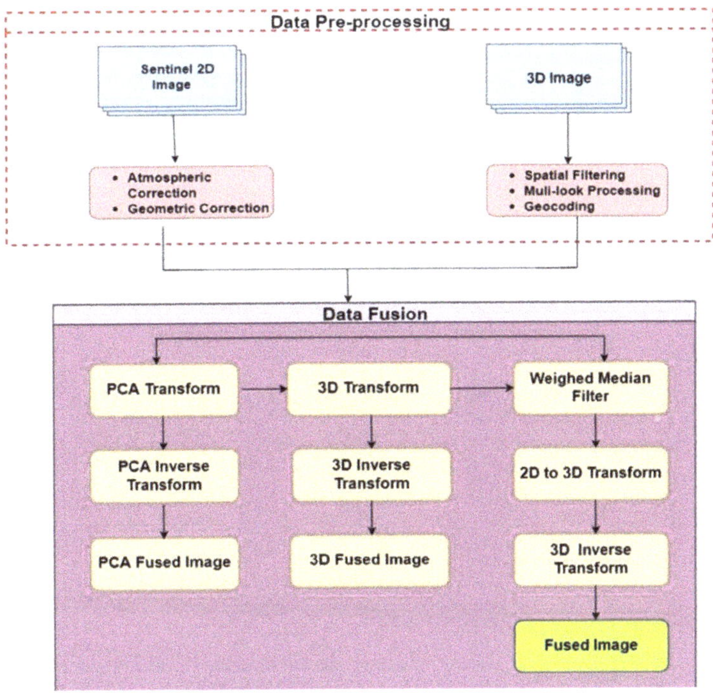

Figure 9. Functioning of hybrid fusion model.

The pixel intensity of the 2D image is multiplied by the weighted intensity values of the 3D image pixels using Equation (4). This task is carried out to convert the 2D pixel intensity values to their corresponding 3D pixel intensity values. Advanced matrix and vector operations that are used for the above-mentioned transformations are discussed in the following paragraphs.

Algorithm 3 describes the steps involved in the proposed hybrid image fusion algorithm. Before starting the main fusion operations, the wavelength intensities and the pixel intensities of the 2D sentinel images are represented as corresponding row and column items of the image matrix using the below-mentioned equation.

$$I_i \times N_j, j, k = I_i, j, k \tag{5}$$

In the next stage, the multispectral 3D images are extracted from the 2D sentinel images by multiplying the wavelength frequencies A and B of panchromatic images with the pixels of the 2D images. This operation is carried out with the help of multiplying Equation (4) by the below-mentioned equation. By carrying out this operation, the pixel values of 2D sentinel images are normalized and converted to 3D image pixel values.

$$E_n = \sum_{k=0}^{A-1} = A_{n,k} W_k \tag{6}$$

Using the new notations, Equation (4) can be rewritten as

$$E_n = AW^t \tag{7}$$

where W is the line vector of size A and components of W_T stand for the action of transposing the image matrix. Though the 2D image pixels are converted to 3D image pixels, the transformation only converts the values; advanced vector projections must be introduced to convert the 2D sentinel image to an actual 3D image. This task is accomplished using Equation (9).

Algorithm 3: Proposed hybrid image fusion algorithm.

```
Input 2D image
B:3D GF-3 image
Assign Variables
A: No of wavelength of 2D image
B: No of wavelength of 3D image
N_A: No of pixels in 2D image
N_B: No of pixels in 3D image
N_F: No of pixels in fused 3D image
Projection of 2D to 3D
For i = 1 to N_A do
Step 1: Compute w_i for 2D image matrix
Step 2: Compute projection of A on B using equation P_{B_k}(E)
End for
Transformation to vector
For j = 1 to N_B
Step 3: Compute E using equation
Step 4: Compute B'using Equation (12)
Step 5: Compute [illegible] using [illegible] and E' with
End For For i = 1 to N_A do
Step 6: Compute B' using equation End for for k = 1 to N_F
Step 7: Compute RGB intensities for the fused panchromatic
Step 8: image projected on a 3D image using Brovey fusion Equation (17)
Output
Step 9: B = Fused Image
```

The general projection of a vector, v, on vector u is performed using the equation

$$P_v = \frac{u.v}{u.u}u \tag{8}$$

$$PB_k(E) = \frac{B_k^T.B}{B_k^T B_k}B_k. \tag{9}$$

where B_k is a column of vector B.

From a statistical point of view, the multispectral image obtained is a combination of any 3D image format along with the weights of the pixel intensities, which are correlated to the pixels of 2D sentinel images [44,45], thus using a projection operation on every pixel of a 2D image. This calculation with a single weight is really a projection onto a wavelength-specific picture, and the entire computation is identical to summing all projections onto all (F) wavelengths. The operations are listed below in Equation (10)

$$\hat{E} = \left[\frac{B_k^T.B}{B_k^T B_k}B_k\right]_k = BB^T E \sum_D^{-1} \tag{10}$$

where $\sum_D$ is a diagonal FXF matrix whose diagonal components are B_T^k and B_k.

Now, we denote $\cap W$ as the line vector of size B, which is equal to $B^T B \sum_D^{-1}$. Now, the 3D multispectral image can be defined as

$$\hat{E} = BW^T \tag{11}$$

By replacing the values of the first wavelength, $\hat{E}$, a modified 3D image is obtained, which is listed in the below equation. When compared with the previous operations, this image matrix is a three-dimensional matrix representing pixel wavelength values, intensity values, and the dimension of the image, respectively.

$$B' = B\hat{W}^T = B\begin{vmatrix} \hat{W} & & \\ 010 & \ldots & 0 \\ 001 & \ldots & 0 \\ & . & \\ & . & \\ & . & \\ 000 & \ldots & 1 \end{vmatrix} \tag{12}$$

The next important stage is acquiring the original 3D image. To obtain the original 3D image, the inverse transformation of the new 3D image obtained using Equation (12) is decomposed using the QR() decomposition method, which postulates two important rules [46].

- Q is an orthogonal matrix, $Q^T Q = I_F$ where I_F is the identity matrix.
- R is an upper triangle matrix, having an upper triangle matrix inverse and $QQ^{-1} = I_F$

Now, the R inverse matrix obtained from the QR() decomposition is multiplied with the modified 3D image obtained using Equation (12) to obtain the original 3D fused image. The equations used to obtain the actual 3D image are discussed below.

$$S = B'R^{-1} \tag{13}$$

An important property of QR() decomposition is that the first column of the matrix S is proportional to that of B', and hence it can still be interpreted as a multispectral image. Now, the first column of the multispectral image obtained from Equation (13) is replaced with the modified 3D image matrix obtained using Equation (12), while all other columns are unchanged. The transformations are illustrated in the below final equations.

$$S' = SW_s^T + AW_A^T and W = \begin{vmatrix} 00 & \dots & 0 \\ 10 & \dots & 0 \\ \cdot & \cdot & \cdot \\ \cdot & \cdot & \cdot \\ \cdot & \cdot & \cdot \\ 00 & . & 1 \end{vmatrix} W_A = \begin{vmatrix} W \\ 0 \end{vmatrix} \quad (14)$$

$$= \begin{vmatrix} W_0 W_1 & \dots & W_{A-1} \\ \cdot & \cdot & \cdot \\ \cdot & \cdot & \cdot \\ \cdot & \cdot & \cdot \\ 00 & . & 0 \end{vmatrix} \quad (15)$$

Since the first column of Equation (15) represents the pixel values of the 2D sentinel image and the other two columns represent the pixel values of 3D images, the final fusion equation can be represented as

$$\hat{B} = S'R \quad (16)$$

4.2.4. Enhanced Brovey Image Fusion Method

Though the multispectral fused image obtained from the proposed fusion method helps to improve the accuracy of the acquired 2D images, it lacks in producing the actual color intensity. In order to improve the RGB color intensities of the fused 3D image, the pixel values of the fused images are integrated with the Brovey fusion model, which computes the mean of individual red, green, and blue intensity values. The obtained mean value is multiplied by the PAN (panchromatic matrix value). The results obtained are finally multiplied with Equation (9) to obtain the full-fledged 3D fused image with the actual RGB color intensities.

$$\begin{aligned} \hat{F} = R_{new} &= \frac{R}{R+G+B} \times PAN(1) \times P_{B_k}(E) \\ G_{new} &= \frac{G}{R+G+B} \times PAN(2) \times P_{B_k}(E) \\ B_{new} &= \frac{B}{R+G+B} \times PAN(3) \times P_{B_k}(E) \end{aligned} \quad (17)$$

4.3. *Proposed GAN Model*

The proposed hybrid image fusion model did not achieve desirable results when integrating 3D point cloud data, and its performance deteriorated when evaluating extensive datasets. Hence, the study has extended its contribution by proposing a versatile GAN model that integrates the proposed feature selection and image fusion models. The proposed GAN model organizes and performs many activities such as organizing the incoming image frames, preprocessing them, and finally fusing them to improve the efficiency and accuracy of the perceived data, which in turn improves the accuracy of decision-making in the AVs.

Additionally, this study proposes optimized kernel functions to select the best-fused image for better decision-making. The new GAN model consists of two discriminators: one is the spatial discriminator and the other the spectral discriminator. After acquiring the images from the sensors, the proposed GAN separates the spatial and spectral information through the discriminators. Advanced Fourier transformation is used as a kernel function to extract the spatial and temporal image information.

After separating them, the GAN model uses two-fold image fusion. First, it uses the previously proposed image fusion model to perform the initial fusion. Then, using the proposed GAN model, additional projection functions are used to convert the fused images obtained from the hybrid image fusion Model to 3D point cloud data. The proposed GAN model uses attention-gate principles and compares the fused images with the source image information. If the difference is minimum then the 3D point cloud data generated in the first-fold fusion are further fused with the source images to obtain the final fused 3D point cloud information. Figure 10 illustrates the overall functionality of the proposed GAN model to convert 2D data to 3D data and fuse them accordingly.

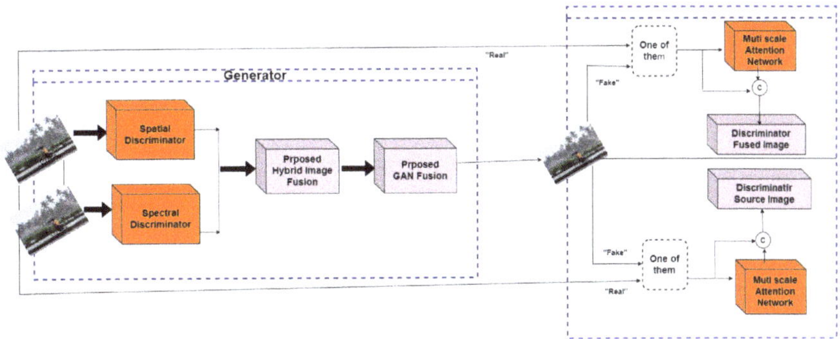

Figure 10. Functioning of proposed GAN model to fuse 3D point cloud data.

The fusion image obtained from Equation (17) is represented in Cartesian coordinates, which have to be converted into 3D projection coordinates using Equation (18) listed below. An additional column for height is introduced and Equation (17) is multiplied with an identity matrix, which converts the 2D image format to pointed 3D projection data. In Equation (18), h, w, and f correspond to the width, height, and focal length of the sensing device, respectively.

$$\begin{pmatrix} x' \\ y' \\ z' \end{pmatrix} = \begin{bmatrix} 1 & 0 & -w/2 \\ 0 & 1 & -h/2 \\ 0 & 0 & 1 \\ 0 & 0 & 1 \end{bmatrix} \begin{bmatrix} x_{00} & x_{01} & x_{02} & x_{03} \\ y_{10} & y_{11} & y_{12} & y_{13} \\ z_{20} & z_{21} & z_{22} & z_{23} \\ x_{00} & x_{01} & x_{02} & x_{05} \end{bmatrix} \begin{bmatrix} f & 0 & w/2 & 0 \\ 0 & f & h/2 & 0 \\ 0 & 0 & 1 & 0 \\ 0 & 0 & 1 & 0 \end{bmatrix} \begin{bmatrix} x \\ y \\ z \end{bmatrix} \tag{18}$$

The proposed GAN model has five layers. The first layer is used for the feature extraction process, the second to fuse the extracted features using the proposed hybrid fusion model, the third to convert the 2D fused image to 3D point cloud data using Equation (21), the fourth to perform the discriminator tasks, and the last layer to perform the final fusion. Spatial features are extracted using the kernel function (19), where $F(i,j)$ is the image information existing in the spatial domain and the exponential term is the basis function corresponding to each pixel value of the image, and temporal features are extracted using the kernel function (20), where the exponential function is the basis of every temporal pixel value of the image information, $F(i,j)$ and ψ is an optimization factor. Following multiple iterations of the training phase, we fine-tuned the factor to an optimized value of 0.65.

$$F(i,j) = \sum_{i=0}^{N-1} \sum_{j=0}^{N-1} f(i,j) e^{-t2\pi(i/N + j/N)} + \psi \tag{19}$$

$$F(i,j) = I/N \sum_{b=0}^{N-1} f(i,j) e^{-t2\pi(b/N)} + \psi \tag{20}$$

The proposed GAN model is trained using Equation (21), where W_s^k is the weight for the kth feature of f_s^k of the s-th pooling scale.

$$W_s^k = \sigma(w_i \sum_{i,j} f_s^k(i,j)) \tag{21}$$

5. Experimental Analysis

Table 3 highlights important software and tools used to implement and evaluate the proposed models. Different categories of comparisons are carried out to evaluate the proposed models, and Table 4 displays various key metrics used to evaluate the proposed and other popular image fusion models.

Table 3. Various tools used for experimental analysis.

S. No	Tools	Version	Description
1	PyCharm	2.7	IDE
2	Python & Libraries (Pandas, NumPy, Open3D, Shapely)	3.10.4	Development of the models
3	MATLAB	R2021a	Developing applications
4	Vehicle perception software system	1.0	To create 3D point cloud vehicle perception
5	Mogo DB	5.0	Database
6	Linux	5.4. 0–26	Operating system
7	Weka	3.0 (GPLv3)	Preprocessing
8	Dataset	nuScenes, GeoTiles (For 3D point cloud data)	To evaluate the proposed image fusion model

Table 4. Key metrics used in the research to evaluate image fusion models.

Metric Name	Purpose
Accuracy (Acc) [47]	To compare accuracy
Efficiency (Eff) [47]	To evaluate the performance of the models
Standard Deviation (SD) [48]	To estimate the contrast
Average Gradient (AG) [47]	To express small detail contrast and texture changes, as well as the sharpness of the image
Spatial Frequency (SF) [47,48]	To measure the overall activity level of the image
Peak Signal-to-Noise Ratio (PSNR) [49]	To compute the visual error between the fused image and the reference image
Correlation Coefficient (CC) [48]	To find the similarity between the reference image and the fused image
Mutual Information (MI) [49]	Is an indicator of how closely the two unrelated factors are related. It measures how knowledge difference between two random factors
Entropy (EN) [49]	Estimates uncertain index

5.1. Performance of Hybrid Image Fusion Model with Other Studies

The proposed model is compared with other studies [50,51] that use hybrid approaches to fuse image data. From the observations, the proposed hybrid image fusion model produces a more accurate fused image than the other studies. From the results portrayed in Figure 11 it is evident that the image clarity and resolution of the proposed image fusion are better than the referred hybrid fusion models.

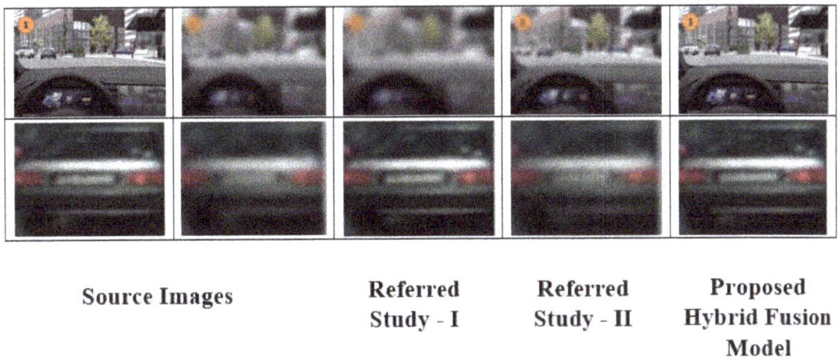

Figure 11. Comparison with other referred models.

Further, the accuracy and efficiency of the proposed hybrid fusion model are compared with other standard fusion models [50,51]. The accuracy of the fused images derived from different models is calculated using Equation (22), and efficiency is estimated by the time taken by the fusion model to fuse the images. The results obtained are portrayed in the below Figures 12 and 13. The results prove that the accuracy and efficiency of the proposed image/video fusion models are better than the other popular image/video fusion models. The proposed hybrid image/video fusion models achieve more than 94% accuracy after fusing the images with minimum resolution and clarity. Further, the proposed hybrid fusion models consume less CPU time to fuse the images over the other referred fusion models.

$$Accuracy = \frac{Pixel_{intensity}(Original)}{Pixel_{intensity}(Fused)} \quad (22)$$

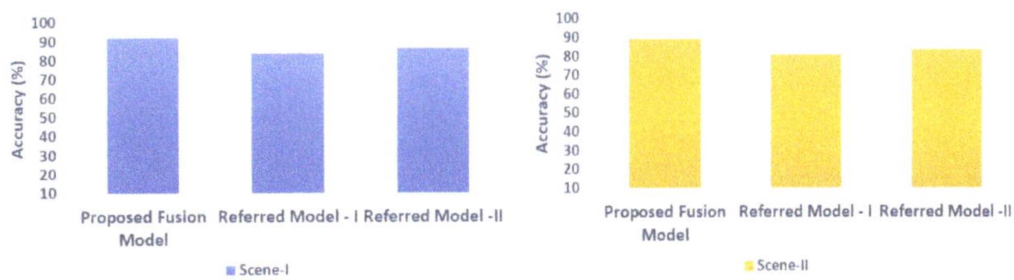

Figure 12. Accuracy comparison with other models.

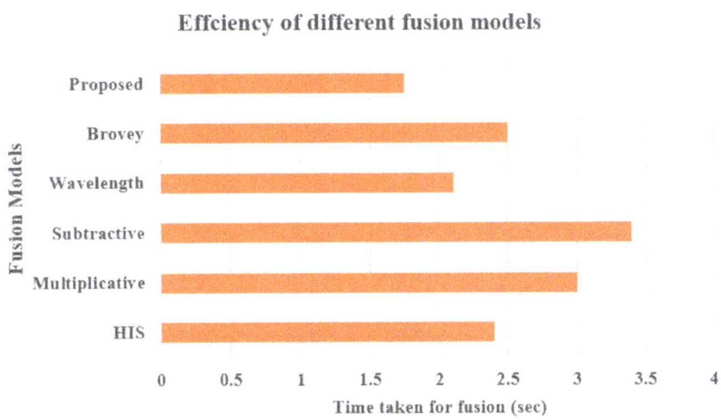

Figure 13. Efficiency comparison.

The final fused images obtained using the proposed and other popular fusion models are evaluated using various metrics including standard deviation, average gradient, spatial frequency, peak signal-to-noise ratio, and correlation cofactor. A sample of three fused images obtained from different models is considered for evaluation. The proposed fusion models perform better than the other popular fusion models. Table 5 portrays the results obtained after evaluation. The impact of the metrics can be visually seen in Figures 14 and 15. The study also evaluated the proposed edge-detection model with other popular edge-detecting models such as KFA (Kalman filter algorithm) and Sobel & Prewitt. Root mean square error (RMSE) values of the images obtained after edge detection are estimated. The lower the RMSE values are, the higher is the accuracy of the constructed image. The accuracy of the proposed edge detection model is better than the other popular edge detection models. RMSE is calculated using Equation (23), and Table 6 illustrates the RMSE values obtained from different edge detection models for different images. The results show that the proposed edge detection model provides a lower value of RMSE, which indicates the improved accuracy in image construction over the other popular edge detection models. Figure 14 illustrates the results obtained after implementing the proposed and other popular edge detection models. From the results, it is evident that the images constructed after implementing the proposed approach are better than other popular edge detection models.

$$RMSE = \sqrt{\frac{1}{MN}\sum_{r=0}^{N-1}[E(r,c) - o(r,c)]^2} \tag{23}$$

In another comparison, advanced machine learning models are used to compare the classification accuracy of the fused images obtained from different fusing models, namely principle component analysis (PCA), weighted median filter-based Gram–Schmidt transform (WMFGS) and Gram–Schmidt (GS). The visual differences between different classification results are portrayed in Figure 15. The CNN model using the random forest (RF) classification model is used to classify the fused images obtained from the four fusion models. The CNN model designed has three layers: one for preprocessing, the next layer for fusion, and the final layer for classification. The proposed feature extraction model and fusion models are converted into appropriate kernel functions and implemented in the first two layers. The random forest classification model is a combination of self-relying decision trees, and decision trees handle alphanumeric data easily when compared with other classification models. From the sample of a thousand fused images obtained from the four fusion models, seven hundred samples are used to train and three hundred samples for testing the RF model. The following steps are carried out to build the infrastructure of the CNN model. The first stage is data acquisition, followed by data preprocessing. Subsequently, the RF classification model using sklearn and NumPy python packages is implemented to train and test the sample fused images obtained by the four fusion models used in this study.

Table 5. Overall performance of proposed and other popular image fusion models for various key metrics.

Images	Model	SD	AG	SF	PSNR	CC
1	GS	305.3385	9.4444	60.0375	89.6709	0.6908
	HIS	[illegible]	[illegible]	[illegible]	[illegible]	[illegible]
	Subtractive	375.7824	9.4150	133.5344	94.5325	0.7107
	Proposed	380.298	10.2148	135.6382	96.2867	0.9472
2	GS	289.3498	9.6105	78.5329	90.0781	0.6040
	HIS	325.7555	9.7290	177.7829	93.0075	0.6077
	Subtractive	339.9461	10.4368	178.6462	93.0360	0.6360
	Proposed	365.4752	11.4976	180.5643	94.2335	0.8235
3	GS	625.3025	10.5363	131.9687	114.0561	0.7434
	HIS	646.8171	10.8526	132.6416	95.5631	0.7313
	Subtractive	657.6024	9.9434	120.6161	95.6224	0.7344
	Proposed	370.2539	11.3145	123.6213	93.25	0.6543

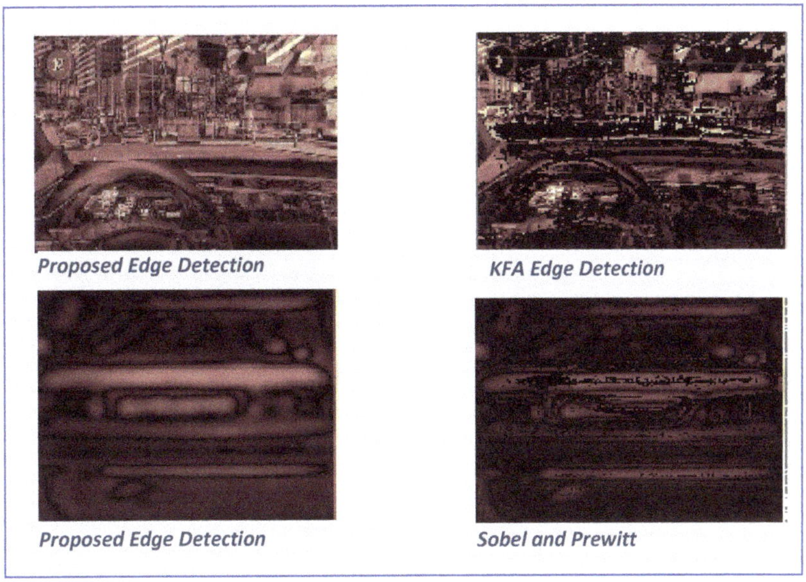

Figure 14. Comparison between edge detection models.

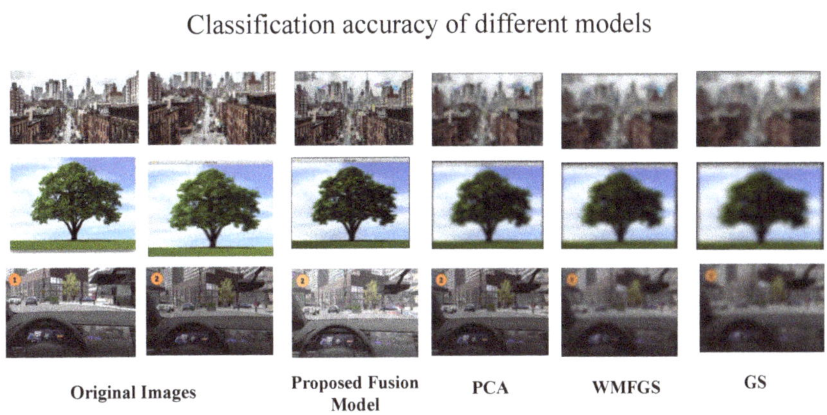

Figure 15. Classification accuracy of different models.

Table 6. RMSE scores of different edge detection models.

Model Name	Image Info	RMSE Value
Proposed	Image 1	0.1330
Proposed	Image 2	0.1295
KFA	Image 1	0.3895
KFA	Image 2	0.2522
Sobel & Prewitt	Image 1	0.4831
Sobel & Prewitt	Image 2	0.3187

Confusion matrix techniques are used to determine the accuracy of classification. From the results obtained, it is evident that the RF model shows more accuracy when it is tested with the fused images obtained from the proposed hybrid image fusion model than with images from the other models. The results obtained are portrayed in Table 7.

Table 7. Accuracy of different image fusion models.

Models	Proposed	PCA	WMFGS	GS
True positives (TP)	15	130	140	120
True negatives (TN)	130	120	120	110
False positives (FP)	12	28	25	40
False negatives (FN)	8	22	15	30
Accuracy (%)	93	83	86	77

5.2. Overall Analysis of Proposed GAN Fusion Model

In the extended work (this study), the proposed GAN model has been included to convert the proposed hybrid image fusion model into a machine learning model to handle huge volumes of image data, specifically point cloud data. The proposed GAN model is an advanced CNN model, which is designed as a single convolutional model layer that includes five main layers. The initial layer is utilized to carry out the process of extracting features, while the second layer is responsible for combining the extracted features using a proposed hybrid fusion model. The third layer converts the resulting 2D fused image into 3D point cloud data using the equation referred to in (21). Following that, the fourth layer is dedicated to performing the tasks of the discriminator. Finally, the last layer is responsible for executing the final fusion. Table 8 depicts the design implementation of the proposed GAN model. The 3D point cloud environment setup was carried out using the following steps. Initially, a vehicle perception system purchased from a third-party vendor was used, which helped to create an advanced 3D point cloud perception system. This cloud-based system was installed in the Ubuntu operating system using advanced Docker commands and Python scripting. Applications were developed using Python scripts and later integrated with the perception system to perform advanced analysis. Both real-time data collected from Velodyne sensors and 3D point cloud data obtained from the GeoTiles dataset were used to evaluate the performance of the proposed GAN and other popular fusion models such as FusionGAN, curvelet transform (CVT), dual-tree complex wavelength transform (DTCWT), Laplacian pyramid transform (LP), lowpass pyramid (LP) and multi-resolution singular value decomposition (MSVD). A sample of 20 images was selected from the GeoTiles dataset to train the GAN model, and a maximum of four images were used to test the GAN model initially.

Table 8. Implementation details of proposed GAN model.

Layer	Discriminator Network					
	I	O	KS	SS	PS	
Layer1 Conv	1	64	4	1	0	LeakyReLU
Layer2 Conv	64	128	4	2	0	LeakyReLU
Layer3 Conv	128	128	4	1	0	LeakyReLU
Layer4 Conv	128	256	4	2	0	LeakyReLU
Layer5 Conv	256	256	4	1	0	LeakyReLU

Figure 16 illustrates the performance of various fusion models for the metrics (MI, EN, SD, and PSNR). A sample of 30 point cloud scenes was taken, which analyzed the performance of various popular fusion models along with our extended GAN model for four metrics, namely MI, EN, SD, and PSNR. From the results, the proposed GAN outperformed other fusion models. Table 9 depicts the CPU time consumed by different fusion models to fuse 2D image data and 3D point cloud data. A sample of 10 images was collected from the nuScenes and GeoTiles dataset to evaluate the efficiency of the fusion models. Once again, the proposed fusion models performed better than the other fusion models. Again, a sample of three 3D point cloud data from the GeoTiles dataset along with their blurred images (since it was difficult to obtain blurred images from the dataset, online applications to blur the selected images were used) was taken to analyze the performance of different fusion models. From the analysis, it was observed that the fusion accuracy of the proposed GAN model was better than other fusion models. Figure 17 depicts the fusion results of various fusion models used to fuse 3D point cloud data sampled from the GeoTiles dataset.

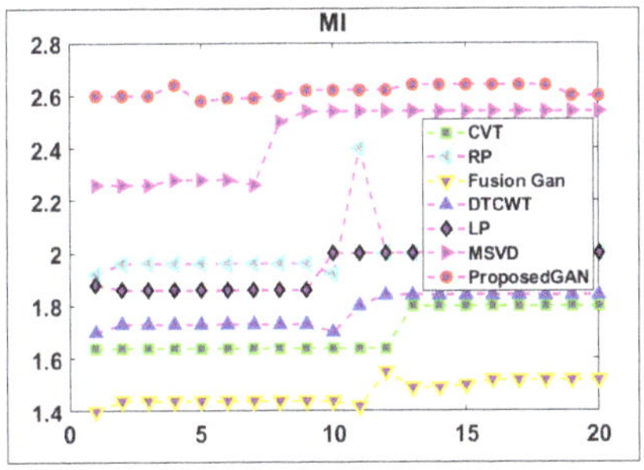

(a) Mutual Information

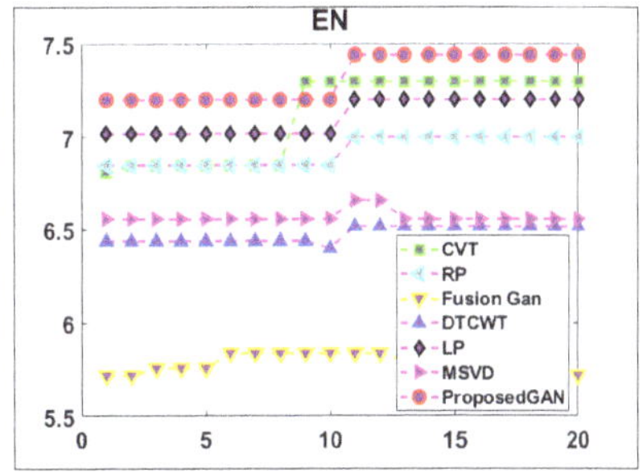

(b) Entropy

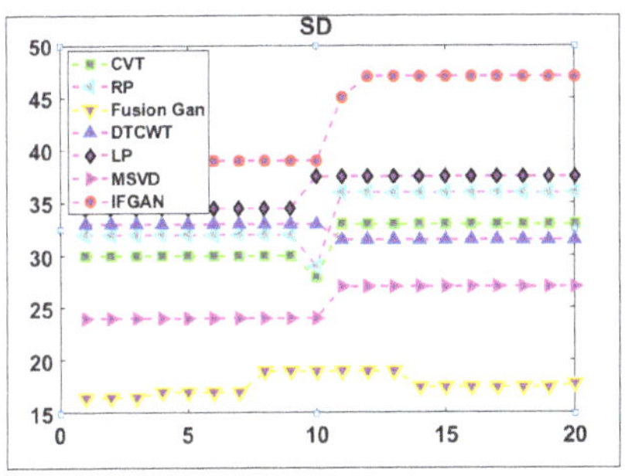

(c) Standard Deviation

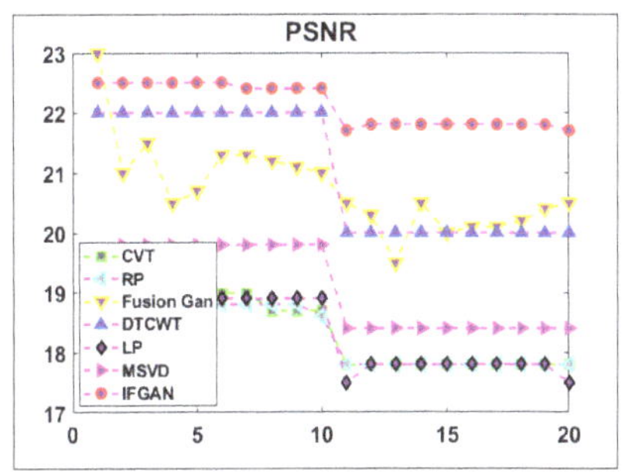

(d) Peal Signal-to-Noise

Figure 16. Comparison between different fusion models for GeoTiles 3D point cloud data.

Table 9. Run-time comparison between different fusion models.

Fusion Model	nuScenes	GeoTiles
CVT	0.6625	0.7742
DTCWT	0.1578	0.1954
LP	0.1426	0.1828
RP	0.1678	0.1938
MSVD	0.1204	0.1621
FusionGAn	0.2575	0.4216
Proposed GAN	0.0067	0.0082

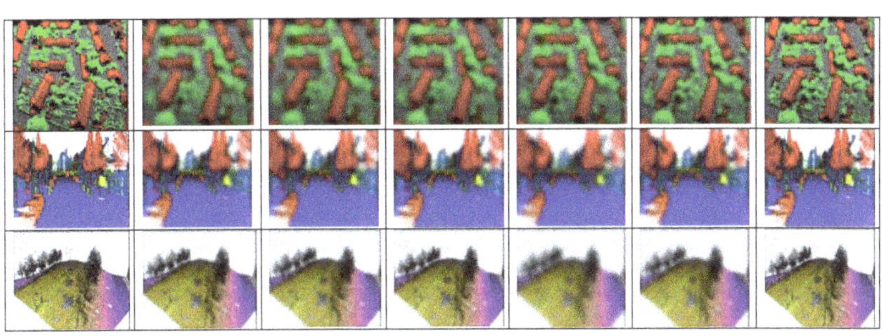

Source Images LP RP MSVD FusionGAN Proposed GAN

Figure 17. Fusion results of various fusion models using GeoTiles 3D point cloud data.

5.3. Limitations of the Proposed Study

The study introduced two fusion models: one for merging 2D and 3D images and another for combining various image types, with a specific focus on 3D point cloud data. The initial hybrid image fusion model encountered significant challenges, including ineffective fusion of 3D point cloud data and difficulties in handling large datasets. These issues were addressed by the extended GAN model. However, the proposed GAN model has its own limitations, such as training instability, which necessitates separate training for generators and discriminators. While the assigned parameters for initializing GAN layers performed well with the GeoTiles dataset, further evaluations are needed to assess the model's ability to fuse images from other datasets. Considerable time was dedicated to optimizing the hyperparameters employed for initializing the model. The study employed a minimal number of images for training and testing the model, and its performance with extensive datasets should be validated in future research.

6. Conclusions

Providing security to customers is a vital task for both AVs and manually driven vehicles. Most of the researchers who propose innovative solutions for AVs have their major focus on providing proper safety and security to the users. This proposal analyzed various reasons for the failure of full-fledged AVs in the current market, which are launched after many trials and research activities. The proposed study found that most of the recently launched AVs failed due to inaccurate and improper decision-making policies, due to their poor perception of their environmental information. Hence, much emphasis has to be given to data preprocessing, namely data cleaning and data fusion of AVs' sensory data. This research suggests a generic data fusion engine to fuse different formats of data, such as text, images, video, and audio, which exhibit multimodal characteristics. The first stage of the research focused heavily on image fusion since most modern sensors, such as LiDAR and Velodyne, collect data in the form of image frames represented as point cloud data. To facilitate this scenario, this study proposed an innovative hybrid image fusion model, which integrates the concepts of the Brovey fusion model and Gram–Schmidt transformation. The suggested image fusion model is robust enough to handle multispectral image data, retain the RGB color intensity of the fused images, and perform image transformation more efficiently. The proposed hybrid image data fusion model was evaluated with the nuScenes dataset and its performance was compared using standard image metrics with other referred and popular data fusion models. The proposed image fusion model provides better results than the other standard image fusion models in terms of accuracy and efficiency.

Despite the proposed hybrid fusion model demonstrating proficiency in handling 2D and 3D data, it encountered difficulties when dealing with 3D point cloud data and managing large volumes of image data. To address these challenges, the study extended its research by introducing an innovative GAN model capable of handling diverse image data types. In addition to conventional metrics, new metrics such as mutual information and entropy were employed to evaluate the performance of the proposed GAN model. Evaluation of the GAN model utilized point cloud data extracted from the GeoTiles dataset. The performance of the proposed GAN model was compared with other popular fusion models that extensively utilize point cloud data. The results revealed that the proposed GAN model outperforms other fusion models. To enhance the proposed fusion model further, future work entails fine-tuning and evaluating it using different datasets to assess its improved performance. Additionally, plans are underway to enhance the GAN architecture by incorporating principles of multiscale attention and advanced discriminators to improve feature selection and data fusion accuracy. Furthermore, efforts will be made to design and develop advanced kernel functions for the various tasks performed by different layers of the GAN model. Further analysis is required to check whether the accuracy of the fused data enhances perception creation for accurate decision-making in AVs. Future work could look into incorporating the video and audio fusion models, which are currently underway, into the proposed framework.

Author Contributions: H.A.I.: Conceptualization, methodology, experimental analysis, original draft preparation, writing, review, editing, and proofreading. H.E.-S.: Conceptualization, supervision, fund acquisition, and proofreading. M.A.K.: Conceptualization, Methodology, Supervision. P.K.: Conceptualization, supervision, review, editing, and proofreading. All authors have read and agreed to the published version of the manuscript.

Funding: This paper was supported by the Emirates Center for Mobility Research of the United Arab Emirates University (grant 31R271) and ASPIRE Award for Research Excellence, Project Reference AARE20-368.

Institutional Review Board Statement: Not Applicable.

Informed Consent Statement: Not Applicable.

Data Availability Statement: This research exclusively uses existing datasets like nuScenes for vehicle scenarios and GeoTiles for 3D point cloud data. No new dataset has been created.

Acknowledgments: This paper was supported by the Emirates Center for Mobility Research of the United Arab Emirates University (grant 31R271) and ASPIRE Award for Research Excellence, Project Reference AARE20-368. Manzoor Khan was involved with this work during his employment with United Arab Emirates University. He has subsequently moved to Nokia Bell Labs.

Conflicts of Interest: The authors declare no conflict of interest. The funders had no role in the design of the study; in the collection, analyses, or interpretation of data; in the writing of the manuscript, or in the decision to publish the results.

References

1. Litman, T. *Autonomous Vehicle Implementation Predictions: Implications for Transport Planning*; Victoria Transport Policy Institute: Victoria, BC, Canada, 2020.
2. Fagnant, D.J.; Kockelman, K. Preparing a nation for autonomous vehicles: Opportunities, barriers and policy recommendations. *Transp. Res. Part Policy Pract.* **2015**, *77*, 167–181. [CrossRef]
3. Mohapatra, H.; Rath, A.K.; Panda, N. IoT infrastructure for the accident avoidance: An approach of smart transportation. *Int. J. Inf. Technol.* **2022**, *14*, 761–768. [CrossRef]
4. Endsley, M.R. Design and evaluation for situation awareness enhancement. In Proceedings of the Human Factors Society Annual Meeting, Anaheim, CA, USA, 1 October 1988; Sage Publications: Los Angeles, CA, USA, 1988, Volume 32, pp. 97–101.
5. Ozguner, U.; Acarman, T.; Redmill, K.A. *Autonomous Ground Vehicles*; Artech House: Norwood, MA, USA, 2011.
6. Terzic, J.; Terzic, E.; Nagarajah, R.; Alamgir, M. Ultrasonic sensing technology. In *Ultrasonic Fluid Quantity Measurement in Dynamic Vehicular Applications*; Springer: Cham, Switzerland, 2013; pp. 11–35.
7. Ignatious, H.A.; Khan, M. An overview of sensors in Autonomous Vehicles. *Procedia Comput. Sci.* **2022**, *198*, 736–741. [CrossRef]
8. Kaur, H.; Koundal, D.; Kadyan, V. Image fusion techniques: A survey. *Arch. Comput. Methods Eng.* **2021**, *28*, 4425–4447.
9. Shahdoosti, H.R. MS and PAN image fusion by combining Brovey and wavelet methods. *arXiv* **2017**, arXiv:1701.01996.
10. Mahanti, N.K.; Pandiselvam, R.; Kothakota, A.; Chakraborty, S.K.; Kumar, M.; Cozzolino, D. Emerging non-destructive imaging techniques for fruit damage detection: Image processing and analysis. *Trends Food Sci. Technol.* **2022**, *120*, 418–438. [CrossRef]
11. Son, D.M.; Kwon, H.J.; Lee, S.H. Visible and Near Infrared Image Fusion Using Base Tone Compression and Detail Transform Fusion. *Chemosensors* **2022**, *10*, 124.
12. Alseelawi, N.; Hazim, H.T.; Salim ALRikabi, H.T. A Novel Method of Multimodal Medical Image Fusion Based on Hybrid Approach of NSCT and DTCWT. *Int. J. Online Biomed. Eng.* **2022**, *18*, 28011. [CrossRef]
13. Chen, D.; Fei, X.; Wang, Z.; Gao, Y.; Shen, X.; Han, T.; Zhang, Y. Classifying Vegetation Types in Mountainous Areas with Fused High Spatial Resolution Images: The Case of Huaguo Mountain, Jiangsu, China. *Sustainability* **2022**, *14*, 13390. [CrossRef]
14. Shahian Jahromi, B.; Tulabandhula, T.; Cetin, S. Real-time hybrid multi-sensor fusion framework for perception in autonomous vehicles. *Sensors* **2019**, *19*, 4357. [CrossRef]
15. Jia, D.; Cheng, C.; Song, C.; Shen, S.; Ning, L.; Zhang, T. A hybrid deep learning-based spatiotemporal fusion method for combining satellite images with different resolutions. *Remote Sens.* **2021**, *13*, 645. [CrossRef]
16. Wang, Y.; Von Der Weth, C.; Zhang, Y.; Low, K.H.; Singh, V.K.; Kankanhalli, M. Concept based hybrid fusion of multimodal event signals. In Proceedings of the 2016 IEEE International Symposium on Multimedia (ISM), San Jose, CA, USA, 11–13 December 2016; pp. 14–19.
17. Malawade, A.V.; Mortlock, T.; Faruque, M.A.A. HydraFusion: Context-Aware Selective Sensor Fusion for Robust and Efficient Autonomous Vehicle Perception. *arXiv* **2022**, arXiv:2201.06644.

18. Zhao, Y.; Huang, B. A hybrid image fusion model for generating high spatial-temporal-spectral resolution data using OLI-MODIS-Hyperion satellite imagery. *World Acad. Sci. Eng. Technol. Int. J. Environ. Chem. Ecol. Geol. Geophys. Eng.* **2017**, *11*, 843–848.
19. Latreche, B.; Saadi, S.; Kious, M.; Benziane, A. A novel hybrid image fusion method based on integer lifting wavelet and discrete cosine transformer for visual sensor networks. *Multimed. Tools Appl.* **2019**, *78*, 10865–10887. [CrossRef]
20. Zhang, X. Multi-focus image fusion: A benchmark. *arXiv* **2020**, arXiv:2005.01116.
21. Kaimaris, D.; Kandylas, A. Small Multispectral UAV Sensor and Its Image Fusion Capability in Cultural Heritage Applications. *Heritage* **2020**, *3*, 1046–1062. [CrossRef]
22. Luo, Y.; He, K.; Xu, D.; Yin, W.; Liu, W. Infrared and visible image fusion based on visibility enhancement and hybrid multiscale decomposition. *Optik* **2022**, *258*, 168914. [CrossRef]
23. Ma, J.; Yu, W.; Chen, C.; Liang, P.; Guo, X.; Jiang, J. Pan-GAN: An unsupervised pan-sharpening method for remote sensing image fusion. *Inf. Fusion* **2020**, *62*, 110–120. [CrossRef]
24. Zhang, H.; Yuan, J.; Tian, X.; Ma, J. GAN-FM: Infrared and visible image fusion using GAN with full-scale skip connection and dual Markovian discriminators. *IEEE Trans. Comput. Imaging* **2021**, *7*, 1134–1147. [CrossRef]
25. Chen, J.; Ding, J.; Yu, Y.; Gong, W. THFuse: An Infrared and Visible Image Fusion Network using Transformer and Hybrid Feature Extractor. *Neurocomputing* **2023**, *527*, 71–82. [CrossRef]
26. Guo, C.; Fan, D.; Jiang, Z.; Zhang, D. MDFN: Mask deep fusion network for visible and infrared image fusion without reference ground-truth. *Expert Syst. Appl.* **2023**, *211*, 118631. [CrossRef]
27. Zhang, G.; Nie, R.; Cao, J.; Chen, L.; Zhu, Y. FDGNet: A pair feature difference guided network for multimodal medical image fusion. *Biomed. Signal Process. Control.* **2023**, *81*, 104545. [CrossRef]
28. Li, J.; Li, B.; Jiang, Y.; Cai, W. MSAt-GAN: A generative adversarial network based on multi-scale and deep attention mechanism for infrared and visible light image fusion. *Complex Intell. Syst.* **2022**, *8*, 4753–4781. [CrossRef]
29. Xu, H.; Ma, J.; Zhang, X.P. MEF-GAN: Multi-exposure image fusion via generative adversarial networks. *IEEE Trans. Image Process.* **2020**, *29*, 7203–7216. [CrossRef]
30. Zhao, Z.; Bai, H.; Zhang, J.; Zhang, Y.; Xu, S.; Lin, Z.; Timofte, R.; Van Gool, L. Cddfuse: Correlation-driven dual-branch feature decomposition for multi-modality image fusion. In Proceedings of the IEEE/CVF Conference on Computer Vision and Pattern Recognition, Vancouver, BC, Canada, 24–31 January 2023; pp. 5906–5916.
31. Wang, X.; Hua, Z.; Li, J. Multi-focus image fusion framework based on transformer and feedback mechanism. *Ain Shams Eng. J.* **2023**, *14*, 101978. [CrossRef]
32. Tiede, P.; Broderick, A.E.; Palumbo, D.C. Variational Image Feature Extraction for the Event Horizon Telescope. *Astrophys. J.* **2022**, *925*, 122. [CrossRef]
33. Liu, Y.; Pu, H.; Sun, D.W. Efficient extraction of deep image features using convolutional neural network (CNN) for applications in detecting and analysing complex food matrices. *Trends Food Sci. Technol.* **2021**, *113*, 193–204. [CrossRef]
34. Liang, N.; Duan, P.; Xu, H.; Cui, L. Multi-View Structural Feature Extraction for Hyperspectral Image Classification. *Remote Sens.* **2022**, *14*, 1971. [CrossRef]
35. Barburiceanu, S.; Terebes, R.; Meza, S. 3D texture feature extraction and classification using GLCM and LBP-based descriptors. *Appl. Sci.* **2021**, *11*, 2332. [CrossRef]
36. Ahmed Bhuiyan, R.; Ahmed, N.; Amiruzzaman, M.; Islam, M.R. A robust feature extraction model for human activity characterization using 3-axis accelerometer and gyroscope data. *Sensors* **2020**, *20*, 6990. [CrossRef]
37. Du, B.; Xiong, W.; Wu, J.; Zhang, L.; Zhang, L.; Tao, D. Stacked convolutional denoising auto-encoders for feature representation. *IEEE Trans. Cybern.* **2016**, *47*, 1017–1027. [CrossRef] [PubMed]
38. El-Sayed, H.; Alexander, H.; Khan, M.A.; Kulkarni, P.; Bouktif, S. DyReT: A Dynamic Rule Framing Engine Equipped With Trust Management for Vehicular Networks. *IEEE Access* **2020**, *8*, 72757–72767. [CrossRef]
39. Gomes, P.R.; Sabuj, H.H.; Uddin, M.A.; Reza, M.T.; Faiz, R.I.; Alam, M.A. A Deep Learning Approach for Reconstruction of Color Images in Different Lighting Conditions Based on Autoencoder Technique. In Proceedings of the IEEE 2021 International Conference on Electronics, Information, and Communication (ICEIC), Jeju, Korea, 31 January–3 February 2021; pp. 1–4.
40. Gai, J.; Tang, L.; Steward, B.L. Automated crop plant detection based on the fusion of color and depth images for robotic weed control. *J. Field Robot.* **2020**, *37*, 35–52. [CrossRef]
41. Kour, H. Analysis on image color model. *Comput. Sci.* **2015**, *4*, 233–235.
42. Zhang, Y.; Liu, Y.; Sun, P.; Yan, H.; Zhao, X.; Zhang, L. IFCNN: A general image fusion framework based on convolutional neural network. *Inf. Fusion* **2020**, *54*, 99–118. [CrossRef]
43. He, C.; Liu, Q.; Li, H.; Wang, H. Multimodal medical image fusion based on IHS and PCA. *Procedia Eng.* **2010**, *7*, 280–285. [CrossRef]
44. Ranchin, T.; Aiazzi, B.; Alparone, L.; Baronti, S.; Wald, L. Image fusion—The ARSIS concept and some successful implementation schemes. *ISPRS J. Photogramm. Remote. Sens.* **2003**, *58*, 4–18. [CrossRef]
45. Dian, R.; Li, S.; Sun, B.; Guo, A. Recent advances and new guidelines on hyperspectral and multispectral image fusion. *Inf. Fusion* **2021**, *69*, 40–51. [CrossRef]
46. Yang, Z.; Diao, C.; Li, B. A robust hybrid deep learning model for spatiotemporal image fusion. *Remote Sens.* **2021**, *13*, 5005. [CrossRef]

47. Metrics. Mterics Used for Evaluation. 2008. Available online: https://academic-accelerator.com/Manuscript-Generator/zh/Performance-Metrics (accessed on 19 July 2008).
48. Jagalingam, P.; Hegde, A.V. A review of quality metrics for fused image. *Aquat. Procedia* **2015**, *4*, 133–142. [CrossRef]
49. Islam, M.N.; Tahtali, M.; Pickering, M. Hybrid Fusion-Based Background Segmentation in Multispectral Polarimetric Imagery. *Remote Sens.* **2020**, *12*, 1776. [CrossRef]
50. Barrero, O.; Perdomo, S.A. RGB and multispectral UAV image fusion for Gramineae weed detection in rice fields. *Precis. Agric.* **2018**, *19*, 809–822. [CrossRef]
51. Talal, T.M.; Attiya, G.; Metwalli, M.R.; El-Samie, A.; Fathi, E.; Dessouky, M.I. Satellite image fusion based on modified central force optimization. *Multimed. Tools Appl.* **2020**, *79*, 21129–21154. [CrossRef]

applied sciences

MDPI

Article

Multi-Tag Fusion Localization Method Based on Geometric Constraints

Zhuojun Liu [1], Wengang Qin [2], Zexing Li [2] and Guofeng Zhou [1,*]

1 College of Engineering, Shanghai Ocean University, Shanghai 201306, China; m210801339@st.shou.edu.cn
2 College of Mechanical and Power Engineering, Shanghai Jiao Tong University, Shanghai 200240, China; wengang.qin@sjtu.edu.cn (W.Q.); lizexing@sjtu.edu.cn (Z.L.)
* Correspondence: gfzhou@shou.edu.cn

Abstract: In environments where Global Navigation Satellite System (GNSS) signals are unavailable, our proposed multi-tag fusion localization method offers a robust solution for the precise positioning of vehicles or robots. During our research, we observed variations in the positioning information estimated from tags located at different positions within the same frame. Our goal was to extract reliable positioning information from this noisy data. By constructing geometric constraints, our method introduces an outlier factor to quantify the differences between tags. After effectively eliminating outliers, we enhanced the Kalman filter framework to accommodate the fusion of data from two or more tags, with the outlier factor dynamically adjusting the observation noise during the fusion process. The experimental results demonstrate that, even under the influence of motion and obstacles, our method maintains position errors within a 3 cm range and orientation errors within 3°. This indicates that our method possesses high positioning accuracy and stability.

Keywords: localization; visual tag; data fusion; geometric constraints; AprilTag

Citation: Liu, Z.; Qin, W.; Li, Z.; Zhou, G. Multi-Tag Fusion Localization Method Based on Geometric Constraints. *Appl. Sci.* **2024**, *14*, 5480. https://doi.org/10.3390/app14135480

Academic Editors: Fabrizio Giulietti, Nadjim Horri and William Holderbaum

Received: 18 May 2024
Revised: 21 June 2024
Accepted: 21 June 2024
Published: 24 June 2024

1. Introduction

Localization is fundamental to vehicle planning, control, and subsequent tasks, essential for the integrity of vehicle functions [1–3]. The most common positioning technology is Global Navigation Satellite System (GNSS); however, not all scenarios offer ample and reliable signals, such as urban canyons, parking lots, tunnels, and indoor areas [4,5]. Visual tag positioning technology has become increasingly prevalent across a spectrum of applications, including robotics [6,7] and Unmanned Aerial Vehicles (UAVs) [8,9], offering innovative solutions for the precise positioning of intelligent transport systems. By employing tags with predefined coordinate information, this technology achieves centimeter-level accuracy in positioning, which is crucial for various mobile platforms, including but not limited to automotive applications.

Despite the low cost and capability for absolute positioning of visual tags [10,11], they still face issues of inaccuracy and instability. During the localization process utilizing tags, the detection quality is significantly influenced by the distance and rotation angle between the camera and the tag [12]. The precision of the system is also impacted by the number and position of the tags within the image. Furthermore, lighting conditions can affect the accuracy of tag detection, particularly for tags like AprilTag [13]. It is crucial that the tag size is substantial enough to offer adequate information while remaining within the camera's field of view [14]. There are three main approaches to address these issues: improving the detection accuracy of individual tags, fusing individual tag data with other sensor information, and increasing the number of tag detections.

To enhance positioning accuracy, some researchers have focused on improving marker detection accuracy. Two processing techniques were proposed by [15], including a method that can filter out very inaccurate detections resulting from partial border occlusion, and a new highly accurate method for edge refinement. Based on observations, ref. [16] proposed improvements include trigonometric yaw angle correction for camera alignment, real-time tag center tracking via a custom yaw-axis gimbal, and a novel pose-indexed probabilistic sensor error model for AprilTag utilizing Gaussian Processes and validated by particle filter tracking. In the process of vehicle movement, ref. [17] found that the marker detection error ranged from 0.028 m to 0.13 m. To reduce pose jitter, they designed a strategy to constrain the inner and outer corner points of the markers, limiting the relative error between markers to within 0.05 m.

Meanwhile, fusing marker information with other data can also improve positioning accuracy. Ref. [13] described an algorithm that improves the pose estimation accuracy of square fiducial markers in difficult scenes by fusing information from RGB and depth sensors. Ref. [18] utilized particle filtering techniques to fuse data from ArUco markers and sonar sensors, achieving effective integration of multi-source sensor information. Ref. [19] introduced a variational Bayesian estimator for mobile robot localization with unknown noise covariance, utilizing tags as landmarks and an onboard camera. By treating the unknown measurement noise covariance as a random matrix following an inverse Wishart distribution, the method simultaneously estimates the robot pose. Ref. [20] proposed an on-manifold extended Kalman filter for tag-based visual–inertial localization, addressing the topological structure of rotation and pose groups and associated uncertainty propagations. Ref. [21] combined the robustness of makers to environmental uncertainty with motion data to minimize the required number of tags. An Iterative Extended Kalman Filter (IEKF) is employed to fuse visual and motion information, and an area detection algorithm is introduced to accelerate AprilTag processing.

While single-tag localization techniques have seen significant advancements, they encounter challenges when marker observation is obstructed or incomplete during vehicle movement [22]. This highlights the necessity for a method that can integrate information from multiple markers, especially in environments with partial occlusions or dynamics. Ref. [23] designed an integrated vision and LIDAR robot localization method using two AprilTag tags for pose estimation, proving the stability of the proposed method over single-tag approaches.

The third category of methods, which we focus on, involves detecting multiple tags simultaneously and fusing their data. For example, Ref. [24] averages observations from multiple markers and applies a constant velocity motion model for Kalman Filtering (KF) but does not adequately account for variability among different markers. Ref. [17] integrates markers into SLAM and applies the 3σ principle to filter outliers, which may be insufficient with fewer than ten markers as it defaults to ranking pose estimates and selecting the median, potentially not reflecting true poses if the observation distribution is skewed. Ref. [25] fuses IMU data with multiple markers using Extended Kalman Filtering (EKF), which can negatively impact localization accuracy when kinematic predictions significantly deviate from actual conditions. The optimization algorithm presented in [26], fusing multiple ArUco marker data based on the Mahalanobis distance, shares conceptual similarities with our approach but also reports trajectory discontinuities due to marker detection errors, especially when markers are distant or partially occluded. In the aforementioned filtering methods, observation noise is often determined by trial and error. However, ref. [27] discovered that the observation noise of pose from planar fiducial markers varies across the measurement range, a fact that must be considered during sensor fusion for a reliable estimate. They present experimental measurements of fiducial markers in real and simulation scenarios for 2D pose estimation and propose analytical functions that approximate the variances of pose estimates.

This paper proposes a multi-tag fusion localization method based on geometric constraints. The approach begins with detecting visual markers to acquire relative pose information between the target and the markers. Following coordinate transformations, the global pose of the target in the world coordinate system is computed. To ensure data accuracy and reliability, geometric constraints are constructed for all observations, effectively eliminating outliers. Subsequently, the filtered and reliable observations are incorporated into a multi-observation fusion framework utilizing a modified KF. This integration facilitates the effective fusion of multiple observations, ultimately leading to precise positioning information.

Our main contributions are as follows:

(1) A linear fusion method using only a camera: This method is simple, low-cost, and avoids complexities and potential inaccuracies associated with motion model-dependent predictions.
(2) Introducing an outlier factor based on geometric constraints: The outlier factor quantifies the variability among different tags, excluding outliers before fusion and dynamically adjusting observation noise during the fusion process, reducing the likelihood of initial noise settings being unreasonable.

The paper consists of several parts. Section 2 introduces common visual tags and provides a detailed description of the AprilTag's detection process. Section 3 describes the outlier handling method before data fusion, focusing on the construction of geometric constraints. Section 4 outlines the fusion approach, utilizing a modified KF for integrating data from multiple tags. Section 5 summarizes the experimental setup and presents the results and analysis. The final section concludes this paper with a summary of the key findings.

2. Visual Tag Process

Visual tags serve as objects in the image that provide key reference points and measurement points, and they have a known size and shape. Common visual tags include ARTag [28], AprilTag [29], ArUco [30], and Stag [31], as shown in Figure 1. Ref. [32] conducted an experimental comparison of these four tags, evaluating them comprehensively in terms of accuracy, recognition rate, and computational cost. The experimental results showed that ARTag has a low computational cost but performs poorly in the detection rate of a single marker; AprilTag, Stag, and ArUco exhibit excellent performance in terms of direction and position; however, AprilTag has a high computational cost and is sensitive to motion blur; Stag and ArUco are sensitive to smaller marker sizes and larger distances; and the computational cost of ArUco increases with the number of markers. Given the diversity of visual tags and their differences in characteristics, it is necessary to weigh the selection of tags based on the needs and scenarios of the actual application.

(a) (b) (c) (d)

Figure 1. Common visual labels: (**a**) ARTag; (**b**) ArUco; (**c**) AprilTag; (**d**) Stag.

2.1. AprilTag3 Detection

In the pursuit of a pose estimation method well-suited for our application, AprilTag3 [33] has been identified as the optimal choice, predominantly attributable to its outstanding capabilities in delivering precise orientation results and elevated detection rates. AprilTag3, a sophisticated iteration within the AprilTag family, stands out for its open-source accessibility, high real-time processing capabilities, and enhanced accuracy.

This advanced version has undergone significant optimization, resulting in a more robust algorithmic framework. Notably, these enhancements have led to substantial gains in computational efficiency and a reduction in memory usage, making AprilTag3 an attractive proposition for resource-intensive applications. Additionally, AprilTag3 supports an innovative format of two-dimensional codes, expanding its utility and applicability.

As depicted in Figure 2, the algorithmic process begins with an image filtering step that downscales the initial image to accelerate subsequent operations. This is followed by adaptive thresholding, which generates a binarized image that clearly delineates the tag boundaries against the background.

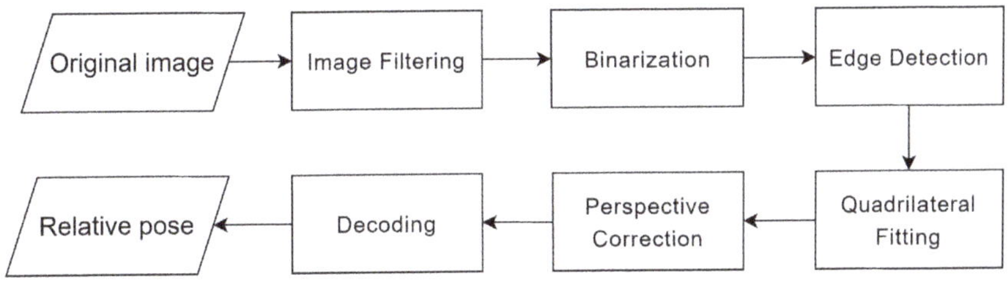

Figure 2. AprilTag3 detection process.

The subsequent edge detection phase employs the union-find algorithm to segment the image into meaningful components, effectively filtering out irrelevant sections and thus enhancing the precision of the detection process. Quadrilateral fitting then identifies potential tags from clustered line segments through a calculated sequence of operations.

To address perspective distortion, a perspective correction step is integrated, utilizing a transformation matrix to reconcile the disparity between the tag's anticipated and actual poses. The final stage involves image sharpening to mitigate blurriness, enabling the clear decoding of the tag ID and the determination of the tag's relative position to the camera.

2.2. Coordinate Transformation

To ascertain the global coordinates of a target within a scene, it is imperative to perform transformations across multiple coordinate systems. This process engages several defined coordinate systems: the world coordinate system (W), tag coordinate system (T), camera coordinate system (C), and the self or robot coordinate system (M), all adhering to the right-hand rule for orientation. As illustrated in Figure 3, the X-axis is represented in red, the Y-axis in green, and the Z-axis in blue.

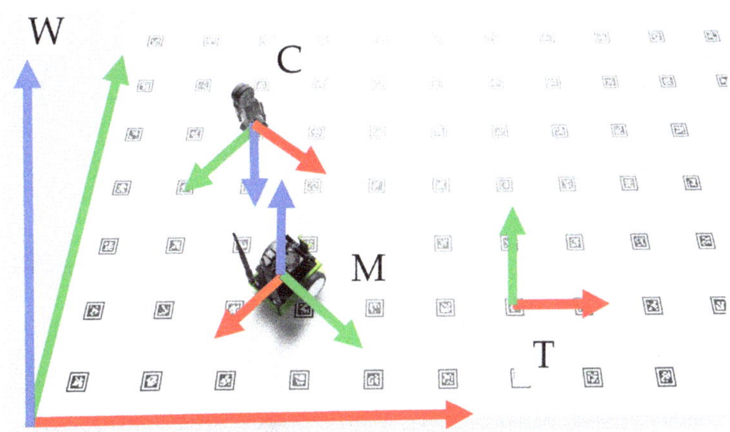

Figure 3. Diagram of the coordinate system.

Firstly, the relative pose between the tag and the camera, denoted as P_C, is transformed into the relative pose between the tag and the target, denoted as P_M:

$$P_M = R_C^M P_C + T_C^M \quad (1)$$

In the equation, R_C^M represents the rotation matrix from the self-coordinate system to the camera coordinate system, and T_C^M is the installation position of the camera. Subsequently, the coordinates of the target P in the tag coordinate system are obtained as P_T:

$$P_T = R_C^T P_M + T_C^T \tag{2}$$

R_C^T is the rotation matrix between the tag coordinate system and the camera coordinate system. During the initial deployment of the tags, each tag is assigned a unique identifier, and the world coordinates of the center of the tag are represented by the matrix $T(id)$:

$$T(id) = [x, y, z]^T \tag{3}$$

Finally, the coordinates of the target P in the world coordinate system are calculated as P_W:

$$P_W = R_T^W P_T + T(\mathrm{id}) \tag{4}$$

In this formula, R_T^W is the rotation matrix from the tag coordinate system to the world coordinate system. The aforementioned rotation matrices R_C^M, R_C^T, R_T^W are derived from the Equation (5):

$$R = R_x R_y R_z \tag{5}$$

R_x, R_y, and R_z, are the rotation matrices calculated from the corresponding angles, ϕ, θ and ψ representing the rotation around the X-axis, Y-axis, and Z-axis, respectively.

$$R_x = \begin{bmatrix} 1 & 0 & 0 \\ 0 & \cos\phi & \sin\phi \\ 0 & -\sin\phi & \cos\phi \end{bmatrix} \tag{6}$$

$$R_y = \begin{bmatrix} \cos\theta & 0 & -\sin\theta \\ 0 & 1 & 0 \\ \sin\theta & 0 & \cos\theta \end{bmatrix} \tag{7}$$

$$R_z = \begin{bmatrix} \cos\psi & \sin\psi & 0 \\ -\sin\psi & \cos\psi & 0 \\ 0 & 0 & 1 \end{bmatrix} \tag{8}$$

An image containing N tags, once processed through tag detection, yields N distinct relative poses. After undergoing coordinate transformation, these relative poses result in N global poses of the target in the world coordinate system at the same moment. These global poses are then utilized as observation values for subsequent fusion steps. The observation values include the two-dimensional coordinates in the world coordinate system and the orientation angle.

3. Construction of Geometric Constraints

When multiple tags are detected in an image, the positioning data extracted from different tags possess varying degrees of uncertainty. This implies that some observed values may have significant errors, and these outliers can substantially affect the final positioning outcome. As depicted in Figure 4, at the same moment, observed values i, j, and k are obtained. Under ideal circumstances, observed values i, j, and k should coincide with the true position. However, due to factors such as location and lighting, deviations occur between the observed values and the actual position. In statistical terms, an outlier is defined as one or several values within a dataset that differ significantly from the rest, constituting abnormal data. This paper constructs geometric constraints to measure the variability among observed values, thereby identifying and eliminating outliers.

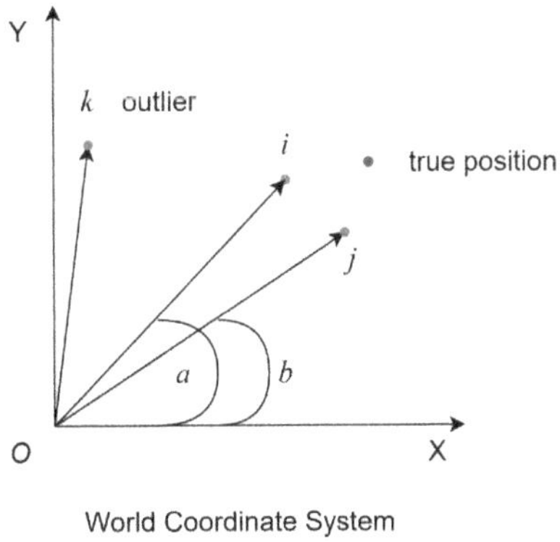

Figure 4. Schematic representation of the geometric constraints.

Consider any two observed values, i and j. The coordinates of observed value i in the world coordinate system are $[x_i, y_i, z_i]$, and the angle it makes with the X-axis is ∠a. The coordinates of observed value j in the world coordinate system are $[x_j, y_j, z_j]$, and the angle it makes with the X-axis is ∠b.

In the XOY plane, the difference in distance between any two observed values, l_{ij}, can be calculated as

$$l_{ij} = \sqrt{(x_i - x_j)^2 + (y_i - y_j)^2} \tag{9}$$

To calculate the angular difference d_{ij}, first determine the angles ∠a and ∠b:

$$\begin{cases} a = \arctan(\frac{y_i}{x_i}) \\ b = \arctan(\frac{y_j}{x_j}) \end{cases} \tag{10}$$

Then, find the difference in angles:

$$d_{ij} = |a - b| \tag{11}$$

If there are a total of N observed values, by combining them in pairs to construct geometric constraints and calculating the corresponding distances l_{ij} and angular differences d_{ij}, $G(N)$ combinations are needed in total. The number of combinations is given by the binomial coefficient:

$$G(N) = C_N^2 = \frac{N!}{(N-2)!2} \tag{12}$$

Each observed value will participate in (N − 1) geometric constraint constructions, resulting in (N − 1) different l_{ij} and d_{ij}. The sum of the distance differences for each observed value with respect to the other observed values is

$$l_i = \sum_{j=1}^{N-1} l_{ij} \tag{13}$$

The sum of the angular differences for each observed value with respect to the other observed values is

$$d_i = \sum_{j=1}^{N-1} d_{ij} \tag{14}$$

An outlier factor, $S(i)$, is introduced and defined using a normalization method based on the maximum and minimum values to describe the variability between observed values from both distance and angular differences:

$$S(i) = k_1 \frac{l_i - \min(l_i)}{\max(l_i) - \min(l_i)} + k_2 \frac{d_i - \min(d_i)}{\max(d_i) - \min(d_i)} \tag{15}$$

In the equations, the weight coefficient k_1, k_2 is used, which must satisfy

$$k_1 + k_2 = 1 \tag{16}$$

$$\begin{aligned} \max(l_i) &= \max[l_1, \ldots, l_i, \ldots, l_{N-1}] \\ \min(d_i) &= \min[d_1, \ldots, d_i, \ldots, d_{N-1}] \end{aligned} \tag{17}$$

4. Fusion of Multiple Observations

In the context of multi-tag localization, the identification and removal of aberrant observations is a fundamental step. Yet, the integration of the subsequent reliable observations from multiple tags is an equally significant challenge. Traditional KF has long been a cornerstone for data fusion, operating under the assumption that the motion of the target conforms to specific patterns, such as uniform velocity or constant acceleration. This assumption facilitates the formulation of kinematic equations that predict system states.

Within this established framework, observational data from detected tags are merged with these predicted states to enhance the accuracy of the localization process. However, the standard KF is limited in its capacity to integrate only two elements: the predicted motion and the actual observation. This limitation becomes particularly evident when attempting to incorporate the wealth of data from multiple reliable observations, a scenario that surpasses the filter's original design.

To address this, a novel approach has been introduced. This approach diverges from the standard method by performing predictive updates based solely on the available observational data. This enhanced method is capable of handling an increased number of observations through an iterative process, which, in turn, refines the precision and stability of the localization system. The flow of this algorithm is illustrated in Figure 5.

At time t, N tags are observed, and after constructing $G(N)$ geometric constraints, n reliable observations are as follows: $\boldsymbol{Z}_1, \boldsymbol{Z}_2, \ldots, \boldsymbol{Z}_n$. The system state at time t is $\boldsymbol{X}_t$, and the system state at the k-th fusion is $\boldsymbol{X}_k$, where $k \in [1, n-1]$. A total of n − 1 fusions are required for each moment. The system state includes the two-dimensional coordinates in the world coordinate system and the orientation angle:

$$\boldsymbol{X}_t = [x, y, \psi]^{\mathrm{T}} \tag{18}$$

$$\boldsymbol{X}_k = [x, y, \psi]^{\mathrm{T}} \tag{19}$$

The relationship between $\boldsymbol{X}_t$ and $\boldsymbol{X}_k$ is given by:

$$\boldsymbol{X}_t = \boldsymbol{X}_k = \boldsymbol{X}_{k-1} + w_{k-1} \tag{20}$$

In this equation, w_k represents the process noise in the fusion, which is normally distributed, $w_k \sim (0, Q)$, where Q is the covariance matrix of the process noise.

The relationship between the observation values and the system state is

$$\boldsymbol{Z}_i = \boldsymbol{H}\boldsymbol{X}_k + v_i \tag{21}$$

Here, $\boldsymbol{H}$ is the observation matrix, and v_i represents the measurement noise of the i-th reliable observation, which is also normally distributed, $v_i \sim (0, \boldsymbol{R})$, $i \in [1, n]$, where $\boldsymbol{R}$ is the covariance matrix of the measurement noise.

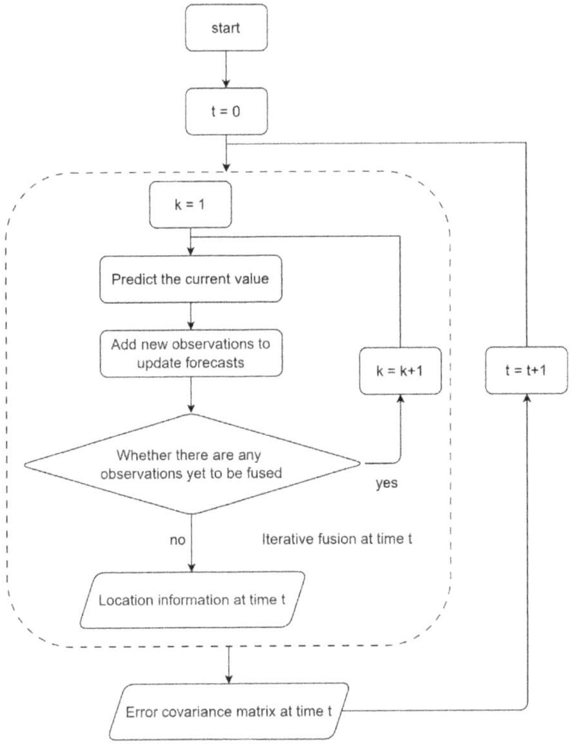

Figure 5. Algorithm flowchart.

At the k-th fusion, where $k \in [1, n-1]$, the fused value from the (k − 1)-th fusion, $\hat{X}_{k-1}$ is used as the current moment's predicted value, $\overline{X_k}$:

$$\overline{X_k} = \hat{X}_{k-1} \tag{22}$$

The initial value, $\overline{X_1}$ is given by:

$$\overline{X_1} = H^{-}Z_1 \tag{23}$$

The prediction error, denoted as $\overline{e}_k$, is calculated as

$$\begin{aligned} \overline{e}_k &= X_k - \overline{X_k} \\ &= X_k - \hat{X}_{k-1} \\ &= X_{k-1} + w_{k-1} - \hat{X}_{k-1} \\ &= e_{k-1} + w_{k-1} \end{aligned} \tag{24}$$

The prediction error covariance matrix $\overline{P}_k$ is computed as

$$\begin{aligned} \overline{P}_k &= E[\overline{e}_k \overline{e}_k^{T}] \\ &= P_{k-1} + Q_{k-1} \end{aligned} \tag{25}$$

where Q is the process noise covariance matrix.

The initial value, $\overline{P}_1$, is given by

$$\overline{P}_1 = P_{t-1} \tag{26}$$

The observation noise is adjusted based on the outlier factor of the observation values:

$$R_k = (S(k) - S(1))R_1 \tag{27}$$

Here, $S(1)$ is the outlier factor of the first reliable observation, and R_1 is the initial observation noise. Since the reliable observations are sorted in ascending order based on the outlier factor, the first reliable observation has the smallest outlier factor value. This implies that a larger outlier factor corresponds to greater observation noise.

The Kalman gain K_k is calculated as

$$K_k = \frac{\overline{P}_k H^{\mathrm{T}}}{H\overline{P}_k H^{\mathrm{T}} + R_k} \tag{28}$$

The new observation value Z_{i+1} is introduced as the k-th observation

$$Z_k = Z_{i+1} \tag{29}$$

The prediction is updated using the Kalman gain and the new observation value:

$$\hat{X}_k = \overline{X}_k + K_k(Z_k - H\overline{X}_k) \tag{30}$$

Update the error:

$$\begin{aligned} e_k &= X_k - \hat{X}_k \\ &= X_k - (\overline{X}_k + K_k(Z_k - H\overline{X}_k)) \\ &= (I - K_k H)(X_k - \overline{X}_k) - K_k v_k \end{aligned} \tag{31}$$

where I is the identity matrix.

Update the error covariance matrix:

$$P_k = E[e_k e_k^{\mathrm{T}}] = (I - K_k H)\overline{P_k} \tag{32}$$

The iteration continues until all reliable observations have been integrated. At the end of the iteration, the final fused value at time t, $\hat{X}_t$ is obtained. At this point, the error covariance matrix will serve as the predicted error covariance matrix for the next moment $t + 1$:

$$\hat{X}_t = \hat{X}_{k=n-1} \tag{33}$$

$$\overline{P}_{t+1} = P_t = P_{k=n-1} \tag{34}$$

5. Experiment

In this section, we detail the experimental approach undertaken to validate the efficacy of our multi-tag fusion localization method. Utilizing the Isaac Sim platform, we have conducted simulation-based experiments that offer a cost-effective and flexible framework for assessing our method. The decision to employ simulation was driven by the constraints of time and budget, while also allowing for a diverse range of scenario configurations through parameter adjustments. This approach enables us to systematically test the robustness and accuracy of our method under controlled conditions. Our primary objectives were twofold: to initially validate the geometric constraints by demonstrating their effectiveness in filtering out outliers and subsequently to assess the overall positioning accuracy and stability of the proposed localization method. These objectives guide the design of our experiments and the analysis of the results obtained.

5.1. Experimental Setup

The Isaac Sim camera model, depicted in Figure 6, generates high-fidelity images for reliable testing. The reliability of the simulation camera was validated through fixed-point tests, confirming its ability to provide trustworthy and high-quality images. Positioned at the rear of the vehicle at a height of 1.15 m, the camera captured 6 to 8 tags per frame at a resolution of 1280 × 720 pixels and a frame rate of 25 fps.

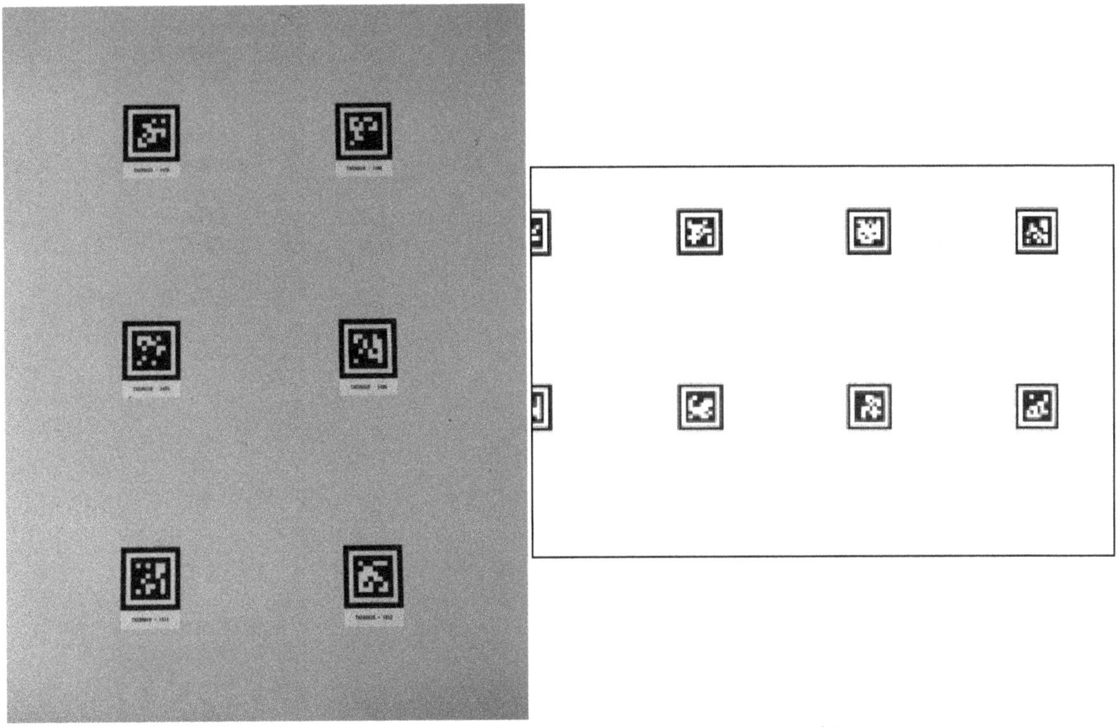

(a) (b)

Figure 6. (**a**) The real camera takes the image; (**b**) The simulation camera takes the image.

This study designed multiple simulation scenarios, as depicted in Figure 7.

Simulation Scenario 1: Visual tags are laid out on the ground. AprilTag's tag36 h 10 series is used for the visual tags, with each tag measuring 0.04 m × 0.04 m and spaced 0.15 m apart. This initial setup establishes a baseline for evaluating the method's accuracy in an unobstructed environment.

Simulation Scenario 2: Adds obstacles based on Scenario 1. The incorporation of obstacles is intended to simulate real-world conditions where line-of-sight to tags may be partially blocked, thereby testing the method's resilience to occlusions.

Simulation Scenario 3: Similar to Scenario 1, but with tags measuring 0.03 m × 0.03 m and spaced 0.1 m apart. Reducing the tag size and adjusting the spacing allows us to examine the method's sensitivity to changes in tag density and visibility range.

Simulation Scenario 4: Adds obstacles based on Scenario 3. Further introducing obstacles with the smaller tags assesses the method's adaptability to challenging conditions with increased occlusion and reduced tag size.

This study conducted a series of six experiments designed to evaluate the effectiveness of geometric constraints and to assess the positioning accuracy and stability of the proposed method, as detailed in Table 1. The simulation software provided the ground truth for the experiments, establishing a reliable benchmark for comparison.

Due to the uniform scaling of the tag models, changes in tag size were accompanied by corresponding adjustments in the spacing between tags. As depicted in Figure 7, the simulation software offered realistic lighting conditions, resulting in varying illumination for tags at different locations. This feature of the simulation software introduced an additional layer of realism to the experiments, as it mimicked the diverse lighting conditions that tags might encounter in real-world scenarios.

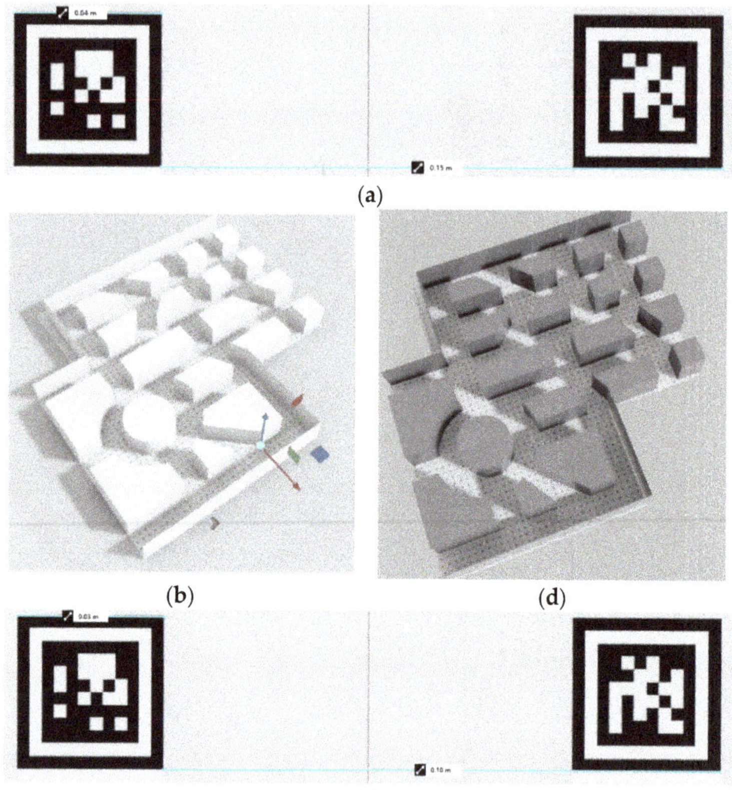

Figure 7. (**a**) Simulation Scenario 1; (**b**) Simulation Scenario 2; (**c**) Simulation Scenario 3; (**d**) Simulation Scenario 4.

Table 1. Experimental setup.

Group	State of Motion	Tag Size	Tag Interval	Whether Obstructions Exist
1	still	0.04 m × 0.04 m	0.15 m	no
2	moving	0.04 m × 0.04 m	0.15 m	no
3	moving	0.04 m × 0.04 m	0.15 m	yes
4	still	0.03 m × 0.03 m	0.1 m	no
5	moving	0.03 m × 0.03 m	0.1 m	no
6	moving	0.03 m × 0.03 m	0.1 m	yes

The inclusion of these factors in the experimental design allowed for a more comprehensive assessment of the proposed method's performance under conditions that closely resemble real-world applications. This attention to detail ensures that the results obtained from the simulation are not only indicative of the method's efficacy but also reflective of its robustness in varying environmental conditions.

5.2. Geometric Constraint Effectiveness

A critical aspect of this research was to demonstrate the effectiveness of geometric constraints in enhancing the accuracy of the positioning system.

5.2.1. Single-Frame Validation

For a demonstrative purpose, we extracted a specific frame from the simulation dataset, capturing both the true and observed positioning values. Within this frame, eight tags were discerned, with the identifiers {1120, 1121, 1136, 1137, 1408, 1409, 1410, 1425}. Implementing geometric constraints, we assigned a consistent weight coefficient of 0.5 to all observations, as per Equation (15), to ascertain the outlier factor for each. The results of these calculations are presented in Table 2.

Table 2. Outlier factor.

ID	Outlier Factor	Whether It Is a Reliable Observation
1121	0	yes
1410	0.04049	yes
1409	0.08078	yes
1137	0.25164	yes
1425	0.67949	no
1408	0.80974	no
1120	0.83909	no
1136	1	no

After sorting the observations by their outlier factors in ascending order, we designated the top 4 observations with the lowest outlier factors as reliable. These reliable observations were identified by the IDs {1121, 1410, 1409, 1137}, while the remaining observations were considered outliers: {1425, 1408, 1120, 1136}.

Figure 8 presents a scatter plot of the observed tag coordinates within the world coordinate system, where each distinct point corresponds to a single observation identified by its unique tag ID. The positioning of these points in proximity to the true values illustrates the precision of our method, demonstrating the successful application of our filtering process in discerning reliable observations from outliers.

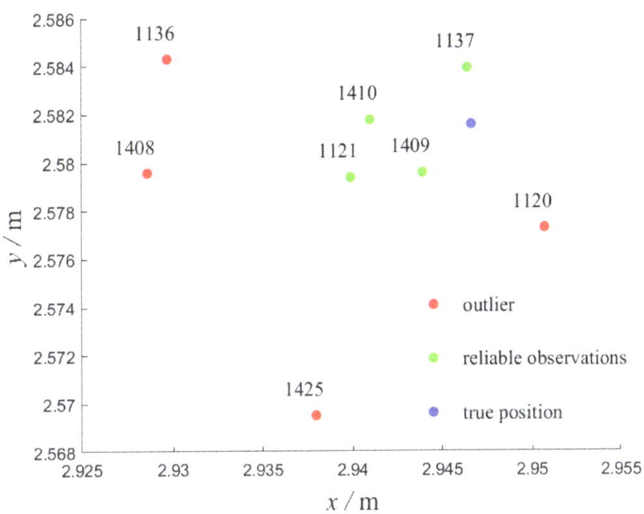

Figure 8. Schematic representation of the observed values.

Furthermore, Table 3 illustrates the variance of the observations before and after the removal of outliers. The variance in the X, Y, and yaw dimensions of the reliable observations is consistently lower than that of all observations combined. This reduction in variance demonstrates that the data become more concentrated and stable after the application of geometric constraints, confirming the reliability of the filtered observations.

Table 3. Comparison of variance before and after excluding outliers.

Dimensionality	The Variance of All Observations	Variance of Reliable Observations
X	0.00715	0.00254
Y	0.00436	0.00184
Yaw	0.01124	0.00846

5.2.2. Process Validation

In this section, we compare the method without geometric constraints and our proposed method over a sustained period, with detailed results presented in Table 4. This comparison allows us to evaluate the impact of geometric constraints on the system's accuracy and stability in a continuous positioning process.

Table 4. Comparative error analysis.

Group		Method Without Geometric Constraints		Our Method	
		Distance Error (m)	Yaw Error (rad)	Distance Error (m)	Yaw Error (rad)
1	Maximum Error	0.02246	0.03816	0.01627	0.01262
	Minimum Error	0.00386	0.00089	0.00273	0.00018
	Average Error	0.01113	0.00989	0.00824	0.00432
2	Maximum Error	0.16475	0.14654	0.02968	0.05063
	Minimum Error	0.00231	0.00014	0.00278	0.00004
	Average Error	0.05813	0.02442	0.01772	0.01298
3	Maximum Error	0.61676	0.10386	0.02858	0.04184
	Minimum Error	0.01283	0.00321	0.00355	0.00036
	Average Error	0.05587	0.05908	0.02294	0.02153
4	Maximum Error	0.05618	0.03842	0.02213	0.02451
	Minimum Error	0.00033	0.00025	0.00033	0.00006
	Average Error	0.01646	0.00837	0.00889	0.00550
5	Maximum Error	0.09544	0.10569	0.02826	0.03641
	Minimum Error	0.00502	0.01421	0.00772	0.00020
	Average Error	0.04035	0.03189	0.01871	0.03478
6	Maximum Error	0.14736	0.10569	0.02624	0.04256
	Minimum Error	0.00148	0.00113	0.00219	0.00027
	Average Error	0.07850	0.03613	0.02611	0.02389

The Method Without Geometric Constraints: This method fused all raw observations. It treated each tag's data equally, applying a uniform and constant level of observation noise throughout the fusion process.

Our Method: The proposed method first culled outliers based on outlier factors, then dynamically adjusted observation noise for each tag's data during fusion, enhancing accuracy and stability.

The table provides a clear comparison of the error metrics between the method without geometric constraints and our proposed method, highlighting the impact of incorporating geometric constraints on positioning accuracy.

For positional error, the method without geometric constraints had an average error within 8 cm, but this error peaked at a maximum of 14 cm during the motion sequence. This variability suggests a potential lack of reliability in positioning accuracy.

In contrast, our method, which applies geometric constraints, showcased superior precision with a maximum average error of 2.611 cm and a maximum error capped at 2.968 cm. This represents a significant improvement in positional accuracy, demonstrating the advantage of using geometric constraints.

When it comes to orientation angle error, the method without geometric constraints exhibited a maximum error of 8.39° and a maximum average error of 3.38°. These figures indicate a higher likelihood of inaccuracy in orientation estimation.

However, our method achieved a maximum error of 2.90° and a maximum average error of just 1.99° in orientation angle estimation. This substantial enhancement in orientation accuracy further underscores the effectiveness of geometric constraints in refining the precision and stability of the positioning system.

We have selected a specific dataset to visually demonstrate the comparative performance between our proposed method and the method without geometric constraints. The visualization focuses on the X, Y, and YAW positioning data for both methods. As depicted in Figure 9, our method exhibits a smoother trajectory, indicative of more stable and precise localization. In contrast, the method lacking geometric constraints shows noticeable spikes and abrupt changes, revealing its susceptibility to inaccuracies.

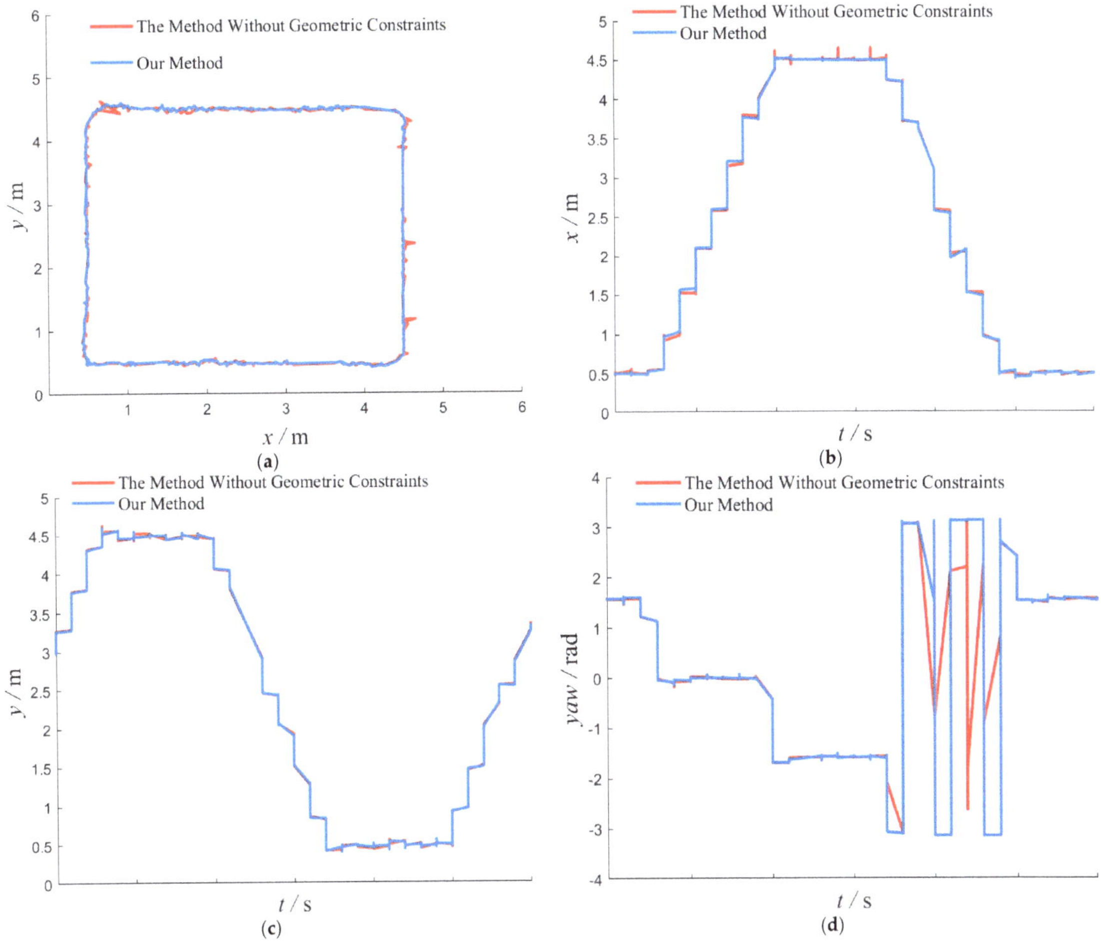

Figure 9. Method comparison with/without geometric constraints. (**a**) Trajectory; (**b**) X; (**c**) Y; (**d**) Yaw.

5.3. Method Performance

In this section, we present graphical representations of our method's performance, focusing on the trajectory and positional data from Groups 2 and 3 as outlined in the experimental setup table, where the positioning information obtained through our method is compared against the true positions provided by the simulation software.

Figure 10 (Group 2): This figure displays the raw observational data for the vehicle's motion as a scatter plot, which clearly shows the substantial dispersion of points due to the presence of outliers. Our method adeptly addresses these irregularities, producing a trajectory that closely mirrors the true path with a notable reduction in fluctuations.

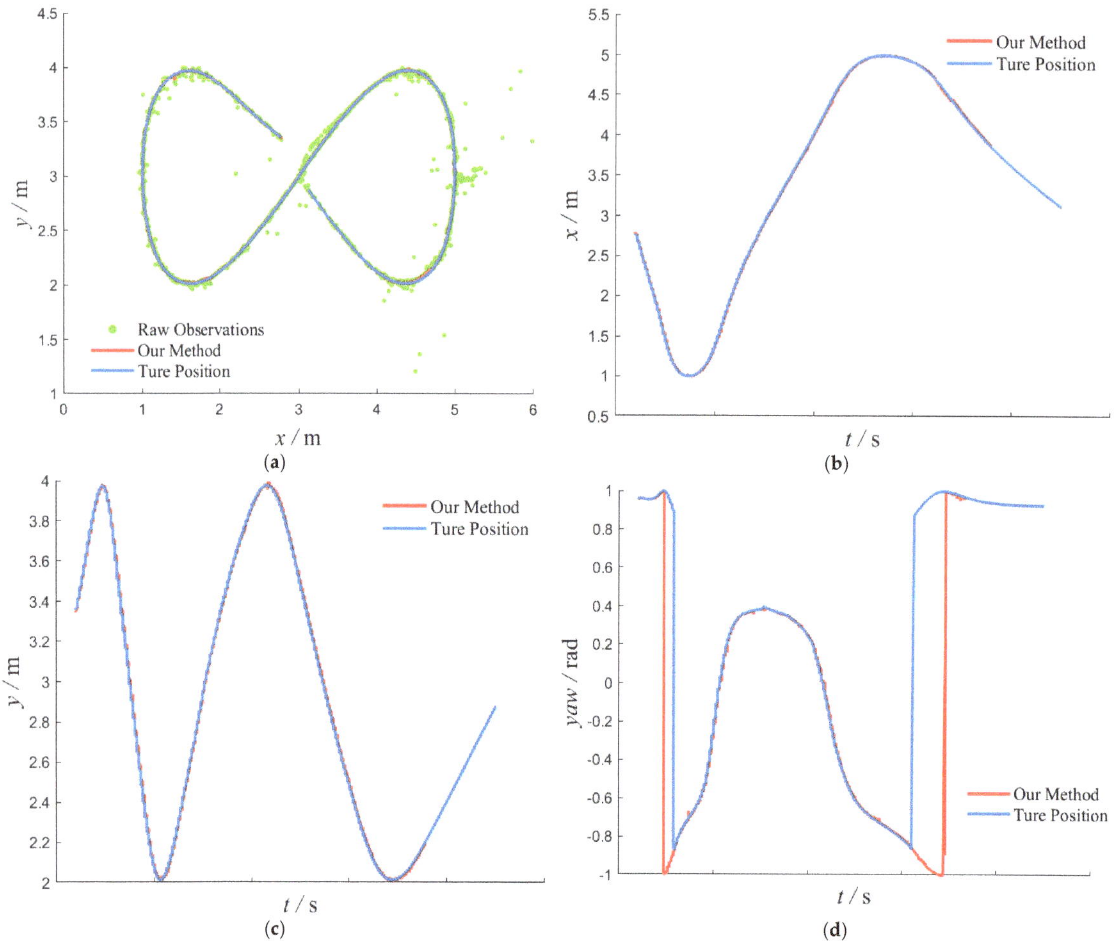

Figure 10. Experimental results of Group 2. (**a**) Trajectory; (**b**) X; (**c**) Y; (**d**) Yaw.

Figure 11 (Group 3): The motion trajectory of the vehicle is portrayed under more complex conditions, where movement is compounded by the presence of obstacles. The increased presence of outliers, particularly at corners where occlusions affect observation quality, is evident. Despite these challenges, our method demonstrates its robustness by producing a trajectory that is significantly smoother and more aligned with the true path compared to the raw data.

Based on the experimental setup outlined in Table 1 and the data presented in Table 4, we conclude the following:

Motion State: The motion state of the vehicle significantly affects positioning accuracy, with moving Groups 2 and 5 showing higher positioning errors compared to the stationary Groups 1 and 4. Our method, however, reduces these errors, bringing the positioning accuracy of moving states close to that of stationary states, indicating its effectiveness in handling motion challenges.

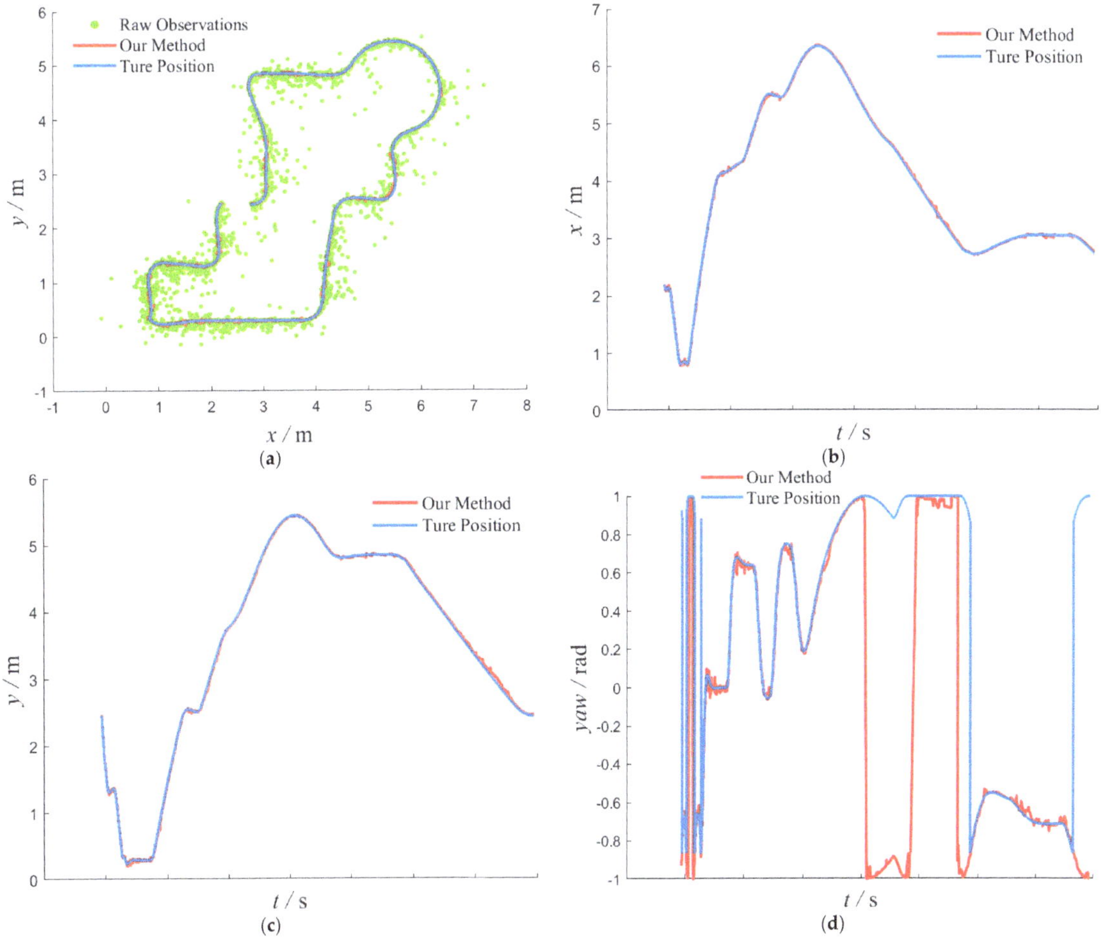

Figure 11. Experimental results of Group 3. (**a**) Trajectory; (**b**) X; (**c**) Y; (**d**) Yaw.

Tag Size: Comparisons across different tag sizes demonstrate that our method maintains low positioning errors regardless of tag size, whether larger (Groups 1 and 2) or smaller (Groups 4 and 5). This finding suggests that our method is compatible and consistent with various tag dimensions.

Obstacle Obstruction: The presence of obstacles in Groups 3 and 6, as opposed to the absence of obstructions in Groups 2 and 5, results in decreased positioning accuracy. However, our method's identification and removal of outliers effectively lessen the impact of obstructions on positioning outcomes.

6. Conclusions

The present study offers a contribution to the field of multi-tag localization by proposing a method that incorporates geometric constraints to address the issues of instability and inaccuracy that may arise in such scenarios.

We have developed a method that calculates an outlier factor based on geometric constraints, which may provide a means to characterize the variability among observational data points from multiple tags. This approach could potentially enhance the accuracy of the positioning process by dynamically adjusting the observation noise for each tag after identifying and removing outliers. Our method also includes an extension to the traditional

KF technique, designed to manage the fusion of data from multiple tags. This extension, which is based solely on observational data for prediction updates, may offer a pathway to bypass the limitations of conventional KF that typically integrate two data sources.

Preliminary results from high-fidelity simulations indicate that our method can achieve a reasonable level of positioning accuracy and stability, with errors maintained within a 3 cm range for position and 3° for orientation. These results are encouraging and suggest that the method may be applicable in environments where GNSS signals are not available or are subject to interference. In our future work, we aim to integrate Z-axis constraints into our localization method to achieve six-degree-of-freedom (6-DOF) positioning. This enhancement will be crucial for high-precision navigation in environments with varying terrain. Additionally, we plan to deploy our method on a real vehicle, where we will tackle and overcome the practical issues encountered in actual deployment scenarios.

Author Contributions: Conceptualization, Z.L. (Zhuojun Liu) and Z.L. (Zexing Li); Data curation, Z.L. (Zhuojun Liu); Formal analysis, Z.L. (Zhuojun Liu); Funding acquisition, W.Q. and G.Z.; Investigation, Z.L. (Zhuojun Liu) and W.Q.; Methodology, Z.L. (Zhuojun Liu); Project administration, W.Q.; Resources, W.Q., Z.L. (Zexing Li), and G.Z.; Software, Z.L. (Zhuojun Liu) and Z.L. (Zexing Li); Supervision, W.Q. and G.Z.; Validation, Z.L. (Zhuojun Liu); Visualization, Z.L. (Zhuojun Liu); Writing—original draft, Z.L. (Zhuojun Liu) and G.Z.; Writing—review and editing, Z.L. (Zhuojun Liu), W.Q. and G.Z. All authors have read and agreed to the published version of the manuscript.

Funding: This research received no external funding.

Institutional Review Board Statement: Not applicable.

Informed Consent Statement: Not applicable.

Data Availability Statement: The original contributions presented in the study are included in the article, further inquiries can be directed to the first author.

Acknowledgments: The authors would like to acknowledge the generous support of Qin Wengang, Li Zexing, and Zhou Guofeng in funding aspects of this work.

Conflicts of Interest: The authors declare no conflicts of interest.

References

1. Chalvatzaras, A.; Pratikakis, I.; Amanatiadis, A.A. A Survey on Map-Based Localization Techniques for Autonomous Vehicles. *IEEE Trans. Intell. Veh.* **2023**, *8*, 1574–1596. [CrossRef]
2. Li, Z.; Wang, Y.; Zhang, R.; Ding, F.; Wei, C.; Lu, J.-G. A LiDAR-OpenStreetMap Matching Method for Vehicle Global Position Initialization Based on Boundary Directional Feature Extraction. *IEEE Trans. Intell. Veh.* **2024**, 1–13. [CrossRef]
3. Zhou, Z.; Wang, Y.; Zhou, G.; Nam, K.; Ji, Z.; Yin, C. A Twisted Gaussian Risk Model Considering Target Vehicle Longitudinal-Lateral Motion States for Host Vehicle Trajectory Planning. *IEEE Trans. Intell. Transp. Syst.* **2023**, *24*, 13685–13697. [CrossRef]
4. Lu, Y.; Ma, H.; Smart, E.; Yu, H. Real-Time Performance-Focused Localization Techniques for Autonomous Vehicle: A Review. *IEEE Trans. Intell. Transp. Syst.* **2022**, *23*, 6082–6100. [CrossRef]
5. Springer, J.; Kyas, M. Autonomous Drone Landing with Fiducial Markers and a Gimbal-Mounted Camera for Active Tracking. In Proceedings of the 2022 Sixth IEEE International Conference on Robotic Computing (IRC), Naples, Italy, 5–7 December 2022; pp. 243–247. [CrossRef]
6. Mantha, B.R.K.; Garcia de Soto, B. Investigating the Fiducial Marker Network Characteristics for Autonomous Mobile Indoor Robot Navigation Using ROS and Gazebo. *J. Constr. Eng. Manag.* **2022**, *148*, 04022115. [CrossRef]
7. Zhang, H.; Zhang, C.; Yang, W.; Chen, C.-Y. Localization and navigation using QR code for mobile robot in indoor environment. In Proceedings of the 2015 IEEE International Conference on Robotics and Biomimetics (ROBIO), Zhuhai, China, 6–9 December 2015; pp. 2501–2506. [CrossRef]
8. Wang, S.; Hu, T. ROS-Gazebo Supported Platform for Tag-in-Loop Indoor Localization of Quadrocopter. In *Intelligent Autonomous Systems 14*; Chen, W., Hosoda, K., Menegatti, E., Shimizu, M., Wang, H., Eds.; Springer International Publishing: Cham, Switzerland, 2017; pp. 185–197. [CrossRef]
9. Ghasemi, A.; Parivash, F.; Ebrahimian, S. Autonomous landing of a quadrotor on a moving platform using vision-based FOFPID control. *Robotica* **2022**, *40*, 1431–1449. [CrossRef]
10. Zhang, Z.; Zhong, L.; Lin, M.; Lin, R.; Li, D. Triangle codes and tracer lights based absolute positioning method for terminal visual docking of autonomous underwater vehicles. *Ind. Robot. Int. J. Robot. Res. Appl.* **2024**, *51*, 269–286. [CrossRef]

11. Kulikov, V.; Spirin, E.; Pikalov, I.; Saramud, M. Methods for determining spatial position and production objects orientation based on the proposed fiducial markers for technical vision system. *Int. J. Adv. Manuf. Technol.* **2024**, *132*, 2549–2562. [CrossRef]
12. López-Cerón, A.; Cañas, J. Accuracy analysis of marker-based 3D visual localization. In Proceedings of the XXXVII Conference on Automation, Madrid, Spain, 7–9 September 2016; pp. 1124–1131. [CrossRef]
13. Jin, P.; Matikainen, P.; Srinivasa, S.S. Sensor fusion for fiducial tags: Highly robust pose estimation from single frame RGBD. In Proceedings of the 2017 IEEE/RSJ International Conference on Intelligent Robots and Systems (IROS), Vancouver, BC, Canada, 24–28 September 2017; pp. 5770–5776. [CrossRef]
14. Szentandrási, I.; Zachariáš, M.; Havel, J.; Herout, A.; Dubská, M.; Kajan, R. Uniform Marker Fields: Camera localization by orientable De Bruijn tori. In Proceedings of the 2012 IEEE International Symposium on Mixed and Augmented Reality (ISMAR), Atlanta, GA, USA, 5–8 November 2012; pp. 319–320. [CrossRef]
15. Kallwies, J.; Forkel, B.; Wuensche, H.-J. Determining and Improving the Localization Accuracy of AprilTag Detection. In Proceedings of the 2020 IEEE International Conference on Robotics and Automation (ICRA), Paris, France, 31 May–31 August 2020; pp. 8288–8294. [CrossRef]
16. Abbas, S.M.; Aslam, S.; Berns, K.; Muhammad, A. Analysis and Improvements in AprilTag Based State Estimation. *Sensors* **2019**, *19*, 5480. [CrossRef] [PubMed]
17. Qin, H.; Shen, G.; Zhou, Y.; Huang, S.; Qin, X.; Xie, G.; Ding, R. Tag-Based Vehicle Visual SLAM in Sparse Feature Scenes. *Automot. Eng.* **2023**, *45*, 1543–1552. [CrossRef]
18. Martínez-Barberá, H.; Bernal-Polo, P.; Herrero-Pérez, D. Sensor Modeling for Underwater Localization Using a Particle Filter. *Sensors* **2021**, *21*, 1549. [CrossRef] [PubMed]
19. Zhang, S.; Shan, J.; Liu, Y. Variational Bayesian Estimator for Mobile Robot Localization With Unknown Noise Covariance. *IEEE/ASME Trans. Mechatron.* **2022**, *27*, 2185–2193. [CrossRef]
20. Kayhani, N.; Zhao, W.; McCabe, B.; Schoellig, A. Tag-based visual-inertial localization of unmanned aerial vehicles in indoor construction environments using an on-manifold extended Kalman filter. *Autom. Constr.* **2022**, *135*, 104112. [CrossRef]
21. Chen, J.; Gao, Y.; Li, S. Real-time Apriltag Inertial Fusion Localization for Large Indoor Navigation. In Proceedings of the 2020 Chinese Automation Congress (CAC), Shanghai, China, 6–8 November 2020; pp. 6912–6916. [CrossRef]
22. Neunert, M.; Bloesch, M.; Buchli, J. An open source, fiducial based, visual-inertial motion capture system. In Proceedings of the 2016 19th International Conference on Information Fusion (FUSION), Heidelberg, Germany, 5–8 July 2016; pp. 1523–1530. Available online: https://ieeexplore.ieee.org/abstract/document/7528064 (accessed on 20 June 2024).
23. Huang, Y.-H.; Lin, C.-T. Indoor Localization Method for a Mobile Robot Using LiDAR and a Dual AprilTag. *Electronics* **2023**, *12*, 1023. [CrossRef]
24. Zhenglong, G.; Qiang, F.; Quan, Q. Pose Estimation for Multicopters Based on Monocular Vision and AprilTag. In Proceedings of the 2018 37th Chinese Control Conference (CCC), Wuhan, China, 25–27 July 2018; pp. 4717–4722. [CrossRef]
25. Kayhani, N.; Heins, A.; Zhao, W.; Nahangi, M.; McCabe, B.; Schoellig, A. Improved Tag-based Indoor Localization of UAVs Using Extended Kalman Filter. Presented at the 36th International Symposium on Automation and Robotics in Construction, Banff, AB, Canada, 21–24 May 2019. [CrossRef]
26. Xu, Z.; Haroutunian, M.; Murphy, A.J.; Neasham, J.; Norman, R. An Underwater Visual Navigation Method Based on Multiple ArUco Markers. *J. Mar. Sci. Eng.* **2021**, *9*, 1432. [CrossRef]
27. Adámek, R.; Brablc, M.; Vávra, P.; Dobossy, B.; Formánek, M.; Radil, F. Analytical Models for Pose Estimate Variance of Planar Fiducial Markers for Mobile Robot Localisation. *Sensors* **2023**, *23*, 5746. [CrossRef] [PubMed]
28. Fiala, M. ARTag, a fiducial marker system using digital techniques. In Proceedings of the 2005 IEEE Computer Society Conference on Computer Vision and Pattern Recognition (CVPR'05), San Diego, CA, USA, 20–25 June 2005; Volume 2, pp. 590–596. [CrossRef]
29. Wang, J.; Olson, E. AprilTag 2: Efficient and robust fiducial detection. In Proceedings of the 2016 IEEE/RSJ International Conference on Intelligent Robots and Systems (IROS), Daejeon, Republic of Korea, 9–14 October 2016; pp. 4193–4198. [CrossRef]
30. Garrido-Jurado, S.; Muñoz-Salinas, R.; Madrid-Cuevas, F.J.; Marín-Jiménez, M.J. Automatic generation and detection of highly reliable fiducial markers under occlusion. *Pattern Recognit.* **2014**, *47*, 2280–2292. [CrossRef]
31. Benligiray, B.; Topal, C.; Akinlar, C. STag: A stable fiducial marker system. *Image Vis. Comput.* **2019**, *89*, 158–169. [CrossRef]
32. Kalaitzakis, M.; Cain, B.; Carroll, S.; Ambrosi, A.; Whitehead, C.; Vitzilaios, N. Fiducial Markers for Pose Estimation: Overview, Applications and Experimental Comparison of the ARTag, AprilTag, ArUco and STag Markers. *J. Intell. Robot. Syst.* **2021**, *101*, 71. [CrossRef]
33. Krogius, M.; Haggenmiller, A.; Olson, E. Flexible Layouts for Fiducial Tags. In Proceedings of the 2019 IEEE/RSJ International Conference on Intelligent Robots and Systems (IROS), Macau, China, 3–8 November 2019; pp. 1898–1903. [CrossRef]

applied sciences

MDPI

Article

Jam Mitigation for Autonomous Convoys via Behavior-Based Robotics

Calvin Cheung, Samir Rawashdeh and Alireza Mohammadi *

Department of Electrical and Computer Engineering, University of Michigan-Dearborn, Dearborn, MI 48128, USA
* Correspondence: amohmmad@umich.edu

Abstract: Autonomous ground vehicle convoys heavily rely on wireless communications to perform leader-follower operations, which make them particularly vulnerable to denial-of-service attacks such as jamming. To mitigate the effects of jamming on autonomous convoys, this paper proposes a behavior-based architecture, called the Behavior Manager, that utilizes layered costmaps and vector field histogram motion planning to implement motor schema behaviors. Using our proposed Behavior Manager, multiple behaviors can be created to form a convoy controller assemblage capable of continuing convoy operations while under a jamming attack. To measure the performance of our proposed solution to jammed autonomous convoying, simulated convoy runs are performed on multiple path plans under different types of jamming attacks, using both the assemblage and a basic delayed follower convoy controller. Extensive simulation results demonstrated that our proposed solution, the Behavior Manager, can be leveraged to dramatically improve the robustness of autonomous convoys when faced with jamming attacks and can be further extended due to its modular nature to combat other types of attacks through the development of additional behaviors and assemblages. When comparing the performance of the Behavior Manager convoy to that of the basic convoy controller, improvements were seen across all jammer types and path plans, ranging from 13.33% to 86.61% reductions in path error.

Keywords: autonomous convoy; behavior-based robotics; jamming

Citation: Cheung, C.; Rawashdeh, S.; Mohammadi, A. Jam Mitigation for Autonomous Convoys via Behavior-Based Robotics. *Appl. Sci.* **2022**, *12*, 9863. https://doi.org/10.3390/app12199863

Academic Editors: Jose Machado, Fabrizio Giulietti, Nadjim Horri and William Holderbaum

Received: 19 August 2022
Accepted: 27 September 2022
Published: 30 September 2022

1. Introduction

Ground vehicle convoys are utilized in both commercial and military applications in order to reduce costs, increase vehicle efficiency, and improve safety for the transport personnel. At the most basic level, a convoy is a group of two or more vehicles, traveling from a set origin to an objective destination under a single leader [1]. From a military perspective, the use of ground vehicle convoys to perform supply operations is an important part of an efficient logistics strategy. While several different transportation methods such as aircrafts, trains, and ships are available, ground vehicles still account for a significant portion of supply and equipment distribution due to the battlefield complexities and the need to embed protective measures such as gun trucks [2]. Similarly, the usage of ground vehicle convoys is vital in the commercial sector, due to the prevalence of paved roads and cost effectiveness when moving large or heavy materials [3].

With the advent of autonomous driving, there has been heavy focus on leveraging autonomous vehicles to improve convoys. Much of the research and development towards autonomous vehicle convoys focus around three iterative employment concepts: minimally manned (MM), partially manned (PM), and fully autonomous (FA) [4]. In a minimally manned system, the lead vehicle in the convoy has a human driver, while the follower vehicles operate autonomously with a safety rider monitoring autonomous performance, ready to take over in the event a fault occurs. In a partially manned system, the lead vehicle convoy has a human driver, but all the follower vehicles are autonomous and unmanned. In a fully autonomous system, all vehicles in the convoy, leader and followers, are unmanned autonomous. In all cases, there are numerous benefits in applying autonomy to ground vehicle convoys. From both the commercial and military perspective, the utilization of autonomous vehicle convoys can reduce fuel consumption, reduce traffic congestion, and improve safety [2].

In all the autonomous vehicle convoy employment concepts, the utilization of wireless communications has been paramount in their development. Two important applications of wireless communications include inter-vehicle communications (IVC) and Global Positioning Systems (GPS). IVC is used to exchange vital information between vehicles in the convoy, including pose, heading, speed, acceleration, and maneuver intentions [5]. This information is used by follower vehicles in maintaining the desired convoy formation along the desired path. In addition, GPS is often used as a vital tool in providing position, navigation, and timing information for each vehicle [4]. In many cases, proper wireless communications are a necessity in the operation of an autonomous convoy, where disruption of the wireless communications system can cause the autonomous convoy to fail. The criticality of proper wireless communications extends beyond autonomous convoys and applies to various other autonomous applications such as surveillance [6], mining, and agriculture [4]. This outsized importance of wireless communications makes it a critical piece of the system to protect from potential cyber-physical attacks, such as jamming.

Jamming is a type of denial-of-service attack that attempts to disrupt or block wireless communications. Jammers interfere with a radio's ability to transmit or receive data, and are the most widely used denial-of-service attack for vehicular networks due to their effectiveness and simplicity [7–9]. The risk of jamming is especially prevalent in military applications, due to electronic warfare systems and strategies employed by adversaries to gain tactical advantages [9]. Given the importance of IVC in autonomous convoys, disruptions caused by jamming could cause instability in the following functionality of the autonomous convoys, or even potentially cause a complete breakdown of the convoy [4]. Since jamming is one of the most simple and effective denial-of-service attacks [10], it is important to understand the attack and the proper ways to defend against it. There have been some limited efforts focused more narrowly on the topic of jamming as it relates to convoys, with a greater body of work focused on investigating the effects of jamming on vehicular ad-hoc networks (VANET) in general [11–13].

On the topic of jamming and its effects of vehicular convoys, there has been some work in developing a framework to measure the impact of the attack and a convoy's resilience. Van der Heijden et al. examined three Cooperative Adaptive Cruise Control convoy controllers and how various attacks such as jamming and data injection impacted their respective performance [7]. Both an attacker model and an evaluation framework were developed for a thorough examination of what sort of attack detection and attack resilience algorithms should be used to resist the effects of attacks on the VANETs. In another convoy specific effort, Hu et al. investigated how stealthy jamming attacks impacted convoy stability [14]. The stealthy jammer performed jamming against a simulated convoy's basic safety message (BSM), which contains information such as vehicle speed, position, and acceleration, over the control channel of dedicated short-range communications (DSRC), but only when the probability of being detected was low. The probability of being detected was determined based on how significantly jamming at a particular time would affect the packet loss ratio of the communications. To detect the stealthy jams, they analyzed received power and transmission delays of messages to differentiate between jamming attacks and normal interference. The effects of jamming on VANETs in general is a more thoroughly explored topic. Various efforts have focused on how jamming affects the transmission of BSMs over DSRC in relation to Forward Collision Warning capabilities [11–13]. These efforts create attack models to quantify the effects of the jamming attacks and come up with different mechanisms to mitigate their effects. Alturkostani et al. looked at vehicle distance with packet delivery ratio to detect jams, causing the system to enter a fail-safe mode [11]. Serageldin et al. utilized redundant channels and data through dissimilar message types to reduce the effectiveness of jams [12]. In a different effort, Serageldin et al.

looked to develop architectures resistant to jamming by using dual and triple-redundant channels at higher power ratings to enable BSM transmission [13]. Other efforts have leveraged more of the Wireless Access in Vehicular Environment standards in trying to detect jams, such as Malebary's work in looking at beacon frequency from roadside units to detect jams and broadcast warnings to other network members in its vicinity [8]. As a whole, prior efforts related to jamming of convoys and VANETS have had a heavy focus on detection, with a response of either broadcasting warnings to convoy vehicles or entering fail-safe modes. Rather than investigating the issue of jam detection, which has been the focus of the previous literature, our proposed solution instead focuses on control-oriented countermeasures when jamming is unavoidable. We aim to develop methods and techniques to allow for continued convoy operation in the face of these denial-of-service attacks, mitigating the undesired outcome of jamming attacks, rather than the jamming itself.

There are a variety of control approaches and architectures utilized in multirobot autonomous convoying. More broadly speaking, multirobot convoying is a subset of formation control, and the proper choice for formation control design is dependent on which factors need to be prioritized, such as formation shape generation time [15], formation robustness during network congestion [16], and flexible prioritization between system performance and control effort [17]. One particularly robust formation control approach for dynamic environments is behavior-based robotics. In behavior-based robotics, robotic control is built as a collection of basic behaviors. These behaviors are modular and run in a concurrent and distributed manner to achieve desired system functionality. The utilization of behavior-based robotics architectures in multirobot teaming has been previously explored in various research efforts seeking to take advantage of its benefits in dynamic environments. Early efforts include DARPA studies on the integration of navigational behaviors and formation behaviors to create human-led robotic teams to be used in different types of task environments [18]. This effort looked to take four robot teams and set them in a line, column, diamond, or wedge using a motor schema behavior approach. The behaviors used were move-to-goal, avoid-static-obstacle, avoid-robot, maintain-formation, and noise. Member vehicles in the robotic team transmitted position information to properly space out appropriately given the formation and method of formation centering. The path error and duration out of formation of the robots for each formation was compared to one another to determine which one had the best performance in different experimental scenarios. Another effort proposed a decentralized method utilizing a formation matrix [19]. In this effort, each robot leveraged a formation matrix, which defined leader follower pairs. This formation matrix allocated robots to specific positions in a formation before the system began operation by using onboard sensors to detect the location of surrounding robots. Once the proper position of each member robot was allocated, the system was able to traverse to a destination avoiding obstacles, while maintaining the desired formation based on what was defined in the matrix. In this approach, a behavior network architecture was used, in which basic behaviors were linked together and either stimulated or suppressed. This stimulation/suppression was based on input vector values such as detected object location. The output vectors of the network defined the robot's path. While this effort did not use communications between robotics in motion, initial allocation of vehicles in the specific starting positions was required for formation matrix setup. In addition, this effort focused on static robotic formations that maintained their shape, making this method less suitable for applications such as a convoy following defined roads. More recent efforts in behavior-based multi-robot teaming have sought to improve performance by optimizing inter-robot communications, rather than eliminating them. One such effort created a distributed framework for multirobot behavior sequencing that allowed teams of robots to adjust to their configuration to meet communication requirements for the different tasks [20]. The effort focused on the idea that coordinated behaviors between multiple robots must be sequenced together in a way that takes inter-robot distances into account to meet information flow constraints due to communications. The constraints translate into specific robot position configurations that need to be met within a finite time for the behaviors to be performed. The framework leveraged finite-time convergence control barrier functions to adjust configurations to meet the communication requirements for different sequences of behaviors. While these efforts focused on utilizing behavior-based control architectures for multi-vehicle teaming, they either heavily rely on IVC throughout their operations or require a manually arranged initial state. Furthermore, the focus on traveling in formation shapes makes the multi-vehicle movement less suitable for convoy path following operations, which is the outcome being sought by industry and military from autonomous convoying. Our efforts focus on the development of a more flexible framework towards convoy path following that utilizes on IVC when available and having the additional capability to activate more robust behaviors not relying on communications when faced with jamming to continue convoy operations. Refer to Table 1 for a summary of the control approaches referenced in this section.

Table 1. Prior work in convoying and formation control.

Title	Authors	Summary
Path Planning for Multiple Unmanned Vehicles (MUVs) Formation Shape Generation Based on Dual RRT Optimization [15]	Gong, Tianhao; Yang, Yu; Song, Jianhui	Perform shape formation generation with multiple unmanned vehicles using rapidly-exploring random trees to plan vehicle paths during formation.
Research on Self-Organizing Behavior-Based UAV Formation Based on Distributed Control [16]	Yunhe, Li; Bo, Li; Xiaowei, Niu	Optimize unmanned aerial vehicle formation through use of detected information with self-organizing behaviors with a quasi-static communication topological structure.
Autonomous Vehicle Convoy Formation Control with Size/Shape Switching for Automated Highways [17]	Jond, Hossein Barghi; Platoš, J; Sadreddini, Zhaleh	Development of a convoy formation control architecture consisting of four closed-form control law algorithms to handle formation size and shape switching.
Behavior-Based Formation Control for Multirobot Teams [18]	Balch, Tucker; Arkin, Ronald	Integrated navigational and formation behaviors to create human-led robotic teams to be used in different types of task environments.
Decentralized behavior-based formation control of multiple robots considering obstacle avoidance [19]	Lee, Giroung; Chwa, Dongkyoung	Proposed a decentralized formation control method utilizing a formation matrix and behavior network architecture.
A Sequential Composition Framework for Coordinating Multirobot Behaviors [20]	Pierpaoli, Pietro; Li, Anqi; Srinivasan, Mohit; Cai, Xiaoyi; Coogan, Samuel; Egerstedt, Magnus	Created a framework for behavior sequencing that allowed teams of robots to adjust their configuration to meet communication requirements for the different tasks.

This paper investigates developing an autonomous convoy system that mitigates the effects of jamming attacks. We aim to create an autonomous convoy that does not inherently rely on any Vehicle-to-Vehicle (V2V) or Vehicle-to-Infrastructure (V2I) network communications by design, creating a highly scalable system that is robust to system error and loss of communications by using only the on-board sensors of each vehicle. A behavior-based robotics architecture is used to facilitate the employment of simpler sensors and reduce the reliance on complex planning with world models.

The contributions of this paper are as follows:

- This paper develops control-oriented countermeasures against jamming attacks on autonomous vehicle convoys that are independent of traditional radio anti-jamming techniques. This allows for layers of protection against jamming at both the network and convoy controller level, improving the robustness and performance of an overall convoy system when confronted with jamming.
- This paper creates an architecture for behavior-based robotics that is uniquely integrated with the Robot Operating System (ROS) framework. This architecture combined a Motor Schema behavior-based robotics with the ROS navigation stack to a depth that had not previously been seen, paving the path for the greater ROS community to leverage behavior-based robotic architectures.
- The jamming-proof motion planning in this paper improves upon the basic Motor Schema approach by integrating it with vector field histogram motion planning, allowing us to avoid the pitfalls of potential field motion planning.

The remainder of the paper is organized as follows: Section 2 provides background on autonomous ground vehicle convoys, jammers, and behavior based robotics; Sections 3 and 4 discusses the materials and methods used, including the behavior-based robotics architecture, algorithms, and design of experiment; Section 4 shows the results to quantify the performance differences between autonomous ground convoy systems when jam mitigation is present; and Section 5 contains the final discussions and conclusions.

2. Background

The purpose of this effort is to develop an autonomous ground vehicle convoy system that mitigates the effects of jamming by utilizing a behavior-based robotics approach. To enable proper exploration of the effort, the following sections provide a brief background on three topics fundamental to our work: autonomous ground vehicle convoys, jammers, and behavior-based robotics.

2.1. Autonomous Ground Vehicle Convoys

Autonomous ground vehicle convoys are composed of a leader vehicle and follower vehicles. Follower vehicles follow the path of the leader at a given offset distance. While there are various levels of autonomy employed by the autonomous ground vehicle convoys, such as MM, PU, and FA [4], the fundamental design of autonomous convoys remains consistent. The leader vehicle sets the path, either through human driving or autonomous navigation, and utilizes V2V communications to send information to the follower vehicles. The information sent varies depending on system design, but typically includes path points, vehicle pose, vehicle speed, and other sensor information to help follower vehicles track their leader. The follower vehicle uses onboard sensors to estimate its own state, and produces a state estimate of its leader using the information received from V2V and various onboard sensors. A path planning system uses the state information and leader path points to generate a motion plan, and the vehicle is actuated accordingly [3]. Figure 1 lays out the high-level architecture as described.

The V2V communications sent between vehicles can contain a broad range of information from a wide suite of sensors. Information relevant to the vehicle's speed and position, such as sensor data (cameras, GPS, Light Detection and Ranging [LiDAR], wheel encoders, etc.), vehicle kinematics and intended maneuvers, would all need to go through the V2V communications to allow follower vehicles to reach the desired speed and formation for leader following [2]. Various V2V communication topologies can be utilized in autonomous ground vehicle convoys, depending on the needs of the system. Figure 2 shows some common leader-follower topologies used, including predecessor following, predecessor-leader following, bidirectional, and bidirectional-leader [21].

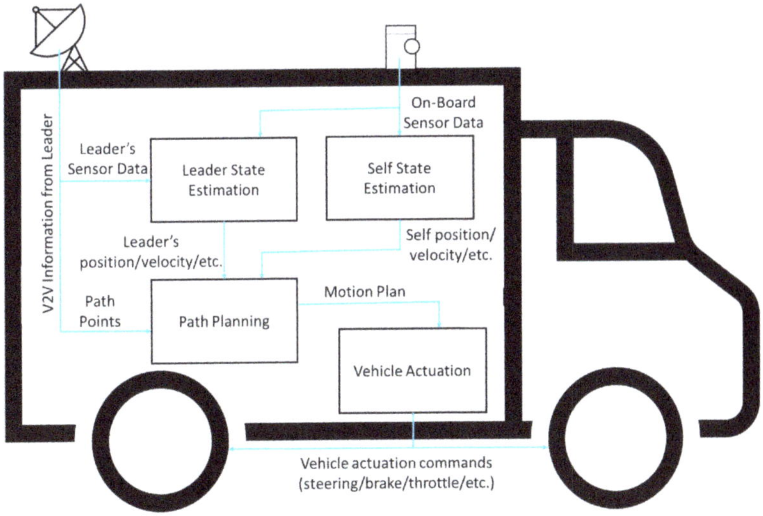

Figure 1. High-level architecture for a follower vehicle in an autonomous ground vehicle convoy.

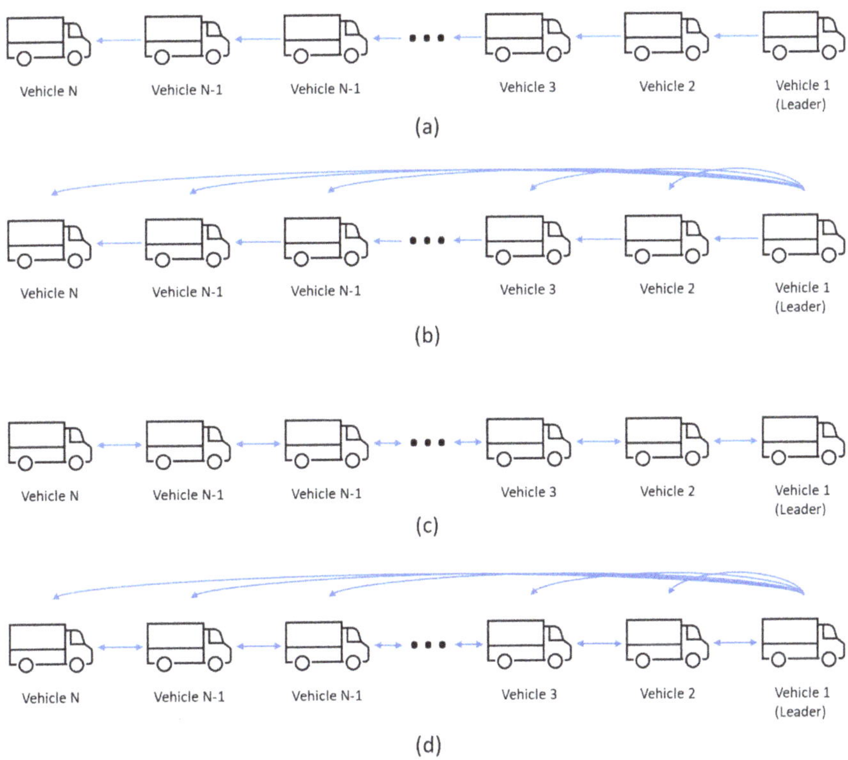

Figure 2. Common leader-follower communication topologies, adapted from [21]: (**a**) predecessor following; (**b**) predecessor-leader following; (**c**) bidirectional; (**d**) bidirectional-leader.

2.2. Jammers

Jamming is an effective and simple radio interference attack that can be used against wireless networks. In its most basic form, jammers emit radio frequency signals to fill a wireless channel to interfere with the sending and receiving of wireless communications. The jammer can work to either prevent the source from transmitting packets or disrupt the receiver from properly recognizing and processing legitimate packets. The end result in either case is the interference of the physical transmission and reception of wireless communications [10]. Different classifications of jammers utilize different techniques in jamming wireless networks. The five most commonly described techniques in literature are as follows in Table 2.

Table 2. Common Jamming Techniques.

Jammer Type	Description
Constant [8,10–12,22]	Continuously emits random noise over a wireless medium to interfere with legitimate communications.
Deceptive [10,12,22]	Periodically inject valid packets in their transmissions to deceive receivers into believing that legitimate messages are being sent.
Random [8,10–12,22]	Operates on jam and sleep periods. They will jam for a time duration of t_J, and sleep for a duration t_S before jamming again. Both t_J and t_S can be either fixed or values random.
Reactive [8,10,12,22]	Continuously monitor a communications channel and only jams when it senses activity.
Intelligent [11,22]	Uses knowledge of the communications protocols they are seeking to jam and analyze the transmissions to target specific messages or message types.

Despite the relative simplicity of basic jammers, such as constant jamming and random jamming, jamming attacks are difficult to defend against. Conventional security mechanisms are ineffective, because attackers can disrupt the entire medium of communications itself rather than having to exploit the traditional areas of confidentiality, authentication, and integrity [10]. Additionally, the deployment of jammers has a low barrier to entry. All radio operations on a wireless communications channel can be completely disrupted by a single device emitting noise at high power, regardless of what security measures are built into the underlying communications protocols. In addition, Software Defined Radios that can be used to launch jamming attacks are becoming increasingly affordable and easy to use [9]. In response to the vulnerabilities presented by jamming, several efforts throughout the years have proposed different anti-jamming methods. While many novel anti-jamming methods exist, the primary mechanisms to mitigate jamming can be broadly classified into larger groupings, with two prominent groupings being filtering and Spread Spectrum (SS) techniques. In filtering techniques, the radio receiver attempts to filter out the jammed signal by various means so that the intended signal can be properly parsed. One method of doing this used beamforming, in which a beamformer performed spatial filtering of spatial samples collected from propagating wave fields [23]. It then separated signals with overlapping frequency content that were delivered from different spatial locations. In a different filtering approach, an adaptive Gaussian filter was used to combat jamming caused by narrowband interference with gaussian white noise and pulsed noise [24]. The adaptive filter used optimal time-frequency localization and variable notch depth in conjunction with a fast Fourier-transform-based correlation to filter both continuous and time-varying narrowband interface. In addition to filtering techniques, SS techniques are often used to combat jamming. With SS techniques, a narrowband information signal uses data-independent, random sequences to spread the signal over a wide band of frequencies, in hopes that the jammer is unable to disrupt the entire frequency band [25]. The receiver can then correlate the signal it received with a copy of the random sequence to decode the data that was meant to be delivered. The two most commonly used SS techniques are direct sequence SS (DSSS) and frequency hopping SS (FHSS) [26]. In DSSS systems, the information signal is spread throughout the entire available bandwidth at the same time. This spreading of the signal causes the energy present at each particular frequency of the bandwidth to be very low and mistaken as noise by jammers that are actively listening for signals to jam. The intended receiver can then reconstruct the signal that it received from all the frequencies in the bandwidth. In FHSS systems, data is sent through a narrowband signal, but the frequency and channel being used to transmit signal rapidly changes. The frequency hopping pattern is known by the sender and intended receiver, allowing for messages to successfully be sent between them. In addition to filtering and spread spectrum techniques, various unique anti-jamming methods have been proposed to protect against the attacks, including exploiting reactive jammer reaction times [27], jammed node isolation [28], and directional antennas [29]. While all these methods can be effective at countering jamming attacks, no solution can offer absolute protection in all scenarios. For this reason, creating robust autonomous convoy solutions that can maintain operation in the face of a jamming attack is important for widespread commercial and military adoption.

2.3. Behavior-Based Robotics

The development of behavior-based robotics was a response to the prevalent control paradigms of the time, which were rooted in function-based, deliberative models [30], as shown in Figure 3a. Under the *deliberative approach*, robotic control relied heavily on world models. Updates to the world model would be determined in a sensing phase, plans based on the model were calculated in a planning phase, and action would then be taken in an acting phase. This approach responded poorly in dynamic environments due to the slow speed of model updates coupled with the need for re-planning [31].

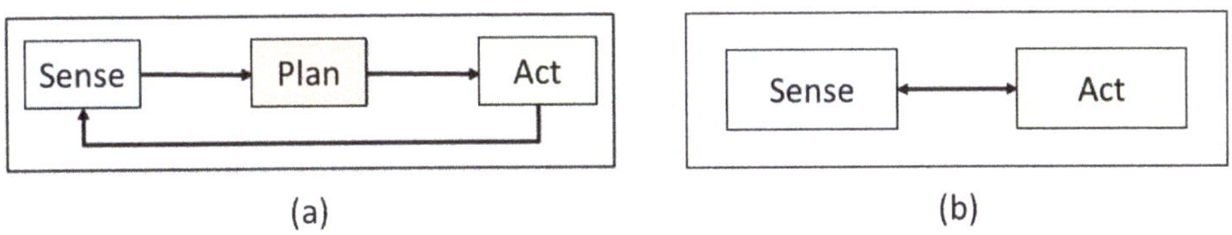

Figure 3. Robotic control approaches: (**a**) deliberative approach; and (**b**) reactive approach [31].

To address the issues of deliberative control, researchers began to develop more *reactive approaches* with faster response times that would be more suited to the dynamic, real-world environments. In these reactive control approaches, there is a tight coupling between sensors and actuators. Control is performed via concurrent condition-action rules so that no complex computations or world models are needed, as shown in Figure 3b. A sensor reads the environment and acts upon the information given the condition-action rules in place. The tight coupling minimizes the need for heavy computation and complex world model [31]. Despite the benefits afforded by reactive control approaches, a purely reactive approach limits an autonomous system's ability to perform complex tasks. While there are various approaches that leverage reactive control with internal states and models, one approach that is particularly well suited to dynamic environments and multi-vehicle control is behavior-based robotics.

In a *behavior-based robotics approach*, robotic control is built as a collection of basic behaviors, with behaviors being defined as something that generates a motor response given some sensory stimulus. These behaviors are modular, and run in a concurrent and distributed manner to achieve desired system functionality. For instance, an autonomous vehicle that moves from its origin to some target location may be concurrently running behaviors such as Move-to-Goal, Avoid-Obstacles, and Stay-on-Path at all times in order to meet the system objective. These different behaviors may be purely reactive, in which they take sensor readings and use rule-based logic to immediately generate some desired motor response, or they can maintain their own state and memory to allow for more complex behaviors. With behaviors being run concurrently, an arbitration mechanism is used to decide on what actions to take, whether it be a single behavior's output or a fusion of multiple behaviors [32]. Arbitration techniques vary based on the specific behavior-based robotics architecture that is being used. Some examples include subsumption architectures [30], motor schemas [33], and circuit architectures [34]. Figure 4 shows an example of how a generalized behavioral-based robotics architecture can be laid out.

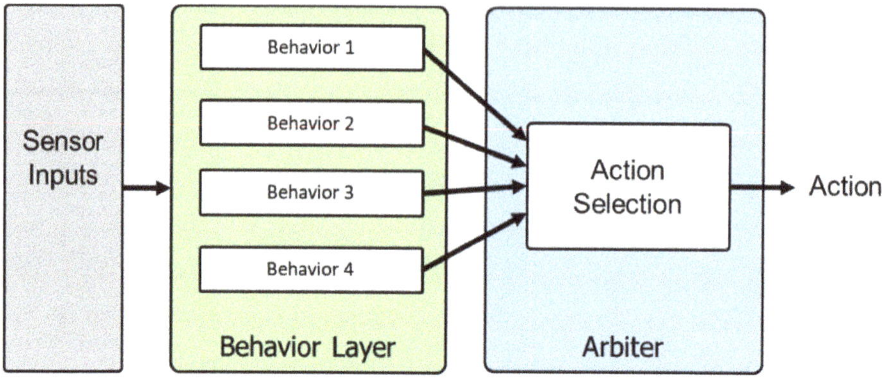

Figure 4. A generalized behavioral-based robotics architecture.

The behavior-based robotics architecture developed for this effort was based on Motor Schemas, a neurological-based approach towards behavioral robotics inspired by schema theory [33]. A schema is a unit of behavior that describes how an agent should react in a given situation and how the reaction can be performed [32]. The concurrent control of many different schemas is used to explain motor behavior. In the Motor Schema approach, each motor schema has an associated perceptual schema that provides the sensor information, which the motor schema uses to generate motor responses per its defined behavior. This is traditionally conducted through potential field response vectors around the robot. Coordination of motor schemas is achieved cooperatively, with no hierarchy of behaviors. Instead, weightings for the motor schemas are set based on a robot's current needs and combined through vector addition. The coordination between different schemas allows for the execution of complex, emergent actions with only a small number of primitive behaviors.

3. Materials and Methods

This following section details the development and testing of the proposed behavior-based autonomous ground vehicle convoy system. A description of the tools leveraged, jamming attacker model, behavior-based robotics architecture, experimental design, and data analysis methods ensue for a thorough discussion on the quantifiable benefits of the proposed system.

3.1. Tools

Development of our autonomous ground vehicle convoy system leveraged the ROS. ROS is a flexible framework for the development of autonomous robotic software, leveraging a collection of libraries, tools, and conventions to facilitate and encourage collaborative software development and reuse [35]. In addition, Gazebo, an open-source 3D robotics simulator, was used in development, due to its integration with ROS and overall community support [36]. Utilizing ROS and Gazebo in our effort allowed us to build upon pre-existing models for our own developmental and simulation environment. To that end, the Clearpath Husky model was chosen as the robotic platform for development.

The Clearpath Husky is a robotic platform that is specialized for research and rapid prototyping [37]. The ROS libraries and Gazebo models are well supported with a large user base, making it an ideal platform to leverage for development [38]. The sensor suite used in this effort includes a GPS, inertial measurement unit, and a forward facing 270° field-of-view LiDAR system. Figure 5 shows a convoy of Clearpath Husky models in a column convoy formation, along with ROS sensor visualization tools showing the 270° field-of-view LMS1xx LiDAR for each vehicle.

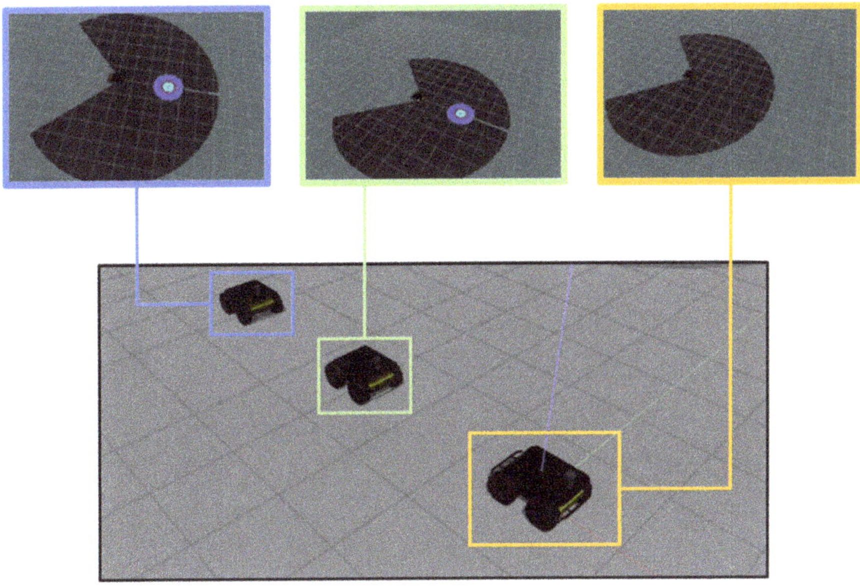

Figure 5. Convoy of Clearpath Husky robots in the Gazebo robotics simulator with each robot's sensor visualization in ROS tools.

3.2. Attacker Model

To develop a jamming attacker model, certain assumptions regarding the network configuration of the robots had to be established beforehand. We assumed that the three Clearpath Husky robots were equipped with wireless radios, allowing them to become mobile nodes. Each node would be able to transmit and receive data packets, forming a generic autonomous vehicular ad-hoc network, similar to what is described in an IEEE 802.11bd network. In addition, each node would be able to leave and rejoin the network without causing the ad-hoc network to fail. Due to the proximity of the robots, we simplified the overall model by assuming that no signal loss or degradation occurs from separation distances between the robots.

The jamming attacker model developed for this effort allows for a choice between simulating a constant jammer and a random jammer. The constant jammer type was chosen due to it being the worst-case jamming attack [12], while the random jammer type allows us to test a greater range of unique jamming scenarios in which follower vehicles end up being partially in the jamming zone. The other types of jammers described are focused on disguising jamming attacks. Since detecting disguised jamming attacks was not a focus of this effort, we focused on constant and random attacker models.

In developing the attacker model, a few key assumptions were made regarding its configuration and capabilities. Firstly, the jammer would be a single node, acting as a single point of jamming in the environment. The jammer was assumed to not be restricted by its power source, meaning it would run continuously for indefinite periods of time without any issue. In addition, it would be frequency unbounded, meaning that the jammer would successfully jam on all frequencies the network communicated on. This maximized the impact of the jamming attacks, making it so we considered packets to be dropped completely when going from the sender to receiver. Implementing jamming in this fashion allowed us to simulate an upper bound on successful denial-of-service attacks [7]. To simplify the jamming model, we assumed that all areas of the jamming zone were equally strong, with no wireless communication being possible anywhere within the jammed area.

The jammer developed for this effort is configurable with the following inputs:

- lat, long—latitudinal and longitudinal coordinates for the center of the jammer;
- r_J—radius of the jamming area, centered at lat, long;
- type of jammer;
- t_J—jamming time for random jammer;
- t_S—sleep time for random jammer.

The latitudinal and longitudinal coordinates for the jammers center and the jamming radius create a cylindrical jamming zone. Figure 6 illustrates the jamming attack model targeted at a robotic convoy.

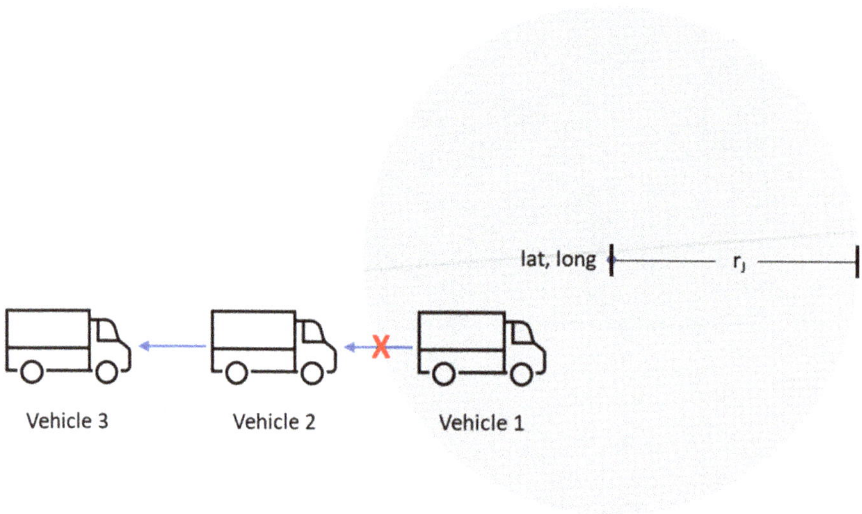

Figure 6. A convoy entering a jamming zone of radius r_J centered at lat, long.

3.3. Behavior Manager

As previously described, a behavior-based robotics architecture uses a collection of basic behaviors to perform robotic control, where behaviors are modular components that generate motor responses given some sensory stimulus which are run in a concurrent and distributed manner. Behavior-based robotic architectures are particularly well-suited to multi-vehicle control in dynamic environments due to scalability, decentralization of control, and tight coupling of rapid sensor readings to real-time path planning [31,39]. The advantages inherent to a behavior-based robotics architecture make it highly well-suited to provide robust protections against jamming for an autonomous convoy. Using the Motor Schema approach as a starting point, we developed a novel behavior-based robotics architecture that uniquely takes advantage of the layered costmap system implicit in the ROS navigation stack. Our behavior-based architecture, henceforth referred to as the Behavior Manager, encompasses behavioral costmaps, behaviors, assemblages, and how they holistically interact. The following sections provide details on the general design of the Behavior Manager, along with the specific implemented system we are proposing for jamming mitigation of autonomous convoys.

3.3.1. Behavioral Costmaps

A core component of the Behavior Manager are the behavioral costmaps. Traditionally, costmaps are two-dimensional grids in which every cell represents a traversibility cost around a robot, based on sensor readings. These costs are used in calculating the optimal path when traversing to a goal by a path planner [31]. ROS utilizes a layered costmap system, in which multiple, separate costmaps are defined for different contexts. Each costmap is considered a layer that is combined to create a master costmap, which is utilized by the path planner [40].

The Behavior Manager leverages ROS' layered costmap system by having each layer represent the costs as defined by a perceptual schema behavior. In essence, every costmap layer exists due to a behavior dictating that a costmap is needed to perform some behavior. These costmap layers, which we call behavioral layers, are utilized both individually and as a combined master costmap by the different behaviors and by the Behavior Manager's path planning system. In our implementation of the behavioral layers, the costmaps are 200 × 200 grids, with every cell containing a value between 0 and 254. The autonomous vehicle is centered at the approximate center of the costmap, at the grid index (100,100). Figure 7 shows a simplified example of the behavioral costmaps.

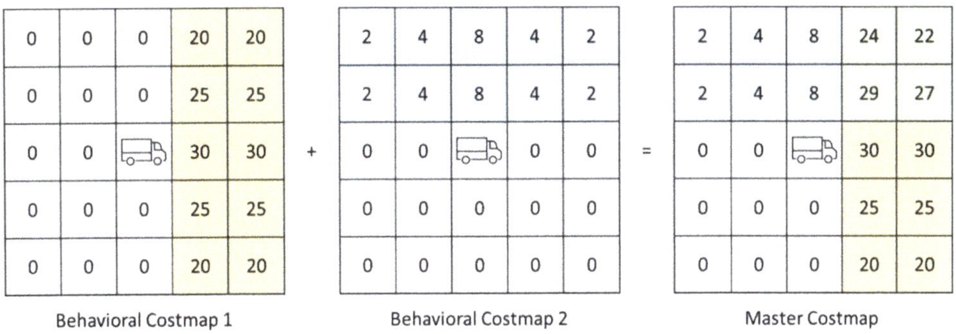

Figure 7. Simplified example of 5 × 5 behavioral costmaps, combined into a master costmap via addition.

3.3.2. Behaviors

In general, behaviors are modular components that generate motor responses given some sensory stimulus which are run in a concurrent and distributed manner [32]. In the Motor Schema approach, the definition of behaviors is expanded, and they are categorized as either motor schemas, which generate motor responses per its defined behavior, or perceptual schemas, which provide sensor information to the other behaviors [33]. The Behavior Manager is based on the Motor Schema approach, and accordingly, categorizes its behaviors as either perceptual schemas or motor schemas.

Behavior Manager perceptual schema behaviors utilize sensor data and generate behavioral costmaps based on the requirements of the behavior. For example, Figure 8 shows how a Keep-Standoff-Distance perceptual schema behavior would work. Figure 8a shows the world state, with the autonomous vehicle in the center, and an obstacle being detected in the upper right-hand corner. Figure 8b shows what the LiDAR readings would be, with a value of 253 being associated with the detected obstacle. Figure 8c shows the final behavioral costmap, with the perceptual schema increasing the costs in the area directly around the obstacle to meet the behavior's goal of keeping a standoff distance between the autonomous vehicle and objects detected by the sensors. Each perceptual schema has a gain that it assigns to its behavioral costmap to be used when combining the layers into a master costmap. Multiple perceptual schemas are run concurrently, creating multiple behavioral costmaps that are used by the greater Behavior Manager in individual behaviors and in overall path planning.

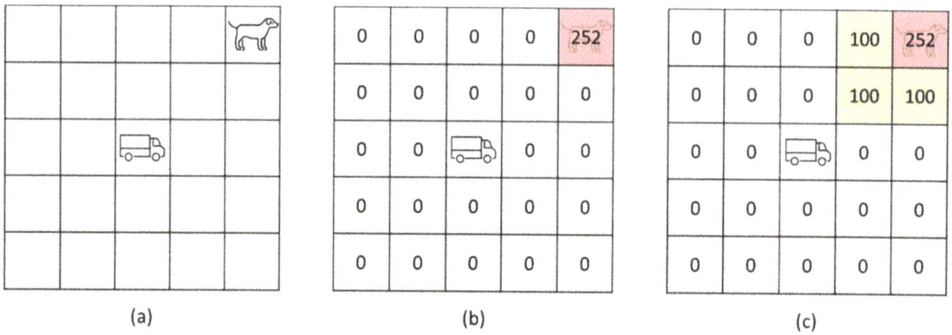

Figure 8. A simplified example of a Keep-Standoff-Distance perceptual schema behavior: (**a**) the world state, with the autonomous vehicle in the center, and an obstacle being detected in the upper right-hand corner; (**b**) costmap with LiDAR readings; (**c**) final behavioral costmap, with the perceptual schema increasing the costs in the area directly around the obstacle.

Behavior Manager motor schemas produce target goals, expressed as two-dimensional Cartesian coordinates **T**, based on the requirements of the behavior. For example, Figure 9 shows two different notional motor schema behaviors. Figure 9a represents a Stay-on-Road behavior. The target goal produced, represented by the star, causes the vehicle to stay on the road. Figure 9b on the other hand, represents a Follow-Leader behavior, with the leader being represented by a dog on the grass. The target goal produced directs the vehicle towards the leader. Each motor schema has a gain that it assigns to the goal that it produces. The goals and associated gains are utilized by the Behavior Manager to determine the final goal point for path planning.

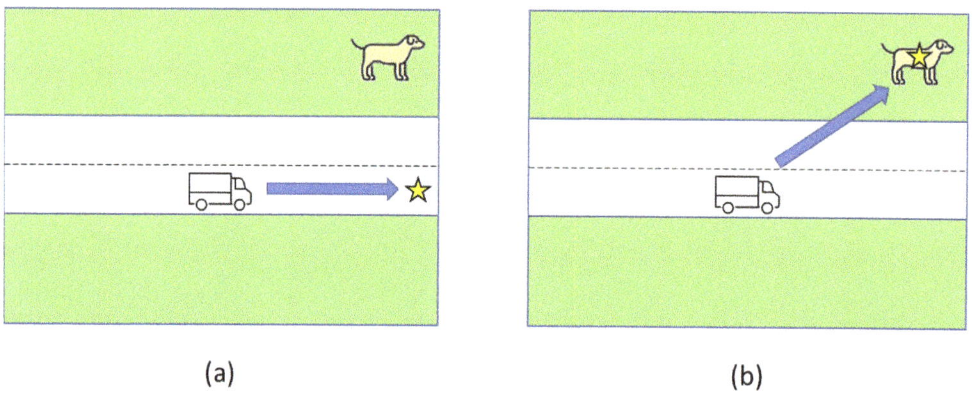

Figure 9. Two examples of motor schema behaviors: (**a**) Stay-of-Path; (**b**) Follow-Leader.

In the development of our autonomous convoy system, we created four behaviors to allow for leader following in the presence of jams: two motor schemas and two perceptual schemas. The behaviors are Move-to-Goal, Maintain-Formation, Avoid-Obstacle-Proximity, and Avoid-Leader-Zone, respectively.

Move-to-Goal

The Move-to-Goal motor schema behavior enables basic goal following. A high-level planner provides a coordinate **B** as an input to the Move-to-Goal behavior. The behavior then publishes the coordinates as the target goal. For our autonomous convoy following system, Move-to-Goal is used by the vehicles to move towards GPS breadcrumbs provided by their respective leaders for a PF convoy approach as follows:

$$\mathbf{T} = \mathbf{B}, \tag{1}$$

where new GPS breadcrumb **B** coordinates are provided periodically.

Maintain-Formation

The Maintain-Formation motor schema behavior mitigates the effects of jamming attacks on autonomous convoy following by generating target goals **T** that are not dependent on a leader's GPS coordinates. Instead, the behavior uses the density-based spatial clustering of applications with noise (DBSCAN) algorithm to find a target goal nearest to the last known valid goal and to continue convoy operations during jamming attacks.

DBSCAN is a non-parametric clustering algorithm that, when given a set of points, groups together points that are closely packed together into different clusters [41]. In the Maintain-Formation behavior, the points provided are from the behavioral costmaps generated by the Avoid-Obstacle-Proximity perceptual schema. For every cluster identified, the center of mass for each cluster is calculated for comparison to the last known valid **T** coordinates. To find the center of mass $\mathbf{C}_i$, where **C** are the coordinates of the center of mass for cluster I, and $i = 1, \ldots, n$ clusters, we considered each cluster as a system of weighted particles P_j, where $j = 1, \ldots, n$. The coordinates for each particle were given by the costmap cell coordinates c_j, with the mass m_j for each particle being defined as the cost for each given coordinate. The formula for $\mathbf{C}_i$ is as follows:

$$\mathbf{C}_i = \frac{1}{M_i} \sum_{j=1}^{n} m_j c_j, \tag{2}$$

where M_i is the total cost of all the cluster points in cluster i, given by:

$$M_i = \sum_{j=1}^{n} m_j. \tag{3}$$

When the autonomous vehicle fails to receive a GPS breadcrumb coordinate **B**, the Behavior Manager determines that a denial-of-service attack is occurring and that communications with the rest of the convoy have been disrupted. The Maintain-Formation behavior will then find the $\mathbf{C}_i$ nearest to the last valid target goal **T**, and set that as **T**. This process is repeated until a new valid **B** is provided, as shown in Algorithm 1.

Algorithm 1 Maintain-Formation

```
if no B received from leader
    C_i = center of mass of clusters 1, . . . , n from DBSCAN(costmap)
    temp_distance = high value placeholder
    for i = 1, . . . , n
        cluster_distance = distance between T and C_i
        if cluster_distance < temp_distance
            T = C_i
            temp_distance = cluster_distance
        end if
    next i
end if
```

Avoid-Obstacle-Proximity

The Avoid-Obstacle-Proximity perceptual schema behavior produces a behavioral costmap that includes costs for objects detected by LiDAR, along with a buffer zone of costs around the objects, in which the cost of the cell decreases as the distance from the object increases, up to a preset radius. The implementation of the buffer zone leverages the inflation layer provided by the ROS Navigation stack [42], which assigns a cost around objects with the following formula:

$$cost = e^{(-1 \times s \times (d-r))}(o-1), \tag{4}$$

where s is a cost scaling factor, d is the distance from the obstacle, r is the robot's radius, and o is the cost to assign a cell that falls within the robot's radius. Refer to Figure 10 for an example of how the buffer zone inflation is represented on a costmap. The buffer zone provided by the Avoid-Obstacle-Proximity behavior helps to prevent the vehicle from getting too close to obstacles, decreasing the chances of collisions.

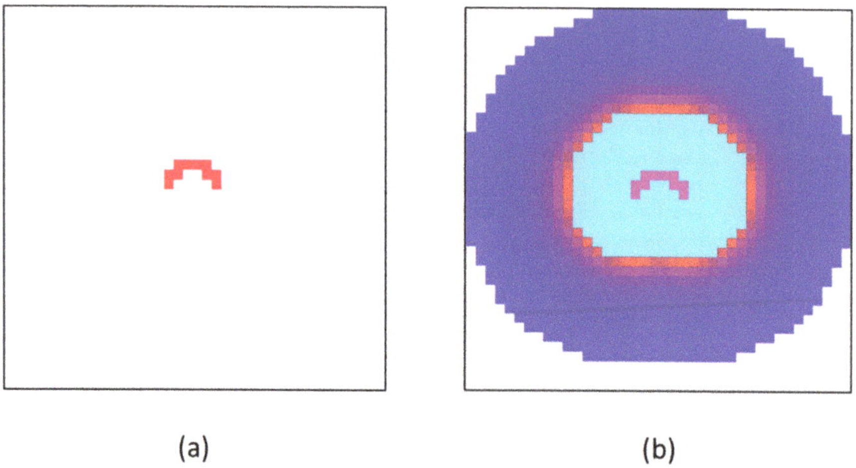

Figure 10. Costmap inflation in the Avoid-Obstacle-Proximity behavior: (**a**) the costmap produced from LiDAR sensing a construction cone; (**b**) the costmap produced from creating a buffer zone of inflated costs around the construction cone.

Avoid-Leader-Zone

The Avoid-Leader-Zone perceptual schema behavior produces a behavioral costmap that inserts a high-cost ring of a preset radius around a robot's leader. This ring creates a zone around the leader that the follower vehicles will not enter, in effect enforcing a following distance between the vehicles. To do this, the behavior takes the position of its leader as an input. If network communications are available, the position is provided wirelessly by the lead vehicle. If the convoy is under a jamming attack, the target goal **T**, as provided by the Maintain-Formation behavior is used instead. The costmap coordinates that comprise the ring are calculated using basic trigonometric functions to obtaining coordinates of a scaled unit circle with an offset center, as follows:

$$x_i = floor\left(\frac{a}{f}\cos\cos\left(\frac{i}{n}2\pi\right)\right) + l_x, \tag{5}$$

$$y_i = floor\left(\frac{a}{f}\sin\sin\left(\frac{i}{n}2\pi\right)\right) + l_y, \tag{6}$$

where $[x_i, y_i]$ are costmap indices for each point in the ring; $i = 1, \ldots, n$ costmap cells, with n representing the number of total number costmap cells the ring should contain; a is the radius of the leader's following distance zone; f is the costmap cell resolution; and $[l_x, l_y]$ is the costmap index of the leader. Figure 11 shows a robot and the leader zone produced by the Avoid-Leader-Zone behavior.

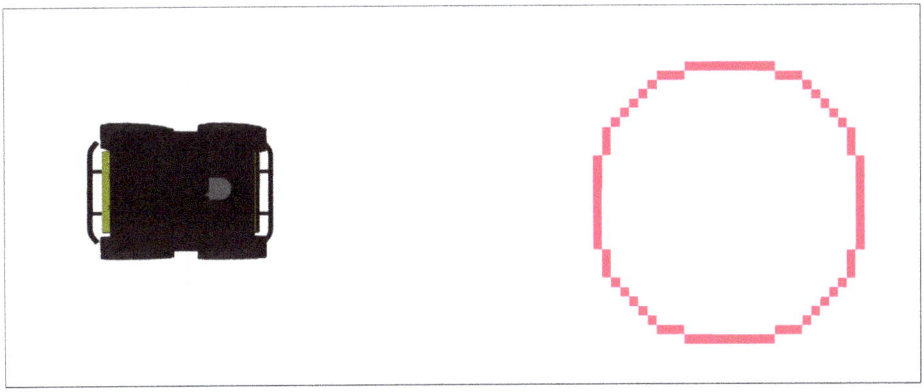

Figure 11. Behavioral costmap created by the Avoid-Leader-Zone perceptual schema behavior.

3.3.3. Assemblages

At a basic level, assemblages are the complex tasks that the robot is trying to accomplish. The assemblages are composed of multiple concurrently running behaviors. The behaviors can be classified as motor schemas, which produce goals to generate motor responses, or perceptual schemas, which generate behavioral costmaps. The behavioral costmaps are used by the motor schema behaviors in goal generation and by the assemblage for navigation. The assemblage acts as a coordinator of the behaviors, weighing and combining them as appropriate for the task at hand. Figure 12 shows the architecture of an assemblage.

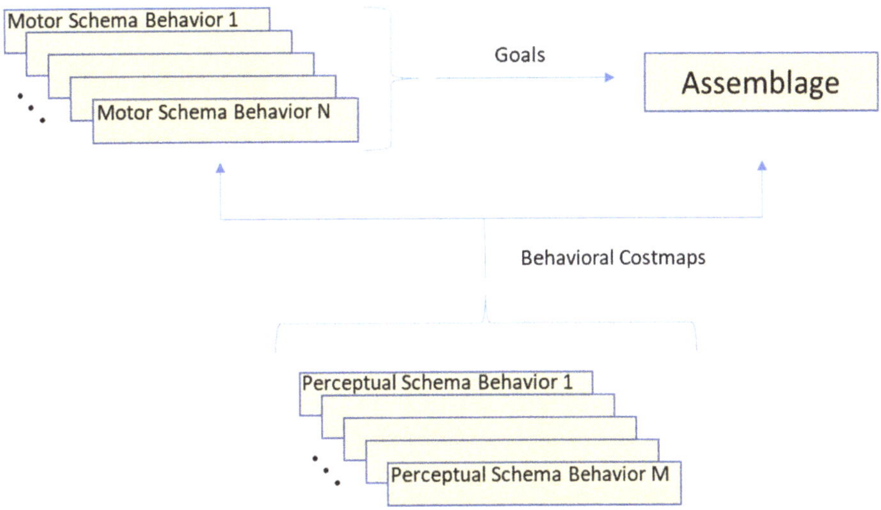

Figure 12. Architecture of an assemblage.

For our effort, we utilized the behaviors described above and developed a Follow-Leader-with-Jam-Mitigation (FLJM) assemblage that performs autonomous convoying with jam mitigation techniques. The architecture of the assemblage and the associated behaviors can be seen in Figure 13.

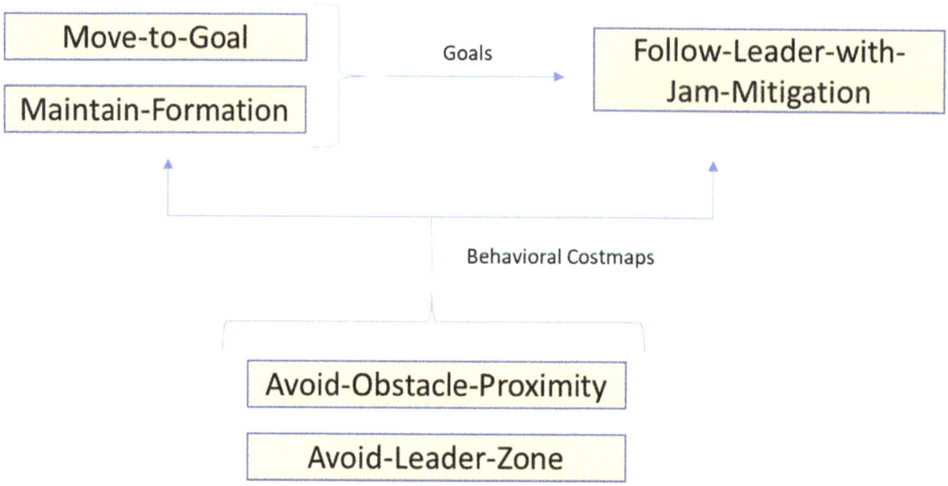

Figure 13. Architecture of the Follow-Leader-with-Jam-Mitigation assemblage.

The basic task of the FLJM assemblage is to perform leader following. An autonomous vehicle running the assemblage will take an assigned leader and await waypoints from them to follow. If the vehicle's network communications are unavailable, either due to a system failure or a denial-of-service attack, it will continue following the moving cluster of costmap points nearest to the last valid waypoint. Once network communications and waypoints are available again, the assemblage will revert to following waypoints. As with all assemblages, the FLJM assemblage is responsible for goal and behavioral costmap combination/selection for all the behaviors. The costmap gains for Avoid-Obstacle-Proximity and Avoid-Leader-Zone are set to be equal, meaning that combining their behavioral costmaps is conducted through addition. Essentially, the costmaps are stacked on top of each other and the sum cost of every cell forms the master costmap. For goal combination, a selection method is utilized. If the system detects valid waypoints from network communications, the **T** provided by Move-to-Goal is used, allowing for waypoint following. If no network communications are received, the **T** provided by Maintain-Formation is used until network communications are restored and waypoints are provided again. Figure 14 characterizes the interactions between behaviors and waypoint transmissions through the network.

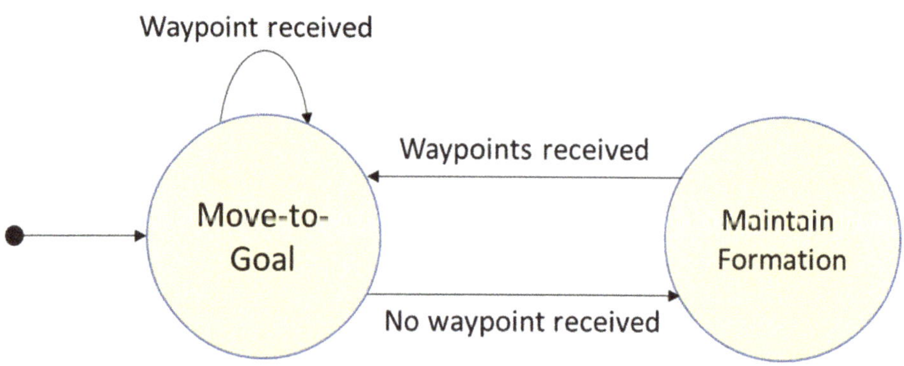

Figure 14. Architecture of the Follow-Leader-with-Jam-Mitigation assemblage.

3.3.4. Path Planning

A key difference between the Behavior Manager and the traditional Motor Schema approach is how motor responses are generated and acted upon. While the Motor Schema approach uses potential field response vectors and navigation, the Behavior Manager leverages layered costmaps with Vector Field Histogram (VFH) path planning. VFH is a path planning technique that was developed specifically to address the inherent limitations of potential field methods, such as U-shaped obstacle traps, oscillation in narrow corridors, or the inability to pass between closely spaced objects [43]. The VFH method represents the obstacles around a robot with certainty values in a Cartesian histogram grid. Data reduction over the histogram grid reduces it to a polar histogram of angular sectors around the robot with associated polar obstacle densities. Based on the desired goal and polar obstacle densities, drive and steering commands are given that best avoid obstacles while making progress towards the goal [44]. VFH was chosen as our path planning technique because of the aforementioned histogram grid. By using the combined layered behavioral costmap as the histogram grid, we naturally integrated costmap-based perceptual schemas with path planning, thereby avoiding potential field methods and their pitfalls.

3.4. Experimental Setup

To test the efficacy of the FLJM assemblage, we compared its performance to that of a basic convoy controller in a Gazebo simulation environment. The Gazebo world used in the experiment contained a flat ground pane without terrain or obstacles and a three-vehicle convoy of Cleapath Husky robots, a four-wheel skid-steer robotic vehicle. The setup was kept minimal to reduce the variables present when testing convoy performance. For the basic convoy controller, we implemented a leader-follower controller that performed "delayed following" with a predecessor-leader following network topology. In a "delayed following" approach, a lead vehicle records the path it drove and relays that information to followers to repeat after some time delay [45]. This approach is recognized as a common leader-follower convoying method [46], making it the ideal baseline for comparison.

The high-level autonomous convoy setup was kept consistent between the basic convoy controller and the FLJM assemblage for an accurate comparison. In both cases, the autonomous convoy was FA and consisted of three vehicles: one convoy leader that drove along preset path plans, and two autonomous followers. The convoy leader precisely followed through given path points, allowing for repeatability in test runs, resulting in better comparisons between the basic convoy controller and the FLJM assemblage. Both unmanned followers would be running either the basic convoy controller or FLJM assemblage to perform autonomous following, depending on which system was being tested at the time. Each vehicle recorded GPS breadcrumbs of its position that it transmitted to its direct follower when wireless network communications were available.

The two preset path plans created for the test runs were a square loop and a roundabout turn, as seen in Figure 15. These paths were chosen to mimic common road formations and driving maneuvers. The square loop tests basic turns, while the roundabout tests the intersections of the same name. In addition, jamming zones were established along the paths to simulate the effects of denial-of-service attacks on the autonomous convoy. Figure 16 overlays the jamming zones along the paths.

To comparatively test the performance of the convoy controllers, multiple test runs with the three vehicle convoy were performed for each jammed path shown in Figure 16. Ahead of each test run, the autonomous followers were configured to use either the basic convoy controller or the FLJM assemblage. Five runs were performed under constant jamming, and an additional five runs were performed under random jamming, with t_J, and t_S set to 10 s and 2 s, respectively. Both the basic convoy controller and the FLJM assemblage were tested in this fashion for each path plan.

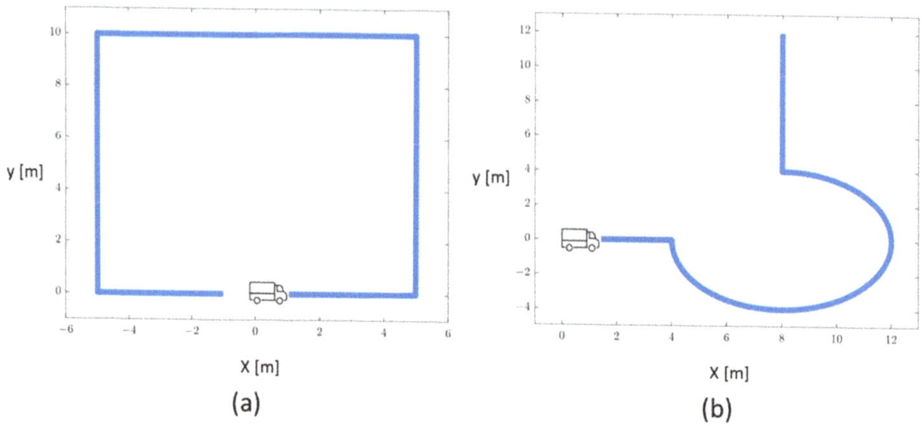

Figure 15. The preset path plans for the lead vehicle: (**a**) square loop and (**b**) roundabout.

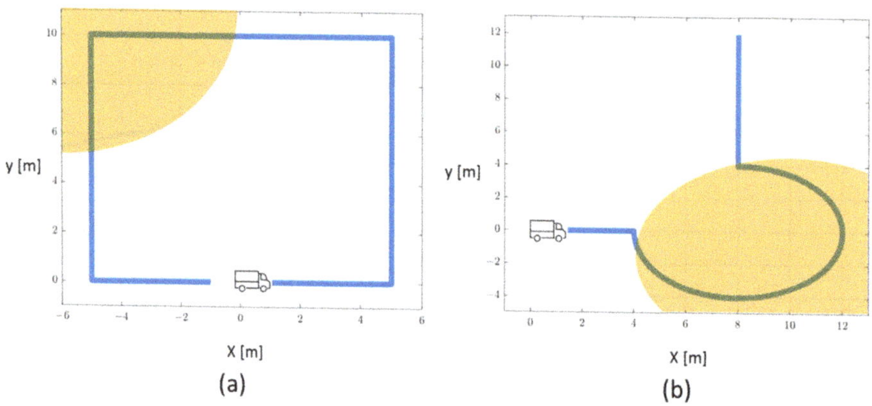

Figure 16. Jamming zones overlaid onto the preset plans for the lead vehicle: (**a**) square loop and (**b**) roundabout.

4. Results

To measure the performance of the convoy controllers, we calculated the mean absolute error (MAE) between the follower vehicles' path and the convoy leader's path. The MAE was chosen as the measurement for evaluation as it has a direct interpretation to the real-world quantities being compared, namely the Euclidian distances between follower points and convoy leader points, in addition to not inflating the penalty for larger errors via squaring, which occurs in other commonly used performance measurements such as root mean squared error [47]. The paths of both the followers and the convoy leader were sampled at a rate of 1000 Hz, respectively creating arrays of position points for each vehicle. The Euclidean distance between the follower position points and corresponding convoy leader positions points were then used to calculate the MAE with the following formula:

$$\text{MAE} = \frac{\sum_{i=1}^{n} |e_i|}{n}, \tag{7}$$

where e is the Euclidean distance between the follower and convoy leader position points, and $i = 1, \ldots, n$, where n is the size of the array e. Table 3 shows the average MAE over five runs for each convoy controller for both the square path and roundabout path when under a constant jamming attack. Figure 17 shows the paths of the five runs taken by the followers, overlaid on the convoy leader's goal points for a visual comparison of performance.

Table 3. Average MAE under a constant jamming attack.

Convoy Configuration	Square (m)	Roundabout (m)
Follower 1 Using Basic Convoy Controller	1.2665	3.1334
Follower 2 Using Basic Convoy Controller	1.9226	3.5205
Follower 1 Using FLJM assemblage	0.4942	0.4197
Follower 2 Using FLJM assemblage	0.7321	0.8452

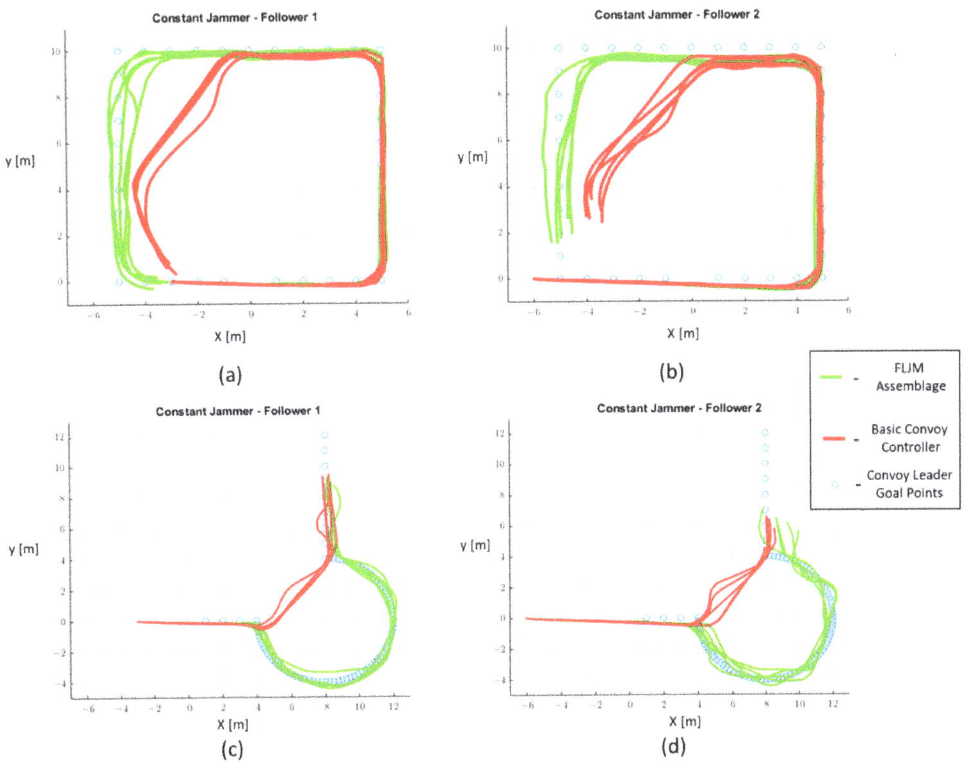

Figure 17. Plot of the paths taken by the followers under constant jamming on the square path for (**a**) follower 1 and (**b**) follower 2; and on the roundabout path for (**c**) follower 1 and (**d**) follower 2.

Likewise, Table 4 shows the average MAE while under a random jamming attack with t_J, and t_S set to 10 s and 2 s, while Figure 18 shows the paths taken for the five runs by the followers, overlaid on the convoy leader's goal points for a visual comparison of performance.

Table 4. Average MAE under a random jamming attack.

Convoy Configuration	Square (m)	Roundabout (m)
Follower 1 Using Basic Convoy Controller	0.6844	1.2637
Follower 2 Using Basic Convoy Controller	1.0028	1.0697
Follower 1 Using FLJM assemblage	0.4680	0.4821
Follower 2 Using FLJM assemblage	0.7654	0.9271

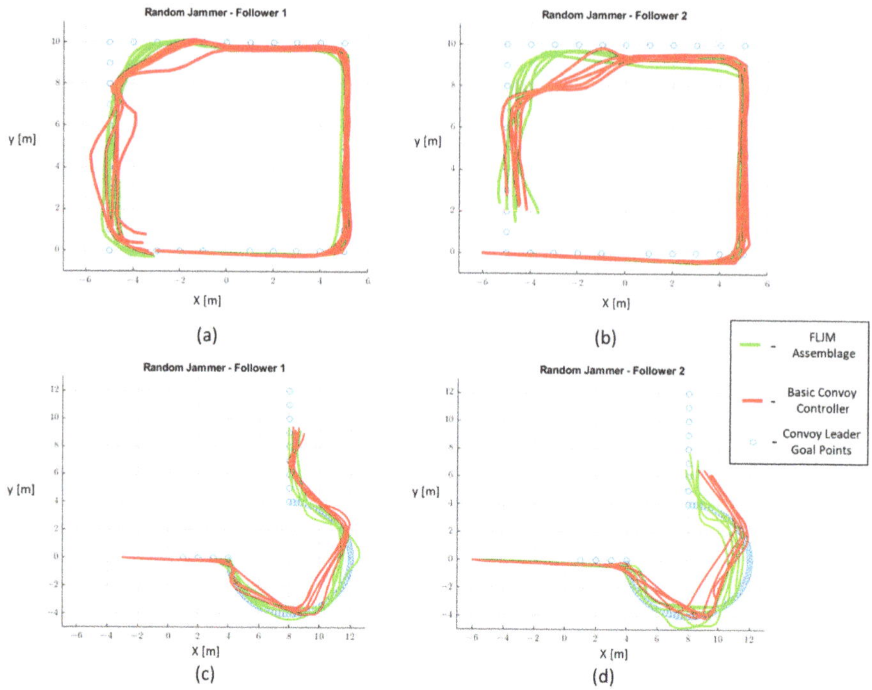

Figure 18. Plot of the paths taken by the followers under random jamming on the square path for (**a**) follower 1 and (**b**) follower 2; and on the roundabout path for (**c**) follower 1 and (**d**) follower 2.

As seen in Table 3, the utilization of the FLJM assemblage significantly improved convoy performance across both path plans when compared to the basic convoy controller under constant jamming. When following the squared loop path, using the FLJM assemblage decreased the average MAE by 60.98% for follower 1 and 61.92% for follower 2. The improvement was even more drastic with the roundabout path, where the average MAE was decreased by 86.61% and 75.99% for follower 1 and follower 2, respectively. These significant improvements were due to the basic convoy controller missing turns that the convoy leader performed in the jamming zones laid out in yellow in Figure 16. As seen in Figure 17, the FLJM assemblage allowed the followers to successfully continue leader-following inside the constant jamming zones, whereas the stoppage of communications prevented the followers from getting proper waypoints when relying solely on basic delayed following.

As seen in Table 4, the utilization of the FLJM assemblage also improved convoy performance across both path plans when compared to the basic convoy controller under random jamming. When following the squared loop path, using the FLJM assemblage decreased the average MAE by 31.62% for follower 1 and 23.67% for follower 2. Improvement was exhibited with the roundabout path as well, where the average MAE was decreased by 61.85% and 13.33% for follower 1 and follower 2, respectively.

While the utilization of the FLJM assemblage yielded similar performance across both the constant and random jammers, the performance improvements when compared to the basic convoy controller were less significant under random jamming attacks. This is due to the nature of random jammers, which alternate between a jamming and sleeping state. Rather than missing all the convoy leader's waypoints in a jamming zone, as in the case of constant jamming, the follower vehicles would periodically receive network transmissions from the jamming zone under random jamming, allowing for occasional transmission of waypoints inside the jammed area. This can be seen when comparing the basic convoy controller paths in Figures 17 and 18. In Figure 17, the basic convoy controller followers stop outside of the jamming zones and only restart following when the convoy leader exits the constant jamming zone. In Figure 18 however, the followers are able to break into the random jamming zone and follow the occasional waypoints they receive when the jammer is sleeping. These results demonstrate that the value of using jamming mitigation techniques increases accordingly with the severity of the jamming attack. As previously stated, constant jamming is considered the worst-case jamming scenario with the most damaging impact, so the potential benefits afforded by using the FLJM assemblage is greater when compared to a random jamming attack, as seen in the experimental results described here.

While the results described here were focused on the constant and random jammer attack model developed for this effort, the performance benefits provided by the FLJM assemblage would be realized even when applied to other attacker models. Various other attacker models have focused on reactive jammers with assumptions of global eavesdropping, zero-delay jamming startup, and unbounded frequency jamming [9,48]. With these attacker models, the FLJM assemblage performance would match that of our constant jammer, given that the reactive jamming occurs instantaneously when entering the jammed areas and disrupts all communications, which closely aligns with our model for constant jamming. Other attacker models have additional parameters for their reactive jammers, such as jamming range, triggering range, and multiple jammer nodes, but they make the same uniform power and omnidirectionality of jamming assumptions we made in our model, resulting in the same cylindrical jamming zones [28]. While the notion of multiple jamming zones introduces coverage areas of different shapes due to overlapping cylinders, the performance improvements from using the FLJM assemblage will remain due to the reactive jammer closely aligning with our constant jamming model. In addition, other attacker models have focused on more dynamic jamming scenarios, such as the jammer being embedded in one of the convoy's follower vehicles [7]. In this attacker model, a follower vehicle in a convoy would switch from a standard network node into a uniform power mobile jammer that prevented messages from arriving at their destination. Assuming the mobile jammer continues following its leader to disguise its adversarial nature, the FLJM assemblage would once again allow for continued convoy operations. In this scenario, the jammed vehicles would remain in the Maintain-Formation behavior for as long as the mobile jammer was activated and switch back to Move-to-Goal upon the deactivation of the jammer. Convoy operations would continue for all vehicles in the convoy. As described in these examples, the benefits of utilizing the FLJM assemblage are generally applicable and has the potential to improve the performance and robustness of autonomous vehicle convoys against a wide range of jamming attacks.

5. Discussion

Jamming attacks are a simple, yet effective type of denial-of-service attack that have the potential to adversely degrade the performance of autonomous convoys. While many efforts have been undertaken to detect the presence of jams, we focused our efforts on finding ways to mitigate a jammer's effects to allow for continued convoy operation. The prevalence of behavior-based robotics approaches in multi-vehicle teaming made it a prime starting point for developing an approach to allow convoy following when confronted with jamming, and lead to the development of the Behavior Manager.

By utilizing the Behavior Manager and creating behaviors and assemblages to mitigate the effects of jamming, we were able to show improved convoy performance with up to an 86.61% reduction in average MAE. This demonstrates that a behavior-based robotics approach can be leveraged to dramatically improve the robustness of autonomous convoys when faced with jamming attacks. Furthermore, the modular nature of the Behavior Manager means that the capabilities can be extended to mitigate the effects of other types of attacks, such as LiDAR spoofing, replay attacks, and RADAR absorption [49], through the development of additional behaviors and assemblages.

A potential future area of research on the usage of behavior-based robotics towards mitigating the effects of attacks is dynamic behavioral weighting. While the usage of Q-learning towards behavior selection has been researched in the past [50], the application of modern learning techniques, such as deep reinforcement learning, towards multi-robot attack mitigation is a novel area that warrants further investigation. By training on datasets that demonstrate human reactions to attacks on manned convoys, the robustness of autonomous convoy systems utilizing behavior-based robotics could be greatly improved on, creating even more stable systems able to handle a broad spectrum of attacks.

Author Contributions: Conceptualization, methodology, software, validation, formal analysis, investigation, resources, data curation, writing—original draft preparation, visualization, project administration: C.C.; writing—review and editing, supervision, funding acquisition: A.M. and S.R. All authors have read and agreed to the published version of the manuscript.

Funding: This research received no external funding.

Institutional Review Board Statement: Not applicable.

Informed Consent Statement: Not applicable.

Data Availability Statement: The data presented in this study are available on request from the corresponding author.

Conflicts of Interest: The authors declare no conflict of interest.

References

1. Murphy, T.J. *Convoy Operations in Afghanistan Handbook*; U. S. Army Combined Arms Center: Fort Leavenworth, KS, USA, 2010.
2. Cheung, C.; Mohammadi, A.; Rawashdeh, S.; Baek, S. Delivery of Healthcare Resources Using Autonomous Ground Vehicle Convoy Systems: An Overview. *Front. Robot. AI* **2021**, *8*, 250. [CrossRef]
3. Nahavandi, S.; Mohamed, S.; Hossain, I.; Nahavandi, D.; Salaken, S.M.; Rokonuzzaman, M.; Ayoub, R.; Smith, R. Autonomous Convoying: A Survey on Current Research and Development. *IEEE Access* **2022**, *10*, 13663–13683. [CrossRef]
4. McKay, S.; Boyer, M.E.; Beyene, N.M. *Automating Army Convoys: Technical and Tactical Risks and Opportunities*; RAND: Santa Monica, CA, USA, 2020.
5. Llatser, I.; Festag, A.; Fettweis, G. Vehicular Communication Performance in Convoys of Automated Vehicles. In Proceedings of the 2016 IEEE International Conference on Communications (ICC), Kuala Lumpur, Malaysia, 22–27 May 2016.
6. Lopes, H.J.; Lima, D.A. Cellular Automata in path planning navigation control applied in surveillance task using the e-Puck architecture. In Proceedings of the 2020 IEEE International Conference on Systems, Man, and Cybernetics (SMC), Toronto, ON, Canada, 11–14 October 2020.
7. van der Heijden, R.; Lukaseder, T.; Kargl, F. Analyzing Attacks on Cooperative Adaptive Cruise Control (CACC). In Proceedings of the 2017 IEEE Vehicular Networking Conference (VNC), Turin, Italy, 27–29 November 2017.
8. Malebary, S. Real-Time Jamming Detection in Vehicular Network. *Int. J. Sci. Res. Innov. Technol.* **2016**, *3*, 159–166.
9. Sciancalepore, S.; Di Pietro, R. Bittransfer: Mitigating Reactive Jamming in Electronic Warfare Scenarios. *IEEE Access* **2019**, *7*, 156175–156190. [CrossRef]
10. Xu, W.; Trappe, W.; Zhang, Y.; Wood, T. The Feasibility of Launching and Detecting Jamming Attacks in Wireless Networks. In Proceedings of the 6th ACM International Symposium on Mobile ad Hoc Networking and Computing-MobiHoc '05, Champaign, IL, USA, 25–27 May 2005.
11. Alturkostani, H.; Chitrakar, A.; Rinker, R.; Krings, A. On the Design of Jamming-Aware Safety Applications in VANETs. In Proceedings of the 10th Annual Cyber and Information Security Research Conference, Oak Ridge, TN, USA, 7–9 April 2015.
12. Serageldin, A.; Alturkostani, H.; Krings, A. On the Reliability of DSRC Safety Applications: A Case of Jamming. In Proceedings of the 2013 International Conference on Connected Vehicles and Expo (ICCVE), Las Vegas, NV, USA, 2–6 December 2013.
13. Serageldin, A.; Krings, A. The Impact of Dissimilarity and Redundancy on the Reliability of DSRC Safety Applications. In Proceedings of the 2014 28th International Conference on Advanced Information Networking and Applications Workshops, Victoria, BC, Canada, 13–16 May 2014.
14. Hu, Y.; Shan, H.; Dutta, R.G.; Jin, Y. Protecting Platoons from Stealthy Jamming Attack. In Proceedings of the 2020 Asian Hardware Oriented Security and Trust Symposium (AsianHOST), Kolkata, India, 15–17 December 2020.
15. Gong, T.; Yu, Y.; Song, J. Path Planning for Multiple Unmanned Vehicles (Muvs) Formation Shape Generation Based on Dual RRT Optimization. *Actuators* **2022**, *11*, 190. [CrossRef]
16. Yunhe, L.; Bo, L.; Xiaowei, N. Research on Self-Organizing Behavior-Based UAV Formation Based on Distributed Control. In Proceedings of the 3rd International Conference on Vision, Image and Signal Processing, Vancouver, BC, Canada, 26–28 August 2019.

17. Jond, H.B.; Sadreddini, Z.; Platoš, J. Autonomous Vehicle Convoy Formation Control with Size/Shape Switching for Automated Highways. *Int. J. Eng.* **2020**, *33*, 2174–2180.
18. Balch, T.; Arkin, R.C. Behavior-Based Formation Control for Multirobot Teams. *IEEE Trans. Robot. Autom.* **1998**, *14*, 926–939. [CrossRef]
19. Lee, G.; Chwa, D. Decentralized Behavior-Based Formation Control of Multiple Robots Considering Obstacle Avoidance. *Intell. Serv. Robot.* **2017**, *11*, 127–138. [CrossRef]
20. Pierpaoli, P.; Li, A.; Srinivasan, M.; Cai, X.; Coogan, S.; Egerstedt, M. A Sequential Composition Framework for Coordinating Multirobot Behaviors. *IEEE Trans. Robot.* **2021**, *37*, 864–876. [CrossRef]
21. Zheng, Y.; Eben Li, S.; Wang, J.; Cao, D.; Li, K. Stability and Scalability of Homogeneous Vehicular Platoon: Study on the Influence of Information Flow Topologies. *IEEE Trans. Intell. Transp. Syst.* **2016**, *17*, 14–26. [CrossRef]
22. Pelechrinis, K.; Iliofotou, M.; Krishnamurthy, S.V. Denial of Service Attacks in Wireless Networks: The Case of Jammers. *IEEE Commun. Surv. Tutor.* **2011**, *13*, 245–257. [CrossRef]
23. Van Veen, B.D.; Buckley, K.M. Beamforming: A Versatile Approach to Spatial Filtering. *IEEE ASSP Mag.* **1988**, *5*, 4–24. [CrossRef]
24. Luo, G.Y. On-Line Wavelet Filtering of Narrowband Noise in Signal Detection of Spread Spectrum System for Location Tracking. *Int. J. Commun. Syst.* **2011**, *25*, 598–615. [CrossRef]
25. Popper, C.; Strasser, M.; Capkun, S. Anti-Jamming Broadcast Communication Using Uncoordinated Spread Spectrum Techniques. *IEEE J. Sel. Areas Commun.* **2010**, *28*, 703–715. [CrossRef]
26. Poisel, R. *Modern Communications Jamming Principles and Techniques*; Artech House: Boston, MA, USA, 2011.
27. Liu, Y.; Ning, P. Bittrickle: Defending against Broadband and High-Power Reactive Jamming Attacks. In Proceedings of the 2012 Proceedings IEEE INFOCOM, Orlando, FL, USA, 25–30 March 2012.
28. Xuan, Y.; Shen, Y.; Nguyen, N.P.; Thai, M.T. A Trigger Identification Service for Defending Reactive Jammers in WSN. *IEEE Trans. Mob. Comput.* **2012**, *11*, 793–806. [CrossRef]
29. Noubir, G. On Connectivity in Ad Hoc Networks under Jamming Using Directional Antennas and Mobility. In *Lecture Notes in Computer Science*; Springer: Berlin/Heidelberg, Germany, 2004; pp. 186–200.
30. Brooks, R. A Robust Layered Control System for a Mobile Robot. *IEEE J. Robot. Autom.* **1986**, *2*, 14–23. [CrossRef]
31. Siciliano, B.; Khatib, O. (Eds.) *Springer Handbook of Robotics*; Springer International Publishing: Cham, Switzerland, 2016.
32. Arkin, R.C. *Behavior-Based Robotics*; MIT Press: Cambridge, MA, USA, 2000.
33. Arkin, R.C. Motor Schema—Based Mobile Robot Navigation. *Int. J. Robot. Res.* **1989**, *8*, 92–112. [CrossRef]
34. Kaelbling, L.P. An Architecture for Intelligent Reactive Systems. *Reason. About Actions Plans* **1987**, 395–410. [CrossRef]
35. Quigley, M.; Conley, K.; Gerkey, B.; Faust, J.; Foote, T.; Leibs, J.; Wheeler, R.; Ng, A. ROS: An open-source Robot Operating System. In Proceedings of the IEEE International Conference on Robotics and Automation (ICRA) Workshop on Open Source Robotics, Kobe, Japan, 12–17 May 2009.
36. Koenig, N.; Howard, A. Design and Use Paradigms for Gazebo, an Open-Source Multi-Robot Simulator. In Proceedings of the 2004 IEEE/RSJ International Conference on Intelligent Robots and Systems (IROS) (IEEE Cat. No.04CH37566), Sendai, Japan, 28 September–2 October 2004.
37. Clearpath Robotics. Robots/Husky-ROS. Available online: http://wiki.ros.org/Robots/Husky (accessed on 10 June 2022).
38. Clearpath Robotics. Husky/husky: Common Packages for the Clearpath Husky. Available online: https://github.com/husky/husky (accessed on 10 June 2022).
39. Soni, A.; Hu, H. Formation Control for a Fleet of Autonomous Ground Vehicles: A Survey. *Robotics* **2018**, *7*, 67. [CrossRef]
40. Lu, D.V.; Hershberger, D.; Smart, W.D. Layered Costmaps for Context-Sensitive Navigation. In Proceedings of the 2014 IEEE/RSJ International Conference on Intelligent Robots and Systems, Chicago, IL, USA, 14–18 September 2014.
41. Ester, M.; Kriegel, H.-P.; Sander, J.; Xu, X. A density-based algorithm for discovering clusters in large spatial databases with noise. In Proceedings of the Second International Conference on Knowledge Discovery and Data Mining, Portland, OR, USA, 2–4 August 1996; pp. 226–231.
42. Macenski, S. costmap_2d/hydro/inflation. Available online: http://wiki.ros.org/costmap_2d/hydro/inflation (accessed on 12 July 2022).
43. Koren, Y.; Borenstein, J. Potential Field Methods and Their Inherent Limitations for Mobile Robot Navigation. In Proceedings of the 1991 IEEE International Conference on Robotics and Automation, Sacramento, CA, USA, 9–11 April 1991.
44. Borenstein, J.; Koren, Y. The Vector Field Histogram-Fast Obstacle Avoidance for Mobile Robots. *IEEE Trans. Robot. Autom.* **1991**, *7*, 278–288. [CrossRef]
45. Vasseur, L.; Lecointe, O.; Dento, J.; Cherfaoui, N.; Marion, V.; Morillon, J.G. Leader-Follower Function for Autonomous Military Convoys. *SPIE Proc.* **2004**, *5422*, 326–337.
46. Zhao, X.; Yao, W.; Li, N.; Wang, Y. Design of Leader's Path Following System for Multi-Vehicle Autonomous Convoy. In Proceedings of the 2017 IEEE International Conference on Unmanned Systems (ICUS), Beijing, China, 27–29 October 2017.
47. Li, X.R.; Zhao, Z. Measures of Performance for Evaluation of Estimators and Filters. *SPIE Proc.* **2001**, *4473*, 530–541.
48. Pietro, R.D.; Oligeri, G. Silence Is Golden: Exploiting Jamming and Radio Silence to Communicate. *ACM Trans. Inf. Syst. Secur.* **2015**, *17*, 1–24. [CrossRef]

49. Chowdhury, A.; Karmakar, G.; Kamruzzaman, J.; Jolfaei, A.; Das, R. Attacks on Self-Driving Cars and Their Countermeasures: A Survey. *IEEE Access* **2020**, *8*, 207308–207342. [CrossRef]
50. Martinson, E.; Stoytchev, A.; Arkin, R. Robot Behavioral Selection Using Q-Learning. In Proceedings of the IEEE/RSJ International Conference on Intelligent Robots and System, Lausanne, Switzerland, 30 September–4 October 2002.

applied sciences

MDPI

Article

Application and Assessment of Cooperative Localization in Three-Dimensional Vehicle Networks

Juan Carlos Oliveros † and Hashem Ashrafiuon *,†

Center for Nonlinear Dynamics and Control, Department of Mechanical Engineering, Villanova University, 800 Lancaster Ave, Villanova, PA 19085, USA

* Correspondence: hashem.ashrafiuon@villanova.edu

† These authors contributed equally to this work.

Abstract: The trajectory planning and control of multi-agent systems requires accurate localization, which may not be possible when GPS signals and fixed features required for SLAM are not available. Cooperative Localization (CL) in multi-agent systems offers a short-term solution that may significantly improve vehicle pose estimation. CL algorithms have been mainly developed and assessed for planar mobile robot networks due to complexities and singularities in three-dimensional (3D) motion. In this paper, we develop the required singularity-free equations and apply and assess an EKF-based CL for 3D vehicle networks. We assess the performance of CL with respect to the number of simultaneous and redundant measurements. We further assess CL performance with only relative position measurements available. Finally, experiments are performed to validate the proposed algorithms. We further investigate the effect of absolute position measurements in CL. **Conclusions:** Cooperative localization is an effective method when applied to 3D vehicle networks. However, CL does not improve localization with only relative position measurements, and thus previously reported results for 2D vehicle models were only effective due to relative orientation measurements. Absolute measurement reduces the overall localization errors much more significantly when there has been CL with prior relative position measurements.

Keywords: cooperative localization; extended Kalman filter; multi-robot navigation; three-dimensional motion; state estimation; autonomous vehicles

Citation: Oliveros, J.C.; Ashrafiuon, H. Application and Assessment of Cooperative Localization in Three-Dimensional Vehicle Networks. *Appl. Sci.* **2022**, *12*, 11805. https://doi.org/10.3390/app122211805

Academic Editors: Nadjim Horri, William Holderbaum and Fabrizio Giulietti

Received: 27 October 2022

Accepted: 16 November 2022

Published: 20 November 2022

1. Introduction

One of the vital technologies required for autonomous navigation is a localization technique that is capable of recognizing the position and orientation of a vehicle or vehicle networks, especially in GPS-denied environments. This problem is further exacerbated by environments that are often unstructured and lack distinct features (such as deserts), contain features that may change over time due to environment, or as the vehicles move to new areas (such as urban settings). Methods such as feature-based Simultaneous Localization and Mapping (SLAM) [1] are not capable of providing accurate localization in these environments. Collaborative systems of unmanned ground and air vehicles show great promise for various applications [2,3]. Cooperative Localization (CL) can be utilized for these systems to significantly reduce the rate of growth of estimation errors until GPS becomes available again. In addition, CL can help reduce the sample rate requirement for various SLAM methods in structured environments, since these methods require extensive computation. Cooperative SLAM techniques [4,5] have shown great promise for precise localization, especially when combined with other techniques such as the Extended Kalman Filter (EKF) [6]. However, the success of these techniques relies on the knowledge of the environment.

CL can utilize several techniques, including Kalman filters [7–9] and particle filters [10,11]. Utilizing these techniques, CL has been suggested as a powerful method to improve pose estimation in vehicle networks capable of inter-agent communication for data sharing [12]. In an early application of CL, an EKF algorithm [13] was developed that distributes vehicle state information while keeping track of the cross-covariance terms. These terms are created by relative measurements of each 2D vehicle's pose. However, this method requires a completely synchronous and interconnected communication network. A decoupled CL algorithm is proposed [14] that eliminates the need for complete and constant between vehicles. Vehicles are assumed to be equipped with the capability of obtaining their own pose (via equations of motion and onboard sensors), measuring the poses of nearby vehicles relative to their own reference frame, and processing this information and communicating with other vehicles. CL allows a group of vehicles to cooperatively approximate an individual vehicle's pose by sharing the relative and absolute pose measurements that each vehicle has obtained.

2. Motivation

Research works on multi-agent systems have focused on 2D vehicle models, even treating vehicles as simple nodes [15]. CL methods have focused primarily on maneuvering in 2D space [16–19] and particularly using wheeled mobile robots [20–22] and vehicles [23] due to the complexities involved in the singularity-free representation of 3D orientation. It is, however, critically important to apply CL to three-dimensional (3D) vehicle models to improve the localization of networks with homogeneous and heterogeneous air, marine, and ground vehicles. Others only consider relative localization without any regard to the global reference frames [24,25]. There is also a CL method [26] that only uses range measurements of a vehicle with access to absolute measurements via GPS and therefore does not take advantage of cross-covariances terms. A common theme in these methods is suffering from singularities due to using Euler angles to represent orientation. EKF-based CL with relative pose measurements to improve the absolute pose of networks with 3D vehicles has remained relatively unexplored due to the intricacies involved in the kinematic representation and linearization of state and measurement equations. Alternative relative localization methods [27] have also been explored. Euler angles are normally used to measure orientation, but are associated with singularities in aggressive vehicle maneuvers with relative or absolute angles reaching 90°. Quaternions, on the other hand, do not have any singularities but are not measurable by any sensors. Hence, they are calculated from the rotation matrix, which may be challenging in the case of relative orientation representation.

3. Related Works

EKFs have commonly been used as the standard technique for state estimation in CL. There are EKF-based methods that achieve localization utilizing stereo-vision and sensor fusion [28] and can be further extended to target tracking [29]. However, these methods require maximizing the overlap of each agent's field of view to enhance localization. Many localization techniques rely on a central system or single/multiple "leader" agents responsible for sensor fusion [30–32]. Distributed or decentralized methods have been investigated [33–35] to ameliorate the computational bottlenecks associated with centralized systems. While these methods can be efficient and effective, they come at the cost of accuracy. A combination of centralized–decentralized CL in ad hoc networks shows great promise in improving localization without prior knowledge of an environment [14,36]. A distributed CL utilizing an EKF with low communication path requirements can also provide centralized-equivalent performance [37]. It is also possible to model the CL problem as a graph model to further utilize additional algorithms [38].

EKF-based CL algorithms are popular due to their ease of execution and computational efficiency. However, linearizing an extremely nonlinear system can lead to inaccuracies and numerical instabilities. Unscented Kalman Filter (UKF)-based methods were introduced in [39] with the advantage of not requiring any linearization. The UKF is deemed to be more robust [40] since it does not rely on the computation of the Jacobian matrix and is an excellent option for self-localization and tracking [41–44]. On the other hand, the UKF has a much higher computation cost compared to the EKF, especially considering the

Cholesky factorizations at every step. As an alternative, the Squared Root UKF [45] does help mitigate the computation cost. However, implementation of UKF-based CL algorithms and simultaneous measurements is very complex and has not been fully explored. Table 1 lists some of these related works along with their advantages and limitations.

Table 1. Limitations of related works.

Algorithm	Advantages	Limitations
SLAM	Obtain localization of multiple vehicles, mapping of an environment	Requires knowledge of the environment and designated landmarks, needs assigned "leaders" and "followers" or a central processing station
Particle Filter	Nonlinear applications, low error in ideal situations, no linearization required	Requires individual agent identification, not robust to false-positive detections
Decentralized EKF with Euler angles	No designated central processing center needed	Prone to singularity issues due to gimbal lock, requires linearization (Jacobian)
Current Method	Singularity-free, computationally effective	Quaternions cannot be measured directly, requires linearization

4. Main Contributions and Novelties

In this paper, we extend a previous conference proceedings publication [46] to include additional subcases, physical experiments, and identifying additional limitations to CL. We extend the centralized EKF-based CL algorithm presented in [13,14] to include 3D vehicle models and assess its performance. Specifically, we utilize quaternions to represent vehicle orientation in order to avoid singularities. For measurements, we introduce quaternions corresponding to relative orientation represented by the rotation matrix and present linearized measurement equations, i.e., the output matrix. We then assess the performance of CL in 3D vehicle networks through a series of Monte Carlo-type simulations. We use a sequential update procedure [47] to assess the effect of simultaneous and redundant measurements on individual vehicles, as well as the overall network. Assuming that absolute orientation can be accurately measured by onboard sensors, we further assess the performance of CL when only relative position measurements are available and investigate the circumstances under which the method could be effective or ineffective. Finally, experiments are preformed to verify the algorithm.

The main contributions and novelties of this publication are outlined as follows:

1. The development and application of singularity-free EKF-based CL for networks of three-dimensional vehicles;
2. The first comprehensive assessment of EKF-based CL with only relative position measurements;
3. The identification of limitations of CL with respect to the number of simultaneous measurements, redundant measurements, and relative position-only measurements;
4. The experimental verification of the EKF CL algorithm's performance and limitations.

The paper is organized as follows. The vehicle model is presented in Section 5. A multi-vehicle system with state and measurement equations and their linearization are presented in Section 6, followed by the CL algorithm in Section 7. Simulations and experiments are presented in Sections 8 and 9, respectively, followed by the concluding remarks in Section 10.

5. Vehicle Model

5.1. Vehicle Pose

Each vehicle is assigned a body-fixed reference frame $X_v Y_v Z_v$. Figure 1 depicts the pose of the vehicle, which is then represented by the position of the origin of $X_v Y_v Z_v$ and its orientation with respect to the inertial reference frame $X_0 Y_0 Z_0$. Hence, the vector $r_0^v = [x_0^v \;\; y_0^v \;\; z_0^v]^\top$ represents the position of the vehicle, and the orthogonal rotation matrix R_0^v that transforms the coordinates of any vector in $X_vY_vZ_v$ to $X_0Y_0Z_0$ determines its orientation. R_0^v can be calculated using quaternions, which are favored over Euler angles since they do not suffer from singularities and require less computational overhead. The quaternion vector of the vehicle reference frame $\bar{q}^v$ can be presented as:

$$\bar{q}^v = \begin{bmatrix} q_0^v \\ q^v \end{bmatrix}, \quad q^v = \begin{bmatrix} q_1^v \\ q_2^v \\ q_3^v \end{bmatrix} \tag{1}$$

where there is redundancy due to the four scalar values representing three rotational degrees of freedom. The constraint relating the quaternions is $\bar{q}^{v^\top}\bar{q}^v - 1 = 0$, indicating that $\bar{q}^v$ is a unit vector.

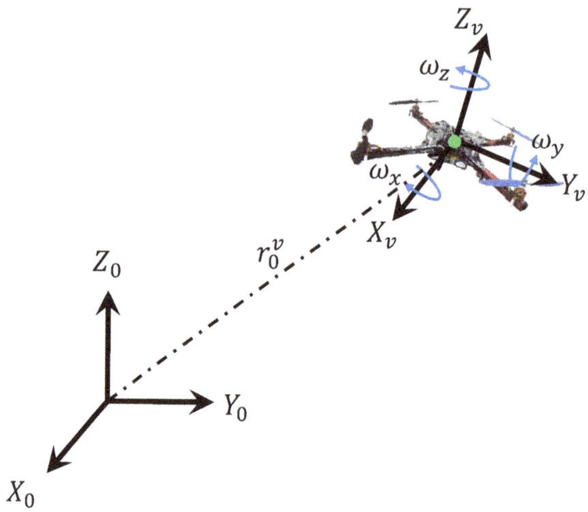

Figure 1. Vehicle 3D Kinematic Model.

The orthogonal rotation matrix is given in terms of the quaternions as [48]:

$$R_0^v(\bar{q}^v) = [2(q_0^v)^2 - 1]I_3 + 2[q^v q^{v^\top} + q_0^v S(q^v)] \tag{2}$$

where $S(q^v)$ defines a skew-symmetric matrix of a vector such that

$$S(q^v) = \begin{bmatrix} 0 & -q_3^v & q_2^v \\ q_3^v & 0 & -q_1^v \\ -q_2^v & q_1^v & 0 \end{bmatrix} \tag{3}$$

and I_3 is the 3×3 identity matrix. Given the rotation matrix in (2), we can calculate $\bar{q}^v$, as described in [46].

3.2. Vehicle Velocity

In the global coordinate frame, $X_0Y_0Z_0$, the vehicle translational velocity vector, $\dot{r}_0^v$, is related to its body-fixed velocities in $X_vY_vZ_v$, $\dot{r}_v^v$, as [49]:

$$\dot{r}_0^v = R_0^v(\bar{q}^v)v^v \tag{4}$$

where $v^v \triangleq \dot{r}_v^v$. The angular velocity vector of a vehicle about its body-fixed reference frame, $\omega^v = [\omega_x^v \ \omega_y^v \ \omega_z^v]^\top$, is used to calculate the time derivative of the quaternion vector, $\dot{\bar{q}}^v$, as:

$$\dot{\bar{q}}^v = \frac{1}{2} G^{v^\top} \omega^v, \quad \omega^v = 2G^v \dot{\bar{q}}^v \tag{5a}$$

$$G^v = \left[-q^v, \quad q_0^v I_3 - S(q^v)\right] \tag{5b}$$

The unit-length constraint equation that relates the time derivatives of the quaternions is also given as $\bar{q}^{v^\top} \dot{\bar{q}}^v = 0$.

5.3. Vehicle State Equations

Considering the single vehicle shown in Figure 1 and as described in the previous section, the pose of the vehicle can be represented by $r_0^v = [x_0^v \ y_0^v \ z_0^v]^\top$ and $\bar{q}^v$. For a kinematic model, we may choose the state vector as a collection of the seven pose variables:

$$\mathrm{x}^v = \begin{bmatrix} r_0^v \\ \bar{q}^v \end{bmatrix} \tag{6}$$

Using (4) and (5), the driftless state equations may be written as:

$$\dot{\mathrm{x}}^v = g^v(\mathrm{x}^v)\mathrm{u}^v \tag{7}$$

where

$$g^v = \begin{bmatrix} R_0^v & 0_{3\times4} \\ 0_{3\times3} & \frac{1}{2}G^{v^T} \end{bmatrix}, \quad \mathrm{u}^v = \begin{bmatrix} v^v \\ \omega^v \end{bmatrix} \tag{8}$$

where $0_{n\times m}$ represents the $n \times m$ zero matrix and the vehicle input vector consists of the body-fixed velocities v^v and ω^v.

The 7×7 Jacobian matrix of state equations, J^v, necessary for EKF linearization is calculated as:

$$J^v = \frac{\partial(g^v(\mathrm{x}^v)\mathrm{u}^v)}{\partial \mathrm{x}^v} = \begin{bmatrix} 0_{3\times3} & \dfrac{\partial \dot{r}_0^v}{\partial \bar{q}^v} \\ 0_{4\times3} & \dfrac{\partial \dot{\bar{q}}^v}{\partial \bar{q}^v} \end{bmatrix} \tag{9}$$

where the nonzero components are:

$$\frac{\partial \dot{r}_0^v}{\partial \bar{q}^v} = \begin{bmatrix} \dfrac{\partial \dot{r}_0^v}{\partial q_0^v} & \dfrac{\partial \dot{r}_0^v}{\partial q^v} \end{bmatrix} \tag{10a}$$

$$\frac{\partial \dot{\bar{q}}^v}{\partial \bar{q}^v} = \frac{1}{2} \begin{bmatrix} 0 & -\omega^{v^\top} \\ \omega^v & -S(\omega^v) \end{bmatrix} \tag{10b}$$

and

$$\frac{\partial \dot{r}_0^v}{\partial q_0^v} = [(4q_0^v)I_3 + 2S(q^v)]v^v \tag{11a}$$

$$\frac{\partial \dot{r}_0^v}{\partial q^v} = 2q^v v^{v^\top} + 2(q^{v^\top} v^v)I_3 - 2q_0^v S(v^v) \tag{11b}$$

It is important to note that several approaches have been proposed to adhere to the unit quaternion constraint during the integration of the state equations [50–52].

6. Multi-Vehicle System Model

6.1. State Equations

In order to formulate the CL algorithm, let us first consider a three-vehicle system shown in Figure 2. In the illustrated scenario, a vehicle designated as b or "base vehicle" obtains a pose measurement of another vehicle designated as t or "target vehicle" relative to its own body-fixed reference frame. Referring to (6)–(8), the system may be represented by the 14 states representing absolute poses of the two vehicles. The state equations are written as:

$$\begin{aligned}
\dot{\mathrm{x}} &= g(\mathrm{x})\mathrm{u} \\
\mathrm{x} &= \begin{bmatrix} r_0^{b\top} & \bar{q}^{b\top} & r_0^{t\top} & \bar{q}^{t\top} \end{bmatrix}^\top \\
\mathrm{u} &= \begin{bmatrix} v^{b\top} & \omega^{b\top} & v^{t\top} & \omega^{t\top} \end{bmatrix}^\top \\
g &= \mathrm{diag}\begin{bmatrix} R_0^b & \frac{1}{2}G^b & R_0^t & \frac{1}{2}G^t \end{bmatrix}
\end{aligned} \tag{12}$$

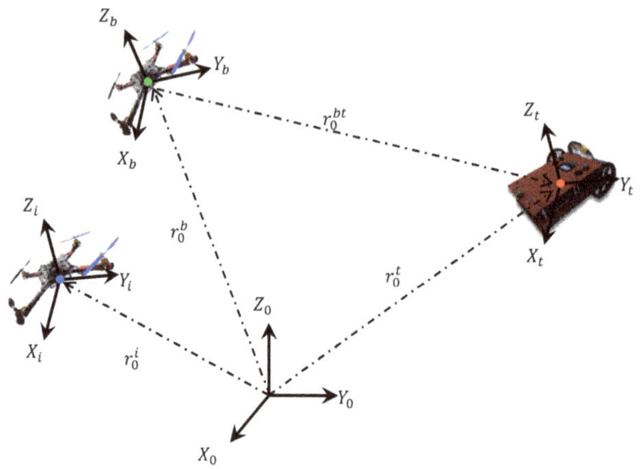

Figure 2. Pose geometry for a two-vehicle network.

The 14×14 Jacobian matrix, J, for the two-vehicle network is simply defined as:

$$J = \mathrm{diag}\begin{bmatrix} J^b & J^t \end{bmatrix} \tag{13}$$

where J^b and J^t are defined for vehicle b and vehicle t according to (9)–(11).

6.2. Measurement Equations

Consider a base vehicle b acquiring a relative pose measurement of a target vehicle t. A problem with quaternions is that they are not kinematic quantities that can be directly measured by any sensor. Thus, we assume the rotation matrix representing the relative orientation of t with respect to b, perhaps derived from Euler angles, is the measured quantity. We then derive the quaternions corresponding to the relative rotation matrix. Similarly, the relative range, bearing angle, and elevation angle of t with respect to b are the quantities normally measured that may be used to determine the relative position. Here, we define the measurement output vector, z, in terms of the relative position vector, r, and the three components of the quaternion vector q, which are determined from the relative orientation matrix:

$$\mathrm{z} = \begin{bmatrix} r \\ q \end{bmatrix}. \tag{14}$$

Note that the q_0 component is not needed since it can be derived from q using the unit-length constraint equation. Using the system states, the relative position r can be written as:

$$r = R_0^{b\top} r_0^{bt} = R_0^{b\top}(r_0^t - r_0^b) = R_0^{b\top} r_0 \tag{15}$$

where $r_0 = r_0^\top - r_0^b$. Quaternions representing the relative orientation are derived from the relative rotation matrix:

$$R(\bar{q}) = R_0^{b\top}(\bar{q}^b)R_0^t(\bar{q}^t) \tag{16}$$

Vector q is then derived from the elements of $R(\bar{q})$ [46].

Next, we linearize the measurement output equations, $z = Hx$ where:

$$H = \begin{bmatrix} H^b & H^t \end{bmatrix} = \begin{bmatrix} -R_0^{b\top} & \frac{\partial r}{\partial \bar{q}^b} & R_0^{b\top} & 0_{3\times 4} \\ 0_{3\times 3} & \frac{\partial q}{\partial \bar{q}^b} & 0_{3\times 3} & \frac{\partial q}{\partial \bar{q}^t} \end{bmatrix}. \tag{17}$$

The term $\frac{\partial r}{\partial \bar{q}^b}$ can be written in terms of its components as $\frac{\partial r}{\partial \bar{q}^b} = \begin{bmatrix} \frac{\partial r}{\partial q_0^b} & \frac{\partial r}{\partial q^b} \end{bmatrix}$. These components can then be defined similar to (11) as:

$$\frac{\partial r}{\partial q_0^b} = [(4q_0^b)I_3 + 2S(q^b)]r_0 \tag{18a}$$

$$\frac{\partial r}{\partial q^b} = 2q^b r_0^\top + 2(q^{b\top} r_0)I_3 - 2q_0^b S(r_0). \tag{18b}$$

Assuming $q_0 \neq 0$, the remaining derivative terms can be shown as:

$$\frac{\partial q}{\partial \bar{q}^b} = \frac{1}{4q_0}\frac{\partial p}{\partial \bar{q}^b} \quad , \quad \frac{\partial q}{\partial \bar{q}^t} = \frac{1}{4q_0}\frac{\partial p}{\partial \bar{q}^t} \tag{19}$$

where p is represented in terms of the elements of the matrix $R(\bar{q})$

$$p = [R_{32} - R_{23} \quad R_{13} - R_{31} \quad R_{21} - R_{12}]^\top, \tag{20}$$

and the components of (19) are derived as:

$$\begin{aligned} \frac{\partial p}{\partial q_0^b} &= 16q_0^b q_0^t q^t + 2[R_0^t - tr(R_0^t)I_3]q^b \\ \frac{\partial p}{\partial q_0^t} &= -16q_0^b q_0^t q^b - 2[R_0^b - tr(R_0^b)I_3]q^t \\ \frac{\partial p}{\partial q^b} &= 2q_0^b[R_0^t - tr(R_0^t)I_3] - 2S(q^b)R_0^{t\top} + 2S(R_0^{t\top} q^b) \\ \frac{\partial p}{\partial q^t} &= -2q_0^t[R_0^b - tr(R_0^b)I_3] + 2S(q^t)R_0^{b\top} - 2S(R_0^{b\top} q^t) \end{aligned} \tag{21}$$

If $q_0 = 0$, components of $\frac{\partial q}{\partial q^b}$ and $\frac{\partial q}{\partial q^t}$ are derived as:

$$\begin{aligned} \frac{\partial q_i}{\partial q_j^b} &= \frac{1}{4q_i}\frac{\partial R_{ii}}{\partial q_j^b}, \quad q_i \neq 0 \\ \frac{\partial q_i}{\partial q_j^t} &= \frac{1}{4q_i}\frac{\partial R_{ii}}{\partial q_j^t}, \quad q_i \neq 0, \quad i,j = 1,2,3 \\ \frac{\partial q_i}{\partial q_j^b} &= \frac{\partial q_i}{\partial q_j^t} = 0, \quad q_i = 0 \end{aligned} \tag{22}$$

Finally, the terms $\frac{\partial R_{ii}}{\partial q_b}$ and $\frac{\partial R_{ii}}{\partial q_t}$ can be derived as:

$$
\begin{aligned}
\frac{\partial R_{ii}}{\partial q_j^b} &= 2R_{0_{ji}}^t q_i^b, \qquad i \neq j \\
\frac{\partial R_{ii}}{\partial q_j^b} &= 2R_{0_{ji}}^t q_i^b + 2\sum_{j=1}^{3} R_{0_{ji}}^t q_j^b, \quad i = j \\
\frac{\partial R_{ii}}{\partial q_j^t} &= 2R_{b_{ji}} q_i^t, \qquad i \neq j \\
\frac{\partial R_{ii}}{\partial q_j^t} &= 2R_{0_{ji}}^b q_i^t + 2\sum_{j=1}^{3} R_{0_{ji}}^b q_j^t, \quad i = j
\end{aligned}
\tag{23}
$$

7. Cooperative Localization

Consider a network of n vehicles numbered 1 to n. A vehicle i, at sample time (k), utilizes absolute measurements to propagate its pose using the discrete form of (7):

$$
\mathrm{x}^i(k+1) = g^i(\mathrm{x}^i(k))\mathrm{u}^i(k) + \gamma^i(\mathrm{x}^i(k))\eta^i(k) \tag{24}
$$

where x^i, u^i, and η^i refer to the state, input, and process noise vectors, respectively, and $g^i(\mathrm{x}^i)$ and $\gamma^i(\mathrm{x}^i)$ are the input and process noise coefficient functions, respectively. Next, we assume that vehicle i in the set $\{1, \cdots, n\}$ measures the relative pose of vehicle, where $j \neq i$. Subsequently, the measurement vector z^{ij} can be written as:

$$
\mathrm{z}^{ij}(k+1) = h^{ij}(\mathrm{x}^i(k), \mathrm{x}^j(k)) + v^i(k) \tag{25}
$$

where $v^i(k)$ represents the measurement noise and $h^{ij} = [r^{ij^\top} \; q^{ij^\top}]^\top$; see (14).

For the EKF procedure, we denote variables at the propagation step with "$-$" and the updated step with "$+$". We also use over "$\hat{}$" for estimated state values. Thus, $\hat{\mathrm{x}}^{i-}$ and $\hat{\mathrm{x}}^{i+}$ indicate the predicted and updated state estimates for vehicle i, respectively. The key to success of EKF-based CL is the cross-covariance [14], which is the correlation term between vehicles i and j due to the relative measurements. Given initial conditions $\mathrm{x}^i(0)$, $P^{ij}(0) = 0_{7\times7}$, and the initial guess of $P^{ii}(0) = I_7$, the prediction step state estimates, covariance matrix, and cross-covariance matrix of vehicles $i = 1, \cdots, n$ at step $k+1$ are given in terms of step k terms as:

$$
\begin{aligned}
\hat{\mathrm{x}}^{i-}(k+1) &= g^i(\mathrm{x}^{i+}(k))\mathrm{u}^i \\
P^{ii-}(k+1) &= J^i P^{ii+} J^{i^\top} + \Gamma_i Q^i \Gamma_i^\top \\
P^{ij-}(k+1) &= J^i P^{ii+} J^{j^\top}
\end{aligned}
\tag{26}
$$

where $j = 1, \cdots, n$, $j \neq i$, $J^i = \frac{\partial g^i}{\partial \mathrm{x}^i}(k)$ is calculated using (9)–(10), $\Gamma^i = \frac{\partial \gamma^i}{\partial \mathrm{x}^i}$, and $Q^i > 0$ is the covariance of process noise.

At the update step, the state vector and (cross) covariance matrices remain unchanged if there are no relative measurements. Let us now assume that a "base" vehicle b takes relative measurements of a "target" vehicle t. Then, the innovation error between the measurement z^{bt} and the estimated output through propagated states is:

$$
\begin{aligned}
e^b(k+1) = &\quad \mathrm{z}^{bt} - h^{bt}(\hat{\mathrm{x}}^{b-}(k+1), \hat{\mathrm{x}}^{t-}(k+1)) \\
\approx &\quad \mathrm{z}^{bt} - \begin{bmatrix} H^b & H^t \end{bmatrix} \begin{bmatrix} \hat{\mathrm{x}}^{b-}(k+1) \\ \hat{\mathrm{x}}^{t-}(k+1) \end{bmatrix}
\end{aligned}
\tag{27}
$$

where $[H^b \ \ H^t]$ is defined in (17). The cross-covariance terms modifies the innovation covariance matrix as follows:

$$\begin{aligned} S_{bt}(k+1) = H^b P^{bb-} {H^b}^\top + H^t P^{tt-} {H^t}^\top \\ + H^b P^{bt-} {H^t}^\top + H^t P^{tb-} {H^b}^\top + R_b \end{aligned} \tag{28}$$

Utilizing (27) and (28), the Kalman gains along with the states and covariance terms for all vehicles $i = 1, \cdots, n$ are updated as follows:

$$\begin{aligned} K^i(k+1) &= [P^{it-} {H^t}^\top + P^{ib-} {H^b}^\top] {S^{bt}}^{-1} \\ \hat{x}^{i+}(k+1) &= \hat{x}^{i-} + K^i e^b \\ P^{ii+}(k+1) &= P^{ii-} - K^i S^{bt} {K^i}^\top \\ P^{ij+}(k+1) &= P^{ij-} - K^i S^{bt} {K^j}^\top \end{aligned} \tag{29}$$

where $j = 1, \cdots, n, \ j \neq i$. The Kalman gain, due to its semi-definite nature, ensures that a decrease in localization error is achieved by any relative measurement, as long as the measurements are not too noisy and the EKF linearization is accurate. The step-by-step procedure for the centralized CL algorithm can be found in [14].

The structure of the EKF-based CL algorithm is illustrated as a diagram in Figure 3 for better understanding of the procedure.

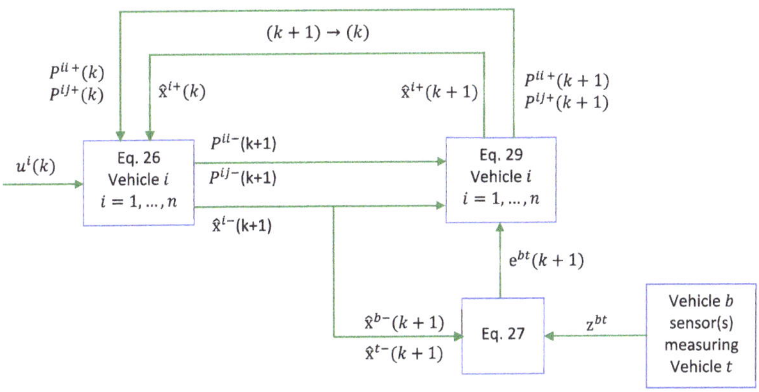

Figure 3. Centralized EKF-based CL diagram.

8. Simulation Results

Given a propagation time step size of Δt, the seven state equations of vehicle i in (4) and (5) can be written in discrete form as:

$$\begin{cases} r_0^i(k+1) = r_0^i(k) + R_0^i(k)\big(v^i(k) + \eta_v^i\big)\Delta t \\ \bar{q}^i(k+1) = \bar{q}^i(k) + \frac{1}{2} G^i(k)\big(\omega^i(k) + \eta_\omega^i\big)\Delta t \end{cases} \tag{30}$$

where η_v^i and η_ω^i are the corresponding components of η^i in (24) representing the input velocity process noise. Similarly, $R_0^i(\bar{q}^i(k))$ and $\frac{1}{2}G^i(\bar{q}^i(k))$ are the components of γ^i in (24).

The relative measurements obtained by all vehicles are polluted by additive Gaussian measurement noise (AWGN) with a standard deviation of 0.05 m for the position states (based on distances between vehicles) and 0.02 for the quaternion states of each vehicle. Vehicles may also be capable of obtaining absolute pose measurements with a standard deviation of 0.1 m for position and 0.02 for the quaternion states. As far as process noise is concerned, the standard deviations are $0.1v^i$ for η_v^i and 1°/s for η_ω^i across all vehicles. Finally, since there is additive noise in both measurement and process steps, after several trial runs, we concluded that about 100 runs in each simulation scenario are required to remove any significant effect stemming from outliers.

All simulations are run for 200 s and use the same input velocities and process and measurement noise properties in order to remove any input bias from the assessments. Input velocities are applied to each vehicle with varying magnitudes, as listed in Table 2. Since vehicles normally have a "forward" velocity input, 3D movement is attained through orientation changes, as a result of input angular velocities. Therefore, linear velocity is only applied in the x-direction. We selected $\Delta t = 0.02$ s, which is sufficiently small for accurate propagation without any noise, given the magnitude of input velocities.

Table 2. Input linear (m/s) and angular (°/s) velocities of vehicles for all simulations.

	$[v_x, v_y, v_z]$ **Linear**	$[\omega_x, \omega_y, \omega_z]$ **Angular**
V^1	−0.30, 0.00, 0.00	0.28, 0.55, 0.55
V^2	−0.35, 0.00, 0.00	0.35, 0.70, 0.70
V^3	0.40, 0.00, 0.00	0.43, 0.85, 0.85
V^4	0.35, 0.00, 0.00	0.50, 1.00, 1.00
V^5	−0.30, 0.00, 0.00	0.58, 1.15, 1.15
V^6	−0.35, 0.00, 0.00	0.65, 1.30, 1.30

To provide a comprehensive assessment, we initially present a test case with a changing measurement scenario. Next, we evaluate the effect of sequentially adding measurements that do not directly involve a given vehicle to show how CL can improve its localization and the localization of the entire network. We then assess the effect of redundant measurements when two or more vehicles simultaneously measure each other's poses. Next, we assess the effectiveness of CL when only relative positions are measured, followed by an investigation of how a single vehicle's absolute measurement can affect localization in this scenario.

8.1. A Test Case

We initially simulate a test case to demonstrate the effectiveness of CL using the first four vehicles listed in Table 2. In the test case, there are no relative measurements for the first 40 s, and therefore, each state is estimated using propagation only. Then, some vehicles fall within the line of sight of other vehicles such that vehicle 1 measures the relative pose of vehicle 2, vehicle 2 measures vehicle 3, and vehicle 3 measures vehicle 4 from $t = 40$ s until $t = 160$ s. These measurements all happen simultaneously. For the final 40 s, all vehicles lose sight of each other, such that there are no relative measurements. However, vehicles 1 and 4 take absolute measurements of their own pose until $t = 200$ s possibly due to the availability of GPS or SLAM for those vehicles.

The actual paths of the four vehicles along with those estimated with and without CL are shown in Figure 4 where significant improvement in the estimation of the path is observed after 40 s when CL is applied. We denote the norm of error in position vector r_0^v as Δr and the norm of error in orientation vector q^v as Δq. Figures 5 and 6 compares the time history of Δr and Δq for the four vehicles with and without CL, along with their 2σ plots. It is clear that there is a significant improvement in the estimation of the pose of all vehicles during CL (from 40 s to 160 s), including vehicle 1, whose relative pose is never measured. In addition, the absolute measurements of vehicles 1 and 4 not only greatly reduce localization errors for those vehicles but also for vehicles 2 and 3 due to their cross-covariance matrices with vehicles 1 and 4. Another observation is that there are discontinuities in state estimations when measurement scenarios are changed, caused by abrupt changes in the covariance matrix, as observed in the 2σ plots in Figures 5 and 6. These discontinuities, however, are not of any concern as long as the algorithm remains numerically stable.

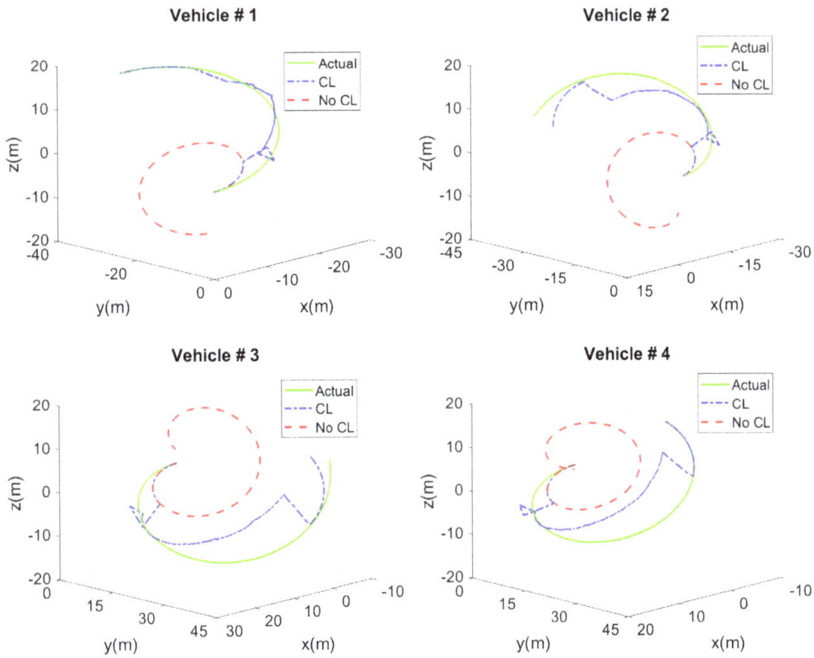

Figure 4. Vehicle actual path versus paths estimated with and without CL in the test case.

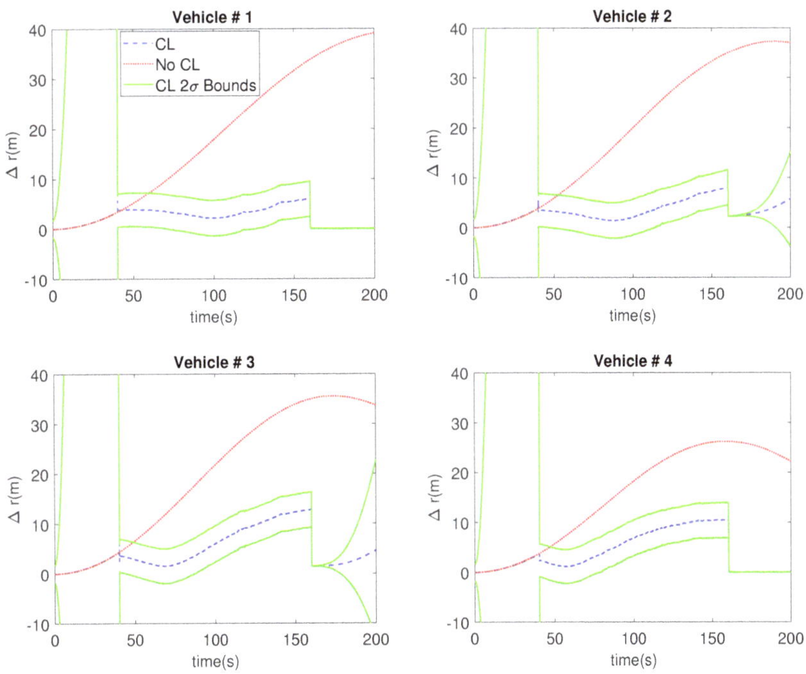

Figure 5. Time history of vehicle position error norm in the test case.

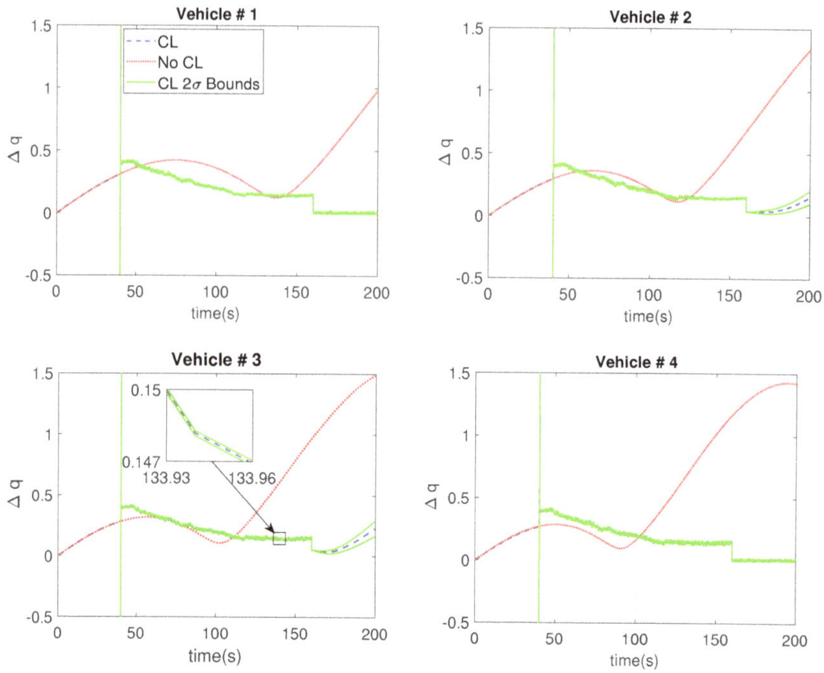

Figure 6. Time history of vehicle orientation error norm in the test case.

8.2. Simultaneous Relative Pose Measurement Assessment

For CL, the number of relative measurements plays a significant role in lowering uncertainty. Therefore, we investigate how the error for a given vehicle (and across all vehicles) is affected by sequentially adding simultaneous measurements by other vehicles. We also assess what happens when vehicles make redundant measurements, such as two vehicles simultaneously taking relative measurements and exchanging information with each other. Hence, we consider two networks, one with all six vehicles, $n = 6$, for assessment of additional simultaneous measurements (Scenario 1), and another with only the first three vehicles, $n = 3$, for the assessment of redundant measurements (Scenario 2). In each scenario, we start with one vehicle taking the relative measurements of another vehicle. In subsequent cases, we continue to add more relative measurements to assess their contribution.

8.2.1. Scenario 1—Effect of Other Vehicle Measurements on Single-Vehicle Localization

The idea for Scenario 1 is to see how other vehicles taking additional simultaneous relative measurements affects the localization of a single vehicle and, consequently, all vehicles in the network. Table 3 shows five cases that were simulated, concentrating on the localization of vehicle 1. In case 1, relative measurements of vehicle 2 are taken by vehicle 1, while in case 2, vehicle 2 simultaneously takes relative measurements of vehicle 3 and so on. In each case, errors are averaged across the 100 simulations.

The mean of the position error norm (Δr) and the orientation error norm (Δq) for vehicle 1 in all 100 simulations are plotted in Figures 7 and 8, respectively, for propagation without measurements and the five CL cases. Both Δr and Δq show that the performance increases as more relative measurements are added, and all CL cases show significant improvement over the propagation. An interesting observation is the significant error reduction in both Δr and Δq of vehicle 1 from Case 1 to Case 2 when vehicle 2 measures the relative pose of vehicle 3. Hence, the overall error for vehicle 1 is reduced even though it is not involved in the additional measurement. When subsequent measurements are added in cases 3 through 5, the localization errors continue to decrease but are relatively marginal. This demonstrates that the beneficial effects of CL may not outweigh its additional computational cost as the number of simultaneous measurements is increased. It should also be noted that the temporary dip in quaternion propagation error in Figure 8 is due to the complete rotation of the vehicles, bringing some components of the quaternion closer to their actual values.

Table 3. Simultaneous measurement descriptions for Scenario 1.

	Simultaneous Measurements ($b \rightarrow t$)
Case 1	$1 \rightarrow 2$
Case 2	$1 \rightarrow 2, 2 \rightarrow 3$
Case 3	$1 \rightarrow 2, 2 \rightarrow 3, 3 \rightarrow 4$
Case 4	$1 \rightarrow 2, 2 \rightarrow 3, 3 \rightarrow 4, 4 \rightarrow 5$
Case 5	$1 \rightarrow 2, 2 \rightarrow 3, 3 \rightarrow 4, 4 \rightarrow 5, 5 \rightarrow 6$

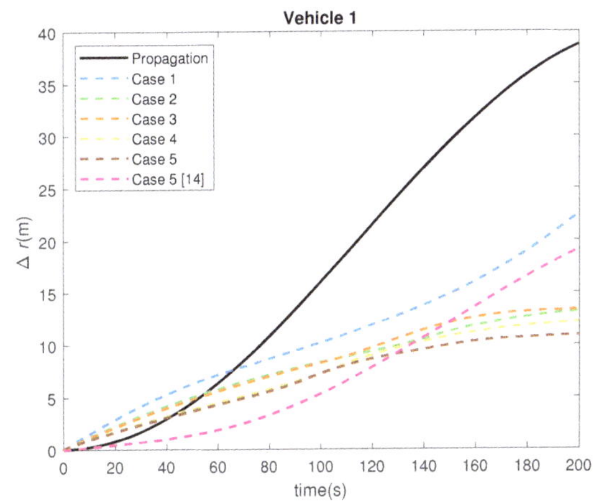

Figure 7. Scenario 1—vehicle 1 mean Δr norm.

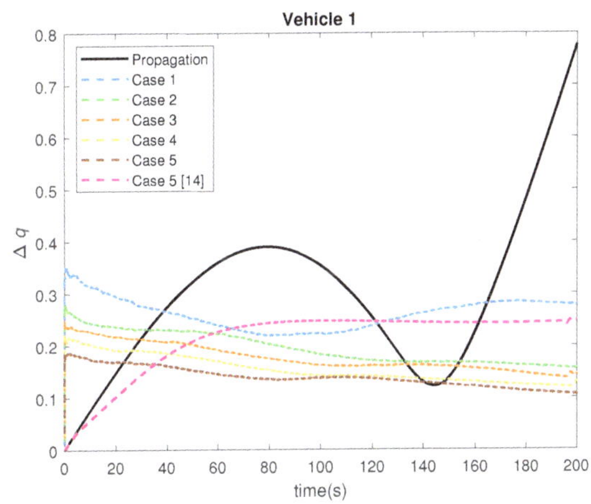

Figure 8. Scenario 1—vehicle 1 mean Δq norm.

Table 4 lists the error norm statistics across the 100 simulations for vehicle 1, which includes statistical terms such as mean and standard deviation for position r and orientation q.

To calculate these terms, we determine the norms of Δr and Δq at each time step of each simulation, and their root means squared are derived over time. Finally, the mean, standard deviation, min, and max are derived across all simulations. In CL, the cross-covariances contribute to lower overall error for vehicles when additional measurements are performed. The statistics again illustrate that the error reduction beyond Case 2 is relatively marginal. However, it also demonstrates that the variability is significantly increased in CL (see the standard deviation of CL cases compared to propagation).

Table 4. Position (m) and quaternion error statistics of *vehicle 1* across 100 simulations for Scenario 1.

	Position				Orientation			
	$e_{r_{mean}}$	$e_{r_{std}}$	$e_{r_{min}}$	$e_{r_{max}}$	$e_{q_{mean}}$	$e_{q_{std}}$	$e_{q_{min}}$	$e_{q_{max}}$
Propagate	2.87	0.01	2.84	2.91	0.05	0.00	0.05	0.05
Case 1	1.99	0.82	0.70	4.48	0.05	0.01	0.02	0.09
Case 2	1.48	0.85	0.38	4.56	0.04	0.02	0.01	0.09
Case 3	1.50	0.76	0.39	4.34	0.03	0.01	0.01	0.08
Case 4	1.33	0.69	0.39	4.34	0.03	0.01	0.01	0.07
Case 5	1.23	0.58	0.24	3.03	0.03	0.01	0.01	0.05
Case 5 [14]	1.27	0.01	1.16	1.19	0.04	0.00	0.03	0.04

We also compared the same statistics across all vehicles. In this case, Δr and Δq are again calculated at each time step of each simulation, and the root mean squared of the means is derived across all time steps and then averaged for all vehicles. Finally, the mean, standard deviation, min, and max are derived across all simulations. The results are listed in Table 5. The data show a similar trend to Table 4, where there is a reduction in the error estimates of both Δr and Δq as simultaneous measurements are added. However, as expected, the reduction in overall localization errors continues to be significant as the number of vehicles that rely solely on propagation decreases. The same increase in variability is also observed in the location of all vehicles.

Table 5. Position (m) and quaternion error statistics across 100 simulations of *all vehicles* for Scenario 1.

	Position				Orientation			
	$e_{r_{mean}}$	$e_{r_{std}}$	$e_{r_{min}}$	$e_{r_{max}}$	$e_{q_{mean}}$	$e_{q_{std}}$	$e_{q_{min}}$	$e_{q_{max}}$
Propagate	14.97	0.02	14.91	15.03	0.43	0.00	0.43	0.44
Case 1	13.30	1.61	10.80	18.37	0.42	0.03	0.37	0.49
Case 2	10.58	2.58	7.16	19.59	0.36	0.04	0.28	0.50
Case 3	9.33	2.77	5.40	19.83	0.30	0.05	0.21	0.44
Case 4	7.82	2.91	3.50	20.95	0.23	0.05	0.13	0.40
Case 5	6.15	2.59	1.67	14.13	0.15	0.06	0.05	0.33
Case 5 [14]	8.20	0.05	8.07	8.32	0.21	0.00	0.20	0.21

To demonstrate the importance of cross-covariances, we repeated the simulations in case 5 but without maintaining the cross-covariance updates as presented in [14]. The error norm statistics in the 100 simulations for vehicle 1 and all vehicles are listed as the last rows of Tables 4 and 5. It is clear that our method performs better for the overall network and the localization of individual vehicles. In addition, we plotted the time history of the error norms Δr and Δq for this method in Figures 7 and 8, respectively. It is observed that while initially the method in [14] performs better, the errors grow at much faster rate.

8.2.2. Scenario 2—Effect of Redundant Pose Measurements

Scenario 2 is added in order to assess simultaneous redundant pose measurements and communications. In this scenario, we only consider the first three vehicles, $n = 3$, while presenting the localization errors across all vehicles. We again assume vehicle 1 measures the relative pose of vehicle 2 as Case 1 and then add measurements until all vehicles simultaneously measure the relative pose of all other vehicles in the network, as listed in Table 6.

Table 6. Simultaneous measurement descriptions for Scenario 2.

	Simultaneous Measurements ($b \to t$)
Case 1	$1 \to 2$
Case 2	$1 \to 2, 2 \to 3$
Case 3	$1 \to 2, 2 \to 3, 3 \to 1$
Case 4	$1 \to 2, 2 \to 3, 3 \to 1, 2 \to 1$
Case 5	$1 \to 2, 2 \to 3, 3 \to 1, 2 \to 1, 3 \to 2$
Case 6	$1 \to 2, 2 \to 3, 3 \to 1, 2 \to 1, 3 \to 2, 1 \to 3$

Table 7 summarizes the mean, standard deviation, minimum, and maximum error statistics for position and orientation estimates across all vehicle. As expected, the initial errors are reduced as measurements are added. However, there is no improvement beyond Case 3. In fact, there seems to be an increase in localization errors beyond Case 4 attributed mainly to computational inaccuracies. Therefore, redundant measurements (i.e., two vehicles simultaneous measuring each other's relative poses and communicating their estimations and statistical properties) can negatively affect CL.

Table 7. Position (m) and quaternion error statistics across 100 simulations of *all vehicles* for Scenario 2.

	Position				**Orientation**			
	$e_{r_{mean}}$	$e_{r_{std}}$	$e_{r_{min}}$	$e_{r_{max}}$	$e_{q_{mean}}$	$e_{q_{std}}$	$e_{q_{min}}$	$e_{q_{max}}$
Propagate	18.08	0.04	17.98	18.16	0.36	0.00	0.36	0.36
Case 1	14.20	2.93	9.95	23.07	0.33	0.05	0.24	0.47
Case 2	10.28	4.76	3.69	24.40	0.23	0.08	0.09	0.50
Case 3	8.44	3.66	2.61	20.94	0.18	0.07	0.06	0.37
Case 4	8.23	4.60	2.56	23.84	0.17	0.07	0.06	0.33
Case 5	10.73	4.57	3.47	23.46	0.20	0.07	0.05	0.39
Case 6	11.33	4.13	4.66	24.18	0.21	0.06	0.10	0.40

8.3. *Relative Position Only Measurement Assessment*

In this section, we assume each vehicle can reliably obtain its own orientation using an IMU and/or a gyroscope and does not measure the relative orientation of other vehicles. Therefore, the only means of CL is through relative position measurements. We use Scenario 2 for the simulations, but only relative position measurements are considered. Table 8 lists the error statistics for only Δr across the 100 simulations. Our first observation is that the inclusion of absolute orientation measurements drastically reduces the overall error across vehicles, since IMUs can provide relatively accurate orientation information. Our second observation is that CL does not achieve statistically significant improvement in any of the cases. Therefore, the additional computational cost of CL does not justify its use in conjunction with the kinematic model presented in this and other works. There were no advantages in applying CL to this set because the equations of motion (4) and (5) and consequently the state equations (7) and (30) do not depend on position variables. We can further conclude that the improvements observed in Section 8.2 are only due to relative orientation measurements. These conclusions have not been previously reported and are valid even when applying CL to 2D vehicle models employed by other researchers [13,14].

Table 8. Mean position error (m) statistics across 100 simulations of *all vehicles* for Scenario 2 with only relative position measurements.

	$e_{r_{mean}}$	$e_{r_{std}}$	$e_{r_{min}}$	$e_{r_{max}}$
Propagate	1.35	0.01	1.32	1.39
Case 1	1.34	0.01	1.31	1.38
Case 2	1.30	0.10	1.25	2.28
Case 3	1.28	0.02	1.24	1.32
Case 4	1.28	0.02	1.25	1.32
Case 5	1.28	0.01	1.25	1.32
Case 6	1.27	0.01	1.25	1.31

8.4. Effect of Absolute Position Measurements in CL

It has previously been reported that measuring the absolute position of one vehicle can reduce the localization errors of other vehicles in CL [14]. Therefore, CL with only relative position measurements may still be useful if a vehicle is able to attain the absolute position measurement via GPS, SLAM, etc. The three-vehicle system of Scenario 2 was utilized to illustrate the effect of absolute measurements. We consider two cases. In one case, the relative positions of vehicles 2 and 3 are measured by vehicles 1 and 2, respectively, for the first 100 s (CL). Afterwards, vehicle 1 measures its own absolute position for a short period of 1 s in addition to the relative measurements. Afterwards, there are no subsequent measurements by any of the vehicles. In the other case, there are no measurements for the first 100 s (no CL), after which vehicle 1 measures its own absolute position. There are no other measurements involved.

The mean error norm Δr across all vehicles and simulations was calculated to be 0.85 ± 0.04 for the CL case and 1.05 ± 0.01 for the no CL case. This clearly indicates that the absolute measurement reduces the overall localization errors much more significantly when there has been CL with prior relative position measurements.

Additional details can be found in Figure 9 where the comparisons between the time histories of the two cases for the three vehicles are plotted. It is clear that CL does not make a significant contribution during the first 100 s. After the first 100 s vehicle 1 measures its own position and its localization error drops to near zero in both cases. However, despite the lack of relative position measurements from 100 s to 200 s, the localization errors for the "CL" case are significantly lower than the "no CL" case for vehicles 2 and 3. This is due to the cross-covariance evolution in the CL algorithm during the first 100 s of the CL case. In conclusion, we propose that when dependable absolute orientation measurements are accessible, CL with relative position measurement does not improve localization errors using the kinematic model. Nevertheless, if any vehicle can measure its absolute position at some point during the experiment, CL can still greatly reduce the position errors of all vehicles.

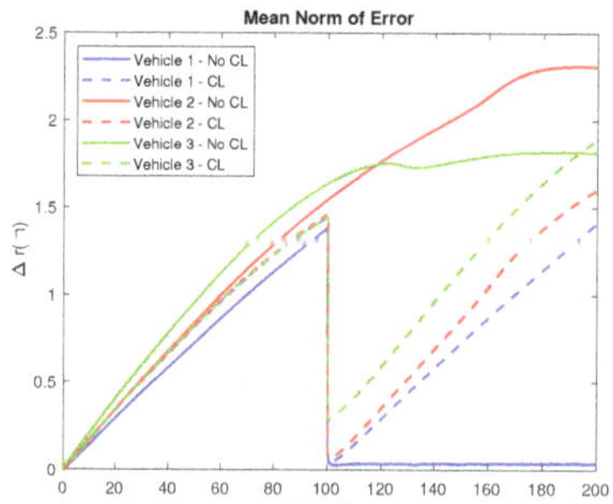

Figure 9. Comparison of error norms when vehicle 1 measures its own position for 1 s at the 100 s mark. Dashed lines indicate CL with relative position measurements for the first 100 s, and solid lines indicate no CL.

9. Experimental Results

We conduct two experiments with three vehicles to verify and evaluate the CL algorithm. The first experiment utilizes relative pose measurements, while the second one only takes relative position measurements. The duration for both experiments is 80 s, and each vehicle follows a predefined path. The actual path data are processed through the CL EKF algorithm as measurements obtained by a vehicle. Each vehicle propagates its own states at a rate of $\Delta t = 0.02$ s. However, relative measurements are obtained every 0.2 s.

Figure 10 shows an overview of one of the vehicles. All three vehicles have the same generic configuration and possess similar capabilities. The vehicles are differential-drive mobile robots with two motors controlled via a dual motor control and a PI controller implemented on an Arduino Duo board using motor encoder measurements. A Raspberry Pi computer performs most of the computation required to navigate the robot autonomously. Relative positions are calculated via LiDAR measurements of distance and bearing angle. However, the relative orientations are measured using a motion capture system that tracks markers, which are otherwise unused. The 0.2 s measurement rate is the time required for the LiDAR to rotate fully and the required data processing.

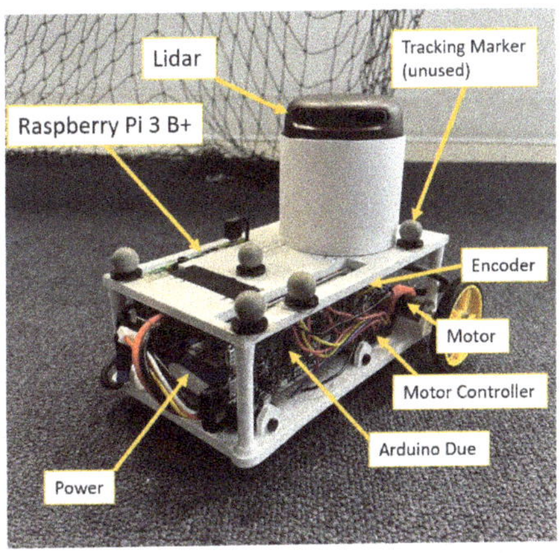

Figure 10. Overview of a mobile robot used for experiment.

9.1. Relative Pose Measurement Experiment

In these experiments, we investigated the effect of CL when relative pose measurements are only available for a very short period. For the first case (no CL), each vehicle propagates its own states for 40 s. Starting at 40 s, vehicle 3 measures its own pose for 1 s, after which there are no other measurements. In the second case (CL), the same scenario is repeated except that between 40 s and 41 s, vehicle 3 simultaneously measures the relative pose of vehicle 2 and 1 in addition to its own pose.

Figure 11 compares the time history of the position error norms of the two cases in this experiment for all three vehicles. As expected the performances for all three vehicles are identical for the first 40 s and vehicle 3's localization is similar throughout. However, despite only 1 s of relative measurements (i.e., five samples), vehicle 1's position estimation is dramatically improved. Vehicle 2, on the other hand, does not initially benefit from CL. However, as time passes, the localization errors start decreasing for CL while increasing for the "no CL" case.

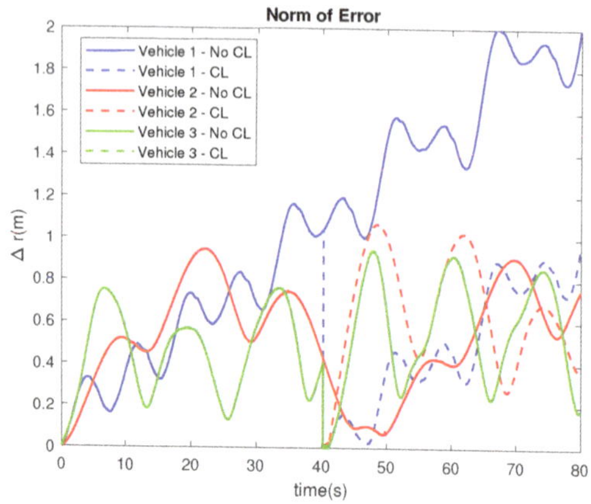

Figure 11. Comparison of error norms when vehicle 3 measures its own pose for 1 s at the 40 s mark. Dashed lines indicate CL with relative and absolute pose measurements starting at 40 s, and solid lines indicate no CL.

9.2. Relative Position Measurement Experiment

Next, we performed experiments with only relative position measurements, similar to the simulations in Section 8.4, where the absolute orientation measurements are available throughout the experiments for all vehicles. In the first experiment (no CL), the three vehicles propagate their position throughout while vehicle 1 measures its absolute position at 40 s for only 1 s. The second experiment (CL) is similar to the first except that vehicles 1 and 2 measure the relative positions of vehicles 2 and 3, respectively, for the first 40 s. Note that vehicle simply propagate their positions for the last 39 s in both cases.

Figure 12 shows the comparisons between the time history of the position error norms of the two experiments for the three vehicles. For the first 40 s, it is evident that CL has a mixed performance, where vehicles 2 and 3 seem to benefit from it while vehicle 1 localization deteriorates. However, after 40 s when vehicle 1 measures its own position for only 1 s, there is a significant drop in estimation errors of vehicles 2 and 3 when compared to the "no CL" case. This experiment confirms that even with intermittent absolute orientation measurements, CL can be very effective in improving localization errors.

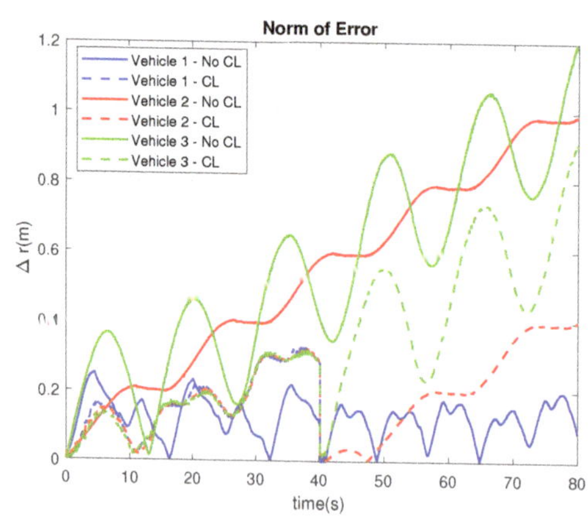

Figure 12. Comparison of error norms when vehicle 1 measures its own position for 1 s at the 40 s mark. Dashed lines indicate CL with relative position measurements for the first 40 s, and solid lines indicate no CL.

10. Conclusions

In this work, we implemented a centralized EKF-based CL algorithm to several networks of 3D vehicles. We build upon a previous publication [46] by adding additional cases, experimental results, and explicit limitations of the proposed CL algorithm. We derived state and relative measurement equations in terms of quaternions and also produced the singularity-free linearized matrices required for EKF. We utilize quaternions to avoid singularities linked to Euler angles. Simulation results demonstrated that CL considerably reduces state estimation errors when relative orientation measurements are available, though the performance does not proportionally improve with additional measurements. We further showed that CL does not provide any notable advantages when redundant measurements are applied. When utilizing CL with only relative position measurements, no improvement is observed, which is attributed to the absence of position variables in the state equations. However, CL with only position measurements can still provide significant improvement in state estimation if absolute measurements by a vehicle in the network becomes available any time during simulation. Furthermore, these conclusions are verified with experimental results that demonstrate the effect that absolute position and pose measurements have on CL. These experimental findings provide additional justification and credibility towards the CL algorithm presented in this paper.

The main limitation of the current algorithm is the use of a simplistic kinematic model that is not realistic for real-world applications. This limitation also leads to the lack of effective localization when only relative positions are employed. Future work involves updating the algorithm to utilize a fully nonlinear dynamic model and applying closed-loop control. There will also be closed-loop control experiments to further evaluate the effect CL has on a vehicle's state estimation.

Author Contributions: All authors contributed to the study conception and design. Material preparation and data collection and analysis were performed by J.C.O. The first draft of the manuscript was written by J.C.O. and H.A. All authors have read and agreed to the published version of the manuscript.

Funding: This research received no external funding.

Institutional Review Board Statement: Not applicable.

Informed Consent Statement: Not applicable.

Data Availability Statement: The datasets generated during and/or analyzed during the current study are available from the corresponding author on reasonable request.

Conflicts of Interest: The authors declare no conflict of interest.

Abbreviations

The following abbreviations are used in this manuscript:

CL	Cooperative Localization
2D	Two-dimensional
3D	Three-dimensional
EKF	Extended Kalman Filter
UKF	Unscented Kalman Filter
GPS	Global Positioning System
SLAM	Simultaneous Localization And Mapping

References

1. Gamini Dissanayake, M.W.M.; Newman, P.; Clark, S.; Durrant-Whyte, H.F.; Csorba, M. A solution to the simultaneous localization and map building (SLAM) problem. *IEEE Trans. Robot. Autom.* **2001**, *17*, 229–241. [CrossRef]
2. Hager, C.; Zarzhitsky, D.; Kwon, H.; Pack, D. Cooperative target localization using heterogeneous unmanned ground and aerial vehicles. In Proceedings of the 2010 IEEE/RSJ International Conference on Intelligent Robots and Systems, Taipei, Taiwan, 18–22 October 2010; pp. 2952–2957. [CrossRef]

3. Minaeian, S.; Liu, J.; Son, Y.J. Vision-Based Target Detection and Localization via a Team of Cooperative UAV and UGVs. *IEEE Trans. Syst. Man Cybern. Syst.* **2016**, *46*, 1005–1016. [CrossRef]
4. Fabresse, F.R.; Caballero, F.; Maza, I.; Ollero, A. Robust Range-Only SLAM for Unmanned Aerial Systems. *J. Intell. Robot. Syst. Theory Appl.* **2016**, *84*, 297–310. [CrossRef]
5. Waniek, N.; Biedermann, J.; Conradt, J. Cooperative SLAM on small mobile robots. In Proceedings of the 2015 IEEE International Conference on Robotics and Biomimetics (ROBIO), Zhuhai, China, 6–9 December 2015; pp. 1810–1815. [CrossRef]
6. Morrison, J.G.; Gávez-López, D.; Sibley, G. Scalable multirobot localization and mapping with relative maps: Introducing MOARSLAM. *IEEE Control Syst. Mag.* **2016**, *36*, 75–85. [CrossRef]
7. Madhavan, R.; Fregene, K.; Parker, L. Distributed cooperative outdoor multirobot localisation and mapping. *Auton. Robot.* **2004**, *17*, 23–41. [CrossRef]
8. Ivanjko, E.; Kitanov, A.; Petrovic, I. Model based Kalman Filter Mobile Robot Self-Localization. In *Robot Localization Map Building*; IntechOpen: London, UK, 2010; pp. 59–90. [CrossRef]
9. Romaniuk, S.; Ambroziak, L.; Gosiewski, Z.; Isto, P. Real time localization system with Extended Kalman Filter for indoor applications. In Proceedings of the 2016 21st International Conference on Methods and Models in Automation and Robotics (MMAR), Miedzyzdroje, Poland, 29 August–1 September 2016; pp. 42–47. [CrossRef]
10. Fox, D.; Burgard, W.; Kruppa, H.; Thrun, S. Probabilistic approach to collaborative multi-robot localization. *Auton. Robot.* **2000**, *8*, 325–344. [CrossRef]
11. Prorok, A.; Martinoli, A. A reciprocal sampling algorithm for lightweight distributed multi-robot localization. In Proceedings of the 2011 IEEE/RSJ International Conference on Intelligent Robots and Systems, San Francisco, CA, USA, 25–30 September 2011; pp. 3241–3247. [CrossRef]
12. Al-Shareeda, M.A.; Manickam, S.; Mohammed, B.A.; Al-Mekhlafi, Z.G.; Qtaish, A.; Alzahrani, A.J.; Alshammari, G.; Sallam, A.A.; Almekhlafi, K. Provably Secure with Efficient Data Sharing Scheme for Fifth-Generation (5G)-Enabled Vehicular Networks without Road-Side Unit (RSU). *Sustainability* **2022**, *14*, 9961. [CrossRef]
13. Roumeliotis, S.I.; Bekey, G.A. Distributed multirobot localization. *IEEE Trans. Robot. Autom.* **2002**, *18*, 781–795. [CrossRef]
14. Kia, S.S.; Rounds, S.; Martinez, S. Cooperative Localization for Mobile Agents: A Recursive Decentralized Algorithm Based on Kalman-Filter Decoupling. *IEEE Control Syst. Mag.* **2016**, *36*, 86–101. [CrossRef]
15. Li, W.; Jelfs, B.; Kealy, A.; Wang, X.; Moran, B. Cooperative Localization Using Distance Measurements for Mobile Nodes. *Sensors* **2021**, *21*, 1507. [CrossRef]
16. Ligorio, G.; Sabatini, A.M. Extended Kalman filter-based methods for pose estimation using visual, inertial and magnetic sensors: Comparative analysis and performance evaluation. *Sensors* **2013**, *13*, 1919–1941. [CrossRef] [PubMed]
17. Lin, J.; Lou, Z.; Zweigel, R.; Abel, D. Cooperative Localization of a Networked Multi-Vehicle System. *IFAC-PapersOnLine* **2019**, *52*, 67–72. [CrossRef]
18. Wu, Y.; Peng, B.; Wymeersch, H.; Seco-Granados, G.; Kakkavas, A.; Garcia, M.H.C.; Stirling-Gallacher, R.A.; Casta, M.H.; Stirling-Gallacher, R.A. Cooperative Localization with Angular Measurements and Posterior Linearization. In Proceedings of the 2020 IEEE International Conference on Communications Workshops (ICC Workshops), Dublin, Ireland, 7–11 June 2020. [CrossRef]
19. Qu, M.; Lu, H. Cooperative Simultaneous Localization and Mapping with Local Measurements in 2D Space. In Proceedings of the 2021 6th International Conference on Automation, Control and Robotics Engineering (CACRE), Dalian, China, 15–17 July 2021; pp. 509–513. [CrossRef]
20. Han, Y.; Wei, C.; Li, R.; Wang, J.; Yu, H. A novel cooperative localization method based on IMU and UWB. *Sensors* **2020**, *20*, 467. [CrossRef] [PubMed]
21. Héry, E.; Xu, P.; Bonnifait, P.; Hery, E.; Xu, P.; Bonnifait, P. Consistent decentralized cooperative localization for autonomous vehicles using LiDAR, GNSS, and HD maps. *J. Field Robot.* **2021**, *38*, 552–571. [CrossRef]
22. Pires, A.G.; Rezeck, P.A.F.; Chaves, R.A.; Macharet, D.G.; Chaimowicz, L. Cooperative Localization and Mapping with Robotic Swarms. *J. Intell. Robot. Syst. Theory Appl.* **2021**, *102*, 47. [CrossRef]
23. Ben, Y.; Sun, Y.; Li, Q.; Cui, W.; Zhang, Q. A Cooperative Localization Algorithm Considering Unknown Ocean Currents for Multiple AUVs. In Proceedings of the 2021 International Conference on Autonomous Unmanned Systems (ICAUS 2021), Changsha, China, 24–26 September 2021; Lecture Notes in Electrical Engineering; Springer Science and Business Media Deutschland GmbH: Berlin/Heidelberg, Germany, 2022; Volume 861, pp. 1379–1387. [CrossRef]
24. Qi, Y.; Zhong, Y.; Shi, Z. Cooperative 3-D relative localization for UAV swarm by fusing UWB with IMU and GPS. *J. Phys. Conf. Ser.* **2020**, *1642*, 012028. [CrossRef]
25. Li, W.; Liu, Y.; Li, X.; Shen, Y. Three-dimensional cooperative localization via space-air-ground integrated networks. *China Commun.* **2022**, *19*, 253–263. [CrossRef]
26. Mayle, M.N.; Sharma, R. Cooperative Localization in a GPS-Limited Environment Using Inter-Vehicle Range Measurements for a System of Multiple, Non-Homogeneous Vehicles. In Proceedings of the AIAA Scitech 2021 Forum, Virtual Event, 11–15 and 19–21 January 2021; AIAA American Institute of Aeronautics and Astronautics, Inc.: Reston, VA, USA, 2021; pp. 1–18. [CrossRef]
27. Wanasinghe, T.R.; George, G.K.; Gosine, R.G. Relative Localization Approach for Combined Aerial and Ground Robotic System. *J. Intell. Robot. Syst. Theory Appl.* **2015**, *77*, 113–133. [CrossRef]

28. Piasco, N.; Marzat, J.; Sanfourche, M. Collaborative localization and formation flying using distributed stereo-vision. In Proceedings of the 2016 IEEE International Conference on Robotics and Automation (ICRA), Stockholm, Sweden, 16–21 May 2016; pp. 1202–1207. [CrossRef]
29. Modalavalasa, N.; Sasi, G.; Rao, B.; Satya Prasad, K.; Ganesh, L.; Kumar, M.N.V.S.S. A new method of target tracking by EKF using bearing and elevation measurements for underwater environment. *Rob. Auton. Syst.* **2015**, *74*, 221–228. [CrossRef]
30. De Silva, O.; Mann, G.K.I.I.; Gosine, R.G. Development of a relative localization scheme for ground-aerial multi-robot systems. In Proceedings of the 2012 IEEE/RSJ International Conference on Intelligent Robots and Systems, Vilamoura-Algarve, Portugal, 7–12 October 2012; pp. 870–875. [CrossRef]
31. Wanasinghe, T.R.; Mann, G.K.I.; Gosine, R.G. Distributed collaborative localization for a heterogeneous multi-robot system. In Proceedings of the 2014 IEEE 27th Canadian Conference on Electrical and Computer Engineering (CCECE), Toronto, ON, Canada, 4–7 May 2014; pp. 1–6. [CrossRef]
32. Wanasinghe, T.R.; Mann, G.K.I.I.; Gosine, R.G. Distributed Leader-Assistive Localization Method for a Heterogeneous Multi-robotic System. *IEEE Trans. Autom. Sci. Eng.* **2015**, *12*, 795–809. [CrossRef]
33. Panzieri, S.; Pascucci, F.; Sciavicco, L.; Setola, R. Distributed multi-robot localization. In *Robotics: Concepts, Methodologies, Tools, and Applications*; IGI Global: Hershey, PA, USA, 2013; Volume 1, pp. 391–406. [CrossRef]
34. Carrillo-Arce, L.C.; Nerurkar, E.D.; Gordillo, J.L.; Roumeliotis, S.I. Decentralized multi-robot cooperative localization using covariance intersection. In Proceedings of the 2013 IEEE/RSJ International Conference on Intelligent Robots and Systems, Tokyo, Japan, 3–7 November 2013; pp. 1412–1417. [CrossRef]
35. Wanasinghe, T.R.; Mann, G.K.I.; Gosine, R.G. Decentralized Cooperative Localization Approach for Autonomous Multirobot Systems. *J. Robot.* **2016**, *2016*, 2560573. [CrossRef]
36. Kia, S.S.; Hechtbauer, J.; Gogokhiya, D.; Martinez, S. Server-Assisted Distributed Cooperative Localization over Unreliable Communication Links. *IEEE Trans. Robot.* **2018**, *34*, 1392–1399. [CrossRef]
37. Wang, L.; Zhang, T.; Gao, F. Distributed cooperative localization with lower communication path requirements. *Rob. Auton. Syst.* **2015**, *79*, 26–39. [CrossRef]
38. Wang, X.; Guo, Y.; Cao, J.; Wu, M.; Sun, Z.; Lv, C. Simplify Belief Propagation and Variation Expectation Maximization for Distributed Cooperative Localization. Appl. Sci. **2022**, *12*, 3851. [CrossRef]
39. Shi, X.; Wang, T.; Huang, B.; Zhao, C. Cooperative multi-robot localization based on distributed UKF. In Proceedings of the 2010 3rd International Conference on Computer Science and Information Technology, Chengdu, China, 9–11 July 2010; Volume 6, pp. 590–593. [CrossRef]
40. Kandepu, R.; Foss, B.; Imsland, L. Applying the unscented Kalman filter for nonlinear state estimation. *J. Process Control* **2008**, *18*, 753–768. [CrossRef]
41. Julier, S.J.; Uhlmann, J.K. Unscented filtering and nonlinear estimation. *Proc. IEEE* **2004**, *92*, 1958. [CrossRef]
42. Ponsa, D.; López, A.; Serrat, J.; Lumbreras, F.; Graf, T. Multiple vehicle 3D tracking using an unscented Kalman Filter. In Proceedings of the 2005 IEEE Conference on Intelligent Transportation Systems, Vienna, Austria, 16 September 2005; pp. 1108–1113. [CrossRef]
43. Huang, G.; Kaess, M.; Leonard, J.J. Consistent unscented incremental smoothing for multi-robot cooperative target tracking. *Robot. Auton. Syst.* **2015**, *69*, 52–67. [CrossRef]
44. Lyu, Y.; Pan, Q.; Lv, J. Unscented transformation-based multi-robot collaborative self-localization and distributed target tracking. *Appl. Sci.* **2019**, *9*, 903. [CrossRef]
45. der Merwe, R.; Wan, E.A.; Van Der Merwe, R.; Wan, E.A. The square-root unscented Kalman filter for state and parameter-estimation. In Proceedings of the 2001 IEEE International Conference on Acoustics, Speech, and Signal Processing, Salt Lake City, UT, USA, 7–11 May 2001; Volume 6, pp. 3461–3464. [CrossRef]
46. Oliveros, J.C.; Ashrafiuon, H. Cooperative Localization of Vehicles in Three-Dimensional Space. In Proceedings of the 2001 IEEE International Conference on Acoustics, Speech, and Signal Processing, Salt Lake City, UT, USA, 7–11 May 2001; Volume 2. [CrossRef]
47. Bar-Shalom, Y.; Willett, P.K.; Tian, X. *Tracking and Data Fusion: A Handbook of Algorithms*; YBS Publishing: London, UK, 2011.
48. Betsch, P.; Siebert, R. Rigid body dynamics in terms of quaternions: Hamiltonian formulation and conserving numerical integration. *Int. J. Numer. Methods Eng.* **2009**, *79*, 444–473. [CrossRef]
49. Fetzer, K.K.L.; Nersesov, S.; Ashrafiuon, H. Nonlinear control of three-dimensional underactuated vehicles. *Int. J. Robust Nonlinear Control* **2020**, *30*, 1607–1621. [CrossRef]
50. Zhao, F.; Van Wachem, B.G. A novel Quaternion integration approach for describing the behaviour of non-spherical particles. *Acta Mech.* **2013**, *3109*, 3091–3109. [CrossRef]
51. Gros, S.; Zanon, M.; Diehl, M. Baumgarte stabilisation over the SO(3) rotation group for control. In Proceedings of the 2015 54th IEEE Conference on Decision and Control (CDC), Osaka, Japan, 15–18 December 2015; pp. 620–625. [CrossRef]
52. Rucker, C. Integrating Rotations Using Nonunit Quaternions. *IEEE Robot. Autom. Lett.* **2018**, *3*, 2979–2986. [CrossRef]

applied sciences

MDPI

Article

UGAVs-MDVR: A Cluster-Based Multicast Routing Protocol for Unmanned Ground and Aerial Vehicles Communication in VANET

Waqar Farooq [1,*], Saif ul Islam [1], Muzzam Ali Khan [2], Saad Rehman [3], Usman Ali Gulzari [1] and Jalil Boudjadar [4,*]

1 Institute of Space Technology, KICSIT, Rawalpindi 46000, Pakistan
2 Department of Computer Science, Quaid-i-Azam University, Islamabad 44000, Pakistan
3 International Institute of Science, Arts & Technology, HITEC University, Gujranwala 52250, Pakistan
4 Department of Electrical & Computer Engineering, Aarhus University, 8200 Aarhus, Denmark
* Correspondence: waqar.farooq@mail.ist.edu.pk (W.F.); jalil@ece.au.dk (J.B.)

Abstract: Unmanned ground vehicles (UGVs) are becoming the foremost part of rescue teams for protecting human lives from severe disasters and reducing human casualties by informing them about the risks ahead, such as mine detection and clearance. In mine detection, a centralized system is required so that the UGVs can communicate with each other efficiently to disseminate the mine detection messages (MDMs) to incoming vehicles of military and civilians. Therefore, in this piece of research, a novel unmanned ground and aerial vehicle (UGAV)-based mine-detection-vehicle routing (MDVR) protocol has been proposed, mainly for the mine detection and clearance teams using a vehicular ad hoc network (VANET). The protocol disseminates the MDMs using UGVs and unmanned aerial vehicles (UAVs) in combination to overcome the limitations of only inter-UGV communication. The proposed protocol performs cluster-based multicast communication in real time using UAVs so that the dynamic mobility of UGVs cannot affect the performance of MDM dissemination. Hence, the proposed scheme is adaptable because any failure in message delivery can cause a high level of destruction. The proposed cluster-based scheme can adapt to any real-time scenario by introducing the level-based cluster-head election scheme (LBCHE), which works concerning its assigned priority for reducing the delay incurred in MDMs dissemination. The simulation of the proposed protocol in the network simulator (NS) shows that the overhead and delay are reduced in MDMs dissemination. At the same time, the throughput, packet delivery ratio, and stability increased compared to the other competing protocols.

Keywords: unmanned ground vehicles; unmanned aerial vehicles; mines; VANET; cluster-head election; multicast communication

Citation: Farooq, W.; Islam, S.u.; Khan, M.A.; Rehman, S.; Gulzari, U.A.; Boudjadar, J. UGAVs-MDVR: A Cluster-Based Multicast Routing Protocol for Unmanned Ground and Aerial Vehicles Communication in VANET. *Appl. Sci.* **2022**, *12*, 11995. https://doi.org/10.3390/app122311995

Academic Editors: Nadjim Horri, William Holderbaum and Fabrizio Giulietti

Received: 19 October 2022
Accepted: 18 November 2022
Published: 24 November 2022

1. Introduction

UGVs and UAVs have an important role in the recent developments in various fields such as rescue operations, monitoring of inaccessible areas, inspection of building construction, monitoring of electrical work, development of intelligent transportation systems, etc. [1,2]. UGVs are being used for detecting buried mines to safeguard the civilian population and military troops [3–5]. According to World Bank and United Nations statistics, hundreds of civilian and military casualties are reported yearly [6]. UGVs are now becoming an essential part of military fleets to meet the challenges of human safety. The efficiency of UGVs can be improved by developing a centralized coordination system based on cluster formation for disseminating the MDMs to incoming vehicles (IVs) successfully, as shown in Figure 1 [7–9]. In this centralized coordinating system, the vehicles share information with the cluster head for making collective decisions within a cluster rather than individual assessments [10]. For example, the movement decisions taken collectively within dangerous remote areas provide safety to mine detection and clearance staff [11,12].

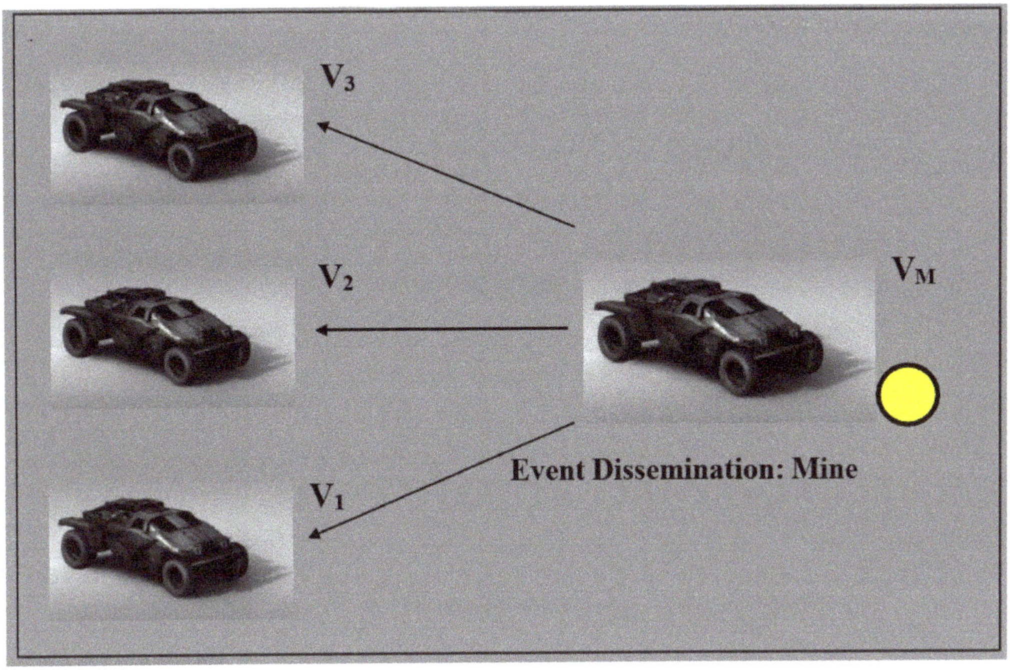

Figure 1. Mine detection and message dissemination to incoming UGVs.

Inter-vehicular communication (IVC) has been successfully implemented within VANET, where messages are transmitted to incoming vehicles (IVs) about the risks ahead [13–15]. In VANET, the vehicles communicate with each other at 5.9 GHz within the range of 300–1000 m using the IEEE-802.11p standard [16–18]. The safety messages and MDM-disseminating schemes have been developed by Farooq et al. [19–21]. However, the defense system analysis (DSA) shows that further work and developments are needed to meet human safety challenges [22]. Therefore, the performance of these protocols is improved by modifying the routing strategy through a combination of UGVs and UAVs. UAVs can maintain a constant altitude of 300 to 1000 m by satisfying the IEEE-802.11p standard range of communication within VANET.

Moreover, communication between UGVs and UAVs was also proposed and implemented efficiently for urban areas in [23–25]. The UAVs can efficiently communicate and disseminate messages from air to air and air to ground, as illustrated in [26,27]. Combining UGVs and UAVs into a single routing scheme results in strong IVC links and an effective central coordination system, which reduces any likelihood of delay or failure in MDM dissemination. Moreover, the proposed scheme can better address the VANET challenges of network fragmentation, dynamic mobility, delay, unstable IVC links, and overhead by reducing the transmission and communication gap of UGVs by using the air-to-ground links of UAVs. The overhead shows the ratio of total control packets sent to total received data packets at the destination, calculated during dynamic vehicular mobility and by varying the size of the cluster.

In the proposed design, the cluster-based scheme is adopted because the mine-detection vehicles need to move in groups to clear the required piece of land of buried mines. Additionally, the UGVs can cooperate better when moving within a single cluster by developing a central coordination system led by the head of the cluster. Furthermore, the overhead is distributed upon several clusters, reducing the delay and increasing the throughput. Besides this, the UGVs can adapt to unpredictable scenarios such as hilly areas, deserts, rough terrains, irregular trajectories, etc. Equally, the multicast approach enhances the proposed protocol performance manifold by increasing the throughput and reducing the overhead by disseminating the message to desired vehicles only, rather than to all incoming vehicles (IVs) [16,28]. Moreover, the multicasting technique increases the packet delivery ratio [29,30].

The distinctiveness of the proposed protocol can also be found in the several real-time applications in which efficient message dissemination is required, such as during landslides, floods, snow falling, fog, etc. In these situations, the UGVs are not enough to achieve the desired goals because of inaccessible areas without the assistance of UAVs. Although the proposed protocol is a small development toward saving human casualties, it will be useful for researchers to enhance its performance and scope within several other applications, as illustrated in the proposed future work. The UGV and UAV combination provides the following advantages:

1. The UGVs maintain a safe distance from the mines, and UAVs eliminate the communication gap between mine-detection and incoming vehicles (IVs).
2. The UAVs provide better connectivity and establish stable IVC links among UGVs through the air-to-ground links where UGVs cannot move together because of rough terrain, irregular trajectories, and hilly areas. Consequently, the network fragmentation in the proposed protocol reduces, resulting in low delay, minute overhead, high throughput, and stable IVC links.

The main motivation for developing the UGAVs-MDVR protocol is to save human life in rescue operations, detect buried mines, monitor inaccessible areas, inspect construction, develop intelligent transportation systems, etc. The proposed protocol works by combining the UGVs and UAVs into a single routing scheme that results in strong IVC links and an effective central coordination system, which reduces any likelihood of delay or failure in MDM dissemination in rescue operations. The cluster-based centralized coordinating system is adopted for taking the collective decision of movement within dangerous remote areas to provide safety to mine detection and clearance staff.

The remainder of the paper is structured as follows. The relevant VANET protocols are reviewed and analyzed in detail in Section 2. The proposed protocol is developed and designed in Section 3, along with the algorithms. Further, in Section 4, the proposed protocol is simulated and compared with the state-of-the-art protocols using a network simulator (NS). Finally, the designed protocol is concluded in Section 5, while future work is suggested in Section 6.

2. Related Work

In this section, the existing VANET routing protocols based on multicast routing are reviewed along with those developed for military purposes. The multicast protocols and their advantages and disadvantages are also analyzed in a Farooq et al. review article [16]. In this survey, the taxonomy of multicast routing protocols is also developed. In addition to this, the reviewed protocols are also classified into multicast routing methods and types.

The clustering for open inter-vehicle communication network (COIN) protocol [31] and cluster-based directional routing protocol (CBDRP) [32] are based on multicast routing protocols developed for urban traffic, but these protocols are unsuitable for real-time applications where the topology changes dynamically and unpredictably. Therefore, these protocols cannot work efficiently for MDM dissemination where the scenarios change rapidly. Additionally, the CBDRP depends upon the geographical position information for disseminating messages, which may be inaccessible at the required time and cause a delay in message transmission [16]. Alternatively, the proposed protocol has been designed to adapt to the vehicle movements in real time to meet the requirements of VANET and to fulfill the safety standards for saving human lives.

The inter-vehicle geocast (IVG) [33] protocol experiences high delay and overhead in remote and hilly areas because of its dependence upon the position information for routing messages. The cached geocast [34,35] protocol causes high network overhead that is due to the maintenance of cache records within the dynamic vehicular mobility. The message transmission scheme of the abiding geocast protocol [34] is based upon the short message duration, which results in low throughput within dense traffic. GVGrid [36] faces a high delay in message transmission because of the formation of complex routes within the congested traffic, creating unstable IVC links.

The mobile just-in-time multicasting protocol (Mobicast) is a spatiotemporary multicast protocol along with the feature of geography in which the messages are transmitted at a specified geographical area and at a given time 't' only. The specified area is called a zone of reference (ZOR). The packet delivery ratio of the protocol is high, but the connection suddenly fails in the case of a change of ZOR. Hence, the protocol is unsuitable for MDM dissemination where the scenario is unpredictable [37].

Dynamic time-stable geocast routing (DTSG) is a geographical and time-stable protocol that keeps messages saved within the desired area for a specified period. The protocol is useful for disseminating short-term information within a specified area. Hence, the protocol's performance remains unaffected by the density of the vehicles using the store-and-carry scheme. On the other hand, the protocol experiences network overhead in managing message saving for delivery in the specified area and in handling the sudden switching between the stability modes [38].

The constrained geocast protocol has been developed for cooperative adaptive cruise control (CACC) for automatic traffic control systems where the future positions of vehicles are estimated for routing decisions. The protocol's performance increases in dense traffic because the future positions can be better estimated. At the same time, the performance decreases in the case of high mobility, where the density decreases [39].

The geocache protocol has been developed to control the density of the vehicles by distributing the load on all possible routes. The incoming vehicles are informed ahead of time by disseminating messages about the density of the vehicles on approaching routes. The performance of the protocol increases by using the cache mechanism for storing the information and sharing it during road congestion, whereas the response time increases in dense traffic [40].

In this paper [41], the communication between military vehicles has been established by proposing the dynamic routing algorithm. This algorithm combines the location information with the intelligent ant colony algorithm to update the routing table efficiently to improve the node's linkages. The protocol updates the dynamic routing tables by consuming less setup time to improve the network's response time. The overhead increases in the case of high mobility because of the maintenance of tables [41].

In this paper [42], the path planning scheme for mobile robots has been introduced to find the path in a dynamic environment. The first feasible path is found in the initial phase, and then the optimal path is found for the robots in a dynamic environment using the proposed algorithm. The overhead and response time increases in managing path planning during dense traffic [42].

In this paper [43], the unmanned aerial vehicle is used to improve the network performance by reducing the latency and signaling overhead and improving the packet delivery ratio by proposing a relay selection algorithm with low complexity. The aerial vehicles communicate with the ground vehicles by using the carry-and-forward scheme [43].

In the protocol proposed in [44], the UAVs are used to increase the efficiency of message dissemination among the UGVs. The main purpose of the scheme is to perform search-and-rescue operations. The protocol uses maps and GPS locations to establish routes and suggest the fastest route for rescue operations [44].

Similarly, the rest of the reviewed multicast protocols [33,45,46] in [16] are also found to be unsuitable for the dissemination of MDMs in real time because of their dependence upon the position information because these are based upon the geocast routing scheme. Hence, the delay incurred while accessing the location information can cause high destruction.

The routing schemes of [41,47] are developed for military vehicle communication without considering their stable IVC links, which is the foremost requirement of MDM dissemination. In ref. [42], the procedure of path planning has been developed, but it cannot produce any central coordination system for creating stable IVC links. Similarly, [48] is also based on the geographical position information for routing, which is inappropriate for our desired application. The protocols [49–51] are also unsuitable for MDM dissemination because of their aim and focus on path planning only for routing messages. The protocols [52–54] are also reviewed and analyzed for message dissemination in VANET and found inappropriate because of the increase in overhead within dense and congested traffic. The analysis of the literature review has been summarized in Table 1:

Table 1. Comparison Summary of Existing Routing Protocols.

Sr. No	Protocols	Geographical	Cluster-Based	Spatiotemporary	Store and Forward Scheme	UGVs or UAVs Enable	Location-Based
1	COIN [31]	N	Y	N	N	N	Y
2	IVG [33]	Y	N	N	Y	N	Y
3	Cached Geocast [34,35]	Y	N	N	Y	N	Y
4	Abiding Geocast [34]	Y	N	Y	Y	N	Y
5	GVGrid [36]	Y	N	N	Y	N	Y
6	Mobicast [37]	Y	N	Y	N	N	Y
7	DTSG [38]	Y	N	N	N	N	Y
8	CBDRP [32]	N	Y	N	Y	N	Y
9	Constrained Geocast [39]	Y	N	N	Y	N	Y
10	Geocache Protocol [40]	Y	N	N	N	N	Y
11	Dynamic Routing [41]	N	N	N	N	N	Y
12	Path Planning [42]	N	N	N	N	N	Y
13	Relay Selection for UAVs [43]	N	N	N	Y	Y	Y
14	Leveraging Communicating UAVs [44]	N	N	N	N	Y	Y

3. Proposed UGAVs-MDVR Protocol

The protocols reviewed in the literature show that these protocols cannot be used for MDM dissemination in real time because these are developed for handling the traffic of urban or highway scenarios. Moreover, applying a single VANET protocol in all scenarios is impossible. Besides this, there is a need to propose a better solution for the VANET challenges of network fragmentation, dynamic mobility, delay, unstable IVC links, and overhead by reducing the transmission and communication gap of UGVs by using the air-to-ground links of UAVs. The overhead shows the ratio of total control packets sent to total received data packets at the destination, calculated during dynamic vehicular mobility and by varying the size of the cluster.

Combining UGVs and UAVs into a single routing scheme for message MDM dissemination is required with strong IVC links and an effective central coordination system to reduce any likelihood of delay or failure in MDM dissemination. Moreover, there is also a need for a scheme in which the mine-detection vehicles can move in groups to clear the required piece of land of buried mines. These vehicles share information in a central coordination system led by the head of the cluster and are able to adapt themselves to unpredictable scenarios.

Hence, our previous work [20,21,55] has been enhanced by introducing UAVs to overcome the existing VANET challenges, as discussed earlier. The basic layout and structure of communication between UGVs and UAVs have been implemented only to analyze the scheme's performance [55]. In the current research, the scheme of cluster formation, cluster-head election, and cluster speed-limit features have been enhanced. The cluster members joining the scheme and cluster merging process have been refined and implemented along with the computation of the speed and direction. Moreover, the scheme for cluster-head election and cluster management has also been improved. Hence, the proposed protocol's architecture and performance have been significantly enhanced.

The hierarchical approach of the proposed protocol reduces the network overhead and delay along with the increase in throughput caused by the logical organization of UGVs in clusters [56]. The hierarchical scheme consists of the following stages:

a. Cluster creation (CC),
b. Level-based cluster-head election (LBCHE),
c. Multicast routing (MRT),
d. Cluster maintenance (CMN)

3.1. Cluster Creation (CC)

The clusters are formed by considering the similar speeds of vehicles because the UGVs used for mine detection move together during mine detection and clearance. The direction is identical for cluster formation; otherwise, the UGV will be dropped from the cluster membership, as given in Algorithm 1 and shown in Equation (1). The cluster creation process is summarized in the following steps, as given in Algorithm 1:

a. Cluster construction,
b. Cluster-head election,
c. Joining process execution of CMs,
d. Dropping off the CMs,
e. Modification of cluster record,
f. Cluster maintaining process.

Algorithm 1. Creation and maintenance functions for clusters

```
TO complete Vw in Φ(n) Implement //all UGVs
IF CL.HD = NOT then //CH not elected
Function cluster_creation(VZ)      //cluster construction
Function CL.HD_election(VZ) //CHE Function
IF INVC_military = TRUE && INVC_direct = SAME //the joining of incoming vehicles by
    considering direction
Function joining_cluster(VZ) //joining as a cluster member
modify_cluster(CR) //Update the cluster Record (CR)
ENDIF
IF Vehicle_CMs(VZ)= CR(modify_stop) //when no change detected
Function drop_veh(VZ) //left CM
modify_cluster(CR)            //modify the CR
ENDIF
ELSE
Function maintain_status_CL.HEAD(VZ) //maintain CH
Function keep_status_CLMN(VZ) //maintain Cluster
Function keep_connection(UAV)
ENDIF
```

$$Cluster\ (C) = \left\{ \begin{array}{c} CMB_1 \in [SPD_{CMB1}|dr_1] \\ CMB_2 \in [SPD_{CMB2}|dr_2] \\ . \\ . \\ CMB_n \in [SPD_{CMBn}|dr_n] \end{array} \right\} \quad (1)$$

where CMB_1, CMB_2, and CMB_3 are cluster members that move with speed SPD_{CMBn} and direction' dr_1'.

3.2. Level-Based Cluster-Head Election (LBCHE)

In LBCHE, a single vehicle is elected from each cluster to manage the cluster members (CMs) and to develop communication with all adjacent clusters. The LBCHE scheme is specified as various levels, which adapt to the dynamic mobility of vehicles and scenarios, such that the IVC links remain stable. The LBCHE scheme is given as follows:

1. The first level has low priority, in which the cluster head (CH) is elected based on the technology or resources of that UGV and is capable of handling the mine-detection function efficiently as specified in [45]. The priority is set as low because the UGVs

are considered completely autonomous in this case. However, the preferences can be changed manually, also according to the desired requirements. The procedure is given in Algorithm 2.

2. The second level has medium priority compared to the first and is designed for scenarios where mines are detected and cleared without human involvement. In this level, the CH is elected based upon the vehicular speed because, in VANET, the speed mismatch causes weak IVC links, as analyzed in a survey of VANET protocols [14]. The speed-based cluster-head election (SBCHE) reduces the switching of CHs and CMs and increases the cluster's stability by reducing continuous maintenance. In addition, the SBCHE scheme is made flexible and adaptable by defining the average speed as the cluster threshold value (CTV) for the CH election. The maximum cluster speed limit (CSL) for UGVs is 60 m/s because the vehicles move slowly within the rough terrain, as shown in Equation (2). The CTV is given in Equation (3).

$$\text{CSL} = \text{maximum (60) m/s} \tag{2}$$

$$\text{CTV} = \text{average (CSL) m/s} \tag{3}$$

The CTV is further improved by introducing speed adjustment factor (SAF) 'u' as shown in Equation (4) so that the UGVs can adapt themselves to any scenario:

$$\text{CTV} = \text{CSL} \pm \text{u} \tag{4}$$

The value of 'u' can be varied to select the desired vehicle, specifically CH. For example, if the value of 'u' is increased, then the CH will be selected from the front vehicles of the cluster since the CTV exceeds the mobility of 30 m/s. Similarly, the CH can be elected from the last vehicles of the cluster by reducing the value of 'u' because the CTV will go below 30 m/s in this case.

3. At the third level, the manual cluster-head election (MCHE) is proposed with high priority to give the major and foremost control to mine disposal staff so that they can handle the situation concerning real-time scenarios and by considering the level of risk. In MCHE, the convoy's leader can elect any vehicle as CH.
4. The fourth level is introduced with UAVs as CH, where it is difficult to maintain IVC links of UGVs because of hilly areas and asymmetrical surfaces. Correspondingly, the UGVs cannot move together in some scenarios because of irregular trajectories where the inter-vehicular distance increases. Consequently, the UAVs are the best possible solution to disseminate the MDMs to incoming vehicles (IVs) by reducing the communication gap through the air-to-ground link, as shown in Figure 2.

The flowchart of the UGAVs-MDVR protocol is shown in Figure 3. The flowchart elaborates on the steps of cluster creation, the level-based cluster-head election scheme, and multicast routing. The incoming cluster member joining and cluster joining process is also demonstrated along with the conditions of similar direction and speed. The cluster merging function is also illustrated in the flowchart, and finally, the multicast record updating function is executed to update the records.

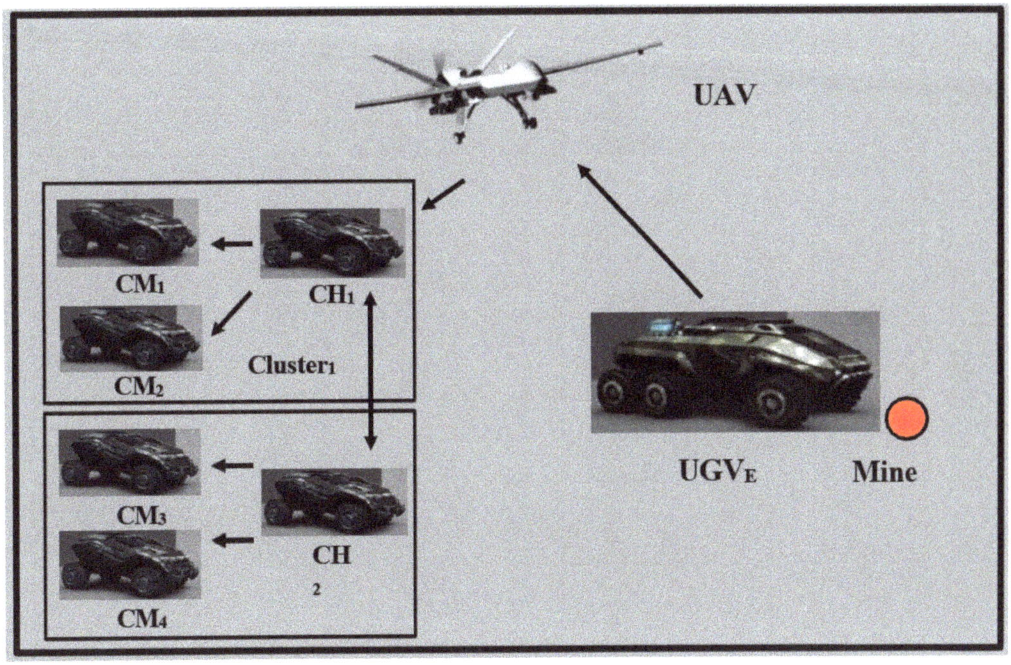

Figure 2. Event message dissemination by UGV to UAV and incoming UGVs.

The cluster election process consists of the following steps, as demonstrated in Algorithm 2 below:

a. Cluster-head election computation,
b. Resources comparison by sharing information with neighboring nodes,
c. Cluster-head election process execution,
d. Maintaining the cluster head and cluster.

Algorithm 2. Election function for clusters

```
TO complete Vw in Φ(n) Implement          //all UGVs
IF CL.HEAD = NOT then //CH not elected
veh_technology _status = find_veh_type(VZ) //find the resources according to mines detection staff
    strategy
Function calculate_spd_dirct_UGVs(VZ)
IF veh_resources_cluster > neighb_vehicles (VZ) //compare UGVs resources with neighboring vehicles
then
CL.HEAD =VZ //CH election
assign vehicle = CL.HD
ELSE IF UGVs(Speed && Direction == SIMILAR) && UGVs(Average speed value)==(Speed Threshold
    value)
then
CL.HEAD =VZ
assign Vehicle = CL.HD
ENDIF
ELSE IF VZ.CLUSTER = CL.HD          //CH maintenance, if elected already
Function keep_ CLMN (VZ)          //maintain CH
Function joining_cluster_member(VZ)
Function leave_cluster_member(VZ)
ELSE
set VZ.CL(Existing State= Tempoary_Elect (UAV) //if no existing head then elect UAV as CH
END IF
```

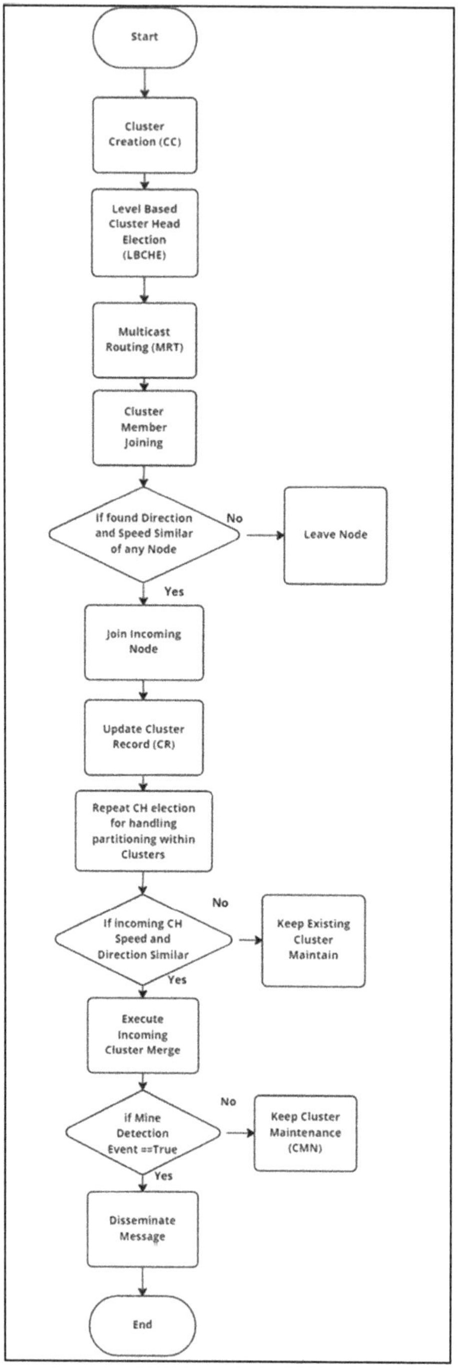

Figure 3. Flowchart of UGAVs-MDVR Protocol.

3.3. *Cluster Multicast Routing (MRT)*

In cluster-based routing, the packets of 512 bytes are transmitted from the source node to its CH in the first step. Then, the CH forwards the packet to the CH of the destination node through intermediate CHs. Hence, the record of all CHs and CMs is maintained within the cluster record (CR), which is updated by transmitting cluster advertisements (CADVs) periodically after every 5 s. The CADVs consist of cluster identification, UGV speed, direction, and sequence number. The sequence numbers are used to differentiate between the old and new advertisements. The MRT is defined in Algorithm 3, whereas the message dissemination scheme for all vehicles or only a specific group of vehicles is illustrated in Algorithm 4. The occurrence of network partitioning can be detected through periodic CADVs. If the CADVs are not received for more than 10 s, then it shows the event of network fragmentation. Therefore, the CH election scheme is executed again to handle the CMs and to establish inter-cluster communication through CHs, as shown in Algorithm 5. The CR is updated continuously to add the record of newly elected CHs.

The cluster record maintenance and updating process are illustrated as follows in Algorithm 3:

a. Speed and direction computation,
b. Sequence number update,
c. Advertisement dissemination,
d. Cluster record update process execution.

Algorithm 3. CR development for UGVs

TO complete V_w in $\Phi(n)$ Implement //all UGVs
Function calculate_spd_dirct_UGVs(V_Z) //speed and direction computation of UGVs
Function modify_vehicle_recent(Identity, spd, dirct, UMDVs)
Function modify_current_time(Identity, UGVs)
Function modify(Identity, UGVs, sequencenos) //modify sequence number
IF time >= 5 **then** //advertisement dissemination periodically
Function distribute_CR(Identity, sqn, spd, dirct,v.time, UGVs, UAVs) //CR dissemination
END IF

Algorithm 4. Mine event message transmission to all or specified incoming vehicles

TO complete V_w in $\Phi(n)$ Implement //all UGVs
IF Defined_event_occur = TRUE && transmit (V_Z) = YES then
Function send_Mine_message(INOMING_ CL.HD)
IF CL.Head _ Mine_message _receive(V_Z) = YES
Function Mine_message (INVC_ CL.HD) // Mine_message transmission to CHs
Function Mine_message (INVC_VC_CMs) //Mine_message transmission to other CMs
modify_cluster_info(CR) //modify the CR
ELSEIF CM_ Mine_message _receive = YES
Function Mine_message (Cluster Members)
modify_vehicles_record(CR) //modify the CR
ENDIF
ELSE
Function Mine_message (desired_ CL.HD) //dissemination of MDMs to desired CH
Function Mine_message (specific_VC_CMs) //dissemination of MDMs to desired CMs
END IF

The CH election process is executed again for re-cluster head election upon network partitioning. The re-election process keeps the network organized for efficiently disseminating the event detection messages irrespective of the sudden and dynamic changes occurring in the network, as shown in Algorithm 5.

Algorithm 5. Repeat CH election for handling partitioning within clusters

```
TO complete V_w Φ(n) Implement                                  //all UGVs
IF multicast(advertisement receive time)> 10 then //missing of advertisement show the partitioning of the
    network
Repeat (election CL.Head, V_Z) //repeat CH election Function
Function send_ advertis_time (V_Z) //advertisements dissemination after five seconds periodically
modify_clust_info(CR) //modify the CR
Function cluster_maintenance(V_Z) //modify the CR for a newly formed cluster
Function maint(current_cluster, new cluster) //maintain newly formed cluster
ELSE
Function send_ advertis (periodic_time_space)
Function keep_ CL.HD(V_Z)
Function CLMN(existing_V_Z)
modify_cluster(CR) //modify the CR
END IF
```

3.4. Cluster Maintenance (CMN)

The CMs join and leave the clusters continuously because of variations in speed, dynamic trajectories, and hilly areas. Thus, the CMN scheme is defined to keep the record of all CMs and updates in CR, as shown in Algorithm 6. The computation of the vehicle's speed and direction is illustrated in Algorithm 7. The CMN scheme specifies the procedure to add the new CMs or drop the missing vehicles, as depicted in Algorithm 8. Similarly, the procedure of cluster merging is also defined in this scheme, as shown in Algorithm 9.

Algorithm 6. Maintenance of cluster record (CR)

```
TO complete V_w in Φ(n) Implement                     //all UGVs
Function send_advertise_UGVs(V_Z)       //Dissemination of advertisement to adjacent vehicles
Function calculate_spd_dirct_UGVs(V_Z) //UGVs speed and direction computation
IF INCOMING(UGVs) = YES && UGVS_dir = SINGLE THEN //for new vehicles
Function add (UGVs)              //CR Update
modify(CR)
END IF
IF nearby(CLUSTER V_Z)= CR(modify_incomplete_miss_data) THEN
Function left (vehicle, CR)
modify_record(cluster, CR)
END IF
```

Algorithm 7. UGVs speed and direction computation

```
TO complete V_w in Φ(n) Implement                     //all UGVs
IF send_advertise_time(t) >= 5 THEN
Function calculate_spd_dirct_UGVs(V_Z) //UGVs speed and direction computation
Function share(UGVs, spd, dirct,V_Z)         //UGVs speed and direction are shared
IF UGVs_dirct = dissimilar THEN
Function left_(CR,V_Z) //drop the vehicle from CR
modify_CR(V_Z, spd, dirct)           //modification of cluster record
END IF
ELSE
Function CR_maintenance(VC_CMs,V_Z)          //the CR maintenance
END IF
```

The cluster member joining function is elaborated as follows and shown in Algorithm 8:

a. Computation of speed and direction for incoming cluster members,
b. Transmission of request by CH to incoming CMs upon similar speed and direction,
c. Transmission CMs acknowledgement,
d. CMs joining delay computation.

Algorithm 8. Cluster joining function upon new CM

```
TO complete V_w in Φ(n) Implement                                  //all UGVs
IF sp(INVC_UGVs) && dir(INVC_UGVs) == UGVs(Vehicles_Ahead) //if IVs speed matches vehicles ahead
CH_send_joining_request(C,cluster_head_ID,T_1) //transmit join request by CH to IV-CM vehicles with its CH_ID at T_1
joining_ack_CM(C,Cluster_member_ID,T_2) //Transmission of CM ack for joining with its CM_ID at T_2
CM_joining_delay(T_2-T_1)                  //Cluster member joining delay calculation
END IF
```

The cluster merging function for the merger of a similar incoming cluster is given as follows and in Algorithm 9 below:

a. Advertisement reception,
b. Speed and direction computation,
c. Incoming CH joining as CM,
d. Incoming CMs joining.

Algorithm 9. Cluster merging function upon incoming cluster

```
TO complete V_w in Φ(n) Implement                                  //all UGVs
IF CLH_ INVC(C) ← received(CADV(CLH_NEXT)) //reception of advertisement by CH from CH ahead
    IF CH_ INVC(speed && direction)==CH_AHEAD(speed && direction) THEN
        CLH_ INVC = CM(CLH_NEXT) //incoming CH becomes member of CH ahead
        CLH_ INVC(CMs) = CMs(CLH_NEXT)
        cluster_merge(CH_ INVC,CLH_NEXT) //cluster head merge with CH ahead
    END IF
END IF
```

The implementation of the proposed protocol has been elaborated concerning the creation of clusters and the cluster-head election process. Then, the cluster members joining and the cluster merging process are illustrated along with the maintenance of the cluster heads and cluster members. Finally, the last step is maintaining the updated cluster record for managing and handling the clusters efficiently. The proposed protocol's execution and performance evaluation are demonstrated in the next section.

4. Performance Evaluation

The performance of the implemented protocol has been evaluated in a network simulator (NS-2.34) by creating vehicular scenarios and establishing wireless IVC links. NS2 is an ideal open-source simulator for the implementation of new protocols, where the network traffic can be visualized graphically [57,58], and these results will become the foundation for developing the comparative analysis of simulation in NS-3 [59] along with the simulation of urban mobility (SUMO) [60] in future work. The object tool command language (OTCL) interpreter is used to create various scenarios according to the requirements of the proposed protocol to analyze its performance. Similarly, multicast communication, packet handling, channel configuration, event scheduling, timer adjustment, etc. are handled in C++ language. Awk scripts are also written in C++ to extract the required information from the trace file to compute packet delivery ratio, throughput, delay, etc.

The performance of the proposed protocol has also been compared to the multicast ad hoc on-demand distance vector (MAODV) [61] and dynamic source routing (DSR) [62] protocols. The performance evaluation is based on the computation of the packet delivery ratio, routing overhead, throughput, cluster stability, and delay with the confidence interval of 95%. The simulation parameters for performance evaluation within NS2 are given in Table 2. The total vehicles considered within a kilometer are 5–20 per kilometer during simulation.

Table 2. UGAVs-MDVR simulation parameters.

Attributes	Values
Maximum Vehicles	45
Vehicles per km	5–20
VANET Range	300–1000 m
Advertisement Interval	5 s
IEEE Protocol	IEEE 802.11p
Phy-Modulation	OFDM
Channel Bandwidth	10 MHz
IEEE 802.11p Data Rate	3–27 Mbps
Frequency IEEE 802.11p	5.850–5.925 GHz
Transmission Mechanism (MAC)	Carrier Sense MA/CA
Maximum Speed	60 km/h
NS Connection category	Transmission Control Protocol
Application Category in NS	File Transfer Protocol
NS Radio Propagation	Two Ray Ground
NS Radio Propagation Aerial	Free Space Model

The MAODV has been selected to analyze and compare the multicasting feature of MAODV with the proposed protocol because the multicast tree is maintained by the group leader in MAODV and can quickly adapt.

Similarly, the node's joining and leaving features can be analyzed by making a comparison with the DSR, which is based upon source routing that establishes the routes on demand. The routes are updated upon the successful delivery of messages at the destination.

In MAODV, the multicast tree is maintained by the group's leader through the transmission of "Hello" messages at regular intervals. The node replies are lost in MAODV because of the movement of intermediate nodes to new positions, increasing the routing overhead.

The DSR is based upon source routing that establishes the routes on demand. The routes are updated upon successful delivery of a message at the destination. DSR floods the packets to all the network nodes and can transmit them to a single node using the address filtration technique, which causes high overhead and delay during dynamic mobility and dense traffic [13,62].

4.1. Overhead

The overhead shows the ratio of total control packets sent to total received data packets at the destination. The overhead is calculated during dynamic vehicular mobility and by varying the size of the cluster. The cluster-based technique and multicast communication approach result in low overhead because of stable IVC links. Moreover, the overhead is distributed among several clusters, which does not affect the overall performance of the protocol, as depicted in Figure 4. The overhead may rise because of cluster maintenance at dynamic mobility. However, it is still low compared to other competing protocols, as shown in Figure 4b. The overhead is less than 4%, even at the cluster size of 18 cm, because of the LBCHE and CMN schemes, as noticed in Figure 4b.

The overhead of the DSR and MAODV rises upon varying the mobility of the vehicles. This behavior shows that the DSR and MAODV are not adapting to dynamic changes in the network. Hence, the number of control packets increases to maintain the network connections, resulting in a sudden increase in overhead.

4.2. Throughput

The throughput illustrates the total data packets delivered to the total simulation time of the scenario, as given in Equation (5) [63]:

$$Throughput = \frac{\sum All\ received\ Packets}{\sum Complete\ Simulation\ Time}(Size\ of\ Packet) \quad (5)$$

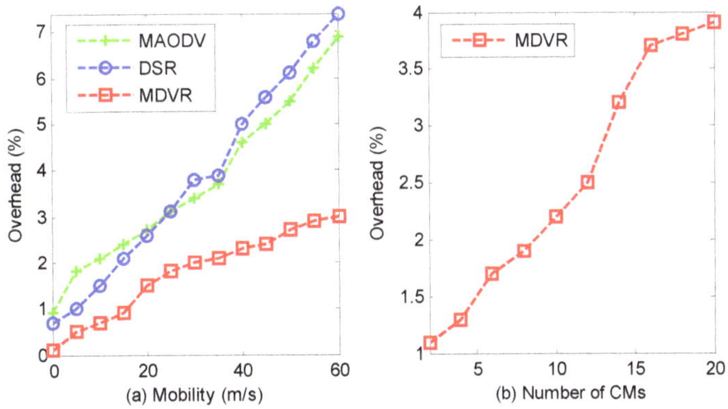

Figure 4. (**a**) UGAVs-MDVR overhead versus mobility; (**b**) UGAVs-MDVR overhead versus the number of CMs.

The throughput of the proposed protocol has been recorded as above 86% at the vehicular mobility of 60 m/s because of strong IVC links, stable cluster formation, and a multicast communication approach. On the other hand, the throughput of MAODV and DSR decreases continuously upon high mobility. The throughput of the proposed protocol is observed to be more than 80%, even at 31 CM per cluster, as depicted in Figure 5.

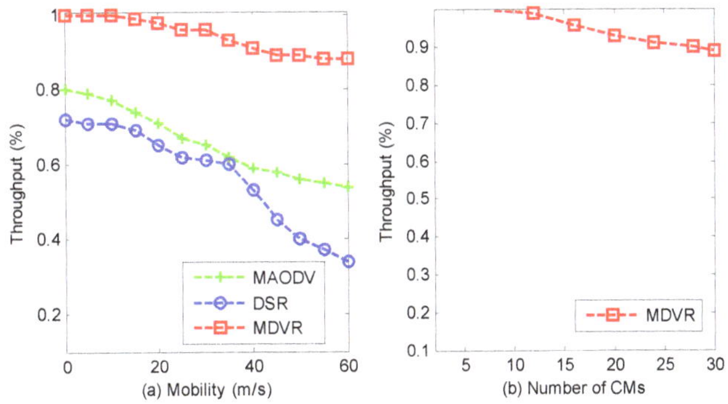

Figure 5. (**a**) UGAVs-MDVR throughput versus mobility; (**b**) UGAVs-MDVR throughput versus the number of CMs.

The throughput of the MAODV and DSR is declining continuously because of a rise in mobility. Hence, the nodes cannot maintain the inter-vehicular connection in dynamic movements and conditions, resulting in network partitioning. Hence, the MDM dissemination is affected by varying vehicle mobility in MAODV and DSR. However, the MDM dissemination is found efficient with stable inter-vehicular links in the proposed protocol because of its ability to cater to vehicles' dynamic and sudden mobility.

4.3. Packet Delay Ratio

The PDR calculation demonstrates the ratio of delivered packets to total transmitted packets as shown in Equation (6) [63]:

$$PDR = \frac{\sum All\ Delivered\ Packets}{\sum All\ Sent\ Packets} \tag{6}$$

The PDR is calculated by varying the size of the cluster and speed of UGVs in various scenarios to analyze the performance of the proposed protocol. The PDR remains above 86%, even at a maximum vehicle speed of 60 m/s, because of a combination of an adaptable

LBCHE scheme and a multicast communication approach, as depicted in Figure 6a. Alternatively, the PDR of MAODV and DSR decrease gradually because of instability caused by dynamic vehicular mobility, as shown in Figure 6a. The performance of the proposed protocol remains unaffected by varying the number of CMs from 6 to 30, as shown in Figure 6b.

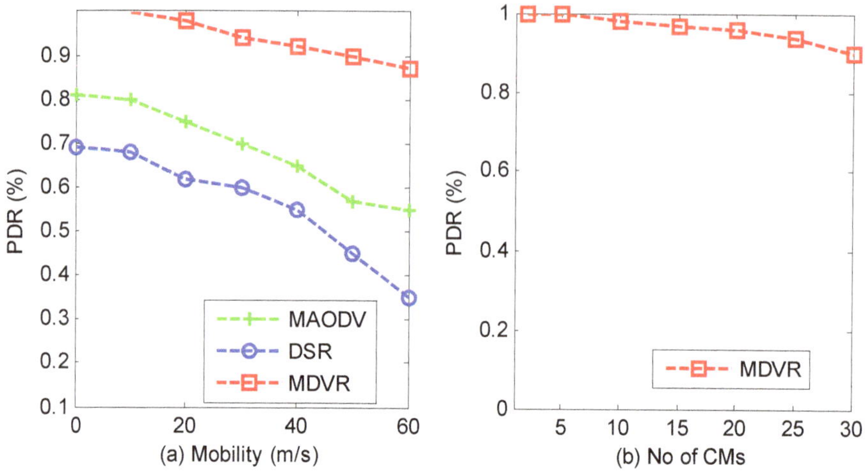

Figure 6. (**a**) UGAVs-MDVR PDR concerning mobility; (**b**) UGAVs-MDVR PDR for the number of CMs.

The simulation results show that the proposed protocol performs efficiently compared to existing schemes even by increasing the mobility of the vehicles.

The packet delivery ratio is declining in MAODV and DSR by increasing the mobility of the vehicles. The MAODV and DSR do not maintain the inter-vehicular links stable upon varying the speed of vehicles and this results in weak inter-vehicular links, which cause network partitioning. The comparison of the proposed protocol illustrates that the important need is to maintain the network links as stable. Stable links are only possible by keeping the network organized and stable irrespective of the network's dynamic changes and mobility variations. The UGAVs-MDVR protocol maintains stable inter-vehicular links by keeping the clusters maintained and organized, irrespective of any dynamic changes in the network. The rapid updating process of the cluster record plays a major role in keeping the cluster organized in the proposed protocol.

4.4. Delay

The delay computation signifies the time difference between the packet dissemination and delivery at the destination, as shown in Equation (7):

$$Delay = \frac{\sum(All\ Received\ Packets - All\ Disseminated\ Packets}{\sum Packets\ Delivered} \quad (7)$$

The delay is calculated by varying the cluster size and the speed of UGVs. The delay is below 4.5 s by increasing the speed to 60 m/s, which signifies the efficiency of LBCHE, CMN, MRT, and stable IVC links, as shown in Figure 7a. Similarly, Figure 7b illustrates that the delay is 4 s at 10 adjacent clusters because of the CMN scheme, which effectively manages the CMs' joining and leaving. On the other side, the delay is observed to be higher upon variation in the speed and cluster size of UGVs.

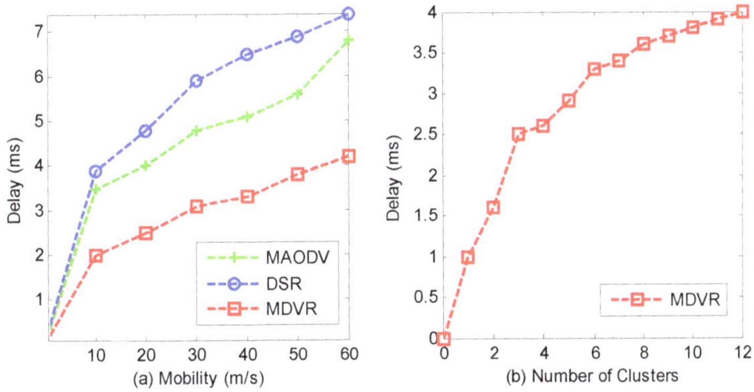

Figure 7. (**a**) UGAVs-MDVR delay with respect to mobility; (**b**) UGAVs-MDVR delay with respect to the number of clusters.

The simulation results show that the MDM dissemination of the proposed protocol is efficient by incurring less delay than MAODV and DSR, even by increasing the mobility of the vehicles. Hence, the MDM dissemination will not be affected by the dynamic mobility of vehicles. The MAODV and DSR are facing delays because of their unstable inter-vehicular links upon the dynamic mobility of the vehicles. The comparison results show that the delay increases upon the rise in mobility, indicating that MDM dissemination will be affected. The major reason for the rise in the delay of MAODV and DSR is the time the network takes to organize and update the inter-vehicular links because of the continuous occurrence of changes in the network. The proposed protocol can adapt itself despite the dynamic mobility of the vehicles, resulting in the efficient delivery of MDMs by incurring less delay compared to the MAODV and DSR.

The delay is also computed with respect to the number of CMs joining within the cluster. The results show that the delay variation is low even by increasing the number of the CMs within clusters because of the strong inter-vehicular communication links.

4.5. Cluster Stability

The stability of IVC links determines the protocol performance at dynamic mobility. Otherwise, the MDMs could not be disseminated successfully. The number of CHs and CMs switching to adjacent clusters shows the stability of a cluster. Only three CH and five CM switches were found at the maximum speed of 60 m/s, as depicted in Figure 8. Besides this, the CH and CM switches are very low at the mobility of less than 50 m/s, which shows the creation of a stable cluster formation.

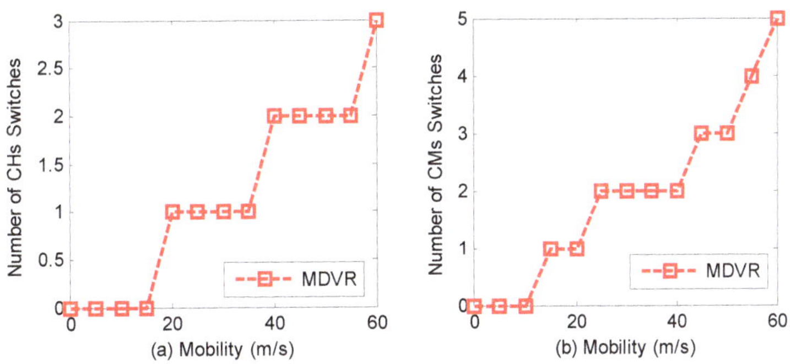

Figure 8. (**a**) UGAVs-MDVR mobility versus the number of CH switches; (**b**) UGAVs-MDVR mobility versus the number of CM switches.

The cluster's stability has been found by varying the mobility of the vehicles. The CH switching and CM switching are less, which illustrates the strong inter-vehicular communication links that result in efficient MDM dissemination among vehicles. The stability of the clusters and the consistent links of the nodes are directly dependent upon the quick organization of the network and the instant updating of the node's records. Therefore, the network remains maintained and consistent for efficient dissemination of MDMs, irrespective of the dynamic mobility of the nodes. Hence, the proposed protocol has achieved a stable and organized network by keeping the rapid update of cluster records, reducing the CH and CM switching.

5. Conclusions

UGVs play a major role in saving human lives, especially during dangerous operations like mine detection and clearance by keeping and maintaining a safe distance from the mines. In this paper, we applied the VANET paradigm among UGVs and UAVs in combination to develop a cluster-based centralized vehicular coordinating system to overcome the deficiencies of our previous work, in which only UGV communication was proposed. The UGVs and UAVs cooperate and assist each other so that the MDM dissemination becomes efficient in all possible real-time scenarios by fulfilling all desired requirements of the mine disposal staff. The stability of UGVs has been achieved by proposing an LBCHE scheme in which the protocol adapts itself to any environment because of the assignment of priorities to each level. The multicast approach combines the unique cluster-based technique to increase the throughput and reduce the delay and overhead.

6. Future Work

In future work, the protocol will be enhanced by introducing the larger architecture of UGV and UAV cooperation, which will be helpful for the rescue teams in a range of other applications to save human lives and to reduce human casualties from severe disasters by informing them about the level of risk ahead. The simulations will also be discussed and analyzed in more detail concerning the UGV density beyond 20 vehicles per kilometer, throughput versus overhead, etc.

Author Contributions: Conceptualization, W.F., M.A.K. and S.R.; methodology, W.F.; software, W.F.; validation, M.A.K., S.u.I., J.B. and S.R.; formal analysis, J.B. and U.A.G.; investigation, W.F., U.A.G. and J.B.; resources, M.A.K., S.R. and J.B.; data curation, W.F.; supervision, M.A.K. and S.R.; project administration, M.A.K., S.R., J.B. and S.u.I., writing—original draft preparation, W.F.; writing—review and editing, S.u.I.; visualization, W.F. and U.A.G. All authors have read and agreed to the published version of the manuscript.

Funding: This research received no external funding.

Institutional Review Board Statement: Not Applicable.

Informed Consent Statement: Not Applicable.

Data Availability Statement: Not Applicable.

Conflicts of Interest: The authors have no conflict of interest to declare.

References

1. Devendiran, R.; Kasinathan, P.; Ramachandaramurthy, V.K.; Subramaniam, U.; Govindarajan, U.; Fernando, X. Intelligent optimization for charging scheduling of electric vehicle using exponential Harris Hawks technique. *Int. J. Intell. Syst.* **2021**, *36*, 5816–5844. [CrossRef]
2. Rajamoorthy, R.; Arunachalam, G.; Kasinathan, P.; Devendiran, R.; Ahmadi, P.; Pandiyan, S.; Muthusamy, S.; Panchal, H.; Kazem, H.A.; Sharma, P. A novel intelligent transport system charging scheduling for electric vehicles using Grey Wolf Optimizer and Sail Fish Optimization algorithms. *Energy Sources Part A Recover. Util. Environ. Eff.* **2022**, *44*, 3555–3575. [CrossRef]
3. Rawashdeh, N.A.; Jasim, H.T. Multi-sensor input path planning for an autonomous ground vehicle. In Proceedings of the 2013 9th International Symposium on Mechatronics and its Applications (ISMA), Amman, Jordan, 9–11 April 2013. [CrossRef]
4. Scheidt, D.; Stipes, J.; Neighoff, T. Cooperating unmanned vehicles. In Proceedings of the AIAA 1st Intelligent Systems Technical Conference, Chicago, IL, USA, 20–22 September 2004. [CrossRef]

5. Han, R.; Shi, J.; Guan, Q.; Banoori, F.; Shen, W. Speed and Position Aware Dynamic Routing for Emergency Message Dissemination in VANETs. *IEEE Access* **2022**, *10*, 1376–1385. [CrossRef]
6. Landmine Contamination, Casualties and Clearance Database. 2016. Available online: http://data.worldbank.org/data-catalog/landmine-database (accessed on 1 January 2017).
7. Abdelfatah, R.I.; Abdal-Ghafour, N.M.; Nasr, M.E. Secure VANET Authentication Protocol (SVAP) Using Chebyshev Chaotic Maps for Emergency Conditions. *IEEE Access* **2022**, *10*, 1096–1115. [CrossRef]
8. De Almeida, T.T.; de Carvalho Gomes, L.; Ortiz, F.M.; Junior, J.G.R.; Costa, L.H.M. Comparative Analysis of a Vehicular Safety Application in NS-3 and Veins. *IEEE Trans. Intell. Transp. Syst.* **2020**, *23*, 620–629. [CrossRef]
9. Duarte, E.K.; Da Costa, L.A.L.F.; Erneberg, M.; De Freitas, E.P.; Bellalta, B.; Vinel, A. SafeSmart: A VANET System for Faster Responses and Increased Safety in Time-Critical Scenarios. *IEEE Access* **2021**, *9*, 151590–151606. [CrossRef]
10. Ali, I.; Lim, H. NameCent: Name Centrality-Based Data Broadcast Mitigation in Vehicular Named Data Networks. *IEEE Access* **2021**, *9*, 162438–162447. [CrossRef]
11. Ardakani, S.P.; Kwong, C.F.; Kar, P.; Liu, Q.; Li, L. CNN: A Cluster-Based Named Data Routing for Vehicular Networks. *IEEE Access* **2021**, *9*, 159036–159047. [CrossRef]
12. Wu, L.; Zhang, L.; Zhou, Q. Event-Based Control and Scheduling of a Platoon of Vehicles in VANETs. *IEEE Access* **2021**, *9*, 166223. [CrossRef]
13. Lim, J.M.-Y.; Chang, Y.C.; Alias, M.Y.; Loo, J. Joint optimization and threshold structure dynamic programming with enhanced priority scheme for adaptive VANET MAC. *Wirel. Netw.* **2016**, *22*, 897–913. [CrossRef]
14. Jan, S.A.; Amin, N.U.; Othman, M.; Ali, M.; Umar, A.I.; Basir, A. A Survey on Privacy-Preserving Authentication Schemes in VANETs: Attacks, Challenges and Open Issues. *IEEE Access* **2021**, *9*, 153701–153726. [CrossRef]
15. Kazi, A.K.; Khan, S.M.; Haider, N.G. Reliable Group of Vehicles (RGoV) in VANET. *IEEE Access* **2021**, *9*, 111407–111416. [CrossRef]
16. Farooq, W.; Khan, M.A.; Rehman, S.; Saqib, N.A. A Survey of Multicast Routing Protocols for Vehicular Ad Hoc Networks. *Int. J. Distrib. Sens. Netw.* **2015**, *11*, 923086. [CrossRef]
17. Lee, S.; Lim, A. An Empirical Study on Ad Hoc Performance of DSRC and Wi-Fi Vehicular Communications. *Int. J. Distrib. Sens. Netw.* **2013**, *9*, 482695. [CrossRef]
18. Chang, B.-J.; Liang, Y.-H.; Huang, Y.-D. Adaptive message forwarding for avoiding broadcast storm and guaranteeing delay in active safe driving VANET. *Wirel. Netw.* **2015**, *21*, 739–756. [CrossRef]
19. Farooq, W.; Khan, M.A.; Rehman, S. A Novel Real Time Framework for Cluster Based Multicast Communication in Vehicular Ad Hoc Networks. *Int. J. Distrib. Sens. Netw.* **2016**, *12*, 8064908. [CrossRef]
20. Farooq, W.; Khan, M.A.; Rehman, S. AMVR: A multicast routing protocol for autonomous military vehicles communication in VANET. In Proceedings of the 4th IEEE International Bhurban Conference on Applied Sciences & Technology (IBCAST), Islamabad, Pakistan, 10–14 January 2017; pp. 699–706. [CrossRef]
21. Farooq, W.; Khan, M.A.; Rehman, S. A cluster based multicast routing protocol for Autonomous Unmanned Military Vehicles (AUMVs) communication in VANET. In Proceedings of the IEEE International Conference on Computing, Electronic and Electrical Engineering (ICE Cube), Quetta, Pakistan, 11–12 April 2016; pp. 42–48. [CrossRef]
22. Delaney, W.; Atkins, R.; Bernard, A.; Boroson, D.; Ebel, D.; Feder, A.; Fleischman, J.; Shatz, M.; Stein, R.; Weiner, S. Broad Views of Defense Systems Analysis. In *Perspectives on Defense Systems Analysis*; MIT Press: Cambridge, MA, USA, 2015; p. 288. Available online: https://ieeexplore.ieee.org/document/7109408 (accessed on 15 September 2022).
23. Oubbati, O.S.; Lakas, A.; Zhou, F.; Güneş, M.; Lagraa, N.; Yagoubi, M.B. Intelligent UAV-assisted routing protocol for urban VANETs. *Comput. Commun.* **2017**, *107*, 93–111. [CrossRef]
24. Bhoi, S.K.; Khilar, P.; Singh, M. A path selection based routing protocol for urban vehicular ad hoc network (UVAN) environment. *Wirel. Netw.* **2017**, *23*, 311–322. [CrossRef]
25. Ahmed, F.; Mohanta, J.C.; Keshari, A.; Yadav, P.S. Recent Advances in Unmanned Aerial Vehicles: A Review. *Arab. J. Sci. Eng.* **2022**, *47*, 7963–7984. [CrossRef]
26. Song, Q.; Zeng, Y.; Xu, J.; Jin, S. A survey of prototype and experiment for UAV communications. *Sci. China Inf. Sci.* **2021**, *64*, 140301. [CrossRef]
27. Ramkumar, D.; Annadurai, C.; Nirmaladevi, K. Continuous authentication consoles in mobile ad hoc network (MANET). *Clust. Comput.* **2019**, *22*, 7777–7786. [CrossRef]
28. Zhang, L.; Gao, D.; Foh, C.H.; Yang, D.; Gao, S. A Survey of Abnormal Traffic Information Detection and Transmission Mechanisms in VSNs. *Int. J. Distrib. Sens. Netw.* **2014**, *10*, 582761. [CrossRef]
29. Khan, M.A.; Ahsan, M.; Shah, G.A.; Sher, M. Multicast Routing Protocols in Wireless Sensor Networks (WSNs). *J. Comput.* **2012**, *4*, 9–17.
30. Mohapatra, P.; Gui, C.; Li, J. Group communications in mobile ad hoc networks. *Computer* **2004**, *37*, 52–59. [CrossRef]
31. Blum, J.; Eskandarian, A.; Hoffman, L. Mobility management in IVC networks. In Proceedings of the IEEE (IV2003) Intelligent Vehicles Symposium, Columbus, OH, USA, 9–11 June 2003. [CrossRef]
32. Song, T.; Xia, W.; Song, T.; Shen, L. A cluster-based directional routing protocol in VANET. In Proceedings of the 12th International Conference on Communication Technology Proceedings, ICCT, Nanjing, China, 11–14 November 2010; pp. 1172–1175. [CrossRef]

33. Joshi, H.; Sichitiu, M.; Kihl, M.; Distributed Robust Geocast Multicast Routing for Inter-Vehicle Communication. 1st WEIRD Workshop on WiMax, Wireless and Mobility. Available online: http://scholar.google.com/scholar?hl=en&btnG=Search&q=intitle:Distributed+Robust+Geocast+Multicast+Routing+for+Inter-Vehicle+Communication#0%5Cnhttp://scholar.google.com/scholar?hl=en&btnG=Search&q=intitle:Distributed+robust+geocast+multicast+routing+for+in (accessed on 5 August 2022).
34. Maihofer, C.; Leinmuller, T.; Schoch, E. Abiding geocast: Time stable geocast for ad hoc networks. In Proceedings of the 2nd ACM International Workshop on Vehicular Ad Hoc Networks, Cologne, Germany, 2 September 2005; pp. 20–29.
35. Maihofer, C.; Eberhardt, R. Geocast in vehicular environments: Caching and transmission range control for improved efficiency. In Proceedings of the IEEE Intelligent Vehicles Symposium, Parma, Italy, 14–17 June 2004. [CrossRef]
36. Sun, W.; Yamaguchi, H.; Yukimasa, K.; Kusumoto, S. GVGrid: A QoS Routing Protocol for Vehicular Ad Hoc Networks. In Proceedings of the 14th IEEE International Workshop on Quality of Service, New Haven, CT, USA, 19–21 June 2006; pp. 130–139. [CrossRef]
37. Chen, Y.-S.; Lin, Y.-W.; Lee, S.-L. A Mobicast Routing Protocol in Vehicular Ad-Hoc Networks. In Proceedings of the EEE Global Telecommunications Conference, Honolulu, HI, USA, 30 November–4 December 2009; pp. 1–6. [CrossRef]
38. Rahbar, H.; Naik, K.; Nayak, A. DTSG: Dynamic time-stable geocast routing in vehicular ad hoc networks. In Proceedings of the 9th IFIP Annual Mediterranean Ad Hoc Networking Workshop, Juan-les-Pins, France, 23–25 June 2010; pp. 1–7. [CrossRef]
39. Wolterink, W.K.; Heijenk, G.; Karagiannis, G. Constrained geocast to support Cooperative Adaptive Cruise Control (CACC) merging. In Proceedings of the 2nd IEEE Vehicular Networking Conference (VNC'10), Jersey City, NJ, USA, 13–15 December 2010; pp. 41–48. [CrossRef]
40. Lakas, A.; Shaqfa, M. Geocache: Sharing and Exchanging Road Traffic Information Using Peer-to-Peer Vehicular Communication. In Proceedings of the 73th IEEE Vehicular Technology Conference, Budapest, Hungary, 15–18 May 2011; pp. 1–7. [CrossRef]
41. Chenghai, W.; Jianjun, Z.; Xiquan, F.; Kaiyu, Q.; Xiangping, L. Research on dynamic routing algorithm of the combat collaboration communication network. In Proceedings of the 2015 27th Chinese Control and Decision Conference (CCDC 2015), Qingdao, China, 23–25 May 2015; pp. 4440–4445. [CrossRef]
42. Ganeshmurthy, M.; Suresh, G. Path planning algorithm for autonomous mobile robot in dynamic environment. In Proceedings of the 3rd International Conference on Signal Processing, Communication and Networking (ICSCN), Chennai, India, 26–28 March 2015; pp. 1–6. [CrossRef]
43. He, Y.; Zhai, D.; Jiang, Y.; Zhang, R. Relay Selection for UAV-Assisted Urban Vehicular Ad Hoc Networks. *IEEE Wirel. Commun. Lett.* **2020**, *9*, 1379–1383. [CrossRef]
44. Oubbati, O.S.; Lakas, A.; Lorenz, P.; Atiquzzaman, M.; Jamalipour, A. Leveraging Communicating UAVs for Emergency Vehicle Guidance in Urban Areas. *IEEE Trans. Emerg. Top. Comput.* **2021**, *9*, 1070–1082. [CrossRef]
45. Kihl, M.; Sichitiu, M.; Ekeroth, T.; Rozenberg, M. Reliable Geographical Multicast Routing in Vehicular Ad-Hoc Networks. In Proceedings of the International Conference on Wired/Wireless Internet Communications, Coimbra, Portugal, 23–25 May 2007; pp. 315–325. [CrossRef]
46. Hassanabadi, B.; Valaee, S. Reliable Periodic Safety Message Broadcasting in VANETs Using Network Coding. *IEEE Trans. Wirel. Commun.* **2014**, *13*, 1284–1297. [CrossRef]
47. Cailleux, L.; Bouabdallah, A. A correspondence model for a future military messaging handling system. *IEEE Commun. Mag.* **2015**, *53*, 46–51. [CrossRef]
48. Preuss, M.; Thomas, S. Wireless, Mesh & Ad Hoc Networks; Military Convoy Location and Situation Awareness. In Proceedings of the 2008 IEEE Sarnoff Symposium, Princeton, NJ, USA, 28–30 April 2008; pp. 1–5. [CrossRef]
49. Suthaputchakun, C.; Ganz, A. Military Inter-Vehicle Communication with Message Priority using IEEE 802.11e. In Proceedings of the MILCOM 2006-2006 IEEE Military Communications Conference, Washington, DC, USA, 23–25 October 2006; pp. 1–5. [CrossRef]
50. Stone, J.; Stratmoen, S.; McCarthy, P.; Collar, S. A practical personal and inter-vehicle cordless communication system for today's mounted and dismounted combat vehicle crewman. In Proceedings of the MILCOM 2000: 21st Century Military Communications, Los Angeles, CA, USA, 22–25 October 2002; Volume 1, pp. 30–34. [CrossRef]
51. Kim, S.-W.; Seo, S.-W. Cooperative Unmanned Autonomous Vehicle Control for Spatially Secure Group Communications. *IEEE J. Sel. Areas Commun.* **2012**, *30*, 870–882. [CrossRef]
52. Chen, H.; Hovareshti, P.; Baras, J.S. Distributed Collaborative Controlled autonomous vehicle systems over wireless networks. In Proceedings of the 8th Mediterranean Conference on Control & Automation (MED), Marrakech, Morocco, 23–25 June 2010; pp. 1695–1700. [CrossRef]
53. Tripathi, B.; Mishra, S.; Garg, S.; Kumar, A.; Kumar, P. An RF relay based control and communication system for Unmanned Ground Vehicle and Micro Air Vehicle. In Proceedings of the 2nd International Conference in Computing for Sustainable Global Development (INDIACom), New Delhi, India, 11–13 March 2015; pp. 406–409.
54. Jeyaprakash, T.; Mukesh, R. A Tactical Information Management System for Unmanned Vehicles Using Vehicular Adhoc Networks. In Proceedings of the 4th International Conference on Intelligent Systems Modelling & Simulation (ISMS), Bangkok, Thailand, 29–31 January 2013; pp. 472–474. [CrossRef]
55. Farooq, W.; Khan, M.A.; Rehman, S.; Saqib, N.A.; Abbas, M. AAGV: A Cluster Based Multicast Routing Protocol for Autonomous Aerial and Ground Vehicles Communication in VANET. In Proceedings of the 15th International Conference on Frontiers of Information Technology (FIT'17), Islamabad, Pakistan, 18–20 December 2017; pp. 315–320. [CrossRef]

56. Khan, M.A.; Shah, G.A.; Ahsan, M.; Sher, M. An efficient and reliable clustering algorithm for wireless sensor actor networks (WSANs). In Proceedings of the 53rd IEEE International Midwest Symposium on Circuits and Systems (MWSCAS), Seattle, WA, USA, 1–4 August 2010; pp. 332–338. [CrossRef]
57. Kevin Fall. The ns Manual-The VINT Project. 2011. Available online: http://www.isi.edu/nsnam/ns/ns-documentation.html (accessed on 1 August 2022).
58. The Network Simulator 2 (NS2). Available online: http://www.isi.edu/nsnam/ns/ (accessed on 9 July 2021).
59. ns-3 | A Discrete-Event Network Simulator for Internet Systems. Available online: https://www.nsnam.org/ (accessed on 18 November 2020).
60. Lopez, P.A.; Behrisch, M.; Bieker-Walz, L.; Erdmann, J.; Flotterod, Y.P.; Hilbrich, R.; Lucken, L.; Rummel, J.; Wagner, P.; Wiebner, E. Microscopic Traffic Simulation using SUMO. In Proceedings of the 2018 21st International Conference on Intelligent Transportation Systems (ITSC), Maui, HI, USA, 4–7 November 2018; pp. 2575–2582. [CrossRef]
61. Kunz, T.; Cheng, E. Multicasting in Ad-Hoc Networks: Comparing MAODV and ODMRP. In Proceedings of the Workshop on Ad hoc Communications, Bonn, Germany, September 2001; pp. 10186–10190.
62. Johnson, D.B.; Maltz, D.A.; Broch, J. DSR: The Dynamic Source Routing Protocol for Multi-Hop Wireless Ad Hoc Networks. In *Ad Hoc Networking*; Addison-Wesley Longman Publishing Co., Inc.: Boston, MA, USA, 2001; pp. 139–172.
63. Verdone, R.; Fabbri, F.; Buratti, C. Maximizing area throughput in clustered wireless sensor networks. *IEEE J. Sel. Areas Commun.* **2010**, *28*, 1200–1210. [CrossRef]

applied sciences

MDPI

Article

Cooperative Roundabout Control Strategy for Connected and Autonomous Vehicles

Chaojie Wang [1], Yu Wang [1] and Srinivas Peeta [1,2,*]

1 School of Civil and Environmental Engineering, Georgia Institute of Technology, Atlanta, GA 30332, USA
2 H. Milton Stewart School of Industrial and Systems Engineering, Georgia Institute of Technology, Atlanta, GA 30332, USA
* Correspondence: peeta@gatech.edu

Abstract: Intersections in the urban network are potential sources of traffic flow inefficiency. Existing intersection control mostly adopts the "cross" flow pattern model, while the use of the roundabout circular flow pattern is rather sparse. Connected and autonomous vehicle (CAV) technologies can enable roundabouts to better compete with traditional intersection designs in terms of performance. This study proposes a roundabout control strategy for CAVs to enhance intersection performance while ensuring vehicle safety. A hierarchical framework is developed to decouple the flow-level performance objective and vehicle-level safety constraints to achieve computational tractability for real-time applications. It entails developing a roundabout flow control model to optimize merge-in flows, a merge-in decision model to generate vehicle passing sequence from the optimal flows, and a virtual platoon control model to achieve safe and stable vehicle operations in a circular roundabout platoon. The performance of the proposed roundabout control strategy is illustrated through numerical studies and compared to existing intersection control methods. Its stability and safety characteristics are also demonstrated.

Keywords: roundabout; optimal control; connected and autonomous vehicle

Citation: Wang, C.; Wang, Y.; Peeta, S. Cooperative Roundabout Control Strategy for Connected and Autonomous Vehicles. *Appl. Sci.* **2022**, *12*, 12678. https://doi.org/10.3390/app122412678

Academic Editors: Nadjim Horri, William Holderbaum and Fabrizio Giulietti

Received: 3 November 2022
Accepted: 7 December 2022
Published: 10 December 2022

1. Introduction

Traffic smoothness is a description of the consistency and invariance of traffic flow. A common symptom of unsmooth traffic is traffic oscillations, which is the "stop-and-go" or "slow-and-fast" flow pattern [1]. Traffic oscillations can result in traffic congestion, thus affecting traffic capacity, reducing traffic safety, and increasing energy consumption. Compared to traffic flow on freeways, urban network traffic has lesser smoothness due to the presence of intersections. The intersection is an essential component of urban road networks. Compared to other road facilities, traffic flow conflicts are more complex at intersections due to the number of different flow directions [2]. Therefore, the efficiency and capacity of an intersection are critical factors for measuring urban transportation system performance. In general, intersections are stop-controlled or signal-controlled. Stop-controlled intersections do not allow multiple vehicles to use the intersection at the same time to ensure safety, which can be inefficient in high-demand cases. Traffic signals separate conflicting traffic flows into different signal phases to enhance the intersection safety. However, the variance and uncertainty in demand decrease the efficiency of signalized intersections. Intersection signal optimization methods, such as [3–5], focus on optimizing the phase lengths to the incoming demands adaptively to reduce inefficient phase time and enhance the overall throughput of the intersection. However, due to the existence of the red phase, traffic has to stop at intersections, leading to limited smoothness.

Compared to signalized intersections, a roundabout has smoother traffic flow characteristics. Roundabouts are circular intersections where road traffic only flows in one direction around a central island. Vehicles only yield to the traffic rather than stop entirely at entry points of the roundabout. Therefore, roundabouts can enhance the smoothness of the intersection. Ref. [6] showed that there is an 89 percent reduction in delays and a 56 percent reduction in vehicle stops after the conversion from signalized intersections to roundabouts. Additionally, roundabouts can improve traffic safety. On the one hand, vehicles will not speed up to try to beat a traffic light because there is no incentive that vehicles can skip a long red phase if they do so. On the other hand, the one-way traffic in the roundabout reduces the possibility of severe T-bone and head-on collisions. Ref. [6] found a 72–80 percent reduction in injury crashes and a 35–47 percent reduction in all crashes in studies of intersections converted from traffic signals to roundabouts. This is because there are only eight conflicting points in a typical four-approach roundabout (without any crossing conflicting points), while a four-approach signalized intersection has 32 conflicting points (16 of them are crossing conflicting points). Compared to merging and diverging conflicting points, crossing conflicting points can entail a higher likelihood of severe accidents because vehicles have higher relative speeds. However, since traffic flows from different directions are not as clearly separated as in signalized intersections, there are more merging and diverging behaviors in roundabouts, which can be challenging for human drivers. To ensure safety, drivers are required to operate vehicles at low speeds when passing through a roundabout. The major causes of collisions in roundabouts are speeding, failing to yield, and improper stopping when proceeding in roundabouts [6]. Further, signalized intersections can adjust the phase lengths of flows from different directions to achieve better system performance, while roundabouts follow the first-come-first-serve rule more strictly as they lack holistic control schemes. In summary, the lack of precision in human drivers' driving behavior can weaken roundabouts' safety advantage, while the oversimplified cooperation rules compared to adaptive traffic signal control scheme can limit the system performance and the application of roundabouts in high-demand situations.

Advances in communication and automation technologies have led to the development of connected and autonomous vehicles (CAVs), which can enable accurate vehicle operations and comprehensive cooperation through vehicle-to-vehicle (V2V) communications. CAVs can address the aforementioned limitations of roundabouts in the context of human-driven vehicles and thereby illustrate the advantages of roundabouts over signalized intersections. Existing studies have shown that CAVs are able to enhance system performance with safety guarantees by incorporating additional real-time information collected using V2V communications into comprehensive control schemes [1,7]. In this context, cooperative adaptive cruise control (CACC) could be leveraged to ensure CAVs' safety while allowing them to pass through smaller-radius roundabouts at higher speeds. The connectivity ability further enables holistic cooperative control strategies to enhance system performance. Several studies have developed model-based [8,9] or machine learning-based [10,11] control strategies for traditional intersection geometries without using traffic signals, which are labeled as signal-free intersection control strategies, to reduce traffic oscillations caused by signalized intersections and enhance smoothness. However, the "cross" flow pattern within the intersection remains the same as in signalized intersections, thereby retaining the safety issues caused by crossing conflicting points. For safety reasons, more constraints are incorporated in signal-free intersection control models, which further constrains the enhancement of system performance under CAV technologies.

While roundabouts can potentially enable better performance in a CAV environment compared to signal-free intersections, few efforts have focused on developing roundabout control strategies. Ref. [12] proposed a fuzzy logic steering control for autonomous vehicles in roundabouts. Ref. [13] developed a parameter-based path generation method for autonomous vehicles in roundabouts. However, these studies focus on the control strategy for a single-vehicle scenario. Other studies [14,15] proposed game-theoretic-based decision-making methods for a two-vehicle scenario in a roundabout. Hence, existing studies address the control problem for specific scenarios and cannot be generalized to high-demand intersections. Moreover, they mainly address safety considerations. However, demand is typically higher at intersections in urban networks, which entails the need for system performance (i.e., smoothness, total delays, and throughput) analysis for intersection control strategies. This motivates the need for a systematic approach to leverage the advantages of roundabouts in the CAV context.

This study seeks to explore the following two research questions. How can CAVs leverage emerging technologies when traveling through intersections with the circular flow pattern to enhance efficiency and safety? Does the circular flow pattern have advantages over the "cross" flow pattern in future CAV environments? The answers to these questions can shed light on future traffic operations and infrastructure design at intersections. Therefore, this study proposes a cooperative roundabout control strategy for CAVs to exploit the advantages of roundabout design to promote performance efficiency of roundabouts while ensuring vehicle safety. As shown in Figure 1, the proposed model is a hierarchical framework that consists of a roundabout flow control model, a merge-in decision model, and a within-roundabout virtual platoon control model. The time horizon of interest is divided into equal time intervals within which the incoming demands of the roundabout are assumed constant. During each time interval τ, the proposed roundabout flow control model solves for the optimal desired merge-in flows (which are the flows merging into the circular flow within the roundabout) to reduce the total estimated waiting time in queues in the current time interval to enhance system performance. Then, the vehicle merge-in control model performs a probabilistic merge-in decision control to achieve the optimal merge-in flows and generate the vehicle passing sequence (vehicles' positions in the virtual platoon) at the same time. Finally, a CACC-based in-roundabout virtual platoon control model is proposed to keep the circular platoon cruising at the desired angular speed in a stable manner while each vehicle maintains a safe distance with the preceding and following vehicles.

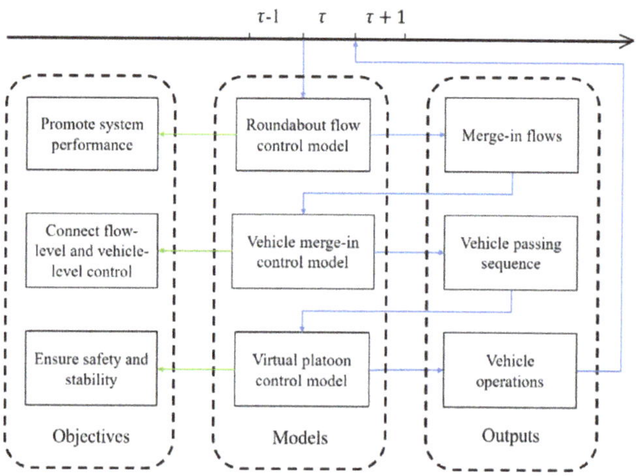

Figure 1. Conceptual structure of the roundabout control strategy.

The rest of the paper is organized as follows. Section 2 presents the three components of the proposed roundabout control strategy. Numerical studies on the effectiveness of the roundabout flow control and virtual platoon control are discussed in Section 3. Section 4 concludes the paper by summarizing contributions and identifying potential future directions.

2. Methodology

2.1. Roundabout Flow Control

Consider a typical roundabout shown in Figure 2, which has four approaches indexed by $i = 1, 2, 3, 4$. The incoming flows on the four approaches are denoted by Q_i, assumed constant during each time period τ, and known at the beginning. In real world, Q_i can be predicted using real-time loop detector data, while the length of τ can be adjusted according to the changing rate of Q_i. The queue length on approach i, $l_i(t)$, is a state variable of t. In each time period, t starts from 0 and ends at $\Delta\tau$ (the time period length). Merge-in flows, $q_i(t)$, are control inputs, which represent the flows merging into the circular flow within the roundabout from approach i. The queue length dynamics can be described as

$$\frac{dl_i(t)}{dt} = Q_i - q_i(t),\ \forall i = 1,\ 2,\ 3,\ 4. \tag{1}$$

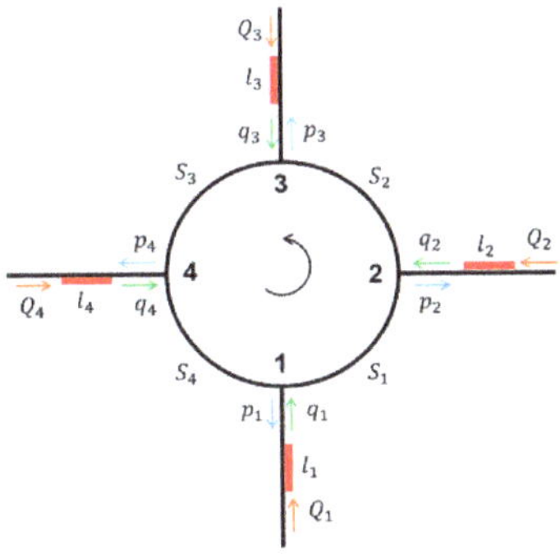

Figure 2. Four-approach roundabout.

Apart from the non-negative constraint, it is also reasonable to have an upper bound $L_{i,max}$ for queue length $l_i(t)$. In applications, $L_{i,max}$ can be decided by the critical queue length that may cause spillback on approach i. Therefore, the state constraints are

$$0 \le l_i(t) \le L_{i,max},\ \forall i = 1,\ 2,\ 3,\ 4. \tag{2}$$

To model the constraints of control inputs q_i, the roundabout segments are introduced and described. As seen in Figure 2, the circular roundabout is separated by four approaches into four arc-shaped segments, S_1, S_2, S_3, and S_4. Since these segments are likely to have similar road design as they belong to the same roundabout, it is assumed that all segments have the same service capacity $\mathcal{Q}_{S,max}$ (i.e., the upper bound of segment flow $\amalg_{S_k}$, $k = 1,\ 2,\ 3,\ 4$). Note that $\amalg_{S_k}$ may consist of vehicles from every approach. For example, vehicles entering from approach 1 heading to approach 3 will drive through S_1 and S_2, and thereby contribute to both $\amalg_{S_1}$ and $\amalg_{S_2}$. Similarly, vehicles entering from approach 2 heading to approach 1 will also drive through S_2, thereby contributing to $\amalg_{S_2}$. Denote flows from approach i heading to approach j as q_{ij} $(i, j = 1,\ 2,\ 3,\ 4)$, which can be represented as $q_{ij} = \eta_{ij} q_i$, where η_{ij} is the proportion of the flow heading to approach j in the merge-in flow from approach i. To describe the relationships between S_k and q_{ij}, define the service tables of four segments as follows (note that this study assumes that vehicles will drive through each segment no more than once when passing through the roundabout). The element in row i and column j of the service table for S_k, a^k_{ij}, is an indicator variable of whether q_{ij} will pass through S_k. For instance, $a^4_{22} = 1$, indicating the U-turn flow from approach 2 will pass through S_4 in the roundabout; while $a^2_{32} = 0$ means vehicles coming from approach 3 heading to approach 2 will not drive through S_2. Therefore, the flow on S_k is

$$\amalg_{S_k}(t) = \sum_{i=1}^{4} \sum_{j=1}^{4} a^k_{ij} \eta_{ij} q_i(t),\ \forall k = 1,\ 2,\ 3,\ 4 \tag{3}$$

Then, the control input constraints can be represented as

$$\amalg_{S_k}(t) \le \mathcal{Q}_{S,max},\ \forall k = 1,\ 2,\ 3,\ 4 \tag{4}$$

$$q_i(t) > 0,\ \forall i = 1,\ 2,\ 3,\ 4 \quad (5)$$

The control objective is

$$\min_{q_1, q_2, q_3, q_4} \int_0^{\Delta\tau} \left(\sum_{i=1}^{4} \frac{Q_i l_i(t)}{q_i(t)} \right) dt \quad (6)$$

where $l_i(t)/q_i(t)$ is the estimated waiting time for vehicles at the end of the queue on approach i, and $Q_i dt$ is the number of incoming vehicles or vehicles joining the queue at the end on approach i. Therefore, the integration can be interpreted as the total estimated waiting time of vehicles arriving from $t = 0$ to $t = \Delta\tau$. Compared to the vehicle-level delay representation, Equation (6) provides an informative and computationally tractable estimation for the total system delay. The complexity of the problem does not increase with the number of vehicles, enabling its use in real-time applications. The optimal control problem defined by (1)~(6) falls in the domain of nonlinear optimal control, which generally does not have analytical solutions. Therefore, a direct collection method [16] is applied to solve it, which converts the optimal control problem into a nonlinear optimization problem by discretizing state variables $l_i(t)$ and control inputs $q_i(t)$ into vectors $\hat{\boldsymbol{l}}_i = [l_i(0), \ldots, l_i(\Delta\tau)]$, $\hat{\boldsymbol{q}}_i = [q_i(0), \ldots, q_i(\Delta\tau)]$.

2.2. Vehicle Merge-in Decision Control

Given the optimal merge-in flows from the last subsection, the intersection controller needs to coordinate vehicles coming from four approaches on when they should merge to achieve the desired merge-in flows. Therefore, a probabilistic merge-in decision control model is developed to determine when the vehicles at the head of queues can start to merge in. In this way, the flow-level control inputs are converted into an optimal vehicle passing sequence, which can be connected to vehicle-level operations.

Let the spare capacity s_k of each segment k denote the difference between the capacity $Q_{S,max}$ and the flow on S_k, that is $s_k(t) = Q_{S,max} - \amalg_{S_k}(t)$. Define S_1, S_2, S_3, and S_4 as the upstream segment of approaches 2, 3, 4, and 1, respectively. Apart from the spare capacity from approach i's upstream segment, there is additional p_i capacity available for vehicles on approach i, which can be described as follows:

$$p_i(t) = \sum_{j=1}^{4} \eta_{ji} q_j(t),\ \forall i = 1,\ 2,\ 3,\ 4, \quad (7)$$

Note that p_i can also be interpreted as the flow that diverges out at approach i (diverge-out flow). Therefore, the total available capacity for the merge-in flow of approach i (suppose its upstream segment is S_k) is $s_k + p_i$. Let P_i denote the probability of the leading vehicle on approach i choosing to merge into the circular flow within the roundabout whenever there is available capacity. Then, the probability P_i should satisfy the following condition to achieve the desired merge-in flows:

$$P_i = \frac{q_i}{Q_{S,max} - \amalg_{S_k} + p_i} \quad (8)$$

where S_k is the upstream segment of approach i. With this condition met, the actual merge-in flow from approach i would be $P_i(s_k + p_i) = q_i$. Following (3) and (7), it is proved that $0 < P_i \leq 1$, which guarantees that our decision control is applicable at all times.

Since the merge-in decision control is probabilistic, the actual merge-in flows given by this model are different from the desired merge-in flows. Therefore, the optimal system performance given by the flow control model is an upper bound to the actual system performance, which is illustrated in our numerical studies as well.

2.3. Vehicle Platoon Control

Once vehicles choose to merge into the roundabout based on the probabilistic decision control model, they will start the merge-in procedure and eventually cruise with the circular platoon until they diverge to the approaches they are heading to. Note that this subsection focuses on longitudinal control therefore does not consider non-holonomic vehicle dynamics or low-level vehicle control such as throttle/steering control. The merge-in procedure consists of a self-adjusting stage and a virtual platoon control stage. Figure 3 illustrates an example of the two stages using the merge-in procedure of vehicle B_1. First, B_1 waits at position W_2 (marked by a red line) when it is at the head of the queue on approach 2. When the decision control model allows it to merge into a spare slot (labeled the matching slot of B_1) in the platoon, B_1 enters the self-adjusting stage. The objective of this stage is to have vehicle B_1 drive through a critical position C_2 (marked by a blue line on approach 2) to enter the virtual platoon control stage at the desired speed v_{des} after some desired time Δt_{des}. The desired speed is also the desired cruising speed of the platoon. The desired time Δt_{des} is estimated as the time it takes for the matching slot to travel to a virtual critical position V_2 (marked by the yellow line on the right part of the roundabout in Figure 3) from the beginning of the self-adjusting stage. The virtual critical position V_2 is a projection of C_2 on the roundabout circle, that is $|M_2V_2| = |M_2C_2|$, where M_2 is a merging point of the roundabout and the on-ramp from approach 2. When vehicle B_1 travels from C_2 to M_2, a virtual vehicle Y_1, which is the projection of B_1, also travels from V_2 to M_2. At the same time, B_1 communicates with vehicles G_1 and G_2, making Y_1 behave like a vehicle communicating with its surrounding vehicles in the circular platoon. This stage is labeled as the virtual platoon control stage. Note that when determining the length of $|M_2C_2|$, $|M_2D_2| \geq |M_2C_2| = |M_2V_2|$ should be ensured, where D_2 is the diverging point of the off-ramp to approach 2. It is assumed that once a diverging vehicle passes the diverging point and drives to the off-ramp, it is no longer a part of the circular platoon (i.e., its movement will not affect surrounding vehicles in the platoon anymore). This guarantees that during the virtual platoon control stage, there will be no real diverging vehicle at the slot where the virtual vehicle projected from a real merge-in vehicle is located. Otherwise, they may have a conflicting influence on the surrounding vehicles since they are at the same slot in the platoon.

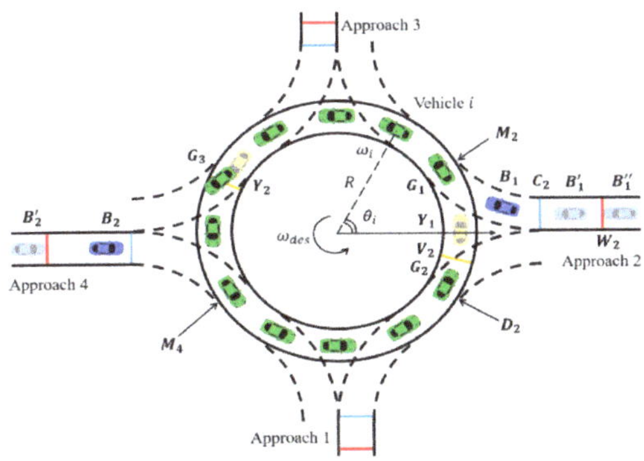

Figure 3. Illustration of self-adjusting stage and virtual platoon control stage in vehicles' merge-in procedure.

The control of a vehicle from approach i in the self-adjusting stage can be formulated as

$$\int_0^{\Delta t_{des}} v(t)dt = |C_iW_i| \tag{9a}$$

$$v(\Delta t_{des}) = v_{des} \tag{9b}$$

$$0 \le v(t) \le v_{max},\ \forall t \in [0,\ \Delta t_{des}] \tag{9c}$$

$$a_{min} \le \frac{dv(t)}{dt} \le a_{max},\ \forall t \in [0,\ \Delta t_{des}] \tag{9d}$$

where v_{max}, a_{min}, a_{max} are the maximum speed, maximum deceleration, and maximum acceleration, respectively. As shown in Figure 4, in a speed-time graph, the formulation seeks for trajectories starting from $(0,0)$ ending at $(\Delta t_{des},\ v_{des})$, whose time integration equals to $|C_iW_i|$. The formulation can have multiple feasible solutions if $S_g \le |C_iW_i| \le S_b$, where S_g and S_b are the blue (right)-shaded and green (left)-shaded areas, respectively, in Figure 4. The blue-shaded area stands for the maximum distance a vehicle can travel within Δt_{des} (the vehicle accelerates at a_{max}, all the way to v_{max}, then maintains the speed until an instant after which it can keep decelerating at a_{min} to arrive at C_i with desired final speed v_{des}). Similarly, the green-shaded area is the lower bound of the distance with the given constraints (9b-d). One feasible strategy is to follow the speed trajectory indicated by the yellow (dash-dot)/red (solid) lines when $|C_iW_i|$ is greater than or equal to/less than $\frac{v_{des}^2}{2a_{max}} + v_{des}(\Delta t_{des} - \frac{v_{des}}{a_{max}})$. Specifically, the yellow dash-dot speed profile indicates that the vehicle accelerates at a_{max}, all the way to a specific speed v', then maintains the speed until an instant after which it can keep decelerating at a_{min} to arrive at C_i with the desired final speed v_{des}. The red solid line shows that the vehicle stops at W_i until moment t', then accelerates at a_{max} until the speed reaches v_{des}, and maintains this speed up to the end Δt_{des}. Both v' and t' can be derived from the constraint (9a).

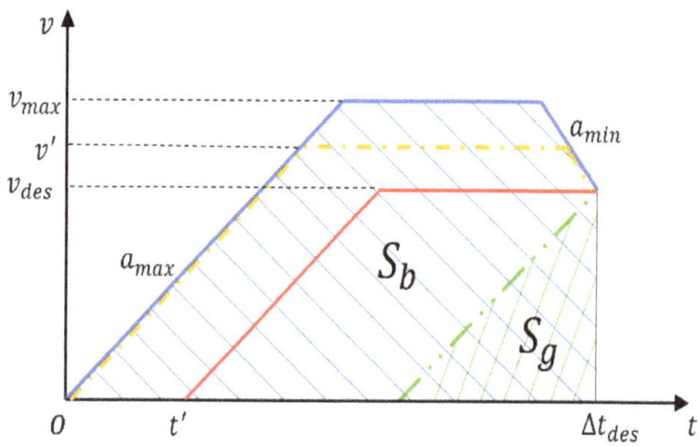

Figure 4. Feasible speed trajectories for vehicles in the self-adjusting stage.

The merge-in procedure will be seamless if the transition of two stages is as precise as described by (9a–d). However, it is possible that when the matching slot arrives at the virtual critical position, the vehicle is not exactly at the critical position at the desired speed because a critical parameter Δt_{des} is estimated and possibly inaccurate when solving (9a–d). For example, in Figure 3, the matching slot of vehicle B_2 arrives at the diverging point with the diverging vehicle G_3 earlier than Δt_{des}. Additionally, it is possible that, at Δt_{des}, G_3 still needs a little more time to get to the off-ramp. Then, B_2 will have passed the critical point when G_3 leaves the platoon. Note that the start time of the virtual platoon control stage is when the available slot arrives at the diverging point, not when the vehicle arrives at the critical point or the blue lines in Figure 3. Therefore, the virtual platoon control should ensure stability to mitigate the initial speed and position errors introduced by virtual vehicles when they enter the virtual platoon control stage.

To ensure stability, the circular platoon in the roundabout is investigated, which consists of real vehicles, virtual vehicles projected from the vehicles on on-ramps at the second merge-in stage, and some spare vehicle slots. Note that here we assume a circular roundabout for simplicity; however, the model is not restricted only to ideal circular roundabouts. The platoon control model can be extended to non-circular roundabouts by mapping the roadways to circular roundabouts with the same road length. Suppose that when the roundabout serves at its maximum capacity, there are N homogeneous vehicles cruising in the roundabout with a desired angular speed $\omega_{des} = v_{des}/R$ (where R is the radius of the roundabout). It implies that the roundabout has N vehicle slots labeled from 1 to N. When there are spare vehicle slots, the circular platoon can be converted into one (if there is only one spare vehicle slot) or more linear vehicle platoons using the spare slots to separate them. Then, existing linear CACC platoon control strategies [17] can be applied.

For a circular platoon with no spare vehicle slots, a symmetric bidirectional control architecture is developed (where predecessor and follower position errors influence the vehicle controller symmetrically). As shown in Figure 3, an angular coordinate θ_n, ω_n is used to describe the position and angular speed of vehicle n (i.e., vehicle matched with vehicle slot n). Each vehicle is modeled as a double integrator. The control inputs for vehicle n in the platoon are assumed to depend only on its speed error $\omega_n - \omega_{des}$ and the relative headway errors between itself and its immediate neighbors (i.e., its predecessor and follower). Denote the positions of the preceding and following vehicles of vehicle n as θ_n^p and θ_n^f, respectively. The desired angular headway is $\Delta = 2\pi/N$. The vehicle dynamics can be described as

$$\frac{d\theta_n(t)}{dt} = \omega_n \tag{10a}$$

$$\frac{d\omega_n(t)}{dt} = -k_v(\omega_n - \omega_{des}) + k_d\left(\theta_n^p - \theta_n - \Delta\right) - k_d\left(\theta_n - \theta_n^f - \Delta\right) \tag{10b}$$

where k_v and k_d are positive constants. To facilitate analysis, a coordinate change is considered using the initial position of slot 1, $\theta_1(0)$, as a reference point:

$$\widetilde{\theta}_n(t) = \theta_n(t) - \omega_{des}t - \theta_1(0) - (n-1)\Delta \tag{11a}$$

$$\widetilde{\omega}_n(t) = \omega_n(t) - \omega_{des} \tag{11b}$$

To describe the dynamics of the circular platoon, define $\widetilde{\boldsymbol{\theta}} := \left[\widetilde{\theta}_1, \widetilde{\theta}_2, \ldots, \widetilde{\theta}_N\right]^T$, $\widetilde{\boldsymbol{\omega}} := [\widetilde{\omega}_1, \widetilde{\omega}_2, \ldots, \widetilde{\omega}_N]^T$. Substituting (11a–b) into (10a–b), the following equation is obtained:

$$\begin{bmatrix} \dot{\widetilde{\boldsymbol{\theta}}} \\ \dot{\widetilde{\boldsymbol{\omega}}} \end{bmatrix} = \begin{bmatrix} \mathbf{0} & \boldsymbol{I} \\ -k_d\boldsymbol{T} & -k_v\boldsymbol{I} \end{bmatrix} \begin{bmatrix} \widetilde{\boldsymbol{\theta}} \\ \widetilde{\boldsymbol{\omega}} \end{bmatrix}, \tag{12}$$

$$\boldsymbol{T} = \begin{bmatrix} 2 & -1 & 0 & \ldots & 0 & -1 \\ -1 & 2 & -1 & 0 & \ldots & 0 \\ 0 & -1 & 2 & -1 & \cdots & 0 \\ \vdots & \ddots & \ddots & \ddots & \ddots & \vdots \\ 0 & \ldots & 0 & -1 & 2 & -1 \\ -1 & 0 & \ldots & 0 & -1 & 2 \end{bmatrix}. \tag{13}$$

Equation (12) indicates that the circular platoon to be controlled is a linear system. The following lemma [18] illustrates the non-negativity of T's eigenvalues.

Lemma 1 [18]. *Given an $n \times n$ circulant matrix C with the form*

$$\boldsymbol{C} = \begin{bmatrix} c_0 & c_{n-1} & \ldots & c_2 & c_1 \\ c_1 & c_0 & c_{n-1} & \cdots & c_2 \\ \vdots & c_1 & c_0 & \ddots & \vdots \\ c_{n-2} & \cdots & \ddots & \ddots & c_{n-1} \\ c_{n-1} & c_{n-2} & \cdots & c_1 & c_0 \end{bmatrix}, \tag{14}$$

the normalized eigenvectors are the Fourier modes, namely,

$$v_j = \frac{1}{\sqrt{n}}\left(1,\ \kappa^j,\ \kappa^{2j},\ \ldots,\ \kappa^{(n-1)j}\right),\ j = 0,\ 1,\ \ldots,\ n-1, \tag{15}$$

where $\kappa = exp\left(\frac{2\pi i}{n}\right)$ and *i is the imaginary unit. The corresponding eigenvalues are then given by*

$$\lambda_j = c_0 + c_{n-1}\kappa^j + c_{n-2}\kappa^{2j} + \ldots + c_1\kappa^{(n-1)j},\ j = 0,\ 1,\ \ldots,\ n-1 \tag{16}$$

Note that $\boldsymbol{T}$ is a $N \times N$ circulant matrix with $c_0 = 2,\ c_1 = c_{N-1} = -1,\ c_i = 0\ (i = 2,\ 3,\ \ldots,\ N-2)$. According to Lemma 1, the jth eigenvalue of $\boldsymbol{T}$ is

$$\lambda_{Tj} = 2 - \kappa^j - \kappa^{(N-1)j},\ j = 0,\ 1,\ \ldots,\ N-1, \tag{17}$$

where $\kappa = \exp\left(\frac{2\pi i}{N}\right)$. Note that $\kappa^{(N-1)j} = -\kappa^j$. Therefore,

$$\lambda_{Tj} = 2 - \kappa^j - \kappa^{-j} = 2 - 2\cos\left(\frac{2\pi j}{N}\right) \geq 0,\ j = 0,\ 1,\ \ldots,\ N-1, \tag{18}$$

Theorem 1 [19]. *The equilibrium states for a general linear time-invariant system* $\dot{x} = Ax$ *are stable if and only if each eigenvalue* λ_A *of* A *satisfies* $\Re(\lambda_A) \leq 0$ *and* $\Re(\lambda_A) < 0$ *if* λ_A *is defective, where* $\Re(\lambda_A)$ *is the real part of* λ_A.

From Theorem 1 (Theorem 5.4 of [19]), the eigenvalues of the system can be described by (12) as follows:

$$\begin{bmatrix} \boldsymbol{0} & \boldsymbol{I} \\ -k_d\boldsymbol{T} & -k_v\boldsymbol{I} \end{bmatrix}\begin{bmatrix} \widetilde{\boldsymbol{\theta}} \\ \widetilde{\boldsymbol{\omega}} \end{bmatrix} = \lambda\begin{bmatrix} \widetilde{\boldsymbol{\theta}} \\ \widetilde{\boldsymbol{\omega}} \end{bmatrix}\left(\lambda^2 + k_v\lambda\right)\widetilde{\boldsymbol{\theta}} = -k_d\boldsymbol{T}\widetilde{\boldsymbol{\theta}} \tag{19}$$

With $\boldsymbol{T}$'s eigenvalues $\lambda_{Tj},\ j = 0,\ 1,\ \ldots,\ N-1$, we have

$$\lambda^2 + k_v\lambda + k_d\lambda_{Tj} = 0\lambda = \frac{-k_v \pm \sqrt{k_v^2 - 4k_d\lambda_{Tj}}}{2} \tag{20}$$

Recall that (18) has shown the non-negativity of λ_{Tj}. Therefore, for $k_v, k_d > 0, \Re(\lambda) \leq 0$ and when $k_v^2 - 4k_d\lambda_T = 0$, $\Re(\lambda) = -k_v/2 < 0$. According to Theorem 1, the equilibrium states of (12) are stable.

3. Numerical Studies

To demonstrate the performance of the proposed CAV control strategy at roundabouts, several numerical studies are conducted for a four-approach single-lane roundabout with $N = 12$ vehicle slots. The desired speed v_{des} is set as 8 m/s (17.9 mph). The service capacity of the roundabout $\mathcal{Q}_{S,max}$ is 60 vehicles/minute. The control time period $\Delta\tau = 5$ min. Note that in real-world applications, the control time period is related to the variation of incoming flows. The principle for the determination of $\Delta\tau$ is that the incoming flows do not change significantly within the control time period because the optimal merge-in flows are solved based on the constant incoming flow assumption.

First, the effectiveness of the roundabout flow control is investigated. The parameters and initial conditions for each approach are shown in Table 1.

Table 1. Experiment settings.

Approach ID	Q_i (Vehicles/min)	$l_i(0)$
1	30	2
2	25	1
3	35	3
4	20	4

For $\eta = [\eta_{ij}]$, the experiment uses the following setting.

$$\eta = \begin{bmatrix} 0.0 & 0.7 & 0.2 & 0.1 \\ 0.2 & 0.0 & 0.6 & 0.2 \\ 0.8 & 0.1 & 0.0 & 0.1 \\ 0.1 & 0.5 & 0.4 & 0.0 \end{bmatrix}. \tag{21}$$

Figure 5 shows the optimal control trajectory of merge-in flows, queue lengths, and segment flows of the system. Five phases (separated by dashed lines) are observed in the 5 min optimal control solution. Within each phase, the control inputs (merge-in flows) remain constant, and the queue lengths change linearly. Figure 5c shows that the maximum capacity constraints for the four segments are all active in the first phase, indicating that the roundabout is operating at its maximum capacity. Each time a queue disappears, the system adjusts the merge-in flows, which demonstrates the adaptiveness of the control strategy. The number of active segment-maximum-capacity constraints decreases with the number of non-zero queues. Because when there is no queue on approach i, the corresponding merge-in flow $q_i = Q_i$. With fewer undetermined control inputs, fewer constraints are active. In the last phase, when there is no queue on four approaches, the merge-in flow q_i on each approach is equal to the incoming flow Q_i. Compared to conventional roundabouts' first-come-first-serve principle, the cooperation among the four approaches is also reflected in the results. For example, the queue on approach 2 grows in the first phase to prioritize vehicles from other approaches with higher demand pressures (i.e., longer queues or higher incoming flows). However, in real-world applications, some approaches may have an upper bound on queue length to avoid spillbacks. Figure 6 illustrates how the queue lengths evolve with/without an upper-bound value of three, on the length of queue on approach 2, l_2. Figure 6b illustrates that after l_2 reaches 3, the system prioritizes vehicles from the other approaches. However, it adjusts the merge-in flows to meet the queue length constraint on approach 2.

Next, the stability performance of the circular platoon control is validated. Note that the merge-in control and the self-adjusting stage are not simulated. It is assumed that the occurrence of merging-in follows a Poisson distribution ($\lambda = 1$), because on average there will be one vehicle merging in when the roundabout operates at its maximum capacity $Q_{S,max} = 60$ vehicles/minute. The initial speed and position disturbances are assumed to follow a uniform distribution $[-1,\ 1]$ m/s and a uniform distribution $[-1,\ 1]$ m, respectively, which are reasonable ranges for speed and position errors in the real world. Figure 7 shows the position errors, headways, and speed errors of 12 vehicles due to disturbances introduced by newly merging vehicles. The equilibrium state of the circular platoon changes each time a new vehicle merges in. For example, in Figure 7a, vehicles' equilibrium states in the platoon move in the same circular direction (counterclockwise in Figure 3) during $t = 30 \sim 46$ s because of continuous positive position errors introduced by new merge-in vehicles. However, it is related more to coordinate shifts and does not influence vehicle safety. Figure 7b shows that vehicles' headways remain in a safe range (the desired time headway is set as 1 s) during the entire simulation. Figure 7c illustrates that all newly introduced speed errors can be mitigated in a short time. The spikes in Figure 7b,c mostly occur on on-ramps because position and speed errors are introduced by virtual vehicles. It allows the system to converge closer to the equilibrium state before real vehicles merge into the circular platoon, which enhances smoothness and safety.

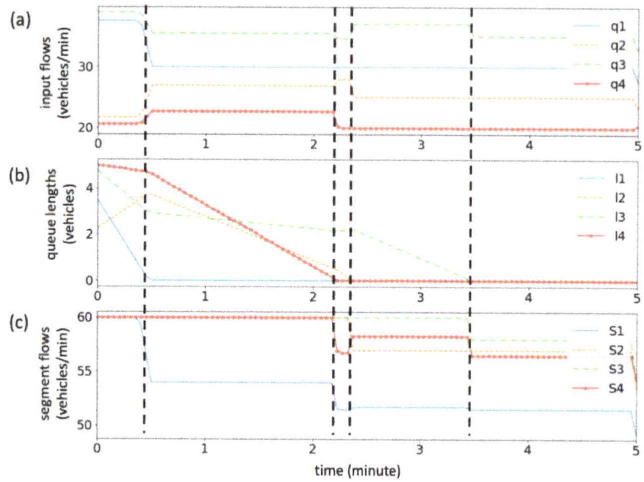

Figure 5. (**a**) Optimal merge-in flows; (**b**) queue lengths; (**c**) segment flows.

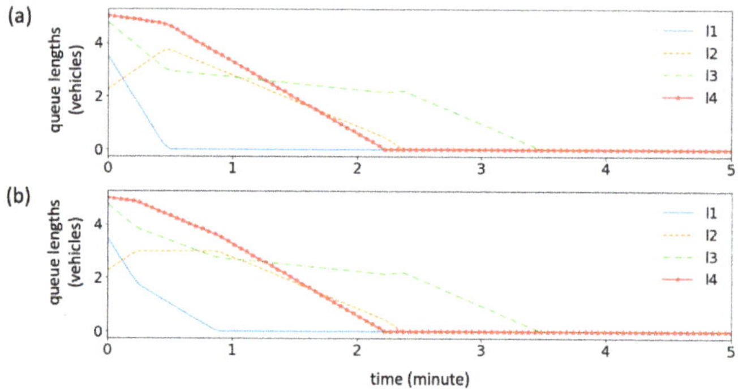

Figure 6. Queue length evolution: (**a**) without queue length upper bound; (**b**) with queue length upper bound.

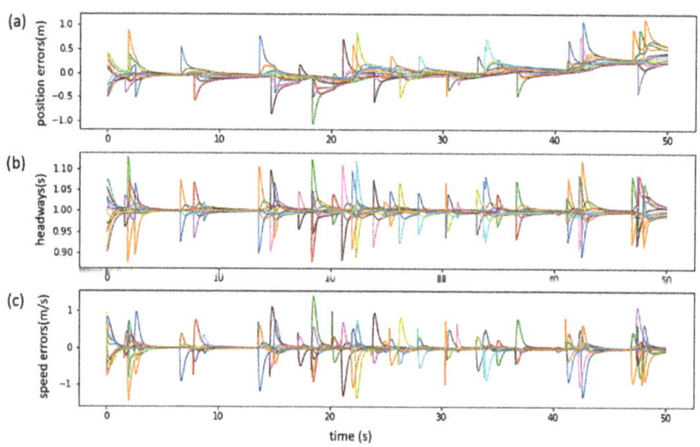

Figure 7. Stability and safety analysis of virtual platoon control: (**a**) position errors; (**b**) headways; (**c**) speed errors.

In summary, the performances of both upper-level flow control and lower-level vehicle control are illustrated separately. Next, these two components are integrated with the proposed vehicle merge-in control and a comprehensive simulation (microscopic simulation) is performed. Figure 8 shows a screenshot of the simulation interface, which illustrates the integration of the proposed control components. The blue dots denote incoming vehicles and platooning vehicles, while the orange ones are vehicles that have passed through the roundabout. In the top-right zoomed-in view, desired positions of platooning vehicles (slot positions) are marked with the slot number labeled. The bottom-right of Figure 8 shows that the speed errors and headway errors of the platooning vehicles in the last 250 steps (the step size is 0.05 s) are updated dynamically. Damping patterns can be observed similar to those in the validation of the platoon stability in Figure 7.

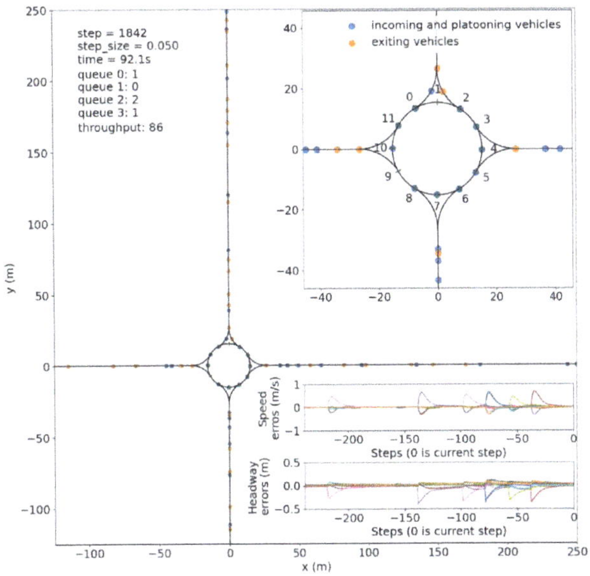

Figure 8. Simulation of the proposed roundabout control strategy.

To compare the overall performance of the proposed roundabout control strategy with other intersection control strategies, the demand pattern associated with the high demand case (case 6) in [9] is used. Table 2 shows that though the demands are different, the proportions of corresponding demands in the two intersection designs are identical (0.75). As the roundabout model uses only a two-lane roadway setting, it is not meaningful to compare its performance using the demand on various approaches of typical four-lane crossroads. To address this issue, and account for the effects of different time headway settings and different numbers of lanes on incoming approaches, a new intersection design efficiency measure labeled as $\rangle$ is proposed, which is the throughput of the intersection divided by the sum of the capacity of all input lanes. The intersection design efficiency quantifies the capability of the intersection design (both flow pattern design and control strategy design) to serve the capacity of all incoming approaches. The proposed unit-free measure is justified as follows. The most efficient intersection design is to use separate facilities (for example, grade-separated roads) to connect all approaches. Thereby, vehicles traveling in different directions will not affect each other, which implies that the intersection can serve the capacity of all incoming approaches (regardless of the number of lanes and how large the capacities are); thus $\rangle = 1$. This "perfect" intersection design serves as an idealized benchmark to evaluate the efficiency of various intersection designs using $\rangle$.

Table 2. Demand settings for performance comparison.

Intersection Design	East and West Bound		North and South Bound	
	Through Demand (veh/h/lane)	Left-Turn Demand (veh/h/lane)	Through Demand (veh/h/lane)	Left-Turn Demand (veh/h/lane)
Two-lane roundabout	1125	450	1125	450
Four-lane crossroad	1500	600	1500	600

Note that in flow-level control, the vehicle-level randomness in the incoming flow is neglected. Additionally, the vehicle merge-in control assumes instantaneous changes of within-roundabout flow due to merge-in flows. Hence, the overall performance of the proposed model can be less efficient than the performance determined from the roundabout flow control component. In the following comparison, the idealized performance associated with the roundabout flow control is used as a benchmark.

The proposed measure enables comparisons between intersection designs with different incoming approach capacities. The results for signal-free intersection control, fully actuated signal control, and fixed-time signal control are from [9] (simulated on intersections with four-lane roads). The proposed roundabout control strategy is simulated for an intersection with two-lane roads (one lane in each direction). The seven-minute (including a two-minute warm-up) simulation recording can be accessed online (https://youtu.be/XEZhPJnr4eg) on 12 September 2022. As shown by Table 3, while the signal-free intersection control strategy from [9] has the highest throughput, its intersection design efficiency is only 0.188. This is because the "cross" flow pattern in conventional intersection design has significant conflicts associated with the flows from different directions, which limits the spatiotemporal occupancy within the intersection. By leveraging CAVs, the roundabouts' circular flow pattern can fully exploit road capacity inside the intersection. The overall performance of the proposed roundabout control yields a high intersection design efficiency, 0.343, which is almost twice that of the signal-free intersection control strategy. Additionally, note that the intersection design efficiency calculated from flow-level control is higher than that of the overall control, indicating that the proposed model suffers mild efficiency loss when converting optimal merge-in flows into vehicle-level controls. This is because the upper-level flow control neglects the vehicle level randomness. As the top-left part of Figure 8 shows, the queue length on approaches 1, 2, and 3 are 1, 2, 1, respectively, while the roundabout flow control components indicates that there should be no queues during the entire time period. Due to vehicle-level randomness, the queue lengths in the comprehensive simulation have small positive values all the time. The solution optimality is impaired as the flow control model addresses the problem in the continuous domain and an additional discretization step is introduced in the vehicle merge-in control. However, solving optimal control problems at the flow level mitigates the computational burden compared to signal-free intersection control in [9] which directly solves the high-dimensional vehicle-level optimal control problem. It is beneficial to sacrifice optimality slightly for real-time computational tractability.

To summarize, the numerical results illustrate that: (i) the roundabout flow control can coordinate the merging flows from different directions to enhance efficiency; (ii) the vehicle platoon control can dampen the oscillations induced by the speed and position errors at the self-adjusting stage to ensure vehicle safety; (iii) the merge-in control is able to integrate the aforementioned two components with slight impairment to the solution optimality; and (iv) the proposed roundabout control strategy shows the advantage of the circular flow pattern over the "cross" flow pattern in terms of the intersection design efficiency.

Table 3. Performance comparison.

Intersection Control Strategy	Proposed Roundabout Flow Control	Proposed Roundabout Control	Signal-Free Intersection Control	Fully Actuated Signal Control
Throughput (trips/min)	105	82.2	120.5	76.5
Lane capacity (veh/lane/min)	60	60	80	80
Number of lanes	4	4	8	8
Intersection design efficiency	0.438	0.343	0.188	0.119

4. Concluding Comments

This study aims to fill a gap in the current literature by illustrating the advantages of roundabouts in a CAV environment. The main contributions of this study are to: (i) develop a hierarchical roundabout control framework to decouple system performance measures and vehicle safety constraints, to relieve the computational burden by significantly decreasing the dimension of the optimal control problem; (ii) investigate the potential advantages of roundabout designs compared to traditional intersection designs to provide insights for future infrastructure design in urban networks in CAV environments; and (iii) extend existing linear CACC models to a circular CACC virtual platoon control model to address the safety and stability considerations for vehicles operating within a roundabout.

This study suggests another potential venue for intersection infrastructure design for CAVs beyond signal-free intersections. This exploratory investigation has some limitations: (i) the proposed model assumes constant incoming flows; therefore real-world dynamics may not be fully captured; (ii) the roadways of the approaches and the roundabout are both assumed to be one-lane roads; so, multi-lane problems such as lane change are not addressed; and (iii) how to adapt human drivers in the proposed control strategy during the transition period is another unexplored issue. Therefore, the insights from this study can provide directions for further research, including: (i) leveraging closed-loop control strategies such as model predictive control to enhance the roundabout flow control performance (i.e., the flow control model in the proposed hierarchical framework can be replaced with more advanced ones to deal with changing incoming flows and measurement delays); (ii) extending the proposed control strategy to multilane roundabouts; and (iii) investigating roundabout control problem in a mixed traffic environment (e.g., enabling larger operational uncertainty for human driven vehicles in models and employing augmented reality or mixed reality technologies to project suggested vehicle movements in roundabouts to assist human drivers).

Author Contributions: Conceptualization, C.W. and S.P.; methodology, C.W.; software, Y.W.; validation, C.W. and Y.W.; formal analysis, C.W.; investigation, C.W.; resources, S.P.; data curation, Y.W.; writing—original draft preparation, C.W.; writing—review and editing, S.P. and Y.W.; visualization, C.W.; supervision, S.P.; project administration, S.P.; funding acquisition, S.P. All authors have read and agreed to the published version of the manuscript.

Funding: This research was supported by the Center for Connected and Automated Transportation (CCAT) Region V University Transportation Center funded by the U.S. Department of Transportation under Award #69A3551747105.

Institutional Review Board Statement: Not applicable.

Informed Consent Statement: Not applicable.

Data Availability Statement: Not applicable.

Conflicts of Interest: The authors declare no conflict of interest.

References

1. Wang, C.; Gong, S.; Zhou, A.; Li, T.; Peeta, S. Cooperative Adaptive Cruise Control for Connected Autonomous Vehicles by Factoring Communication-Related Constraints. *Transp. Res. Part C Emerg. Technol.* **2020**, *113*, 124–145. [CrossRef]
2. Levin, M.W.; Rey, D. Conflict-Point Formulation of Intersection Control for Autonomous Vehicles. *Transp. Res. Part C Emerg. Technol.* **2017**, *85*, 528–547. [CrossRef]
3. Li, Z.; Elefteriadou, L.; Ranka, S. Signal Control Optimization for Automated Vehicles at Isolated Signalized Intersections. *Transp. Res. Part C Emerg. Technol.* **2014**, *49*, 1–18. [CrossRef]
4. Eom, M.; Kim, B.I. The Traffic Signal Control Problem for Intersections: A Review. *Eur. Transp. Res. Rev.* **2020**, *12*, 50. [CrossRef]
5. Jiang, X.; Zhang, J.; Wang, B. Energy-Efficient Driving for Adaptive Traffic Signal Control Environment via Explainable Reinforcement Learning. *Appl. Sci.* **2022**, *12*, 5380. [CrossRef]
6. Retting, R.A.; Mandavilli, S.; McCartt, A.T.; Russell, E.R. Roundabouts, Traffic Flow and Public Opinion. *Traffic Eng. Control* **2006**, *47*, 268–272.
7. Eisenman, S.; Josselyn, J.; List, G.; Persaud, B.; Lyon, C.; Robinson, B.; Blogg, M.; Waltman, E.; Troutbeck, R.J. *Operational and Safety Performance of Modern Roundabouts and Other Intersection Types*; No. Project SPR C-01-47; New York State Department of Transportation: New York, NY, USA, 2004; Volume 37.
8. Malikopoulos, A.A.; Cassandras, C.G.; Zhang, Y.J. A Decentralized Energy-Optimal Control Framework for Connected Automated Vehicles at Signal-Free Intersections. *Automatica* **2018**, *93*, 244–256. [CrossRef]
9. Mirheli, A.; Tajalli, M.; Hajibabai, L.; Hajbabaie, A. A Consensus-Based Distributed Trajectory Control in a Signal-Free Intersection. *Transp. Res. Part C Emerg. Technol.* **2019**, *100*, 161–176. [CrossRef]
10. Tran, D.Q.; Bae, S.-H. Proximal Policy Optimization through a Deep Reinforcement Learning Framework for Multiple Autonomous Vehicles at a Non-Signalized Intersection. *Appl. Sci.* **2020**, *10*, 5722. [CrossRef]
11. Tran, D.Q.; Bae, S.-H. Comprehensive Automated Driving Maneuvers under a Non-Signalized Intersection Adopting Deep Reinforcement Learning. *Appl. Sci.* **2022**, *12*, 9653. [CrossRef]
12. Rastelli, J.P.; Peñas, M.S. Fuzzy Logic Steering Control of Autonomous Vehicles inside Roundabouts. *Appl. Soft Comput. J.* **2015**, *35*, 662–669. [CrossRef]
13. González, D.; Pérez, J.; Milanés, V. Parametric-Based Path Generation for Automated Vehicles at Roundabouts. *Expert Syst. Appl.* **2017**, *71*, 332–341. [CrossRef]
14. Banjanovic-Mehmedovic, L.; Halilovic, E.; Bosankic, I.; Kantardzic, M.; Kasapovic, S. Autonomous Vehicle-to-Vehicle (V2V) Decision Making in Roundabout Using Game Theory. *Int. J. Adv. Comput. Sci. Appl.* **2016**, *7*, 292–298. [CrossRef]
15. Tian, R.; Li, S.; Li, N.; Kolmanovsky, I.; Girard, A.; Yildiz, Y. Adaptive Game-Theoretic Decision Making for Autonomous Vehicle Control at Roundabouts. In Proceedings of the IEEE Conference on Decision and Control, Miami, FL, USA, 17–19 December 2018; Volume 2018.
16. Beiner, L.; Paris, S.W. Direct Trajectory Optimization Using Nonlinear Programming and Collocation. *J. Guid. Control Dyn.* **1987**, *10*, 338–342. [CrossRef]
17. Wang, Z.; Wu, G.; Barth, M.J. A Review on Cooperative Adaptive Cruise Control (CACC) Systems: Architectures, Controls, and Applications. In Proceedings of the IEEE Conference on Intelligent Transportation Systems, Maui, HI, USA, 4–7 November 2018; Volume 2018.
18. Davis, P.J. *Circulant Matrices*; Wiley: New York, NY, USA, 1979.
19. Johan Åström, K.; Murray, R.M. *Feedback Systems: An Introduction for Scientists and Engineers*; Princeton University Press: Princeton, NJ, USA, 2010.

applied sciences

MDPI

Article

Design and Implementation of an Energy-Efficient Vehicle Platoon Control Algorithm Using Prescribed Performance and Extremum Seeking Control

Andreas Katsanikakis and Charalampos P. Bechlioulis *

Department of Electrical and Computer Engineering (ECE), University of Patras, 26504 Rio, Greece; up1047668@upnet.gr
* Correspondence: chmpechl@upatras.gr

Abstract: Platooning has emerged as a promising approach to enhancing the fuel efficiency of vehicles, but determining the inter-vehicular distance that achieves the minimum consumption remains a challenge. In this article, an algorithm is proposed that employs extremum seeking control integrated with the prescribed performance control technique to find the optimal inter-vehicular distance. The algorithm utilizes the predecessor-following architecture to track the desired distance while minimizing the estimated aerodynamic drag coefficient to seek the optimal value. To estimate the coefficient, an observer is designed. Simulation results are presented to demonstrate the effectiveness of the approach. The proposed algorithm exhibits a significant improvement over existing methods that do not incorporate prescribed performance. Consequently, our scheme provides a valuable contribution to the field of platooning and paves the way for future research directions.

Keywords: platoon; optimal distance; energy efficiency; extremum seeking; prescribed performance control; aerodynamic drag

Citation: Katsanikakis, A.; Bechlioulis, C.P. Design and Implementation of an Energy-Efficient Vehicle Platoon Control Algorithm Using Prescribed Performance and Extremum Seeking Control. *Appl. Sci.* **2023**, *13*, 5650. https://doi.org/10.3390/app13095650

Academic Editors: Fabrizio Giulietti, Nadjim Horri and William Holderbaum

Received: 7 April 2023
Revised: 30 April 2023
Accepted: 2 May 2023
Published: 4 May 2023

1. Introduction

The transportation sector is under pressure to reduce greenhouse gas emissions and develop more energy-efficient systems, especially when it comes to heavy-duty vehicles. These vehicles, according to official reports, Ref. [1] cover approximately 25% of CO_2 emissions and 5% of energy consumption. Reducing fuel consumption is a major concern in order to improve efficiency. The scientific community indicates that platooning seems to be a huge solution to the problem. Recent studies [2] show that vehicle platooning can significantly improve fuel consumption when minimizing aerodynamic drag, a major source of energy loss in transportation, and reduce emissions, while at the same time providing safety on the road by reducing the potential for accidents due to human error. In addition, vehicle platooning can also increase the capacity of existing roads by reducing the space required between vehicles. This means that more vehicles can travel on the same stretch of road, reducing congestion and improving traffic flow. As explicitly explained in [3,4], the aerodynamic drag that develops between following vehicles varies with the inter-vehicular distance. Reducing the gap between two or more vehicles moving in close proximity significantly improves the efficiency of the system. This is because the objects can take advantage of the reduced pressure zone created by the object in front of it, which can help reduce turbulence and lower drag. Therefore, in order to minimize the aerodynamic effect, it is necessary to find the optimal distance value.

It must be stated, though, that accurate aerodynamics modelling is not an easy task [4] because the airflow around a vehicle can be very complex. Some common factors that affect the flow of air are speed and direction, the shape of a vehicle, and the compressibility of the air. Concluding, equations that model these behaviours are highly complicated and difficult to solve analytically, even for simple geometries. The aforementioned studies attempt to approach and observe air behaviour through experimentation. The experimental results are extracted under several constraints since measured values can be affected by uncertainties, such as wind tunnel calibration errors, sensor inaccuracies, and uncertainties in the properties of the test environment. The aforementioned statements are also discussed in [5] , where authors review the literature about the benefits of platooning in energy savings. The work describes several ways to optimize the efficiency and a detailed focus is given in the aerodynamics concerning drag reductions. The authors propose two different ways to reduce the aerodynamic drag in a platoon and they conclude that it plays a significant role in energy consumption, suggesting that further studies should focus on developing ways to reduce it; however, they underline that due to the complex aerodynamic nature, conclusions are not always safe to conduct.

1.1. Motivation and Contributions

Our inspiration is to discover new innovative and effective control strategies that can enhance the performance of complex systems, such as platoons. The main aim of this work is to develop a solution for optimizing vehicle platoon distance that can be implemented in real-world transportation systems. The proposed control algorithm should allow the platoon to seek and track, throughout the whole road course, the optimal inter-vehicular distance in order to achieve the minimum aerodynamic drag, thus improving the energy consumption of the total system. It is noteworthy that the optimal distance is not always the smaller feasible one; for that reason, the algorithm should be in place to identify online proper factors. In order to estimate the aerodynamic drag and find the optimal distance, an extremum seeking scheme that continuously seeks the extremum of a performance metric is proposed; in particular, the estimation of the drag coefficient, which is extracted by an observer. This approach is integrated with a robust prescribed performance control algorithm [6] that ensures the correct tracking of the desired distance. Prescribed performance is particularly useful when designing control algorithms for platoon systems because it allows the designer to explicitly specify desirable performance specifications. This can lead to more effective and efficient control techniques that can meet the desired performance requirements under various operating conditions and ensure collision avoidance while limiting problems that may occur due to sensor limitations.

Extremum seeking control is a model-free optimization method that is used to optimize the performance of a system by continuously seeking and tracking the extremum (maximum or minimum) of a performance metric. Studies of the method have been carried out for over half a century now with the most iconic ones [7,8] where stability proofs and optimization solutions for several complex non-linear problems are presented. The key advantage of ESC is that it does not require a mathematical model of the system being controlled. Instead, it relies on direct measurements of the system's output and input signals to estimate the objective function and adjust the system parameters accordingly. ESC has been successfully applied to a wide range of applications, including power electronics, renewable energy systems, air crafts, and process control. However, like any control method, it has its own set of limitations and challenges that must be carefully considered when designing and implementing an ESC system. Fewer studies can be found for ESC when concerning platoons systems since the idea is still in the early stages. Eventually, the outcome of this work should be a well-set designed novel ESC algorithm that can effectively balance the trade-offs between minimizing drag force, thus improving fuel efficiency and carbon footprint and maintaining the stability and safety of the platoon system.

Concluding, it is worth mentioning that the ultimate goal of the proposed control algorithm is its future implementation in various types of real-world platooning systems, including semi-autonomous and fully autonomous vehicles. The implementation of such a system can have a significant impact on the transportation sector, reducing greenhouse gas emissions, fuel consumption, and traffic congestion, while improving the safety and efficiency of the system. Additionally, the development of this control algorithm can pave the way for future research and innovation in the field of platooning systems and extremum seeking control.

1.2. Related Works

Vehicle platooning has been a subject of interest for several years in the transportation research community. As stated above, the focus has shifted towards improving the energy efficiency of platooning systems, which has led to a surge in research on this topic. In this section, some of the relevant work that has been carried out in the area of our discussion is reviewed.

A number of studies have investigated the benefits of platooning in terms of fuel consumption and greenhouse gas emissions. For instance, in [9] it is demonstrated that platooning can reduce fuel consumption by up to 13.6% and travel time in connected and automated vehicles. Additionally, in [10], the authors conduct experiments using computational fluid dynamics (CFD) to explore the issue for different platoon formulations and they conclude that for specific formulations, energy savings can reach up to 10.3%, and at the same time aerodynamic reductions can be reduced up to 24%. To achieve these benefits, several techniques have been proposed for optimizing platooning systems. One common approach is to use adaptive cruise control. In [11], the authors investigate the effect of that control method in terms of stability, safety, and traffic flow behaviour. A pioneering approach of that method, cooperative adaptive cruise control (CACC) [12], which is a variant of adaptive cruise control, allows vehicles to communicate with each other to coordinate their movements. The ecological approach of CACC has been shown to improve fuel efficiency and reduce emissions by minimizing the distance between vehicles while maintaining a safe following distance. Another approach is to use predictive control algorithms to optimize platoon performance. In [13], a model predictive control (MPC) is implemented and compared with other current methods. The algorithm uses a model of the platoon dynamics to predict the future behaviour of the vehicles and adjust the control inputs accordingly. Another MPC method that considers energy reduction utilizing the air-drag coefficient, similar to the approach of this study, is proposed in [14]. Despite the fact that these strategies show great potential, both ACC and MPC methods require knowledge of the dynamics of the system.

In recent years, several studies have focused on using dynamic programming and machine learning techniques to optimize platooning systems. Dynamic programming in [15] helps the control system to explore the optimal trajectory that reduces fuel consumption based on road topography maps, and the final scheme is stated as a look-ahead control strategy. In [16], a reinforcement learning-based algorithm for vehicle platooning is proposed that provides a solution to the longitudinal control problem considering efficiency policies. The algorithm uses a deep neural network to learn the optimal control policy for the platoon, which is then used to adjust the control inputs in real time. These methods are quite promising; however, they require high-end process systems and several restraints have been made throughout their experimental implementations. A different approach is provided in [17], where a control method is developed to provide a solution for the platoon rooting problem in a real-world metropolitan network using small predefined inter-vehicular distances to minimize fuel consumption. Concerning the extremum-seeking approach in [18], the authors developed an ESC method to reduce aerodynamic drag on a single truck by utilizing the parameters of a wind deflector installed on the roof of the tractor cab. Although this study is not focused on platooning, the patent can be adopted for the case in future works.

Overall, these studies demonstrate the potential benefits of platooning in reducing energy consumption and greenhouse gas emissions and highlight the importance of developing efficient control strategies for platooning systems. In this work, we propose an extremum seeking control-based algorithm that can be used to optimize the inter-vehicular distance in platooning systems, with the aim of minimizing the aerodynamic drag and improving the energy efficiency of the platoon. The algorithm does not require the knowledge of system dynamics and enforces collision avoidance and connectivity maintenance employing the prescribed performance control technique. Our study will be directly compared to [19], where an improved ESC is proposed to reduce aerodynamic drag in real time for a platoon of electric vehicles. The authors in that work developed an ESC scheme that is integrated with a sliding mode controller to track the desired inter-vehicular distance. In addition, a different observer, based on the high-gain observation technique, was proposed.

2. Problem Formulation

The dynamics of a platoon with N vehicles are given by:

$$\begin{aligned} \dot{p}_i &= v_i \\ m_i\, \dot{v}_i &= f_i + u_i \end{aligned} \tag{1}$$

for $i = 1, \ldots, N$. The system is approached with second-order nonlinear dynamics where p_i, v_i indicate the position and the velocity of each vehicle, respectively. Additionally, m_i is the mass of each vehicle, f_i resembles the aerodynamic drag resistance force, and u_i is the vehicle acceleration force. In this study, other forces such as rolling resistance, gradient resistance, and so on are considered much smaller, and thus they are neglected. According to the aerodynamics of platooning when considering fuel consumption [2], drag can be modeled by:

$$F_{d_i} = \frac{1}{2}\, C_{d_i}(d_i)\, \rho\, A_i\, v_{r_i}^2 \tag{2}$$

where C_{d_i} is the drag coefficient related to the inter-vehicular distance d_i, ρ is the air density, A_i is the frontal area of the vehicle, and $v_{r_i} = v_i - v_{\text{air}}$ is the relative velocity of the vehicle with respect to the airflow v_{air}. Combining Equations (1) and (2) and considering that $d_i = p_{i-1} - p_i$, it can be deduced that

$$m_i\, \dot{v}_i = -\frac{1}{2}\, C_{d_i}(d_i)\, \rho\, A_i\, |v_{r_i}| v_{r_i} + u_i \tag{3}$$

For the system, the force u_i is the control input, whereas m_i, $C_d(d_i)$ are considered unknown, thus rendering f_i uncertain.

The aim of the proposed algorithm is to seek the optimal rigid formation with the extremum seeking technique by minimizing the estimation of the aerodynamic drag coefficient, thus approximating the unknown f_i, and to ensure that this formation will be settled with the prescribed performance control protocol. For the drag coefficient estimation, an extended state observer is proposed. Additionally, rigid formation is represented geometrically by the gaps $\Delta_{i-1,i}$, for $i = 1, \ldots, N$ between consecutive vehicles, where $\Delta^*_{i-1,i} > 0$ is the optimal distance, which extremum seeking calculates, between the $(i-1)$-th and the i-th vehicle. Eventually, the algorithm exports the optimal force that should be applied to each vehicle in order to achieve the minimum drag. The closed-loop system diagram given in Figure 1 provides a better understanding of the above.

Therefore, the proposed scheme requires:

- Tracking the desired optimal distance with the prescribed performance control protocol.
- Seeking the optimal desired distance with extremum seeking control.
- Estimating the air drag coefficient.

In the following section, a detailed description of the aforementioned modules is provided.

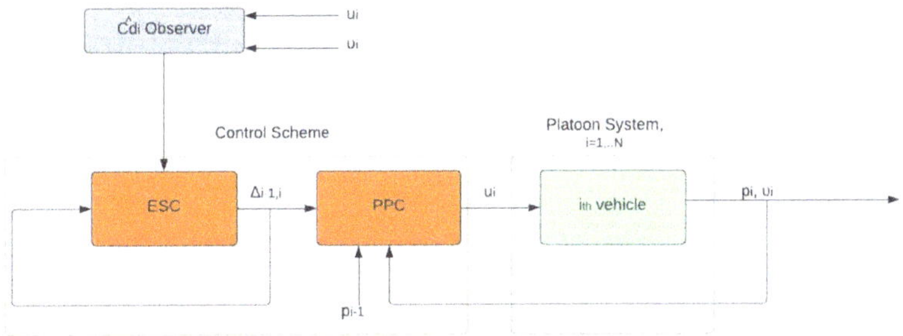

Figure 1. The closed-loop system after the control design for the i-th vehicle with the PPC and ESC controllers and the drag coefficient observer.

3. Extremum Seeking with Prescribed Performance Control

3.1. Prescribed Performance Control

In this work, the leader-following architecture, for the design of the prescribed performance controller that tracks the desired distance, is adopted. The efficiency of this architecture is fully described and tested both in simulation and in an experimental environment in the author's previous work [6].

An important aspect of the PPC controller is that it maintains the inter-vehicular distance $(p_{i-1} - p_i)$ between boundaries that ensure collision avoidance while limiting connectivity failures due to sensor constraints. Let, $\Delta_{col} < p_{i-1} - p_i < \Delta_{con}, i = 1, \ldots, N$. Therefore, for the desired distance $\Delta_{col} < \Delta_{i-1,i} < \Delta_{con}$, where Δ_{col},Δ_{con} are artificially selected positive constants for the collision avoidance and connectivity constraints, respectively. The initial state of the platoon should also meet this assumption. Hence, the control law u_i is designed as:

$$u(\xi_v, t) \equiv \begin{bmatrix} u_1(\xi_{v_1}, t) \\ . \\ . \\ . \\ u_N(\xi_{v_N}, t) \end{bmatrix} = -\, k_v\, (\rho_v(t))^{-1}\, r_v(\xi_v)\, \epsilon_v(\xi_v) \tag{4}$$

where $k_v > 0$ and $r_v(\xi_v)$, $\epsilon_v(\xi_v)$ are:

$$r_v(\xi_v) = \text{diag}(\left[\ \tfrac{2}{(1+\xi_{v_i})(1-\xi_{v_i})}\ \right]_{i=1,\ldots,N}) \tag{5}$$

$$\epsilon_v(\xi_v) = \left[\ \ln(\tfrac{1+\xi_{v_i}}{1-\xi_{v_i}}),\quad \ldots,\quad \ln(\tfrac{1+\xi_{v_N}}{1-\xi_{v_N}})\ \right]^T \tag{6}$$

To determine the normalized error vector ξ_v, we must first define the velocity error vector as $e_v \equiv [e_{v_1}, \ldots, e_{v_N}]^T = vs. - v_d(\xi_p, t)$, where $vs. \equiv [v_1, \ldots, v_N]^T$ and $v_d(\xi_p, t)$ is the vector that contains the reference velocities based on the desired distance. Subsequently, the velocity performance function $\rho_{v_i}(t)$, $i = 1, \ldots, N$ must be properly chosen in order to satisfy $\rho_{v_i}(0) > |e_{v_i}(0)|$, $i = 1, \ldots, N$. Additionally,

$$\xi_v(e_v, t) \equiv [\xi_{v_1}(e_{v_1}, t), \ldots, \xi_{v_N}(e_{v_N}, t)]^T \equiv (\rho_v(t))^{-1}\, e_u \tag{7}$$

with $\rho_v(t) = \text{diag}([\rho_{v_i}(t)]_{i=1,\ldots,N})$

The intermediate control law that formulates the reference velocity vector $v_d(\xi_p, t)$ at the kinematic level is selected as:

$$v_d(\xi_p, t) \equiv \begin{bmatrix} v_{d_1}(\xi_{p_1}, t) \\ \cdot \\ \cdot \\ \cdot \\ u_{d_{N-1}}(\xi_{p_{N-1}}, t) \\ u_{d_N}(\xi_{p_N}, t) \end{bmatrix} = k_p\,(\rho_p(t))^{-1}\, r_p(\xi_p)\, \epsilon_p(\xi_p) \tag{8}$$

Similarly, $k_p > 0$ and $r_p(\xi_p), \epsilon_p(\xi_p)$ are:

$$r_p(\xi_p) = \mathrm{diag}\left(\left[\frac{\frac{1}{\underline{M}_{p_i}} + \frac{1}{\overline{M}_{p_i}}}{(1+\frac{\xi_{p_i}}{\underline{M}_{p_i}})\,(1-\frac{\xi_{p_i}}{\overline{M}_{p_i}})} \right]_{i=1,\ldots,N} \right) \tag{9}$$

$$\epsilon_p(\xi_p) = \left[\ln\left(\frac{1+\frac{\xi_{p_1}}{\underline{M}_{p_1}}}{1-\frac{\xi_{p_1}}{\overline{M}_{p_1}}}\right), \quad \ldots, \quad \ln\left(\frac{1+\frac{\xi_{p_N}}{\underline{M}_{p_N}}}{1-\frac{\xi_{p_N}}{\overline{M}_{p_N}}}\right) \right]^T \tag{10}$$

where, $\underline{M}_{p_i}, \overline{M}_{p_i}$ are positive parameters that are related to collision and connectivity boundaries. For the normalized position error vector ξ_p, we define the position error between two neighboured vehicles $e_{p_i}(t) = p_{i-1}(t) - p_i(t) - \Delta_{i-1,i}, \quad i = 1, \ldots, N$ and properly choose function $\rho_{p_i}(t)$ to achieve the prescribed performance objectives. Thus,

$$\xi_p(e_p, t) \equiv [\xi_{p_1}(e_{p_1}, t), \ldots, \xi_{p_N}(e_{u_N}, t)]^T \equiv (\rho_p(t))^{-1}\, e_p \tag{11}$$

with $\rho_p(t) = \mathrm{diag}([\rho_{p_i}(t)]_{i=1,\ldots,N})$

Finally, performance function is designed as:

$$\rho_{p_i}(t) = (1 - \frac{\rho_\infty}{\max(\underline{M}_{p_i}, \overline{M}_{p_i})})\exp(-lt) + \frac{\rho_\infty}{\max(\underline{M}_{p_i}, \overline{M}_{p_i})} \tag{12}$$

with l, ρ_∞ positive parameters. Finally, for the parameters $\underline{M}_{p_i}, \overline{M}_{p_i}$ we set:

$$\begin{aligned} \underline{M}_{p_i} &= L_i - \Delta_{col} \\ \overline{M}_{p_i} &= \Delta_{con} - H_i \end{aligned} \tag{13}$$

where L_i, H_i are artificially selected positive defined parameters. Proper selection should ensure that $\underline{M}_{p_i}, \overline{M}_{p_i} > 0$. Further analysis will not be given here, whereas the proof of convergence and stability can be found in [6].

3.2. Extremum Seeking Control

Extremum seeking control is a feedback control method that seeks to optimize a system without the knowledge of the model and maintain it to the optimal operating point by adjusting iteratively its parameters to find the optimal value that maximizes or minimizes a given objective function. There are several variations of ESC that differ in the specific method used to estimate the objective function and compute the gradient. These methods include perturbation-based ESC, harmonic ESC, and sliding mode ESC, among others. Each method has its own advantages and limitations and is typically chosen based on the specific requirements and characteristics of the system to be controlled. In this case, in order to explore the parameter space and find the optimal value the system is perturbed with a sinusoidal excitation signal. Adopting the methods proposed in [8], a single parameter ESC design for a non-linear plant is proposed in Figure 2.

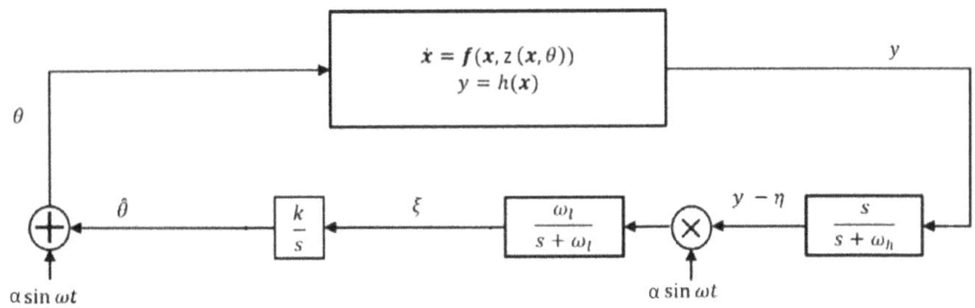

Figure 2. Single parameter basic ESC design with the perturbation signal.

After extensive analysis and with respect to our platoon system of N vehicles, the dynamics of the system can be summarized as:

$$\dot{x}_i = f(x_i, z_i(x_i, \hat{\theta}_i + a_i \sin \omega_i t)) \tag{14}$$

$$\dot{\hat{\theta}}_i = k_i \xi_i \tag{15}$$

$$\dot{\xi}_i = -\omega_{l_i} \xi_i + \omega_{l_i}(y_i - n_i) a_i \sin \omega_i t \tag{16}$$

$$\dot{n}_i = -\omega_{h_i} + \omega_{h_i} y_i, \qquad \text{for } i = 1, \ldots N. \tag{17}$$

where $\dot{x}_i$ is the state and θ_i is the equilibrium of the system. Moreover, z_i is a known control law parameterized by θ_i and $a_i, \omega_i, \omega_{l_i}, \omega_{h_i}, k_i, a_i(0), \hat{\theta}_i(0)$ are tuning parameters with the low-frequency components signal stated as n_i. In our case, $x_i = [p_i, u_i]^T$, $\theta_i = \Delta_{i-1,i}$, with the optimal point $\theta_i^* = \Delta_{i-1,i}^*$, z_i, is the control law from the PPC part and $y_i = h(x_i)$ is the objective function. The target is to deploy an objective function that approximates the aerodynamic drag. Studies in [3,20] show that the relation of the C_d coefficient with the inter-vehicular distance, as Equation (3) shows, cannot be measured and thus in [19] an observer is proposed. Hence, $y_i = h(x_i)$ can be a function that contains that value. Since the goal is to minimize the drag, y should be replaced by $-y$, yet the design is the same; thus, $y_i = -f(\hat{C}_{d_i})$. The reasoning and proof of that point can be found [8]. Additionally, a quick inspection of the main assumptions should be conducted to prove the stability of the ESC.

1. There exists a smooth function $l : \mathbb{R} \to \mathbb{R}^n$ such that $f(x_i, a_i(x_i, \theta_i)) = 0$ if and only if $x_i = l(\theta_i)$. The assumption is satisfied considering $\theta_i = \Delta_{i-1,i}$, $l(\Delta_{i-1,i}) = \Delta_{i-1,i}$.
2. For each $\theta_i \in \mathbb{R}$, the equilibrium $x_i = l(\theta_i)$ of the system is locally exponentially stable. The stability proof of the PPC controller proves it.
3. There exists $\theta_i^\star = \Delta_{i-1,i}^* \in \mathbb{R}$ such that

$$(h \circ l)'(\theta_i^*) = 0$$
$$(h \circ l)''(\theta_i^*) > 0 \tag{18}$$

Observing the graph results in [3,20] there seems to be a unique point between two consecutive vehicles in a platoon where the aerodynamic drag coefficient, that has a continuous profile, has the lowest value. That statement proves the above two properties.

At this point, the main idea of the design has been completed. However, it should be highlighted that the use of sinusoidal perturbation in the aforementioned scheme results in undesired steady-state oscillations, which exhibit unnecessary platoon actions that can lead not only to increased fuel consumption but also to drive discomforts and eventually a risk of collision. In [21], an ESC with attenuated steady-state oscillation (ESCWAO) is proposed and the authors in [19] adopt the method for a similar purpose. The ESCWAO is designed to reduce the magnitude of the dither signal as the estimated gradient approaches zero.

This approach can effectively reduce steady-state oscillations and improve the performance of the ESC for platoons. The specific design of the ESCWAO may vary depending on the particular application and the characteristics of the system being controlled. However, in general, the approach involves modifying the gain of the dither signal based on the estimated gradient. By adapting the magnitude of the dither signal to the estimated gradient, the ESCWAO can effectively reduce oscillations and improve the performance of the controller. Figure 3 shows that the ESCWAO design is similar to the standard ESC with the difference that there is not a constant amplitude a but an adaptive law is proposed for the decay in excitation signal amplitude. Moreover, the demodulated signal after high-pass filtering is multiplied without the amplitude.

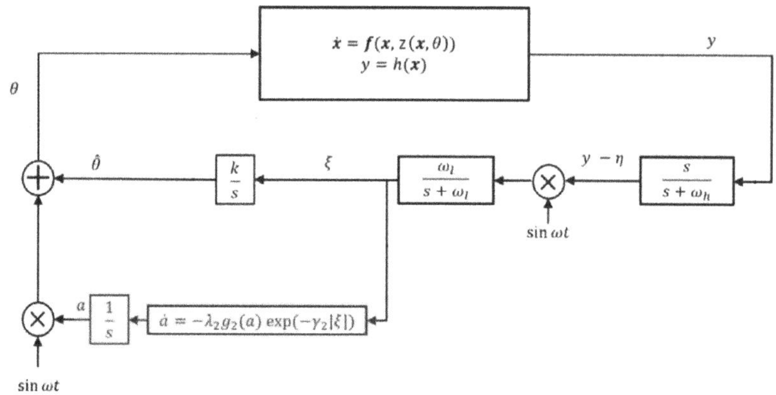

Figure 3. Single parameter ESCWAO design. The previously constant a signal is now replaced by the scheme of the dithered signal.

Dynamic analysis of the system is identical to standard ESC and the dithered signal is added to the system as indicated in Equation (14):

$$\dot{a}_i = -\lambda_{2_i} g_2(a_i) \exp(-\gamma_{2_i}|\xi_i|), \quad a_i(t_0) = a_{i_0} > 0 \tag{19}$$

where, $\lambda_{2_i}, \gamma_{2_i}, a_{i_0}, \ for \ i = 1, \ldots N$, are positive design parameters.

For the stability proof of the proposed algorithm, the same assumptions as stated in standard ESC are followed, whereas for the extra dithered signal, further analysis can be found in [21]. ESCWAO requires parameters $\omega, \ \omega_l, \ \omega_h, k, \ \lambda_2, \ \gamma_2, a(0), \ \hat{\theta}(0)$ to be fine-tuned. Henceforth, the reduction in steady-state oscillations achieved by the ESCWAO leads to efficient design. However, it is important to note that proper tuning of the ESCWAO is necessary to achieve the desired performance, which sometimes is a labour-intensive process. More specifically, filter cut-off frequencies need to be lower than the frequency of the perturbation signal and the adaption gain k value should also be small to achieve the desired time scales of the system [8] and properly estimate the optimal value. Additionally, the γ_2 parameter determines the rates of decay of the dither signal. Lowering the value will lead to bigger decays and the results will be similar to the standard ESC. Thus, at the beginning of the tuning process, the parameter should be sufficiently small in order to detect at which point the ESCWAO converges to the optimal point. Afterwards, the designer should increase the value and maintain it at a point that ensures faster decay.

3.3. Air-Drag Coefficient Observer ($\hat{C}_{d_i}$)

Estimating the drag coefficient C_d is essential in order to approximate the aerodynamic force. An extended state observer is designed according to [22]. Considering that our system is one-dimensional, Equation (3) can be written as :

$$\dot{v}_i = -\frac{C_{d_i}(d_i)\rho A_i |v_{r_i}| v_{r_i}}{2m_i} + \frac{u_i}{m_i} \tag{20}$$

$$y = v_i$$

for $i = 1, \ldots, N$, where $-\frac{C_{d_i}(d_i)\rho A_i |v_{r_i}| v_{r_i}}{2m_i}$ denotes the unknown system function. Based on the non-linearity of our system, we design the following extended state observer:

$$\dot{\hat{v}}_i = z + \frac{u_i}{m_i} + \frac{v_i - \hat{v}_i}{\epsilon} \tag{21}$$

$$\dot{z} = \frac{v_i - \hat{v}_i}{\epsilon^2} \tag{22}$$

for a small positive gain ϵ. Subsequently, we shall estimate the drag coefficient as:

$$\hat{C}_{d_i} = -\frac{2m_i z}{\rho A_i |v_{r_i}| v_{r_i}} \tag{23}$$

4. Simulation Results and Discussion

The algorithm proposed in this study was tested in MATLAB software. Simulations were conducted on a laptop equipped with an AMD Ryzen 5 4600H processor, 8 GB of RAM, and a 512 GB solid-state drive. To demonstrate the algorithm's feasibility, a platoon of $N = 4$ vehicles was considered with a leader ($i = 0$), where $m_i = [1200, 700, 800, 900]$ kg, for $i = 1, \ldots, 4$. The aerodynamic drag profiles for the platoon members were similar to the experimental measurements from [20]. The authors highlight that for inter-vehicular distances below 3m, drag has non-constant and complex behaviour. The measured values used to create the drag profiles are shown in Figure 4. Spline fitting was performed on these values in order to smooth the profile functions and comply with the conditions as stated in Equation (18). Furthermore, we assumed zero wind speed ($v_{r_i} = v_i$), $\rho = 1.293$ kg m^{-3} and $A_i = [7, 2.5, 3, 3.5]$ m^2, for $i = 1, \ldots, 4$.

The leading vehicle was set to have a continuously differentiable velocity profile as $v_0 = 18 + 0.1\sin(\frac{2\pi t}{100}) - 0.05\sin(\frac{2\pi t}{50})$. Concerning the PPC design, we set the collision and connectivity constraints as $\Delta_{Col} = 2.5$ m, $\Delta_{Con} = 20$ m, and $L_i = H_i = 10$ m. To establish predefined transient and steady-state performance, we selected $l = 0.1$, $\rho_\infty = 0.07$. Hence, we set the performance functions $\rho_{p_i}(t) = (1 - \frac{0.07}{0.95L_i})\exp(-0.1t) + \frac{0.07}{0.95L_i}$ and $\rho_{v_i}(t) = 2|e_{v_i}(0)|\exp(-0.1t) + 0.3$ for $i = 1,2,3,4$, obeying the initialization restrictions. Finally, the control gains were selected as $k_p = 1$ and $k_v = 100$.

The gains for the observers were set as $\epsilon_i = 10^{-4}$ to estimate each vehicle's drag coefficient. For the ESCWAO part, we chose the objective functions $y_i = [-400\hat{C}_{d_1}^2, -500\hat{C}_{d_2}^2, -700\hat{C}d_3{}^2, -800\hat{C}_{d_4}^2]$ and the parameters for the extremum seeking controllers were tuned as $\omega_{h_1} = 3.2$, $\omega_{h_2} = 2.9$, $\omega_{h_3} = \omega_{h_4} = 2.7$, $\omega_{l_1} = 0.645$, $\omega_{l_2} = \omega_{l_3} = 0.6$, $\omega_{l_4} = 0.85$, and $\omega_i = 1, k_i = 0.7$, $\gamma_{2_i} = 6$, $g_2(a_i) = a_i$, $a_i(0) = 4$, for $i = 1,2,3,4$. Moreover, $\lambda_{2_i} = 0.2$, $i = 1,2,3,4$. We assumed that the initial position of the vehicles was $p_0(0) = 0$, $p_1(0) = -10$, $p_2(0) = -20$, $p_3(0) = -30$, $p_4(0) = -40$ and the desired $\Delta_{0,1}(0) = \hat{\theta}_1(0) = 17$, $\Delta_{i-1,i}(0) = \hat{\theta}_i(0) = 15$, for $i = 2,3,4$. It should be noted that the initial values complied with all restrictions.

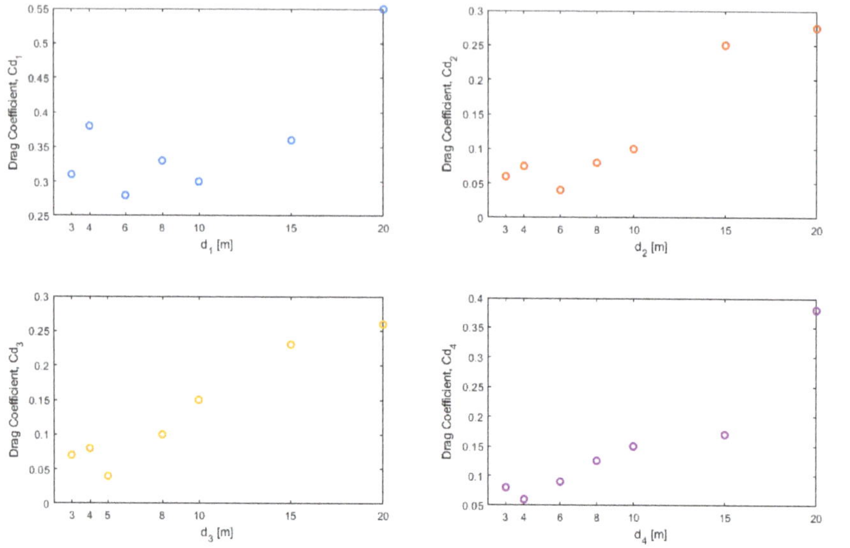

Figure 4. The real drag coefficient relation to the inter-vehicular distance for each follower vehicle in the platoon.

A long simulation time was chosen $t = 800$ s to allow the ESCWAO to converge to the optimal point and to prove that whereas ESCWAO optimized the distance, PPC works efficiently. The computational time of the algorithm in our hardware was 40.42 s, indicating that the implementation worked sufficiently in our setup. Figure 4 shows that for the first two vehicles in the platoon, the lowest drag coefficient was achieved when the inter-vehicular distances reached 6 m, for the third vehicle at 5 m, and for the fourth vehicle at 4 m; thus, each vehicle's ESC controller was expected to converge around these values, respectively. Although in [20] the authors conclude that for all vehicles in the platoon, the optimal distance is 6m, in that simulation, we slightly changed these values for the third and the fourth vehicle in order to investigate whether the algorithm could correctly track and maintain different optimal distances. Furthermore, the initial distances were intentionally selected far from these values to demonstrate the robustness and efficiency of the controllers. Simulation results are shown in the figures below.

Figure 5 shows that the ESCWAO controllers converged correctly to the desired values for all the vehicles in a short time, and for most of the simulation, oscillations at the steady state disappeared, and hence the efficiency was not affected negatively. However, it is clear that in the zoomed plots in the same figure, oscillations in the transient state can be momentarily abrupt. The observers also show promising performance Figure 6, where it is noticeable that the estimations of the drag coefficients are very close to the actual optimal values as stated before in Figure 4. This is a reasonable outcome since it is proved that if the estimations are correct and if fine-tuning has been conducted to the parameters of the ESCWAO part, the algorithm should be able to seek and detect any change in the coefficients that leads to the optimal distance.

Figure 7 shows the evolution of the position errors $e_{p_i}(t)$, $i = 1, 2, 3, 4$, along with the corresponding performance functions. It seems that the algorithm not only can maintain the formation of the vehicles despite the lack of knowledge of their dynamic models but also is in place to track the desired distances even when the ESCWAO have not converged yet and formulate correctly all the vehicles along the simulation with prescribed transient and steady-state performance. The zoomed plots in the same figure provide a better insight into the last statement.

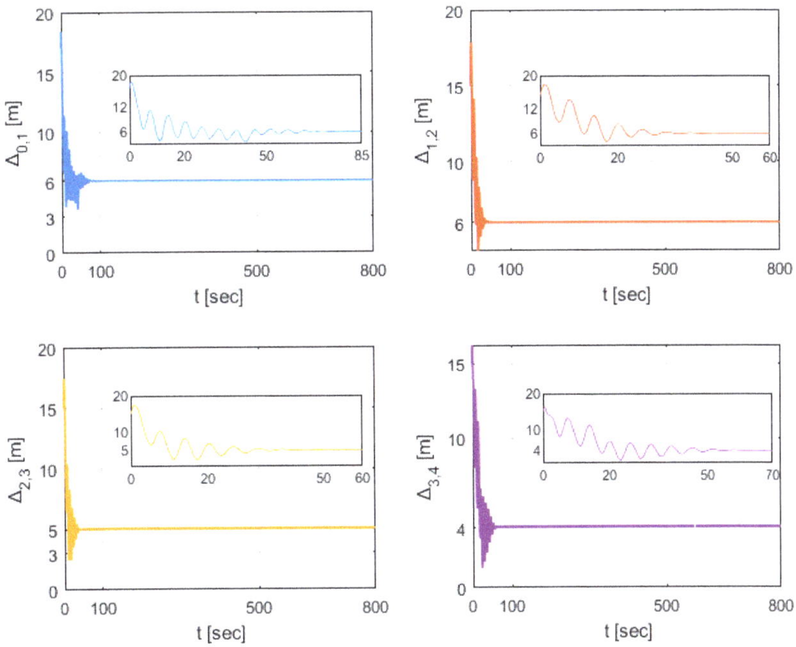

Figure 5. The convergence of the desired distance, $\Delta_{i-1,i}$, for each platoon member under ESCWAO with zoomed plots to clarify the transient state.

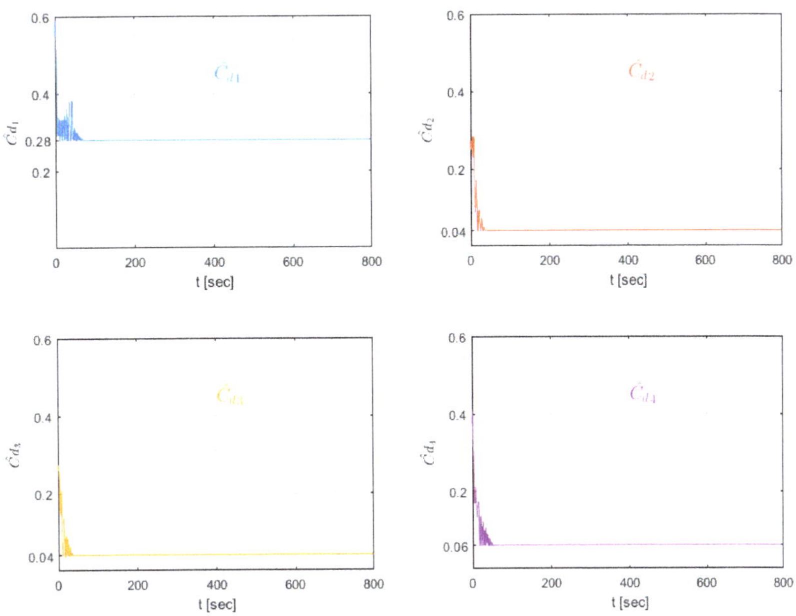

Figure 6. The change in the observed drag coefficient throughout the simulation for each vehicle.

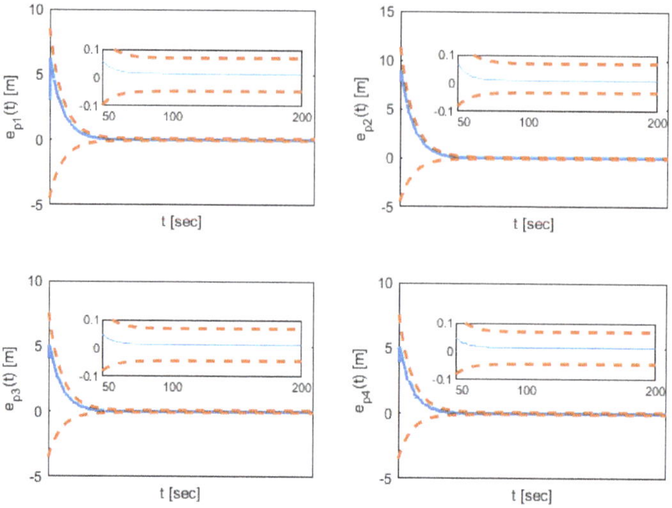

Figure 7. The position errors $e_{pi} = p_{i-1} - p_i - \Delta_i$ illustrated in blue solid lines along with the corresponding performance bounds depicted in red dashed lines. The zoomed plots clarify the transition from the transient to the steady state and the convergence to zero.

Comparative Studies

A quite interesting point is to investigate what happens in the algorithm in terms of energy efficiency. The force vector u_i is used to calculate the energy of each vehicle and eventually the platoon's system in total. Specifically, $Energy_i = \int_0^{800} u_i^2(\tau)\, d\tau,\ i = 1,2,3,4.$ Employing the same system as before, we compare the final energy values for the plain PPC (fixed inter-vehicular distances), the proposed PPC with ESCWAO, and the basic ESC with PPC, as designed in Figure 2. The setup is the same as before: in the plain PPC we set the desired distance $\Delta_{i-1,i} = 8,\ i = 1,2,3,4$ and in the basic ESC with PPC we set $a_i = 1$, for $i = 1,2,3,4$. Figure 8 shows the comparative results. The total energy that the system requires in the plain PPC is almost double the energy required by the proposed ESCWAO one. Moreover, PPC with a basic ESC system energy requirement is also high and for the first two vehicles in the platoon is individually higher even than the plain ESC indicating that PPC with basic ESC is not worthy at all. It should be noted that the first vehicle behind the leading one requires in all cases more energy than the others due to the fact that its frontal area is set far bigger, demonstrating a larger truck, but still, in our case, the energy is less.

Further investigation of the energy efficiency is provided in Table 1. We implemented the algorithm proposed in [19] for a system of two vehicles neglecting rolling and gradient resistance. For the leading vehicle, we adopted the same velocity profile as before, and for the following vehicle, the second spline-fitted profile for the drag resistance (C_{d_2}) is shown in Figure 4. For the following vehicle $m = 1200$ kg, $A = 2.5$ m^2, $p = 1.28$ kg m^{-3}, and the initial distance between the two vehicles was 20m in both cases. Additionally, the sliding mode controller's (SMC) and high-gain observer parameter values were set the same as the authors suggested in their work. In our case, for the PPC scheme, $k_p = 1, k_v = 149$, $\Delta_{Col_1} = 2.5, \Delta_{Con_1} = 20, H_1 = L_1 = 10$ use the same performance functions as in the previous section and for our observer $\epsilon_1 = 10^{-3}$.

Concerning the extremum seeking scheme, in the implementation of the ESCWAO with the sliding mode controller $\omega_h = 1.1, \omega_l = 0.4, \omega = 1, k = 0.6, \lambda_2 = 0.2$, $\gamma_2 = 6, g(a) = a, a(0) = 4, y = -100\hat{C_d}^2$, and $\hat{x}_d = \hat{\theta}(0) = 17$. In our case, the parameters are set $\omega_{h_1} = 3.27, \omega_{l_1} = 0.5, \omega_1 = 1, k_1 = 0.7, \lambda_{2_1} = 0.5, \gamma_{2_1} = 6, g(a_1) = a_1, a_1(0) = 4$, $y_1 = -700\hat{C_d}^2$, and $\Delta_{0,1}(0) = \hat{\theta}_1(0) = 17$. Following the same energy calculations for a simulation time $t = 800$ s, it is clear that energy demanding in both cases is close, although in SMC, implementation is slightly less. The important fact is that the computational time in our proposed scheme is almost half the time needed for the computational time of the SMC implementation. Hence, the algorithm proposed in this research provides a different aspect of minimizing the aerodynamic drag force while considering collision and connectivity issues without risking higher energy losses or higher hardware demands.

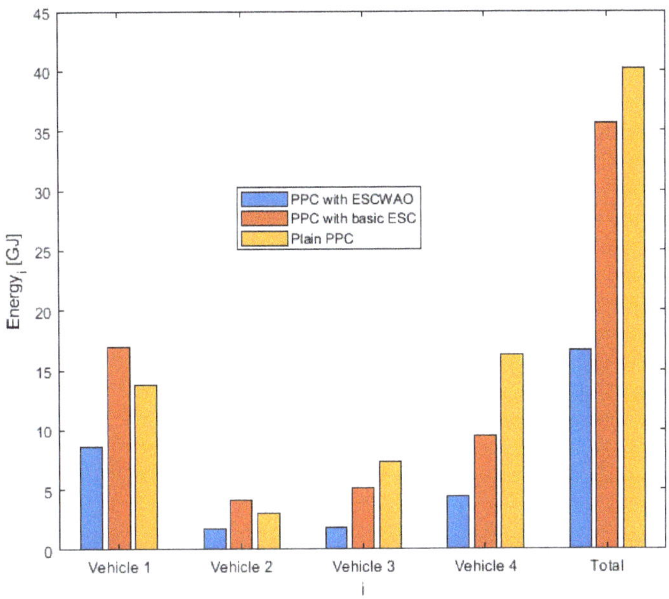

Figure 8. Required energy that each vehicle and the platoon system demand in total after the simulation, under the three different control cases.

Table 1. The energy demands and computational time comparison for the two different implementations.

	Energy [GJ]	Computational Time [s]
SMC with ESCWAO	0.677	5.61
PPC with ESCWAO	0.586	12.17

5. Conclusions and Future Works

The above study suggests an energy-efficient extremum seeking algorithm without steady-state oscillations embedded in the prescribed performance control technique in a vehicle platoon system to minimize the aerodynamic drag force between consecutive vehicles and help the system safely maintain a desired formation while avoiding connectivity issues. Simulation results are quite promising since the goal is achieved and there is no need for a high-end hardware setup in the first place. The algorithm indeed seems to optimize the system in terms of energy demands. However, we should take into consideration that several constraints have been set for the system, and in a real environment, several complex conditions take part in energy consumption. To ensure experimental feasibility, various tests should be conducted to investigate potential solutions for various problems that are caused by uncertainties in the model. Additionally, drag behaviour should be examined under different weather/road conditions to extract richer data and provide a more realistic approach to the drag profiles. Reassured that the algorithm can work under the aforementioned issues, the next immediate goal is to conduct platoon experiments to verify its actual potential. One main objective is to find out whether the oscillations in the transient state, as shown before in Figure 7, can highly affect driving comfort or not. On the other side, the last statement shows that this work can immediately contribute to studies for unmanned vehicles, where comfort issues are not taken into consideration.

Further research can explore how the algorithm can be implemented in real-world transportation systems to improve performance and fuel efficiency. Additionally, research can focus on refining the algorithm to accommodate for uncertainties, such as wind tunnel calibration errors and sensor inaccuracies, that may affect the experimental observations of drag coefficients. In this work, only the longitudinal problem is considered and the vehicle state is obtained directly, which is impossible in a real scenario. Future research can focus on the case. Adopting methods using Kalman filters is a common solution to these problems. For instance, in [23], Kalman filter techniques are proposed to estimate the velocity errors and improve the observations of the heading vector in a complex autonomous vehicle dynamic model. Another interesting study [24] proposes Kalman filter techniques to estimate roll and pitch for vehicles in a platoon system using only the planar kinematic model. These robust works in total rely on the estimated speed, position, and orientation of the vehicles. Another common issue in the real application is the system delays due to different modules which are not considered in our control design. In [25,26], different solutions are proposed to solve the issue for a variety of sensors, both in unmanned independent systems and platoon formations. Adapting the insights of these works in our proposed algorithm can open the way to utilize it in a real platoon scenario. Lastly, a significant challenge is to develop an ESC scheme that considers more than one parameter to improve the accuracy of the optimization, such as energy consumption based on the engine torque. Finally, research can also delve into how to optimize platooning systems considering different weather conditions, altitudes, and other environmental variables that are not considered in this work.

Author Contributions: Conceptualization, C.P.B.; methodology, A.K. and C.P.B.; software, A.K.; validation, A.K.; formal analysis, A.K.; investigation, A.K.; writing—original draft preparation, A.K.; writing—review and editing, C.P.B.; visualization, A.K.; supervision, C.P.B.; project administration, C.P.B. All authors have read and agreed to the published version of the manuscript.

Funding: This research received no external funding.

Institutional Review Board Statement: Not applicable.

Informed Consent Statement: Not applicable.

Data Availability Statement: Not applicable.

Conflicts of Interest: The authors declare no conflict of interest.

Abbreviations

The following abbreviations are used in this manuscript:

PPC	Prescribed Performance Control
SMC	Sliding Mode Control
ESC	Extremum Seeking Control
ESCWAO	Extremum Seeking Control with Attenuated Oscillation
MPC	Model Predictive Control
ACC	Adaptive Cruise Control

References

1. IEA. *The Future of Trucks*; IEA: Paris, France, 2017; p. 167. [CrossRef]
2. Alam, A.; Besselink, B.; Turri, V.; MåRtensson, J.; Johansson, K.H. Heavy-Duty Vehicle Platooning for Sustainable Freight Transportation: A Cooperative Method to Enhance Safety and Efficiency. *IEEE Control. Syst. Mag.* **2015**, 35, 34–56. [CrossRef]
3. Vohra, V.; Wahba, M.; Akarslan, G.; Ni, R.; Brennan, S. An Examination of Vehicle Spacing to Reduce Aerodynamic Drag in Truck Platoons. In Proceedings of the 2018 IEEE Vehicle Power and Propulsion Conference (VPPC), Chicago, IL, USA, 27–30 August 2018; Volume 2018. [CrossRef]

4. Robertson, F.; Bourriez, F.; He, M.; Soper, D.; Baker, C.; Hemida, H.; Sterling, M. An experimental investigation of the aerodynamic flows created by lorries travelling in a long platoon. *J. Wind. Eng. Ind. Aerodyn.* **2019**, *193*, 103966. [CrossRef]
5. Pi, D.; Xue, P.; Wang, W.; Xie, B.; Wang, H.; Wang, X.; Yin, G. Automotive platoon energy-saving: A review. *Renew. Sustain. Energy Rev.* **2023**, *179*, 113268. [CrossRef]
6. Verginis, C.K.; Bechlioulis, C.P.; Dimarogonas, D.V.; Kyriakopoulos, K.J. Robust Distributed Control Protocols for Large Vehicular Platoons With Prescribed Transient and Steady-State Performance. *IEEE Trans. Control. Syst. Technol.* **2018**, *26*, 299–304. [CrossRef]
7. Krstic, M. Performance improvement and limitations in extremum seeking control. *Syst. Control. Lett.* **2000**, *39*, 313–326. . [CrossRef]
8. Ariyur, K.; Krstic, M. *Real-Time Optimization by Extremum-Seeking Control*; John Wiley & Sons: Nashville, TN, USA, 2003.
9. Rios-Torres, J.; Malikopoulos, A.A. Automated and Cooperative Vehicle Merging at Highway On-Ramps. *IEEE Trans. Intell. Transp. Syst.* **2017**, *18*, 780–789. [CrossRef]
10. Kaluva, S.T.; Pathak, A.; Ongel, A. Aerodynamic Drag Analysis of Autonomous Electric Vehicle Platoons. *Energies* **2020**, *13*, 4028. [CrossRef]
11. Swaroop, D.; Rajagopal, K. A review of constant time headway policy for automatic vehicle following. In Proceedings of the ITSC 2001. 2001 IEEE Intelligent Transportation Systems Proceedings (Cat. No.01TH8585), Oakland, CA, USA, 25–29 August 2001; pp. 65–69. [CrossRef]
12. Zhai, C.; Chen, X.; Yan, C.; Liu, Y.; Li, H. Ecological Cooperative Adaptive Cruise Control for a Heterogeneous Platoon of Heavy-Duty Vehicles With Time Delays. *IEEE Access* **2020**, *8*, 146208–146219. [CrossRef]
13. van Lierop, M. Model Predictive Control for Vehicle Platooning: A Practical Comparison against Traditional Methods. Master's Thesis, Delft University of Technology, Delft, The Netherlands, 2022.
14. Caiazzo, B.; Coppola, A.; Petrillo, A.; Santini, S. Distributed Nonlinear Model Predictive Control for Connected Autonomous Electric Vehicles Platoon with Distance-Dependent Air Drag Formulation. *Energies* **2021**, *14*, 5122. [CrossRef]
15. Turri, V.; Besselink, B.; Mårtensson, J.; Johansson, K.H. Fuel-efficient heavy-duty vehicle platooning by look-ahead control. In Proceedings of the 53rd IEEE Conference on Decision and Control, Los Angeles, CA, USA, 15–17 December 2014; pp. 654–660. [CrossRef]
16. Desjardins, C.; Chaib-draa, B. Cooperative Adaptive Cruise Control: A Reinforcement Learning Approach. *IEEE Trans. Intell. Transp. Syst.* **2012**, *12*, 1248–1260. [CrossRef]
17. Larson, J.; Munson, T.; Sokolov, V. *Coordinated Platoon Routing in a Metropolitan Network*; Society for Industrial and Applied Mathematics: Philadelphia, PA, USA, 2016; pp. 73–82. [CrossRef]
18. Papageorgiou, G.; Barden, J.; Whidborne, J.F.; Garry, K.P. Extremum Seeking Control for Truck Drag Reduction. In Proceedings of the 2018 UKACC 12th International Conference on Control (CONTROL), Sheffield, UK, 5–7 September 2018; pp. 114–119. [CrossRef]
19. Su, Z.; Chen, P. Extremum Seeking Control-based Control Framework for Electric Vehicle Platooning. *IFAC-PapersOnLine* **2022**, *55*, 123–128. [CrossRef]
20. Davila, A.; Aramburu, E.; Freixas, A. *Making the Best Out of Aerodynamics: Platoons*; SAE International: Warrendale, PA, USA, 2013. [CrossRef]
21. Bhattacharjee, D.; Subbarao, K. Extremum seeking control with attenuated steady-state oscillations. *Automatica* **2021**, *125*, 109432. [CrossRef]
22. Guo, B.Z.; Zhao, Z.L. On the convergence of an extended state observer for nonlinear systems with uncertainty. *Syst. Control. Lett.* **2011**, *60*, 420–430. [CrossRef]
23. Xia, X.; Hashemi, E.; Xiong, L.; Khajepour, A. Autonomous Vehicle Kinematics and Dynamics Synthesis for Sideslip Angle Estimation Based on Consensus Kalman Filter. *IEEE Trans. Control. Syst. Technol.* **2023**, *31*, 179–192. [CrossRef]
24. Liu, W.; Xia, X.; Xiong, L.; Lu, Y.; Gao, L.; Yu, Z. Automated Vehicle Sideslip Angle Estimation Considering Signal Measurement Characteristic. *IEEE Sensors J.* **2021**, *21*, 21675–21687. [CrossRef]
25. Liu, W.; Quijano, K.; Crawford, M.M. YOLOv5-Tassel: Detecting Tassels in RGB UAV Imagery With Improved YOLOv5 Based on Transfer Learning. *IEEE J. Sel. Top. Appl. Earth Obs. Remote. Sens.* **2022**, *15*, 8085–8094. [CrossRef]
26. Han, X.; Xu, R.; Xia, X.; Sathyan, A.; Guo, Y.; Bujanovia, P.; Leslie, E.; Goli, M.; Ma, J. Strategic and tactical decision-making for cooperative vehicle platooning with organized behavior on multi-lane highways. *Transp. Res. Part C Emerg. Technol.* **2022**, *145*, 103952. [CrossRef]

applied sciences

MDPI

Article

A New HEV Power Distribution Algorithm Using Nonlinear Programming

Jooin Lee [1] and Hyeongcheol Lee [2,*]

[1] Department of Electrical Engineering, Hanyang University, Seoul 04763, Republic of Korea
[2] Department of Electrical and Biomedical Engineering, Hanyang University, Seoul 04763, Republic of Korea
* Correspondence: hclee@hanyang.ac.kr; Tel.: +82-22-220-1685

Abstract: An equivalent consumption minimization strategy (ECMS) is one of the most powerful and practical ways to improve the fuel efficiency of hybrid electric vehicles (HEVs). In an ECMS, it is important to determine the optimal equivalent factor to reach a global optimal solution. The optimal equivalent factor is determined by driving conditions. Previous studies have used an adaptive ECMS (A-ECMS) to determine the appropriate equivalent factor according to changing driving conditions. An A-ECMS adjusts the equivalent factor by controlling the battery's state of charge (SOC) to follow a reference SOC trajectory. It is therefore critical to identify a reference SOC trajectory that reflects real-world driving conditions. These conditions, which are composed of the HEV's nonlinear dynamics and complex constraints, can be formulated into a nonlinear optimal control problem (NOCP). Here, we propose applying nonlinear programming (NLP) to an A-ECMS. The NLP-based ECMS algorithm can be divided into two parts: the use of an NLP to solve an NOCP to obtain the reference SOC trajectory and the application of an NLP solution (the result of the first part) to an A-ECMS. Simulation results demonstrate that the proposed NLP-based ECMS closely resembles a global optimal solution for dynamic programming in a relatively brief calculation time.

Keywords: nonlinear programming (NLP); adaptive equivalent consumption minimization strategy (A-ECMS); reference state of charge (SOC) trajectory; energy management strategy (EMS); transmission-mounted electric device (TMED)

Citation: Lee, J.; Lee, H. A New HEV Power Distribution Algorithm Using Nonlinear Programming. *Appl. Sci.* **2022**, *12*, 12724. https://doi.org/10.3390/app122412724

Academic Editors: Nadjim Horri, William Holderbaum and Fabrizio Giulietti

Received: 15 November 2022
Accepted: 8 December 2022
Published: 12 December 2022

1. Introduction

As global warming accelerates, environmental regulations are being introduced and enforced around the world. In the United States, the National Highway Traffic Safety Administration has developed corporate average fuel economy standards that require an average fuel efficiency of 23.2 km/L by 2025. In Europe, the Euro 7 regulations for vehicle emissions also require nitrogen oxide emission to be lowered to 40–120 mg/kWh by 2025 [1]. In response to strong environmental regulations, the automobile industry is making great efforts to electrify the vehicle powertrain.

Hybrid vehicles that provide propulsion power from two or more sources are typical examples of electrified powertrains. Hybrid electric vehicles (HEVs) are generally powered by both an internal combustion engine (ICE) and an electric motor (EM). HEV can reduce fuel consumption by operating ICE and EM in high-efficiency areas. When combined with an ICE-based powertrain, an EM can support the ICE in low-speed sections that are associated with low efficiencies and can regenerate energy during braking. As a result, an effective energy management strategy (EMS) that reflects driving and vehicle conditions is important to improve the efficiency of a HEV.

The study of EMSs of HEVs can be classified into two main categories. The first comprises rule-based EMSs, in which rules based on the experience of engineers determine the power distribution ratio between the ICE and EM. The advantage of a rule-based EMS is its intuitiveness. However, due to its innate rigidity, it is impossible to design a rule that anticipates every possible driving situation, and fuel efficiency may deteriorate in situations that are not reflected in the rule [2,3]. In addition, a rule-based EMS requires considerable time and resources as intensive parameter-tuning is required to anticipate diverse driving situations [4]. The second category comprises optimization-based EMSs, which use optimization theories to determine the power distribution ratio between the ICE and EM. An optimization-based EMS can determine the power distribution ratio closest to a global optimal solution. However, such a strategy can be complicated and require lengthy computational times due to complex optimization problems.

Various studies have been conducted on optimization-based EMSs because the hardware performance of the hybrid control unit (HCU) has been increased and fuel consumption can be minimized by using a global optimal solution in various driving situations. In an optimization-based EMS, an optimal control problem (OCP) can be defined and calculated by using optimization theory. Depending on how it is defined, or which optimization theory is applied, an OCP can be divided into offline and online optimization algorithms.

An offline optimization algorithm produces a global optimal solution that can minimize fuel consumption in all expected driving conditions, assuming that all conditions are known in advance. The most representative algorithm is dynamic programming (DP), which is based on the principle of optimality proposed by Richard Bellman. After splitting the complex problem into several simpler subproblems, the solution of the subproblems is sequentially obtained in a recursive manner, so that a complete algorithm obtains the global optimal solution. Although DP can identify a global optimal solution, it is difficult to apply such solutions to a real-world HCU because DP requires lots of computational resources. A global optimal solution obtained by DP is used primarily in offline analysis in the development of other EMS options, such as rule-based designs or as machine-learning training data [5].

Online optimization produces a solution that can minimize fuel consumption for a given driving condition in a limited time. In general, it can be applied to a real-world HCU by reducing the computational load by using a simplification method (e.g., linearization) or local optimization. Pontryagin's minimum principle (PMP), model predictive control (MPC), and equivalent consumption minimization strategy (ECMS) are representative algorithms. A PMP performs local optimization by applying a numerical method to a Hamiltonian function, which considers the battery's state of charge (SOC) and the fuel consumption characteristics of the ICE [6–8]. A PMP needs to efficiently design the Hamiltonian function to reduce computational loads. An MPC obtains an optimal solution for a given driving condition on a finite horizon. For an infinite horizon, an OCP is defined by designing cost functions and constraints that represent the characteristics of the system. A numerical solution can be obtained by applying a solver, such as quadratic programming and nonlinear programming (NLP) to a defined OCP [9,10]. An MPC needs to design system models and constraints carefully to reduce computational loads and create a coping mechanism for infeasible regions. Recently, research on the use of nonlinear-MPC (NMPC) by NLP solver has also been actively conducted. NMPC directly determines the torque of the engine and motor. NLP solvers used in NMPC include the sequential quadratic programming (SQP) method [11,12], mixed-integer nonlinear programming (MINLP) method [13], and the interior point method [14]. NMPC is one of the most efficient and powerful online optimization methods. When designing an NMPC, two analyses are most important. The first is an analysis for model simplification to reduce long computational time. The second is an analysis of the infeasible region to design a strategy for the infeasible solution. An ECMS derives a solution that minimizes equivalent energy consumption, which is the equivalent energy consumption in the form of a cost function that reflects fuel energy and electrical energy. This cost function defines the equivalent factor, which confirms the relative importance of electrical energy and fuel energy, according to driving conditions. The equivalent factor is a therefore an important tuning factor when designing an ECMS. A traditional ECMS uses an equivalent factor with a fixed value [15,16]. However, the equivalent factor can be sensitive to driving conditions [17–20].

An adaptive-ECMS (A-ECMS) has been proposed to regulate the equivalent factor according to changes in driving conditions. An A-ECMS adjusts the equivalent factor so that the SOC can follow the reference SOC trajectory while considering driving conditions. To obtain the reference SOC trajectory, researchers have applied particle swarm optimization (PSO) [21–23], approximate DP [24–26], and neural networks (NN) [27–31].

An A-ECMS is one of the most suitable EMS options to use in real-world HCUs [32–36]. However, A-ECMS designs have some drawbacks. First, computational time must be minimized to operate in a real-world HCU. An A-ECMS compares the control input candidates presented in a complex solution space to determine the power distribution ratio. Because the calculation of many control input candidates should reflect the characteristics of the HEV, system models and constraints need to be efficiently designed for computation. If the system model and constraints are too complex or poorly designed, the computational time increases significantly. Determining the reference SOC trajectory accurately and rapidly is also challenging. An A-ECMS adjusts the equivalent factor according to the reference SOC trajectory. If the accuracy of the reference SOC trajectory is insufficient, fuel consumption cannot be minimized. Even an accurate reference SOC trajectory cannot be used for an A-ECMS if it requires a long computational time. Only accurate and rapid calculation of a reference SOC trajectory can maximize the performance of an A-ECMS.

In this paper, an NLP-based ECMS is proposed. NLP-based ECMS does not directly determine the torque of the engine and motor. The equivalent factor is determined by NLP, and the torque of the engine and motor is determined by the A-ECMS scheme using the equivalent factor. Therefore, the OCP size of the NLP can be reduced, and stable control of the vehicle is possible, even if the NLP has derived an infeasible solution. As mentioned above, it is important to obtain the reference SOC trajectory accurately and rapidly, and NLP can provide rapid and accurate calculations. To reduce NLP computation time, an optimal system model and constraints for an HEV with a P2 structure (TMED) was designed. Results of simulations showed that the proposed NLP-based ECMS closely resembled the global optimal solution produced by DP in a relatively short calculation time.

The main contributions of this paper are as follows. First, a suitable required torque preprocessing method for a P2 structure is proposed. In this process, the required torque demanded by the ICE and EM according to the change in gear ratio is calculated. Secondly, a method for calculating the reference SOC trajectory using NLP is proposed. System models, constraints, and cost functions were designed to reflect the characteristics of the HEV with a P2 structure and driving conditions. To reduce the computational time, the NLP model was simplified. Finally, an A-ECMS that can efficiently follow the reference SOC trajectory is proposed as an optimization-based EMS resembling the global optimal solution of DP.

This paper consists of the following structure. Section 2 presents a system model of a HEV with a P2 structure, Section 3 presents analysis result of the NLP-based ECMS, Section 4 discusses the significance of our findings, and Section 5 is discussion of the result.

2. System Model

In this paper, the NLP-based ECMS was applied to HEVs with a P2 structure, which includes a four-cylinder ICE, a permanent-magnet synchronous motor (PMSM), a lithium-ion battery, and a five-speed automatic transmission. The configuration of the P2 hybrid powertrain is shown in Figure 1. An HEV with a P2 powertrain was simulated by using Autonomie [37], a forward-looking simulation tool developed by Argonne. The NLP-based ECMS algorithm was designed to use Autonomie's data. In addition, an Autonomie simulation environment was used to analyze the performance of the NLP-based ECMS. Table 1 lists the specifications of the major components of the Autonomie model.

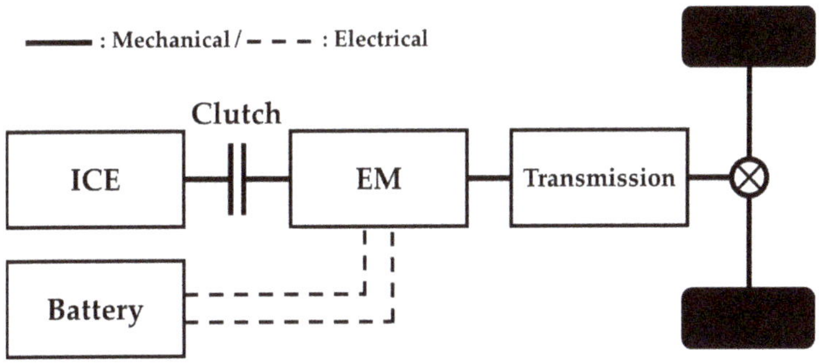

Figure 1. Structure of an HEV with a P2 powertrain.

Table 1. Specification of HEV.

Component	Parameter	Value
Engine (4-cylinder)	Displacement	2.2 [L]
	Maximum power	85.0 [kW]
Motor (PMSM)	Maximum power	29.2 [kW]
Battery (Lithium-ion)	Capacity	7.03 [Ah]
	Nominal voltage	324 [V]
Transmission (5-speed automatic)	Gear ratio	2.563/1.552/1.022/0.727/0.52
Final drive	Gear ratio	4.438
Wheel	Wheel radius	0.30115 [m]
Chassis	Mass	1680 [kg]

2.1. Internal Combustion Engine

An EMS is designed to minimize fuel consumption of an HEV. The ICE of an HEV is modeled as quasistationary, represented by a brake-specific fuel consumption (BSFC) map configured by using the ICE's steady-state test data, as shown in Figure 2. A BSFC map is formulated as a function of torque (T_{eng}) and speed (w_{eng}), as shown in Equation (1). The relationship between BSFC and the power (P_{eng}) is shown in Equation (2). Q_{LHV} is the lower heating value. We have

$$bsfc = f(T_{eng}, w_{eng}) \tag{1}$$

$$P_{eng} = bsfc \times Q_{LHV}. \tag{2}$$

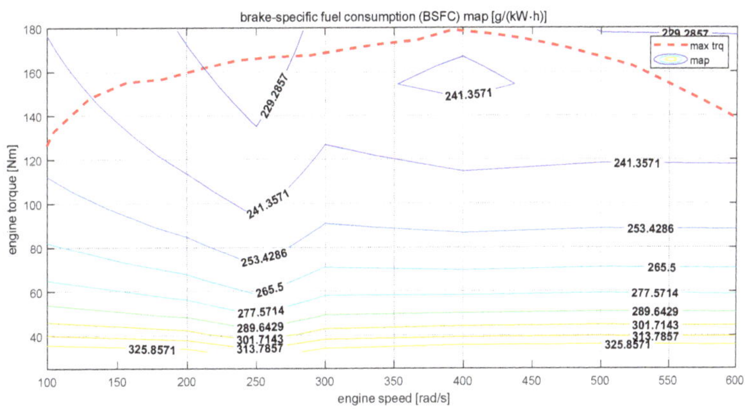

Figure 2. Brake-specific fuel consumption (BSFC) map.

The quasistationary model cannot consider the transient response of the ICE. The transient response of an ICE adversely affects fuel consumption. The transient response of an ICE has complex dynamics, which are difficult to reflect in an EMS accurately. The transient response of the ICE is simplified as a first-order delay function to reflect it in the EMS, as shown in Equation (3). As a result, when the input torque ($T_{e,cmd}$) is determined, the actual torque (T_{eng}) passes through the low-efficiency area of the BSFC map:

$$\dot{T}_{eng} = \frac{1}{\tau_{eng}}\left(T_{e,cmd} - T_{eng}\right). \quad (3)$$

2.2. Electrical Motor

An HEV is being driven while maintaining the SOC at a certain level. The SOC is determined by the current of the EM, when excluding the influence of accessories. The current of the EM is determined by power. The power of the EM was represented in a quasistationary model in the form of a map. The power map was expressed as a function of torque (T_{mot}) and speed (w_{mot}), as shown in Equation (4). The power of an EM also has a nonlinear characteristic, as shown in Figure 3. We have

$$P_{mot} = f(T_{mot},\ w_{mot}) \quad (4)$$

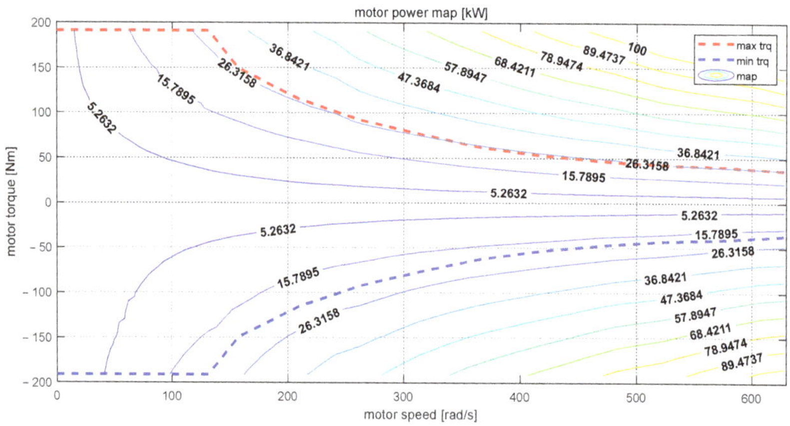

Figure 3. Electric motor power map.

Because the power map of an EM is also quasistationary, it cannot consider a transient response. A transient response is simplified as a first-order delay function, as shown in Equation (5). The time delay (τ_{eng}, τ_{mot}) should be decided by reflecting how quickly the EM is compared to the ICE. As a result, the input torque ($T_{m,cmd}$) passes through the low-efficiency area of the power map when the actual torque is determined. We have

$$\dot{T}_{mot} = \frac{1}{\tau_{mot}}\left(T_{m,cmd} - T_{mot}\right). \quad (5)$$

2.3. Battery

The battery is a complex nonlinear system that is affected by temperature, voltage, and SOC. It is difficult to build a model that accurately represents the battery's dynamic characteristics in an EMS because the formula is complicated and takes a long time to compute. Therefore, the internal resistance model, which is one of the control-oriented models, was applied to the EMS. This model calculates the battery SOC by defining the equivalent circuit expressed by internal resistance (R_{int}), open-circuit voltage (V_{oc}), and

EM power (P_{mot}). The formula is shown in Equation (6). In this formula, R_{int} and V_{oc} are functions of SOC:

$$I_{mot} = \frac{V_{oc} - \sqrt{V_{oc}^2 - 4 \cdot R_{int} \cdot P_{mot}}}{2 \cdot R_{int}}. \quad (6)$$

The SOC of the battery can be calculated integrating the current used by the EM, as shown in Equation (7). This formula is calculated by the initial SOC (SOC_{init}) and battery capacity (Q_{bat}):

$$SOC = SOC_{init} - \frac{1}{Q_{bat}} \cdot \int I_{mot}(t)dt. \quad (7)$$

3. NLP-Based ECMS Algorithm

In this paper, the NLP-based ECMS algorithm, a combination of NLP and an A-ECMS, is proposed. A conceptual diagram of the NLP-based ECMS algorithm is shown in Figure 4. Based on the reference SOC trajectory obtained by NLP, the A-ECMS operates to minimize fuel consumption of the HEV. The process of performing the NLP-based ECMS algorithm is as follows. First, the required torque demanded by ICE and EM is calculated according to the driving condition. The gear ratio must be properly considered when calculating the required torque, because the gear ratio of the P2 powertrain is a discrete variable. Secondly, the reference SOC trajectory corresponding to the required torque through NLP is calculated. NLP can cause long calculation times and infeasible solutions if the system model and constraints are poorly designed. Thirdly, the A-ECMS determines the control input so that the measured SOC can follow the reference SOC trajectory.

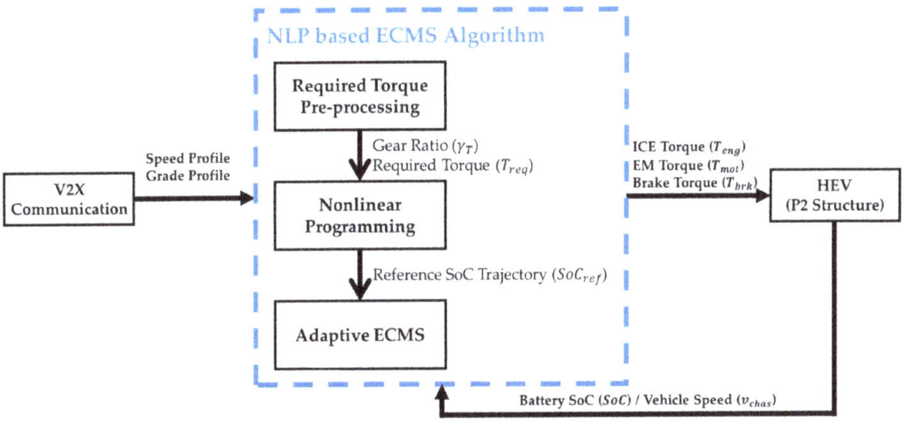

Figure 4. Conceptual diagram of the NLP-based ECMS algorithm.

3.1. Required Torque Preprocessing Algorithm

The required torque preprocessing algorithm determined the required torque used to calculate the reference SOC trajectory in NLP. The required torque preprocessing algorithm consisted of three steps. In first step, the speed was calculated along with the acceleration of the wheel by using the vehicle-speed and road-grade profiles. Inertial force, grade resistance force, air resistance force, rolling resistance torque, and final drive efficiency were reflected. In the second step, the gear ratio trajectory was determined by using the gear shift rule, which is a function of the speed and acceleration determined in the first step. A map for gear shift rules is provided in Figure 5. In third step, the required torque was determined by the gear ratio and transmission efficiency. The algorithm calculated the gear ratio and the torque which will be distributed to the EM and ICE.

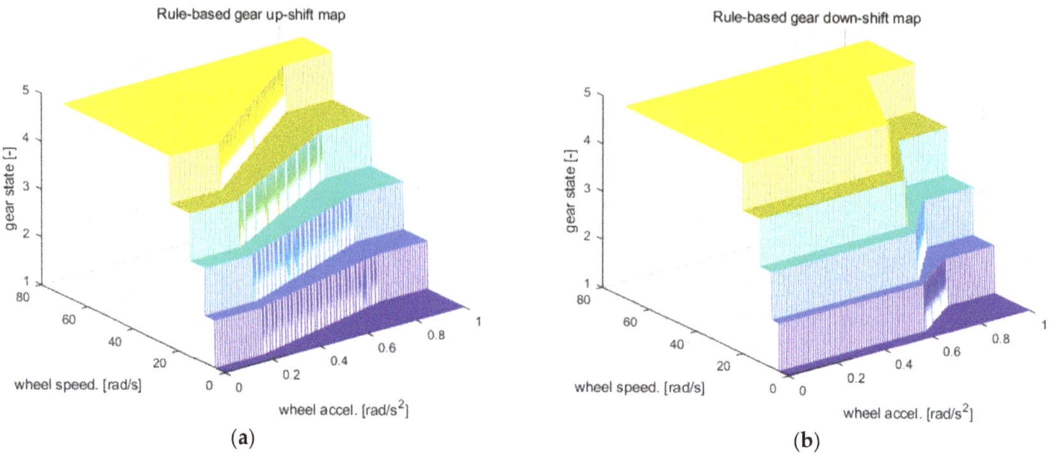

Figure 5. Gear shift rule. (**a**) Upshift map. (**b**) Downshift map.

3.2. Nonlinear Programming

The NLP-based ECMS algorithm calculated the reference SOC trajectory. NLP then calculated the solution for a nonlinear optimal control problem (NOCP) by using a solver. The NOCP was derived from the vehicle model and driving conditions. The IPOPT solver calculated the solution for the NOCP. The solution for the NOCP, the result of IPOPT, was the reference SOC trajectory. The more accurately and efficiently the NOCP was defined, the higher the quality of the solution provided by NLP derived from the solver [38–42].

The NOCP can be explained in three parts. The first part involves a nonlinear system dynamics model. The state vector, input vector, and known-disturbance vector of the HEV model is shown in Equation (8). The system dynamics equations of the components are represented by state space by using vectors, as shown in Equation (9). As part of the simplification work to accelerate the NLP computation, the speed of the EM was treated as a known disturbance. This is because the solution of the NLP, which was used indirectly in the form of the SOC, was determined to satisfy the required torque derived from the EM speed. We have

$$x = \begin{bmatrix} T_{eng} \\ T_{mot} \\ SOC \end{bmatrix},\ u = \begin{bmatrix} T_{cmd,\ eng} \\ T_{cmd,\ mot} \end{bmatrix},\ \text{and } d = w_{mot} = w_{ref} \tag{8}$$

$$\dot{x} = \begin{bmatrix} \dot{T}_{eng} \\ \dot{T}_{mot} \\ S\dot{O}C \end{bmatrix} = \begin{bmatrix} \frac{1}{\tau_{eng}} \cdot \left(-T_{eng} + T_{cmd,\ eng}\right) \\ \frac{1}{\tau_{mot}} \cdot \left(-T_{mot} + T_{cmd,mot}\right) \\ -\frac{1}{Q_{bat}} \cdot \left(\frac{V_{oc} - \sqrt{V_{oc}{}^2 - 4 \cdot R_{int} \cdot T_{mot} \cdot w_{mot}}}{2 \cdot R_{int}}\right) \end{bmatrix} = f(x,u,d). \tag{9}$$

The second part involves nonlinear constraints. The system model was used in NLP to predict how the HEV would behave within the time horizon. That is, the system model operated as a nonlinear constraint when the NLP calculated the solution. The NLP-based ECMS includes constraints of the time horizon by applying the system model to a fourth order Runge–Kutta method, which is a discretization method. The fourth-order Runge–Kutta method is shown in Equation (10). We have

$$\begin{aligned} x[k+1] &= x[k] + \tfrac{T_{step}}{6}\cdot(k_1 + 2\cdot k_2 + 2\cdot k_3 + k_4) \\ k_1 &= f(x[k],\ u[k],\ d[k]) \\ k_2 &= f(x[k] + 0.5\cdot T_{step}\cdot k_1,\ u[k],\ d[k]) \\ k_3 &= f(x[k] + 0.5\cdot T_{step}\cdot k_2,\ u[k],\ d[k]) \\ k_4 &= f(x[k] + T_{step}\cdot k_3,\ u[k],\ d[k]). \end{aligned} \tag{10}$$

The Runge–Kutta method, which requires a relatively large amount of computation, was applied rather than the Euler method, which has a small amount of computation, because the step time (T_{step}) can be relatively large. A large step time can predict longer intervals using the same number of receiving horizons. Equation (10) can be reformulated to equality constraints, as shown in Equation (11). The first nonlinear constraint was derived from a system dynamics model. We have

$$g_1(x) = x[k+1] - \left\{ x[k] + \frac{T_{step}}{6}\cdot(k_1 + 2\cdot k_2 + 2\cdot k_3 + k_4) \right\} = 0. \tag{11}$$

If the described Runge–Kutta method is used, the initial value of the state vector is important because it is accumulated from the initial value. That means the initial values of the state vector are also equality constraints. The initial values of engine torque, motor torque, and SOC were set as shown in Equation (12). Because the SOC must be situated within a certain range at the end of driving, the final value was also important for the SOC. The final value of the SOC was also set to an equality constraint as in Equation (13). We have

$$\begin{cases} g_2(x) = x_1(0) = T_{eng0} = 0 \\ g_3(x) = x_2(0) = T_{mot0} = 0 \\ g_4(x) = x_3(0) = SOC_0 = 0.6 \end{cases} \tag{12}$$

$$g_5(x) = x_3\left(T_{fin}\right) = SOC_{fin}. \tag{13}$$

The maximum/minimum torque is determined by the speed (w_{ref}). Because the speed (w_{ref}) is a known disturbance, the maximum/minimum torque can be known in advance and is an inequality constraint, as shown in Equation (14):

$$\begin{cases} g_6(x) \to f_{eng,min}\left(w_{ref}\right) \le T_{eng} \le f_{eng,max}\left(w_{ref}\right) \\ g_7(x) \to f_{mot,min}\left(w_{ref}\right) \le T_{mot} \le f_{mot,max}\left(w_{ref}\right) \end{cases}. \tag{14}$$

The third part involves a cost function. The algorithm includes a term for fuel consumption and a term for tracking the required torque by using the sum of ICE torque (T_{eng}) and EM torque (T_{mot}). The cost function of an NLP-based ECMS is shown in Equation (15). Q_{fuel} is the weight cost of fuel consumption ($cost_{fuel}$), and Q_{trq} is the weight cost of the reference torque ($cost_{trq}$). We have

$$cost = Q_{fuel}\cdot cost_{fuel} + Q_{trq}\cdot cost_{trq}. \tag{15}$$

The cost of fuel consumption can be calculated by using Equation (16). Surface-fitting for a fuel-consumption rate map was performed to include fuel consumption in the cost function of NLP. To derive a polynomial with an appropriate degree, various models were analyzed as shown in Figure 6. We have

$$cost_{fuel} = sum\left(p_{20}\cdot T_{eng}^2 + p_{10}\cdot T_{eng} + p_{11}\cdot T_{eng}\cdot w_{ref} + p_{01}\cdot w_{ref}\right). \tag{16}$$

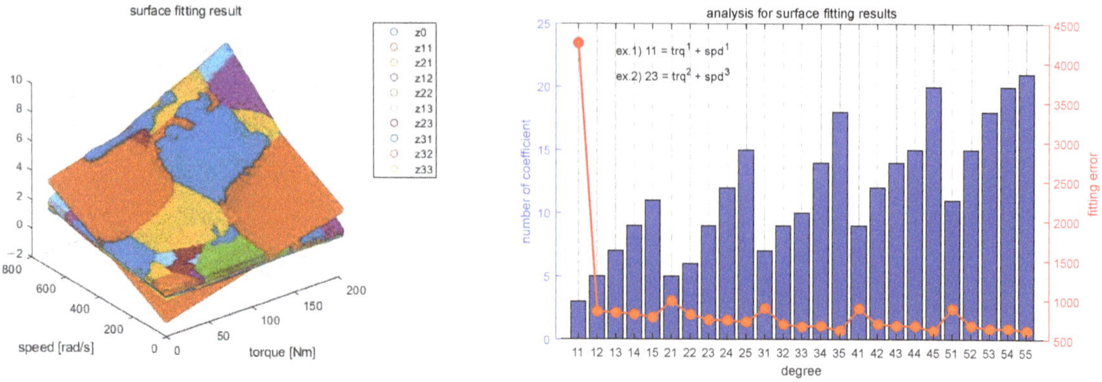

(a) **(b)**

Figure 6. Surface fitting results for a fuel-consumption rate map. (**a**) surface fitting result with various degree. (**b**) Fitting error according to the degree of polynomial.

The cost for tracking the required torque using the sum of ICE torque (T_{eng}) and EM torque (T_{mot}) can be calculated by using Equation (17). The required torque is the reference torque given to the ICE and EM. The cost function was designed to use reference torque, not speed, to decrease the computational time by reducing the size of the system model. We have

$$cost_{trq} = sum\left\{\left(T_{eng} + T_{mot} - T_{ref}\right)^2\right\}. \tag{17}$$

The constraint and cost functions were defined in the form of an OCP for receding horizons. The defined problem was solved by an NLP solver. The IPOPT solver, which is a rapid and powerful NLP solver, was used for the NLP-based ECMS algorithm.

3.3. *Adaptive Equivalent Consumption Minimization Strategy*

The NLP-based ECMS algorithm was controlled by an A-ECMS based on the reference SOC trajectory, which is the result of NLP. An A-ECMS computes the cost function for input grids and obtains the optimal power distribution ratio, which minimize the cost function. A schematic diagram of A-ECMS consisting of five steps is shown in Figure 7.

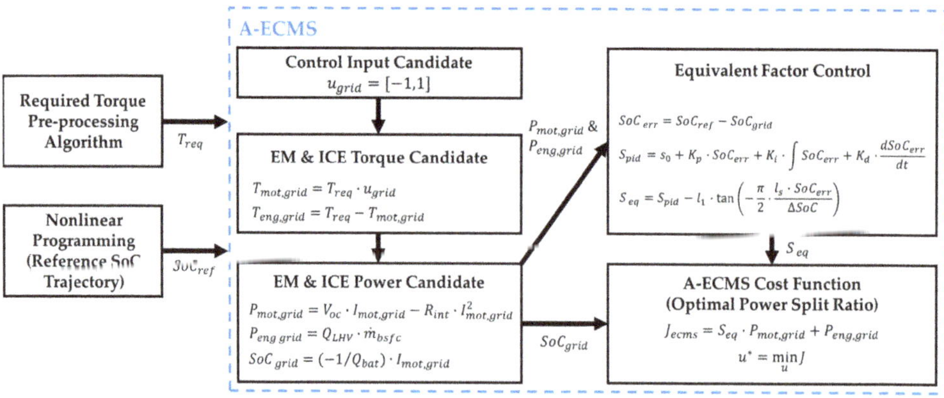

Figure 7. Conceptual diagram of A-ECMS in NLP-ECMS.

The first step determined the number of grids for the control input candidates (u_{grid}) for the power distribution ratio, as described by Equation (18). The A-ECMS in this paper used 201 grids (N = 201). We have

$$u_{grid} = u_{N\times 1} = [-1,\ 1]. \tag{18}$$

The second step calculated the torque candidate of the ICE and EM. The required torque (T_{req}), calculated by a preprocessing algorithm, was distributed to the EM torque candidates ($T_{mot,\ grid}$) by the input grid. The engine torque candidates ($T_{eng,\ grid}$) were based on the difference between the required torque and EM torque candidates. The formulas are found in Equations (19) and (20). In this process, the maximum/minimum torque of the EM and the ICE was considered as shown in Equation (21):

$$T_{mot,\ grid} = T_{req}\cdot u_{grid} \tag{19}$$

$$T_{eng,\ grid} = T_{req} - T_{mot,grid} \tag{20}$$

$$subj.\ to. \begin{cases} T_{mot,min} \le T_{mot,grid} \le T_{mot,max} \\ T_{eng,min} \le T_{eng,grid} \le T_{eng,max} \end{cases}. \tag{21}$$

The third step calculated the power candidate of the ICE and EM. The ICE power candidates ($P_{eng,\ grid}$) were obtained from the BSFC ($\dot{m}_{bsfc}$) and lower heating value (Q_{LHV}). The EM power candidates ($P_{mot,\ grid}$) were obtained from the internal resistance (R_{int}), open-circuit voltage (V_{oc}), and EM current (I_{mot}). Internal resistance and open-circuit voltage are functions of the battery's SOC, and the EM current can be calculated by using Equation (22). SOC candidates were calculated from the EM power candidates because the EM power candidates affected the SOC. The formulas are as follows Equations (23)–(25):

$$I_{mot,grid} = \left(V_{oc} - \sqrt{V_{oc}{}^2 - 4\cdot R_{int}\cdot T_{mot,grid}\cdot w_{mot}}\right) \Big/ 2\cdot R_{int} \tag{22}$$

$$P_{mot,\ grid} = V_{oc}\cdot I_{mot,grid} - R_{int}\cdot I^2_{mot,grid} \tag{23}$$

$$P_{eng\ grid} = Q_{LHV}\cdot \dot{m}_{bsfc}\left(T_{eng,grid},\ w_{eng}\right) \tag{24}$$

$$SOC_{\ grid} = (-1/Q_{bat})\cdot I_{mot,grid}. \tag{25}$$

The fourth step calculates the equivalent factor (S_{eq}), which was used to equalize the power candidates of the ICE and EM. This process determines whether to use power from the EM or the ICE. An A-ECMS regulates the equivalent factor, so that the battery's SOC follows the reference SOC trajectory, which can be calculated by NLP. If the battery's SOC exceeded the reference trajectory, the equivalent factor was decreased. If the battery's SOC was less than reference trajectory, the equivalent factor was increased. This is a form of a proportional–integral–derivative controller (S_{pid}), as shown in Equations (26) and (27). Equation (28) was used to prevent the equivalent factor from changing rapidly [33,38]. We have

$$SOC_{err} = SOC_{ref} - SOC_{grid} \tag{26}$$

$$S_{pid} = S_0 + K_p\cdot SOC_{err} + K_i\cdot \int SOC_{err} + K_d\cdot \frac{dSOC_{err}}{dt} \tag{27}$$

$$S_{eq} = S_{pid} - l_1\cdot tan\left(-\frac{\pi}{2}\cdot\frac{l_s\cdot SOC_{err}}{\Delta SOC}\right). \tag{28}$$

The final step calculates the cost function and derives the optimal input, which is the power-split ratio between the ICE and EM that minimizes the cost function from the candidates. The cost function (J_{ecms}) of an A-ECMS reflecting the equivalent factor is described by Equation (29). The power candidates of the EM were equalized by using the equivalent factor as the power candidates of the ICE. We have

$$J_{ecms} = S_{eq} \cdot P_{mot,grid} + P_{eng,grid}. \quad (29)$$

An A-ECMS is a quasistationary method that uses an input grid. Because the transient response of the ICE was not considered, the ICE torque changed rapidly or was turned on and off frequently. To prevent this problem, Equations (30) and (31) were used to reflect the cost function. Equation (30) is a cost function for engine torque change in previous (T_{eng}) and current steps ($T_{eng,grid}$). Equation (31) is the cost of starting an engine that was in EV mode in the previous step (u_{cmd}):

$$J_{trq} = \sqrt{\left(T_{eng,grid} \cdot w_{eng} - T_{eng} \cdot w_{eng}\right)^2} \quad (30)$$

$$J_{str} = \left(u_{grid} \neq 1\right) \cdot (u_{cmd} = 1). \quad (31)$$

In addition, the torque candidates of the ICE and EM are constrained by maximum/minimum torque conditions. If the engine torque and motor torque do not satisfy the required torque because of the constraints, the vehicle cannot follow the speed profile. To prevent this problem, Equation (32) was applied to reflect the cost function by using the difference between the power components (EM and ICE) and the required power:

$$J_{ref} = abs\left\{ \left(T_{req} - T_{mot} - T_{eng}\right) \cdot w_{gb} \right\}. \quad (32)$$

The cost function of A-ECMS (J) is shown in Equation (33), and the optimal solution of A-ECMS (u^*) can be obtained from Equation (34). Each term that constitutes the cost function must be tuned by using a weight (W_1, W_2, W_3). When setting the weight, it is important to prevent contamination of the original cost function (J_{ecms}) with the ICE power candidate, EM power candidate, and equivalent factor. We have

$$J = J_{ecms} + W_1 \cdot J_{trq} + W_2 \cdot J_{str} + W_3 \cdot J_{ref} \quad (33)$$

$$u^* = \underset{u_{grid}}{minimize}\ J. \quad (34)$$

4. Results

In this paper, the performance of an NLP-based ECMS was validated in a simulation environment by using Autonomie, a forward-looking simulation tool developed by Argonne. An HEV model (P2 powertrain) and an NLP-based EMS were connected in a Simulink environment, as shown in Figure 8. The FTP-75 driving cycle was used to carry out the simulation and verify the performance. Simulations were run on a desktop computer equipped with an AMD Ryzen 5 5600X 6-Core Processor (3.70 GHz) and 16 GB of RAM.

Dynamic programming was used to compare the performance of the NLP-based ECMS. As mentioned above, DP was used to calculate a global optimal solution. That is, an optimal solution that can minimize fuel consumption for a given speed profile. If NLP-based ECMS ensures the HEV follows the optimal trajectory, the result of the NLP-based ECMS will get close to that of DP.

In this paper, the reductions in fuel consumption in the NLP-based ECMS were compared with the results of a rule-based EMS, DP, and DP-based ECMS. The computational times of each EMS were also compared.

Section 4.1 describes the DP model, which is a comparative group, and analyzes the results of the DP model. Section 4.2 analyzes the performance of NLP's reference SOC trajectory calculation. Section 4.3 analyzes the performance of the NLP-based ECMS by using the HEV vehicle model of Autonomie.

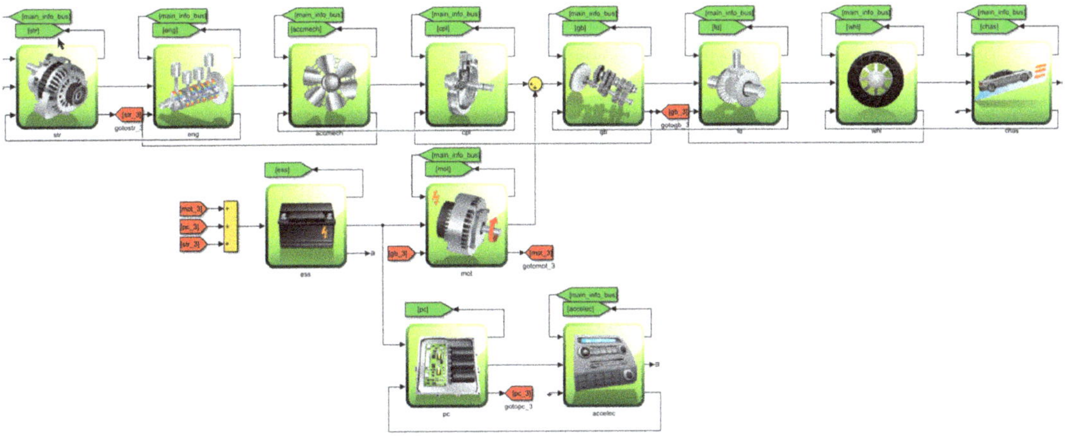

Figure 8. Autonomie simulation environment.

4.1. Comparison Group (Global Optimal Solution/Dynamic Programming)

First, we analyzed the results of deriving the reference SOC trajectory of the NLP-based ECMS. Because an NLP-based ECMS derives the reference SOC trajectory by using NLP, the results of NLP and DP could be compared. The DP setting as a comparison target is shown in Table 2. The input grid involves the power-split ratio, which ranges from −1 to 1. The state grid is for the SOC, which ranges from 0.3 to 0.9. As the number of input grids and state grids increased, the calculation time increased. As the step time decreased, the calculation time increased significantly.

Table 2. Settings for comparison groups (DP).

Case	Input Grid	State Grid	Step Time	Calculation Time
0 (reference)	201	601	0.1 [s]	2481.9 [s]
1	201	61	0.1 [s]	210.7 [s]
2	21	601	0.1 [s]	210.1 [s]
3	21	61	0.1 [s]	89.2 [s]
4	201	61	1.0 [s]	20.4 [s]
5	21	601	1.0 [s]	20.8 [s]
6	21	61	1.0 [s]	8.8 [s]

As shown in Figures 9 and 10, the DP result changed significantly according to the setting for the input grid, the state grid, and the step time. Figure 9 displays the results of DP calculations for the SOC trajectory, which differs greatly depending on the case. Figure 10 depicts the result of DP calculations for ICE torque. If the grid and step time were set incorrectly, the calculated engine torque would include ripple.

Case 0 is the DP result with the largest grid and the smallest step time set. Case 0 had the longest calculation time because of the setting, but the solution was the closest to the global optimal. Case 0 was therefore used as the reference result, which is an ideal result that minimizes fuel consumption during the speed profile. Case 0 cannot be applied to the ECMS because it took too much time to calculate. Accordingly, Case 2, Case 5, and Case 6, which required relatively short computation times, were further compared. The results of Case 2, Case 5, and Case 6 were applied to the ECMS to implement a DP-based ECMS and compared it with the NLP-based ECMS proposed in this paper. Case 2, Case 5, and Case 6, which have relatively short calculation times, were also included in the comparison group. The results of Case 2, Case 5, and Case 6 were applied to the ECMS to design a DP-based ECMS and compared them with the NLP-based ECMS proposed in this paper.

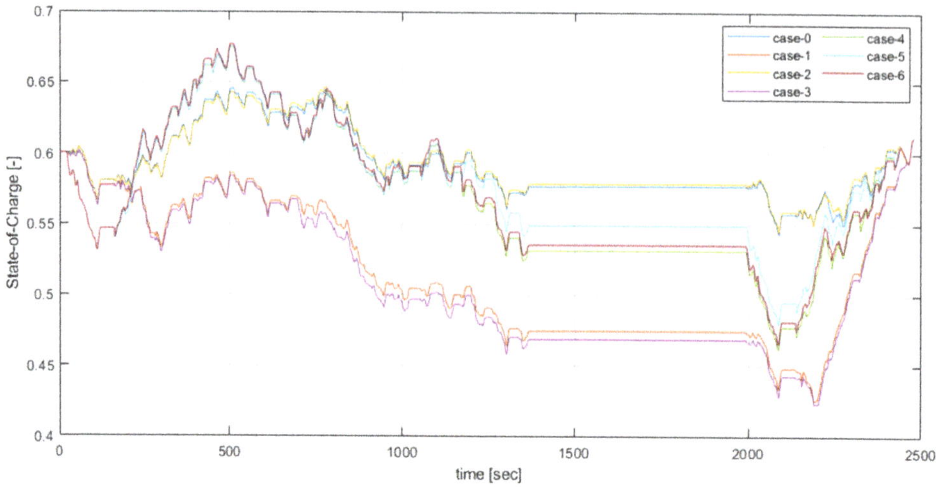

Figure 9. Comparison of DP results (1): state of charge.

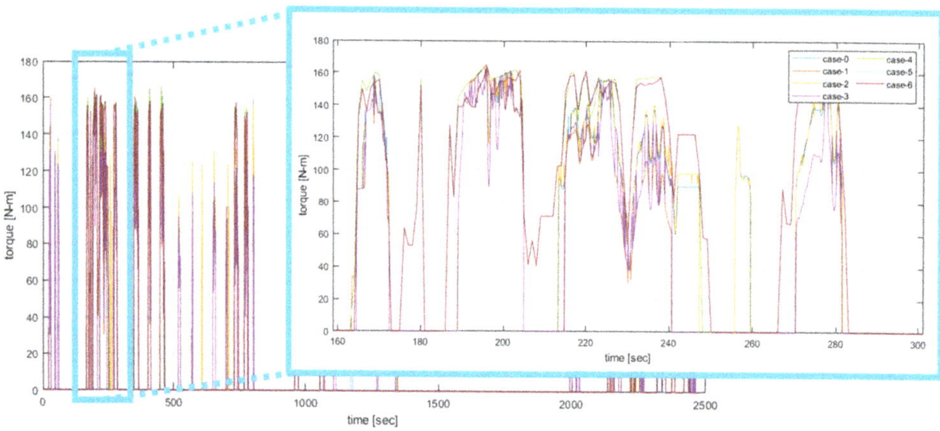

Figure 10. Comparison of DP results (2): engine torque.

4.2. Reference SOC Trajectory

The NLP-based ECMS calculated the reference SOC trajectory by using NLP. Accuracy and computational time were important factors. Accuracy indicates how close the results of NLP were to the global optimal solution.

The results of NLP are compared with DP models in the comparison group. Table 3 compares the NLP and DP results. The calculation time is the time spent calculating the reference SOC trajectory. The sum of error is the difference between the global optimal solution and the reference SOC trajectory obtained by DP and NLP. As a result, NLP was the quickest and the most accurate to calculate the reference SOC trajectory.

Case 0, Case 2, Case 5, and Case 6 are results for the reference framework obtained by the DP model. Case 0 takes the longest time because it has the largest number of input/state grids and the smallest step size. Because the most accurate DP model is used, Case 0 is used as a reference because it is the closest result to a global optimal solution.

Table 3. Comparison of NLP results and DP results (1): calculation time and sum of error.

Case	Input Grid	State Grid	Step Time	Calculation Time	Sum of Error
0 (reference)	201	601	0.1 [s]	2481.9 [s]	0
2	21	601	0.1 [s]	210.1 [s]	45.187
5	21	601	1.0 [s]	20.8 [s]	508.71
6	21	61	1.0 [s]	8.8 [s]	717.21
NLP	-	-	-	2.6 [s]	156.77

In order to be applied to DP based ECMS, the DP model of Case 2, Case 5, and Case 6 has been simplified. Fewer input/state girds and larger step sizes were used, which can reduce the computation time. Because the model is simplified, however, the results were moved away from the global optimal solution. The cost function of the DP is the fuel consumption of the HEV as shown in Equation (1). The cost function used in the DP of each case and the system model (the equations, parameters, maps, etc. shown in Section 2) are also identical.

The NLP case is a result obtained by NLP. The cost function and system models of NLP and DP are different. The cost function and system model of NLP are shown in Section 3.2. However, because the target vehicle is same, same vehicle parameters are used as shown in Section 2.

The tendency of the reference SOC trajectory is important because it determines the charging or discharging of the EM. As shown in Figure 11, the reference SOC trajectory obtained by NLP was similar to the global optimal solution. As shown in Figure 12, the states of ICE ($x1$) and EM torque ($x2$) were found by the NLP input to be first-order delay functions. Additionally, the states of ICE and EM torque were limited by constraints on the maximum/minimum ICE and EM torque.

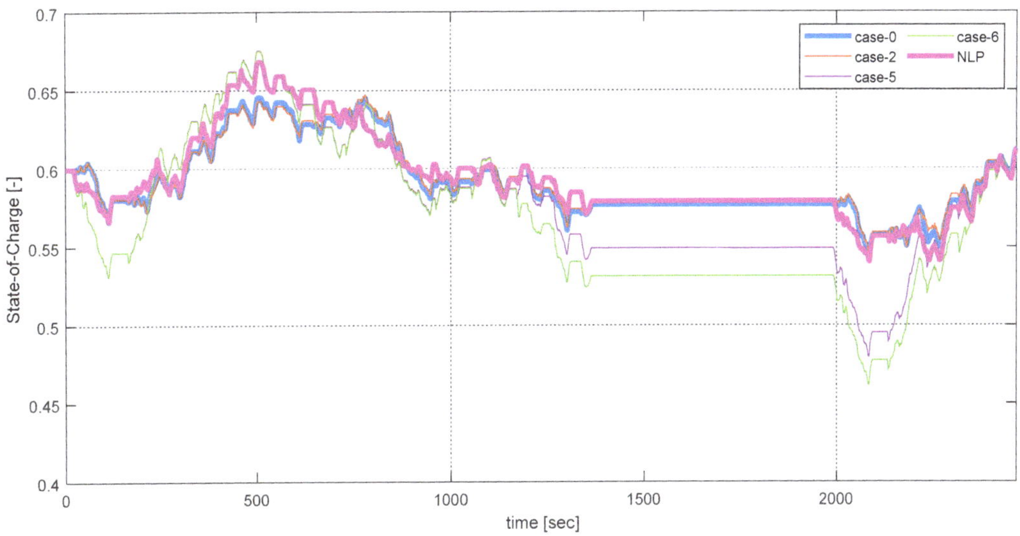

Figure 11. Comparison of NLP results to DP results (2): reference SOC trajectory.

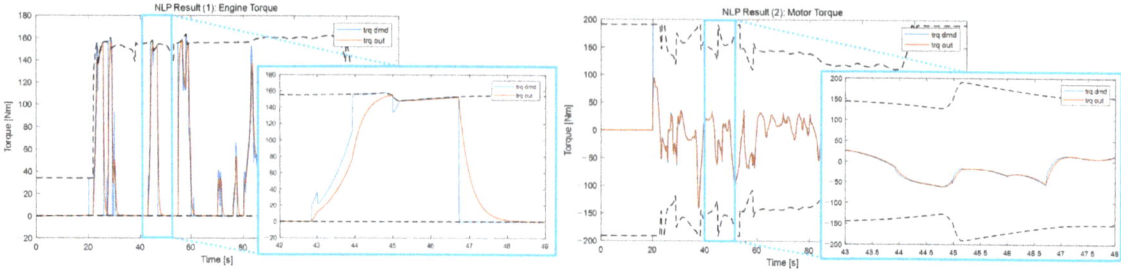

(**a**) (**b**)

Figure 12. Result of NLP: (**a**) ICE torque; (**b**) EM torque. (blue = state/red = input/black = maximum/minimum torque constraints.

4.3. NLP-Based ECMS

Figure 13 depicts the results of the simulation by applying the NLP-based ECMS to Autonomie's HEV model. The NLP-based ECMS properly determined the ICE torque and EM torque so that the HEV follows the speed profile. In addition, the NLP-based ECMS appropriately determined the charging and discharging mode of the EM, so that the battery SOC follows the reference SOC trajectory calculated by NLP. An error occurred between the engine torque and motor torque and command because of the influence of clutch and transmission.

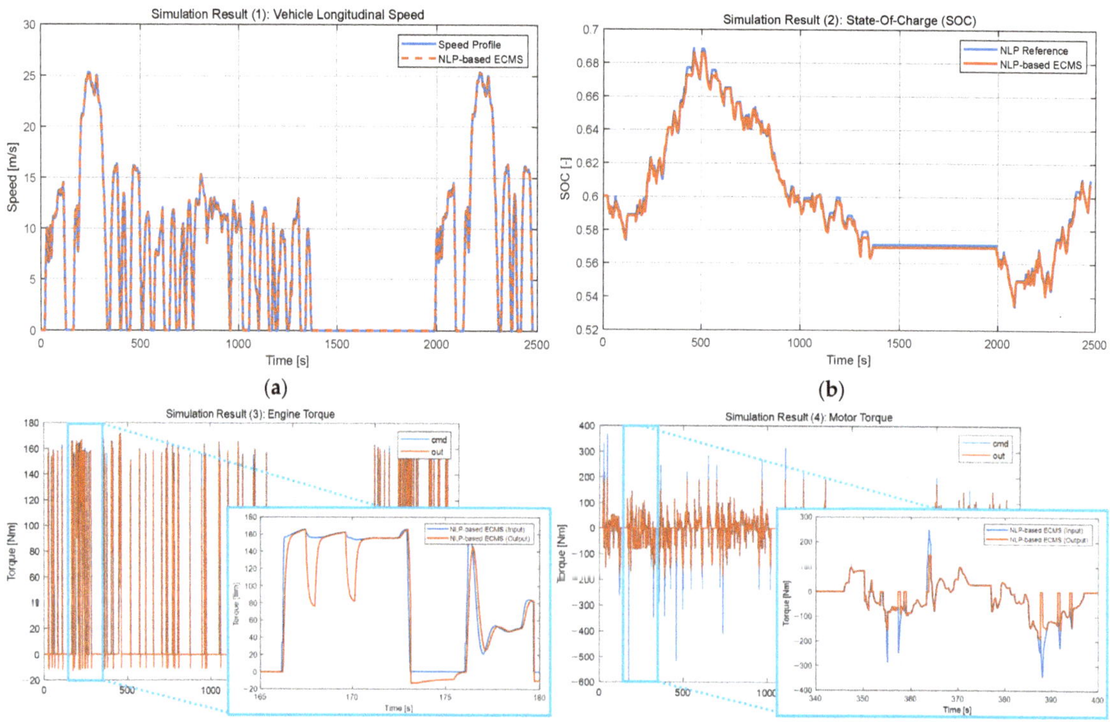

(**a**) (**b**) (**c**) (**d**)

Figure 13. Autonomie (P2 HEV) simulation result with an EMS (NLP-based ECMS): (**a**) vehicle speed; (**b**) state of charge; (**c**) ICE torque; (**d**) EM torque.

Table 4 reveals that the NLP-based ECMS reduced fuel consumption in a manner similar to the Case-0 DP-based ECMS with the shortest simulation time. The Case-0 DP-based ECMS is the closest to the global optimal solution because the A-ECMS used the most accurate DP result. An NLP-based ECMS is an EMS that optimally controls HEV.

Table 4. Analysis of Autonomie simulation results: Fuel Improvement and Simulation Time.

Case	Fuel Consumption	Fuel Improvement	Final SOC	Simulation Time
DP	0.512 [kg]	–	–	–
Ruel-based	0.595 [kg]	0.0 [%]	0.61 [-]	73.79 [s]
0	0.527 [kg]	11.4 [%]	0.61 [-]	2553.18 [s]
2	0.527 [kg]	11.4 [%]	0.61 [-]	281.49 [s]
5	0.533 [kg]	10.4 [%]	0.61 [-]	90.39 [s]
6	0.536 [kg]	9.9 [%]	0.61 [-]	81.00 [s]
NLP	0.531 [kg]	10.8 [%]	0.61 [-]	74.18 [s]

5. Discussion

An NLP-based ECMS that can determine an appropriate equivalent factor for an A-ECMS was proposed. The NLP-based ECMS used a reference SOC trajectory to control the equivalent factor. The NOCP was designed to calculate the reference SOC trajectory. The NOCP includes ICE torque, EM torque, and battery SOC as state variables, and the system models for these variables were applied in a simplified form to increase calculation speed. The NOCP also includes a cost function, which was designed through surface-fitting to the fuel-consumption-rate map. The solution closest to the global optimal with a short computation time through NLP-based ECMS was preferred. In addition, the proposed method was able to respond to infeasible solutions that occurred in NLP by using a strategy involving the equivalent factor. The performance of the proposed method, which was verified in an Autonomie simulation environment, was superior to those of a DP-based ECMS and rule-based EMS in terms of calculation time and fuel consumption.

The proposed NLP-based ECMS completed its tasks rapidly in a desktop-computer simulation. However, the hybrid or electric control unit of an actual vehicle likely has far fewer computational resources. To apply an NLP-based ECMS to a real-world environment, a solver that can be applied to embedded system is therefore required. In this paper, the possibility that NLP-based ECMS can reduce fuel consumption close to the global optimal solution (DP result) is confirmed. However, empirical research on NLP-based ECMS is also important. For empirical verification, it is necessary to compare NLP-based ECMS with other online optimization approaches, such as PSO [21–23] or NN-based method [27–31]. We are conducting a follow-up study for empirical verification of NLP-based ECMS. In conclusion, follow-up studies are required to apply NLP-based ECMS to embedded systems. NLP-based ECMS includes NLP, which is a challenge when applied to embedded systems. To apply NLP-based ECMS to embedded systems, two representative follow-up studies are needed.

The first is to improve the computational speed and computational stability of NLP. To improve the computational performance of NLP, it is necessary to simplify OCP. Simplifying OCP means softening constraints on road information (speed profile, grade profile, etc.) or vehicle models (degree of surface fitting polynomial for engine/motor map). It is important to determine the appropriate prediction/control horizon through this process. Improving computational stability means considering the infeasible region of the NLP. NLP can derive infeasible solutions because of the exceptional driving situation, like a cut-in.

The second is to develop a NLP solver suitable for embedded systems. Various studies have been actively conducted on the research subject [43–45], such as SQP or interior-point method. NLP requires a very large amount of computation because it uses numerical and recursive methods. Therefore, it is necessary to optimize the amount of computation through solution space analysis. The solution space is determined by an operation characteristic or a driving situation of the HEV.

The proposed NLP-based ECMS does not consider clutch and transmission effects. Because the simulation results show that clutch and transmission effects degrade the performance of the proposed method, the influence of clutch and transmission must be considered when defining the OCP in NLP. A MINLP method may be used because the clutch and transmission involve mathematically discrete variables.

In this paper, previously supplied driving conditions were assumed. However, it is impossible to accurately anticipate actual driving conditions. An EMS must therefore reflect changing driving conditions, which depend on traffic conditions.

Author Contributions: Formal analysis, J.L.; project administration, J.L. and H.L.; software, J.L.; validation, J.L.; writing—original draft preparation, J.L.; writing—review and editing, J.L. and H.L.; supervision, H.L. All authors have read and agreed to the published version of the manuscript.

Funding: This research received no external funding.

Institutional Review Board Statement: Not applicable.

Informed Consent Statement: Not applicable.

Data Availability Statement: Not applicable.

Acknowledgments: This work was supported by the Autonomous Driving Development Innovation Program (20014476, Development of Mixed Reality-based Autonomous Driving Parts and System Evaluation Technology) funded by the Ministry of Trade, Industry & Energy (MOTIE, Korea).

Conflicts of Interest: The authors declare no conflict of interest.

References

1. Wang, Y.; Huang, Z. Optimization-based energy management strategy for a 48-V mild parallel hybrid electric power system. *J. Energy Resour. Technol.* **2020**, *142*, 052002.
2. Larsson, V.; Johannesson, L.; Egardt, B.; Lassson, A. Benefit of route recognition in energy management of plug-in hybrid electric vehicles. In Proceedings of the 2012 American Control Conference (ACC), Montréal, QC, Canada, 27–29 June 2012; pp. 1314–1320.
3. Overington, S.; Rajakaruna, S. Review of PHEV and HEV operation and control research for future direction. In Proceedings of the 2012 3rd IEEE International Symposium on Power Electronics for Distributed Generation Systems (PEDG), Aalborg, Denmark, 25–28 June 2012; pp. 385–392.
4. Lee, J.; Park, B. Development and evaluation of a cooperative vehicle intersection control algorithm under the connected vehicles environment. *IEEE Trans. Intell. Transp. Syst.* **2012**, *13*, 81–90. [CrossRef]
5. Zeng, Y.; Cai, Y.; Kou, G.; Gao, W.; Qin, D. Energy management for plug-in hybrid electric vehicle based on adaptive simplified-ECMS. *Sustainability* **2018**, *10*, 2060. [CrossRef]
6. Hou, C.; Ouyang, M.; Xu, L.; Wang, H. Approximate Pontryagin's minimum principle applied to the energy management of plug-in hybrid electric vehicles. *Appl. Energy* **2014**, *115*, 174–189. [CrossRef]
7. Mechichi, O.; Trigui, R.; El Amraoui, L. Adaptive λ-Control Strategy for Plug-In HEV Energy Management Using Fast Initial Multiplier Estimate. *Appl. Sci.* **2022**, *12*, 10543. [CrossRef]
8. Tormos, B.; Pla, B.; Bares, P.; Pinto, D. Energy Management of Hybrid Electric Urban Bus by Off-Line Dynamic Programming Optimization and One-Step Look-Ahead Rollout. *Appl. Sci.* **2022**, *12*, 4474. [CrossRef]
9. Borhan, H.; Vahidi, A.; Phillips, A.M.; Kuang, M.L.; Kolmanovsky, I.V.; Di Cairano, S. MPC-based energy management of a power-split hybrid electric vehicle. *IEEE Trans. Control Syst. Technol.* **2011**, *20*, 593–603. [CrossRef]
10. Vu, T.M.; Moezzi, R.; Cyrus, J.; Hlava, J.; Petru, M. Parallel Hybrid Electric Vehicle Modelling and Model Predictive Control. *Appl. Sci.* **2021**, *11*, 10668. [CrossRef]
11. Zhang, B.; Xu, F.; Shen, T. A Real-Time Energy Management Strategy for Parallel HEVs with MPC. In Proceedings of the 2019 IEEE Vehicle Power and Propulsion Conference (VPPC), Hanoi, Vietnam, 14–17 October 2019; pp. 1–5.
12. Shivappriya, S.N.; Karthikeyan, S.; Prabu, S.; Pérez de Prado, R.; Parameshachari, B.D. A Modified ABC-SQP-Based Combined Approach for the Optimization of a Parallel Hybrid Electric Vehicle. *Energies* **2020**, *13*, 4529. [CrossRef]
13. Yang, L.; Wang, W.; Yang, C.; Du, X.; Zhang, W. Online mixed-integer optimal energy management strategy for connected hybrid electric vehicles. *J. Clean. Prod.* **2022**, *374*, 133908. [CrossRef]
14. Liu, H.; Liu, R.; Xu, R.; Han, L.; Ruan, S. Hierarchical energy management strategy considering switching schedule for a dual-mode hybrid electric vehicle. *Proc. Inst. Mech. Eng. Part D J. Automob. Eng.* **2022**, *236*, 938–949. [CrossRef]
15. Onori, S.; Serrao, L.; Rizzoni, G. *Hybrid Electric Vehicles: Energy Management Strategies*; Springer: Cham, Switzerland, 2016.

16. Simmons, K.; Guezennec, Y.; Onori, S. Modeling and energy management control design for a fuel cell hybrid passenger bus. *J. Power Sources* **2014**, *246*, 736–746. [CrossRef]
17. Zhang, P.; Wu, X.; Du, C.; Xu, H.; Wang, H. Adaptive equivalent consumption minimization strategy for hybrid heavy-duty truck based on driving condition recognition and parameter optimization. *Energies* **2020**, *13*, 5407. [CrossRef]
18. Lin, W.; Zhao, H.; Zhang, B.; Wang, Y.; Xiao, Y.; Xu, K.; Zhao, R. Predictive Energy Management Strategy for Range-Extended Electric Vehicles Based on ITS Information and Start–Stop Optimization with Vehicle Velocity Forecast. *Energies* **2022**, *15*, 7774. [CrossRef]
19. Zanelli, A.; Servetto, E.; De Araujo, P.; Vankayala, S.N.; Vondrak, A. Numerical Assessment of Auto-Adaptive Energy Management Strategies Based on SOC Feedback, Driving Pattern Recognition and Prediction Techniques. *Energies* **2022**, *15*, 3896. [CrossRef]
20. Kommuri, N.K.; McGordon, A.; Allen, A.; Truong, D.Q. A Novel Adaptive Equivalence Fuel Consumption Minimisation Strategy for a Hybrid Electric Two-Wheeler. *Energies* **2022**, *15*, 3192. [CrossRef]
21. Li, T.; Cui, W.; Cui, N. Soft Actor-Critic Algorithm-Based Energy Management Strategy for Plug-In Hybrid Electric Vehicle. *World Electr. Veh. J.* **2022**, *13*, 193. [CrossRef]
22. Wang, Z.; Jiao, X. Optimization of the powertrain and energy management control parameters of a hybrid hydraulic vehicle based on improved multi-objective particle swarm optimization. *Eng. Optim.* **2021**, *53*, 1835–1854. [CrossRef]
23. Oubelaid, A.; Taib, N.; Nikolovski, S.; Alharbi, T.E.; Rekioua, T.; Flah, A.; Ghoneim, S.S. Intelligent speed control and performance investigation of a vector controlled electric vehicle considering driving cycles. *Electronics* **2022**, *11*, 1925. [CrossRef]
24. Pérez, W.; Tulpule, P.; Midlam-Mohler, S.; Rizzoni, G. Data-Driven Adaptive Equivalent Consumption Minimization Strategy for Hybrid Electric and Connected Vehicles. *Appl. Sci.* **2022**, *12*, 2705. [CrossRef]
25. Wang, W.; Cai, Z.; Liu, S. Design of Real-Time Control based on DP and ECMS for PHEVs. *Math. Probl. Eng.* **2021**, *2021*, 6667614. [CrossRef]
26. Han, J.; Park, Y.; Kum, D. Optimal adaptation of equivalent factor of equivalent consumption minimization strategy for fuel cell hybrid electric vehicles under active state inequality constraints. *J. Power Sources* **2014**, *267*, 491–502. [CrossRef]
27. Sun, X.; Xue, M.; Cai, Y.; Tian, X.; Jin, Z.; Chen, L. Adaptive ECMS Based on EF Optimization by Model Predictive Control for Plug-in Hybrid Electric Buses. *IEEE Trans. Transp. Electrif.* **2022**. *early access*. [CrossRef]
28. Chen, Z.; Liu, Y.; Zhang, Y.; Lei, Z.; Chen, Z.; Li, G. A neural network-based ECMS for optimized energy management of plug-in hybrid electric vehicles. *Energy* **2022**, *243*, 122727. [CrossRef]
29. Han, L.; Jiao, X.; Zhang, Z. Recurrent neural network-based adaptive energy management control strategy of plug-in hybrid electric vehicles considering battery aging. *Energies* **2020**, *13*, 202. [CrossRef]
30. Zhang, B.; Meng, Q.; Wu, J.; Ni, Y. An Adaptive Adjustment Method of Equivalent Factor Considering Speed Predict Information. *World Electr. Veh. J.* **2021**, *12*, 211. [CrossRef]
31. Shanmuganathan, J.; Victoire, A.A.; Balraj, G.; Victoire, A. Deep Learning LSTM Recurrent Neural Network Model for Prediction of Electric Vehicle Charging Demand. *Sustainability* **2022**, *14*, 10207. [CrossRef]
32. Musardo, C.; Rizzoni, G.; Guezennec, Y.; Staccia, B. A-ECMS: An adaptive algorithm for hybrid electric vehicle energy management. *Eur. J. Control* **2005**, *11*, 509–524. [CrossRef]
33. Sivertsson, M. Adaptive control using map-based ECMS for a PHEV. *IFAC Proc. Vol.* **2012**, *45*, 357–362. [CrossRef]
34. Onori, S.; Serrao, L.; Rizzoni, G. Adaptive equivalent consumption minimization strategy for hybrid electric vehicles. In Proceedings of the Dynamic Systems and Control Conference, Cambridge, MA, USA, 12–15 September 2010; pp. 499–505.
35. Gu, B.; Rizzoni, G. An adaptive algorithm for hybrid electric vehicle energy management based on driving pattern recognition. In Proceedings of the ASME International Mechanical Engineering Congress and Exposition, Chicago, IL, USA, 5–10 November 2006; pp. 249–258.
36. Kessels, J.T.; Koot, M.W.; Van Den Bosch, P.P.; Kok, D.B. Online energy management for hybrid electric vehicles. *IEEE Trans. Veh. Technol.* **2008**, *57*, 3428–3440. [CrossRef]
37. AUTONOMIE. Available online: https://vms.taps.anl.gov/ (accessed on 26 October 2022).
38. Balazs, A.; Morra, E.; Pischinger, S. Optimization of electrified powertrains for city cars. *SAE Int. J. Altern. Powertrains* **2012**, *1*, 381–394. [CrossRef]
39. CasADi. Available online: https://web.casadi.org/ (accessed on 26 October 2022).
40. IPOPT. Available online: https://coin-or.github.io/Ipopt/ (accessed on 26 October 2022).
41. Biegler, L.T.; Zavala, V.M. Large-scale nonlinear programming using IPOPT: An integrating framework for enterprise-wide dynamic optimization. *Comput. Chem. Eng.* **2009**, *33*, 575–582. [CrossRef]
42. Pirnay, H.; López-Negrete, R.; Biegler, L.T. Optimal sensitivity based on IPOPT. *Math. Program. Comput.* **2012**, *4*, 307–331. [CrossRef]
43. Jabr, R.A.; Coonick, A.H.; Cory, B.J. A primal-dual interior point method for optimal power flow dispatching. *IEEE Trans. Power Syst.* **2002**, *17*, 654–662. [CrossRef]
44. Debert, M.; Yhamaillard, G.; Ketfi-herifellicaud, G.A. Predictive energy management for hybrid electric vehicles-Prediction horizon and battery capacity sensitivity. *IFAC Proc. Vol.* **2010**, *43*, 270–275. [CrossRef]
45. Dueri, D.; Açıkmeşe, B.; Scharf, D.P.; Harris, M.W. Customized real-time interior-point methods for onboard powered-descent guidance. *J. Guid. Control Dyn.* **2017**, *40*, 197–212. [CrossRef]

applied sciences

MDPI

Article

Longitudinal Control Strategy for Connected Electric Vehicle with Regenerative Braking in Eco-Approach and Departure

Rolando Bautista-Montesano [1], Renato Galluzzi [1], Zhaobin Mo [2], Yongjie Fu [2], Rogelio Bustamante-Bello [1] and Xuan Di [2,3,*]

[1] School of Engineering and Sciences, Tecnólogico de Monterrey, Calle del Puente 222, Col. Ejidos de Huipulco, Tlalpan, Ciudad de México 14380, Mexico; rolando.bautista@tec.mx (R.B.-M.); renato.galluzzi@tec.mx (R.G.); rbustama@tec.mx (R.B.-B.)

[2] Department of Civil Engineering and Engineering Mechanics, Columbia University, New York, NY 10027, USA; zm2302@columbia.edu (Z.M.); yf2578@columbia.edu (Y.F.)

[3] Center for Smart Cities, Data Science Institute, Columbia University, New York, NY 10027, USA

* Correspondence: sharon.di@columbia.edu

Abstract: The development of more sustainable urban transportation is prompting the need for better energy management techniques. Connected electric vehicles can take advantage of environmental information regarding the status of traffic lights. In this context, eco-approach and departure methods have been proposed in the literature. Integrating these methods with regenerative braking allows for safe, power-efficient navigation through intersections and crossroad layouts. This paper proposes rule- and fuzzy inference system-based strategies for a coupled eco-approach and departure regenerative braking system. This analysis is carried out through a numerical simulator based on a three-degree-of-freedom connected electric vehicle model. The powertrain is represented by a realistic power loss map in motoring and regenerative quadrants. The simulations aim to compare both longitudinal navigation strategies by means of relevant metrics: power, efficiency, comfort, and usage duty cycle in motor and generator modes. Numerical results show that the vehicle is able to yield safe navigation while focusing on energy regeneration through different navigation conditions.

Keywords: connected electric vehicle; regenerative braking; eco-approach and departure

Citation: Bautista-Montesano, R.; Galluzzi, R.; Mo, Z.; Fu, Y.; Bustamante-Bello, R.; Di, X. Longitudinal Control Strategy for Connected Electric Vehicle with Regenerative Braking in Eco-Approach and Departure. *Appl. Sci.* **2023**, *13*, 5089. https://doi.org/10.3390/app13085089

Academic Editors: Fabrizio Giulietti, Nadjim Horri and William Holderbaum

Received: 12 March 2023
Revised: 14 April 2023
Accepted: 17 April 2023
Published: 19 April 2023

1. Introduction

Currently, climate change and pollution are problems that are being addressed by the United Nations Sustainable Development Goals. These global issues directly affect transportation systems [1]. The goal of achieving net-zero emissions has led to the implementation of stricter government policies for vehicle manufacturers and has paved the way for vehicles powered through alternative energy sources. The automotive sector has invested more money than any other sector in research and development (R&D) in the clean energy innovation sector [2]. Longitudinal control has seen the development of various state-of-the-art algorithms, such as physics-informed deep learning [3] and causal discovery [4,5], with a greater focus on energy efficiency in light of the growing use of EVs. Electric vehicles (EVs) are currently the subject of new designs, battery research, and energy-efficient automated navigation strategies [6]. Longitudinal energy-efficient navigation strategies that extend the driving range for EVs with partial or incomplete information from its environment are currently open research opportunities.

Automated vehicles (AVs) have been introduced to the navigation environment during the past two decades [7]. Sensors provide internal data of the vehicle and the environment. They can be classified into internal (proprioceptive) and external (exteroceptive) state sensors [8]. Internal sensors measure values of the dynamic system such as linear and angular speeds and accelerations, forces, and wheel loads. Inertial measurement units (IMU), encoders, and global navigation satellite system (GNSS) are the most common proprioceptive devices. External sensors retrieve data from the environment, such as distance to the surrounding elements, or light intensity, e.g., cameras, radars, LiDARs, and ultrasonic sensors [9]. Sensor data are then processed to compute the internal states of the vehicle and to perceive its surroundings. Such perception data require filtration to ensure a reliable performance for low- and high-level systems [10]. Kalman filtering and its variations are the most common techniques used for data fusion to estimate ego states and the behavior of external elements [11]. The motion of AVs is controlled by a navigation system that should select suitable lateral and longitudinal actions. High dynamic environments represent a challenge for this technology [12].

Connected vehicles (CVs) are considered as a key element for the constantly evolving autonomous driving and cooperative driving automation fields [13]. They can potentially increase safety and traffic efficiency by sharing information with surrounding vehicles and infrastructure [14] with the goal of implementing a collaborative strategy to optimize traffic. Centralized and decentralized architectures have been proposed to test their performance in terms of communication, intersection control, and vehicle speed management [15]. An enhanced traffic strategy would limit the use of throttle and braking. Consequently, fuel usage, reduced emissions, and energy regeneration can be exploited during navigation [16]. However, state estimation and macroscopic vehicle control is an open research area to ensure safe ego navigation that does not completely rely on communication data.

Eco-approach and departure (EAD) is a longitudinal guidance strategy for CVs navigating an urban signalized or unsignalized [17] intersection environment [18]. The main goal of EAD is to optimize energy, travel time, and comfort by controlling the speed profile of AVs. EAD strategies can face four scenarios in signalized intersections. The cruise control (CC) scenario traverses the intersection at a constant speed. Speed-up denotes when the vehicle increases its speed to cross the road junction. Coast-down with a full stop decreases the AV speed until it halts at the intersection during a red light. Then, it continues moving. Lastly, glide gradually slows down and crosses the intersection without completely stopping [19]. EAD has been tested on internal combustion engine vehicles (ICEV), leading to savings that can range from 5% to 10% for high entry speed and 7% to 26% for low entry speed [20]. EAD on CVs relies on signal phase and timing (SPaT) information. SPaT is a series of messages based on vehicle-to-everything (V2X) communication systems as established by SAE International J2735 standard [21]. It is intended to be a guideline for transportation authorities and the automotive industry to control and communicate within traffic systems [22]. SPaT provides details on the states of the traffic lights, timing data, intersection location, and road layout. Data are transmitted at a 10 Hz rate. SPaT can provide information to driver assistance systems to prevent accidents and to avoid nonessential throttle or braking [23]. Most EAD approaches rely on communications between vehicles and their surrounding environment: pedestrians, vehicles, infrastructure, network, and devices [24]. However, in realistic urban scenarios, not all the information may be available during navigation [25]. Developments of EAD for EVs assume full connectivity and access to SPaT information and consider vehicle queues and battery discharge prediction.

The vehicle models used to validate EAD strategies are typically kinematic or simplified longitudinal dynamic models of ICEVs [26]. However, a powertrain model of the vehicle provides deeper insight on how power consumption and regeneration relate to vehicle dynamics, road characteristics [27], battery charging, discharging, and aging [28], and vehicle motion [29]. Power-efficient longitudinal control for AVs in urban traffic can be achieved when an EAD system merges SPaT data with a valid model of the vehicle, facilitating its migration into a physical platform [30]. Through electric drives, EVs are capable of restoring electrical power to their battery by means of regenerative braking systems (RBS) [31]. These systems retrieve mechanical power from the vehicle and convert it into electricity while reducing the speed of the vehicle during navigation [32]. Power savings and regeneration depend on speed, traveled distance, vehicle [33], and motor features [34].

Algorithms developed to generate target velocity profiles for EAD are known as green light optimal speed advisory (GLOSA) systems [35]. They determine ideal longitudinal speed setpoints to optimize power, travel time, and comfort [36]. Implementations of these systems are based on several approaches. For instance, discontinuous functions are employed to follow recommended speed profiles that would reduce fuel consumption and travel time [19]. Other approaches use rule-based systems. They have been deployed on physical platforms using separate strategies for green, yellow, and red lights [20]. This approach has also been extended to other scenarios where a vehicle can have access to V2X, and whether it finds other vehicles in the environment [24]. On the other hand, fuzzy control systems have been coupled with sliding control to manage energy in EVs and use an RBS [31]. Correspondingly, a dynamic programming approach considers passenger preferences regarding time efficiency, comfort, and fuel usage based on a dataset. It is then deployed in a simulator to validate its performance [37]. Another example is the use of evolutionary methods, where a genetic algorithm is employed to optimize the braking energy recovery efficiency. The study demonstrates its better performance compared with inexperienced and expert drivers [34]. The use of human-like reasoning has led into the use of neural networks. Fuel economy and reduced emissions were achieved by a neural network that uses SPaT data, inter-vehicle communications, and onboard sensing [25]. In hybrid approaches, dynamic programming is used to optimize the longitudinal speed of the vehicle considering energy efficiency. Model predictive control is used to compute the short horizon reactive behavior [26]. Finally, reinforcement learning has also been employed to reduce energy consumption while having a explainable decision process. This approach was validated using microscopic simulators [38]. Table 1 presents a summary of related works.

Artificial intelligence (AI) has been widely used for AVs, where the main goal is to recreate human behavior and reasoning for decision-making by being able to handle large amounts of incomplete or inaccurate data [39]. AI approaches can be classified into logic, heuristic, approximate reasoning, and human-like methods. We will focus on logic-based and approximate reasoning approaches. Logic-based approaches are expert systems that solve specific tasks based on a knowledge base. Their low-computational-cost architecture makes them attractive for predictive and reactive planning. Rule-based systems relate observations with actions. Specific rules aim to capture all the potential states in which the vehicle can be found. Rule explosion is a potential drawback. Therefore, the cause–effect relationships should be carefully considered. They can be considered as one of the most simple forms of AI [40]. Their usage has spread for critical safety applications [41]. Approximate reasoning AI techniques aim to mimic human reasoning. This approach differs from the logic-based ones in their non-Boolean knowledge base representation. Fuzzy logic can express knowledge as a set of Boolean rules that linguistically model a system with vagueness. These types of systems are robust in the case of nonlinear, imprecise, and uncertain data. One of their advantages is that they can be verified by human experts due to their design nature [42]. However, their design methodology is still an area in need of improvement. Fuzzy control systems have been actively used as regenerative braking systems (RBS) in EAD scenarios due to their performance in power recovery and safety [43]. Both approaches use low computational resources, which makes them suitable candidates to implement in physical platforms due to real-time limitations.

Table 1. Longitudinal navigation performance comparison.

Navigation Algorithm	Vehicle Type	Metric
Consensus motion control algorithm [17]	ICEV	MOVES [44] 20% time reduction, 23.7% fuel consumption reduction
Longitudinal speed guidance [18]	ICEV	MOVES 3–5% energy reduction
GlidePath [19]	ICEV	CMEM [45] 17% fuel consumption reduction, up to 64% time reduction
Rule-based [20]	ICEV	Physical platform 6% energy savings
GLOSA [22]	ICEV	21% emission reduction, 121% time increment
Mixed-integer linear programming [23]	Electric bus	Up to 40% energy reduction
Cooperative eco-driving system [24]	ICEV	MOVES 7% energy reduction, 59% emissions reduction
Prediction-Based EAD [25]	ICEV	4.0% energy savings, 4.0–41.7% emissions reduction
Dynamic programming with MPC [26]	Four-wheel-drive EV	15–20% energy reduction
EAD along signalized corridors [27]	ICEV	12–28% fuel savings
Sequential quadratic programming with MPC [28]	EV	10.33% extended battery life
FIS-based EAD/RBS (this work)	EV	101.96% power regenerated than rule-based approach

EAD systems have been used as part of low- and high-level navigation strategies for fuel consumption and optimal trajectory planning [46]. Their impact extends to traffic and surrounding elements (passenger vehicles, trucks, and micromobility actors) to reduce emissions [47]. Typical test scenarios are signalized intersections with one crossroad. The vehicles used to test the developed systems are ICEV and, recently, hybrid architectures [48].

1.1. Modeling Framework

The approach presented in this work is illustrated in the block diagram of Figure 1. It depicts a connected EV (CEV) that acquires information s from the environment and performs longitudinal and lateral actions. The parallel closed-loop scheme implements a lateral pure pursuit controller [49] and one of two longitudinal control options: rule-based or fuzzy inference system (FIS) control systems. The presented approach aims to solve the EAD problem by implementing a coupled GLOSA–RBS system. The simulations and numerical analysis of this work demonstrate that power consumption during navigation is reduced. It also seeks to show that both longitudinal approaches have high motoring and regenerative efficiency during navigation.

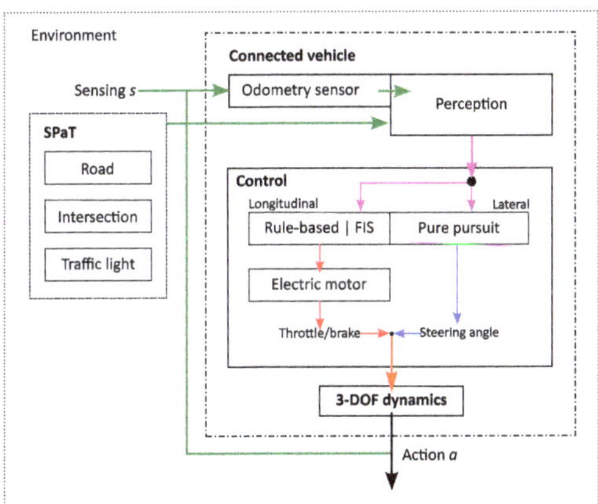

Figure 1. AV navigation block diagram. Sensor and SPaT data is represented by a green arrow, processed perception information in pink, longitudinal control in red, lateral control in blue, navigation inputs in orange and actions in black.

The proposed model uses a hierarchical approach (Figure 1). First, odometry data are retrieved by the internal sensors of the vehicle. It is assumed that the internal states of the vehicles are calculated through sensor fusion techniques and that proprioceptive data are always available and noise-free. Simultaneously, surrounding infrastructure provides SPaT data, which are assumed to be noise-free. Noisy measurements, data loss, and partial observations will be covered in future research. Then, perception module fuses the data to obtain pose information of the vehicle. Next, the pose is fed to the control module, which contains a rule-based or a FIS control systems for longitudinal actions and a pure pursuit controller for lateral actions. Finally, the control signals are sent to the three-degree-of-freedom (3-DOF) dynamics module of the CEV.

1.2. Contributions of This Paper

The major contributions of this paper are listed as follows:

1. We demonstrate the feasibility of implementing rule- and a fuzzy inference system-based longitudinal navigation control systems on a pure electric platform. In both cases, power can be regenerated in the battery with appropriate design considerations. Previous efforts are mostly based on an internal combustion engine and hybrid vehicles, with little or no focus on electric power consumption and regeneration. These techniques are implemented for a B-class electric vehicle. The fuzzy inference system outperforms the rule-based strategy by a magnitude of two in terms of power regeneration. A benchmark experiment is performed in a crossroad intersection environment with a traffic light. Both strategies implement an eco-approach and departure strategy based on a coupled green light optimal speed advisory/regenerative braking system.
2. Lateral, longitudinal, and rolling resistance forces are considered in the dynamic model. They directly impact on the navigation performance and power consumption and regeneration. Works on vehicle navigation usually rely on simplified kinematic models where no forces are considered. Longitudinal and lateral vehicle dynamics provide realistic behavior in terms of powertrain effort and vehicle motion. This differs from most state-of-the-art efforts, where dynamics are partially or totally neglected.
3. A fully characterized electric machine that matches the vehicle powertrain is often neglected in the literature. Ideal actuators are a common choice. Efficiencies are assumed to be ideal or constant. From an energy standpoint, the vehicle propulsion electric motor is represented by its loss map. This allows computing a realistic bidirectional power demand and conversion efficiency.

In summary, the SPaT data usage, sensor fusion, longitudinal and lateral controllers, electric motor, and dynamic vehicle models add realism to the connected automated vehicle control system.

The paper is organized as follows: Section 2 presents the theoretical framework. The rule set and fuzzy inference system used for longitudinal control can be found in Section 3. Section 4 explains the rationale for the experiment design. It also contains the results of the experiment. Finally, Section 5 concludes this work and proposes future research directions. For notation clarity, a nomenclature list is provided at the end of this paper.

2. Powertrain Modeling

2.1. Vehicle Dynamic Model

The CEV is modeled as a 3-DOF bicycle model, as illustrated in Figure 2. Right and left wheels on the same axle are lumped into one element, both for the front and rear. The resulting rear wheel is fixed, whereas the front one can steer at an angle δ. This angle is input for the vehicle dynamic model. The center of gravity C, mass m, polar moment of inertia I_{zz} with respect to the z axis, and wheelbase L of the vehicle are constant parameters. The wheelbase is the distance between axles, thus $L = l_f + l_r$.

The dynamic model of the vehicle is governed by the following equations:

$$m\ddot{y} = -m\dot{x}\dot{\psi} + F_{yf} + F_{yr} \tag{1}$$
$$I_{zz}\ddot{\psi} = l_f F_{yf} - l_r F_{yr} \tag{2}$$
$$m\ddot{x} = m\dot{y}\dot{\psi} + F_{xf} + F_{xr} \tag{3}$$

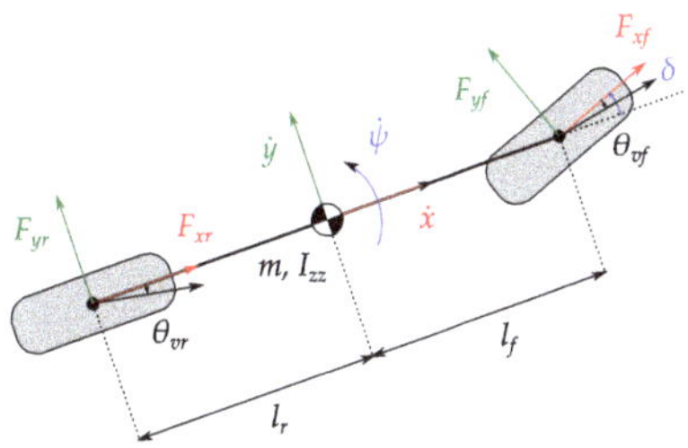

Figure 2. The 3-DOF bicycle CEV model.

These expressions show the coupled dynamic behavior between the longitudinal (x), lateral (y), and yaw (ψ) DOFs. The motion of the vehicle is mainly affected by the longitudinal force on the front (motoring) axle (F_{xf}). Moreover, external lateral forces on the front (F_{yf}) and rear (F_{yr}) axles are further contributions to vehicle dynamics. Roll, pitch, and vertical dynamics are assumed to be uncoupled from this representation. Aerodynamic contributions can be neglected even at the maximum speed of the class-B-sized vehicle (14 m/s). cby the powertrain propulsion minus the rolling resistance. On the rear axle, the longitudinal force is only given by the rolling resistance term:

$$F_{xf} = \frac{T_m \tau_{gb}}{r_t} - \frac{F_c}{2}\text{sign}\,(\dot{x}) \tag{4}$$
$$F_{xr} = -\frac{F_c}{2}\text{sign}\,(\dot{x}) \tag{5}$$

where T_m is the electric motor torque, τ_{gb} the gear box gain, r_t the tire radius, $\dot{x}$ the longitudinal speed of the vehicle, and F_c the Coulomb friction. Aerodynamic effects are neglected in the longitudinal model due to the low velocity values expected in urban environments ($\dot{x} < 60\,\text{km/h}$) [50]. The propulsion force is another input of the vehicle model. The cornering stiffness of the front and rear tires ($C_{\alpha f}$ and $C_{\alpha r}$) contribute to the lateral forces:

$$F_{yf} = 2C_{\alpha f}\left(\delta - \theta_{vf}\right) \tag{6}$$
$$F_{yr} = 2C_{\alpha r}(-\theta_{vr}) \tag{7}$$

In addition, tire velocity angles are given by

$$\theta_{vf} = \arctan\left(\frac{\dot{y} + l_f\dot{\psi}}{\dot{x}}\right) \tag{8}$$
$$\theta_{vr} = \arctan\left(\frac{\dot{y} - l_r\dot{\psi}}{\dot{x}}\right) \tag{9}$$

To favor faster computation, tire velocity angles were linearized assuming small angle deviations:

$$\theta_{vf} \approx \frac{\dot{y} + l_f\dot{\psi}}{\dot{x}} \tag{10}$$
$$\theta_{vr} \approx \frac{\dot{y} - l_r\dot{\psi}}{\dot{x}} \tag{11}$$

The nonlinear model in Equations (1)–(3) describe the lateral dynamics [51]. The global expressions are obtained by substituting Equations (10) into (6), and (11) into (7). Finally, resulting Equations (6) and (7) are substituted into (1) and (2) to yield the following expressions:

$$m\ddot{y} = -m\dot{x}\dot{\psi} + 2\left[C_{\alpha f}\left(\delta - \frac{\dot{y} + l_f\dot{\psi}}{\dot{x}}\right) + C_{\alpha r}\frac{l_r\dot{\psi} - \dot{y}}{\dot{x}}\right] \tag{12}$$

$$I_{zz}\ddot{\psi} = 2\left[l_f C_{\alpha f}\left(\delta - \frac{\dot{y} + l_f\dot{\psi}}{\dot{x}}\right) - l_r C_{\alpha r}\frac{l_r\dot{\psi} - \dot{y}}{\dot{x}}\right] \tag{13}$$

The parameters for the vehicle model used in this study were extracted from the vehicle templates in CarSIM™ of a B-class hatchback vehicle in electric powertrain configuration, front-wheel drive. The values are listed in Table 2. The presented model is used to reproduce the dynamic behavior of the CEV in the rule-based and FIS GLOSA/RBS framework. This representation has been extensively used due its accurate vehicle dynamic behavior [52].

Table 2. B-class hatchback electric vehicle model parameters.

Parameter	Symbol	Value	Unit
Distance from COG to front axle	l_f	1.04	m
Distance from COG to rear axle	l_r	1.56	m
Mass	m	1377	kg
Yaw polar moment of inertia	I_{zz}	1343.1	kgm^2
Front tire cornering stiffness	$C_{\alpha f}$	12	kN/rad
Rear tire cornering stiffness	$C_{\alpha r}$	11	kN/rad
Total rolling resistance force	F_c	5	N

2.2. Motor Model

We use a look-up table containing the power losses of the machine for different torque and shaft speed combinations to describe the electric machine behavior. This approach is advantageous for the following reasons:

1. The loss map gives means for a generic representation of the machine, which is able to fit any motor used in powertrain systems.
2. Focus is given to conversion efficiency through motoring and regenerative quadrants.
3. Power losses can be accurately characterized through electromagnetic finite-element models without the need to represent the machine through a lumped-parameter model.
4. Electromagnetic dynamics are represented through a first-order low-pass filter with a time constant of 50 ms.

The electric machine loss map used in this case is illustrated in Figure 3.

Assuming perfect wheel rolling (no slip) and an ideally efficient gearbox, the powertrain mechanical power can be defined either through vehicle variables or through motor variables, i.e.,

$$P_m = F_{xf}\dot{x} - P_{rr} = T_m\omega_m - P_{rr} \tag{14}$$

where T_m is the motor torque, ω_m is its angular speed, and P_{rr} is the vehicle dissipated power due to rolling resistance.

During operation, the motor will attempt to impose a torque—positive or negative—on the vehicle based on the longitudinal control's request. On the other hand, the vehicle will impose angular velocity on the motor shaft. Part of the mechanical power in Equation (14) will have an impact on the electrical domain of the motor, as it will drive the absorption or regeneration of electrical power, defined as:

$$P_e = V_{dc}i_{dc} \tag{15}$$

with V_{dc}, i_{dc} being the voltage and current of the battery.

In motoring mode, the efficiency of the powertrain is defined as

$$\eta_{mot} = \frac{P_m}{P_e} = \frac{P_m}{P_m + P_{loss}(\omega_m, T_m)} \tag{16}$$

where $P_{loss}(\cdot)$ represents the loss map function of the machine evaluated at a specific operating point.

Conversely, the powertrain has a regenerative efficiency given by

$$\eta_{reg} = \frac{P_e}{P_m} = 1 - \frac{P_{loss}(\omega_m, T_m)}{P_m} \tag{17}$$

Note that in any instance, a generic efficiency term must guarantee $\eta < 1$.

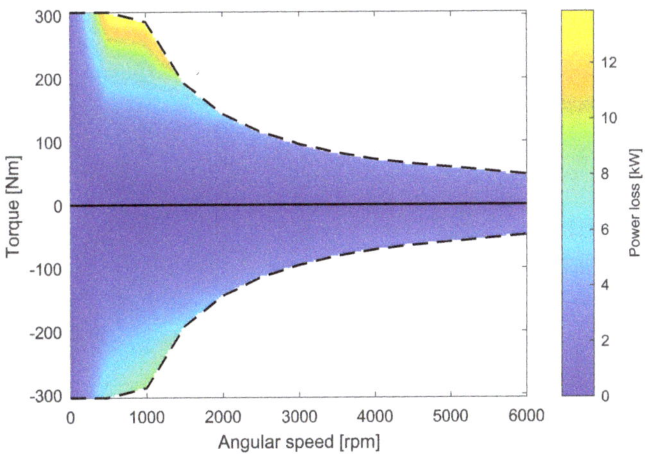

Figure 3. Electric machine loss map. Dashed lines represent the torque–speed characteristic limit. The solid line separates the motoring quadrant (positive torque) from the regenerative quadrant (negative torque).

3. Control Methods

This work proposes a hierarchical control system to optimize a CEV power consumption during EAD using GLOSA/RBS. Two strategies are presented to solve the longitudinal navigation problem: a rule-based and a FIS strategy. Both control methods retrieve information from the SPaT data and the vehicle location and current longitudinal speed. Then, a scenario is identified with the current status of the traffic light. The distance to the intersection and current longitudinal speed are used to determine a motor torque setpoint.

3.1. Rule-Based Controller

The rule-based motor torque setpoint generator is an empiric solution to the GLOSA/RBS problem. Initially, the SPaT data providing the traffic light status are retrieved from the surrounding infrastructure in the intersection. Then, the rule set is divided into three scenarios that depend on the status of the traffic light (L_s): green, yellow, and red. Intersection-crossing or free-driving are the states that the CEV can be in. The selected motor torque (τ) ensures that the vehicle navigates in the environment without crossing the intersection during the red light. The navigation strategy can achieve power restoration by switching between motoring $(\tau > 0)$, regenerative $(\tau < 0)$, or neutral $(\tau = 0)$ behavior of the motor while keeping the longitudinal speed $(\dot{x})$ under the road's speed limit $(\dot{x}_{lim})$ and maintaining comfort. In summary, the driving scenario that the CEV faces depends on L_s,

$\dot{x}$ and the pose (X) to produce a desired τ. The algorithm for the rule-based controller is Algorithm 1.

Algorithm 1 Rule-based controller

```
Input: X, ẋ, L_s
Output: τ
  if L_s is green then
     if X ≥ d_a then
        if ẋ ≥ ẋ_lim then τ ← −τ_min
        else if ẋ < 0 then τ ← τ_min
        elseτ ← (2/3)τ_max
        end if
     else if d_a < X < d_d then
        if ẋ ≥ ẋ_lim then τ ← 0
        else if 0 < ẋ < ẋ_lim then τ ← τ_max
        elseτ ← (1/2)τ_max
        end if
     elseτ ← (2/3)τ_max
     end if
  else if L_s is yellow then
     if X ≥ d_a then
        if ẋ ≥ ẋ_slow then τ ← −(2/3)τ_max
        else if 0 < ẋ < ẋ_slow then τ ← −τ_min
        elseτ ← τ_min
        end if
     else if d_a < X < d_d then
        if ẋ ≥ ẋ_slow then τ ← −(5/6)τ_max
        else if 0 < ẋ < ẋ_slow then τ ← −τ_min
        elseτ ← (1/6)τ_max
        end if
     elseτ ← (2/3)τ_max
     end if
  else if L_s is red then
     if X ≥ d_a then
        if ẋ ≥ ẋ_slow then τ ← −(1/4)τ_max
        else if ẋ ≤ 0 then τ ← τ_min
        elseτ ← (1/10)τ_max
        end if
     else if d_a < X < d_d then
        if ẋ ≥ ẋ_slow then τ ← −τ_max
        else if 0 < ẋ < ẋ_slow then τ ← −τ_min
        elseτ ← (1/6)τ_max
        end if
     elseτ ← (2/3)τ_max
     end if
  end if
```

3.1.1. Green Light

The road is divided into three segments: enter, approach, and departure zones. The first segment, *enter*, extends from the beginning of the road up to an approaching distance (d_a). When the CEV is located in this area, it can have three behaviors. In the first case, the vehicle slowly increases its speed by applying a small torque τ_{min} if the speed is in the $0 < \dot{x} < \dot{x}_{lim}$ interval, where $\dot{x}_{lim}$ is the speed limit of the road. In the second case, if the speed limit is surpassed, a small negative torque $-\tau_{min}$ is used to decrease the velocity. This event encourages power regeneration. Negative speeds are avoided by applying a $\frac{2}{3}\tau_{max}$ torque. The third case avoids using negative values for $\dot{x}$. The second road segment, *approach*, is defined in the $d_a < d < d_d$ interval, where d_d is the beginning of the departing zone or the point after the traffic light in the intersection. As this segment is very close to the crossroad, the CEV will focus on crossing the intersection and reaching the departing zone by applying a τ_{max} torque. Navigating to a higher speed than the limit is avoided by applying a null torque. Negative speeds are avoided by applying a $\frac{1}{2}\tau_{max}$ torque. In the final road segment, *departure*, the vehicle is allowed to increase its throttle by $\frac{2}{3}\tau_{max}$ to cross the intersection and continue navigating.

3.1.2. Yellow Light

The general goal of the navigation strategy during a yellow light is to reduce speed to avoid crossing the intersection during the red light and applying aggressive braking maneuvers. There are three different behaviors in the *enter* interval. The first is to slowly reduce the longitudinal speed of the CEV by applying a $-\tau_{min}$ torque if the current velocity is in the $0 < \dot{x} < \dot{x}_{slow}$ range, where $\dot{x}_{slow}$ is the minimum speed in which the CEV can navigate on the road. If the vehicle is navigating faster than $\dot{x}_{slow}$, a $-\frac{2}{3}\tau_{max}$ torque is applied. Negative speeds are avoided by applying a τ_{min}. The *approach* segment implements similar behaviors to those in the *enter* segment. The main difference are that a $-\frac{5}{6}\tau_{max}$ braking torque that is applied for speeds greater than $\dot{x}_{slow}$, and a $\frac{1}{6}\tau_{max}$ is applied to avoid negative speeds. The vehicle is allowed to increase its throttle by $\frac{2}{3}\tau_{max}$ after crossing the intersection in the *departure* zone. The staged braking strategy during this traffic light status induces a regenerative state for CEV navigation.

3.1.3. Red Light

The strategy of the rule-based control system is to avoid braking to a full stop during red lights at road intersections. The CEV faces a scenario of maximum power regeneration after the speed has been reduced during the yellow light strategy. If the CEV is in the first navigational stage, any longitudinal speed greater than $\dot{x}_{slow}$ causes a $-\frac{1}{4}\tau_{max}$ throttle. A $\frac{1}{10}\tau_{max}$ throttle is applied to keep the vehicle moving in the $0 < \dot{x} < \dot{x}_{slow}$ interval. A full stop and negative velocities are avoided by applying a τ_{min} torque. The *approach* road interval strategy applies a τ_{max} maneuver for speeds greater than $\dot{x}_{slow}$. A $-\tau_{min}$ torque is used to reduce speeds in the range $0 < \dot{x} < \dot{x}_{slow}$. Full stop and negative speeds are avoided with a $\frac{1}{6}\tau_{max}$. The vehicle increases its throttle torque to $\frac{2}{3}\tau_{max}$ once it is in the *departure* zone.

3.2. *Fuzzy Inference System*

Fuzzy set theory is used to specify how a crisp measurement fits into a linguistic description. It is considered a generalization of classic set theory, as it represents uncertainty and memberships. Fuzzy logic maps a degree of membership of a system to a fuzzy set using classic Boolean logic. However, it resembles human-like reasoning with the usage of linguistic variables and decision-making processes with uncertainty [53]. FISs rely on a previously defined knowledge base coming from an expert. They evaluate a statement that relates an antecedent with an implication. These types of systems can represent mathematical complex systems with a simple set of rules model with fuzzy sets [54].

Longitudinal control with FIS require the previous design and validation of appropriate membership functions (MFs). Related research has proposed the usage of uniformly distributed triangular MFs to model the normalized error as input and identical triangular functions with shoulder functions for the brake/throttle as output [55]. However, an uneven distribution and form of MF is proposed to model inter-vehicle distance and the speed error [56]. Similarly, a set with a high number of unevenly distributed MFs was used to capture the behavior of the longitudinal model of a vehicle in [57]. Those guidelines provided insight on how to design the longitudinal FIS-based controller. Three FIS were deigned, one for each state of the traffic light. The green light FIS uses the simplest set of MFs and rules. The yellow FIS and red FIS have a greater number of linguistic variables for the inputs. Even and uneven distributions for the MFs were used to better represent the behavior of every physical variable.

The following MFs are used to represent the physical variables involved in the longitudinal navigation control system: $\dot{x}$, (X), and τ. L_s is not fuzzified, as the light status is a Boolean value. Inputs $\dot{x}$ and X are modeled with Gaussian, Gaussian combination, S, and Z MFs. The longitudinal speed of the CEV is normalized to a $[-1, 1]$ interval. Long-term motor braking can cause reverse motion in the vehicle. Hence, positive and negative values are considered. The maximum speed $\dot{x}_{max}$ of the 3-DOF model is not reachable in urban environments. Thus, the speed limit of the road is used as higher and lower boundaries with an ε tolerance such that

$$\dot{x}_{norm} = \frac{\dot{x}}{(1+\varepsilon)\dot{x}_{lim}} \tag{18}$$

The relative location (X_r) of the CEV respect the intersection (X_I) is also normalized in a $[-1, 1]$ range. The normalization distance of the location is either the SPaT converage (d_{SPaT}) range or the maximum distance in which a visual sensor can detect the traffic light status (d_{sensor}).

$$X_r = \frac{X_I - X}{\max(d_{SPaT}, d_{sensor})} \tag{19}$$

The output τ uses triangular, linear S, and Z MFs. The maximum motor torque τ_{max} is used to normalize τ in a $[-1, 1]$ interval. Defuzzification is performed using the Mamdani method.

$$\tau_{norm} = \frac{\tau}{\tau_{max}} \tag{20}$$

Three different FISs are used to generate the desired longitudinal behavior of the CEV, one for each of the traffic L_s: green, yellow, and red. A multiplexer selects the appropriate system according to the data provided by SPaT.

3.2.1. Green Light

The location X_r of the CEV is represented as three linguistic variables: *entering*, *approaching*, and *departing* (Figure 4a). *Entering* is a Z-shaped MF that covers from the beginning of the road up to a fifth of the road before the intersection. *Approaching* is represented as a two-sided Gaussian MF that runs from 0.4 to a value slightly greater than zero. Finally, *Departing* uses a S-shaped MF that starts a road length before the intersection and ends with the road. Four variables model the velocity $\dot{x}_{norm}$ of the CEV: *reverse*, *slow*, *normal*, and *fast* (Figure 4b). *Reverse* is a Z-shaped MF that goes from the negative speed limit up to zero in an open interval. *Slow* uses a Gaussian MF representation centered at a tenth of $\dot{x}_{norm}$. A *normal* generalized bell-shaped MF describes speeds that are greater than *slow* and less than the speed limit. Any velocity that is near or greater than the speed limit is part of the S-shaped MF *fast*. The motor torque τ_{norm} is described as three variables: *reverse*, *neutral*, and *forward* (Figure 4c). *Reverse* covers all negative torques up to zero. A steep linear Z-shaped MF is employed to express this variable. An unbalanced triangular MF centered on zero depicts the *neutral* torque. The negative part of the MF has a pronounced slope, while the positive ends in a fifth. This behavior was chosen to promote null or slow positive torque. The *forward* variable is a S-shaped MF with slow growth starting from zero.

A set of five rules models the desired behavior during the navigation with L_s = green (Algorithm 2). The main goal is to take advantage of the current status of the light and navigate toward the intersection in the smallest amount of time. Special conditions are considered to avoid navigating backwards or surpassing the speed limit of the road.

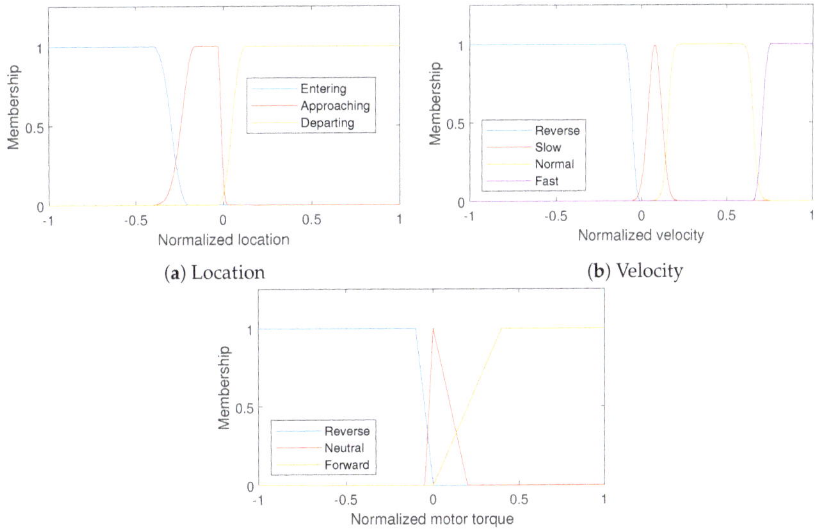

(**a**) Location (**b**) Velocity (**c**) Motor torque

Figure 4. Green light fuzzy sets.

Algorithm 2 Green light FIS knowledge base

Input: $X_r, \dot{x}_{norm}$
Output: $\_norm$
  if $X_r \rightarrow$ entering & $\dot{x}_{norm} \rightarrow$ fast **then**
    $\tau_{norm} \leftarrow$ reverse
  else if $X_r \rightarrow$ entering & $\dot{x}_{norm} \rightarrow \neg$ fast **then**
    $\tau_{norm} \leftarrow$ forward
  else if $X_r \rightarrow$ approaching & $\dot{x}_{norm} \rightarrow$ fast **then**
    $\tau_{norm} \leftarrow$ reverse
  else if $X_r \rightarrow$ approaching & $\dot{x}_{norm} \rightarrow \neg$ fast **then**
    $\tau_{norm} \leftarrow$ forward
  else if $X_r \rightarrow$ departing **then**
    $\tau_{norm} \leftarrow$ forward
  end if

3.2.2. Yellow Light

The same three linguistic variables employed for the X_r set in the green light are used for the yellow light: *entering*, *approaching*, and *departing* (Figure 5a). No changes are made on the type of MF for each variable. However, their limits are changed. The *entering* zone is reduced to a half of its original size. On the other hand, the *approaching* zone is extended three times. Similarly, all the variables of $\dot{x}_{norm}$ in the green light set are reused (Figure 5b). The limits and properties of the *slow*, *normal*, and *fast* MF are updated. *Slow* and *normal* MFs are narrowed and displaced to the left without touching a zero value. *Fast* is just shifted to the left. *Reverse* remains unchanged. The three original MFs of τ_{norm} are preserved. *Neutral* and *forward* continue as they were originally defined. *Reverse* is shifted to the left. A new triangular MF with symmetric boundaries is introduced to add a *soft reverse* variable. This new MF is designed to perform braking and regenerative actions for the CEV.

Nine rules describe the designed navigation navigation behavior for L_s = yellow (Algorithm 3). The main goal is to greatly reduce the speed of the CEV according to its pose and velocity without reaching a full stop, as a part of the RB/EAD strategy. During this maneuver, power can be regenerated because of the performance the electric motor. Crossing the intersection during the yellow light should be avoided. However, traversing the crossroad is encouraged if the speed is high and safety constraints are compromised. Reverse and over the speed limit navigation situations are avoided through specific rules.

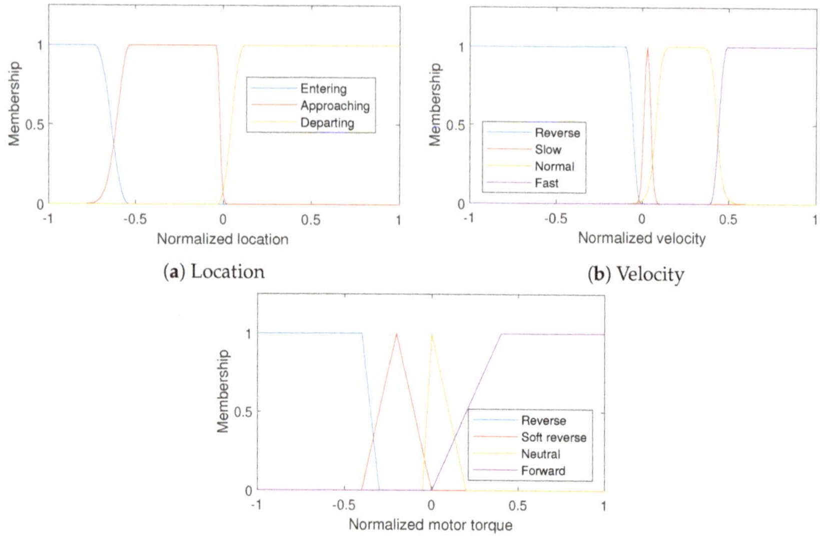

(**a**) Location (**b**) Velocity

(**c**) Motor torque

Figure 5. Yellow fuzzy sets.

Algorithm 3 Yellow light FIS knowledge base

Input: $X_r, \dot{x}_{norm}$
Output: τ_{norm}
  if $X_r \rightarrow$ entering $\&\dot{x}_{norm} \rightarrow$ reverse **then**
    $\tau_{norm} \leftarrow$ forward
  else if $X_r \rightarrow$ entering $\&\dot{x}_{norm} \rightarrow$ slow **then**
    $\tau_{norm} \leftarrow$ neutral
  else if $X_r \rightarrow$ entering $\&\dot{x}_{norm} \rightarrow$ normal **then**
    $\tau_{norm} \leftarrow$ soft reverse
  else if $X_r \rightarrow$ entering $\&\dot{x}_{norm} \rightarrow$ fast **then**
    $\tau_{norm} \leftarrow$ reverse
  else if $X_r \rightarrow$ approaching $\&\dot{x}_{norm} \rightarrow$ reverse **then**
    $\tau_{norm} \leftarrow$ forward
  else if $X_r \rightarrow$ approaching $\&\dot{x}_{norm} \rightarrow$ slow **then**
    $\tau_{norm} \leftarrow$ neutral
  else if $X_r \rightarrow$ approaching $\&\dot{x}_{norm} \rightarrow$ normal **then**
    $\tau_{norm} \leftarrow$ soft reverse
  else if $X_r \rightarrow$ approaching $\&\dot{x}_{norm} \rightarrow$ fast **then**
    $\tau_{norm} \leftarrow$ reverse
  else if $X_r \rightarrow$ departing **then**
    $\tau_{norm} \leftarrow$ forward
  end if

3.2.3. Red Light

The X_r and $\dot{x}_{norm}$ fuzzy sets for the L_s = red are identical to the ones presented in L_s = yellow (Figure 5a,b). The same criteria are considered to fuzzify both crisp values and require no additional modification. Nonetheless, an extra linguistic variable $softforward$ is added to τ_{norm} (Figure 6). The full throttle is not needed during L_s = red in *entering* or *approaching* zones. Therefore, this rule is needed for smoothly navigating toward the intersection. Additionally, it is easier for the motor to switch to *neutral* or *soft reverse*. The *forward* rule is shifted to the right without further modification.

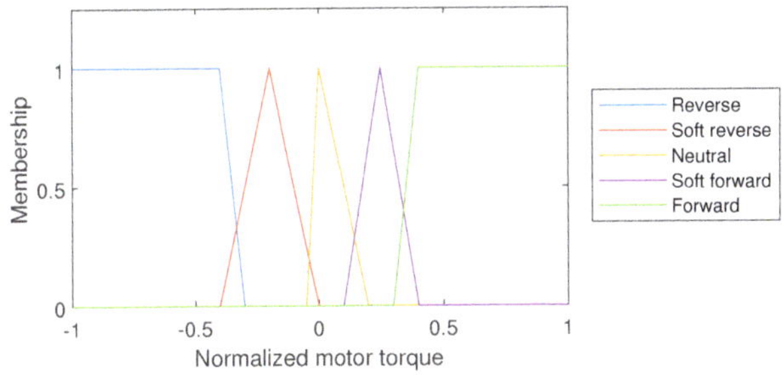

Figure 6. Red light fuzzy sets.

A nine-rule knowledge base is used for L_s = red (Algorithm 4). The goal of this system it to navigating toward the intersection while regenerating the most amount of power. Constantly reducing the speed to a minimum without reaching full stop guarantees the complement of RB/EAD. This strategy is valid for the *entering* and *approaching zones*. During the *entering* stage, the speed is slowly reduced. In contrast, the braking rate is increased in the *approaching* zone to avoid crossing the intersection. Specific rules are designed to evade reverse navigation because of the motor regenerative behavior.

Algorithm 4 Red light FIS knowledge base

Input: $X_r, \dot{x}_{norm}$
Output: τ_{norm}
  if $X_r \rightarrow$ entering $\&\dot{x}_{norm} \rightarrow$ reverse **then**
    $\tau_{norm} \leftarrow$ forward
  else if $X_r \rightarrow$ entering $\&\dot{x}_{norm} \rightarrow$ slow **then**
    $\tau_{norm} \leftarrow$ soft forward
  else if $X_r \rightarrow$ entering $\&\dot{x}_{norm} \rightarrow$ normal **then**
    $\tau_{norm} \leftarrow$ neutral
  else if $X_r \rightarrow$ entering $\&\dot{x}_{norm} \rightarrow$ fast **then**
    $\tau_{norm} \leftarrow$ soft reverse
  else if $X_r \rightarrow$ approaching $\&\dot{x}_{norm} \rightarrow$ reverse **then**
    $\tau_{norm} \leftarrow$ forward
  else if $X_r \rightarrow$ approaching $\&\dot{x}_{norm} \rightarrow$ slow **then**
    $\tau_{norm} \leftarrow$ neutral
  else if $X_r \rightarrow$ approaching $\&\dot{x}_{norm} \rightarrow$ normal **then**
    $\tau_{norm} \leftarrow$ soft reverse
  else if $X_r \rightarrow$ approaching $\&\dot{x}_{norm} \rightarrow$ fast **then**
    $\tau_{norm} \leftarrow$ reverse
  else if $X_r \rightarrow$ departing **then**
    $\tau_{norm} \leftarrow$ forward
  end if

4. Experiments

4.1. Experiment Design

It is assumed that the CEV can retrieve information from the environment with SPaT data or can obtain the status of the traffic light of the intersection with a visual sensor. The internal states are obtained through an odometer. The vehicle uses said data to compute a motor torque setpoint for a longitudinal navigation strategy to achieve RB/EAD. Two strategies were developed: a rule- and a FIS-based strategy. The torque goes into an electric motor, the powertrain, and finally into the 3-DOF vehicle.

4.2. Environment

The street layout is a right-hand, three-lane, one-direction, and no-slope crossroad with a main and secondary roads (Figure 7). Both roads are 300 m long and intersect at the middle. The lanes are 4 m wide. The boundaries of the roads are delimited by solid white lines. A traffic light is located at the middle of the crossroad. Crosswalks are available at the road intersection. It is assumed that the whole environment has SPaT data access without communication interruptions or data integrity threats. The CEV is meant to navigate in a straight path from its initial pose X_i to the goal pose X_f. The initial conditions of the CEV, traffic light, and simulation parameters are shown in Table 3.

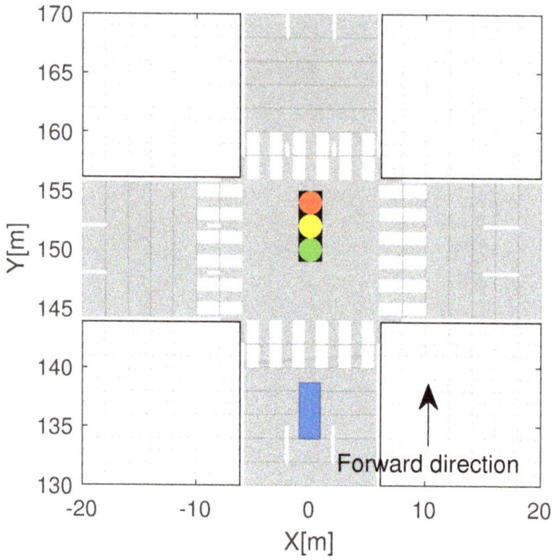

Figure 7. Road layout.

Table 3. Environment parameters.

Parameter	Notation	Default Setting
Initial pose [m]	$X_i = [x_i, y_i]$	[0, 0]
Initial steering angle [rad]	δ	0
Initial velocity [m/s]	$\dot{X}_i = [\dot{x}_i, \dot{y}_i]$	min: [0, 0] max: [13, 0]
Goal pose [m]	$X_f = [x_f, y_f]$	[300, 0]
Green light duration [s]	t_{Lg}	16
Yellow light duration [s]	t_{Ly}	4
Red light duration [s]	t_{Lr}	20
Initial light	L_s	$L_s \in [t_{Lg}, t_{Ly}, t_{Lr}]$
Vehicle sample time [s]	t_s	0.001
Rules and FIS sample time [s]	t_{Ds}	0.1

4.3. Integrated Longitudinal Navigation

Algorithm 5 depicts the integrated longitudinal control system for the implementation of GLOSA/RBS/EAD. The results of said algorithm are presented in terms of navigation success rate and power regeneration. Both generated control methods are based on empiric and expert knowledge. A fixed-step solver was selected with different sampling times for the physical elements (t_s) and the control methods (t_{Ds}) (Table 3). The rule- and FIS-based controllers were tested and validated in a variety of initial speeds and initial light status time. The complete performance data of the rule- and FIS-based longitudinal navigation systems can be found in Table 4.

The rule-based system successfully completed 96.43% of the tests. The cases in which the intersection was not successfully crossed were due to extreme situation such as the following: The vehicle was traveling at very high speeds at the *approaching* zone of the intersection and the status of traffic light was *green*. Then, there was not enough road to decrease the longitudinal speed and make a stop during the sequence change to *yellow* and *red*. The fuzzy system had a similar success rate to the rule-based, 95.9%. The performance in terms of speed and acceleration of both control systems is comparable. Likewise, navigation complies with comfort as the acceleration never surpasses the $[-9.6, 4.905]$ m/s^2 interval. Completion time is consistent with the the longitudinal velocity and throttle.

Algorithm 5 Integrated data acquisition and control algorithm for the connected electric vehicle

Initialize the control methods mentioned in Section 3
Define ego vehicle and environment
Start V2X communication
for $t = 0, 1, 2, \ldots, T$ **do**
    Retrieve SPaT data
    Retrieve pose data
    if navigation = rules **then**
        Compute selected τ_{norm}
    else if navigation = FIS **then**
        Compute X_r and $\dot{x}_{norm}$
        Fuzzify crisp values
        Use knowledge base according to L_s
        Defuzzify linguistic value
        Compute selected τ_{norm}
    end if
    Apply a τ_{max} gain
    Input the τ setpoint to the electric motor
    Apply the motor torque to the vehicle powertrain
end for

Table 4. Navigation system performance.

Method	Velocity [m/s]	Acceleration [m/s^2]	Completion Time [s]	Success Rate [%]
Rule-based	min: 0.025 max: 22.46 avg: 12.22	min: −3.02 max: 3.01 avg: 0.28	33.32	96.43
FIS-based	min: 0.092 max: 23.83 avg: 13.08	min: −4.51 max: 3.74 avg: 0.39	27.76	95.9

Three samples of the performed tests on rule- and FIS-based controllers are presented: one for L_s = green with starting time $t = 0$ s and initial speed $\dot{x}_i = 0$ m/s (Figure 8), another one for L_s = yellow with starting time $t = 0$ s and initial speed $\dot{x}_i = 5$ m/s (Figure 9), and, finally, for L_s = red with starting time $t = 15$ s and initial speed $\dot{x}_i = 11$ m/s (Figure 10). The samples are evaluated with the pose X and velocity $\dot{x}$ metrics. The pose plot depicts the world-based road in meters for x against time. It also shows the position of the intersection and the current status L_s of the traffic light. The velocity plot illustrates the velocity behavior of the CEV through time. Each test is described in the next paragraphs.

4.3.1. Test 1: Green Light Starting From Rest

It can be noted from Figure 8a how the rule- and FIS-based systems approach the intersection and successfully cross during L_s = green before its change to its next state. From Figure 8b, it can be seen how the vehicle starts with an initial velocity of $\dot{x}_i = 0$ m/s. The rule-based system increases its speed until $t \approx 7$ s. Then, it adopts a CC strategy with small increments of speed to navigate the *approach* zone. Speed is kept constant for an interval of ∼10 s. After the junction is passed, the vehicle increments its speed. The FIS-based controller adopts a more aggressive strategy as it constantly increments its speed until $t \approx 10$ s. Finally, it navigates at a constant speed for ∼3 s until overtaking the crossroad and increments its speed.

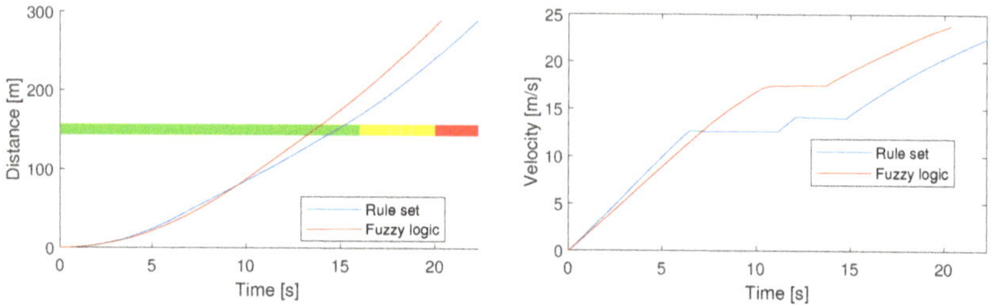

(**a**) Location (**b**) Velocity

Figure 8. Performance metrics for test 1.

4.3.2. Test 2: Yellow Light Starting from a Slow Speed

Figure 9a shows how both longitudinal control systems approach the intersection and successfully cross during L_s = green. In both cases, the CEV adopts an RBS/EAD strategy to avoid crossing during the L_s = yellow or L_s = red status. The vehicle starts with an initial velocity of $\dot{x}_i = 5$ m/s (Figure 9b). The rule-based system initially decreases its speed. Then, a CC behavior is used for $t \approx 22$ s. This approach is maintained until the CEV detects a change of L_s to green. Then, after the junction is passed, the vehicle increments its speed. The FIS-based controller adopts a different strategy. Initially, it increments its speed. Then, when the *approach* zone is reached it decreases its speed until it is close do zero. When L_s = green, it constantly increments its speed, crosses the intersection, and continues on its path.

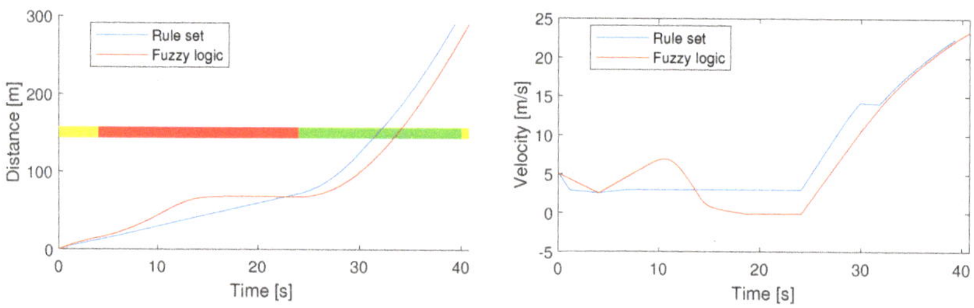

(**a**) Location (**b**) Velocity

Figure 9. Performance metrics for test 2.

4.3.3. Test 3: Red Light Starting from a High Speed

Both longitudinal control systems approach the intersection and successfully cross during L_s = green (Figure 10a). The CEV adotps an RBS/EAD strategy to avoid crossing during the L_s = red status. The vehicle starts with an initial velocity of $\dot{x}_i = 11$ m/s (Figure 10b). The rule-based system initially decreases its speed to $v \approx 7$ m/s. Then, the CEV detects a change of L_s to green. The vehicle gradually increments its speed during the *approach* phase. Small CC behaviors are adopted during this stage. When it crosses the intersection, it increases its speed to continue navigating. The FIS-based controller adopts a less conservative strategy. First, it applies CC until the light change. Next, it increments its speed, goes across the intersection, and continues on its path.

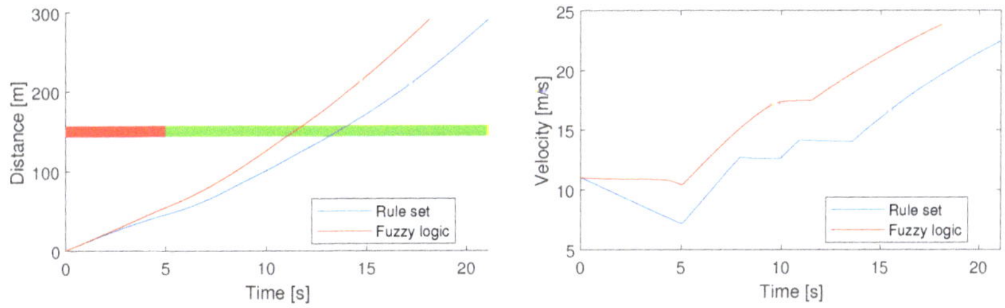

(**a**) Location (**b**) Velocity

Figure 10. Performance metrics for test 3.

4.4. Power Consumption and Regeneration

Related work has focused on pure ICEV or hybrid vehicles. They measure the performance of their EAD systems based on fuel savings and time to completion. We are proposing techniques for CEV. Thus, the power efficiency of the rule- and FIS-based systems is an additional metric used to evaluate the performance of the proposed EAD controllers. The following parameters have been chosen to provide better insight: average power ($\overline{P}$), braking/motoring power ratio ($\overline{P}_r$), and the motoring (η_{mot}) and regenerative (η_{reg}) efficiencies.

The average power is an algebraic mean of the instantaneous power P in a time t. The constant sampling time t_s (Table 3) allows this computation to be valid.

$$\overline{P} = \frac{1}{T}\sum_{t=0}^{T} P_t \tag{21}$$

The regeneration/motoring power ratio provides information on how many instants of the drive cycle were destined to reduce or increment the speed of the CEV and how much power could be restored.

$$\overline{P}_r = \frac{\overline{P}_{brake}}{\overline{P}_{mot}} \tag{22}$$

Power consumption and restoration depend on the initial conditions of the tests and the rationale used to design the longitudinal control system. The performance of the rule- and FIS-based systems can be found in Tables 5 and 6, respectively. They relate the initial conditions of L_s, t_i, and $\dot{x}_i$ with the mechanical and electrical average motoring ($\overline{P}_{mot}$) and regeneration ($\overline{P}_{reg}$) power with their corresponding $\overline{P}_r$ and their efficiencies η_{reg} and η_{mot}.

The rule-based longitudinal system has consistent performance when it enters the intersection at the beginning of the green light $t = (0, 10], L_s \in$ green. Regardless of the initial speed, the $\overline{P}_{mot}$ of the system is consistent in both mechanical and electrical systems. On the other hand, $\overline{P}_{reg}$ decreases uniformly as the speed increases because of the adopted strategy of crossing the junction as soon as possible. It can be noted that η_{mot} increases linearly with $\dot{x}_i$, whereas η_{reg} remains constant. A similar behavior can be found in the FIS-based system. However, $\overline{P}_r$ is almost doubled at low speeds, increased by a half on medium speeds and greater than 15% at high speeds when compared with the rule-based performance. The motoring and regeneration efficiencies show an almost identical operation to the ones in the other system but with higher values.

For $t = [10, 16], L_s \in$ green segment or late green at the traffic light, there is constant power consumption at medium and high initial speeds and lower power consumption for slow speed in the rule-based system. That behavior is found to be inverse for the FIS. Slow and medium speeds show the constant and higher use of power for high velocity. Regeneration of power proportional to $\dot{x}_i$ can be spotted in both controllers. The FIS regenerates from two to three times the power of the rule-based system. Up to 60% of the consumed electrical power can be restored by the CEV. The η_{mot} value is almost constant in both systems, with higher efficiency in the FIS, while η_{reg} increases in both cases, with better power usage for the FIS.

Table 5. Rule-based controller power performance.

L_s Time Interval [s]	$\dot{x}_i$ Interval [m/s]	Mechanical			Electrical			η_{mot} [%]	η_{reg} [%]
		$\overline{P}_{mot}$ [kW]	$\overline{P}_{reg}$ [kW]	$\overline{P}_r$ [%]	$\overline{P}_{mot}$ [kW]	$\overline{P}_{reg}$ [kW]	$\overline{P}_r$ [%]		
(0, 10)	[0, 5)	21.66	6.38	29.36	26.92	4.91	18.12	80.44	77.05
	[5, 10)	22.54	5.46	24.25	27.32	4.21	15.24	82.50	77.03
	[10, 13]	21.57	4.68	21.68	25.58	3.62	13.95	84.31	77.44
[10, 16]	[0, 5)	15.79	3.00	18.99	19.76	2.23	11.28	79.90	74.60
	[5, 10)	19.35	4.60	24.16	24.22	3.48	14.55	79.89	75.55
	[10, 13]	19.92	5.36	27.87	24.90	4.11	17.14	80.01	76.65
(16, 20]	[0, 5)	10.62	0.33	3.19	12.86	0.25	1.99	82.57	75.99
	[5, 10)	15.32	2.69	17.50	18.57	2.03	10.86	82.51	75.55
	[10, 13]	19.21	5.79	30.95	23.69	4.68	20.24	81.08	80.80
(20, 30]	[0, 5)	11.99	0.53	4.44	14.51	0.42	2.89	82.58	78.80
	[5, 10)	18.88	3.03	15.76	22.87	2.48	10.65	82.57	81.90
	[10, 13]	24.46	6.50	26.54	30.01	5.49	18.28	81.50	84.45
(30, 40]	(0, 5)	15.11	0.82	5.44	18.31	0.64	3.49	82.51	77.76
	[5, 10)	23.32	3.82	16.17	28.11	3.14	11.02	82.98	82.18
	[10, 13]	26.12	7.39	27.96	31.04	6.35	20.22	84.14	85.94
Average		19.06	4.02	19.61	23.24	3.20	12.66	81.96	78.78

Table 6. Fuzzy logic-based controller power performance.

L_s Time Interval [s]	$\dot{x}_i$ Interval [m/s]	Mechanical			Electrical			η_{mot} [%]	η_{reg} [%]
		$\overline{P}_{mot}$ [kW]	$\overline{P}_{reg}$ [kW]	$\overline{P}_r$ [%]	$\overline{P}_{mot}$ [kW]	$\overline{P}_{reg}$ [kW]	$\overline{P}_r$ [%]		
(0, 10)	[0, 5)	21.48	12.78	59.13	25.76	10.61	40.35	83.38	83.04
	[5, 10)	23.96	9.78	37.77	28.15	8.23	26.90	85.10	84.15
	[10, 13]	21.69	6.26	26.58	25.10	5.29	19.32	86.43	84.54
[10, 16]	[0, 5)	16.10	5.19	34.47	19.51	3.97	21.68	82.53	76.55
	[5, 10)	16.74	10.35	61.85	20.22	8.14	40.03	82.79	78.72
	[10, 13]	18.27	17.21	90.77	21.88	13.86	60.57	83.48	80.55
(16, 20]	[0, 5)	15.04	5.02	33.39	18.03	3.65	20.25	83.43	72.71
	[5, 10)	15.02	6.66	44.29	17.97	4.97	27.65	83.60	74.65
	[10, 13]	14.94	8.68	58.38	17.90	6.65	37.35	83.45	76.65
(20, 30]	[0, 5)	17.20	8.20	47.36	20.52	6.07	29.36	83.80	74.01
	[5, 10)	18.38	8.91	48.36	21.98	6.59	29.89	83.61	73.91
	[10, 13]	19.82	8.08	41.23	23.77	6.13	26.08	83.35	73.97
(30, 40]	[0, 5)	19.38	2.03	10.04	22.28	1.59	6.89	84.85	78.16
	[5, 10)	21.42	6.96	32.29	25.04	5.42	21.39	85.56	77.91
	[10, 13]	23.79	7.27	30.20	27.82	5.85	20.65	85.52	80.52
Average		18.88	8.22	43.76	22.39	6.46	28.55	84.05	78.13

A new behavior is found during the yellow light or $t = (16, 20], L_s \in$ yellow for the rule-based system. Power consumption and regeneration increase linearly with $\dot{x}_i$. Immediate CC navigation causes a very small $\overline{P}_r$ for low speeds. On the other hand, medium and high speeds require an initial brake to avoid crossing the intersection during $L_s =$ red and encourage CC. This implies an increasing $\overline{P}_r$. The FIS controller has a constant power consumption, but a growing regeneration rate as in the rule-based system. $\overline{P}_r$ is ten times greater during slow, more than double for medium, and almost double for high $\dot{x}_i$. The motoring efficiency is kept constant for each system with a higher value for the FIS. The efficiency for regeneration power is consistent for low and medium speeds and greater for high speeds for the rule-based system, whereas the FIS continuously increases.

The rule-based system has a continuously increasing power consumption and regeneration during the initial red light phase or $t = (20, 30]$, $L_s \in$ red. Slow initial speeds cause little regeneration due to CC as in $L_s \in$ yellow. Similarly, the faster the vehicle starts navigating, the greater a negative throttle should be applied to start CC, causing the increment of $\overline{P}_r$ from below 5% to above one-fifth of the electric motoring power. The FIS power consumption increments slowly, but the regeneration is almost constant. This implies a $\overline{P}_r$ of almost one-third of the consumed electric power. The value of η_{mot} is very near in the rule- and FIS controllers, whereas there is a higher efficiency in η_{reg} of the rule-based system.

Finally, the rule-based controller exhibits a similar behavior to the previous stage for the late red light or $t = (30, 40]$, $L_s \in$ red. This applies to $\overline{P}_r$, η_{mot} and η_{reg}. The FIS has a different performance. This time, its power consumption as well as the regeneration increment slowly. However, a small reduction in the restored power can be noted at high speeds. Motoring and regeneration efficiencies are higher when compared with the previous stage but smaller than for the rule-based ones.

The difference in the success rate of both navigation systems is not significant. Both are close to 96%. The average mechanical motoring power of the rule-based controller and the FIS is almost identical. The rule-based employs slightly more power. Nonetheless, the power regeneration and ratio performance of both systems exhibit different behaviors. The FIS regenerated electrical power outperforms the one of the rule-based by 101.96%. Consequently, the regeneration/motoring power ratio is 125.55% higher.

5. Conclusions and Future Research

In this work, two longitudinal navigational strategies were developed for implementing RBS on EAD while navigating signalized intersections with a traffic light. In the virtual environment, internal data are acquired through odometry and external data through SPaT coming from V2X communications. The data are fed into a perception module that provides location information to a CEV controller. The controller implements a lateral pure pursuit algorithm. A rule- or FIS-based longitudinal controller determines the setpoint for an electric motor that outputs a torque that is implemented by the powertrain. Said torque is then transformed into positive or negative throttle. Finally, it is applied into a dynamic model of an ego vehicle. This model provides a realistic representation of the plant. The proposed algorithms implement a longitudinal navigation strategy that account for speed limit compliance, comfort, and power efficiency and regeneration. The numerical results enable the possibility of implementation of both systems into a physical system for controlled tests.

Navigation of CEV at urban intersections with power management and restoration is a field with open research opportunities. After simulation, the numerical results show that the proposed systems can be employed as a benchmark for future researchers to validate new systems in terms of SPaT data, longitudinal velocity and acceleration, comfort, time to completion, success rate, consumed and regenerated power, and motoring and regeneration efficiency. Therefore, the open research questions along these lines are:

1. Multiple intersection environments to alter the state of charge of a battery during sustained navigation.
2. Background vehicles to increase the complexity of the problem and help better understand the safety, comfort, and power management challenges.
3. The use of machine learning, artificial intelligence, or predictive control techniques that use the proposed longitudinal navigation systems as a benchmark and design rationale.

4. Enhancing vehicles with visual and inertial sensors to cover non-SPaT areas. Sensor fusion techniques are required to tackle data availability, thereby expanding the problem into a partially observable Markov decision process.
5. The integration of the proposed approach within more complex energy management and control strategies, where multiple vehicle functionalities and behaviors coexist.
6. Safe and eco-approach and departure navigation in highly dynamic situations, such as lane changing, merging, overtaking, or obstacle avoidance maneuvers.

Author Contributions: Conceptualization, R.B.-M., R.G. and X.D.; methodology, R.B.-M., R.G., Z.M., Y.F., R.B.-B. and X.D.; software, R.B.-M.; validation, R.B.-M.; formal analysis, R.B.-M. and R.G.; investigation, R.B.-M., R.G. and X.D.; resources, X.D.; data curation, R.B.-M.; writing—original draft preparation, R.B.-M.; writing—review and editing, R.B.-M., R.G., Z.M., Y.F., R.B.-B. and X.D.; visualization, R.B.-M. and R.G.; supervision, X.D.; project administration, R.B.-B. and X.D.; funding acquisition, X.D. All authors have read and agreed to the published version of the manuscript.

Funding: Bautista-Montesano, and Galluzzi are supported by CIMB and the Data Science Hub at Tecnológico de Monterrey, and the AWS Cloud Credit for Research. Bautista-Montesano is funded by Tecnológico de Monterrey—Grant No. A00996397, Consejo Nacional de Ciencia y Tecnología (CONACYT) under the scholarship 679120. Di is supported by NSF CPS-2038984.

Conflicts of Interest: The authors declare no conflict of interest.

Abbreviations

The following abbreviations are used in this manuscript:

AI	Artificial intelligence
AV	Automated vehicle
CC	Cruise control
CV	Connected vehicle
CEV	Connected electric vehicle
EAD	Eco-approach and departure
EV	Electric vehicle
FIS	Fuzzy inference system
GLOSA	Green light optimal speed advisory
ICEV	Internal combustion engine vehicles
MF	Membership functions
RBS	Regenerative braking system
SPaT	Signal phase and signal
V2X	Vehicle to everything communication

References

1. United-Nations. *The Sustainable Development Goals Report*; UN: New York, NY, USA, 2022.
2. Sudhakar, S.; Sze, V.; Karaman, S. Data Centers on Wheels: Emissions From Computing Onboard Autonomous Vehicles. *IEEE Micro* **2023**, *43*, 29–39. [CrossRef]
3. Mo, Z.; Shi, R.; Di, X. A physics-informed deep learning paradigm for car-following models. *Transp. Res. Part C Emerg. Technol.* **2021**, *130*, 103240. [CrossRef]
4. Ruan, K.; Zhang, J.; Di, X.; Bareinboim, E. Causal Imitation Learning via Inverse Reinforcement Learning. In Proceedings of the Eleventh International Conference on Learning Representations, Kigali, Rwanda, 1–5 May 2023.
5. Ruan, K.; Di, X. Learning Human Driving Behaviors with Sequential Causal Imitation Learning. In Proceedings of the AAAI Conference on Artificial Intelligence, Virtual, 22 February–1 March 2022; Volume 36, pp. 4583–4592.
6. International-Energy-Agency. *Clean Energy Innovation*; International-Energy-Agency: Paris, France, 2020.
7. Chen, G.; Wang, F.; Li, W.; Hong, L.; Conradt, J.; Chen, J.; Zhang, Z.; Lu, Y.; Knoll, A. NeuroIV: Neuromorphic Vision Meets Intelligent Vehicle Towards Safe Driving with a New Database and Baseline Evaluations. *IEEE Trans. Intell. Transp. Syst.* **2022**, *23*, 1171–1183. [CrossRef]
8. Yeong, D.J.; Velasco-Hernandez, G.; Barry, J.; Walsh, J. Sensor and Sensor Fusion Technology in Autonomous Vehicles: A Review. *Sensors* **2021**, *21*, 2140. [CrossRef] [PubMed]
9. Woo, A.; Fidan, B.; Melek, W.W. Localization for Autonomous Driving. In *Handbook of Position Location*; Wiley: Hoboken, NJ, USA, 2019; pp. 1051–1087. [CrossRef]

10. Giacalone, J.P.; Bourgeois, L.; Ancora, A. Challenges in aggregation of heterogeneous sensors for Autonomous Driving Systems. In Proceedings of the 2019 IEEE Sensors Applications Symposium (SAS), Sophia Antipolis, France, 11–13 March 2019. [CrossRef]
11. Kim, T.; Park, T.H. Extended Kalman Filter (EKF) Design for Vehicle Position Tracking Using Reliability Function of Radar and Lidar. *Sensors* **2020**, *20*, 4126. [CrossRef] [PubMed]
12. Claussmann, L.; Revilloud, M.; Gruyer, D.; Glaser, S. A Review of Motion Planning for Highway Autonomous Driving. *IEEE Trans. Intell. Transp. Syst.* **2020**, *21*, 1826–1848. [CrossRef]
13. Liu, W.; Hua, M.; Deng, Z.; Huang, Y.; Hu, C.; Song, S.; Gao, L.; Liu, C.; Xiong, L.; Xia, X. A Systematic Survey of Control Techniques and Applications: From Autonomous Vehicles to Connected and Automated Vehicles. *arXiv* **2023**. [CrossRef]
14. Montanaro, U.; Dixit, S.; Fallah, S.; Dianati, M.; Stevens, A.; Oxtoby, D.; Mouzakitis, A. Towards connected autonomous driving: Review of use-cases. *Veh. Syst. Dyn.* **2018**, *57*, 779–814. [CrossRef]
15. Guanetti, J.; Kim, Y.; Borrelli, F. Control of connected and automated vehicles: State of the art and future challenges. *Annu. Rev. Control* **2018**, *45*, 18–40. [CrossRef]
16. Guan, J.; Chen, G.; Huang, J.; Li, Z.; Xiong, L.; Hou, J.; Knoll, A. A Discrete Soft Actor-Critic Decision-Making Strategy with Sample Filter for Freeway Autonomous Driving. *IEEE Trans. Veh. Technol.* **2023**, *72*, 2593–2598. [CrossRef]
17. Wang, Z.; Han, K.; Tiwari, P. Digital Twin-Assisted Cooperative Driving at Non-Signalized Intersections. *IEEE Trans. Intell. Veh.* **2022**, *7*, 198–209. [CrossRef]
18. Williams, N.; Wu, G.; Closas, P. Impact of positioning uncertainty on eco-approach and departure of connected and automated vehicles. In Proceedings of the 2018 IEEE/ION Position, Location and Navigation Symposium (PLANS), Monterey, CA, USA, 23–26 April 2018. [CrossRef]
19. Altan, O.D.; Wu, G.; Barth, M.J.; Boriboonsomsin, K.; Stark, J.A. GlidePath: Eco-Friendly Automated Approach and Departure at Signalized Intersections. *IEEE Trans. Intell. Veh.* **2017**, *2*, 266–277. [CrossRef]
20. Hao, P.; Wu, G.; Boriboonsomsin, K.; Barth, M.J. Eco-Approach and Departure (EAD) Application for Actuated Signals in Real-World Traffic. *IEEE Trans. Intell. Transp. Syst.* **2019**, *20*, 30–40. [CrossRef]
21. Committee, V.C.T. *V2X Communications Message Set Dictionary*; SAE International: Warrenda, PA, USA, 2020. [CrossRef]
22. Wágner, T.; Ormándi, T.; Tettamanti, T.; Varga, I. SPaT/MAP V2X communication between traffic light and vehicles and a realization with digital twin. *Comput. Electr. Eng.* **2023**, *106*, 108560. [CrossRef]
23. Jin, K.; Li, X.; Wang, W.; Hua, X.; Long, W. Energy-optimal speed control for connected electric buses considering passenger load. *J. Clean. Prod.* **2023**, *385*, 135773. [CrossRef]
24. Wang, Z.; Wu, G.; Barth, M.J. Cooperative Eco-Driving at Signalized Intersections in a Partially Connected and Automated Vehicle Environment. *IEEE Trans. Intell. Transp. Syst.* **2020**, *21*, 2029–2038. [CrossRef]
25. Ye, F.; Hao, P.; Qi, X.; Wu, G.; Boriboonsomsin, K.; Barth, M.J. Prediction-Based Eco-Approach and Departure at Signalized Intersections with Speed Forecasting on Preceding Vehicles. *IEEE Trans. Intell. Transp. Syst.* **2019**, *20*, 1378–1389. [CrossRef]
26. Dong, H.; Zhuang, W.; Chen, B.; Yin, G.; Wang, Y. Enhanced Eco-Approach Control of Connected Electric Vehicles at Signalized Intersection with Queue Discharge Prediction. *IEEE Trans. Veh. Technol.* **2021**, *70*, 5457–5469. [CrossRef]
27. Wu, G.; Hao, P.; Wang, Z.; Jiang, Y.; Boriboonsomsin, K.; Barth, M.; McConnell, M.; Qiang, S.; Stark, J. Eco-Approach and Departure along Signalized Corridors Considering Powertrain Characteristics. *SAE Int. J. Sustain. Transp. Energy Environ. Policy* **2021**, *2*, 1–16. [CrossRef]
28. Jia, Y.; Luo, G.; Zhang, Y. Development of optimal speed trajectory control strategy for electric vehicles to suppress battery aging. *Green Energy Intell. Transp.* **2022**, *1*, 100030. [CrossRef]
29. Wang, Q.; Dong, H.; Ju, F.; Zhuang, W.; Lv, C.; Wang, L.; Song, Z. Adaptive Leading Cruise Control in Mixed Traffic Considering Human Behavioral Diversity. *arXiv* **2022**. [CrossRef]
30. Dong, H.; Zhuang, W.; Ding, H.; Zhou, Q.; Wang, Y.; Xu, L.; Yin, G. Event-driven Energy-efficient Driving Control in Urban Traffic for Connected Electric Vehicles. *IEEE Trans. Transp. Electrif.* **2022**, *9*, 99–113. . [CrossRef]
31. Mei, P.; Karimi, H.R.; Yang, S.; Xu, B.; Huang, C. An adaptive fuzzy sliding-mode control for regenerative braking system of electric vehicles. *Int. J. Adapt. Control Signal Process.* **2021**, *36*, 391–410. [CrossRef]
32. Saiteja, P.; Ashok, B.; Wagh, A.S.; Farrag, M.E. Critical review on optimal regenerative braking control system architecture, calibration parameters and development challenges for EVs. *Int. J. Energy Res.* **2022**, *46*, 20146–20179. [CrossRef]
33. Mello, E.F.; Bauer, P.H. Energy-Optimal Speed Trajectories between Stops. *IEEE Trans. Intell. Transp. Syst.* **2020**, *21*, 4328–4337. [CrossRef]
34. Li, N.; Yang, J.; Jiang, J.; Hong, F.; Liu, Y.; Ning, X. Study on Speed Planning of Signalized Intersections with Autonomous Vehicles Considering Regenerative Braking. *Processes* **2022**, *10*, 1414. [CrossRef]
35. Qian, G.; Fu, T.; Sun, L. Research on the fuel consumption conservation potential of ADAS on passenger cars. *E3S Web Conf.* **2021**, *268*, 01035. [CrossRef]
36. Simchon, L.; Rabinovici, R. Real-Time Implementation of Green Light Optimal Speed Advisory for Electric Vehicles. *Vehicles* **2020**, *2*, 35–54. [CrossRef]
37. Yan, Y.; Han, D.; Shen, T.; Wang, Z.; Wang, J.; Yin, G. Velocity Trajectory Planning of Electric Vehicles with Consideration of the Passenger's Individual Preferences. In Proceedings of the 2022 6th CAA International Conference on Vehicular Control and Intelligence (CVCI), Nanjing, China, 28–30 October 2022. [CrossRef]

38. Jiang, X.; Zhang, J.; Wang, B. Energy-Efficient Driving for Adaptive Traffic Signal Control Environment via Explainable Reinforcement Learning. *Appl. Sci.* **2022**, *12*, 5380. [CrossRef]
39. Claussmann, L.; Revilloud, M.; Glaser, S.; Gruyer, D. A study on al-based approaches for high-level decision making in highway autonomous driving. In Proceedings of the 2017 IEEE International Conference on Systems, Man, and Cybernetics (SMC), Banff, AB, Canada, 5–8 October 2017. [CrossRef]
40. Grosan, C.; Abraham, A. Rule-Based Expert Systems. In *Intelligent Systems Reference Library*; Springer: Berlin/Heidelberg, Germany, 2011; pp. 149–185. [CrossRef]
41. Li, N.; Chen, H.; Kolmanovsky, I.; Girard, A. An Explicit Decision Tree Approach for Automated Driving. In *Volume 1: Aerospace Applications. Advances in Control Design Methods, Bio Engineering Applications, Advances in Non-Linear Control, Adaptive and Intelligent Systems Control, Advances in Wind Energy Systems, Advances in Robotics, Assistive and Rehabilitation Robotics, Biomedical and Neural Systems Modeling, Diagnostics, and Control, Bio-Mechatronics and Physical Human Robot, Advanced Driver Assistance Systems and Autonomous Vehicles, Automotive Systems*; American Society of Mechanical Engineers: Tysons, VA, USA, 2017. [CrossRef]
42. Magdalena, L. Fuzzy Rule-Based Systems. In *Springer Handbook of Computational Intelligence*; Springer: Berlin/Heidelberg, Germany, 2015; pp. 203–218. [CrossRef]
43. Guo, J.; Li, W.; Wang, J.; Luo, Y.; Li, K. Safe and Energy-Efficient Car-Following Control Strategy for Intelligent Electric Vehicles Considering Regenerative Braking. *IEEE Trans. Intell. Transp. Syst.* **2022**, *23*, 7070–7081. [CrossRef]
44. Wang, Z.; Wu, G.; Scora, G. MOVESTAR: An Open-Source Vehicle Fuel and Emission Model based on USEPA MOVES. *arXiv* **2020**. [CrossRef]
45. An, F.; Barth, M.; Norbeck, J.; Ross, M. Development of Comprehensive Modal Emissions Model: Operating Under Hot-Stabilized Conditions. *Transp. Res. Rec. J. Transp. Res. Board* **1997**, *1587*, 52–62. [CrossRef]
46. Hao, P.; Boriboonsomsin, K.; Wang, C.; Wu, G.; Barth, M. Connected Eco-approach and Departure System for Diesel Trucks. *SAE Int. J. Commer. Veh.* **2021**, *14*, 217–227. [CrossRef] [PubMed]
47. Oswald, D.; Hao, P.; Williams, N.; Barth, M. Development of an Innovation Corridor Testbed for Shared Electric Connected and Automated Transportation. In *A Research Reportfrom the National Center for Sustainable Transportation*; National Center for Sustainable Transportation: Davis, CA, USA, 2021. [CrossRef]
48. Han, J.; Karbowski, D.; Rousseau, A. State-Constrained Optimal Solutions for Safe Eco-Approach and Departure at Signalized Intersections. In Proceedings of the ASME 2020 Dynamic Systems and Control Conference, Virtual, 5–7 October 2020; American Society of Mechanical Engineers: New York, NY, USA, 2020. [CrossRef]
49. Coulter, R.C. *Implementation of the Pure Pursuit Path Tracking Algorithm*; Technical Report CMU-RI-TR-92-01; Carnegie Mellon University: Pittsburgh, PA, USA, 1992.
50. Genta, G.; Morello, L. *The Automotive Chassis: Volume 2: System Design*; Mechanical Engineering Series; Springer International Publishing: Cham, Switzerland, 2020.
51. Jazar, R.N. *Vehicle Dynamics: Theory and Application*; Springer: New York, NY, USA, 2008. [CrossRef]
52. Miloradović, D.; Glisovic, J.; Stojanovic, N.; Grujic, I. Simulation of vehicle's lateral dynamics using nonlinear model with real inputs. In Proceedings of the IOP Conference Series: Materials Science and Engineering, Kazimierz Dolny, Poland, 21–23 November 2019; IOP Publishing: Bristol, UK, 2019; Volume 659, p. 012060. [CrossRef]
53. Zadeh, L. Fuzzy sets. *Inf. Control* **1965**, *8*, 338–353. [CrossRef]
54. Ponce-Cruz, P.; Ramirez-Figueroa, F.D. *Intelligent Control Systems with LabVIEW™*; Springer: London, UK, 2010. [CrossRef]
55. Cabello, F.; Acuna, A.; Vallejos, P.; Orchard, M.E.; del Solar, J.R. Design and validation of a fuzzy longitudinal controller based on a vehicle dynamic simulator. In Proceedings of the 2011 9th IEEE International Conference on Control and Automation (ICCA), Santiago, Chile, 19–21 December 2011. [CrossRef]
56. Mattas, K.; Botzoris, G.; Papadopoulos, B. Safety aware fuzzy longitudinal controller for automated vehicles. *J. Traffic Transp. Eng. (Engl. Ed.)* **2021**, *8*, 568–581. [CrossRef]
57. Baz, R.; Majdoub, K.E.; Giri, F.; Taouni, A. Self-tuning fuzzy PID speed controller for quarter electric vehicle driven by In-wheel BLDC motor and Pacejka's tire model. *IFAC-PapersOnLine* **2022**, *55*, 598–603. [CrossRef]

applied sciences

MDPI

Article

Torque Vectoring Control Strategies Comparison for Hybrid Vehicles with Two Rear Electric Motors

Henrique de Carvalho Pinheiro [1], Massimiliana Carello [1] and Elisabetta Punta [2,*]

[1] Department of Mechanical and Aerospace Engineering (DIMEAS), Politecnico di Torino, 10129 Torino, Italy; henrique.decarvalho@polito.it (H.d.C.P.); massimiliana.carello@polito.it (M.C.)
[2] National Research Council of Italy, Institute of Electronics, Computer and Telecommunication Engineering, CNR-IEIIT, 10129 Torino, Italy
* Correspondence: elisabetta.punta@cnr.it

Featured Application: The methodology presented in this paper can be applied to the design and evaluation of torque vectoring systems in hybrid electric vehicles. Comparison of different controllers is helpful in making design choices in these applications.

Abstract: In today's automotive industry, electrification is a major trend. In-wheel electric motors are among the most promising technologies yet to be fully developed. Indeed, the presence of multiple in-wheel motors acting as independent actuators allows for the implementation of innovative active systems and control strategies. This paper analyzes different design possibilities for a torque vectoring system applied to an originally compact front-wheel drive hybrid electric vehicle with one internal combustion engine for the front axle and two added electric motors integrated in the wheels of the rear axle. A 14 degrees of freedom vehicle model is present o accurately reproduce the nonlinearities of vehicle dynamic phenomena and exploited to obtain high-fidelity numerical simulation results. Different control methods are compared, a PID, an LQR, and four different sliding mode control strategies. All controllers achieve sufficiently good results in terms of lateral dynamics compared with the basic hybrid version. The various aspects and features of the different strategies are analyzed and discussed. Chattering reduction strategies are developed to improve the performance of sliding mode controllers. For a complete overview, control systems are compared using a performance factor that weighs control accuracy and effort in different driving maneuvers, i.e., ramp and step steering maneuvers performed under quite different conditions ranging up to the limits.

Keywords: hybrid electric vehicles; vehicle dynamics; in-wheel motors; torque vectoring; direct yaw control; PID; LQR; sliding mode control

Citation: de Carvalho Pinheiro, H.; Carello, M.; Punta, E. Torque Vectoring Control Strategies Comparison for Hybrid Vehicles with Two Rear Electric Motors. *Appl. Sci.* **2023**, *13*, 8109. https://doi.org/10.3390/app13148109

Academic Editor: Mihailo Ristic

Received: 16 May 2023
Revised: 3 July 2023
Accepted: 9 July 2023
Published: 12 July 2023

1. Introduction

Electrification is a consolidated reality in the automotive sector. The use of electric motors (EMs) can have an impact on the energy efficiency of vehicles [1], but also create several interesting possibilities in terms of control. Especially when considering architectures with in-wheel motors (IWMs), the presence of EMs allows for the adoption of layouts with multiple power sources, a fact that would be unimaginable in vehicles with only internal combustion engines (ICEs). IWMs are widely discussed in the literature; however, owing to some important shortcomings, they are not yet an industry standard [2].

Direct yaw control (DYC) techniques such as torque vectoring (TV) are perfectly suited to electric vehicles' (EVs) layouts with multiple motors, as they rely on the imbalance between the torques applied to the left and right sides of the vehicle.

The first technical applications based on this concept were implemented with the use of asymmetric brake action, actuated differentials, as well as rear axle steering systems; however, all these strategies have important limitations, which are in turn mitigated by the action of EMs. The TV system can guarantee improvements in terms of dynamic response. In particular, the control objectives in terms of lateral dynamics are to increase the maximum lateral acceleration of the vehicle, to extend the linear response region of the steering system, and to improve the responsiveness of the steering system [2].

Figure 1 shows the graph of the input steering angle (δ) versus lateral acceleration (a_y) and how the curve is modified by the TV system action. The graph changes according to the achievement of the TV control goals.

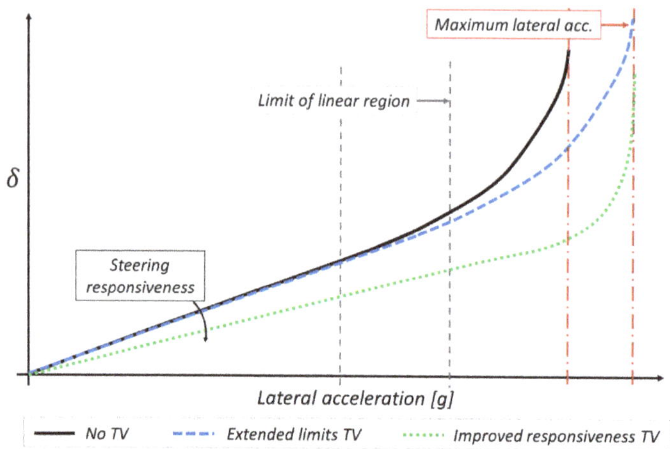

Figure 1. Input steering angle (δ) versus lateral acceleration (a_y). The graph is changed by TV.

Some control strategies particularly suitable to tackle the TV systems issue are present in the relevant control literature, ranging from proportional, integral, derivative (PID) controllers [2,3] to model predictive control (MPC) [4–6] and sliding mode control (SMC) [7–13]. Although, in recent years, there has been a sharp increase in the number of new contributions that consider the TV control problem, there are still gaps in this area, particularly when considering hybrid electric vehicles (HEVs) for which torque allocation controllers usually address the issue of power management [1,14], instead of lateral dynamics. There are few examples of general comparison papers [15].

This paper considers as a case study an A-segment passenger vehicle containing an ICE in the front-wheel drive (FWD) layout, to which two independent EMs are added in each wheel of the rear axle, thus creating an all-wheel drive (AWD) architecture. The two EMs added to the rear axle are used to implement a DYC strategy. The P4 parallel hybrid vehicle (also called "parallel through the road") is intended to both improve the handling and dynamic response of the HEV as well as give it better acceleration performance and enable better fuel efficiency management.

A first study on the subject was presented in [15], where a comparison between two controllers was performed, namely a PID strategy and a first-order sliding mode (FOSM). The two methods proved to be adequate for the set objectives in terms of overall handling, however, the FOSM exhibited pronounced chattering, which resulted in unacceptable vibrations that adversely affected the performance. In this paper, the PID control remains unchanged, while two different strategies are proposed to reduce the chattering in the FOSM, namely, the introduction of a low-pass filter and the replacement with a continuous approximation of the sign function are considered. To complete the discussion and comparison, a full-state linear-quadradic regulator (LQR) and two second-order sliding mode (SOSM) algorithms, the twisting algorithm [16,17] and the suboptimal algorithm [18], are also presented.

The 14 degrees of freedom (DoF) model of the vehicle is described in Section 2, while Section 3 presents the control strategies that have been considered. The main obtained results are discussed in Section 4. Finally, Section 5 provides some concluding remarks on the presented results and future developments.

2. Vehicle Modelling and Maneuvers

2.1. Vehicle Dynamics

The model used to represent the dynamic response of the vehicle is a 14 DoF model widely exploited and presented in the literature [19,20], an application of which can be found in [21].

The derivation of the 14 DoF model and its complete equations is beyond the scope of this paper. The 14 DoF of the model are briefly introduced below, corresponding to the scheme in Figure 2.

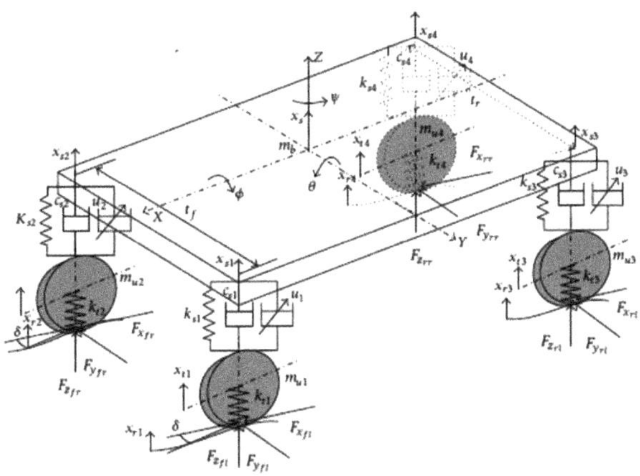

Figure 2. Scheme and variables of the 14 DoF vehicle model.

- 6 DoF describe the free body motion of the vehicle sprung mass;
- 2 DoF for each of the four wheels represent the generalized position with respect to the body of the vehicle ζ (derived from the suspension kinematics) and the longitudinal slip of the tire.

The actual implementation of the model is performed using the CarRealTime software (CRT) from VI-Grade. CRT's interface allows a complete description of the studied vehicle, including the characteristics of its many subsystems (steering, brakes, suspensions, body, powertrain, and so on). Being so, it is possible to analyze a fully non-linear response, especially when it comes to the limit driving conditions, with superior precision when compared with lower DoF mathematical models. The studied vehicle is based on the physical features of the Fiat 500, as shown in Table 1.

The vehicle model relies on custom damper, spring, and suspension curves; as far as tires are concerned, the appropriate Pacejka models are implemented, widely used in professional vehicle dynamics simulations [22]. The 14 DoF model, while suitable for the complete description of the vehicle's motion and dynamics, is rather inconvenient during the implementation and theoretical calculations related to the control systems, owing to its complexity and inherent non-linearity. For that purpose, a simpler model is used in the preliminary phases of system setup, namely a 2 DoF bicycle model [19,20].

Table 1. Parameters of the vehicle [15].

Property	Symbol	Value
Total mass	M	1006 kg
Wheelbase	l	2.3 m
Distance between CoG and front axle	a	0.805 m
Distance between CoG and rear axle	b	1.495 m
Track	t	1.413 m
CoG height	h_{cg}	0.537 m
Yaw inertia	I_z	965.6 kg·m^2
Tire radius (no load)	r	0.291 m
Lateral stiffness (front tires)	C_f	21,094 N/rad
Lateral stiffness (rear tires)	C_r	14,556 N/rad
Motor peak torque	T_{EM}	103 N·m
Motor peak power	P_{EM}	25 kW

This widely known model considers the automobile as a single-track vehicle (collapsing the contribution of left–right wheels); Figure 3. It is valid under certain conditions such as small steering angles and linear range of the tires and it neglects some contributions such aerodynamic forces, suspension compliance, and load transfer.

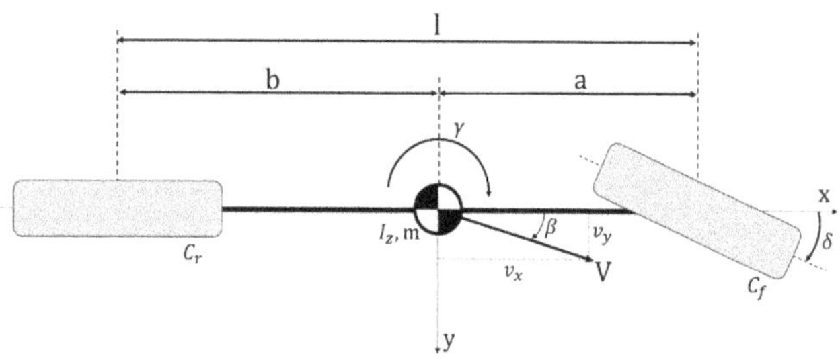

Figure 3. Schematic representation of the 2 DoF bicycle model.

In the bicycle model, the 2 DoF used to describe the behavior of the vehicle are the side slip angle β and yaw rate γ. By mild computations, which are not reported for the sake of brevity, the equations of motion can be written in a state-space representation as follows:

$$\dot{x} = Ax + Bu$$
$$x = \binom{\beta}{\gamma},\ A = \begin{pmatrix} \frac{C_f + C_r}{m v_x} & \frac{aC_f - bC_r}{m {v_x}^2} - 1 \\ \frac{aC_f - bC_r}{I_z} & \frac{a^2 C_f + b^2 C_r}{I_z v_x} \end{pmatrix},\ u = \binom{\delta}{M_z},\ B = \begin{pmatrix} -\frac{C_f}{m v_x} & 0 \\ -\frac{aC_f}{I_z} & \frac{1}{I_z} \end{pmatrix} \quad (1)$$

where the state vector x consists of the variables β and γ, while the input vector u is made up of the steering angle at the wheels δ and the external yaw moment M_z. The elements of the matrices A and B are calculated as functions of the physical parameters of the vehicle (as in Table 1) and of v_x, which is the longitudinal velocity of the vehicle.

Notice that the used 2 DoF bicycle model has M_z as one of the two inputs. This variable is not commonly present in the standard representation of the model. Its addition is particularly well suited for describing the effects of the TV, in which a yaw moment is generated as a consequence of left/right torque imbalance (dimension not present in the "mono-track" approach of the bicycle model).

2.2. Model and Torque Allocation of the Hybrid Powertrain

The components of the hybrid powertrain (PWT), as well as the control logic that defines the distribution between ICE and EMs, are shown in Figure 4, where each block represents a key element of the control system. This layout is presented in detail in [15], where a similar model is used for a preliminary analysis of the parallel hybrid P4 layout in the Fiat 500. The central idea of the system is that the imbalance between torques applied in the left and right wheel will create a yaw moment M_z on the vehicle, which will be the main control action.

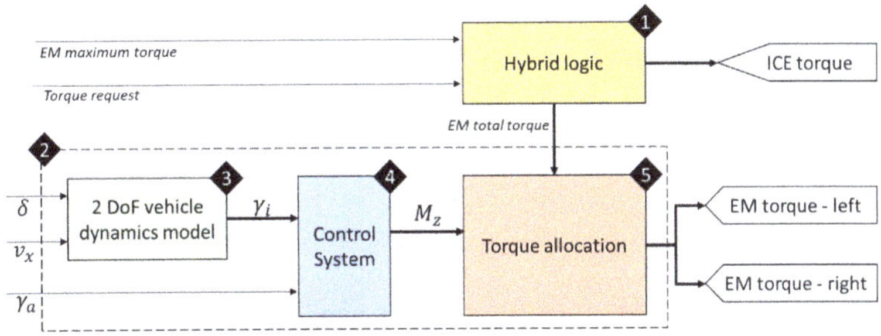

Figure 4. Simulink model organized in different blocks and subsystems [15].

The interface chosen is a co-simulation environment between MATLAB/Simulink and VI-CRT. As displayed in Figure 4, the control system receives the torque request and EM limitations from the plant and, with the block "hybrid logic", (1) decides on the best split between the front ICE and the rear EMs. The ICE torque request is directly sent to the plant again, while the overall EM torque request passes through the block "torque vectoring" (2), which will determine the split between left and right commands. Block (2) is then composed of three internal logic subsystems: the "bicycle model" block (3) calculates the *ideal yaw rate* using the 2 DoF handling model to estimate the yaw rate that a neutral vehicle would present in a given status of steering angle and velocity; then the block "control system" (4) obtains the reference yaw rate and compares it to the real yaw rate coming from the plant observation and, through the various proposed logics, defines the desired control action in terms of *torque bias;* finally, in the final "torque allocation" block (5), the total EM torque and the torque bias requests are combined to calculate the effective left and right torque requests for the EMs [15].

The ideal yaw rate is calculated by targeting the neutral understeer behavior characterized by an understeer coefficient K_u with zero value. The ideal yaw rate can then be determined as follows:

$$\frac{\gamma}{\delta} = \frac{\frac{v_x}{l}}{1 + K_u v_x} \rightarrow \gamma_i = \delta \cdot \frac{v_x}{l} \tag{2}$$

The EM model is energetic, based on the maximum torque and power curves supplied by the manufacturer, as well as the efficiency map. Regarding the battery, a simple saturation strategy is put in place to restrain the current based on the series/parallel configuration of the cells and their maximum discharge rates and nominal voltage. The characteristics of the E-PWT are considered fixed for all values of state of charge and no thermal features are considered.

Some implementations of TV systems are based on the application of braking action. However, these strategies are less efficient because the braking power is dissipated (rather than regenerated). Being able to control each EM individually, this paper does not consider the actuation of the braking system as a control parameter.

After the definition of the individual torque requests, they are sent to CRT each with a timestep of 0.001 s, depicting all significant dynamic quantities. This timestep is also relevant to the dynamic response of the control system; all systems respond with the same overall frequency, even if, in a real application, some differences in refresh times and horizon would be expected.

2.3. Steering Maneuver

Let us now introduce the maneuvering conditions to which the vehicle is subjected to validate and compare the different control strategies under examination.

The first steering maneuver considered is ramp steering, in which the virtual driver linearly increases the steering wheel angle (SWA) at a slow rate and at the same time tries to keep the speed. Ramp steering requires a full range of lateral accelerations and limited throttle input to maintain the target speed, so it is suitable for TV validation and for assessing the quasi-static lateral behavior of vehicles.

The parameters used for ramp steering maneuver are shown in Table 2.

Table 2. Ramp steering maneuver parameters.

Property	Value
Longitudinal velocity	15 m/s
Initial steering angle	0°
SWA increment rate θ	8°/s
Turning direction	Left
Total maneuver time	25 s
Steering start	1 s
Steering end	22 s
Total maneuver time	25 s
Transmission gear	Fixed 3rd
Turning direction	Left

At the start of the simulation and during the first 1 s, the steering is held at zero to allow the solver to stabilize. The final steering angle target is set above 1.0 g and above the expected road and tire grip limits; therefore, all configurations must be saturated and exhibit slip behavior by the end of the steering phase that occurs at 18 s. The simulation continues for another 2 s to fully show the nonlinear response of the vehicle.

Ramp steering shows a quasi-static lateral response; therefore, a more dynamic maneuver is chosen to highlight the transient behavior of the controlled vehicle, i.e., **step steering**. In this standard open loop maneuver, the driver quickly passes from a null steering condition to a fixed value, as described in Table 3. The step steering is repeated in two conditions: the first one far from the limit conditions with a steering angle of 50° (equivalent to a lateral acceleration of around 0.5 g) and the second one with a higher steering (80°) condition in which the actuators are saturated and the tires near the maximum grip.

Table 3. Step steering maneuver parameters.

Property	Value
Longitudinal velocity	15 m/s
Initial steering angle	0°
Final SWA	50° and 80°
Turning direction	Left
Total maneuver time	5 s
Steering start	1 s
Step duration	1 s
Total maneuver time	5 s
Transmission gear	Fixed 3rd
Turning direction	Left

Figure 5 shows the imposed steering angle over time, in both the step steering and ramp steering maneuvers. The chosen maneuvers are open-loop, meaning that the driver inputs are predetermined and do not depend on the response of the system, which is particularly useful for a direct comparison of controllers.

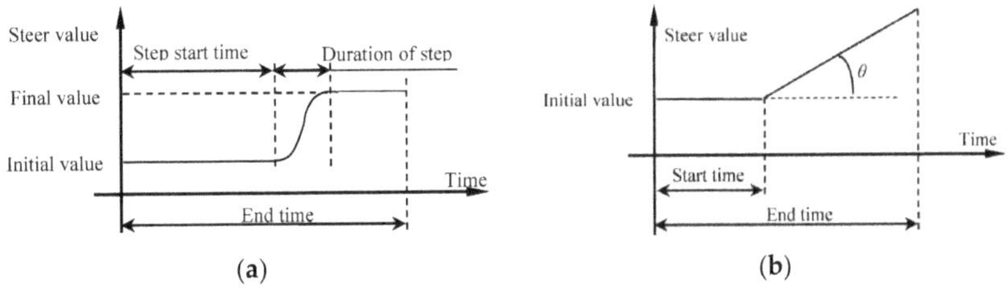

Figure 5. Step steering (**a**) and ramp steering (**b**) maneuver definition.

The steering angle is used in (2) to compute the variation in the ideal yaw rate that serves as a reference for the control systems.

3. Control Systems and Performance Evaluation

The control strategies to be implemented and used to carry out the analysis and comparison are defined and briefly described below.

Four different control methodologies are considered, namely, PID, FOSM, LQR, and SOSM. The individual control strategies (apart from the LQR, which is a full-state control strategy) are implemented in a classic feedback control loop (Figure 6) based on the error between the actual and ideal yaw rate. The control action u in the specific case is the yaw moment induced by the torque bias between the left and right wheels.

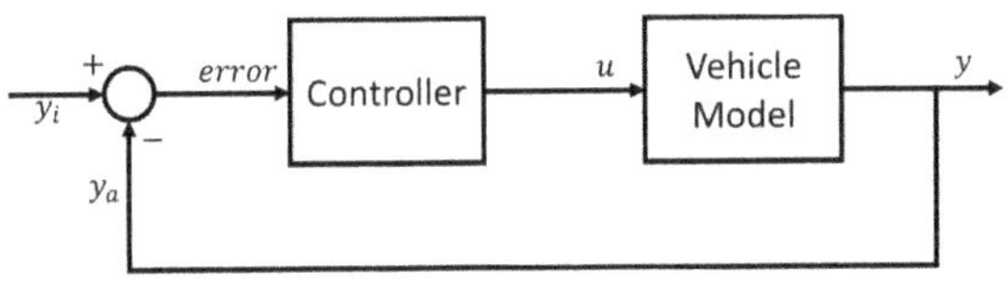

Figure 6. Feedback control block diagram.

The HEV baseline corresponds to the system without torque bias, and it is obtained by setting the control signal to zero. The HEV baseline is without the TV active and is indicated as "OFF" in the following graphs. It is important to underline that, in the TV OFF condition, the rear motors are not idle, but are operated according to a P4 hybrid logic. In this case, the vehicle presents an AWD layout, with a 50% rear torque split for each side as in an open differential.

For all controllers, a minimum steering angle threshold of $\delta_{min} = 5{\cdot}10^{-4}$ rad at the wheels is required to activate the TV system, to eliminate undesired oscillations in straight-line operation.

3.1. PID Controller

The PID strategy is a classic control, widely used in various sectors and applications, and represents the main control reference because its application on direct yaw control systems in the automotive sector is consolidated. The PID designs the control signal according to the following law (3):

$$u = K_P(\gamma_i - \gamma_a) + K_I\frac{1}{s}(\gamma_i - \gamma_a) + K_D\frac{F}{1 + F\frac{1}{s}}(\gamma_i - \gamma_a) \quad (3)$$

where the PID parameters K_P, K_I, K_D, and F are the gains of the proportional, integral, and derivative components, as well as the filtering frequency of the derivative term, respectively; γ_i is the ideal reference yaw rate; and γ_a is the actual yaw rate.

The used parameters values, established by a standard tuning phase evaluating the output performance in the different maneuvers to find an optimized performance, are presented in Table 4.

Table 4. PID controller parameters.

Property	Unit	Value
K_P	-	40
K_I	-	10
K_D	-	0.01
F	Hz	100

As shown by [22], gain scheduling strategies are important during maneuvers with wide speed ranges; however, this is not the case of the proposed study. Furthermore, the work of [23] shows that adaptive PID controllers in DYC systems presents just marginal improvements with respect to fixed gain PID; therefore, a simpler fixed gain control is presented.

3.2. First-Order Sliding Mode (FOSM)

The FOSM controllers [24] are well suited to address the control of nonlinear uncertain systems [25]. In the design of an FOSM controller, a fundamental step is the definition of the sliding surface [24], $S = 0$, on which the trajectories of the controlled system must be steered in a finite time. For the considered TV control problem, the following sliding output (4) is defined:

$$S = \gamma_i - \gamma_a \tag{4}$$

The FOSM control law is then designed according to $u = -K{\cdot}sign(\gamma_i - \gamma_a)$, where the control gain $K > 0$ is a suitably chosen positive constant. The application of the FOSM control guarantees that the sliding surface $S = 0$ is reached in a finite time.

Consider the 2 DoF system (1) [19,20]. Reaching the sliding surface in a finite time, i.e., $S(t) \rightarrow 0$ in a finite time, guarantees the stability according to Lyapunov's theory as well as the achievement of the control objective for system (1).

The straightforward implementation of this SMC is presented in [15], performed by applying symmetric sign function. The work shows that, even though the simple FOSM can achieve the overall lateral performance for the system, it creates unacceptable vibrations to the system. The work also describes the effects of a *delay* and a *hysteresis* band, concluding that they are of small influence.

To attenuate the chattering effect and create suitable FOSM controllers for the proposed problem, two strategies were tested and are presented in this paper.

3.2.1. FOSM with a Low-Pass Filter

The first proposed technique is the inclusion of a low-pass filter after the sign function. As observed in [15], the output signal of the FOSM assumes a high-frequency oscillation behavior to maintain the steady-state control of the vehicle, thus the presence of a low-pass filter with a given dynamic response is expected to eliminate the discontinuity and improve stability and, on the other hand, to create a slower response owing to its inherent delay.

The implementation of the FOSM with the low-pass filter strategy is achieved through a relay function, imposing a zero-hysteresis system saturated at 100% of the maximum available torque bias. The sign function is placed in series with a low pass filter with time constant equal to 1.2. The control law passing through the transfer function with an overall gain of 0.8 is given by (5):

$$u = -\frac{0.8}{1.2s + 1}{\cdot}sign(\gamma_i - \gamma_a) \tag{5}$$

3.2.2. FOSM with Continuous Approximation of the Sign Function

The second strategy studied to reduce the chattering and vibration problems created by the FOSM is the substitution of the sign function with an equivalent expression that is properly continuous. As proposed by [26] and implemented by [27] in relevant application, the following function (6) is used:

$$cont(S) = \frac{S}{|S| + \varnothing} \tag{6}$$

where S is the sliding output as in (4) and ∅ is a positive variable. Note that, as ∅ tends to zero, the function tends to the original sign function, while for high values of ∅, the continuous function tends to zero, thus regulating the value of the constant can help improve the sharpness/smoothness of the control output.

It is also possible to consider an adaptive method for the definition of ∅ as in (7):

$$\varnothing = \varnothing_0 + |\eta|, \ (\eta \approx \int S\, dt) \tag{7}$$

This strategy was tested in the studied system and no relevant advantage was observed, so the results refer to the simpler case with $\varnothing = 2.5$ and no adaptive term.

3.3. Linear-Quadratic Regulator (LQR)

The linear quadratic regulator strategy is part of the optimal control theory, in which the control law is defined to minimize a cost function. This function is typically defined through a quadratic cost function as in (8):

$$J = \int_{t_0}^{t_1} \left(x^T Q x + u^T R u + 2x^T N u \right) dt \tag{8}$$

where x is the state vector; u is the control vector; and Q, R, and N are the weight matrixes related to the system states, to the control inputs, and to the cross-contribution of x and u, respectively. The control law is defined by (9):

$$u = -Kx \tag{9}$$

where K is the matrix that minimizes (8), defined by solving the continuous time Riccati differential equation.

The control law is like the pole-placement strategy, but instead of choosing the ideal poles of the system (not always a straightforward task), the challenge is to define relevant matrixes Q, R, and N to balance the state error and control effort. For the studied system, the used values of Q and R are as in (10), while no cross-contribution was included (N is null):

$$R = \begin{bmatrix} 1 \times 10^5 & 0 \\ 0 & 1 \end{bmatrix}, \ Q = 1 \times 10^6 \begin{bmatrix} 1 & 0 \\ 0 & 1000 \end{bmatrix} \tag{10}$$

As the system state-space matrixes A and B are dependent on the vehicle's speed, as defined in (1), so will be the results of K after the cost function minimization. Therefore, for the practical implementation of the LQR, the K matrix was calculated beforehand with a dedicated MATLAB script for a wide range of discrete velocities (from 1 to 100 m/s) and, during the simulations, the coefficients of the control were linearly interpolated from the closest neighbors of the actual vehicle's velocity for each time step. This approach allowed for good precision of the control system without the requirement of excessive computational power.

Differently from the previous control systems PID and SMC, the LQR is a full-state control, meaning that the control gains and strategy use as input not a single error signal, but rather assume the availability of all of the relevant internal states of the plant. In the case of the DYC problem using the 2 DoF bicycle model, this means knowing not only the yaw rate γ_a as before, but in addition being able to observe the side-slip angle β. The side-slip angle is a variable notoriously hard to directly measure, so its availability for an actual control system would depend on the estimation of the variable based on other output signals. This estimation is not the focus of this paper, so it will be assumed that β is available with sufficiently high precision, as it is demonstrated to be achievable by some works present in the literature [28,29].

3.4. Second-Order Sliding Mode (SOSM)

The last control system category considered in this paper is the second-order sliding mode controllers. This kind of strategy follows similar principles to the FOSM, maintaining its typical features of robustness, but avoiding some of the main problems related to the chattering, which renders it suitable for evaluation in the case of DYC.

In the case of the SOSM, the definition of the sliding surface S can follow the same formulation as (4), being the difference between the real and target values of the yaw rate. However, in the second-order system, the proposed algorithms are such that not only the sliding output S tends to 0 in a finite time, but also its first time derivate $\dot{S}$ [27]. Two algorithms are proposed to accomplish this condition.

3.4.1. SOSM Twisting Algorithm

The first algorithm proposed is the so-called twisting algorithm. This classical approach [16,17], implemented in relevant application [23], can effectively steer both S and $\dot{S}$ to zero in a finite time.

The control law of the twisting algorithm [23], is defined by (11):

$$\dot{u} = \begin{cases} -\alpha_m sign(S(t)), \ when \ S.\dot{S} \leq 0 \\ -\alpha_M sign(S(t)), \ when \ S.\dot{S} > 0 \end{cases} \tag{11}$$

where α_m and α_M are control coefficients with $\alpha_M > \alpha_m > 0$.

To properly control the system, it is important to assure that α_m and the ratio α_M/α_m are sufficiently high.

In the present DYC, the values obtained are $\alpha_m = 5.6$ and $\alpha_M = 64.1$ after tuning the system. The fact that the control law (11) defines $\dot{u}$ instead of u itself means that the typical discontinuities of the SMC systems are displaced to the derivative of the control input, not its direct value, theoretically reducing the issues related to the vibration induced by the FOSM.

3.4.2. SOSM Suboptimal Algorithm

The second algorithm that satisfies the conditions of the SOSM is the suboptimal algorithm [18]. In this version of the sliding mode, the control law is defined as (12):

$$\dot{u} = -k_r \cdot sign\left[S(t) - \frac{1}{2}S(t_{Mk})\right] \tag{12}$$

where t_{Mk} is defined as the last moment in which $\dot{S} = 0$. For the practical application of the simulation environment, in which the timesteps are small yet discrete, t_{Mk} is updated when there is a change of sign in the variation in S between two subsequent timesteps: S(t) − S(t − 1). The simulation begins with $t_{Mk} = t_{M0} = 0$, and thus $S_1(t_{M0})$ equals the initial value of $(\gamma_i - \gamma_a)$.

The choice of the coefficient k_r must be sufficiently high to guarantee stability [23]. In the studied application, the chosen value to satisfy such a condition is $k_r = 28.8$. To further reduce some of the residual chattering, known as the "ringing effect", especially during the step steering maneuvers, the sign function was substituted with the continuous equivalent with a value of $\varnothing = 4.5$, as in (5).

3.5. Performance Evaluation

To properly compare the results of the control systems in an objective way, a performance indicator was applied. This **performance factor (PF)** is composed of three indexes:

- The **control penalty (CP)** is related to the effort of the controller in applying the control law, meaning that control systems that require more torque bias to stabilize the vehicle must be negatively evaluated.
- The **error penalty (EP)** directly evaluates the distance between the actual and target yaw rate during the maneuver, in such a way that less precise controllers are the most penalized.
- The **timed error penalty (TEP)** also takes into consideration the evolution of the error in time, meaning that controllers can afford a higher overshoot or initial displacement relative to the ideal behavior, if they are able to quickly eliminate the error. Steady-state error is heavily penalized in this indicator.

The three indicators are then normalized with respect to the reference value obtained by the original PID in the first step steering maneuver and combined through a weighted sum to compose a single performance indicator, as in (13):

$$\begin{cases} CP = \int |u(t)|dt \\ EP = \int |(\gamma_i(t) - \gamma_a(t))|dt \\ TEP = \int t|(\gamma_i(t) - \gamma_a(t))|dt \\ PF = 0.4CP + 0.4EP + 0.2TEP \end{cases} \tag{13}$$

4. Simulation Results

Simulations have been carried out for the case where no TV is applied and for all cases where the different control methods previously introduced and taken into consideration are applied, namely, PID, LQR, and the four versions of SMC strategies.

In the following section, the graphs of the obtained results are presented, which include control signals, vehicle state, and driver inputs labelled according to the following:

- The baseline HEV without TV (**OFF**);
- The PID controller (**PID**);
- The FOSM controller with the low-pass filter (**FOSM lowpass**);
- The FOSM controller with continuous function (**FOSM continuous**);
- The LQR (**LQR**);
- The SOSM controller with the twisting algorithm (**SOSM twisting**);
- The SOSM controller with the suboptimal algorithm (**SOSM suboptimal**).

The analysis of the graphic interface is useful for the interpretation of the results and for quick debugging and has been performed as the first step of the post processing. A pronounced understeering behavior of the baseline HEV could be found in all maneuvers; instead, all controlled versions are much closer together and can follow the desired yaw rate for much longer, confirming the overall good performance achieved by the control systems proposed. Let us now consider the main results of each maneuver.

4.1. 50° Step Steering

Figure 7 shows the graph of the right-side EM torque where it is possible to grasp the effect of the different control systems.

The baseline plot shows the necessary torque requested to maintain the target speed during the maneuver, in this case applied in a balanced manner between left and right, while the other plots display the baseline contribution plus the bias.

FOSM lowpass shows the sharpest response after the threshold, quickly saturating the actuator at the beginning of the step and establishing a persistent oscillatory behavior, with a continuous signal but with a non-continuous derivative. This is an improvement with respect to the simple FOSM [15] with discontinuous control, but nonetheless it is expected to see a rough effect in terms of vibrations on the vehicle with such aggressive variations.

The actuation shape displayed by the PID, LQR, and FOSM continuous controllers are quite similar, with a first peak value followed by a stabilized steady-state contribution. The main difference between them is the intensity of the transient value, highest in the LQR and lowest in the PID. This difference is directly related to the optimized gains for each controller, which should balance promptness and stability.

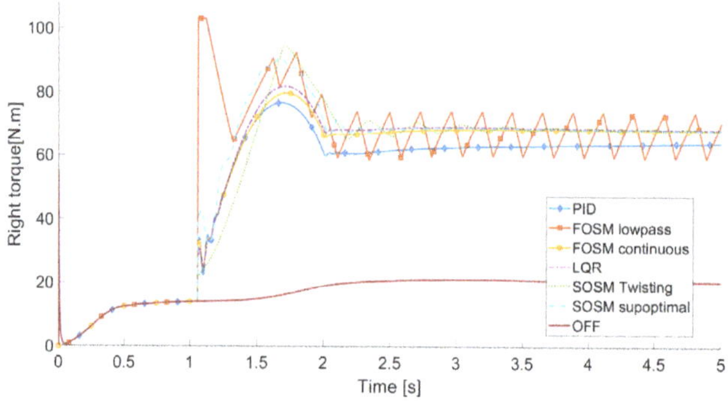

Figure 7. The right-side EM torque during the 50° step steering.

The response of SOSM twisting is the slowest, achieving its peak only after 0.7 s, while SOSM suboptimal presents a prompt actuation after the beginning of the step. Both second-order SMC systems display sufficiently smooth and stable control outputs, as expected.

Regarding the main goal of the control systems, Figure 8 shows the yaw rate of the vehicle during the step steering and the error relative to the calculated ideal yaw rate.

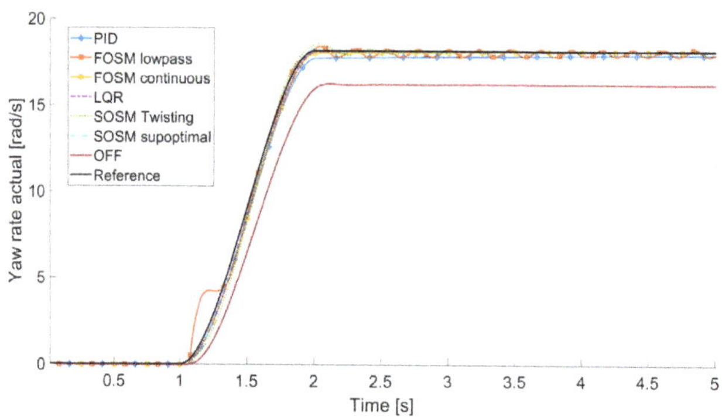

(**a**)

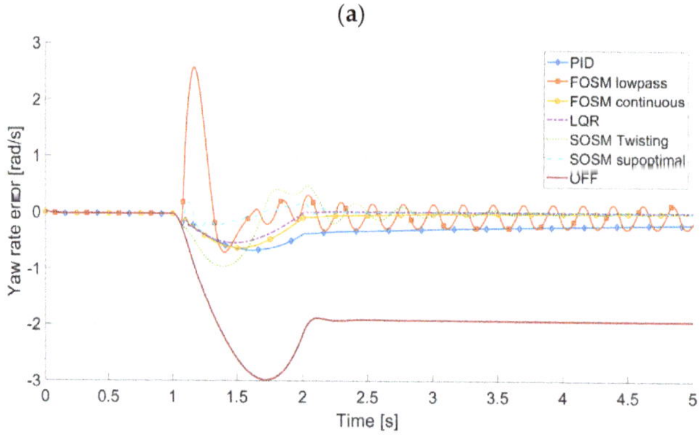

(**b**)

Figure 8. Yaw rate (**a**) and yaw rate error (**b**) in the 50° step steering.

The relations between the control action and the yaw rate outcome are clear. FOSM lowpass presents a high overcontrol peak and a persistent vibration throughout the maneuver, while PID, FOSM continuous, and LQR display a similar shape of the response, even if it can be noted that the PID controller fails to guarantee a zero error in the 4 s after the step. SOSM twisting is relatively slow to respond and to stabilize while SOSM suboptimal is very precise, with the lowest maximum yaw error.

Figure 9 shows the lateral acceleration of the vehicle and its relation to the imposed steering angle, as in the goals presented in Figure 1.

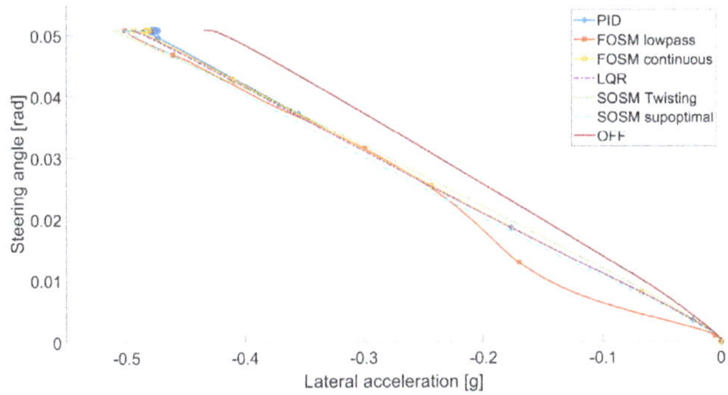

Figure 9. Steering angle vs. lateral acceleration during the 50° step steering.

As expected, the baseline vehicle displays the lowest lateral acceleration in each steering condition. Apart from FOSM lowpass, all the other controllers create a similar outcome, with smooth and linear curves. The controller with the biggest impact on the understeering behavior is SOSM suboptimal, generating the most responsive dynamics; however, the difference between it and the second-best results (LQR) is of just shy of 3% lateral acceleration (at 0.04 rad/s steering) versus a 22% increase relative to the baseline.

4.2. 80° Step Steering

As opposed to the previous step steering where the controllers were able to achieve the yaw rate goal without (almost) never saturating the actuators, in a more challenging condition, this fact shall not hold any longer.

The equivalent results of torque and yaw rate can be observed in Figures 10 and 11, respectively.

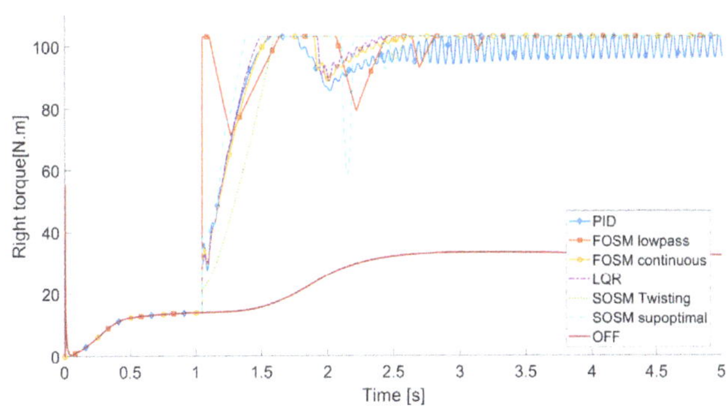

Figure 10. Torque in the right-side EM during the 80° step steering.

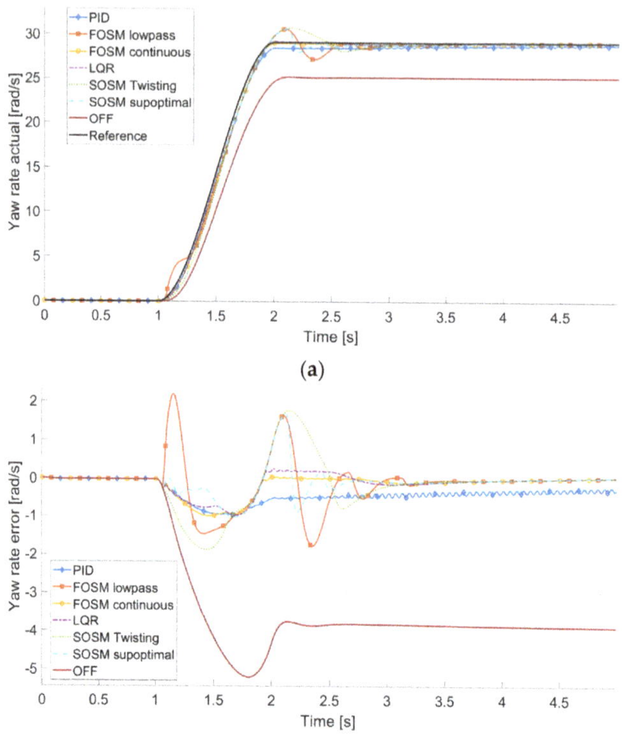

(a)

(b)

Figure 11. Yaw rate (**a**) and yaw rate error (**b**) in the 80° step steering.

As predicted, the controllers take the right EM to its maximum torque value in all configurations around 1.5 s. Unlike the 50° case, in this condition, the most disturbed controller is the PID, displaying important oscillations after the completion of the steering action, most likely due to the integrative term combined with the saturated condition. FOSM lowpass instead reduced its vibrational behavior in the steady-state phase in the absence of the changing sign of the error signal. Once again, SOSM suboptimal is very responsive and quickly raises its torque level after the step input; however, this same promptness also affects its behavior after the saturation period, with a pronounced reduction around 2.2 s.

The best performing control systems in this case are LQR and the FOSM continuous, displaying a low maximum error and no significant overshoot or intense actuation vibrations.

All controllers except the PID successfully take the error to zero in the 5 s period. The PID controller continues to reduce the error very slowly, but the actuation vibrations jeopardize its performance.

4.3. Ramp Steering

The last maneuver studied is ramp steering, highlighting the steady-state response of the vehicle under a wide range of lateral accelerations.

As shown in Figure 12, the control systems present a first overshooting and slight vibrations (due to the initial threshold) followed by a linear ascending phase and a saturation around 12.5 s. Once again, the FOSM lowpass controller behaves poorly regarding the actuation smoothness, presenting quite accentuated vibrations before saturation.

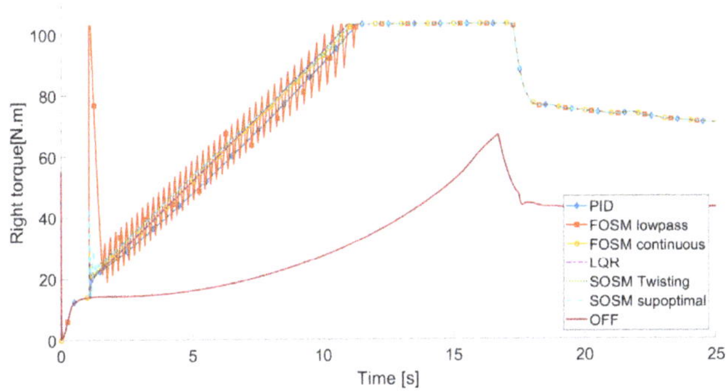

Figure 12. Torque in the right-side EM during ramp steering.

Figure 13 displays the yaw rate and its error, while Figure 14 shows a zoomed-in section of the error plot between 0 and 14 s.

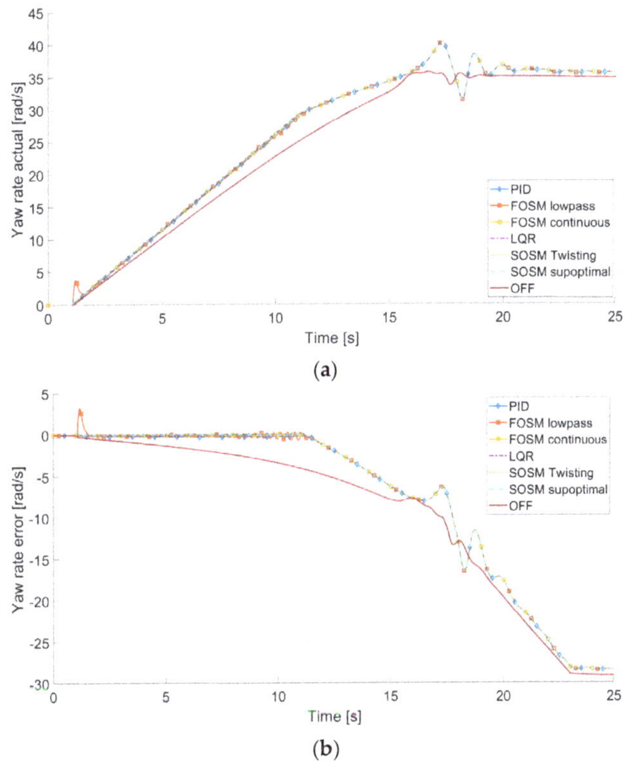

Figure 13. Yaw rate (**a**) and yaw rate error (**b**) in ramp steering.

The controlled vehicles clearly change their behavior and follow quite well the yaw goal for the first part of the maneuver. After the saturation of the actuators, the controlled vehicles see their behavior move closer to the baseline version, as it is no longer possible to supply the required yaw moment to create a perfectly neutral vehicle.

At around 16 s, the baseline vehicle achieves its maximum lateral performance, as shown by the yaw rate peaking, corresponding to the reduction in the requested torque presented before. The same peak effect is displayed at around 17 s for all controlled systems, at a yaw rate of 40.3 rad/s vs. the 35.5 rad/s baseline.

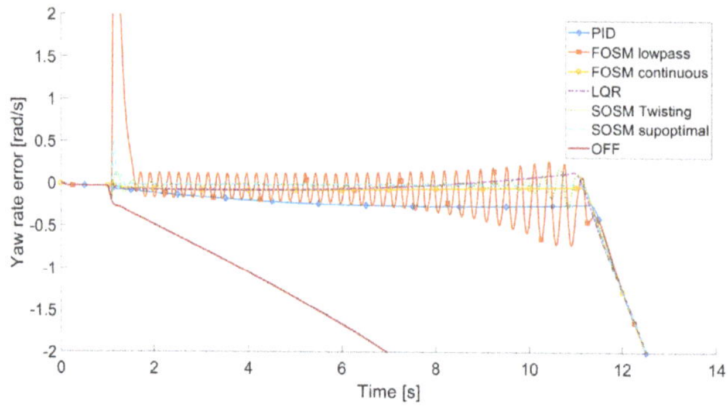

Figure 14. Yaw rate error detail: zoomed-in region before saturation.

After this peak, a highly non-linear response is observed, as expected, as the tires are no longer in their preferred working zone. It is hard to draw meaningful conclusions from the results after the tire saturation because the mechanical models tend to be much less precise.

A closer look at the first section of the maneuver allows for a more interesting discussion of the controller.

The plots confirm some previously stated conclusions, such as the high sensitivity and vibrational response of the FOSM lowpass controller; the quick settling time and prompt response of SOSM suboptimal; the PID controller with good transient behavior but finding it difficult to eliminate small steady-state error; the slower but smooth response of the SOSM twisting algorithm; and the similar smooth response of LQR and FOSM continuous.

Lastly, Figure 15 presents the lateral acceleration and steering angle.

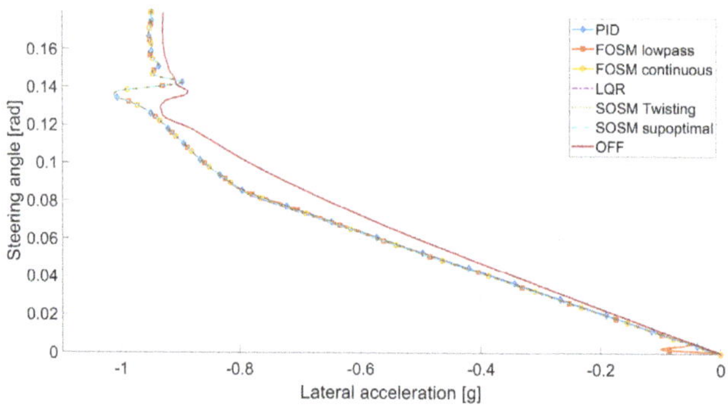

Figure 15. Steering angle vs. lateral acceleration during ramp steering.

In Figures 9 and 15, the advantage of all control systems over the baseline vehicle is clear. Higher maximum lateral acceleration and more responsive behavior can be observed in the plot. It is interesting to observe that the linearity goal is not completely achieved; in fact after the saturation of the EMs at 0.8 g, the slope of the plot changes significantly.

As can be noted in Figure 9, SOSM suboptimal exhibited more responsive behavior. In this case, the SOSM twisting and LQR controllers are much closer and have a higher lateral acceleration in several zones of the plot. This suggests that the SOSM suboptimal controller performs better in transient situations but gives no significant advantage in less dynamic conditions.

In [15], the constant radius corner condition was studied. In this paper, both of the presented maneuvers are open loop.

Finally, Table 5 shows, for comparison purposes, the numerical results obtained by the control systems considered in the three conditions studied and with reference to the proposed PF.

Table 5. Numerical comparison of the control systems in the three conditions studied using the proposed PF.

Controller	Step 50°	Step 80°	Ramp Steering
OFF	3.740	7.258	58.543
PID	1.000	1.469	35.981
FOSM lowpass	0.926	1.577	35.445
FOSM continuous	0.641	1.009	35.011
LQR	0.608	1.104	35.042
SOSM twisting	0.728	2.084	34.776
SOSM suboptimal	0.525	1.491	34.802

The time considered for the step steering maneuvers is the total simulation time, while for ramp steering, the PF was calculated until 17 s of simulation, because, after that, the systems behave similarly, and the simulation precision is lower.

It is interesting to notice that the PID controller and FOSM lowpass generally display a lower level of performance than the other controllers, while SOSM twisting presents a good response in two simulations, but very poor performance in the 80° step. FOSM continuous, LQR, and SOSM suboptimal are the most consistent throughout the tested conditions.

5. Conclusions

In this paper, a comprehensive study of TV systems for an HEV based on different control methodologies was addressed. Analysis and comparison of the performances obtained were conducted using PID, LQR, and different SMC strategies for the TV applied to an FWD HEV with two electric motors integrated in the rear wheels. All the considered control architectures proved to be suitable for achieving the DYC objectives corresponding to increasing lateral acceleration and improving vehicle responsiveness. The extension of the steering linear region is conditioned by the saturation of the Ems, which generates the torque bias.

The chattering issue was encountered when plain FOSM was applied and it was successfully reduced by the proposed techniques, especially when the continuous FOSM algorithm was used. The best controllers with respect to the proposed PF were continuous FOSM, LQR, and suboptimal SOSM. Interestingly, the SMC with properly tuned control gains achieved performances very similar to those obtained by LQR, which theoretically should provide important advantages, being a full-state controller. The application of the SMC strategies with a continuous approximation of the sign proved to be particularly effective in terms of reducing the chattering as well as complexity of the controller, proving to be a valid option for the DYC.

Future work will be devoted to the development of TV systems for HEVs based on innovative and compliant SMC strategies, such as Simplex-based algorithms [25]. Attention will be paid to the inclusion of strategies based on MPC or machine learning, as well as to the experimental and DiL validation [30] of the obtained results.

Author Contributions: Conceptualization, H.d.C.P., M.C. and E.P.; methodology, H.d.C.P., M.C. and E.P.; software, H.d.C.P.; validation, H.d.C.P., M.C. and E.P.; formal analysis, H.d.C.P., M.C. and E.P.; investigation, H.d.C.P., M.C. and E.P.; resources, M.C. and E.P.; data curation, H.d.C.P.; writing—original draft preparation, H.d.C.P.; writing—review and editing, H.d.C.P., M.C. and E.P.; visualization, H.d.C.P.; supervision, M.C. and E.P. All authors have read and agreed to the published version of the manuscript.

Funding: This research received no external funding.

Institutional Review Board Statement: Not applicable.

Informed Consent Statement: Not applicable.

Data Availability Statement: Not applicable.

Acknowledgments: The authors want to acknowledge the support of the VI-Grade Italy team during the execution of the work and extend thanks for the availability of the simulation software used in its development.

Conflicts of Interest: The authors declare no conflict of interest.

References

1. Carello, M.; Bonansea, P.; D'Auria, M. Driveline Optimization for a Hybrid Electric City Vehicle to Minimize Fuel Consumption. In Proceedings of the SAE 2014 World Congress & Exhibition, Detroit, MI, USA, 8–10 April 2014. SAE Technical Paper 2014-01-1090. [CrossRef]
2. Pinheiro, H.D.C.; Messana, A.; Sisca, L.; Ferraris, A.; Airale, A.G.; Carello, M. Torque Vectoring in Electric Vehicles with In-wheel Motors. In *Advances in Mechanism and Machine Science*; Springer International Publishing: Cham, Switzerland, 2019; pp. 3127–3136. [CrossRef]
3. Assadian, F.; Hancock, M. A comparison of yaw stability control strategies for the active differential. In Proceedings of the IEEE International Symposium on Industrial Electronics, Dubrovnik, Croatia, 20–23 June 2005; Volume 1, pp. 373–378. [CrossRef]
4. Siampis, E.; Velenis, E.; Longo, S. Predictive rear wheel torque vectoring control with terminal understeer mitigation using nonlinear estimation. In Proceedings of the 2015 54th IEEE Conference on Decision and Control (CDC), Osaka, Japan, 15–18 December 2015; pp. 4302–4307. [CrossRef]
5. Siampis, E.; Velenis, E.; Longo, S. Model Predictive torque vectoring control for electric vehicles near the limits of handling. In Proceedings of the 2015 European Control Conference (ECC), Linz, Austria, 15–17 July 2015; pp. 2553–2558. [CrossRef]
6. Hu, X.; Chen, H.; Li, Z.; Wang, P. An Energy-Saving Torque Vectoring Control Strategy for Electric Vehicles Considering Handling Stability Under Extreme Conditions. *IEEE Trans. Veh. Technol.* **2020**, *69*, 10787–10796. [CrossRef]
7. Goggia, T.; Sorniotti, A.; De Novellis, L.; Ferrara, A. Torque-vectoring control in fully electric vehicles via integral sliding modes. In Proceedings of the 2014 American Control Conference, Portland, OR, USA, 4–6 June 2014; pp. 3918–3923. [CrossRef]
8. Canale, M.; Fagiano, L.; Ferrara, A.; Vecchio, C. Vehicle Yaw Control via Second-Order Sliding-Mode Technique. *IEEE Trans. Ind. Electron.* **2008**, *55*, 3908–3916. [CrossRef]
9. Wang, Z.; Wang, Y.; Zhang, L.; Liu, M. Vehicle Stability Enhancement through Hierarchical Control for a Four-Wheel-Independently-Actuated Electric Vehicle. *Energies* **2017**, *10*, 947. [CrossRef]
10. Xu, X.; Xiong, L.; Feng, Y. Torque Vectoring Control for Handling Improvement of 4WD EV. *Adv. Mater. Res.* **2013**, *765–767*, 1893–1898. [CrossRef]
11. Chae, M.; Hyun, Y.; Yi, K.; Nam, K. Dynamic Handling Characteristics Control of an in-Wheel-Motor Driven Electric Vehicle Based on Multiple Sliding Mode Control Approach. *IEEE Access* **2019**, *7*, 132448–132458. [CrossRef]
12. Tota, A.; Lenzo, B.; Lu, Q.; Sorniotti, A.; Gruber, P.; Fallah, S.; Velardocchia, M.; Galvagno, E.; De Smet, J. On the Experimental Analysis of Integral Sliding Modes for Yaw Rate and Sideslip Control of an Electric Vehicle with Multiple Motors. *Int. J. Automot. Technol.* **2018**, *19*, 811–823. [CrossRef]
13. Zehetner, J.; Horn, M. Vehicle dynamics control with torque vectoring and active rear steering using sliding mode control. *IFAC Proc. Vol.* **2007**, *40*, 1–8. [CrossRef]
14. Carello, M.; Pinheiro, H.D.C.; Longega, L.; Di Napoli, L. Design and Modelling of the Powertrain of a Hybrid Fuel Cell Electric Vehicle. *SAE Tech. Pap. Ser.* **2021**, 3, 2878–2892. [CrossRef]
15. de Carvalho Pinheiro, H.; Punta, E.; Carello, M.; Ferraris, A.; Airale, A.G. Torque Vectoring in Hybrid Vehicles with In-Wheel Electric Motors: Comparing SMC and PID control. In Proceedings of the 2021 IEEE International Conference on Environment and Electrical Engineering and 2021 IEEE Industrial and Commercial Power Systems Europe (EEEIC/I CPS Europe), Bari, Italy, 7–10 September 2021; pp. 1–6. [CrossRef]
16. Levant, A. Sliding order and sliding accuracy in sliding mode control. *Int. J. Control* **1993**, *58*, 1247–1263. [CrossRef]
17. Levantovsky, L.V. Second order sliding algorithms: Their realization. In *Dynamics of Heterogeneous Systems*; Institute for System Studies Moscow: Moscow, Russia, 1985; pp. 32–43.
18. Bartolini, G.; Ferrara, A.; Usai, E. Applications of a Sub-Optimal Discontinuous Control Algorithm for Uncertain Second Order Systems. *Int. J. Robust Nonlinear Control* **1997**, *7*, 299–319. [CrossRef]
19. Milliken, W.F.; Milliken, D.L.; Metz, L.D. *Race Car Vehicle Dynamics*; Society of Automotive Engineers: Warrendale, PA, USA, 1995.
20. Genta, G.; Morello, L. *The Automotive Chassis: Volume 1–2*; Springer Science & Business Media: Berlin/Heidelberg, Germany, 2008.
21. Setiawan, J.D.; Safarudin, M.; Singh, A. Modeling, simulation and validation of 14 DOF full vehicle model. In Proceedings of the International Conference on Instrumentation, Communication, Information Technology, and Biomedical Engineering, Bandung, Indonesia, 23–25 November 2009; pp. 1–6. [CrossRef]

22. Pacejka, H.B. *Tyre and Vehicle Dynamics*, 2nd ed.; Butterworth-Heinemann: Oxford, UK, 2006.
23. De Novellis, L.; Sorniotti, A.; Gruber, P.; Pennycott, A. Comparison of Feedback Control Techniques for Torque-Vectoring Control of Fully Electric Vehicles. *IEEE Trans. Veh. Technol.* **2014**, *63*, 3612–3623. [CrossRef]
24. Utkin, V.I. *Sliding Modes in Control and Optimization*; Springer: Berlin/Heidelberg, Germany, 1992.
25. Bartolini, G.; Punta, E. Multi-Input Sliding Mode Control of Nonlinear Uncertain Non-Affine Systems with Mono-Directional Actuation. *IEEE Trans. Autom. Control* **2014**, *60*, 393–403. [CrossRef]
26. Harashima, F.; Hashimoto, H.; Maruyama, K. Practical robust control of robot arm using variable structure system. In Proceedings of the 1986 IEEE International Conference on Robotics and Automation Proceedings, San Francisco, CA, USA, 7–10 April 1986; pp. 532–539. [CrossRef]
27. Menhour, L.; Lechner, D.; Charara, A. Sliding mode control to design a driver model for vehicle steering: Experimental validation and stability evaluation of sideslip motion. *IFAC Proc. Vol.* **2010**, *43*, 7–12. [CrossRef]
28. De Martino, M.; Farroni, F.; Pasquino, N.; Sakhnevych, A.; Timpone, F. Real-time estimation of the vehicle sideslip angle through regression based on principal component analysis and neural networks. In Proceedings of the 2017 IEEE International Systems Engineering Symposium (ISSE), Vienna, Austria, 11–13 October 2017; pp. 1–6. [CrossRef]
29. Chindamo, D.; Gadola, M.; Bonera, E.; Magri, P. Experimental Comparison of The Two Most Used Vehicle Sideslip Angle Estimation Methods for Model-Based Design Approach. *J. Phys. Conf. Ser.* **2021**, *1888*, 012006. [CrossRef]
30. de Carvalho Pinheiro, H.; Russo, F.; Sisca, L.; Messana, A.; De Cupis, D.; Ferraris, A.; Airale, A.G.; Carello, M. Active Aerodynamics through Active Body Control: Modelling and Static Simulator Validation. In Proceedings of the ASME 2020 International Design Engineering Technical Conferences and Computers and Information in Engineering Conference, Virtual, 17–19 August 2020. [CrossRef]

applied sciences

MDPI

Article

Design and Control of a Shape Memory Alloy-Based Idle Air Control Actuator for a Mid-Size Passenger Vehicle Application

Pacifique Turabimana [1], Jung Woo Sohn [2,*] and Seung-Bok Choi [3,4,*]

1 Department of Aeronautics, Mechanical and Electronic Convergence Engineering, Graduate School, Kumoh National Institute of Technology, Daehak-Ro 61, Gumi 39177, Republic of Korea; 20215056@kumoh.ac.kr

2 School of Mechanical System Engineering, Kumoh National Institute of Technology, Daehak-Ro 61, Gumi 39177, Republic of Korea

3 Department of Mechanical Engineering, The State University of New York, Korea (SUNY Korea), 119 Songdo Moonhwa-Ro, Yeonsu-Gu, Incheon 21985, Republic of Korea

4 Department of Mechanical Engineering, Industrial University of Ho Chi Minh City (IUH), 12 Nguyen Van Bao Street, Go Vap District, Ho Chi Minh City 70000, Vietnam

* Correspondence: jwsohn@kumoh.ac.kr (J.W.S.); seungbok.choi@sunykorea.ac.kr (S.-B.C.)

Abstract: The idle air control actuator is an important device in automotive engine management systems to reduce fuel consumption by controlling the engine's idling operation. This research proposes an innovative idle air control (IAC) actuator for vehicle applications utilizing shape memory alloy (SMA) technology. The proposed actuator leverages the unique properties of SMAs, such as the ability to undergo large deformations upon thermal activation, to achieve precise and rapid controls in the air intake of automotive engines during idle conditions. The actuator structure mechanism consists of an SMA spring and an antagonistic spring made from steel. The design process utilizes both numerical and analytical approaches. The SMA spring is electrically supplied to activate the opening process of the actuator, and its closing state does not need electricity. However, the PID controller is used to control the applied current, which reduces the time taken by the actuator to achieve the actuation strokes. It shows good operability within multiple numbers of operation cycles. Additionally, the performance of the designed actuator is evaluated through mathematical algorithms by integrating it into the engine's air intake system during idle operating conditions. The results demonstrate the effectiveness of the SMA-based actuator in achieving rapid control of the air intake through bypass, thereby improving engine idle conditions.

Keywords: shape memory alloy; automotive engine; actuator; phase transformation; idle speed

Citation: Turabimana, P.; Sohn, J.W.; Choi, S.-B. Design and Control of a Shape Memory Alloy-Based Idle Air Control Actuator for a Mid-Size Passenger Vehicle Application. *Appl. Sci.* **2024**, *14*, 4784. https://doi.org/10.3390/app14114784

Academic Editors: Fabrizio Giulietti, Nadjim Horri and William Holderbaum

Received: 7 May 2024
Revised: 24 May 2024
Accepted: 28 May 2024
Published: 31 May 2024

1. Introduction

Internal combustion engines (ICEs) have been the dominant power source for automotive applications over the centuries. A relentless pursuit of efficiency and performance has marked the evolution of ICEs over the years [1]. One critical component in optimizing the performance of an internal combustion engine is the idle air control device/actuator, which controls the engine's air inlet and maintains optimal combustion characteristics during idle conditions. A well-performed idle air control actuator contributes to smoother idling and plays a significant role in fuel economy, emission reduction, and overall engine idling performance. Nowadays, researchers are interested in developing automotive engine actuators using quick response, weightless, and controlled materials like shape memory alloy (SMA) materials to offer a potential solution to the challenges associated with mechanical or traditional actuators. Traditional idle air control valves utilize electromagnetic actuators or stepper motors. They have served their purpose well but face limitations of response time, complexity, and electric energy consumption.

Shape memory alloys (SMAs) are unique smart materials that can regain their original shape when subjected to a specific thermal stimulus, such as heat [2,3]. This property has been harnessed in various engineering applications, from biomedical to aerospace actuators. SMAs are classified into different categories based on the alloying types but the most widely known and used is Nickel Titanium (NiTi) Shape Memory Alloy, known as Nitinol. Its shape memory effect (SME) and superelasticity make it suitable for various applications, including in the automotive sector. NiTi shape memory alloys are available on the market in different forms including cold-rolled strips and tubes [4–6], wires [7,8], sheets or plates [9,10], foils [11,12], and springs [10,13,14]. In different complicated actuators that use composite materials, shape memory alloy wires are embedded in the composite to improve their working ability, such as fast response in shape-changing of the structure, mechanical behaviour, and impactful actuation [15,16]. They are used in automotive applications but many are also widely applied in robotic systems [17,18]. The shape memory alloys in the tube form operate at higher temperatures than their wired counterparts [19]. The hollow form of the tube provides the cooling channel of these shape memory alloy-based actuators [19,20]. The cooling time of SMA is reduced and the life span of the actuators increases. Plate-based shape memory alloys are used in aerospace actuators for the vibration and stiffness control of flexible structures [21]. In the automotive industry, they are applied in adaptive shock control actuators [22,23]. The actuators with improved stroke and fast actuation responses use spring-based shape memory alloys [13,24]. SMA spring actuators are mostly applied in the automotive sector. They perform at the constant strain within many actuation cycles and can be deformed up to 200% of their length with low mechanical loads [25,26]. Moreover, the coil spring increases the amount of restoring energy to reduce the consumed energy during actuator operation. An actuator can have two SMA springs or a wire and spring combination, but one element serves as a biasing material [27].

Several researchers are interested in working on SMAs for automotive applications to modernise actuators in this sector. They are used to simulate or experiment to find the results of a proposed designed actuator. None have been used to evaluate their operability in the complete system, which represents the innovation of this work. Song et al. [28] used a shape memory alloy with a Nitinol type to design and control a variable exhaust nozzle actuator. The configuration has SMA wire and a bias spring, but the area is reduced by 40% from the existing mechanical one. Fawcett et al. [29] proposed an emergency brake using an SMA material in its operation. They performed only an experimentation approach during the design. The brake generates high torque and does not require electricity in off-mode conditions. Sathish et al. [30] introduced the two-way SMA spring applied in passenger vehicle suspension to replace the traditional one by improving vehicle comfort and performance. Their experiment results show a significant reduction in the sprung mass displacement from various excitation frequencies. Nissele et al. [31] designed and manufactured an SMA-based active aerodynamic for automotive applications. The surface is smooth and self-controlled with a good arrangement compared to the traditional ones. Pedro et al. [32] performed both theoretical simulations and experiments for designing an SMA wire-based mechanism for aeronautical applications. They investigated the generated torque used for flap actuation. Venkata et al. [33] conducted a research study to morph automotive fender skirts that use composite material embedded with SMA wires. The design process includes mathematical modelling and experiments. The findings describe the reduction in the actuation time of 95% with a 62% weight reduction compared to the dome-shaped steel fender skirt. Pacifique et al. [13] proposed an active thermostat that uses an SMA spring to activate its operations. The design covers mathematical modelling and simulation, which show an engine overheating reduction and low-pressure fluctuation in the cooling system. Tiegang et al. [34] proposed a novel non-embedded, adjustable, and flexible SMA actuator to be applied in the exhaust nozzle to vary its area. The study includes both theoretical and experimental approaches.

This research aims to design the SMA-based Idle Air Control (IAC) actuator for vehicle application. The IAC actuator closes or opens the air path through the bypass for engine idling conditions from its pintle's stroke. The design process includes numerical analysis with COMSOL Multiphysics software and analytical methods with mathematical modelling and simulation using MATLAB software (R2023b). The SMA spring activates the actuator from its phase transformations using electric current. A PID controller and limiters are used for the output responses to reduce the time taken for thermal stimuli. The actuator's operations are evaluated based on the practical spark ignition engine idle-condition input parameters using MATLAB Simulink software (R2023b). The model is simple and weightless, with the rapid opening of the idle air path at a low consumed electric current of only 2 A.

2. Materials and Methods

2.1. Structural Configuration of Proposed IAC Actuator

The structure of the idle air control (IAC) actuator identifies the external physical appearance and internal parts arrangement with their shapes. Figure 1a shows a completely assembled actuator. It has three main parts with different shapes and colours. The bottom part is in rubber colour and the middle part with chromate colour are fixed parts to hold the internal components. The remaining one is a part of the IAC actuator to be fixed into the throttle body. Figure 1b shows the 14 components of a shape memory alloy-based idle air control actuator. Each one has its role in the operation of the actuator. The pintle and rod are the moving components based on the state of the actuator. The stability of the rod holder comes from the vertical support, which assembles it with the outer housing and the fixture plate.

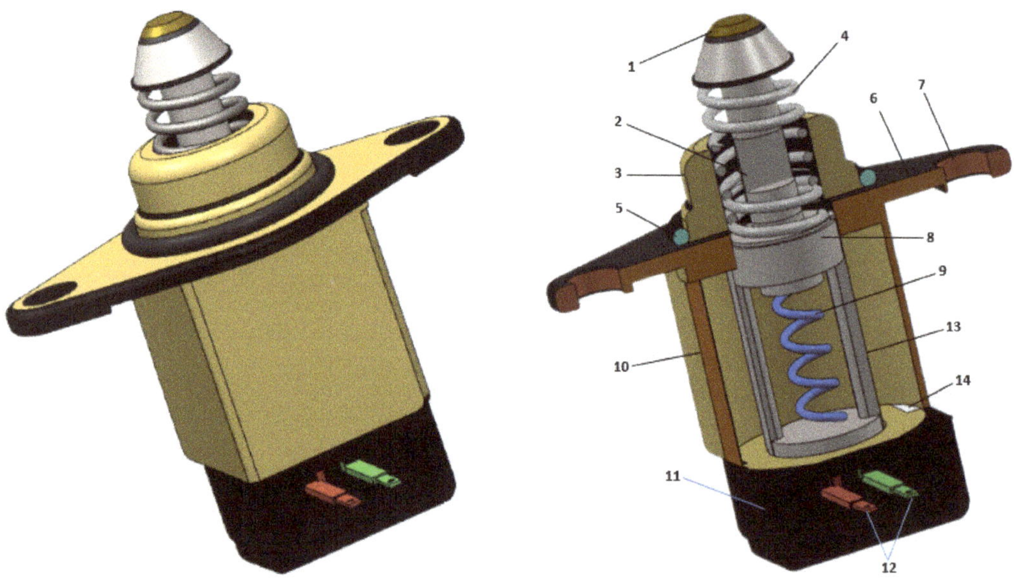

(a) (b)

Figure 1. Structure of proposed idle air control actuator. (**a**) Complete 3D assembled model of the actuator; (**b**) sectioned actuator with all main components (1. Pintle, 2. Rod, 3. Inner housing, 4. Bias spring, 5. O ring, 6. Fixture plate, 7. Bolts-holder, 8. Rod holder, 9. SMA spring, 10. Outer housing, 11. Connecting box, 12. Electric connectors, 13. Vertical support, and 14. Cooling slot).

The fixture plate supports the actuator onto the throttle body through the bolt holder and bolts with nuts. O ring protects the actuator from leakage of air in the bypass. Electric connectors in the connecting box relate to the electric wires from the engine electronic control module and supply the electric current to the SMA spring. This spring is made from a Nickel-and-Titanium (NiTi) alloy with 52% of Ni and 48% of Ti. The reference specifications of the SMA spring are from the released datasheet of Nexmetal Corporation, Sheridan, WY, USA, for the NiTi SMA coil spring [35]. The rod holder, bias spring, and rod are made from stainless steel. The fixture plate and vertical support are made of low-carbon steel. Inner and outer housings are made from grey cast iron. Together with the cooling slot, they facilitate a high rate of heat transferred to the ambient from the SMA spring during cooling time. The connecting box is made of ductile cast iron. Using cast iron in these components increases their ability to work outside with different temperatures and vibrations from the engine. Apart from the simple structure of the actuator, it has a low weight of 17.46 g. However, two springs (SMA spring and Bias spring) shown in Figure 2 are the main components to manage this IAC actuator's operation.

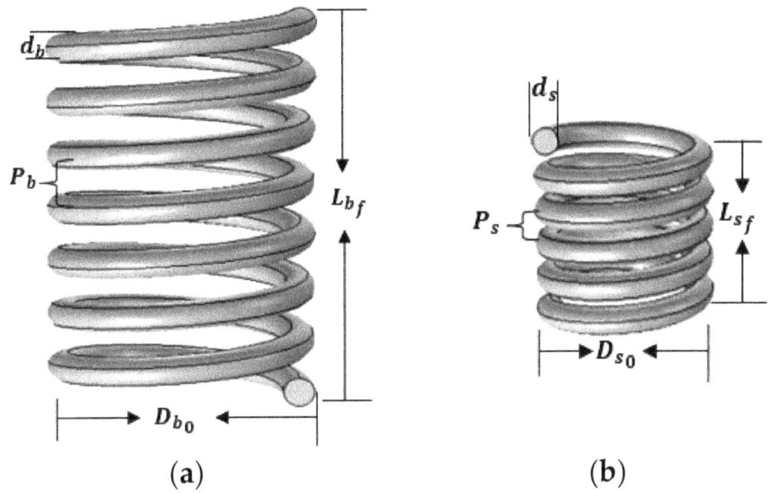

Figure 2. Schematic structure of springs in normal state. (**a**) Bias spring; (**b**) SMA spring (*d*: wire diameter, *P*: vertical pitch, *D*: coil diameter, and *L*: free length of the coil).

The bias spring, shown in Figure 2a, is designed to work under compression. Its parameters are indicated by a subscript b ($._b$). Figure 2b illustrates an SMA spring. It is designed to work under mechanical extension. Its parameters have a subscript s ($._s$) to differentiate them from those of a bias spring. The wire diameter (d_b) was selected from the stainless steel wire available on the market. It is the basic element used to calculate other spring parameters as expressed in Equation (1). It is an equation of modelling modified from a study conducted by Turabimana et al. [13]. The parameters of the SMA spring were identified based on force (F_b) that should extend to the full displacement of the SMA spring at its martensite phase.

$$F_b = \frac{8D_{b_0}}{\tau_m \pi d_b^3} C_f \quad (1)$$

where F_b is a mechanical driving force to extend the SMA alloy spring, C_f is a shear stress correction factor of the spring and it is calculated based on the spring index (i_s) in Table 1. τ_m is a maximum shear stress of the spring, which defines the spring's working ability [36]. This stress for coil spring from stainless steel is 515 MPa. The two springs are designed using the same theory of coil spring, but the working conditions and parametric relationship are different. The parameters of an SMA spring stand to resist F_b and can generate the force that compresses the bias spring during the austenite phase. The parameters of both springs are listed in Table 1.

Table 1. Specifications of the spring.

Parameter	Value	
	Bias Spring ($._b$)	SMA Spring ($._s$)
Wire diameter (d)	2 mm	1.7 mm
Free length of coil (L_f)	23 mm	15.5 mm
Coil diameter (D_0)	16 mm	10.2 mm
Spring index (i_s)	8	6
Generated force (F)	2.25 N	2.416 N
Number of active coils (N_c)	7	5

2.2. *Actuation Mechanism of SMA-Based IAC Actuator*

2.2.1. Actuation Requirements of Proposed Actuator

The actuation mechanism of the proposed SMA-based IAC actuator starts with setting the operational requirements. Those requirements are constraints and specifications related to the environmental and working conditions of the actuator to achieve its final purpose. The constraints are within the bounds of the characteristics of the SMA spring listed in Table 2. The specifications are settled based on the existing idle air control valves of the internal combustion engine. The SMA spring is the operational active component of the actuator and activates its operating states. The specifications include power that allows for heating the SMA spring in a reasonable time, actuation time that defines the heating and cooling cycles, working stroke and temperature to the full opening of the actuator, and repeatability to confirm many possible operating numbers of cycles. Each specification or constraint has its function in the actuator's operation to keep its life span long with good operability.

Table 2. Properties of SMA spring.

Property	Value
Maximum allowable stress (σ_0)	690 Nmm^{-2}
Minimum allowable stress (σ_i)	70 Nmm^{-2}
Operating number of cycles (N_{cy})	10^5
Operating temperature	0–60 °C
Ambient temperature (T_e)	25 °C
Density (ρ_s)	6.45 gmm^{-6}
Poison's ratio (r_s)	0.33
Electric resistivity_Martensite (ρ_{eM})	8.7×10^{-4} Ωmm
Electric resistivity_Austenite (ρ_{eA})	7.2×10^{-4} Ωmm
Specific heat (c_p)	0.2 $Cal(g^oC)^{-1}$
Martensite young's modulus (E_M)	33,500 Nmm^{-2}
Austenite young's modulus (E_A)	62,500 Nmm^{-2}
Coefficient of thermal expansion (C_{th})	11×10^{-6} K^{-1}
Maximum pulling force (F_{smax})	4 N

2.2.2. Working Principle of SMA Based Actuator

The operating principle of the proposed IAC actuator is to close and open the idle air path (bypass) in the throttle body. The SMA spring manages the operation mechanism through its phases such as the martensite and austenite phases. The martensite phase plays a role in the closing state and the austenite phase acts in the opening state of the actuator.

In the closing state, the bias spring extends the SMA spring mechanically with its force (F_b), as shown in Figure 3a. The force (F_b) expressed in Equation (1) is transformed according to the operating principle of the SMA-based IAC actuator and expressed by Equation (2). This is a pseudo-elastic phenomenon of SMA spring. Then, the rod assembled with the SMA spring moves into the rod holder and causes the pintle to displace. Within this state, environmental air passes through the throttle valve from entering the engine intake manifold. The stress of the SMA spring induced by mechanical force (F_b) should be less than the maximum allowable working stress (σ_0) in Table 2. However, the contrary operation is the opening state of the actuator, as shown in Figure 3b.

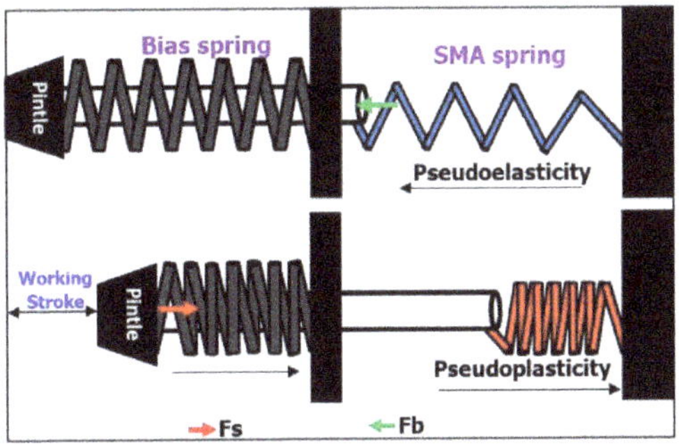

Figure 3. Operational mechanism of the actuator: (**a**) closing state; (**b**) opening state.

In the opening state of the IAC actuator, the SMA spring should return to its original position, but generate force (F_s) to compress the bias spring. The generated force F_s is expressed by Equation (2). It depends on the fraction volume (ϑ) and the SMA spring displacement (S). This force should be able to suppress the mechanical driving force from the bias spring (Fb). The process is a pseudoplastic phenomenon of the SMA spring. With a compressed bias spring, the pintle with the rod follows the motion of the SMA spring and moves in the opposite direction to open the air flows into the bypass of the throttle body. The thermal mechanical behaviour of the SMA alloy spring is the key parameter to activate the opening state, which will be defined in the phase transformations and control techniques in the next sections.

$$\begin{gathered} F_b = \mathrm{F_p} + F_t + K_s x \\ F_s = \frac{\pi d_s^3}{6D_s}\left\{\left(\frac{d_s}{\pi D_s^2 N_{sc}} G_s\right) x_s + (\eta_1 + \eta_2 G_s)(\vartheta_M - \vartheta_A)\right\} = \mathrm{F_p} + F_t + K_b x \end{gathered} \tag{2}$$

where, $\eta_{1,2}$, are the parameters associated with the stress–strain curve. $\vartheta_{M,A}$, stands for the martensite fraction volume and Austenite fraction volume. $d_s, D_s, N_{sc}, G_s, K_s$, are the SMA spring's wire diameter, coil mean diameter, number of active coils, shear modulus, and spring constant, respectively. K_b is a bias spring constant, x represents spring the displacement or stroke achieved by the pintle, and F_p stands for the force that comes from the pressure difference between the inlet and outlet of the bypass. F_t represents the transient flow force. F_t is in the function of pintle dimensions and air characteristics, as expressed in Equation (3):

$$F_t = \mathrm{L} C_A \pi D_p sin\varphi_c \sqrt{2\rho_A \Delta_p \frac{dx}{dt}} \tag{3}$$

where, L, C_A, D_p, φ_c, ρ_A, Δ_p stand for the flowing length, air flow coefficient, pintle diameter, pintle's conic angle, intake air density, and air pressure difference (pressure of air into the intake manifold and pressure entering the bypass idle air path).

2.3. Analytical Modelling of the Proposed IAC Actuator

Mathematical modelling of the SMA spring is necessary to describe two operability states of the proposed actuator using scientific expressions. From the closing to opening state, the shape memory effect (SME) acts undergo the converted current and SMA models to fulfil the actuation of the IAC actuator, as shown in Figure 4. The function that defines the electric current (I) in the converter is dependent on the SMA spring's size (A) and holding temperature circulating into the spring ($\mathrm{T_w}$) [37]. This function is expressed as follows:

$$I = \mathrm{F}(\mathrm{T_w}, A) \tag{4}$$

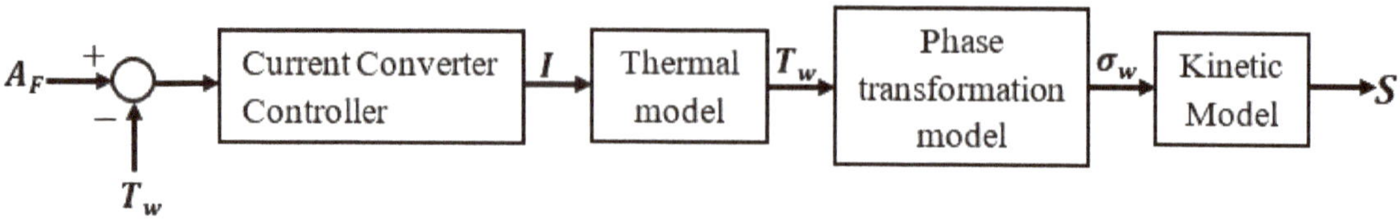

Figure 4. Shape memory effect (SME) analytical model to operate the actuator.

The desired temperature of the actuator's pseudo-plasticity is the final austenite temperature of the SMA spring (A_F). The converter uses this temperature as the input of the current function to predict the required electric current to supply the SMA spring. The system uses a proportional integrated derivative (PID) controller to achieve fast smooth current output responses. However, the controller input is the output signal from the current function. The scientific behind of this controller is the tuning technique of its gains to the significant values. In this work, the parameterisation of the PID controller is performed with a genetic algorithm (GA) method to obtain the preference controller gains ($K_{P,I,D}$). The objective function to give the optimum gain is a function of the length of the integrated time absolute error (ITAE). The variables are three, the lower boundaries are [0 0 0], and the upper boundaries are [50 50 50]. The system output displays K and fitness values. After using the optimum gains, the output signals from PID pass through the current limiter. To prevent saturation output current signals, we opt to use a saturation limiter with upper and lower boundaries [38]. The output signals of the current converter are shown in Figure 5.

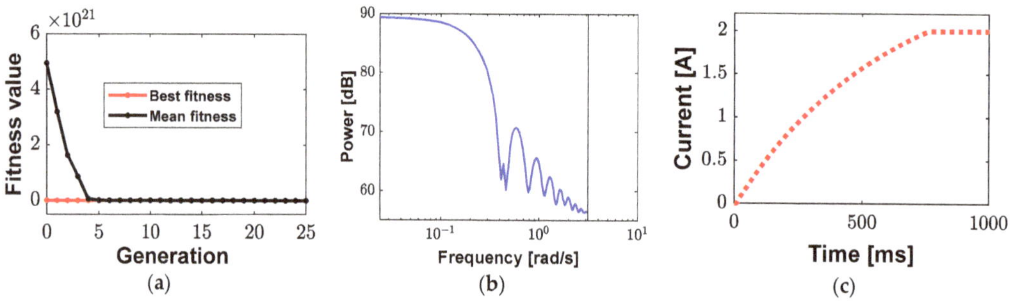

Figure 5. Current converter's characteristics. (**a**) PID optimization fitness values (Best: 504.366, Mean: 1.8951×10^{15}); (**b**) power spectrum of integrated time absolute error (ITAE); and (**c**) output current response.

Figure 5a shows the best fitness value of the PID controller's optimization process achieved within 25 generations. The best fitness value and the last minimum length value of the integrated time absolute error (ITAE) are the key parameters to confirm the optimum gains of the PID controller. Figure 5b shows the frequency of ITAE during controller optimization. The recorded optimum controller gains (K) are [1.03 0.64 0.31]. The controller gain values are important in the controller outputs, accuracy, and time response. Figure 5c shows the electric current output response. The electric current of 2 A is achieved after 700 ms from multiple simulation duty cycles. This current is enough to heat the SMA spring and becomes the input of the SMA models to activate the operation of the SMA-based actuator. The required SMA models to determine the appropriate responses of the actuator include thermal, constitutive, and kinetic models. The models are arranged consecutively to use controlled electric current as the principal input [7,39]. The SME links the models to operate. However, SME is activated by heating the SMA spring at the final required temperature, as described by the thermal model. The thermal model is expressed by Equation (5). For the actuator to perform those two antagonistic states, the heat transfer of the SMA spring is followed by a heating cycle (T_{hc}) for opening and a cooling cycle (T_{cc}) for the actuator's closing process. The temperature response (T_w) from this model becomes the input of the phase transformation model. The temperature (T_w) should be between the final martensite temperature (M_F) and the final austenite temperature (A_F) in Table 3.

$$
\begin{aligned}
\rho_s V_s c_p \tfrac{dT_w}{dt} &= I^2 R - h_s \tfrac{\pi D_s^2}{4}(T_w - T_e) \ \textit{(Heat transfer in SMA spring)} \\
T_{hc} &= T_e + \frac{I^2 R}{h_s \frac{\pi D_s^2}{4}}\left(1 - e^{-\left(\frac{4h_s}{D_s \rho_s c_p}\right)t}\right) \ \textit{(Heating cycle)} \\
T_{cc} &= T_e + (T_w - T_e)e^{-\left(\frac{4h_s}{D_s \rho_s c_p}\right)t} \ \textit{(Cooling cycle)}
\end{aligned}
\tag{5}
$$

where ρ_s, V_s, c_p, I, R, h_s, t, T_e stand for the SMA spring's density, total volume, specific heat, electric current, electric resistance, convection coefficient, operational time, and ambient temperature, respectively.

Table 3. Phase transformation characteristics.

Parameter	Value	Reference
Martensite starting temperature (M_S)	38 °C	
Martensite final temperature (M_F)	26.5 °C	[40]
Austenite starting temperature (A_S)	42 °C	
Austenite final temperature (A_F)	50.5 °C	
Slope of Austenite limit curve (β_A)	7.2 $\text{Nmm}^{-2}\text{K}^{-1}$	
Slope of martensite limit curve (β_M)	6.92 $\text{Nmm}^{-2}\text{K}^{-1}$	
Maximum transformation strain (ε_{tr})	2.3%	[41]

The volume fraction (ϑ) defines the process of the transformation model. From the volume fraction, this model is divided into three classes such as the martensite (M) to the austenite (A) phase, clarifying the austenite fraction volume (ϑ_A), austenite (A) to martensite (M) phase to define the martensite fraction volume (ϑ_M), and twinned (M_{tw}) to detwinned (M_{dt}) martensite phase to derive the detwinned martensite fraction volume (ϑ_{dt}). During the phase transformation process, we set the transformation boundaries of martensite (ϑ_M) and austenite fraction volume (ϑ_A), with a maximum (ϑ_M) of 1 and 0 for (ϑ_A), as shown in Figure 6. Equations (6) and (7) express the phase transformation processes of volume fraction (ϑ). This equation is established based on the phase transformation equation model of the SMA-driven compliant rotary actuator demonstrated by M. Daisuke et al. in 2018 [42] and the operating condition of the proposed actuator. The induced stress during the austenite fraction should be less than the stress of the martensite fraction. However, both stresses could cause the operational strain below the transformation strain (ϵ_{tr}).

$$
\begin{aligned}
\vartheta_M &= \vartheta_{Ai}\left(\tfrac{1}{2}\left(cos\tfrac{\pi(\delta T_M - \delta\sigma_M)}{M_S - M_F} + 1\right)\right)^{z_a} \ \textit{(A to M transformation process)} \\
\vartheta_A &= (\vartheta_{Mi} - 1)\left(\tfrac{1}{2}\left(cos\tfrac{\pi(\delta T_A - \delta\sigma_A)}{A_S - A_F} + 1\right)\right)^{z_b} + 1 \ \textit{(M to A transformation process)} \\
\vartheta_{dt} &= (1 - \vartheta_A)(\vartheta_{dti} - 1)\left(\tfrac{1}{2}\left(cos\tfrac{\pi\delta\sigma}{\nabla\sigma} + 1\right)\right)^{z_c} + (1 - \vartheta_A) \ (M_{tw} \ \textit{to} \ M_{dt} \ \textit{transformation process})
\end{aligned}
\tag{6}
$$

where,

$$
\begin{gathered}
z_a = \left(1 + \left(\tfrac{\delta T_M - \delta\sigma_M}{M_S - M_F}\right)\left(\tfrac{\alpha_1 - 1}{\alpha_1}\right)\right)\alpha_1, \\
z_b = \left(1 - \left(\tfrac{\delta T_A - \delta\sigma_A}{A_F - A_S}\right)\left(\tfrac{\alpha_2 - 1}{\alpha_2}\right)\right)\alpha_2, \\
z_c = \left(1 - \tfrac{\delta\sigma(\alpha_3 - 1)}{\nabla\sigma\alpha_3}\right)\alpha_3, \\
\nabla\sigma = \sigma_0 - \sigma_i, \ \delta\sigma = \sigma_w - \sigma_i, \ \delta\sigma_M = \tfrac{\sigma_w}{\beta_M}, \ \delta\sigma_A = \tfrac{\sigma_w}{\beta_A}, \\
\delta T_M = T_w - M_s, \ \delta T_A = T_w - A_S
\end{gathered}
\tag{7}
$$

where A_S, A_F, M_S, M_F are the starting austenite temperature, final austenite temperature, starting martensite temperature, and final martensite temperature. These parameters are listed in Table 3. $\alpha_{1,2,3}$ represent the crystal variable speed coefficients of nitinol, β_A, β_M are slope limit curves of austenite and martensite, and the other mathematical symbols are defined in Tables 2 and 3. The twinned crystals' components are separated during detwinned martensite. However, introducing the stresses in the transformation process of the M_{tw} to M_{dt} defines the deformation of the SMA spring in the martensitic condition. The kinetic model is used to find the total stroke that is moved by the pintle. It is defined by the displacement of the SMA spring from the axial pitch. This displacement includes both the change in the axial pitch of the coils (Δ_P) and the initial pitch (P_s) of the SMA spring. Hence, this kinetic model determines the change in vertical pitch (Δ_P) of each coil in a reasonable working strain of spring wire. The product of (Δ_P) with the number of active coils (N_{sc}) results in the total displacement of the spring if it is added or deducted from the original vertical pitch of the SMA spring. The change in axial pitch of the coils (Δ_P) is a function of working stress, working strain, SMA spring size and the SMA spring's generated force, as expressed by the first term of Equation (8):

$$S = \left(P_s + \left(4\frac{F_s P_s \varepsilon_w}{\pi D_s^2 \sigma_w}\right)\right) N_{sc} \tag{8}$$

where S represents the total stroke moved by the actuator's pintle, ε_w stands for the working strain, and σ_w is the working stress. Other variables were defined in the previous paragraphs or in Appendix A at the end part of the article. The working strain (ε_w) used in this model is from the strain due to elasticity, strain induced by thermal, and induced strain by phase transformation. It is expressed as follows:

$$\begin{gathered} \varepsilon_w = \varepsilon_i \frac{\sigma_w}{E(\vartheta_A, \vartheta_{dt})} - \frac{\sigma_i}{E(\vartheta_{Ai}, \vartheta_{dti})} + \xi(\vartheta_{dt} - \vartheta_{dti}) + \varnothing(\vartheta_A)T_w - \varnothing(\vartheta_{Ai})T_e \\ E(\vartheta_A, \vartheta_{dt}) = \mathrm{E_{tw}} + (E_A - E_{tw})\vartheta_A + (E_{dt} - E_{tw})\vartheta_{dt} \\ \varnothing(\vartheta_A) = \varnothing_\mathrm{M} + (\varnothing_\mathrm{A} - \varnothing_\mathrm{M})\vartheta_A \end{gathered} \tag{9}$$

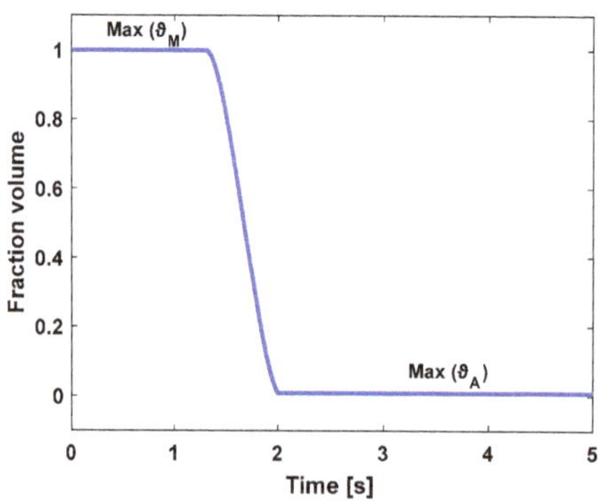

Figure 6. Maximum reference of fraction volume in phase transformation.

In the above, ξ, $\varnothing$, ε_i are the transformation modulus, temperature coefficient from the material's resistance, and initial strain of the SMA spring, respectively. E_A, E_{tw}, E_{dt}, $E(\vartheta_A, \vartheta_{dt})$ represent the austenite young's modulus, twinned young's modulus, detwinned young's modulus, and young's modulus in austenite and detwinned martensite phase transformations, respectively. Equations (4)–(9) were used to develop the MATLAB algorithms that define the operation of the actuator in the analytical mean. The model-based system design approach (Simulink (R2023b)) was adopted and linked with those algorithms because of the multidomain models of the proposed actuator. Simulation outputs are presented in the results section.

2.4. Numerical Analysis of SMA-Based IAC Actuator

The design process of the actuator requires different verification, evaluation, and validation from one analysis approach to another to obtain a well-operating actuator at the final stage [43]. In this section, we used finite element analysis to validate the analytical simulations in the previous section. Both approaches were performed to open and close the air flowing in the bypass of the throttle body for idle engine speed. The numerical analysis uses COMSOL Multiphysics software (R2023b) to parameterise, geometrise, and simulate the IAC actuator. The analysis was arranged to fulfil the opening and closing operations of the actuator described in Figure 3. Table 4 identifies the numerical simulation workflow from the start to the computation of the outputs but highlights the different settings between the closing and opening state of the actuator.

Table 4. Numerical simulation workflow in COMSOL Multiphysics.

Model Builder and Setting	IAC Actuator's Operating Mode	
	Closing State	Opening State
Physics interface	Structure Mechanics/Multibody dynamics	Structure Mechanics/Joule Heating and Thermal Expansion
Study	Stationary	Stationary
Physics constraints	Rigid fixed domain (supports), Prescribed displacement, (Assigned on the Pintle as the desired displacement), and Boundary load (F_b used as pressure and assigned on the bias spring)	Fixed constraint (supports), Temperature (T_e), Ground (negative connector), Electric terminal (positive connector linked with the SMA spring).
Meshing	Free tetrahedral extra fine	Free tetrahedral extra fine
Solver configurations	Stationary nonlinear solver with segregated steps	Stationary nonlinear solver with segregated steps

The model builder uses different physics due to the two operating modes of the actuator. Multibody dynamics is used for the mechanical constraints to simulate the actuator in the closing state while the opening state uses joule heating and thermal expansion physics. After confirming the geometry component with full physics settings, the materials are assigned to the concerned actuator components, as identified in Section 2.1. Table 5 lists the properties of materials assigned to the model, excluding the NiTi properties, which are listed in Tables 2 and 3. When inserting the NiTi input parameters/properties, the basic and phase transformation properties were separated from the martensite and austenite properties.

Table 5. Materials and their properties for the model.

Property	Material				
	Stainless Steel 405 (Bias Spring, Rod Holder, and Rod)	Low Carbon Steel 1008 (Fixture Plate and Vertical Support)	Grey Cast Iron (Housings)	Ductile Cast iron (Connection Box)	Nitrile Rubber (O Ring)
Young's modulus	190 Gpa	200 Gpa	124 Gpa	172 Gpa	0.855 MPa
Poison's ratio	0.3	0.29	0.26	0.26	0.43
Density	7800 kg/m^3	7872 kg/m^3	7150 kg/m^3	7200 kg/m^3	1000 kg/m^3
Thermal conductivity	16.2 W/mK	16.2 W/mK	53.3 W/mK	32.3 W/mK	0.33 W/mK
Specific heat capacity	500 J/kgK	450 J/kgK	490 J/kgK	506 J/kgK	1900 J/kgK
Coefficient of thermal expansion	9.9×10^{-6}/K	1.32×10^{-5}/K	9×10^{-6}/°C	11.6×10^{-6}/°C	0.48/°C

In the meshing process of the model, the mesh element size was 0.005 mm, with a free tetrahedral extra fine that uses a geometric shape order of quadratic serendipity, as shown in Figure 7a. The electric current uses a terminal type of current with 2 A. It is assigned on a positive connector to supply the SMA spring for the thermal effect of the actuator operation. Figure 7b shows how the current input causes the electric potential induced into the actuator's components during the computation of the actuator's opening state solutions. The SMA spring experienced a greater electric potential of 5.12 V and crossed close to the positive connector. The induced temperature to initiate the shape memory effect in the actuator is shown in Figure 7c. In the SMA spring also, the highest temperature of 54.9 °C was recorded. Numerical solutions will be presented in the Results section.

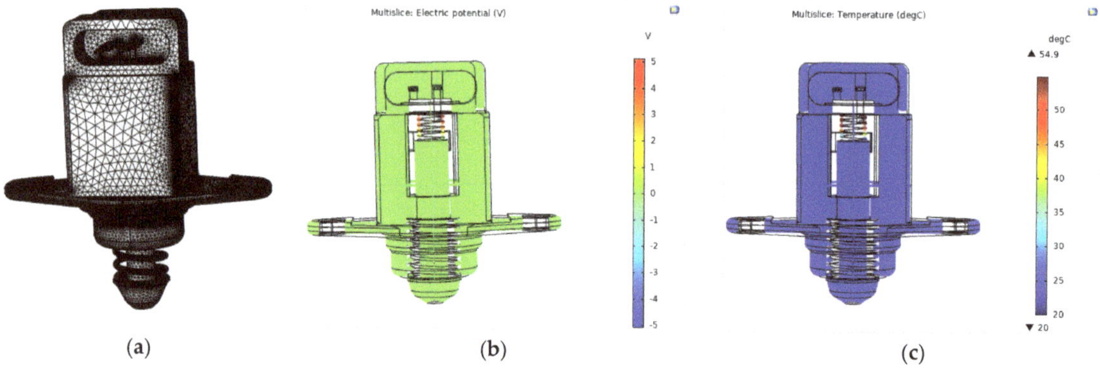

Figure 7. Numerical modelling process of the IAC actuator. (**a**) Generated meshing of IAC actuator model, (**b**) induced electric potential from the input current flowing into the actuator, and (**c**) temperature in the actuator to activate shape memory effect.

2.5. Performance Evaluation of the Proposed IAC Actuator

In this section, we will evaluate the operability of the designed SMA-based IAC actuator through MATLAB (R2023b) simulation using mathematical algorithms and Simulink. We do not conduct experiments for the proposed actuator at this phase. Still, the same examinations will be performed in this section by integrating the designed actuator into the intake manifold and comparing the engine outputs with the real practical operated engines and literature experimental results. The proposed IAC actuator is assembled into the structure of the throttle body to control the flow of bypass air to the intake manifold through the displacement of its pintle, as shown in Figure 8.

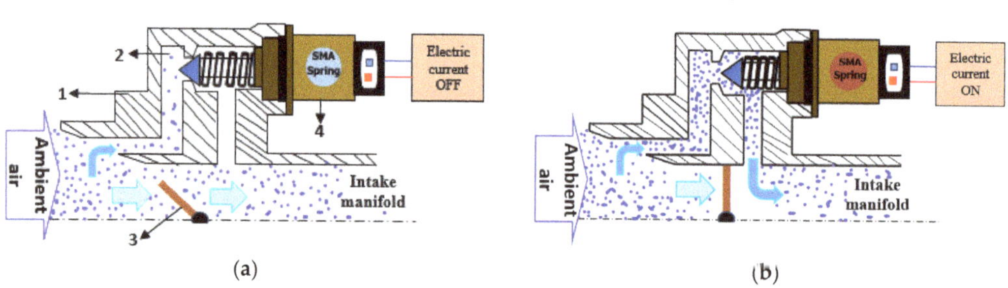

Figure 8. Simplified structural representation of engine intake manifold with throttle body holding an SMA-based IAC actuator. (**a**) Air flowing in the main path with the opening throttle valve and bypass closed by the SMA-based IAC actuator; (**b**) air passes through the bypass path opened by the SMA-based IAC actuator and full closed throttle valve. (1. Throttle body, 2. Bypass, 3. Throttle valve butterfly, 4. SMA-based IAC actuator).

The SMA-based actuator closes the bypass, and fresh air from the environment enters the intake manifold through the throttle valve, as shown in Figure 8a. The actuator is in a closed state, and there is no supply of electric current to the SMA spring. This state of the actuator always occurs whenever the engine is turned off, or the driver presses the accelerator pedal. Figure 8b shows the fresh air flowing through the bypass to enter the intake manifold with an opening actuator. The throttle valve butterfly is in full close with an angle of 90°. This state of the actuator always occurs during engine idling conditions. The dynamic characteristics of air through the bypass and their corresponding engine idling characteristics are expressed in Equations (10)–(13). The main inputs of the equations forming the algorithms in this section include the actual parameters for the spark ignition engine from previous research with experiments [44–46] and practical engines [47,48]. The triggering signal from the SMA-based IAC actuator is the opening and closing strokes that vary with time (S).

$$\dot{m}_{\mathrm{A}}(\mathrm{t}) = \mathrm{D_p}\sin(\varphi_c)S(t)\frac{P_{atm}(t)}{(RT_{atm}(t))^{\frac{1}{2}}}\left(\frac{2P_{in}(t)}{P_{atm}(t)}\left(1-\frac{P_{in}(t)}{P_{atm}(t)}\right)\right)^{\frac{1}{2}} \tag{10}$$

where $\dot{m}_{\mathrm{A}}$, T_{atm}, P_{atm}, P_{in}, R are the air mass flow rate through the pintle of the SMA-based IAC actuator, ambient temperature, atmospheric pressure, air pressure in the intake manifold, and air constant with a value of 287.05 (J/kg°C). The opening or closing stroke ($S(t)$) of the actuator describes the air mass flow rate through the bypass ($\dot{m}_{\mathrm{A}}$) before reaching the intake manifold. The longer the opening stroke of the actuator, the more flowing air enters the intake manifold. This air mass flow rate was used to define the other characteristics of the engine in idle mode.

$$\frac{d}{dt}P_{in}(\mathrm{t}) = \frac{RT_{in}(t)}{V_{in}}\left(\dot{m}_{\mathrm{A}}(t) - \frac{\dot{m}_{en}(t)}{1+\frac{1}{\lambda(t)s_t}}\right) \tag{11}$$

In the above, T_{in}, V_{in}, $\dot{m}_{en}$, λ represent the temperature in the intake manifold, the volume of the intake manifold, the air mass flow rate that passes through and induced by the engine, and the air–fuel ratio. s_t is the stochiometric air–fuel ratio, with a value of 14.7 for the spark ignition engine.

$$\dot{m}_{en}(t) = \frac{P_{in}(t)}{2RT_{in}(t)}\nu_e(P_{in}(t),\omega_{en}(t))V_p\frac{\omega_{en}(t)}{2\pi} \tag{12}$$

where ν_e, V_p, ω_{en} are the volumetric efficiency, piston displacement volume, and engine speed.

$$T_{en}(t) = \left(\eta\frac{H_l}{V_p}\frac{4\pi}{\lambda s_t}\frac{\frac{\dot{m}_{en}}{1+\frac{1}{\lambda(t)s_t}}\left(t-\frac{2\pi}{\omega_{en}(t)}\right)}{\omega_{en}\left(t-\frac{2\pi}{\omega_{en}(t)}\right)}\right) - \left(\frac{\eta_f 4\pi}{V_p}\right) - (P_{en}(t)-P_{in}(t))\frac{V_p}{4\pi} \tag{13}$$

where T_{en}, η, η_f, H_l stand for the torque produced by the engine during the idle operating mode, ignition parameter, piston's friction coefficient, and heating due to ignition. The engine modelling equation completed by Gharib et al. [47] and the input parameters form the basis of the equation used to compute the idle engine torque output. Then, we simulated the idle engine speed in revolutions per minute using the generated engine torque and the engine's load torque. The product of the idle engine speed in radians per second and torque gives the total power output of the engine in watts [49].

3. Results and Discussion

This section presents the findings obtained from the methods discussed in the previous section. It includes the results of analytical analysis, numerical simulations, and integration of the SMA-based IAC actuator into the automotive engine's intake manifold. Therefore, the discussions will bear the results in line with the IAC actuator's operations.

3.1. Analytical Results

Analytical simulation results were obtained based on the equations in the analytical method. Figure 9a shows the extension displacement of the shape memory alloy spring from the input mechanical driving force (F_b). The extension displacement of the SMA spring can be the same as the closing stroke moved by the pintle of the IAC actuator during the closing state. The SMA spring keeps stretching as the input force increases. With the maximum force of 2.39 N, the displacement reached 6.26 mm. However, a coil spring (bias spring) pulled another coil spring (SMA spring), and the displacement curve became rough because the coils were not stretching equally.

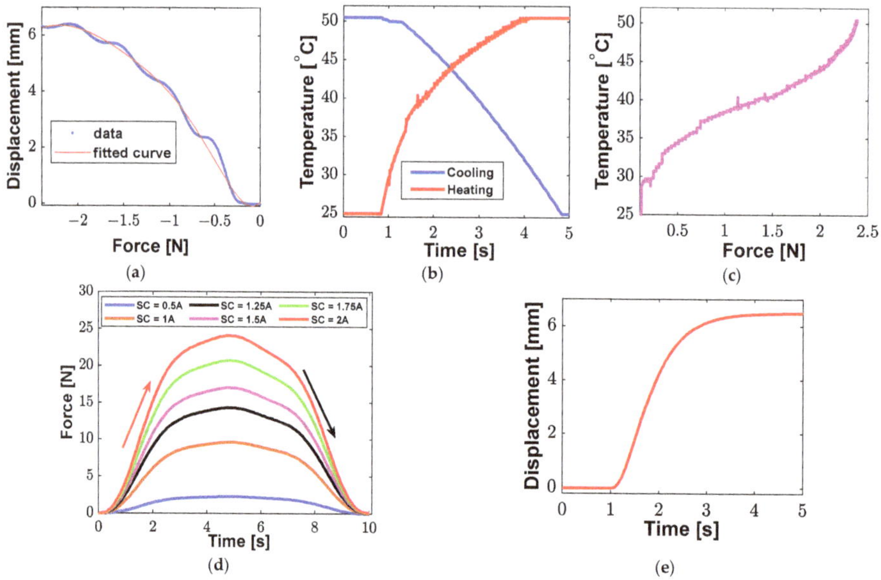

Figure 9. Shape memory effect outputs in the operating mechanism of the proposed IAC actuator. (**a**) Mechanical force versus the extension displacement of proposed actuator caused by bias spring; (**b**) heating and cooling cycles with used time to the final stage; (**c**) generated force of SMA spring to pull the bias spring with heating temperature; (**d**) output generated force of SMA spring produced by different supplied electric current; (**e**) displacement of SMA spring to regain its normal position.

The heating and cooling cycles of the SMA spring are shown in Figure 9b. Before 1 s, the heating cycle curve rises from the ambient temperature (25 °C) to the final austenite temperature (50.5 °C), but the final one was achieved at 3.8 s. The main input here was the controlled current (I) from the current converter. The heating cycle remained active when the actuator opened until the SMA spring's current supply stopped. The choice of a maximum heating temperature of 50.5 °C prevents an actuator from overheating and high electricity consumption. It reduces the time needed to cool it down for another operational state. However, the cooling cycle curve starts to fall after 1.2 s of off-current supply. It takes 4.75 s to complete cooling the SMA spring from the maximum heating temperature to the ambient temperature of 25 °C. It implies that the time required to cool the SMA spring is more than its heating time. Figure 9c shows the generated force of the SMA spring used to compress the bias spring and regain its position. The generated force is zero when the temperature of the heating cycle is between 25 °C and 30 °C. The increasing temperature provides the rise of the generated force. With maximum heating temperature, the spring generates its maximum force of 2.46 N. However, the variation in output force generated by the SMA spring depends on the supplied current, as shown in Figure 9d. This force increases proportionally with the electric current supplied to the SMA spring increases. The current of 2 A provides the highest generated force of 2.46 N. The rising curve of force is active every time the current is on and starts descending directly after the supply current is off for all force curves. Figure 9e shows the displacement of the SMA spring under the electrical supply. The extended coils of the SMA spring return to their original positions when it heats up, causing the spring to regain its original length. In 3.8 s, the maximum settling displacement of 6.5 mm is reached. This displacement of the SMA spring from the input current can be the same as the opening stroke moved by the actuator's pintle.

The fraction volume curve that defines the transformation of the SMA spring is shown in Figure 10a. According to this curve, the SMA spring's operability ranges from 0.01 to 0.14 for the austenite fraction volume and 1 to 0.5 for the martensite fraction volume. However, no fraction volume curve passes the settled maximum fraction volume of martensite and austenite. With more repeating cycles, the required temperature for phase transformation is lower. The applied maximum temperature is 54 °C instead of the 50.5 °C to test the performance ability of the SMA spring in the actuator. However, the fraction volume curve confirms the ability of this SMA spring to perform well at such applied temperature without fail. Figure 10b shows the hysteresis curves of stress and strain for both extension and SMA regaining its initial position. The maximum mechanical stress with the SMA spring under extension is 452 Mpa. It is below the maximum allowable working stress (σ_0) of 690 Mpa. The corresponding extension strain to complete the cycle is 0.0219. The regaining process of the SMA spring pulls the bias spring, which causes it to have a maximum stress of 347 Mpa with a strain of 0.016 from the thermal effect. The strength operation of the SMA spring in this actuator is advantageous to the design to prevent overstressing and overstraining of the actuator to increase its life span. The operating with repeatability of the actuator is shown in Figure 10c,d. The actuator performs 100 operating cycles in 5 s with a time interval of 0.0001. They are very congested but indicate that the stress and strain curves of the 1st cycle are the same as the ones recorded in the 100th cycle. The recorded strain from multiple cycles is below the transformation strain in the material properties. The output stress at the 100th cycle is also less than the maximum allowable stress of the SMA spring. They designate the high repeatability and long-life operation of the SMA-based actuator [50].

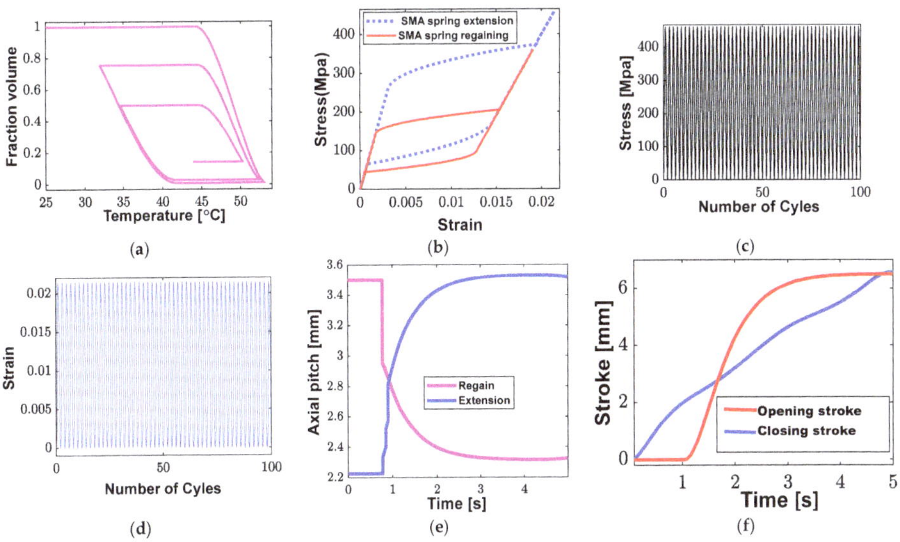

Figure 10. Effect of shape memory alloy in actuation mechanism of the proposed actuator. (**a**) Fraction volume of SMA spring concerning temperature taken during phase transformation; (**b**) hysteresis curve of stress–strain from the mechanical extension of SMA spring and stress–strain curve during spring regains its initial position; (**c**) stress of SMA within multiple numbers of operating cycles; (**d**) operating strain of SMA spring from a multiple number of cycles; (**e**) the change in vertical pith to the extension and regaining position of the SMA spring for closing and opening states of the actuator; (**f**) closing and opening strokes of the proposed IAC actuator.

Figure 10e shows the change in vertical pitches of the coils of the SMA spring. The vertical pitch of each coil changes to extend during the closing state of the actuator. It starts from the initial pitch and extends until the final displacement of the SMA spring. The displacement from a coil covers the initial vertical pitch and its change. The extension starts from the initial pitch of 2.23 mm and ends at 3.5 mm. Regaining axial pitch starts and the SME boosts to activate the SMA spring to the opening state of the actuator. The pitch is changed in the opposite direction from the extension but does not reach the starting point of the extension stage, because of the residual resistance from the bias spring. The final extension pitch becomes the starting point of the regaining pitch, and the final regaining pitch is 2.3 mm. The change in vertical pitch for extension is 1.27 mm, while the vertical pitch for regaining is 1.2 mm. Figure 10f shows the IAC actuator's closing and opening strokes. The maximum stroke of both states is 6.5 mm, but achieved at different times. The closing stroke curve is not smooth and takes a full simulation time of 5 s to complete. The opening curve from zero to the maximum stroke is smooth and was achieved in 3 s of simulation. The maximum closing stroke is the sum of the total changes of axial pitches and the initial pitch, and the maximum opening stroke is the total change in the axial pitches for regaining. The maximum stroke was settled based on the existing operating space of the pintle from previous research on idle engine speed [51–53]. The quick opening of the actuator is required to facilitate the idle speed activation in a short time.

3.2. Numerical Simulations

The numerical simulation results were recorded when the convergence plot was fully generated with the final minimum estimation error of the solution process for the selected nonlinear solver, as shown in Figure 11. It includes both physical interfaces in Table 4 that identify the operational states of the proposed actuator. Although the starting and ending iteration numbers for the operational states differ, they both end with the same estimation error of 10^{-3}. The opening state ends at the 186th iteration number and the closing state starts at the previous state's last iteration number and ends at the 489th iteration number.

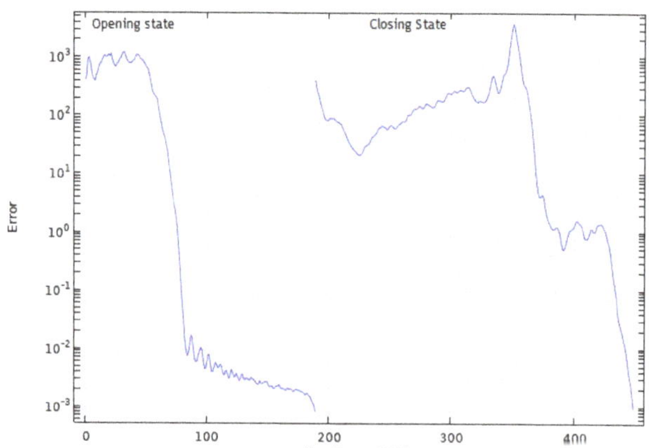

Figure 11. Solver's convergence plot of the estimation errors.

Figure 12 shows the numerical results of both the actuator's opening and closing operation states. They use the internal structure of the actuator that defines its operational mechanism in Figure 3. The external structure is hidden to prevent the use of results with multi-sliced graphs of results to present the internal structure, which can cause some confusion in analysis.

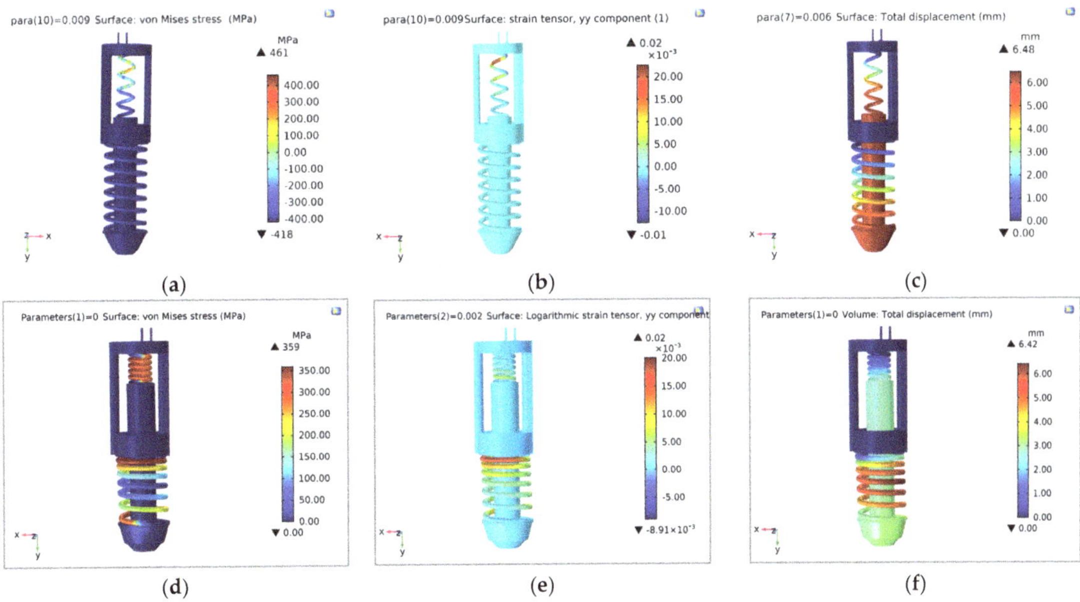

Figure 12. Numerical analysis results of the SMA-based IAC actuator's closing and opening operating mechanisms. (**a**) Induced stress of the actuator during closing state; (**b**) induced strain of the actuator during closing state; (**c**) closing displacement of the actuator; (**d**) stress induced by the actuator during opening state; (**e**) induced strain of the actuator during opening state; and (**f**) the actuator opening displacement.

The extension of the SMA spring activates the closing position of the actuator. It allows the rod and pintle to move in a downward position. However, the stretching of the SMA spring experiences a maximum stress of 461 MPa, while the other parts of the actuator are less stressed, as shown in Figure 12a. It is noticed that there is a difference of 9 MPa between the numerical and analytical maximum stresses. The difference between the numerical output stress and allowable maximum stress of the SMA spring is 229 MPa. It defines the operability of the actuator for a long time without overstressing. Figure 12b shows the strain of the actuator during the closing state with a maximum strain of the actuator that is recorded on the SMA spring. This strain was below the transformation train prescribed in the SMA properties, with a strain difference of 0.3% between them. Figure 12c shows the displacement that is moved by the pintle to close the IAC actuator. The red colour that indicates the maximum displacement appears on the parts which play a role in the pintle's movements. The red colour on the bias spring comes from the settings chosen to keep the axial pitch of this spring and this spring is identified as the force F_b during simulation. The maximum closing displacement of the actuator is 6.48 mm, which indicates a difference of 0.02 mm from the analytical simulation. Figure 12d shows the stress of the actuator during the opening process. The maximum stress is 359 MPa, which appears in the springs. The SMA spring compresses the bias spring to regain its initial position and pulls the rod with a pintle to move upward, which leads to high stress like a bias spring. The color bar of stress clearly shows the bias spring as a part of the actuator that experiences stress. This stress is the lowest compared to the closing stress and the permissible working stress of the SMA spring. The strain from the actuator's opening state is shown in Figure 12e. Based on a predefined colourmap of the actuator's strain for the opening state, the bias spring shows the highest strain, and the SMA spring follows it. However, the recorded maximum strain of 2% is lower than the allowable working strain of the spring made from steel, which is between 15% and 20% [54,55]. This strain is also below the maximum strain transformation of the SMA spring. Figure 12f shows the displacement of the actuator to the opening state. The maximum displacement is 6.42 mm, defined by the SMA spring moving upward with the rod and pintle to regain its initial position after compressing the bias spring. There is a difference of 0.06 mm between the closing and opening displacements from the residual energy of the SMA spring after mechanical force application. The coils' axial pitches of the bias spring were changed, which caused the bias spring to be recorded as the highly displaced part of the actuator. The moving parts, like the rod and pintle in this position, move back to their initial positions. During the design, the desired actuator's stroke was 6 mm. The decimal values outside of the desired stroke are the operation tolerance. The nature of each analysis method causes apparent differences between numerical and analytical results [56,57]. In the analytical analysis, the solutions were exact based on the inputs we inserted into the equations for forming and material properties. However, numerical analysis provides approximate solutions based on the inputs, material properties assigned, and others already in COMSOL Multiphysics, as well as a nonlinear solver with high computing capability.

3.3. Performance Evaluation Outputs

The results from the performance ability of the designed IAC actuator in the engine's air intake system are presented here. They are compared to previous studies with experiments and practical engines operating in idle mode. This section also verifies the key parameters and inputs from the practical engines, which may reduce the cost of experiments and manufacturing the proposed IAC actuator. The air mass flow rate through the bypass when the IAC actuator opens is shown in Figure 13a. It describes the amount of air that flows into the intake manifold per defined time. The air mass flow rate in a steady state is 2.76×10^{-3} kg/s after 6 s of simulation. The air mass flow rate is a triggering characteristic of idle engine operation. The IAC actuator's stroke defines the amount of this air mass flow rate. Figure 13b shows the air pressure through the intake manifold. It is a critical parameter after the air mass flow rate to characterise the intake manifold of an engine operating in idling mode. It is steadily sated after 6 s with an air pressure of 6.06×10^4 Pa. The air mass flow rate variation determines how air pressure changes in the intake manifold. The intake manifold's air pressure is higher with the SMA-based IAC actuator than in the literature study from reference [53], and the air pressure difference in these results is 1.56×10^4 Pa.

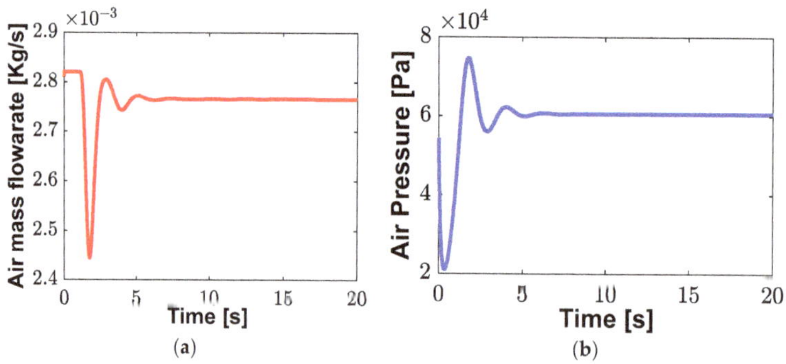

Figure 13. Intake manifold characteristics of idling engine. (**a**) Air mass flow rate through the bypass to the intake manifold; (**b**) pressure of air flows in the intake manifold during engine operation in idle mode.

Figure 14a shows the output torque and power of an engine that operates in idling mode. Engine torque is the first output performance characteristic to measure in the simulation. The maximum idling torque is 31.08 Nm. This torque response is steadily stated at 5.3 s with 25 Nm. However, the steady state of output power is 16.5 Kw at the same time as the torque. The maximum simulated idle engine power is 19.78 Kw. The results demonstrate an improvement in engine idle torque of 1.24% and power of 1.236% compare with the experimental results from the previous works in the literature [48,51].

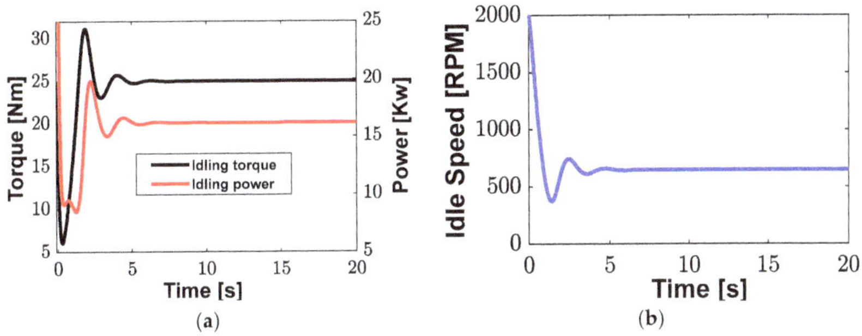

Figure 14. Performance characteristics of an engine running in idle mode. (**a**) Idling engine torque and power output; (**b**) output idling engine speed.

Figure 14b shows the speed of the running engine in idle mode. The output speed curve after 5.3 s of simulation is in a steady state with 650 *RPM*, and the maximum idling speed is 743.8 RPM. All simulated steady-state values are between the maximum and minimum idle performance output characteristics of the practical operated engine. Furthermore, the enhancement of engine idling speed is also shown by comparing with the experimental results from previous research, as cited in references [47,48,51,52,58]. This means the performance improvement of the engine during idling conditions with the designed IAC actuator is achieved. The engine idling characteristics concerning the IAC actuator's performance are shown in Figure 15. As the actuator's opening stroke increases, the engine's idling torque and speed increase, but in different ways. The maximum idling torque of 31.08 Nm is attained with an opening stroke level of 3.8–4.1 mm and a corresponding idling speed of 540.6 RPM. At an opening stroke level of 5.975 mm, the engine reaches its maximum idling speed of 743.8 RPM, and its corresponding idling torque of 24.7 Nm. With this IAC actuator, the engine cannot stall from the torque disturbance or speed variations. Thus, the operating performance of the proposed SMA actuator contributes to the engine's good idling characteristics.

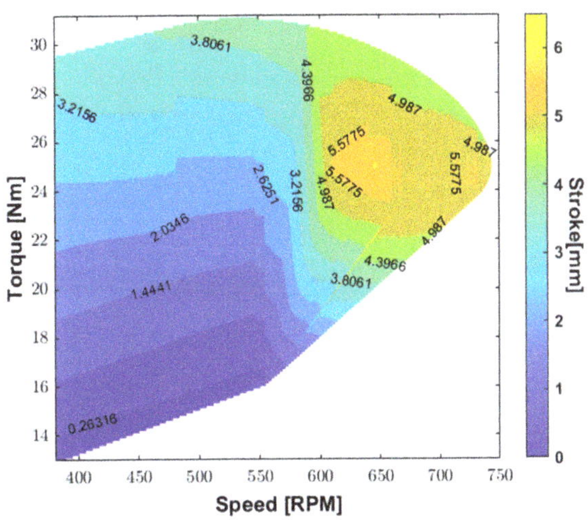

Figure 15. Effect of the SMA-based IAC actuator's operating performance on the engine's idling output characteristics.

4. Conclusions

In this study, a novel SMA-based idle air control actuator for automotive engines was proposed to improve the engine's idling conditions. The innovative idle air control actuator uses shape memory alloy (SMA) smart material as a new technology to activate and control the idling mode of the engine. The inherent characteristics of SMAs, particularly reversible phase transformations in response to thermal changes, were harnessed to create an IAC actuator that is not only simple in structure and compactness but also possesses energy efficiency actuation capabilities with only 2 A of supplied electric current. Through mathematical modelling in analytical analysis and numerical analysis using COMSOL Multiphysics software (R2023b), it was demonstrated that the SMA-based IAC actuator effectively modulates air intake during engine idle mode. This level of idling control contributes significantly to the overall idling efficiency of the engine, addressing key challenges in achieving good performance during idling conditions. The successful control strategy with the PID controller in SMA spring models further enhances the adaptability of the SMA-based IAC actuator's full opening state in 3 s and attaining the engine's idling speed of 743.8 RPM. The operability of the actuator at low stress within multiple cycles permits SMA material to be a viable and robust solution for the demanding conditions of real-world vehicle applications. Finally, the looking forward ongoing research will focus on refining the actuator's design with optimizing control not based on the existing actuators but on the engine's idling conditions, multiple experiments, and practical implementation of this actuator in a real spark ignition engine to validate the simulation results and address any potential challenges associated with its integration into diverse vehicle platforms.

Author Contributions: Conceptualization, P.T., J.W.S. and S.-B.C.; software, P.T.; formal analysis, P.T.; investigation, P.T.; writing—original draft preparation, P.T.; writing—review and editing, S.-B.C. and J.W.S.; supervision, J.W.S. All authors have read and agreed to the published version of the manuscript.

Funding: This work was supported by the National Research Foundation of Korea (NRF) grant funded by the Korea government (Ministry of Science and ICT, MSIT) (No. 2023R1A2C1007973). This research was supported by the MSIT (Ministry of Science and ICT, MSIT), Korea, under the Grand Information Technology Research Center support program (IITP-2024-2020-0-01612) supervised by the IITP (Institute for Information & communications Technology Planning & Evaluation).

Data Availability Statement: Data is contained within the article.

Conflicts of Interest: The authors declare no conflicts of interest.

Appendix A. Mathematical Variables and Their Descriptions

Variable	Description	Variable	Description
$\cdot_b$	Indication of Bias spring	A	SMA spring's area
$\cdot_s$	Indication of SMA spring	ρ_s	Density of SMA spring
C_f	Shear stress correction factor	V_s	The total volume of SMA spring
D_{b_0}	The outer coil diameter of the bias spring	c_p	Specific heat
T_{cc}	Cooling cycle	R	The electric resistance of the SMA spring
d_b	The wire diameter of the bias spring	h_s	Convective coefficient
i_s	Spring index	ι	Operational time
ι_m	Spring's maximum shear stress	T_e	Ambient temperature
$\mathrm{F_p}$	Force from the pressure difference between the inlet and outlet of the bypass	F_b	A mechanical driving force to extend the SMA spring
K_s	Spring constant of SMA spring	T_{hc}	Heating cycle
K_b	Spring constant of bias spring	A_F	Austenite final temperature
F_s	Generated force by SMA spring to compress the bias spring	A_S	Austenite starting temperature
d_s	SMA spring's wire diameter	M_F	Martensite final temperature
D_s	SMA spring's coil mean diameter	M_S	Martensite stating temperature
N_{sc}	SMA spring's number of active coils	ϑ_A	Austenite fraction volume
F_t	Transient flow force	ϑ_M	Martensite fraction volume

x	Spring displacement	ϑ_{dt}	Detwinned martensite fraction volume
$\eta_{1,2}$	Parameters associated with the stress–strain curve	$\varnothing$	Temperature coefficient from material's resistance
ε_w	SMA spring working strain	G_s	SMA spring's shear modulus
ϑ_1	Tensile detwinned martensite	Δ_P	Change in vertical pitch for spring
ϑ_2	Compressive detwinned martensite	$\alpha_{1,2,3}$	Cristal variable speed coefficients of Nitinol
ϵ_{tr}	Transformation strain	ρ_A	Density of air to flow in the intake manifold
Δ_p	Air pressure difference	β_A	Slope of Austenite limit curve
L	Air Flowing length	β_M	Slope of martensite limit curve
C_A	Air flow coefficient	ϑ_{Mi}	Initial martensite fraction volume
D_p	Pintle diameter	ϑ_{Ai}	Initial austenite fraction volume
φ_c	Pintle's conic angle	ϑ_{dti}	Initial detwinned fraction volume
ξ	Transformation modulus	σ_i	Minimum allowable stress
I	Electric current	σ_0	Maximum allowable stress
T_w	Temperature in the SMA spring	ε_i	Minimum strain
S	Actuator's defined stroke	P_s	Initial vertical pitch of the coil spring
$\dot{m}_A$	Air mass flow through pintle of SMA-based IAC actuator	$\dot{m}_{en}$	Air mass flow rate passes through the IAC actuator and induced by the engine
T_{atm}	Ambient temperature	V_{in}	Volume of the intake manifold
P_{atm}	Atmospheric pressure	T_{in}	Temperature in the intake manifold
P_{in}	Air pressure in the intake manifold	λ	Air fuel ratio
R	Air constant	s_t	Stochiometric air-fuel ratio
T_{en}	Engine torque	v_e	Volumetric efficiency
η	Ignition parameter	V_p	Piston displacement volume
η_f	Piston's friction coefficient	ω_{en}	Engine speed
H_l	Heating due to ignition	E_A	Austenite young's modulus
E_{tw}	Twinned young's modulus	E_{dt}	Detwinned young's modulus

References

1. Chakirov, R.; Vagapov, Y. Rapid Control Prototyping Platform for the Design of Control Systems for Automotive Electromechanical Actuators [Internet]. *IFAC Proc. Vol.* **2010**, *43*, 646–651. [CrossRef]
2. Battaglia, A.R.M.; Sellitto, A.; Giamundo, A.; Visone, M. Shape Memory Alloys Applied to Automotive Adaptive Aerodynamics. *Materials* **2023**, *16*, 4832. [CrossRef] [PubMed]
3. Dezaki, M.L.; Bodaghi, M.; Serjouei, A.; Afazov, S.; Zolfagharian, A. Adaptive reversible composite-based shape memory alloy soft actuators. *Sens. Actuators A Phys.* **2022**, *345*, 113779. [CrossRef]
4. Purohit, R.; Patel, K.K.; Gupta, G.K.; Rana, R.S. Development of Ni-Ti Shape Memory Alloys through Novel Powder Metallurgy Route and Effect of Rolling on their properties. *Mater. Today Proc.* **2017**, *4*, 5330–5335. [CrossRef]
5. Kanada, T.; Enokizono, M. A slant type shape memory alloy. *J. Magn. Magn. Mater.* **2000**, *215–216*, 573–575. [CrossRef]
6. Benafan, O.; Gaydosh, D.J. Scale-up of NiTiHf shape memory alloy tubes with high torque capability. *Smart Mater. Struct.* **2019**, *28*, 085035. [CrossRef]
7. Copaci, D.; Blanco, D.; Moreno, L.E. Flexible shape-memory alloy-based actuator: Mechanical design optimization according to application. *Actuators* **2019**, *8*, 63. [CrossRef]
8. Song, G.; Kelly, B.; Agrawal, B.N. Active position control of a shape memory alloy wire actuated composite beam. *Smart Mater. Struct.* **2000**, *9*, 711–716. [CrossRef]
9. Fann, K.J.; Su, J.Y. Cold Forming of Ni-Ti Shape Memory Alloy Sheet. *IOP Conf. Ser. Mater. Sci. Eng.* **2018**, *317*, 012079. [CrossRef]
10. Costanza, G.; Tata, M.E. Shape Memory Alloys for Aerospace, Recent Developments, and New Applications: A Short Review. *Materials* **2020**, *13*, 1856. [CrossRef]
11. Prendota, W.; Goc, K.; Strączek, T.; Yamada, E.; Takasaki, A.; Przewoźnik, J.; Radziszewska, A.; Uematsu, S.; Kapusta, C. Properties of NiTi shape memory alloy micro-foils obtained by pulsed-current sintering of Ni/Ti foils. *Metals* **2019**, *9*, 323. [CrossRef]
12. Tomus, D.; Tsuchiya, K.; Inuzuka, M.; Sasaki, M.; Imai, D.; Ohmori, T.; Umemoto, M. Fabrication of shape memory TiNi foils via Ti/Ni ultrafine laminates. *Scr. Mater.* **2003**, *48*, 489–494. [CrossRef]
13. Turabimana, P.; Sohn, J.W.; Choi, S.B. A Novel Active Cooling System for Internal Combustion Engine Using Shape Memory Alloy Based Thermostat. *Sensors* **2023**, *23*, 3972. [CrossRef]
14. Spaggiari, A.; Spinella, I.; Dragoni, E. Design equations for binary shape memory actuators under arbitrary external forces. *J. Intell. Mater. Syst. Struct.* **2013**, *24*, 682–694. [CrossRef]
15. Bhaskar, J.; Sharma, A.K.; Bhattacharya, B.; Adhikari, S. A review on shape memory alloy reinforced polymer composite materials and structures. *Smart Mater. Struct.* **2020**, *29*, 073001. [CrossRef]
16. Mizzi, L.; Hoseini, S.F.; Formighieri, M.; Spaggiari, A. Development and prototyping of SMA-metamaterial biaxial composite actuators. *Smart Mater. Struct.* **2023**, *32*, 035027. [CrossRef]
17. Serrano, D.; Copaci, D.S.; Moreno, L.; Blanco, D. SMA based wrist exoskeleton for rehabilitation therapy. In Proceedings of the 2018 IEEE/RSJ International Conference on Intelligent Robots and Systems (IROS), Madrid, Spain, 1–5 October 2018; pp. 2318–2323. [CrossRef]
18. Hamid, Q.Y.; Hasan, W.Z.W.; Hanim, M.A.A.; Nuraini, A.A.; Hamidon, M.N.; Ramli, H.R. Shape memory alloys actuated upper limb devices: A review. *Sens. Actuators Rep.* **2023**, *5*, 100160. [CrossRef]

19. Liu, Q.; Ghodrat, S.; Jansen, K.M.B. Modelling and mechanical design of a flexible tube-guided SMA actuator. *Mater. Des.* **2022**, *216*, 110571. [CrossRef]
20. Jiang, S.; Zhang, Y.; Zheng, Y.; Li, C. Deformation mechanism of hot spinning of NiTi shape memory alloy tube based on FEM. *J. Wuhan Univ. Technol. Mater. Sci. Ed.* **2012**, *27*, 811–814. [CrossRef]
21. Sohn, J.W.; Ruth, J.S.; Yuk, D.G.; Choi, S.B. Application of Shape Memory Alloy Actuators to Vibration and Motion Control of Structural Systems: A Review. *Appl. Sci.* **2023**, *13*, 995. [CrossRef]
22. Liu, R.; Zhang, C.; Ji, H.; Zhang, C.; Qiu, J. Training, Control and Application of SMA-Based Actuators with Two-Way Shape Memory Effect. *Actuators* **2023**, *12*, 25. [CrossRef]
23. Hao, L.; Qiu, J.; Ji, H.; Nie, R. Numerical analysis on shape memory alloy–based adaptive shock control bump. *J. Intell. Mater. Syst. Struct.* **2018**, *29*, 3055–3066. [CrossRef]
24. John, S.; Hariri, M. Effect of shape memory alloy actuation on the dynamic response of polymeric composite plates. *Compos. Part A Appl. Sci. Manuf.* **2008**, *39*, 769–776. [CrossRef]
25. Koh, J.S. Design of shape memory alloy coil spring actuator for improving performance in cyclic actuation. *Materials* **2018**, *11*, 2324. [CrossRef] [PubMed]
26. Soother, D.K.; Daudpoto, J.; Chowdhry, B.S. Challenges for practical applications of shape memory alloy actuators. *Mater. Res. Express* **2020**, *7*, 073001. [CrossRef]
27. Bhatt, N.; Gurung, H.; Soni, S.; Singla, A. Effect of biasing conditions on the performance of a SMA spring actuator under thermo-mechanical loading. *Mech. Adv. Mater. Struct.* **2022**, *29*, 4599–4613. [CrossRef]
28. Song, G.; Ma, N.; Lee, H.J.; Arnold, S. Design and control of a proof-of-concept variable area exhaust nozzle using shape-memory alloy actuators. *Smart Mater. Struct.* **2007**, *16*, 1342–1347. [CrossRef]
29. Fawcett, J.N.; Burdess, J.S. Application of a shape memory alloy in a smart thermal clutch or brake. *Proc. Inst. Mech. Eng. Part L J. Mater. Des. Appl.* **1999**, *213*, 21–26. [CrossRef]
30. Palaniappan, S.K.; Rathanasamy, R.; Chellamuthu, S.; Pal, S.K. Development of Shape Memory Alloy Based Quarter Car Suspension System. *J. Mech. Eng. Res.* **2020**, *3*, 21–24. [CrossRef]
31. Hübler, M.; Nissle, S.; Gurka, M.; Breuer, U. Active aerodynamic components for automotive applications—FRP spoiler with integrated SMA actuation. *ICCM Int. Conf. Compos. Mater.* **2015**, *2015*, 1–6. [CrossRef]
32. Leal, P.B.C.; Savi, M.A. Shape memory alloy-based mechanism for aeronautical application: Theory, optimization and experiment. *Aerosp. Sci. Technol.* **2018**, *76*, 155–163. [CrossRef]
33. Chillara, V.S.C.; Headings, L.M.; Tsuruta, R.; Itakura, E.; Gandhi, U.; Dapino, M.J. Shape memory alloy–actuated prestressed composites with application to morphing automotive fender skirts. *J. Intell. Mater. Syst. Struct.* **2019**, *30*, 479–494. [CrossRef]
34. Sazonov, E.; Daoud, W.A. Grand Challenges in Wearable Electronics. *Front. Electron.* **2021**, *2*, 668619. [CrossRef]
35. Shimoga, G.; Kim, T.H.; Kim, S.Y. An intermetallic niti-based shape memory coil spring for actuator technologies. *Metals* **2021**, *11*, 1212. [CrossRef]
36. Ishii, T. Design of shape memory alloy (SMA) coil springs for actuator applications [Internet]. In *Shape Memory and Superelastic Alloys*; Woodhead Publishing Limited: Cambridge, UK, 2011; pp. 63–76. [CrossRef]
37. An, L.; Huang, W.M.; Fu, Y.Q.; Guo, N.Q. A note on size effect in actuating NiTi shape memory alloys by electrical current. *Mater. Des.* **2008**, *29*, 1432–1437. [CrossRef]
38. Mozhi, G.T.; Dhanalakshmi, K.; Choi, S.B. Design and Control of Monolithic Compliant Gripper Using Shape Memory Alloy Wires. *Sensors* **2023**, *23*, 2052. [CrossRef] [PubMed]
39. Grigorie, T.L.; Botez, R.M. A Self-Tuning Intelligent Controller for a Smart Actuation Mechanism of a Morphing Wing Based on Shape Memory Alloys. *Actuators* **2023**, *12*, 350. [CrossRef]
40. Pillai, R.R.; Murali, G.; Gopal, M. Modeling and Simulation of a Shape Memory Alloy Spring Actuated Flexible Parallel Manipulator. *Procedia Comput. Sci.* **2018**, *133*, 895–904. [CrossRef]
41. Qian, H.; Li, H.; Song, G.; Guo, W. A constitutive model for superelastic shape memory alloys considering the influence of strain rate. *Math. Probl. Eng.* **2013**, *2013*, 248671. [CrossRef]
42. Kittinanthapanya, R.; Sugahara, Y.; Matsuura, D.; Takeda, Y. Development of a novel SMA-driven compliant rotary actuator based on a double helical structure. *Robotics* **2019**, *8*, 12. [CrossRef]
43. Turabimana, P. Design of Automobile Concrete Pump. *IOSR J. Mech. Civ. Eng.* **2019**, *15*, 1–5. [CrossRef]
44. Meng, L.; Luo, J.; Yang, X.; Zeng, C. Intake air mass observer design based on extended Kalman filter for air-fuel ratio control on SI engine. *Energies* **2019**, *12*, 3444. [CrossRef]
45. Amaral, A.A.D.C.; Saidatin, N.; Mahmud, R.; Maulana, H.S. Experimental Study of Spark Ignition Engine Performance under Different Electronic Control Units, Hole Injectors and Spark Plugs. *J. Mech. Eng. Sci. Innov.* **2023**, *3*, 22–29. [CrossRef]
46. Wong, P.K.; Tam, L.M.; Li, K.; Vong, C.M. Engine idle-speed system modelling and control optimization using artificial intelligence. *Proc. Inst. Mech. Eng. Part D J. Automob. Eng.* **2010**, *224*, 55–72. [CrossRef]
47. Gharib, A.R.; Gharib, M.R.; Mousavi, S.A.S.; Gorkani, M.R. Control of Throttle Valve in idle speed condition. In Proceedings of the 2010 International Conference on Modelling, Identification and Control, ICMIC, Okayama, Japan, 17–19 July 2010; pp. 394–399.
48. Shamekhi, A.M.; Taghavipour, A.; Shamekhi, A.H. Engine idle speed control using nonlinear multiparametric model predictive control. *Optim. Control Appl. Methods* **2020**, *41*, 960–979. [CrossRef]

49. Nikzadfar, K.; Noorpoor, A.; Shamekhi, A.H. Design of an optimal idle speed controller for a turbocharged diesel engine using fuzzy logic method. *J. Mech. Sci. Technol.* **2012**, *26*, 2325–2336. [CrossRef]
50. Hashemi, Y.M.; Kadkhodaei, M.; Mohammadzadeh, M.R. Fatigue analysis of shape memory alloy helical springs. *Int. J. Mech. Sci.* **2019**, *161–162*, 105059. [CrossRef]
51. Di Cairano, S.; Yanakiev, D.; Bemporad, A.; Kolmanovsky, I.V.; Hrovat, D. Model predictive idle speed control: Design, analysis, and experimental evaluation. *IEEE Trans. Control Syst. Technol.* **2012**, *20*, 84–97. [CrossRef]
52. Yin, F.; Sun, J.; Wang, W. *Engine Idle Speed Control Research Status and Development Trend*; no. Icmmcce; Atlantis Press: Berlin/Heidelberg, Germany, 2015; pp. 1884–1888. [CrossRef]
53. Shulga, E.; Lanusse, P.; Airimitoaie, T.B.; Maurel, S.; Trutet, A. Model Predictive Control of engine intake manifold pressure with an uncertain model. *IFAC-PapersOnLine* **2021**, *54*, 335–340. [CrossRef]
54. Wei, H.; Chen, Y.L.; Yu, W.; Su, L.; Wang, X.; Tang, D. Study on corrosion resistance of high-strength medium-carbon spring steel and its hydrogen-induced delayed fracture. *Constr. Build. Mater.* **2020**, *239*, 117815. [CrossRef]
55. Nov, Z.; Salvetr, P.; Kotous, J.; Moty, P.; Gokhman, A. Enhanced Spring Steel's Strength Using Strain Assisted Tempering. *Materials* **2022**, *15*, 7354. [CrossRef] [PubMed]
56. Mirzaeifar, R.; Desroches, R.; Yavari, A. A combined analytical, numerical, and experimental study of shape-memory-alloy helical springs. *Int. J. Solids Struct.* **2011**, *48*, 611–624. [CrossRef]
57. Zhang, B.; Zeng, S.; Tang, F.; Hu, S.; Zhou, Q.; Jia, Y. Experimental and numerical analysis of the mechanical properties of a pretreated shape memory alloy wire in a self-centering steel brace. *Processes* **2021**, *9*, 80. [CrossRef]
58. Yildiz, Y.; Annaswamy, A.M.; Yanakiev, D.; Kolmanovsky, I. Spark-ignition-engine idle speed control: An adaptive control approach. *IEEE Trans. Control Syst. Technol.* **2011**, *19*, 990–1002. [CrossRef]

applied sciences

MDPI

Article

High-Efficiency Navigation of Nonholonomic Mobile Robots Based on Improved Hybrid A* Algorithm

Zhaozhan Chi [1], Zhenhan Yu [2], Qianyu Wei [3], Qiancheng He [1], Guangxian Li [1,4,*] and Songlin Ding [4,*]

1 School of Mechanical Engineering, Guangxi University, Nanning 530004, China
2 School of Public Policy and Management, Guangxi University, Nanning 530004, China
3 School of Journalism and Communication, Guangxi University, Nanning 530004, China
4 School of Engineering, RMIT University, Victoria 3083, Australia
* Correspondence: liguangxian@gxu.edu.cn (G.L.); songlin.ding@rmit.edu.au (S.D.)

Abstract: With the development of automation technologies, autonomous robots are increasingly used in many important applications. However, precise self-navigation and accurate path planning remain a significant challenge, particularly for the robots operating in complex circumstances such as city centers. In this paper, a nonholonomically constrained robot with high-precision navigation and path planning capability was designed based on the Robot Operating System (ROS), and an improved hybrid A* algorithm was developed to increase the processing efficiency and accuracy of the global path planning and navigation of the robot. The performance and effectiveness of the algorithm were evaluated by using randomly constructed maps in MATLAB and validated in a practical circumstance. Local path planning and obstacle avoidance were carried out based on the model predictive control (MPC) theory. Compared with the conventional A* + DWA (dynamic window approach) method, the average searching time was reduced by 12.62~24.5%, and the average search length was reduced by 9.25~9.5%. In practical navigating tests, the average search time was reduced by 18~24%, and the average search length was reduced by 10.3~12%, while the overall path was smoother. The results demonstrate that the improved algorithm can enable precise and efficient navigation and path planning of the robot in complex circumstances.

Keywords: nonholonomic mobile robots; self-navigation; mapping; path planning; hybrid A* algorithm

Citation: Chi, Z.; Yu, Z.; Wei, Q.; He, Q.; Li, G.; Ding, S. High-Efficiency Navigation of Nonholonomic Mobile Robots Based on Improved Hybrid A* Algorithm. *Appl. Sci.* **2023**, *13*, 6141. https://doi.org/10.3390/app13106141

Academic Editors: Oscar Reinoso García, Fabrizio Giulietti, Nadjim Horri and William Holderbaum

Received: 29 March 2023
Revised: 11 May 2023
Accepted: 12 May 2023
Published: 17 May 2023

1. Introduction

With the development of artificial intelligent and automation technologies, mobile robots with self-navigation capabilities are gradually replacing conventional machines such as manually operated vehicles. However, the working areas and traffic conditions for autonomous mobile robots could be complex, e.g., in the central business district. To work efficiently in these environments, an autonomous robot should have three functions [1]: (1) global planning capability under different working conditions, which enables the robot to plan a feasible global path efficiently according to the local surroundings; (2) tracking and controlling capability to ensure high precision in the movement based on the path planning; and (3) rapid response to environmental constraints and planning of new feasible paths in a time when the local environments are changed due to the passing-by vehicles or pedestrians [2].

The planning of the path can be divided into global path planning and local path planning. Global path planning is designing the route of the robot in a fully perceived environment. It is the process of constructing a two-dimensional or three-dimensional map of the known environment with specific constraints [3] and target information [4]. For the global path planning, two classical searching methods are often used to detect the positions and areas of obstacles: (1) searching algorithms based on random samplings, such as the probabilistic road-map (PRM) [5], rapidly exploring random tree (RRT) method [6], and RRT algorithm based on convolutional neural network (CNN) [7]; and (2) graph search-based algorithms, such as the Dijkstra [8] and extended Dijkstra algorithms [9]. The Dijkstra algorithm is characterized by its ability to guarantee that the obtained path will be the shortest. However, it needs to scan all the nodes in the map, and the number of nodes is generally immense. This results in a significant increase in computation and a potential decrease in solution rate, while also consuming significant memory space.

To increase the scanning/searching efficiency, A* algorithm was proposed by using a heuristic function as the search-guided approach [10,11]. The mathematical principle of the A* algorithm is presented in Equation (1). To determine the priority of the next node $f(n)$ in the scanning/searching process, two cost functions, $g(n)$ and $h(n)$, are used in the equation: $g(n)$ represents the cost from the current node to the start node of the path, and $h(n)$ represents the estimated cost from the next node to the end node of the path.

$$f(n) = g(n) + h(n) \tag{1}$$

Based on the A* algorithm, some improved algorithms were developed recently. For example, Harabor and Grastien developed the jump point search (JPS) algorithm [12] by storing nodes in the open list to improve search efficiency. Li et al. proposed self-adaptive learning particle swarm optimization (PSO) with different learning strategies [13] to improve the search capability of PSO. With the development of nonholonomically constrained robots, path planning becomes more and more important for robots, such as the Ackermann mobile robot chassis. However, the A* algorithm is not effective in planning path for nonholonomically constrained mobile platforms. When planning the paths for such robots, post-processing and smooth optimization processes are required before the application of the paths because such robots are only feasible to paths that satisfy their kinematic constraints, and they are not capable of tracking all the continuous paths. As shown in Figure 1, the Ackermann robots cannot traverse angles greater than 90° (such as the red robot in Figure 1a) in a limited range when confronting such obstacles (the blue grid) in real-world scenarios. Therefore, flexible path planning should be used to avoid potential collisions (Figure 1b).

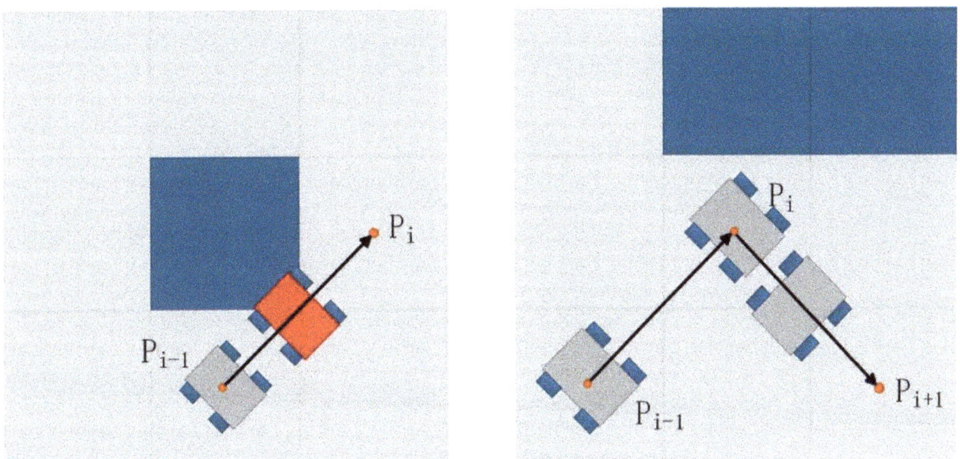

(a) (b)

Figure 1. The collision schematic diagram. (**a**) The collision of Ackermann chassis due to the limitation of transverse angle. (**b**) Avoiding collision via flexible path planning.

In addition to the global path planning, local path planning is also critical when the working environment keeps changing dynamically. The local dynamic path planning is mainly divided into two approaches [14]. The first approach involves the detection of new obstacles in the environment and obtaining space information. The robot then conducts replanning at the navigation planning level to obtain a new collision-free path that is sent to the underlying tracking controller for executing movement. The typical methods in this category include the artificial potential field (APF) method, grid planning method, and others. However, these methods require replanning and outputting control commands, resulting in a large amount of computation, and cannot meet the real-time requirements. The other approach to local path planning at the motion control level involves considering the nonholonomic-constraint characteristics of the robot. The primary concept is to establish an optimal control problem in the unit-time step when facing new obstacles and solve the optimization problem while obtaining the optimal trajectory and the corresponding control input. The primary representative methods of this approach include the dynamic window approach (DWA) [15], time-elastic belt (TEB) approach [16], and local planning based on model predictive control (MPC) [17] which is widely used for avoiding collisions [18]. Yi Gan et al. proposed the A1-0 Bg-RRT algorithm to reduce computational time in complex environments [19]. Zhang et al. proposed the algorithm for ships with INS-aided shipborne GNSS, which considered the measurement update from one-dimensional speed [20]. Zhang et al. introduced the Bézier curve method to smooth the routes [21]. However, when applied to nonholonomic constraints, these algorithms show slow navigation speed and can generate non-optimal paths.

Nonholonomic mobile robots such as Ackermann architecture mobile robots, single-wheeled inverted pendulum robots and three-wheeled omnidirectional mobile robots have fewer controlled variables than state variables during motion, which can easily lead to motion slipping problems. How to control nonholonomic mobile robots to reach the target location and face the target direction are a challenging task, and extensive research has been conducted in this area. For example, Karl Worthmann et al. [22] proposed a predictive control scheme based on tailored nonquadratic stage cost to address this issue. Dang et al. [23] proposed a novel evaluation function based on the Theta-star search algorithm. The new function can generate waypoints with an empirically averaged distance to obstacles. In 2020, Li et al. [24] proposed a graph-based multi-agent path planner. The planner can effectively generate optimal collision-free trajectories for multiple nonholonomic mobile robots in environments with multiple obstacles. Johnson et al. [25] proposed the dynamic motion planning networks (Dynamic MPNet) algorithm. They modified the planning network to achieve real-time planning for nonholonomic robot motion planning problems, instead of simply relaxing constraints. By adjusting the steering of the robot through its path planner or driver, Gonon et al. [26] developed the method to assist navigation in dense crowds.

However, although these path planning algorithms can accommodate nonholonomic constraints in robots, many inherent issues exist, e.g., suboptimal operating speed, failure in producing smoother paths with higher accuracy, and the randomly occurred excessive searches. This is because the heuristic function $h(x)$ of the conventional A* algorithm is generally the Euclidean distance, which traverses a very large number of unnecessary nodes in planning the path and significantly limits the efficiency of navigation. To optimize the path planning, an improved hybrid A* algorithm was proposed in this paper with the intension to improve the performance of autonomous navigation of mobile robots with nonholonomic chassis. The Chebyshev distance [27] was used to weight the heuristic function in the global path planning, and the local path planning algorithm was improved as an MPC local planner. The performance of the algorithm was tested in MATLAB and validated in practical navigation experiments, and the conventional A* algorithm was used as comparison to demonstrate the effectiveness of the new algorithm. By comparing the paths planned for maps with different obstacles, it was found that shorter and smoother paths were generated via the improved A* algorithm due to the application of the Chebyshev distance, which increased the efficiency and accuracy of the overall path planning. The results of this study provide a practical method for the efficient autonomous navigation of robots with the nonholonomic-constraint mobile platform, e.g., the pit robot with Ackermann chassis.

2. Path Planning for Nonholonomic Robots

2.1. Map Construction and Robot Positioning

Simultaneous localization and mapping (SLAM) [28], which includes laser SLAM, visual SLAM, and multi-sensor fusion SLAM according to the type of sensors employed, was the commonly used method in the construction of maps for different working environments [29]. In this study, laser SLAM was chosen as the localization and mapping method due to its higher accuracy, faster speed, and wide-ranging capabilities with the using of LiDAR. The sensor emits laser signals to perceive the surrounding environment, and the obstacles reflected back to the receiver by the laser are used to construct the map based on the sizes, shapes, and positions of the obstacles. In plotting the map, the geometric mapping approach, the topological mapping approach, and the occupancy grid mapping approach are the commonly used methods. In this study, the occupancy grid map approach, which plots the map by small squares, was employed. The number of squares represents the size of the map, and the probability of the obstacle in a square is higher when the number is larger. Then, the plotted map was meshed, and the values in the range of 0–255 were defined to each cell presenting different occupied characters (vacancy, obstacles, etc.). The computer uses this feature to discriminate obstacles. A two-dimensional map was constructed digitally via the G-mapping, which is a widely used SLAM algorithm in ROS. This algorithm is an improvement upon the Rao-Blackwellized particle filter (RBPF) approach, which prioritizes positioning before map construction.

2.2. Robot Positioning

The positions of a robot can be determined via the SLAM algorithm which utilizes the probability distribution approach; particles are used to represent an estimation of the robot's position based on sensor measurements. However, the algorithm could generate a very large number of particles, leading to high computational load and time delay. Furthermore, the SLAM algorithm is less effective in addressing the robot kidnapping problem, i.e., when the robot is suddenly moved to a different location or experiences tire skidding. As a result, the positioning of robot was implemented via the AMCL algorithm, which utilizes a particle filter to calculate the robot's position according to the data from both the odometer and LiDAR sensors. Moreover, AMCL can be applied in solving both global and local localization problems to reduce the workload required for subsequent navigation algorithm. With regard to robot kidnapping, AMCL incorporates weights determined by the information from short-term observation and long-term observation. These weights enable more reasonable particle number and time allocation, as expressed in Equations (2) and (3), respectively:

$$\mathrm{w}_{\mathrm{long\_t+1}} = \mathrm{w}_{\mathrm{long\_t}} + a_{\mathrm{long}}\left(\mathrm{w}_{\mathrm{avg}} - \mathrm{w}_{\mathrm{long\_t}}\right) \tag{2}$$

$$\mathrm{w}_{\mathrm{short\_t+1}} = \mathrm{w}_{\mathrm{short\_t}} + a_{\mathrm{short}}\left(\mathrm{w}_{\mathrm{avg}} - \mathrm{w}_{\mathrm{short\_t}}\right) \tag{3}$$

where w_i is the observation average function, a_i is the average weight coefficient determined by the observed information, i has three states: long-term, short-term, and average. In the AMCL algorithm, the probability of increasing particle number is related to the two weight coefficients. The weights were updated by iteration, and the values were typically within the range of 0.0010~0.01. The probability of increasing the particle number can be calculated using the following equation:

$$\mathrm{P} = \max\left\{0.0, 1.0 - \frac{\mathrm{w}_{\mathrm{long}}}{\mathrm{w}_{\mathrm{short}}}\right\} \tag{4}$$

It can be seen from Equation (4) that the increase in the particle number is related to the value of $\frac{\mathrm{w}_{\mathrm{long}}}{\mathrm{w}_{\mathrm{short}}}$. When the weight of short-term information is less than that of long-term information, which indicates that the robot has deviated from the predetermined trajectory, the system will increase the number of random particles. At this stage, the robot's self-positioning capability can be achieved more quickly and accurately by matching a larger number of particles. The visualized process of the robot's self-positioning process via AMCL algorithm is illustrated in Figure 2. Through multiple updates and resampling of the particle set, the particle filter can adjust the weight and position of the particles to reflect the probability distribution of the robot's position. The location where particles appear more frequently is where the robot is more likely to be located.

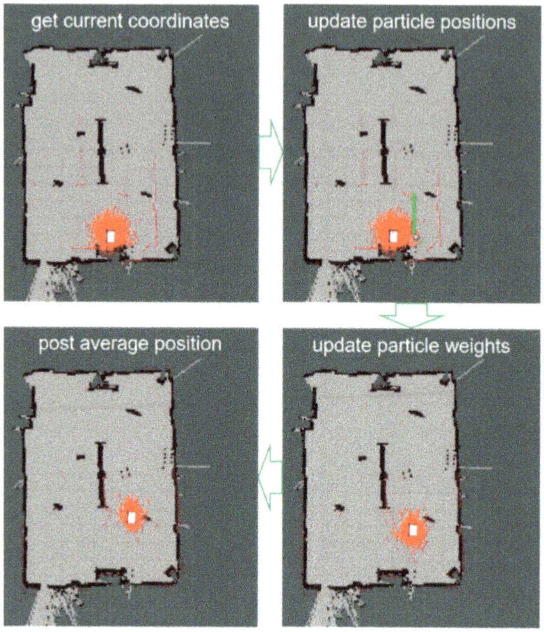

Figure 2. Robot self-positioning.

2.3. *Path Planning Based on Improved Hybrid A* Algorithm*

2.3.1. Grid Environment Modeling

The planning of paths was determined according to the positions of vacancies and obstacles on the constructed maps. As introduced in Section 2.1, grid mapping approach was used for the construction of the map, and the map was meshed into grids with certain numbers presenting starting point, end point, obstacles, security areas, and other relevant information. As shown in Figure 3, the black regions in the grid map represent obstacle areas, while the white regions represent areas which are available for travel, and both areas can be updated in real time according to the positions of obstacles and vacancies when the environment is changed.

2.3.2. Global Path Planning Based on Improved Hybrid A* Algorithm

Robot path planning consists of global path planning and local path planning. As mentioned before, the processing efficiency of conventional A* algorithm is limited when the areas of the obstacles are complex. Therefore, the hybrid A* algorithm [30] was used for the global path planning. The algorithm improved the specification of the heuristic function and the searching method to achieve better results. For the heuristic search-based path planning, the effectiveness of the heuristic function $h(x)$ is critical. In conventional A* algorithm, $h(x)$ is usually the Euclidean distance between the two nodes (x_n, y_n) and (x_g, y_g), as presented in Equation (5).

$$h(x) = \sqrt{(x_n - x_g)^2 + (y_n - y_g)^2} \tag{5}$$

However, this approach causes the traversal of unnecessary nodes and an increase in the redundancy of the path planning when the distance calculated via $h(x)$ is smaller than the actual path. To address the limitations of the Euclidean-distance heuristic function, Chebyshev distance d_q, which was expressed in Equation (6), was applied. The improved algorithm is presented in Equation (7).

$$d_q = max\{|x_n - x_g|, |y_n - y_g|\} \quad (6)$$

$$(\mathrm{n}) = \mathrm{g}(\mathrm{n}) + \mathrm{w}\cdot h(\mathrm{n}) \quad (7)$$

$$\mathrm{w} = \exp\left(\frac{1}{\alpha \mathrm{d_n} + \beta}\right) \quad (8)$$

$$d_n = \frac{max\{|x_n - x_s|, |y_n - y_s|\}}{max\{|x_g - x_s|, |y_g - y_s|\}} \quad (9)$$

Value d_n in Equation (9) is the ratio of the Chebyshev distance from the starting point of the current node to the full Chebyshev distance. Value w is the weighting coefficient, which dynamically weights the heuristic function. Values (x_s, y_s) are the horizontal and vertical coordinates of the starting point, (x_g, y_g) are the horizontal and vertical coordinates of the target point, and (x_n, y_n) are the horizontal and vertical coordinates of the current node. The weighting coefficient α (typically 5~9) is used to obtain better results, depending on the distribution of obstacles in the map. The obstacle ratio β represents the proportion of the area occupied by obstacles to the total area of the map. By weighting the heuristic function with the Chebyshev distance (Equation (8)), a balance between planning efficiency and effectiveness can be achieved; it reduced the gap between the heuristic function $h(x)$ and the actual path length, the number of nodes traversed, and eventually the search time.

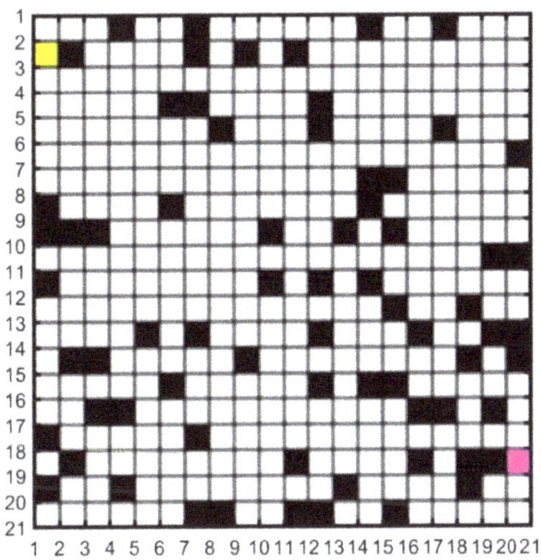

Figure 3. A virtual map generated via the grid mapping approach: the dark squares are obstacles; yellow and purple squares are the starting and end points, respectively.

The smoothing of the path is necessary for the vehicle (the Ackermann-steered mobile platform) and the position of each target point, especially the position of the vehicle at the starting and ending points on the path, which should be appropriately designed in order to minimize the total length of the path. The determination of the position of the vehicle at each point is presented in Figure 4. If the angle formed by a target vertex and its two adjacent points is less than 90° (Figure 4), the scheme of entering and exiting the opposite direction with an angular bisector position will be adopted. On the other hand, if the angle is greater than 90° (Figure 4), the one-way entry–exit method with the stop position being the vertical line of the angular bisector line will be adopted. This approach allows for smoother movement and efficient transverse of the vehicle.

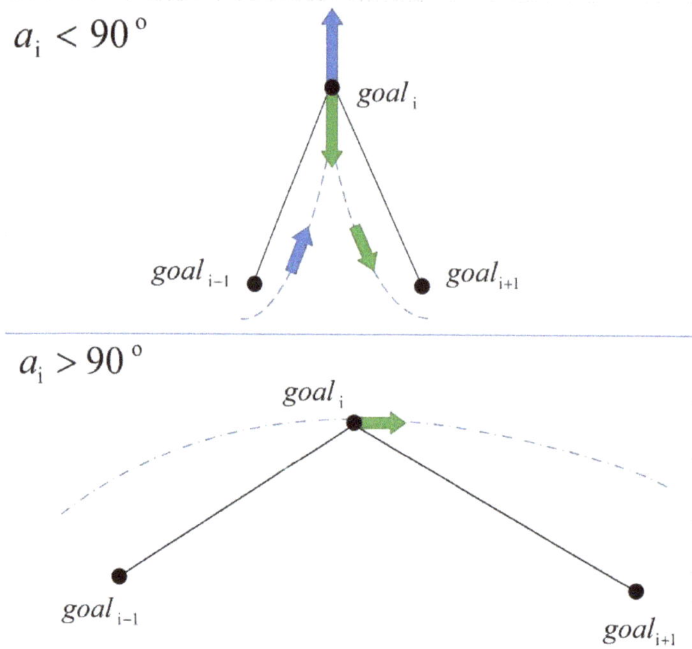

Figure 4. Robot motion path diagram in global planning.

To improve the searching efficiency and meanwhile satisfy the condition of the heuristic function, Reeds-Shepp curve was applied at the end of the global path planning process. The Reeds-Shepp curve is a cost-optimized path curve that has been proven to be optimal for nonholonomically constrained robots in a general-purpose environment free from obstacles [31]. During the global path planning, the improved hybrid A* algorithm with Chebyshev distance attempted to generate a Reeds–Shepp curve from the current node to the target node if a node was expanded during the loop search. If such a curve was generated between the two points without encountering any obstacles, the curve would be combined with the result in the closed list to generate the optimal path. The adoption of Reeds–Shepp curve significantly increased the efficiency of the algorithm.

The flowchart depicting the final global planning algorithm is presented in Figure 5. Figure 6 shows the pseudocode of the global path planning algorithm in the improved hybrid A* algorithm.

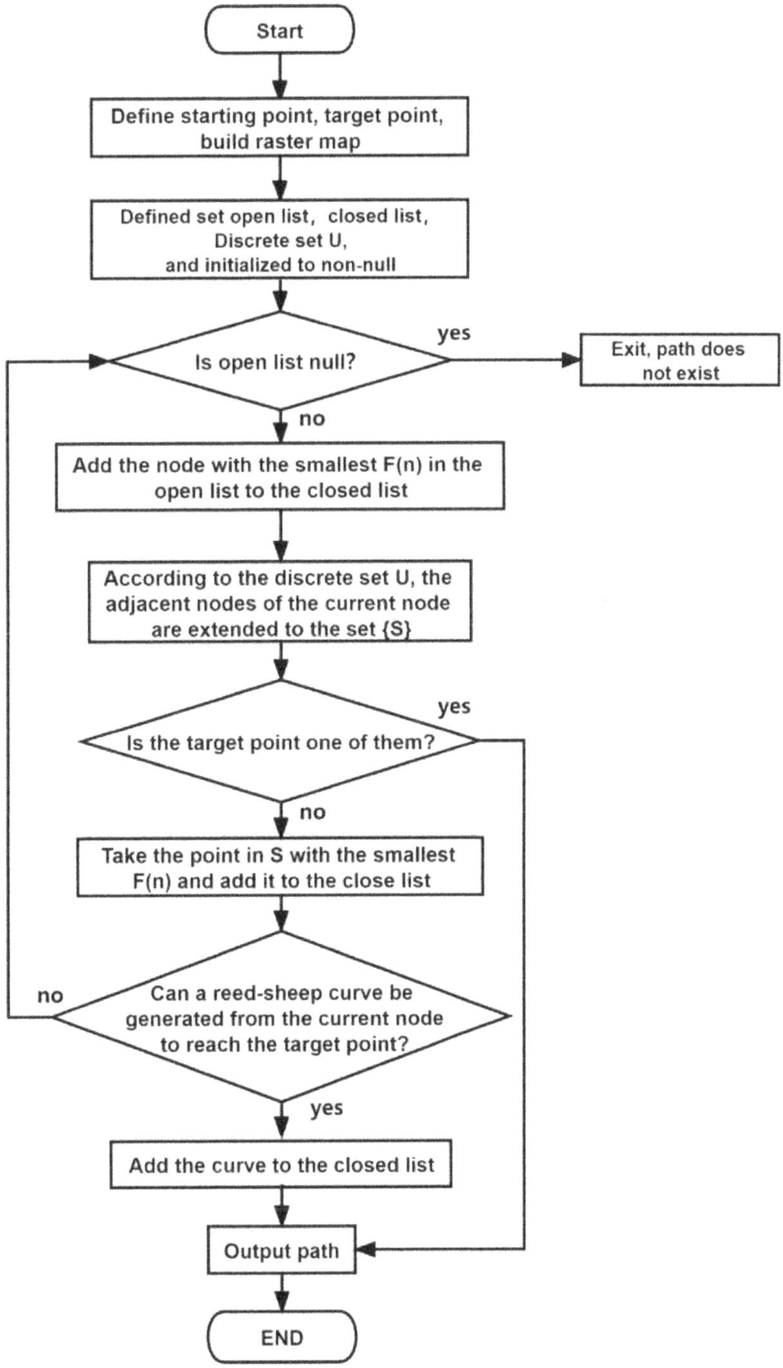

Figure 5. Flowchart of global path planning algorithm.

```
input:
Cumulative cost function g(n), Heuristic function h(n), Dynamic weighting coefficient w, Origin state S_t, Target point state S_e, Child nodes extend the discrete space set {U}
Output:
Optimal globally planned path node sequence closed list
initialization:
A matrix set is built to store all nodes to be expanded   open list←{S_t}
Build a heuristic function for each node in the map {H}
Initializes the cumulative cost function for all nodes g(S_t)←{0}
Child nodes extend the discrete space set {U}←{u}&{w}
A collection of adjacent nodes {S}
Start of algorithm:
While openlist ≠ φ do
    The F(n) minimum cost node in the open list is displayed
    Expand adjacent nodes according to discrete space set
  If  the target point belongs to a set of adjacent nodes, return closed list;
        break;
  END If
    For i=1 to size(closed list) do
      if g(S_n)=infinite
        Push node S_n into the closed list
      END If
     END For
    If  S_n can be connected to S_e by a reed-sheep curve
     RES=reed-sheep(S_n, S_e, R_min)
    return {closed list+RES};break;
    END IF
  End while
Return {s}
```

Figure 6. Pseudocode of global path planning algorithm.

2.3.3. Local Path Planning Based on MPC Theory

Global path planning is generally constructed for one time at the initial of the path planning. In practice, the environment could be dynamically changed during the operation of the robot. Therefore, it is necessary to update the local path according to the changed environment and incorporate such local path planning into the total path planning process [32]. The real-time update of the environment changes (e.g., new obstacles or changes in the existed trajectory) can be realized by the MPC-based local path planning. MPC is a specialized form of control that generates the current control action by solving an open-loop optimal control problem within a finite time horizon [33,34]. It provides a more adaptive and robust control strategy for mobile robots, ensuring safe and efficient navigation in dynamic environments. In the improved hybrid A* algorithm, the total cost $f(n)$ of a node n is expressed as $f(n) = g(n) + h(n)$, where $g(n)$ and $h(n)$ are the actual cost from the start of the path to node n and the estimated cost from node n to the goal. Specifically, $g(n)$ is the stage cost determined by the current state and controlling input, which is calculated using the following formula: $g(n) = g(\text{parent}) + c(\text{parent}, n)$, where $g(\text{parent})$ is the actual cost from the start of the path to the parent of node n, and c(parent, n) is the actual cost from the parent to node n. Value $h(n)$ is the terminal cost according to the predicted future state, and it is obtained by using a heuristic cost estimate function that estimates the cost from node n to the goal. As a result, it can be seen that the optimal control via MPC is directly influenced by $g(n)$ and $h(n)$. With regard to Ackermann steering chassis, the MPC is particularly suitable for local path planning because it can effectively handle the nonholonomic constraints that are unique to the chassis. As shown in Figure 7, MPC can plan a path that takes into account these constraints and also has a feedforward system that includes a prediction function, allowing for dynamic obstacle prediction during local path planning.

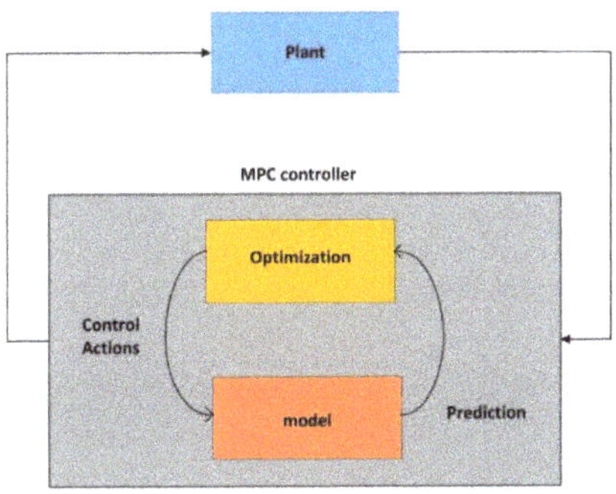

Figure 7. Schematic diagram of the MPC principle.

In the MPC strategy, the constraint conditions of dynamic obstacles are fully considered, reflecting the optimization of the robot motion state and prediction of possible collision situations between the robot and dynamic obstacles at future time steps. Adjustments are made based on these constraints, ensuring safe and efficient navigation. At a given sampling moment, the geometric visualization of the entire constraint condition of dynamic obstacles is illustrated in Figure 8.

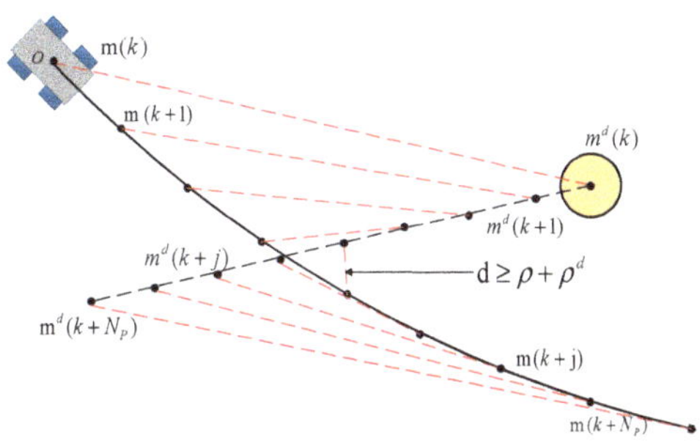

Figure 8. Local path planning diagram based on MPC.

The mathematical expression of the obstacle constraint on the robot position space $m(k)$ at the current time k can be obtained. Equations (10) and (11) show the criteria of collision elimination of a robot when confronting the static obstacle and the dynamic obstacle with constant motion state: the distance d between the center points of the two obstacles should be larger than the sum of the expansion radius in all predicted states.

$$\|m(k+j) - m^{os}\|_2 \geq r^{os} + \rho, \forall j = \{0, 1, 2, ..., N_p\} \quad (10)$$

$$\left\|m(k+j) - m^{d}(k+j)\right\|_2 \geq \rho^{d} + \rho, \forall j = \{0, 1, 2, ..., N_p\} \quad (11)$$

where r^{os} and ρ^d are the expansion radius for the static and dynamic obstacles, respectively, j represents a certain time, m^{os} and m^d represent the position space of static obstacle and dynamic obstacle, m represents the spatial position of the robot, and ρ is the expansion radius of the robot.

3. Simulation and Experiment

3.1. Simulation of Map Construction

To evaluate the effectiveness and efficiency of the improved hybrid A* algorithm, a computer simulation was carried out via MATLAB 2018b. The hardware/software environments are listed in Table 1. The following reasonable assumptions were considered in the simulation:

(1) The robot can perform turns within the reaction time, and the turning angle does not exceed its maximum turning angle while moving in the map, without considering factors such as friction and other resistances.

(2) The robot's orthographic projection area occupies one quarter of a grid. In a map with an obstacle ratio of 0.2, the value of variable α is 5, and the value of variable β is 0.2. In a map with an obstacle ratio of 0.3, the value of variable α is 6, and the value of variable β is 0.3.

These assumptions fit the definition of our model. The Chebyshev distance from start to finish is 8 m.

Grid maps with obstacle proportions of 0.2 and 0.3 (the ratio of the obstacle number to the total number of grids) was used as the comparative experiments. The simulation process is the random generation of a grid graph by MATLAB using the A* algorithm and improved algorithm for path planning. Four experiments were carried out on each map with different ratio of obstacles, and the time and length of each planned path were calculated. The grid map used in the simulation was 20 × 20, with a size of 0.5 m × 0.5 m per grid. The information of the hardware is listed in Table 1. The model of the GPU used in this study is NVIDIA GTX1650 with 896 CUDA cores and the core frequency of 1485 MHz. The SLAM and relevant machine-learning algorithms were processed via CUDA toolkit, and the processing efficiency is directly determined by the core frequency and core number of the GPU. Since GPUs have a large number of cores, which allow for better computation of multiple parallel processes, they can process multiple computations simultaneously. Therefore, the calculation running on a high-performance GPU can have a significantly shorter time compared to those running on the CPU.

Table 1. Hardware environment for simulation.

Operating System	CPU	RAM	GPU
64 bit Windows 10	Intel(R) Core (TM) i5-9300H	8.0 GB	GTX1650

3.2. Physical Mock-Ups and the Controlling System

A robot which consisted of an Ackermann steering chassis, a three-degree-of-freedom mechanical arm, and an actuator was fabricated (Figure 9a). Specifically, the chassis was responsible for tasks such as target navigation and path planning, while the mechanical arm performed track planning assisted with an auxiliary actuator and camera-based target positioning. The Ackermann chassis was equipped with LiDAR, a binocular camera, a control box, a manipulator, a motor, and a drive which facilitated the autonomous driving and operation of the robot. The controlling system of the autonomous robot was made up by a upper computer and a lower computer, as shown in Figure 9b. The path planning algorithm was stored and performed in the upper computer which consisted of a PC terminal with an Intel (R) Core (TM) i5-9300H CPU @ 2.30 GHz processor and a microcontroller with the Jetson Nano demo board. The PCD terminal performed decision-making and information processing, whereas the Jetson Nano controlled the radar and the camera and facilitated communication between the components. The lower computer comprised two STM32 demo boards. The STM32 F103 received commands sent by the upper computer and controlled the movement of chassis. This demo board was also responsible for information collection from odometer and responded back to the upper computer for navigation. The STM32 F405, along with the related drive, received commands from the upper computer to control the mechanical arm according to the track.

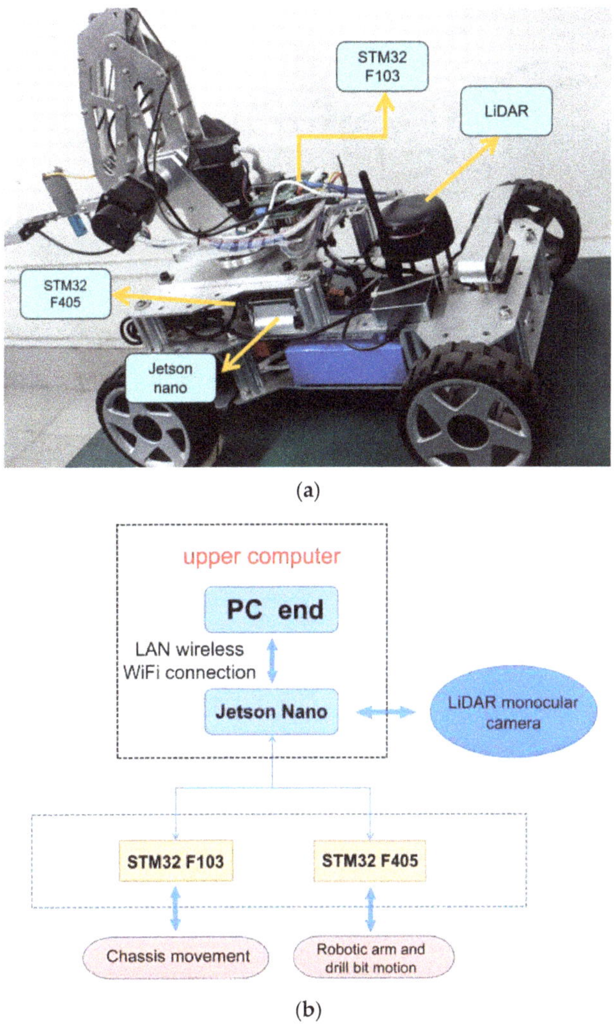

Figure 9. (**a**) The assembly of the robot. (**b**) The working scheme of the controlling system.

3.3. Navigation Experiment

The practical navigation experiment was conducted at a 10 m × 10 m testing site, and cardboard boxes and chairs were allocated within the site to imitate the obstacles (Figure 10). In the experiment, the obstacle ratio of the whole map is 0.15, so β is 0.15, α is 5, and the Chebyshev distance d_q from the starting point to the end point is 7 m.

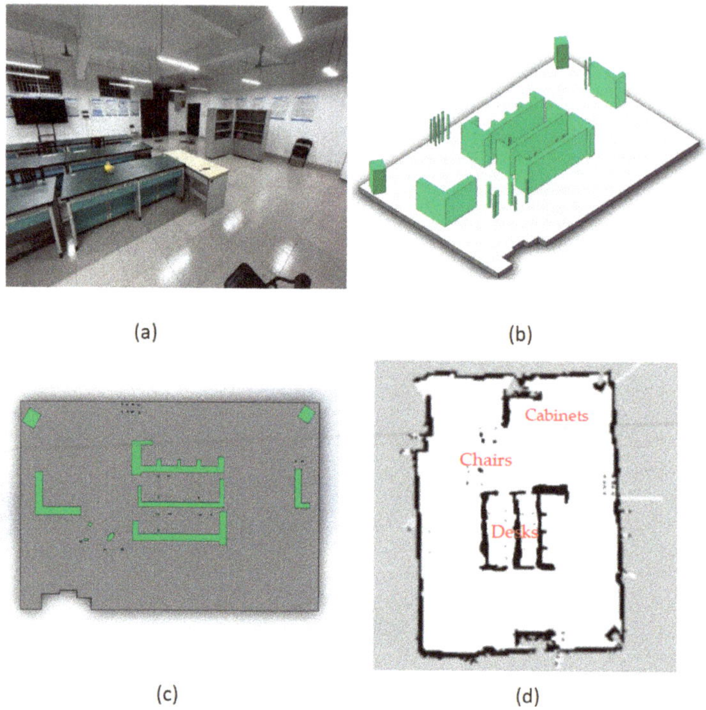

Figure 10. The practical environment. (**a**–**c**) are for navigation experiment and the constructed map. (**d**) practical environment.

We used SLAMTEC A2 radar to build the map (Table 2). The accuracy of the radar is 1 cm to 5 cm, the scanning radius is 12 m, and the scanning frequency is 5 Hz to 15 Hz. The global map was established with the data collected via LiDAR and processed by the mapping algorithm. It can be seen that the constructed map is roughly the same as the practical environment, with a relative error of about 1%. The Rviz visualization interface was launched to verify the normality of map information, robot model, LiDAR, and other relevant data. Moreover, the planned path, such as global and local paths, can be visualized in the Rviz interface to monitor the results of the tests. After the generation of the path according to the global map, various speed instructions and other relevant data such as steering angle and rotating speed of the front wheel were generated in real-time. The autonomous navigation system in the upper computer constantly received and released topics (i.e., data and instructions) in the ROS communication mode. The lower computer received relevant instructions, performed the necessary tasks, and ultimately completed the navigation task.

Table 2. Specific parameters of the radar.

Model Number	Accuracy	Scanning Radius	Scanning Frequency	Sampling Frequency	Angular Resolution
SLAMTEC A2	1–5 cm	12 m	5–15 Hz	16 k	0.9°

4. Results and Discussions

Figure 11 shows the simulation results of mapping and path planning in MATLAB. The results of the improved hybrid A* algorithm were evaluated and compared to those of the traditional A* algorithm. The obstacle ratios of the maps used in the simulation were 0.2 and 0.3, respectively. The blue lines in the figures represent the paths planned by the conventional A* algorithm, and the red lines stand for those from the improved hybrid A* algorithm. It was found that the conventional A* algorithm searched the concave obstacles by scanning the eight surrounding nodes. This increased the search time and distance and reduced the algorithm's efficiency, although it did not result in the local minimum problem. Moreover, the sharp corners of the paths generated by the conventional A* algorithm were not suitable for nonholonomically constrained Ackermann chassis. In contrast, the path generated by the improved hybrid A* algorithm was shorter and smoother. This significantly reduced the total length of the path as well as the turning radius at the boundary of the obstacles, facilitating the movement of the Ackermann chassis.

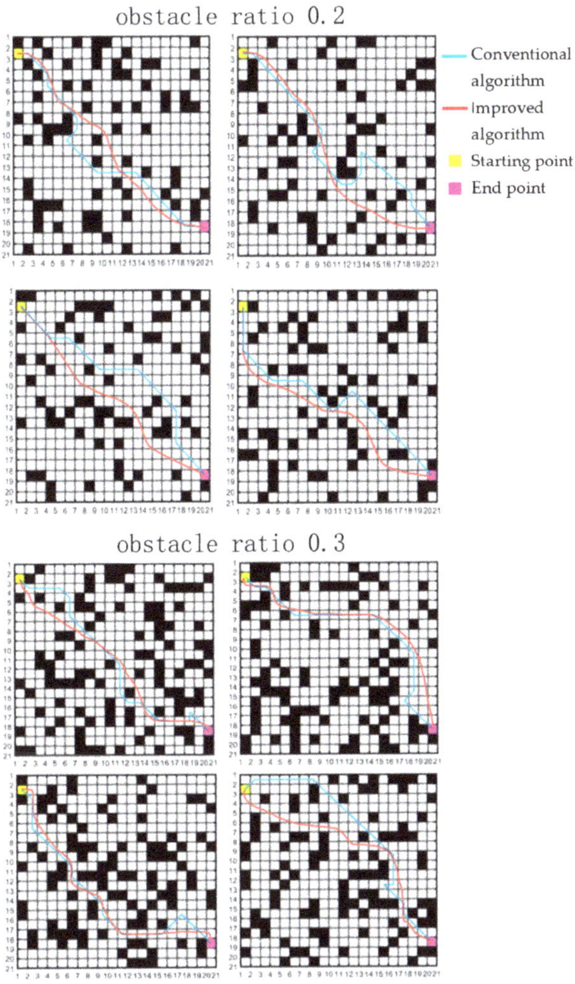

Figure 11. MATLAB simulation experimental results.

Figure 12 shows the difference between the A* algorithm and improved algorithm in the simulation experiments. It can be seen that the path charted by the A* algorithm possesses a large corner, displaying a tendency to collide with the boundary of the obstacles. This problem was caused by the algorithm's manner of node expansion in eight directions without considering motion constraints and the actual size of the robot. This makes the A* algorithm unsuitable for actual navigation. Conversely, the path planned by the improved algorithm is notably smoother with a small corner, presenting a more feasible performance than that of the conventional A* algorithm. This is because the application of the Chebyshev distance optimized the node expansion when the transverse angle was larger than 90°.

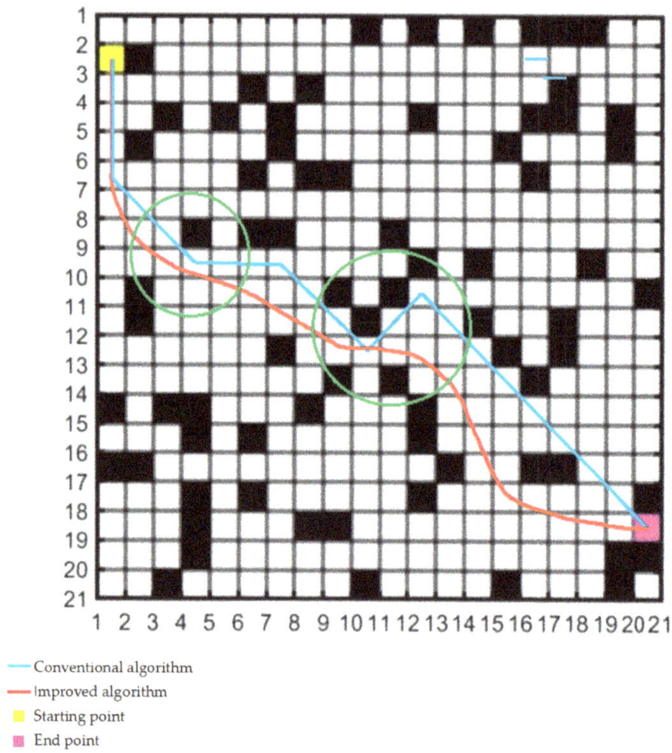

Figure 12. The difference between A* algorithm and improved algorithm in simulation experiments.

The performance of both algorithms was evaluated using a grid map at the two obstacle ratios. The comparison of the improved algorithm and the traditional algorithm is shown in Figure 13. The values of the average searching length and time consumption of the two algorithms are listed in Tables 3 and 4. The results obtained from the simulation and experiment demonstrate that the improved algorithm is advantageous in the path selection. Specifically, the average searching length with the improved hybrid A* algorithm was about 9% less at both obstacle ratios of 0.2 and 0.3. This is because the improved hybrid A* algorithm avoids the generation of nodes that were generated by the conventional A* algorithm. The smoother path generated via the improved hybrid A* algorithm contributed to the overall increase in the efficiency of the algorithm, which was reflected by the average searching time. As listed in the tables, the average search times using the improved hybrid A* algorithm were 12.62% and 24.5% when the obstacle rates were 0.3 and 0.2, respectively. This indicates the better efficiency of the improved algorithm, but the time reduction became less significant when the obstacle rate was increased.

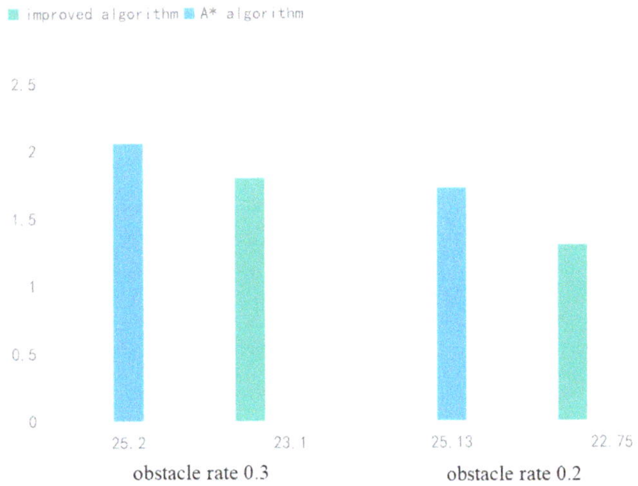

Figure 13. Comparison of the improved algorithm and the traditional algorithm.

Table 3. Comparison between the algorithms at the obstacle rate of 0.3.

Indicators	A* Algorithm	Improved Hybrid A* Algorithm
Path state	Unsmoothed	Smooth and continuous
Average searching length	25.2	23.1 (9.25% reduction)
Average searching time	2.06 s	1.8 s (12.62% reduction)

Table 4. Comparison between the algorithms at the obstacle rate of 0.2.

Indicators	A* Algorithm	Improved Hybrid A* Algorithm
Path state	Unsmoothed	Smooth and continuous
Average searching length	25.13	22.75 (9.5% reduction)
Average searching time	1.722 s	1.3 s (24.5% reduction)

Figure 14 shows a map of passable and impassable areas and obstacles. In this context, the black regions represent impassable areas and obstacles, whereas the gray regions signify passable areas. The starting point and end point for the comparison experiment remain the same. The path planned by the A* algorithm is shown in Figure 14a, while the path planned by the improved hybrid A* algorithm is shown in Figure 14b. It is evident from the image that the improved hybrid A* algorithm effectively circumvents the robot's entry into concave areas while searching for the optimal path. This approach results in a shorter path length, smoother trajectory, and reduced likelihood of collisions. In practical navigating tests, the average time of search was reduced by 18~24%, and the average length of search was reduced by 10.3~12%. Accounting for the positioning error of the robot, the present study suggests that the improved algorithm generates superior paths compared to that of the A* algorithm.

The improved algorithm uses the Chebyshev distance to weight the heuristic function, and under the guidance of global programming; it avoids the concave area of obstacles, improves the redundancy and inefficiency of the traditional A* algorithm, and can quickly generate an optimal curve in the early and middle stages.

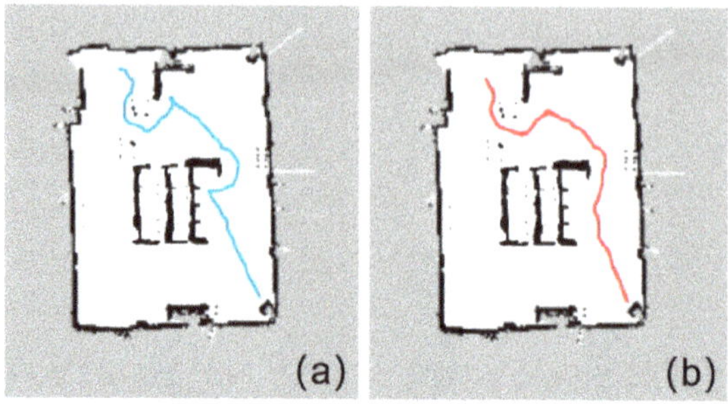

Figure 14. Path generated via different algorithms in the practical navigation experiment: (**a**) conventional A* algorithm and (**b**) improved hybrid A* algorithm.

5. Conclusions and Further Work

The nonholonomic robot with autonomous navigation was developed in this paper. The improved hybrid A* algorithm with Chebyshev weighting was applied in the path planning process to increase the efficiency of the navigation. The development of the self-path planning method was conducted in the following processes: (1) the construction of maps via the Robot Operating System (ROS), (2) the self-positioning of the robot via the adaptive Monte Carlo (AMCL) algorithm, (3) the global path planning via the improved hybrid A* algorithm, and (4) the local path planning via the model predictive control (MPC) strategy. By testing the performance of the algorithm in the simulation and practical environments, it was found that the improved hybrid A* algorithm outperformed the conventional A* algorithm in both search time and search length. These improvements make it an ideal choice for mobile chassis with nonholonomic constraints. Compared with the conventional A* algorithm, the average search time was reduced by 12.62~24.5%, and the average search length was reduced by 9.25~12% at different obstacle rates. In the practical navigating tests, the average time of search was reduced by 18~24%, and the average length of search was reduced by 10.3~12%, while the overall path was smoother. These results demonstrate the superior performance of the improved algorithm in various working environments.

Consideration for moving obstacles is an inevitable aspect of future work. With the evolution and increasing accessibility of autonomous navigation technologies, the rapid detection and avoidance of mobile objects become an imperative concern [35]. Future research focus will be on the development of new sensors, construction of more accurate obstacle-prediction models, and implementation of faster path-planning algorithms. Furthermore, applying artificial intelligence and deep learning in navigation algorithms can result in more precise prediction and avoidance of moving obstacles. The widespread implementation of these technologies will advance the maturity and expansion of robotics.

Author Contributions: Conceptualization, Z.C. and G.L.; methodology, Z.C.; validation, G.L.; investigation, Z.C. and Q.H.; resources, Z.Y. and Q.W.; data curation, Q.H.; writing—original draft, Z.C.; writing—review & editing, G.L. and S.D.; visualization, Z.Y., Q.W. and G.L.; supervision, G.L. and S.D.; funding acquisition, S.D. All authors have read and agreed to the published version of the manuscript.

Funding: This research received no external funding.

Institutional Review Board Statement: Not applicable.

Informed Consent Statement: Not applicable.

Data Availability Statement: No new data were created or analyzed in this study. Data sharing is not applicable to this article.

Conflicts of Interest: The authors declare no conflict of interest.

Nomenclature

Symbol	Definition
$f(n)$	Priority of the next node
$h(n)$	Estimated cost from the next node to the end
$g(n)$	Cost from the current node to the start
w_{long}	Long-term observation average function
w_{short}	Short-term observation average function
a_{long}	Average weight coefficient of the long-term observation information
a_{short}	Average weight coefficient of the short-term observation information
w_{avg}	Long-term observation average function
w_i	Observation average function
w	Weighting coefficient
α	Distribution coefficient
β	Obstacle ratio
d_n	Ratio of Chebyshev distances
d_q	Chebyshev distance
x_n, y_n	Horizontal and vertical coordinates of the current node
x_g, y_g	Horizontal and vertical coordinates of the target point
x_s, y_s	Horizontal and vertical coordinates of the starting point
r^{os}	Expansion radius of static obstacle
ρ^d	Expansion radius of dynamic obstacle
m^{os}	Position space of static obstacle
m^d	Position space of dynamic obstacle
m	Spatial position of the robot
ρ	Expansion radius of the robot

References

1. Gul, F.; Rahiman, W.; Alhady, S.S.N. A comprehensive study for robot navigation techniques. *Cogent Eng.* **2019**, *6*, 1632046. [CrossRef]
2. He, X.; Yang, H.; Hu, Z.; Lv, C. Robust Lane Change Decision Making for Autonomous Vehicles: An Observation Adversarial Reinforcement Learning Approach. *IEEE Trans. Intell. Veh.* **2023**, *8*, 184–193. [CrossRef]
3. Behringer, R.; Muller, N. Autonomous road vehicle guidance from Autobahnen to narrow curves. *IEEE Trans. Robot. Autom.* **1998**, *14*, 810–815. [CrossRef]
4. Skog, I.; Handel, P. In-Car Positioning and Navigation Technologies—A Survey. *IEEE Trans. Intell. Transp. Syst.* **2009**, *10*, 4–21. [CrossRef]
5. Kavraki, L.E.; Svestka, P.; Latombe, J.C.; Overmars, M.H. Probabilistic roadmaps for path planning in high-dimensional configuration spaces. *IEEE Trans. Robot. Autom.* **1996**, *12*, 566–580. [CrossRef]
6. Cui, R.; Li, Y.; Yan, W. Mutual Information-Based Multi-AUV Path Planning for Scalar Field Sampling Using Multidimensional RRT*. *IEEE Trans. Syst. Man Cybern. Syst.* **2016**, *46*, 993–1004. [CrossRef]
7. Wang, J.; Chi, G.; Li, C.; Wang, C.; Meng, M. Neural RRT*: Learning-Based Optimal Path Planning. *IEEE Trans. Autom. Sci. Eng.* **2020**, *17*, 1748–1758. [CrossRef]
8. Yao, Y.F.; He, N.; Zhang, M. The Application of Internet of Things in Robot Route Planning Based on Multisource Information Fusion. *Comput. Intell. Neurosci.* **2022**, *2022*, 1707259. [CrossRef]
9. Jiang, J.R.; Huang, H.W.; Liao, J.H.; Chen, S.Y. Extending Dijkstra's Shortest Path Algorithm for Software Defined Networking. In Proceedings of the 2014 16th Asia-Pacific Network Operations and Management Symposium (APNOMS), Hsinchu, Taiwan, 17–19 September 2014; IEEE: Piscataway, NJ, USA, 2014.
10. Chang, J.R.; Jheng, Y.H.; Chang, C.H.; Lo, C.H. An Efficient Algorithm for Vehicle Guidance Combining Dijkstra and A* Algorithm with Fuzzy Inference Theory. *J. Internet Technol.* **2015**, *16*, 189–200. [CrossRef]
11. Yang, D.; Xu, B.; Rao, K.; Sheng, W. Passive Infrared (PIR)-Based Indoor Position Tracking for Smart Homes Using Accessibility Maps and A-Star Algorithm. *Sensors* **2018**, *18*, 332. [CrossRef]
12. Harabor, D.; Grastien, A. Online Graph Pruning for Pathfinding on Grid Maps. In Proceedings of the Twenty-Fifth AAAI Conference on Artificial Intelligence, San Francisco, CA, USA, 7–11 August 2011; Volume 2.
13. Li, G.; Chou, W. Path planning for mobile robot using self-adaptive learning particle swarm optimization. *Sci. China Inf. Sci.* **2017**, *61*, 052204. [CrossRef]
14. Li, J.C.; Ran, M.P.; Wang, H.; Xie, L.H. MPC-based Unified Trajectory Planning and Tracking Control Approach for Automated Guided Vehicles. In *Proceedings of the 2019 IEEE 15th International Conference on Control and Automation (ICCA), Edinburgh, UK, 16–19 July 2019*; IEEE: Piscataway, NJ, USA, 2019; pp. 374–380.

15. Fox, D.; Burgard, W.; Thrun, S. The Dynamic Window Approach to Collision Avoidance. *IEEE Robot. Autom. Mag.* **1997**, *4*, 23–33. [CrossRef]
16. Wu, J.; Ma, X.; Peng, T.; Wang, H. An Improved Timed Elastic Band (TEB) Algorithm of Autonomous Ground Vehicle (AGV) in Complex Environment. *Sensors* **2021**, *21*, 8312. [CrossRef] [PubMed]
17. Rasekhipour, Y.; Khajepour, A.; Chen, S.K.; Litkouhi, B. A Potential Field-Based Model Predictive Path-Planning Controller for Autonomous Road Vehicles. *IEEE Trans. Intell. Transp. Syst.* **2017**, *18*, 1255–1267. [CrossRef]
18. Ji, J.; Wang, H.; Ren, Y. *Path Planning and Tracking for Vehicle Collision Avoidance in Lateral and Longitudinal Motion Directions*; Synthesis Lectures on Advances in Automotive Technology; Springer: Cham, Switzerland, 2020; Volume 4, pp. 1–152. [CrossRef]
19. Gan, Y.; Zhang, B.; Ke, C.; Zhu, X.F.; He, W.M.; Ihara, T. Research on Robot Motion Planning Based on RRT Algorithm with Nonholonomic Constraints. *Neural Process. Lett.* **2021**, *53*, 3011–3029. [CrossRef]
20. Zhang, Z.Y.; Li, Y.; Wu, P.; Liu, Z.C.; Xiao, J.J.; Zhu, W.H. An INS-aided MASS autonomous navigation algorithm considering virtual motion constraints and the leeway and drift angle. *Ocean Eng.* **2023**, *272*, 113790. [CrossRef]
21. Zhang, Y.F.; Wen, Y.Y.; Tu, H.Y. A Method for Ship Route Planning Fusing the Ant Colony Algorithm and the A* Search Algorithm. *IEEE Access* **2023**, *11*, 15109–15118. [CrossRef]
22. Worthmann, K.; Mehrez, M.W.; Zanon, M.; Mann, G.K.I.; Gosine, R.G.; Diehl, M. Model Predictive Control of Nonholonomic Mobile Robots without Stabilizing Constraints and Costs. *IEEE Trans. Control Syst. Technol.* **2016**, *24*, 1394–1406. [CrossRef]
23. Dang, C.V.; Ahn, H.; Lee, D.S.; Lee, S.C. A Path Planning Method Based on Theta-star Search for Non-Holonomic Robots. In Proceedings of the 2022 Joint 12th International Conference on Soft Computing and Intelligent Systems and 23rd International Symposium on Advanced Intelligent Systems (SCIS&ISIS), Ise, Japan, 29 November–2 December 2022; pp. 1–6.
24. Li, J.; Ran, M.; Xie, L. Efficient Trajectory Planning for Multiple Non-Holonomic Mobile Robots via Prioritized Trajectory Optimization. *IEEE Robot. Autom. Lett.* **2021**, *6*, 405–412. [CrossRef]
25. Johnson, J.J.; Li, L.; Liu, F.; Qureshi, A.H.; Yip, M.C. Dynamically Constrained Motion Planning Networks for Non-Holonomic Robots. In Proceedings of the 2020 IEEE/RSJ International Conference on Intelligent Robots and Systems (IROS), Las Vegas, NV, USA, 24 October 2020–24 January 2021; pp. 6937–6943.
26. Gonon, D.J.; Paez-Granados, D.; Billard, A. Reactive Navigation in Crowds for Non-Holonomic Robots with Convex Bounding Shape. *IEEE Robot. Autom. Lett.* **2021**, *6*, 4728–4735. [CrossRef]
27. De Oliveira, A.L.; Binder, B.J. Discrete Manhattan and Chebyshev pair correlation functions in k dimensions. *Phys. Rev. E* **2020**, *102*, 012130. [CrossRef] [PubMed]
28. Smith, R.; Cheeseman, P. On the Representation and Estimation of Spatial Uncertainty. *Int. J. Robot. Res.* **1987**, *5*, 56–68. [CrossRef]
29. Kang, Y.; Song, Y.; Song, Y.; Yan, D.; Li, D. Square-Root Cubature Kalman Filter and Its Application to SLAM of an Mobile Robot. *Robot* **2013**, *35*, 186–193. [CrossRef]
30. Dolgov, D.; Thrun, S.; Montemerlo, M.; Diebel, J. Practical search techniques in path planning for autonomous driving. *Int. Symp. Comb. Search SoCS* **2008**, *1001*, 18–80.
31. Reeds, J.A.; Shepp, L.A. Optimal Paths for a Car That Goes Both Forwards and Backwards. *Pac. J. Math.* **1990**, *145*, 367–393. [CrossRef]
32. He, X.; Liu, Y.; Lv, C.; Ji, X.; Liu, Y. Emergency steering control of autonomous vehicle for collision avoidance and stabilisation. *Veh. Syst. Dyn.* **2018**, *57*, 1163–1187. [CrossRef]
33. Shim, T.; Adireddy, G.; Yuan, H. Autonomous vehicle collision avoidance system using path planning and model-predictive-control-based active front steering and wheel torque control. *Proc. Inst. Mech. Eng. Part D J. Automob. Eng.* **2012**, *226*, 767–778. [CrossRef]
34. Kayacan, E.; Kayacan, E.; Ramon, H.; Saeys, W. Distributed nonlinear model predictive control of an autonomous tractor–trailer system. *Mechatronics* **2014**, *24*, 926–933. [CrossRef]
35. He, X.; Lou, B.; Yang, H.; Lv, C. Robust Decision Making for Autonomous Vehicles at Highway On-Ramps: A Constrained Adversarial Reinforcement Learning Approach. *IEEE Trans. Intell. Transp. Syst.* **2023**, *24*, 4103–4113. [CrossRef]

applied sciences

Article

Global Path Planning for Differential Drive Mobile Robots Based on Improved BSGA* Algorithm

Ming Yao [1], Haigang Deng [2,*], Xianying Feng [1], Peigang Li [1], Yanfei Li [1] and Haiyang Liu [1]

[1] School of Mechanical Engineering, Shandong University, Jinan 250061, China; ym1601065487@163.com (M.Y.); fxying@sdu.edu.cn (X.F.); pgli@vip.sina.com (P.L.); yanfei-li@foxmail.com (Y.L.); 18943653361@163.com (H.L.)
[2] School of Instrumentation Science and Engineering, Harbin Institute of Technology, Harbin 150001, China
* Correspondence: hgdeng_hit@163.com

Abstract: The global path planner is an important part of the navigation system for autonomous differential drive mobile robots (DDMRs). Aiming at the problems such as long calculation time, large number of search nodes, and poor smoothness of path when A* is applied to global path planning, this study proposes an improved bidirectional search Gaussian-A* (BSGA*) algorithm. First, the Gaussian function is introduced to realize the dynamic weighting of the heuristic function, which reduces the calculation time. Secondly, the bidirectional search (BS) structure is adopted to solve the problem of nodes' repeated search when there are large obstacles between the starting point and the target point. Finally, a multi-layer turning point filter strategy is proposed to further smooth the path. In order to verify the performance of the improved BSGA* algorithm, experiments are carried out in simulation environments with the size of 15 × 15 and 30 × 30, respectively, and compared with the five common global path planning algorithms including ant colony optimization (ACO), D* lite algorithm, and genetic algorithm (GA). The results show that the improved BSGA* algorithm has the lowest calculation time and generates the shortest and smoothest path in the same environment. Finally, the program of the improved BSGA* algorithm is embedded into the LEO ROS mobile robot and two different real environments were built for experimental verification. By comparing with the A* algorithm, Dijkstra algorithm, ACO, D* lite algorithm, and GA, the results show that the improved BSGA* algorithm not only outperforms the above five algorithms in terms of calculation time, length, and total turning angle of the generated paths, but also consumes the least time when DDMR drives along the generated paths.

Keywords: differential driven mobile robots; global path planning; improved BSGA* algorithm; bidirectional structure; Gaussian function; turning point filtering strategy

Citation: Yao, M.; Deng, H.; Feng, X.; Li, P.; Li, Y.; Liu, H. Global Path Planning for Differential Drive Mobile Robots Based on Improved BSGA* Algorithm. *Appl. Sci.* **2023**, *13*, 11290. https://doi.org/10.3390/app132011290

Academic Editors: Nadjim Horri, William Holderbaum and Fabrizio Giulietti

Received: 5 September 2023
Revised: 11 October 2023
Accepted: 12 October 2023
Published: 14 October 2023

1. Introduction

DDMRs are widely used in earthwork construction, firefighting, the military, and other fields because of their ability to operate in high-risk, high-intensity environments [1–3]. At present, most DDMRs rely on manual control, which requires strict manipulation skills and experience from the operators. This way of long-term stable operation is difficult, and due to the relatively harsh working environment, it often brings certain risks to the operators. Therefore, the autonomous navigation systems of DDMRs have become a research hotspot [4,5]. Global path planning is the basis of an autonomous navigation system, which can pre-plan a safe and feasible path according to the environmental information collected by the sensors in advance to guide their moving process [6].

Global path planning, as a kind of static planning (also called offline planning), refers to planning an optimal collision-free path from the starting point to the target point for the robot under the consideration of certain evaluation criteria [7]. It is evaluated in three main ways. (1) Optimality: it can use path length, number of turns, driving time, and other standards to evaluate the superiority of the path. (2) Authenticity: the planned path should meet the driving requirements of the mobile robot. (3) Integrity: the path should connect the starting point and the target point continuously, and achieve the avoidance of obstacles [8].

Many algorithms have been applied to global path planning for mobile robots, including the A* algorithm [9], rapidly exploring random tree (RRT) algorithm [10], ant colony algorithm (ACO) [11], genetic algorithm (GA) [12], etc. On this basis, many scholars have carried out extensive and in-depth research. The RRT algorithm is suitable for continuous domain search because it does not require the discretization of a continuous space during planning [13]. However, this algorithm focuses more on finding feasible paths rather than reducing costs, thus the costs of the path tend to be high [14]. Wang [15] proposed a kinematic constrained bi-directional RRT algorithm (KB-RRT*) for path planning. By introducing kinematic constraints, KB-RRT* can avoid unnecessary tree growth, but the global path obtained based on the KB-RRT* algorithm is still a suboptimal solution. ACO is a heuristic algorithm that simulates the foraging behavior of ant colonies. Although the algorithm can obtain the optimal path, its convergence speed is slow and it is easy to fall into the local optimal. Hou et al. [16] proposed an improved ACO algorithm by adopting the communication mechanism of ants using their antennae to contact each other in nature, and applied the algorithm to mobile robots. However, the algorithm still has problems in calculation efficiency. GA can generate a large number of new individuals when the crossover probability is large, thereby improving the global search scope, but there is a precocious phenomenon and the results are unstable. Tuncer A et al. [17] proposed an improved GA using optimized mutation operators, which simultaneously checked all idle nodes near the mutation nodes, selected nodes according to the fitness value of paths rather than the direction of mutation nodes, and obtained a higher average fit than other methods. However, the paths are too close to the obstacles and are less safe. The Dijkstra algorithm [18] is a single-source path search algorithm that expands outward from the start point until it reaches the target and obtains the shortest path. With the increase in environmental information, the Dijkstra algorithm will consume a lot of computing time. P. E. Art et al. [19] proposed the A* algorithm in 1968 by combining the Dijkstra algorithm with the BFS algorithm. The A* algorithm greatly improves the search efficiency by introducing a heuristic function and the planned path is optimal. In addition, the path planned by the A* algorithm based on the grids has a sudden change in turning angle, while the DDMRs can turn in place by the speed difference between motors on both sides, so the A* algorithm is more suitable for the global path planning of the DDMRs. However, the A* algorithm still has the problems of long calculation time, poor smoothness, and low safety when facing a complex environment.

To solve the problem of the long calculation time of the A* algorithm, many scholars improve the algorithm structure or optimize the heuristic function. Zhang et al. [20] proposed an improved A* algorithm containing a bi-directional sector expansion and a variable-step search strategy, thus improving the computational efficiency. Liu et al. [21] adopted the Delaunary triangulation algorithm to deal with complex obstacles, took the generated Voronoi points as the preferred pathfinding nodes, then designed the dynamic fusion pathfinding algorithm (DFPA) based on the Delaunary triangulation algorithm and the improved A* algorithm. Jiang H et al. [22] introduced a cosine factor into the heuristic function of the A* algorithm to optimize the search direction, and adopted a search strategy that synchronizes the starting point and the target point to realize the path planning for the electric disinfection vehicle, which has higher computational efficiency, but does not consider the kinematic model. For the problem of poor path smoothness, scholars optimize the A* algorithm by improving the heuristic function or adjusting the search step size. Liu et al. [23] designed a global path yaw angle based on the relationship between the real-time position and global path, then introduced it into the heuristic function of the A* algorithm to improve the path smoothness. Tang et al. [9] filter the nodes in the closed list to avoid the irregular path. Meanwhile, a cubic B-spline curve is adopted to smooth the path and improve the stability in turning. Li et al. [24] introduced the jump-point search strategy into

A* algorithm and effectively reduced the number of turning points by ignoring unnecessary nodes. For the problem of low security, some scholars have introduced the influence of obstacles into the A* algorithm. Zhang et al. [25] converted the distance between the mobile robot and the obstacles into the time cost, so as to optimize the cost function and improve the safety. Sang et al. [26] narrow the search scope by imposing constraints of maximum search distance and maximum path length, then maintain a safe distance by reducing the search points near the obstacles. Cui [27] predicted the future movement of the target using Bernstein basis polynomials combined with the information of the obstacle distribution around the target, then utilized the A* algorithm to search for a safe tracking path. Finally, in the case of dynamic constraints, a quadratic programming method is used to optimize the tracking path to improve the smoothness. Dang [28] proposed an Adaptive Back-stepping Hierarchical Sliding Mode Control (ABHSMC) scheme for three-wheeled mobile robots (3WMRs) based on RBF neural networks. By aggregating all uncertain components in specific vectors and estimating using an RBF neural network, the effect of uncertainties will be minimized. Then, combining it with the TEB local planner and A* global planner improves the navigation and obstacle avoidance performance.

Although these studies have optimized the performance of the A* algorithm to some extent, the direct application to DDMRs are still one-sided, because, in addition to the factors mentioned above, the global path planning of DDMRs also needs to consider the total driving time. The former is related to the setting of the heuristic function, algorithm structure, and processor performance, while the latter is related to the motion state, length, and total turning angle of the path. Therefore, the kinematic model of DDMR is established to analyze the motion state on the turning point. On this basis, Gaussian functions are firstly used to dynamically adjust the weight ratio of the heuristic functions, which effectively reduces the calculation time. Secondly, the BS structure is introduced into the A* algorithm, which solves the problems of repeated node search when there is a large area of obstacles between the starting point and the target point. Finally, a multi-layer turning nodes filtering strategy is proposed to reduce the total turning angle and improve the smoothness of the path. The simulation results show that the improved BSGA* algorithm has higher efficiency than the A* algorithm, GA, and ACO algorithm, and can plan a more reasonable driving route. The LEO ROS robot is used for experimental verification. The experimental results show that it has better results than the five common global path planning algorithms.

2. Materials and Methods

This section introduces the kinematic modeling of DDMR, the process of environment modeling using the grid method, and the basic principle of the A* algorithm.

2.1. Analysis of Turning Motion of DDMR

Figure 1 shows the analysis of DDMR's turning motion. The rectangular coordinate system XOY is established, the coordinates of DDMR are set as (x, y), the linear velocity is v, and the angular velocity is ω. The width of DDMR is B, the turning center is ICR, and the turning radius is R. Because DDMR has nonholonomic constraints [29], its motion can be described only through linear velocity v and angular velocity ω [30].

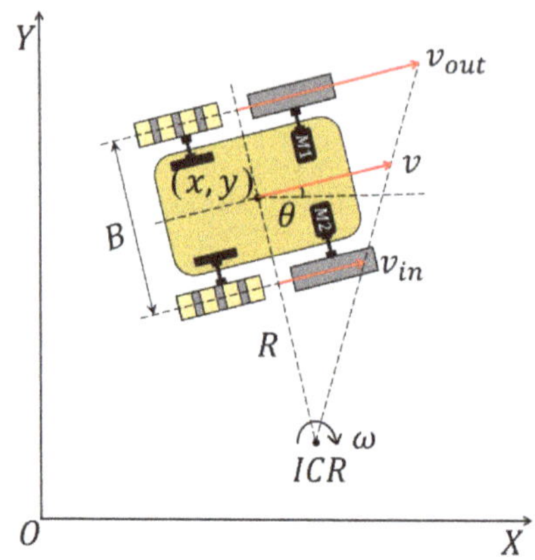

Figure 1. Analysis of DDMR's turning motion.

The controller makes the DDMR reach the expected motion state by controlling the velocities of the drive motors on both sides, the linear velocities of the drive wheels on both sides can be expressed as follows:

$$\begin{cases} v_{in} = (R + B/2)\omega \\ v_{out} = (R - B/2)\omega \end{cases} \tag{1}$$

where v_{in} and v_{out} are, respectively, the inner linear velocity and outer linear velocity of DDMR. From Equation (1), the linear velocity of the center of mass can be expressed as follows:

$$\mathrm{v} = \omega R = (v_{in} + v_{out})/2 \tag{2}$$

Based on Equations (1) and (2), the turning radius R and angular velocity ω of DDMR can be expressed as follows:

$$\begin{cases} \omega = (v_{out} - v_{in})/B \\ R = [B(v_{out} + v_{in})]/[2(v_{out} - v_{in})] \end{cases} \tag{3}$$

According to Equation (3), when v_{out}=$-v_{in}$, $R = 0$, DDMR can realize in place turning. Assuming that the angular velocity of DDMR is constant in the turning process, the turning angle θ is proportional to the turning time Δt. Many studies [31] have lower driving time by reducing the number of turns, but this is not comprehensive. The direct factor affecting the turning time is the total turning angle, and there is no direct functional correspondence between it and the number of turns. Since the DDMR realizes turning through the speed difference of the drive motors arranged on both sides of the body, when the total turning angle is too large, it will aggravate the wear of the drive motors, conveyor belts, and other components, and even reduce the service life. Secondly, the turning of the DDMR is a relatively slow process, which includes three decomposition actions: deceleration, turning, and acceleration, and consumes more time than that of the straight motion. Therefore, reducing the total turning angle can effectively shorten the time required for the DDMR to drive along this path. So the total turning angle should be used as the evaluation index to reduce the turning time.

2.2. Grid Modeling

Based on the driving characteristics of DDMR that can be turned in place, the grid method is used to establish the environment map. The work space is projected as a two-dimensional plane and discretized into a square grid of uniform, continuous, and non-intersecting. According to the environmental information, the grids with obstacle projection can be set as obstacle grids, represented by 1; the grids without obstacle projection can be set as free grids, represented by 0. A Cartesian coordinate system is established in the grid map, with the lower left corner as the origin of coordinates, the horizontal axis for the X axis, with values increasing from left to right. The vertical axis for the Y axis, with values increasing from bottom to top. The position of each grid is represented by the coordinate $p(x_i, y_j)$ in the upper right corner of the grid, where $i, j = 1, 2, 3 \ldots\ldots, n$.

The environmental information in the grid maps will be different from the actual maps, which mainly depends on the size of the grids. If the grids are too large, the resolution of the environment maps will be reduced, which may make the planned paths deviate from the theoretical optimal paths seriously, or even fail to search for feasible paths. If the grids size are too small, the resolution of the environment maps will be too high, which will occupy a lot of computational resources. In order to simplify the planning problem, DDMR is regarded as a square with side length $l_{edge} = 1$, and the side length l of the grid is also set as 1. In order to improve the safety of the path, DDMR should keep away from obstacles during the driving process. Therefore, the obstacles are puffed up as shown in Figure 2, the grids with only partial obstacles are also set as obstacle grids.

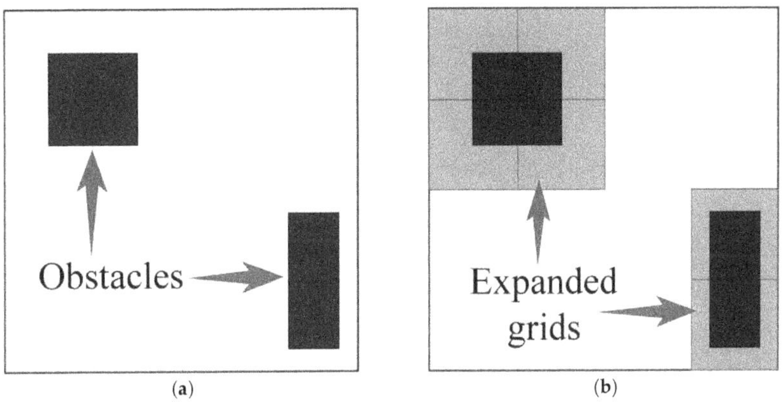

Figure 2. Obstacle puffing process. They should be listed as follows: (**a**) Before puffing process; (**b**) After puffing process.

2.3. Procedure of A* Algorithm

As a heuristic path planning algorithm, the A* algorithm is used to find the optimal path in the static environment [32]. This algorithm searches from the starting point according to the predetermined search strategy, calculates the actual cost of each feasible node around the current node to the starting point and the heuristic cost to the target point, and selects the node with the minimum total cost as the next extended node. The algorithm ends when the extended node is overlapped with the target point [33]. The core of the A* algorithm lies in the cost function, which is as follows:

$$\mathrm{f}(n) = g(n) + h(n) \tag{4}$$

where $f(n)$ represents the total cost from the starting point (X_s, Y_s) to the target point (X_e, Y_e) through the current point (X_n, Y_n); $g(n)$ represents the actual cost from the starting point (X_s, Y_s) to the current point (X_n, Y_n); and $h(n)$ represents the heuristic cost from the current point (X_n, Y_n) to the target point (X_e, Y_e). The traditional A* algorithm adopts the 4-neighborhood search strategy. Because DDMRs can realize in-place turning, it can be reasonably considered to complete the diagonal movement of the grids. In order to make the path generated by the A* algorithm more realistic, the search neighborhood is extended to 8 neighborhoods, as shown in Figure 3, with 8 turning angles $(0°, 45°, -45°, 90°, -90°, 135°, -135°, 180°)$. There are mainly three heuristic functions, including Euclidean distance function [34], Chebyshev distance function [35], and Manhattan distance function [36]. They can be expressed, respectively, as follows:

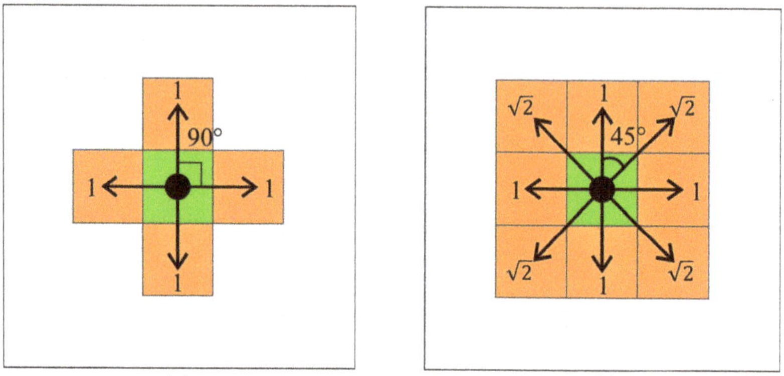

(a) (b)

Figure 3. Extended search neighborhood. They should be listed as: (**a**) 4 search neighborhood; (**b**) 8 search neighborhood.

Euclidean distance function:

$$h(n) = \sqrt{(X_e - X_n)^2 + (Y_e - Y_n)^2} \tag{5}$$

Chebyshev distance function:

$$h(n) = |X_e - X_n| + |Y_e - Y_n| \tag{6}$$

Manhattan distance function:

$$h(n) = \max(|X_e - X_n|, |Y_e - Y_n|) \tag{7}$$

For the heuristic function, if $h(n)$ is the same as the actual cost $P(n)$ from the (X_n, Y_n) to the (X_e, Y_e), all nodes extended by the A* algorithm are on the optimal path, it will not expand any remaining unarticulated points, and the search speed is the fastest at this time, but this is difficult to achieve. For the Manhattan distance, its value is often less than $P(n)$, although the optimal path can be found, the search speed will decrease. For the Chebyshev distance, its value is often greater than $P(n)$, although it searches faster, there is no guarantee that the optimal path can be found. Since the 8-neighborhood search strategy is adopted in this study, $h(n)$ obtained by using Euclidean distance is closer to $P(n)$, which ensures that the search speed can be improved on the basis of finding the optimal path. Therefore, Euclidean distance is adopted as the heuristic function.

The A* algorithm divides all extended nodes into two sets:

OpenList: Stores all extension nodes to be detected in the search process, and sorts them according to the $f(n)$.

CloseList: Stores all detected extension nodes in the search process to prevent repeated search.

On the basis of the above concepts, the steps of the A* algorithm are shown in Figure 4.

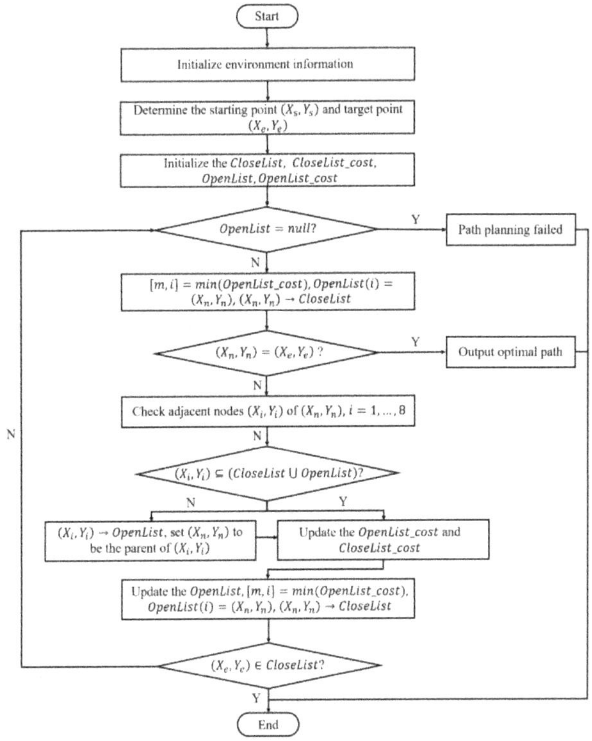

Figure 4. Program flow of A* algorithm.

3. Improved BSGA* Algorithm

The traditional A* algorithm is improved to design a new global path planner, whose inputs include the values of grids, the starting point, the target point, and the attitude of the DDMR. The global path calculated by the improved BSGA* algorithm is divided into a series of waypoint coordinates as the outputs of the algorithm, which are relied upon to generate the linear velocity and angular velocity to guide the DDMR to move to the target point. In this planner, the constraints include kinematic constraints, that is, the maximum velocity of 0.6 m/s and the maximum angular velocity of $\pi/6$ rad/s. The environmental constraints, that is, the side lengths of the grids and the locations of the obstacles. Cost constraints: the sum of the actual costs and heuristic costs of each grid.

3.1. Improved Heuristic Functions

The heuristic function determines the quality of the A* algorithm [37]. As shown in Figure 5, the A* algorithm, using Euclidean distance as the heuristic function, can search a feasible path, but there are too many search nodes and the calculation time is too long. Therefore, this study addresses the above shortcomings by improving the heuristic function. When $h(n) = 0$, the A* algorithm is transformed into a Dijkstra algorithm, and a large number of nodes will be searched. Although the total cost in the planning process is the same as the actual cost, it is inefficient [38]. When $g(n) = 0$, the A* algorithm is transformed into the Breadth-First Search (BFS) algorithm, which has high planning efficiency but not an optimal path [39]. In order to make the path planning process both efficient and qualitative, $h(n)$ should usually occupy a larger weight in the beginning of the planning, so that the algorithm can search near the target point more quickly. At the end of the planning process, $g(n)$ should occupy a larger weight, so as to avoid repeated searches of nodes that may occur near the target point and reduce the total turning angle of paths. Since Euclidean distance is used as the heuristic function in this study, $h(n)$ tends to be slightly smaller than the actual cost from the current node to the target node. As the current node gradually approaches the target node, this gap will become smaller and smaller. In the optimal case, $h(n)$ is always the same as the actual cost, so the algorithm can always select the optimal node. But this is often difficult to achieve.

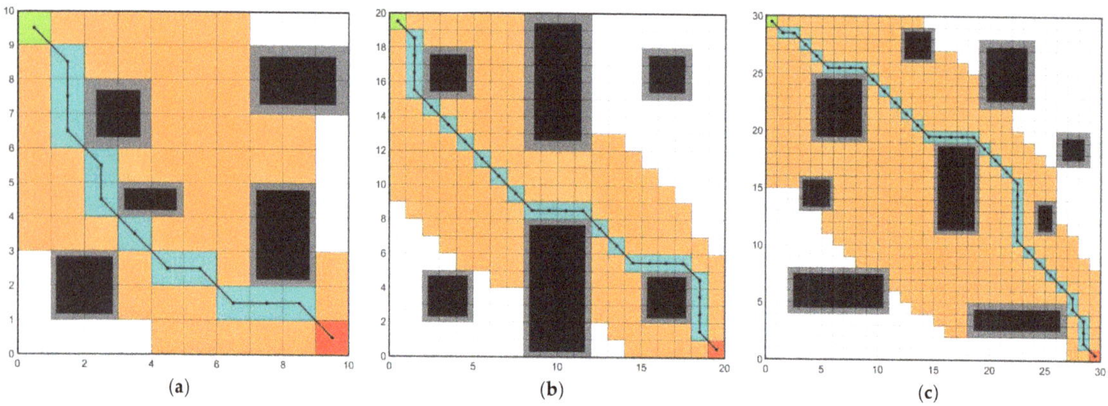

Figure 5. Simulation results of traditional A* algorithm under different map environments. (**a**) Map size is 10 × 10. (**b**) Map size is 20 × 20. (**c**) Map size is 30 × 30.

Therefore, this study combined with the Gaussian function to construct the attenuation coefficient $D(n)$ for dynamically adjusting the weight ratio of the heuristic function; the heuristic function was rewritten as the following equation:

$$\begin{cases} f(n) = g(n) + \mathrm{h}(n)/D(n) \\ D(n) = \frac{1}{\sigma\sqrt{2\pi}} e^{\frac{-(h(n)-\mu)^2}{2\sigma}} \end{cases} \quad (8)$$

Let $\mu = 0$ and $\sigma = 1/\sqrt{2\pi}$. At the beginning of the algorithm, the current node is far from the target point, at this time $D(n) \to 0^+$, so the weight ratio of $h(n)$ increases, thus improving the search efficiency. As the algorithm runs, the current node gradually approaches the target point, $D(n)$ gradually increases, and the weight ratio of $h(n)$ gradually decreases to reduce the total turning angle of the path.

In order to verify the performance of the Gaussian–A* (GA*) algorithm, Matlab2020a software was used to write the algorithm program. The Intel workstation running the program was equipped with an i9-10900 processor, with 64 GB RAM and 3.70 GHz main frequency. The subsequent simulation was also run based on this workstation. As shown in Figures 5 and 6, 50 executions were carried out on grid maps with sizes of 10 × 10, 20 × 20 and 30 × 30, respectively.

Table 1 shows the states represented by the different colored grids. It can be seen that, compared with the A* algorithm, the path generated by the GA* algorithm has fewer turning points and the number of nodes searched is also less. According to Table 2, it can be seen that, on the 10 × 10 grid map, the calculation time, the number of search nodes, and the total turning angle of the GA*s algorithm are, respectively, reduced by 64.43%, 40.00%, and 37.5% relative to the A* algorithm. On the 20 × 20 grid map, the calculation time, number of search nodes, and the total turning angle of the GA*s algorithm are, respectively, reduced by 80.98%, 56.38%, and 25% relative to the A* algorithm. On the 30 × 30 grid map, these three were reduced by 84.09%, 70.91%, and 41.67%, respectively. Although the length of the path generated by the GA* algorithm is not much different from that of the A* algorithm, it has higher efficiency, and with the gradual expansion of the map, this difference becomes more and more obvious. In addition, the GA* algorithm can generate smoother paths, which is more conducive to DDMR driving.

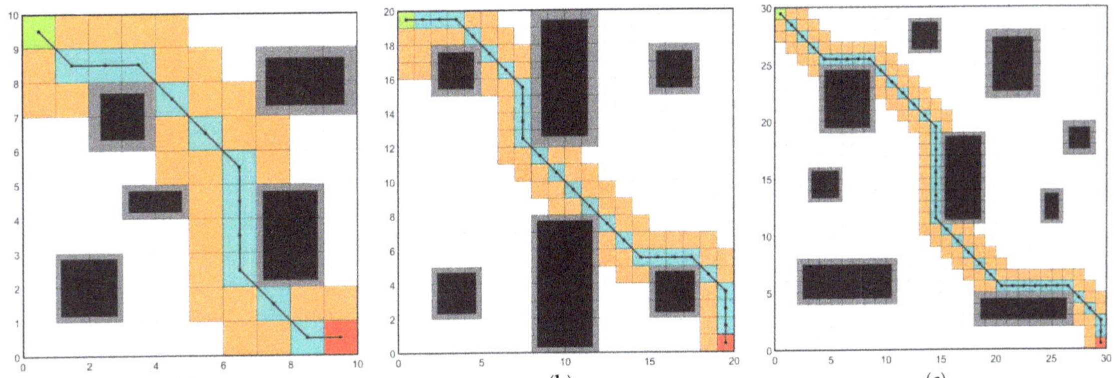

Figure 6. Simulation results of traditional GA* algorithm under different map environments. (**a**) Map size is 10 × 10. (**b**) Map size is 20 × 20. (**c**) Map size is 30 × 30.

Table 1. Grid state correspondence table.

No	Color	State of Grid
1	White	Free grid
2	Black	Obstacle grid
3	Grey	Expanded grid
4	Green	Starting point
5	Red	Target point
6	Orange	Searched point
7	Blue-green	Path point

Table 2. Comparison of GA* algorithm and traditional A* algorithm on different maps; the values of average calculation time are reported as mean ± standard deviation for 50 executions.

Map Size	Algorithm	Average Calculation Time/ms	Number of Searched Nodes	Total Angle/°	Path Length
10 × 10	A*	574.94 ± 12.97	65	360	14.49
	GA*	204.51 ± 2.17	39	225	14.49
20 × 20	A*	3119.03 ± 41.53	188	360	30.39
	GA*	593.38 ± 2.76	82	270	30.39
30 × 30	A*	14,194.60 ± 75.32	440	540	45.70
	GA*	2258.70 ± 9.34	128	315	46.87

3.2. Bidirectional Search Structure

Although the GA* algorithm obviously improves the computational efficiency and smoothness of the path, the planning effect in some special cases still needs to be improved. As shown in Figure 7a,b, when there is a large obstacle between the starting point and the target point, DDMR needs to bypass the obstacle to reach the target point, so the A* algorithm and GA* algorithm will search in the direction that deviate from the starting point and the target point at the beginning process. The numbers in the upper-left, lower-left, and lower-right corners of the grids in Figure 7 correspond to the $f(n)$, $g(n)$, and $h(n)$ of the current node, respectively. The arrow points to the parent point of the node where the arrow tail is located. As can be seen from Figure 7a, $f(n)$ gradually increases in the process of avoiding obstacles. However, after avoiding obstacles, $g(n)$ is too large due to the long path, so $f(n)$ is even higher than $f(n)$ that corresponds to the nodes at the beginning of the search process. Therefore, in the later stage of the A * algorithm, many nodes near the starting point will be repeatedly searched. As shown in Figure 7b, this phenomenon has been improved somewhat in the GA* algorithm but still exists. To address the above problem, this study, using a BS structure to improve the GA* algorithm, which is named the BSGA* algorithm.

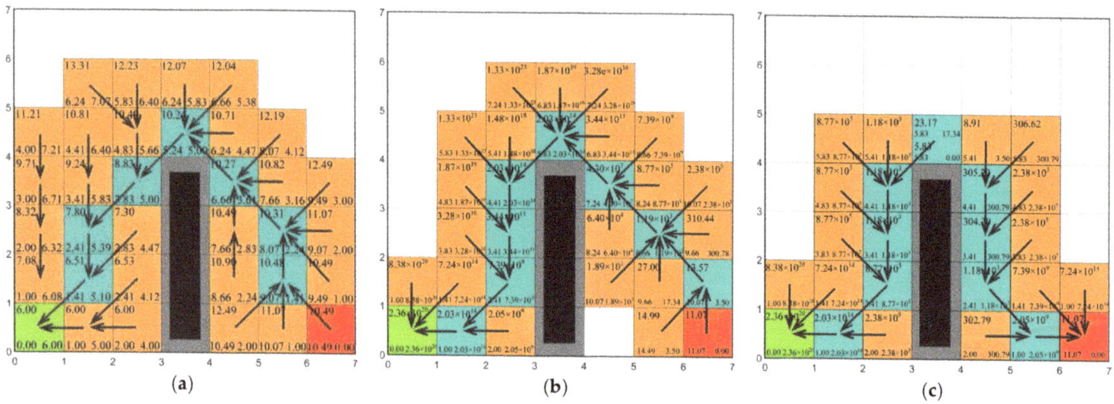

Figure 7. Comparison of the cost during the calculation process. (**a**) A*. (**b**) GA*. (**c**) BSGA*.

The BSGA* algorithm searches from the starting point and the target point simultaneously, where the nodes to be searched in the forward search process beginning from the starting point are stored in $OpenList\_forward$, and the nodes to be searched in the reverse search process beginning from the target point are stored in $OpenList\_backward$. The forward and reverse search processes are roughly the same as that of GA*. The difference is that the target point of the forward search is $Current\_back$, which is the current node of the reverse search, and the cost function is shown in Equation (9). The target point of the reverse search is $Current\_forward$, which is the current node of the forward search, and the cost function is shown in Equation (10). The two search processes are carried out alternately and both adopt the eight neighborhood search strategy until the current nodes of the two search processes meet.

$$\begin{cases} f_{forward}(n_f) = g_{forward}(n_f) + \frac{h_{forward}(n_f)}{D_{forward}(n_f)} \\ D_{forward}(n_f) = \frac{1}{\sigma\sqrt{2\pi}} e^{\frac{-(h_{forward}(n_f)-\mu)^2}{2\sigma}} \end{cases} \tag{9}$$

$$\begin{cases} f_{backward}(n_b) = g_{backward}(n_b) + \frac{h_{backward}(n_b)}{D_{backward}(n_b)} \\ D_{backward}(n_b) = \frac{1}{\sigma\sqrt{2\pi}} e^{\frac{-(h_{backward}(n_b)-\mu)^2}{2\sigma}} \end{cases} \tag{10}$$

where n_f is the current node of forward search; n_b is the current node of the reverse search; $g_{forward}(n_f)$ is the actual cost from the starting point to n_f; $h_{forward}(n_f)$ is the Euclidean heuristic cost of forward search from n_f to n_b; and $D_{forward}(n_f)$ is the dynamic attenuation coefficient of the forward heuristic function. $g_{backward}(n_b)$ is the actual cost of the target point to n_b; $h_{backward}(n_b)$ is the Euclidean heuristic cost of the next reverse search from n_b to the n_f; and $D_{backward}(n_b)$ is the dynamic attenuation coefficient of the backward heuristic function.

The pseudo code for the BSGA* algorithm is as follows (Algorithm 1).

Algorithm 1

Initialization
Puff up obstacles in the grid map
Determine the starting point (X_s, Y_s) and target point (X_e, Y_e)
$(X_s, Y_s) \rightarrow CloseList\_forward$, $(X_e, Y_e) \rightarrow CloseList\_backward$, set $CloseList\_forward\_cost = 0$, $CloseList\_backward\_cost = 0$
$Child\_nodes\_forward \rightarrow OpenList\_forward$, $Child\_nodes\_backward \rightarrow OpenList\_backward$
$Flag = 1$
$while = Flag$
    If $(Child\_nodes\_forward \subset OpenList\_forward) \cap (Child\_nodes\_backward \subset OpenList\_backward) \neq 1$
    Add nodes to the corresponding set
    Calculate the $f_{forward}$ of all nodes in $OpenList\_forward$, min $\_n_f \rightarrow CloseList\_forward$, set $n_f = current\_forward$, $f_{forward}(n_f) \rightarrow CloseList\_forward\_cost$
    Calculate the $f_{backward}$ of all nodes in $OpenList\_backward$, min $\_n_b \rightarrow CloseList\_backward$, set $n_b = current\_backward$, $f_{backward}(n_b) \rightarrow CloseList\_backward\_cost$
    $if\ current\_forward = current\_backward$
        Flip the elements in $CloseList\_backward$ and add them to $CloseList\_forward$
        $Cost = CloseList\_forward\_cost(end) + CloseList\_backward\_cost(end)$
        Generate the optimal path
        $Flag = 0$
    $else$
        Non-obstacle sub-nodes $current\_forward \rightarrow OpenList\_forward$
        Non-obstacle sub-nodes $current\_backward \rightarrow OpenList\_backward$
    $end\ if$
$end\ while$

As shown in Figure 7c, the BS structure is introduced to dynamically switch and adjust the starting point and target point in the search process, so that the current node always keeps the trend of being close to the target point. In this process, $f(n)$ decreases rapidly until the two paths meet and falls to the lowest value. The values of the upper and lower levels of the encounter grids represent the total cost, actual cost, and heuristic cost of forward search and reverse search at this node, respectively. The introduction of the BS structure effectively solves the problem of repeatedly searching nodes when avoiding obstacles and reaching the target point, and improves the efficiency of search.

In order to verify the performance of the BSGA* algorithm, as shown in Figures 8–10, the A* algorithm, GA* algorithm, and BSGA* algorithm were executed 50 times on the maps of 10 × 10, 20 × 20, and 30 × 30, respectively. It can be seen that, in the case of large obstacles between the starting point and the target point, the BSGA* algorithm searches fewer nodes compared to the A* algorithm and the GA* algorithm.

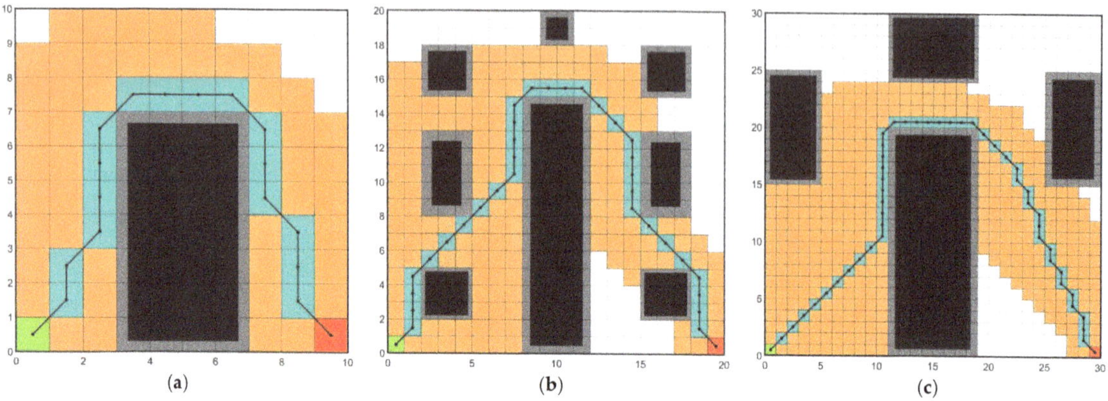

Figure 8. Performance of A* algorithm with a large area of obstacles: (**a**) Map size is 10 × 10. (**b**) Map size is 20 × 20. (**c**) Map size is 30 × 30.

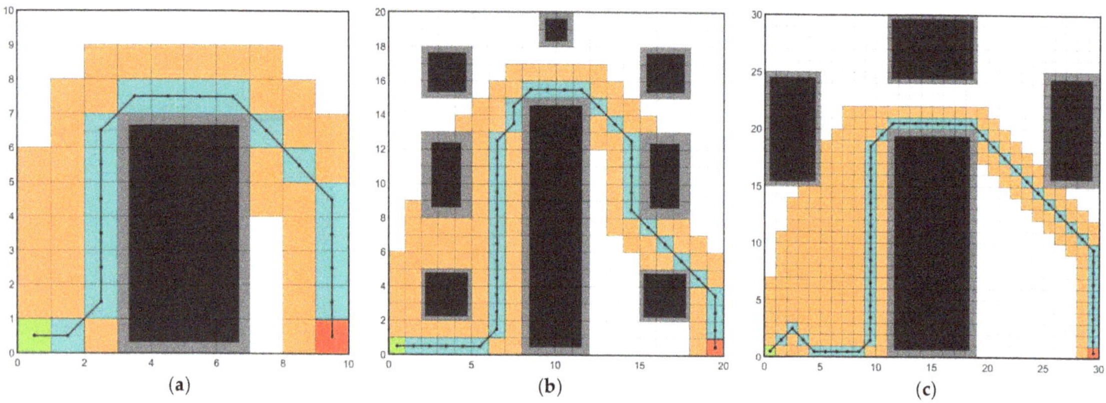

Figure 9. Performance of GA* algorithm with a large area of obstacles: (**a**) Map size is 10 × 10. (**b**) Map size is 20 × 20. (**c**) Map size is 30 × 30.

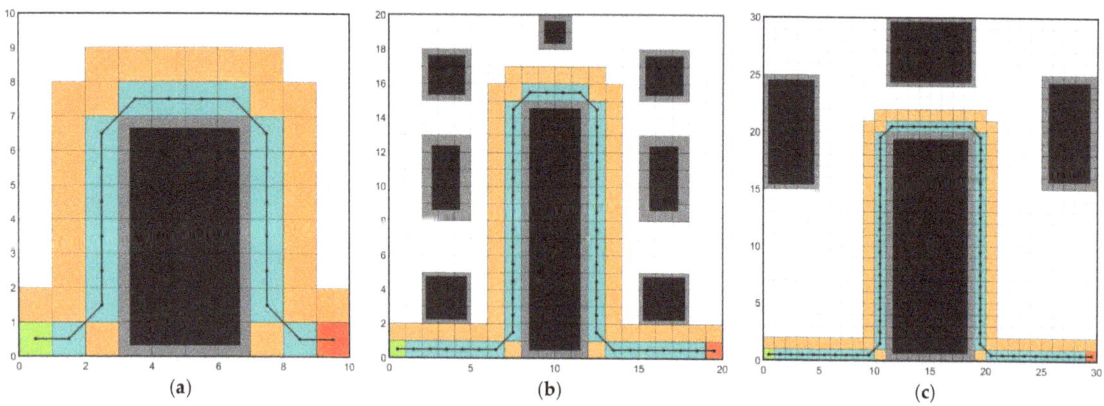

Figure 10. Performance of BSGA* algorithm with a large area of obstacles: (**a**) Map size is 10 × 10. (**b**) Map size is 20 × 20. (**c**) Map size is 30 × 30.

According to Table 3, it can be seen that, on the 10 × 10 grid map, compared with the A* and GA* algorithms, the calculation time of BSGA* is reduced by 61.44% and 21.04% and the number of search nodes is reduced by 28.13% and 9.80%, respectively, but there is no significant advantage in the total turning angle. On the 20 × 20 grid map, compared with the A* algorithm and GA* algorithm, the calculation time of BSGA* is reduced by 59.48% and 41.25%, the number of search nodes is reduced by 45.16% and 28.17%, and the total turning angle is reduced by 20% and 20%, respectively. On the 30 × 30 grid map, compared with the A* algorithm and GA* algorithm, the calculation time of BSGA* is reduced by 80.94% and 69.73% and the number of search nodes is reduced by 64.80% and 49.64%, respectively. The A* algorithm makes several turns in the path after bypassing the obstacles, and the total turning angle rises rapidly, while the total turning angle of the BSGA* algorithm and GA* algorithm is relatively more stable. Although the BSGA* algorithm has obvious advantages over the GA* and A* algorithms in search efficiency, the path length will be slightly longer and the advantages in total turning angle are not obvious.

Table 3. Comparison of the paths generated by the A*, GA*, and BSGA* algorithms under different grid maps, the values of average calculation time are reported as mean ± standard deviation for 50 executions.

Map Size	Algorithm	Average Calculation Time/ms	Number of Searched Nodes	Total Angle/°	Path Length
10 × 10	A*	574.94 ± 12.27	64	450	19.49
	GA*	280.79 ± 4.48	51	270	20.07
	BSGA*	221.71 ± 2.42	46	360	20.66
20 × 20	A*	1593.84 ± 18.83	186	450	39.63
	GA*	1099.21 ± 7.24	142	450	42.56
	BSGA*	645.77 ± 7.24	102	360	46.66
30 × 30	A*	6192.52 ± 38.80	392	810	56.11
	GA*	3898.71 ± 27.79	274	405	62.46
	BSGA*	1180.20 ± 4.77	138	360	66.66

3.3. Multi-Layer Turning Point Filtering Strategy

In order to further reduce the total turning angle and improve smoothness, this study proposed an optimization strategy based on turning point filtering. The specific procedures are as follows:

(1) Store the path points $Grid\_node_i(i = 1,2,\ldots)$ obtained by BSGA* algorithm in the $CloseList\_Path$ and store the obstacles and environment boundaries in the $Obstacle\_nodes$. Store the $Start\_node = (X_s, Y_s)$ and the $target\_node = (X_e, Y_e)$ in the Opt_path.

(2) Calculate the slope k_1 and k_2 of $Start\_node$ and $Grid\_node_1$, $Start\_node$ and $Grid\_node_2$ respectively. If $k_1 = k_2$, then $Grid\_node_1$ is not a turning point, continue to calculate the slope k_i of $Start\_node$ and $Grid\_node_i$ in turn until $k_1 \neq k_i$, then $Grid\_node_{i-1}$ is the turning point. $Grid\_node_i$ will be stored in the set of $Turn\_nodes$, then repeat the above process from $Grid\_node_i$ until it reaches the target point.

(3) Connect the previous node $Grid\_node_{i-2}$ of the $Grid\_node_{i-1}$ with the next node $Grid\_node_i$. If the connection between them passes through the obstacle, then $Grid\_node_{i-1}$ will be stored in Opt_path. If the connection does not pass through the obstacle, $Grid\_node_{i-1}$ will be deleted in $Turn\_nodes$. Judge all the turning points in $Turn\_nodes$ in turn according to this step.

(4) Repeat steps (2) and (3) until the connection between the previous node and the next node of any of the turns in $Turn\_nodes$ passes through the obstacle and connects the nodes in Opt_path to generate the optimized path, then the algorithm stops.

To verify the effectiveness of the turning point filtering strategy, as shown in Figures 11 and 12, 50 executions were conducted, respectively, to compare the A*, BSGA*, and improved BSGA* algorithms under grid maps with sizes of 15 × 15 and 30 × 30. The purple nodes in Figures 11c and 12c are the filtered turning points. As can be seen from Table 4, the calculation time of improved BSGA* is reduced by 62.45% compared with the A* algorithm and improved by 13.79% compared with the BSGA* algorithm in the 15 × 15 grid map. This is because improved BSGA* is implemented by re-performing the turning point filtering on the paths planned by the BSGA* algorithm, so the calculation time is slightly longer. Compared with the A* and BSGA* algorithms, the total turning angle of paths planned by improved BSGA* is reduced by 89.76% and 86.34%, respectively, the path length is reduced by 1.13. In the 30 × 30 grid map, compared with A* and BSGA*, the calculation time of improved BSGA* is reduced by 92.28% and that of BSGA* is improved by 7.26%, the total turning angle of the path is reduced by 87.89% and 81.83%, the path length is reduced by 2.33, respectively. Therefore, it can be seen that the improved BSGA* path has better smoothness and is more suitable for the driving process of the DDMR.

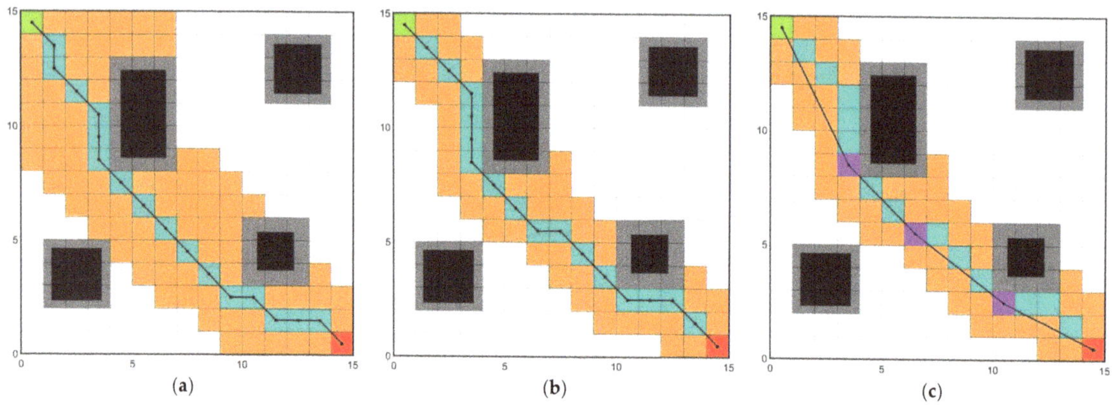

Figure 11. Performance of turning point filtering strategies in a 15 × 15 map: (**a**) A*. (**b**) BSGA*. (**c**) Improved BSGA*.

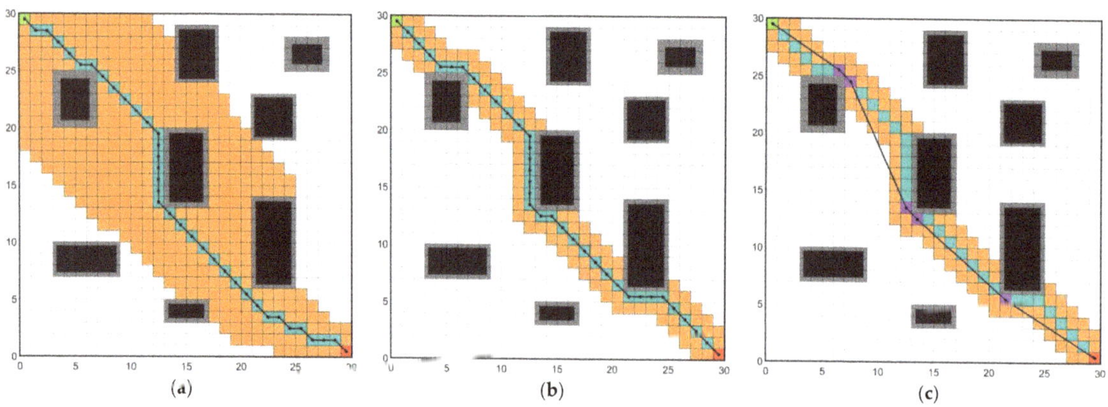

Figure 12. Performance of turning point filtering strategies in a 30 × 30 map: (**a**) A*. (**b**) BSGA*. (**c**) Improved BSGA*.

Table 4. Comparison of paths generated by A*, BSGA*, and improved BSGA* algorithms under 15 × 15 and 30 × 30 grid maps; the values of average calculation time are reported as mean ± standard deviation for 50 executions.

Map Size	Algorithm	Average Calculation Time/ms	Number of Searched Nodes	Total Angle/°	Path Length
15 × 15	A*	975.49 ± 16.29	95	360	21.55
	BSGA*	321.92 ± 7.27	63	270	21.55
	Improved BSGA*	366.31 ± 8.35	63	36.87	20.42
30 × 30	A*	12,316.77 ± 99.37	388	540	44.52
	BSGA*	886.42 ± 12.69	131	360	44.52
	Improved BSGA*	950.80 ± 14.47	131	65.42	42.19

4. Simulation and Experiments

4.1. Performance Comparison of Different Algorithms

In order to verify the performance of the improved BSGA* algorithm, the 15 × 15 and 30 × 30 grid maps constructed in Section 3.3 were used to compare the improved BSGA* algorithm with the ACO algorithm [40,41], D* lite algorithm [42], and GA [43]. The ACO algorithm [44] uses pheromone as the medium to transmit the information learned by ants in the process of moving, so that ants can move to the area with high pheromone concentration spontaneously. The parameter settings of the ACO algorithm are shown in Table 5. The GA [45] retains individuals according to their fitness by simulating the biological evolution process and updates the population through genetic and mutation operations. The parameters settings of GA are shown in Table 6. The D* lite algorithm uses the same heuristic function as the A* algorithm to guide the search [46], but the difference is that the D* lite algorithm first takes the target point as the starting point for reverse search to obtain the shortest path information, and can use the previously obtained information to reduce the search scope when it is necessary to re-plan during a later process.

Table 5. Setting the parameters of ant colony algorithm.

Parameters	Value
Maximum number of iterations	200
Number of ants	50
The expectation heuristic factor	1.5
The pheromone heuristic factor	7
The evaporation coefficient of pheromone	0.3
Initial amount of pheromone concentration	1
The constant coefficient	1

Table 6. Setting of genetic algorithm parameters.

Parameters	Value
Population size	500
Number of generations	200
Crossover probability	0.8
Mutation probability	0.2

Figure 13a,b shows the paths generated by the above-mentioned various algorithms on the 15 × 15 and 30 × 30 grid maps, respectively. It can be seen that, compared with the ACO algorithm, GA, and the D* lite algorithm, the path generated by the improved BSGA* algorithm is smoother and the total turning angle is smaller. It is worth noting that, due to the randomness of the paths generated by the ACO algorithm and GA, the results of each execution may be different. For better comparison, these three algorithms were executed

50 times, respectively, on the above two grid maps; the results are shown in Table 7. It can be seen that, on the 15 × 15 grid map, the improved BSGA* algorithm is 69.84%, 5.81%, and 85.97% lower than the ACO algorithm, 69.76%, 92.50%, and 6.20% lower than the GA, and 89.45%, 5.26%, and 86.34% lower than the D* lite algorithm in terms of calculation time, path length, and total turning angle, respectively. On the 30 × 30 grid map, the improved BSGA* algorithm is 93.74%, 5.45%, and 80.87% lower than the ACO algorithm, 74.34%, 92.64%, and 5.30% lower than the GA, and 86.82%, 5.30%, and 79.23% lower than the D* lite algorithm in these three aspects, respectively. From the above data, it can be seen that the improved BSGA* algorithm has a superior performance compared to the ACO algorithm, GA, and the D* lite algorithm. Therefore, the paths generated by the improved BSGA* algorithm are more suitable for DDMRs.

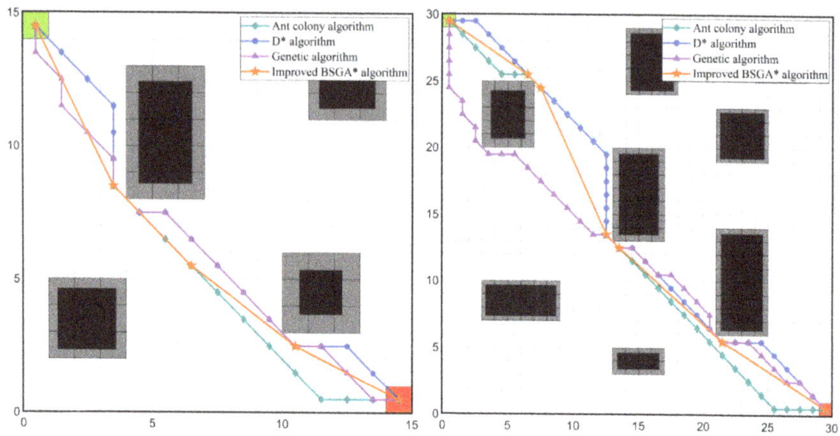

(a) (b)

Figure 13. Comparison of paths generated by ACO, GA, D* lite, and improved BSGA* algorithms under different maps: (**a**) Map size is 15 × 15; (**b**) Map size is 30 × 30.

Table 7. Comparison of paths generated by ACO, GA, D* lite, and improved BSGA* algorithms under 15 × 15 and 30 × 30 grid maps; the values of average calculation time are reported as mean ± standard deviation for 50 executions.

Map Size	Algorithm	Average Calculation Time/ms	Total Angle/°	Path Length
15 × 15	ACO	1214.73 ± 89.46	262.80 ± 78.65	21.68 ± 0.23
	GA	1211.47 ± 156.62	491.40 ± 21.75	21.77 ± 0.45
	D* lite	3473.67 ± 69.46	270.00	21.55
	Improved BSGA*	366.31 ± 8.35	36.87	20.42
30 × 30	ACO	15,190.43 ± 376.30	342.00 ± 106.49	44.62 ± 0.27
	GA	3705.64 ± 338.03	888.30 ± 141.11	45.77 ± 0.77
	D* lite	7217.43 ± 84.26	315.00	44.55
	Improved BSGA*	950.80 ± 14.47	65.42	42.19

Figures 14 and 15 are the column graphs and error bars drawn according to the above comparison experiments. Since both the D* lite algorithm and improved BSGA* algorithm are heuristic algorithms, the optimal paths generated in each execution are the same, so the standard deviations of path length and total turning angle are 0 in 50 executions. The paths generated by the ACO algorithm and the GA have randomness. As can be seen from Figures 14 and 15, after enough iterations, although the path generated by the ACO algorithm and GA is close to the improved BSGA* algorithm in length, the standard deviation of total turning angle, however, is larger due to the randomness of path point selection. This means that both algorithms are not stable enough in terms of path smoothing. In contrast, the paths generated by the improved BSGA* algorithm are not only smoother, but also the optimal path generated is unique and performs more stably when the grid map, starting point, and target point are determined.

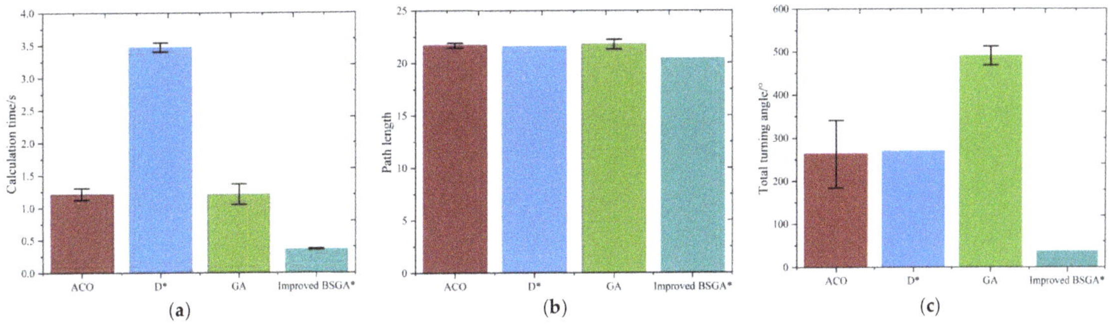

Figure 14. Column graphs and error bars of executing time, path length, and total turning angle in 15 × 15 map: (**a**) Executing time; (**b**) Path length; (**c**) Total turning angle.

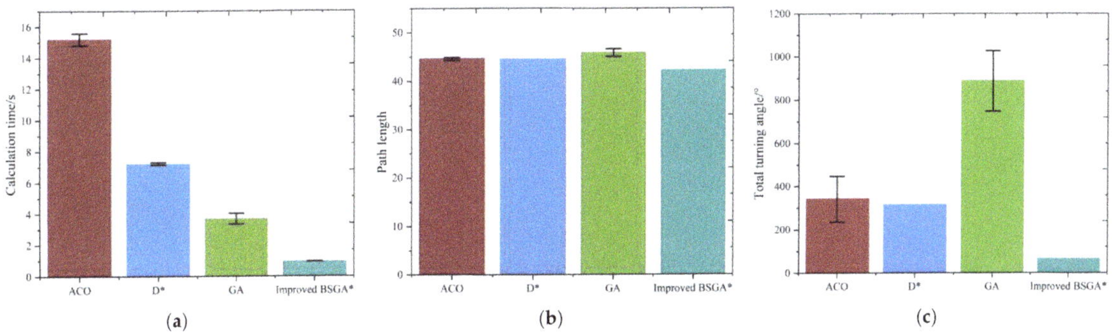

Figure 15. Column graphs and error bars of executing time, path length, and total turning angle in 30 × 30 map: (**a**) Executing time; (**b**) Path length; (**c**) Total turning angle.

4.2. Comparison in Real Environments

In order to verify the effectiveness of the improved BSGA* algorithm in real environments, a LEO ROS mobile robot is selected for the path-planning experiment. As shown in Figure 16, rubber wheels are adopted as the front wheels and two DC three-phase brushless motors are equipped, which can realize the differential control of both sides of the driving wheels. The rear wheels use omnidirectional wheels to achieve turning. The robot is equipped with a US-015 ultrasonic module, which can detect obstacles with a measuring accuracy of 0.1 cm + 1%. In addition, the robot uses EAI G1 LIDAR to test the surrounding environment, obtains obstacle information based on the triangulation principle, constructs environment maps using the *Gmapping* SLAM algorithm, saves them with *map_server*, and generates grid maps through the *map_2d* function package. The AMCL algorithm is used to determine the initial position and pose of the robot before navigation.

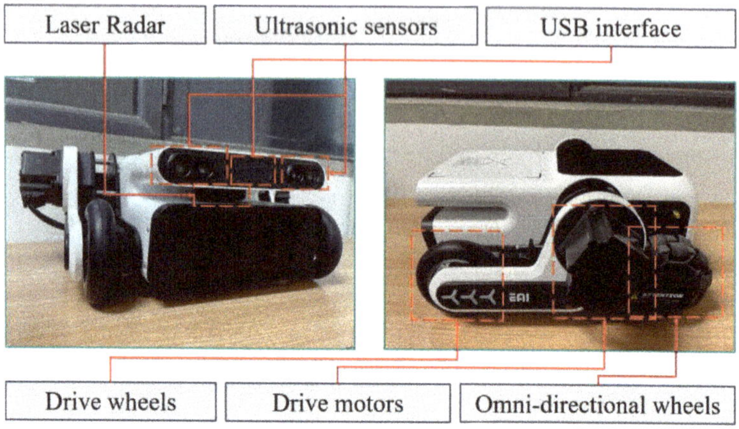

Figure 16. Main structure of the LEO ROS robot.

The global path-planning algorithms embedded in the LEO ROS mobile robots exist in the form of plug-ins in the *move_base* package. By creating and configuring the parameter file *base_global_planner_param.yaml* writing the *GlobalPlanner* class, the selection of different global path-planning algorithms can be realized. In order to apply the improved BSGA* algorithm to the robot, *BSGAstar_global_planner* plug-in is written and added into the *move_base* package based on the current *GlobalPlanner*, and the algorithm was implemented by modifying and configuring the parameter file *move_base_params.yaml*.

The LEO ROS robot is 360 mm in length, 455 mm in width, and 160 mm in height. The LIDAR is 100 mm from the ground. Based on the above size, 360 × 260 × 40 mm cardboard boxes are used to build the experimental environment, as shown in Figure 17. The edge of the square experimental environment consists of 15 cardboard boxes placed side by side. Figure 18a shows environment 1, the starting point and the target point are in a diagonal position, and there are three obstacle areas randomly placed in the environment. Figure 17b shows environment 2, there is a large area of obstacles between the starting point and the target point. The environmental map was constructed using the *Gmapping* SLAM algorithm in the ROS system and displayed on the *rivz* platform in real time. The size of the grid constructed by the robot was set to 360 mm × 360 mm. As the construction of environments is affected by many factors, such as the smoothness of driving, it is difficult to be completely consistent with the actual environment.

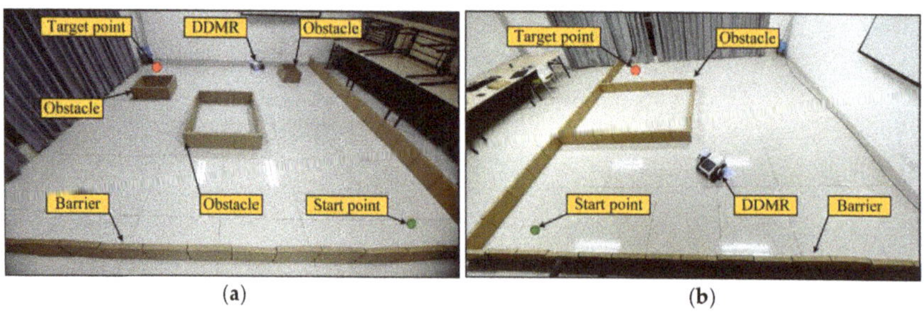

Figure 17. Experimental environments: (**a**) Environment 1; (**b**) Environment 2.

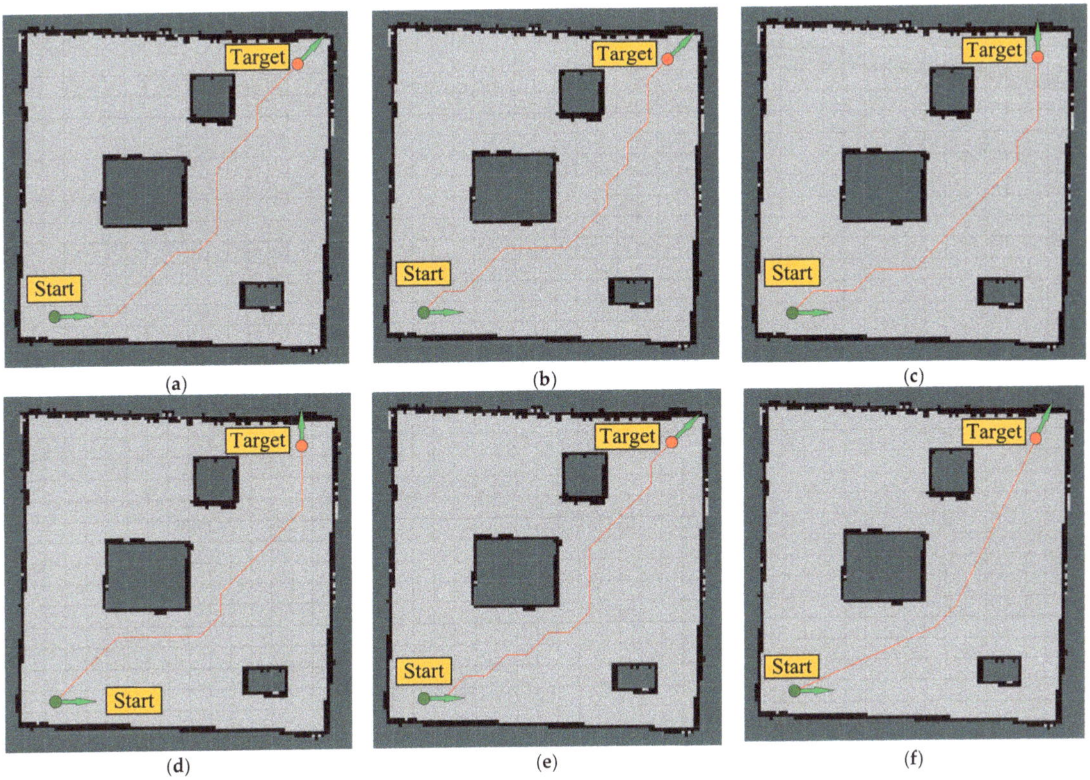

Figure 18. Comparison of paths generated by Dijkstra, A*, and improved BGA* algorithms in environment 1: (**a**) Dijkstra; (**b**) A*; (**c**) D* lite; (**d**) ACO; (**e**) GA; (**f**) Improved BSGA*.

In order to verify the performance of the improved BSGA* algorithm proposed in this study, it is compared with the Dijkstra algorithm, A* algorithm, D* lite algorithm, ACO algorithm, and GA, respectively, in two environments. The above algorithms were executed 10 times in both environments, respectively. Figure 18 shows the paths generated by the above six algorithms in environment 1 and Figure 19 shows the paths generated by the above six algorithms in environment 2. The green arrow in the figure represents the current orientation of the robot, and the red line segment represents the driving path. It can be seen that the paths generated by the improved BSGA* algorithm are smoother in both maps, mainly because the redundant turning points can be removed by using a multi-layer turning point filtering strategy.

Tables 8 and 9 show the comparison of the calculation times of six algorithms and the driving time of the LEO ROS robot in environments 1 and 2.

Table 8. The comparison of the path-planning time of six algorithms and travel time of the LEO ROS robot in environment 1; the values of average calculation time are reported as mean ± standard deviation for 10 executions.

Algorithm	Average Calculation Time/ms	Actual Driving Time/s
Dijkstra	1134.99 ± 12.73	74.82 ± 0.53
A*	981.96 ± 19.28	101.60 ± 0.35
D* lite	4480.39 ± 138.67	54.76 ± 0.35
ACO	3940.42 ± 87.47	58.88 ± 0.41
GA	1049.15 ± 51.95	110.66 ± 0.48
Improved BSGA*	299.69 ± 13.21	35.02 ± 0.28

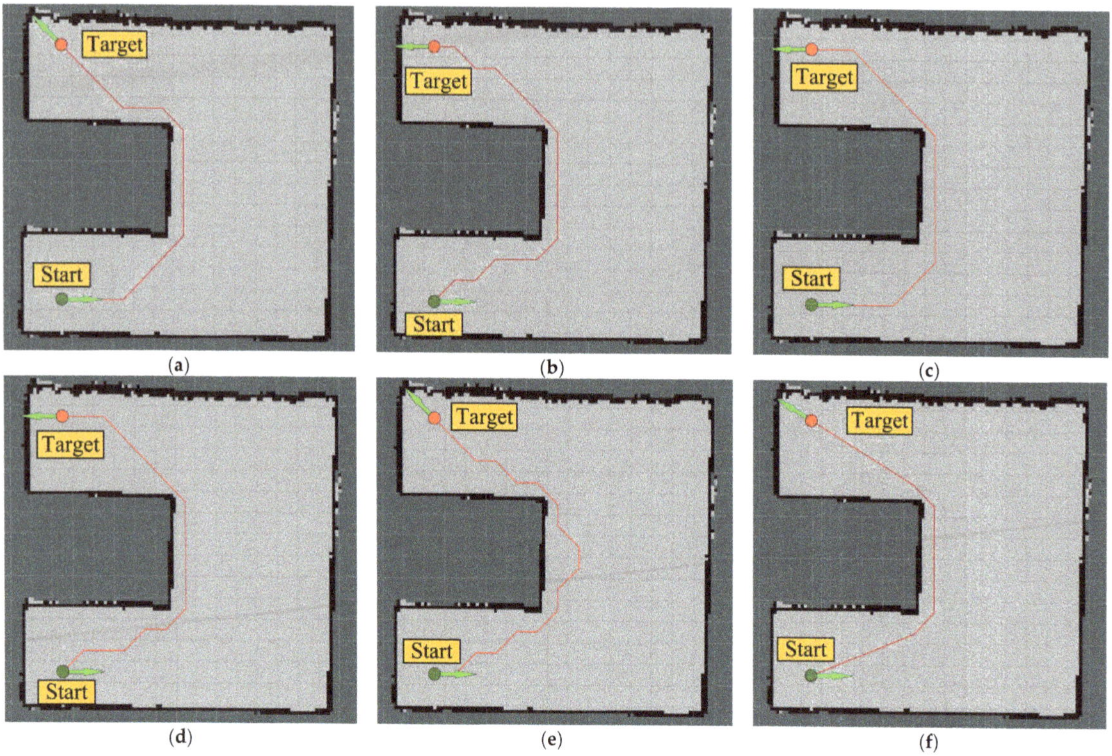

Figure 19. Comparison of paths generated by Dijkstra, A*, and improved BGA* algorithms in environment 2: (**a**) Dijkstra; (**b**) A*; (**c**) D* lite; (**d**) ACO; (**e**) GA; (**f**) Improved BSGA*.

Table 9. The comparison of the path-planning time of six algorithms and travel time of the LEO ROS robot in environment 2; the values of average calculation time are reported as mean ± standard deviation for 10 executions.

Algorithm	Average Calculation Time/ms	Actual Driving Time/s
Dijkstra	1454.97 ± 13.16	60.28 ± 0.34
A*	626.82 ± 8.14	93.61 ± 0.54
D* lite	4976.84 ± 114.65	78.14 ± 0.38
ACO	4426.67 ± 245.95	77.39 ± 0.51
GA	1940.80 ± 138.87	146.24 ± 0.62
Improved BSGA*	324.91 ± 8.02	52.88 ± 0.31

It can be obtained that, compared with the other five algorithms in environment 1, the calculation times of the improved BSGA* algorithm are reduced by 73.60%, 69.48%, 93.31%, 92.39%, and 71.43%, respectively, in turn. In environment 2, the corresponding data are 77.67%, 48.17%, 93.47%, 92.66%, and 83.26%, respectively, in turn. This is because the Gaussian function is used to dynamically weigh the heuristic function of the A* algorithm, and the BS search structure is applied to the A* algorithm, which can effectively reduce the redundant search nodes. From Table 9, it can be concluded that, compared with the other five algorithms in environment 1, the driving time of the LEO ROS robot using the improved BSGA* algorithm is reduced by 53.19%, 65.53%, 36.05%, 40.52%, and 68.35%, respectively, in turn. In environment 2, the corresponding data are 12.28%, 43.51%, 32.33%, 31.67%, and 63.84%, respectively, in turn. This is due to the fact that the LEO ROS robot takes relatively more time during the turning process, whereas the improved BSGA* algorithm generates smoother and shorter paths, thus effectively reducing the driving time. Therefore, in different types of environments, the improved BSGA* algorithm shows better results than the other five algorithms, which is beneficial for improving the service life of DDMRs, reducing motor loss, and improving driving efficiency.

5. Conclusions

In this paper, an improved BSGA* algorithm is proposed based on the traditional A* algorithm and the kinematic analysis of DDMRs. By introducing a Gaussian function to dynamically weigh the heuristic function and optimizing it using the BS structure, the algorithm performance in the face of different types of obstacles is effectively improved. In addition, a multi-layer turning point filtering strategy is proposed to reduce the redundant nodes and improve the smoothness of the path. The simulation results show that the improved BSGA* algorithm has a better performance than the ACO algorithm, D* lite algorithm, and GA in different maps. Finally, the improved BSGA* algorithm is applied to the LEO ROS mobile robot, and two different environments are built. The experimental results show that the improved BSGA* algorithm is more suitable to provide global path guidance for DDMRs than the Dijkstra algorithm, A* algorithm, ACO algorithm, D* algorithm, and GA. As a static planning method, the improved BSGA* cannot deal with dynamic obstacles, so it should be combined with a local obstacle avoidance algorithm to ensure the safety of driving in future work.

Author Contributions: Conceptualization, M.Y.; Methodology, M.Y.; Software, M.Y.; Investigation, M.Y.; Writing—original draft, M.Y.; Writing—review and editing, M.Y., H.D. and X.F.; Funding acquisition, H.D.; Resources, X.F.; Project administration, X.F.; Form analysis, P.L.; Data curation, Y.L.; Supervision, H.L.; Visualization, H.L.; Validation, H.L. All authors have read and agreed to the published version of the manuscript.

Funding: This research was funded by the Key Technology Research and Development Program of Shandong, grant number 2022CXGC010101 and the Key Technology Research and Development Program of Shandong, grant number 2019JZZY010443.

Institutional Review Board Statement: Not applicable.

Informed Consent Statement: Not applicable.

Data Availability Statement: Not applicable.

Conflicts of Interest: The authors declare that they have no known competing financial interest or personal relationship that could have appeared to influence the work reported in this paper.

References

1. Stefek, A.; Pham, T.V.; Krivanek, V.; Pham, K.L. Energy Comparison of Controllers Used for a Differential Drive Wheeled Mobile Robot. *IEEE Access* **2020**, *8*, 170915–170927. [CrossRef]
2. Felix-Rendon, J.; Bello-Robles, J.C.; Fuentes-Aguilar, R.Q. Control of differential-drive mobile robots for soft object deformation. *ISA Trans.* **2021**, *117*, 221–233. [CrossRef]
3. Xie, H.; Zheng, J.; Chai, R.; Nguyen, H.T. Robust tracking control of a differential drive wheeled mobile robot using fast nonsingular terminal sliding mode. *Comput. Electr. Eng.* **2021**, *96*, 107488. [CrossRef]
4. Haider, M.H.; Wang, Z.; Khan, A.A.; Ali, H.; Zheng, H.; Usman, S.; Kumar, R.; Bhutta, M.U.M.; Zhi, P. Robust mobile robot navigation in cluttered environments based on hybrid adaptive neuro-fuzzy inference and sensor fusion. *J. King Saud Univ. Comput. Inf. Sci.* **2022**, *34*, 9060–9070. [CrossRef]
5. Raikwar, S.; Fehrmann, J.; Herlitzius, T. Navigation and control development for a four-wheel-steered mobile orchard robot using model-based design. *Comput. Electron. Agric.* **2022**, *202*, 107410. [CrossRef]
6. Bai, Y.; Zhang, B.; Xu, N.; Zhou, J.; Shi, J.; Diao, Z. Vision-based navigation and guidance for agricultural autonomous vehicles and robots: A review. *Comput. Electron. Agric.* **2023**, *205*, 107584. [CrossRef]
7. Jian, Z.; Zhang, S.; Chen, S.; Nan, Z.; Zheng, N. A Global-Local Coupling Two-Stage Path Planning Method for Mobile Robots. *IEEE Robot. Autom. Lett.* **2021**, *6*, 5349–5356. [CrossRef]
8. Mandloi, D.; Arya, R.; Verma, A.K. Unmanned aerial vehicle path planning based on A* algorithm and its variants in 3d environment. *Int. J. Syst. Assur. Eng. Manag.* **2021**, *12*, 990–1000. [CrossRef]

9. Tang, G.; Tang, C.; Claramunt, C.; Hu, X.; Zhou, P. Geometric A-Star Algorithm: An Improved A-Star Algorithm for AGV Path Planning in a Port Environment. *Ieee Access* **2021**, *9*, 59196–59210. [CrossRef]
10. Liao, B.; Wan, F.; Hua, Y.; Ma, R.; Zhu, S.; Qing, X. F-RRT*: An improved path planning algorithm with improved initial solution and convergence rate. *Expert Syst. Appl.* **2021**, *184*, 115457. [CrossRef]
11. Miao, C.; Chen, G.; Yan, C.; Wu, Y. Path planning optimization of indoor mobile robot based on adaptive ant colony algorithm. *Comput. Ind. Eng.* **2021**, *156*, 107230. [CrossRef]
12. Sarkar, R.; Barman, D.; Chowdhury, N. Domain knowledge based genetic algorithms for mobile robot path planning having single and multiple targets. *J. King Saud Univ. Comput. Inf. Sci.* **2022**, *34*, 4269–4283. [CrossRef]
13. Lavalle, S.M. *Planning Algorithms*; Cambridge University Press: Cambridge, UK, 2006; ISBN 9780511546877.
14. Guo, J.; Xia, W.; Hu, X.; Ma, H. Feedback RRT* algorithm for UAV path planning in a hostile environment. *Comput. Ind. Eng.* **2022**, *174*, 108771. [CrossRef]
15. Wang, J.; Li, B.; Meng, M.Q.H. Kinematic Constrained Bi-directional RRT with Efficient Branch Pruning for robot path planning. *Expert Syst. Appl.* **2021**, *170*, 114541. [CrossRef]
16. Hou, W.; Xiong, Z.; Wang, C.; Chen, H. Enhanced ant colony algorithm with communication mechanism for mobile robot path planning. *Robot. Auton. Syst.* **2022**, *148*, 103949. [CrossRef]
17. Tuncer, A.; Yildirim, M. Dynamic path planning of mobile robots with improved genetic algorithm. *Comput. Electr. Eng.* **2012**, *38*, 1564–1572. [CrossRef]
18. Dijkstra, E.W. A note on two problems in connexion with graphs. *Numer. Math.* **1959**, *1*, 269–271. [CrossRef]
19. Hart, P.E.; Nilsson, N.J.; Raphael, B. A Formal Basis for the Heuristic Determination. *IEEE Trans. Syst. Sci. Cybern.* **1968**, *2*, 28–29.
20. Zhang, Z.; Jiang, J.; Wu, J.; Zhu, X. Efficient and optimal penetration path planning for stealth unmanned aerial vehicle using minimal radar cross-section tactics and modified A-Star algorithm. *ISA Trans.* **2023**, *134*, 42–57. [CrossRef]
21. Liu, Z.; Liu, H.; Lu, Z.; Zeng, Q. A Dynamic Fusion Pathfinding Algorithm Using Delaunay Triangulation and Improved A-Star for Mobile Robots. *IEEE Access* **2021**, *9*, 20602–20621. [CrossRef]
22. Jiang, H.; Sun, Y. Research on global path planning of electric disinfection vehicle based on improved A* algorithm. *Energy Rep.* **2021**, *7*, 1270–1279. [CrossRef]
23. Liu, H.; Zhang, Y. ASL-DWA: An Improved A-Star Algorithm for Indoor Cleaning Robots. *IEEE Access* **2022**, *10*, 99498–99515. [CrossRef]
24. Li, J.; Liao, C.; Zhang, W.; Fu, H.; Fu, S. UAV Path Planning Model Based on R5DOS Model Improved A-Star Algorithm. *Appl. Sci.* **2022**, *12*, 11338. [CrossRef]
25. Zhang, H.; Li, M.; Yang, L. Safe Path Planning of Mobile Robot Based on Improved A* Algorithm in Complex Terrains. *Algorithms* **2018**, *11*, 44. [CrossRef]
26. Sang, H.; You, Y.; Sun, X.; Zhou, Y.; Liu, F. The hybrid path planning algorithm based on improved A* and artificial potential field for unmanned surface vehicle formations. *Ocean. Eng.* **2021**, *223*, 108709. [CrossRef]
27. Cui, S.; Chen, Y.; Li, X. A Robust and Efficient UAV Path Planning Approach for Tracking Agile Targets in Complex Environments. *Machines* **2022**, *10*, 931. [CrossRef]
28. Dang, S.T.; Dinh, X.M.; Kim, T.D.; Xuan, H.L.; Ha, M. Adaptive Backstepping Hierarchical Sliding Mode Control for 3-Wheeled Mobile Robots Based on RBF Neural Networks. *Electronics* **2023**, *12*, 2345. [CrossRef]
29. Latombe, J.C. *Robot Motion Planning*; Springer: Berlin/Heidelberg, Germany, 1991.
30. Klančar, G.; Matko, D.; Blažič, S. A control strategy for platoons of differential drive wheeled mobile robot. *Robot. Auton. Syst.* **2011**, *59*, 57–64. [CrossRef]
31. Li, Y.; Jin, R.; Xu, X.; Qian, Y.; Wang, H.; Xu, S.; Wang, Z. A Mobile Robot Path Planning Algorithm Based on Improved A* Algorithm and Dynamic Window Approach. *IEEE Access* **2022**, *10*, 57736–57747. [CrossRef]
32. Duchoň, F.; Babinec, A.; Kajan, M.; Beňo, P.; Florek, M.; Fico, T.; Jurišica, L. Path Planning with Modified a Star Algorithm for a Mobile Robot. *Procedia Eng.* **2014**, *96*, 59–69. [CrossRef]
33. Hong, Z.; Sun, P.; Tong, X.; Pan, H.; Zhou, R.; Zhang, Y.; Han, Y.; Wang, J.; Yang, S.; Xu, L. Improved A-Star Algorithm for Long-Distance Off-Road Path Planning Using Terrain Data Map. *ISPRS Int. J. Geo-Inf.* **2021**, *10*, 785. [CrossRef]
34. Zhao, H.; Zhang, B.; Sun, J.; Yang, L.; Yu, H. Spot-welding path planning method for the curved surface workpiece of body-in-white based on a memetic algorithm. *Int. J. Adv. Manuf. Technol.* **2021**, *117*, 3083–3100. [CrossRef]
35. Liu, C.; Mao, Q.; Chu, X.; Xie, S. An Improved A-Star Algorithm Considering Water Current, Traffic Separation and Berthing for Vessel Path Planning. *Appl. Sci.* **2019**, *9*, 1057. [CrossRef]
36. Flores-Caballero, G.; Rodriguez-Molina, A.; Aldape-Perez, M.; Villarreal-Cervantes, M.G. Optimized Path-Planning in Continuous Spaces for Unmanned Aerial Vehicles Using Meta-Heuristics. *IEEE Access* **2020**, *8*, 176774–176788. [CrossRef]
37. Li, C.; Huang, X.; Ding, J.; Song, K.; Lu, S. Global path planning based on a bidirectional alternating search A* algorithm for mobile robots. *Comput. Ind. Eng.* **2022**, *168*, 108123. [CrossRef]
38. Soltani, A.R.; Tawfik, H.; Goulermas, J.Y.; Fernando, T. Path planning in construction sites: Performance evaluation of the Dijkstra, A∗, and GA search algorithms. *Adv. Eng. Inform.* **2002**, *16*, 291–303. [CrossRef]
39. Ryu, H.; Chung, W.K. Local map-based exploration using a breadth-first search algorithm for mobile robots. *Int. J. Precis. Eng. Manuf.* **2015**, *16*, 2073–2080. [CrossRef]

40. Dorigo, M.; Gambardella, L.M. Ant colony system: A cooperative learning approach to the traveling salesman problem. *Ieee Trans. Evol. Comput.* **1997**, *1*, 53–66. [CrossRef]
41. Dorigo, M.; Maniezzo, V.; Colorni, A. Ant system: Optimization by a colony of cooperating agents. *IEEE Trans. Syst. Man Cybern. Part B Cybern.* **1996**, *26*, 29–41. [CrossRef]
42. Stentz, A. Optimal and Efficient Path Planning for Partially-Known Environments. In Proceedings of the 1994 IEEE International Conference on Robotics and Automation, San Diego, CA, USA, 8–13 May 1994; pp. 3310–3317.
43. Sugihara, K.; Smith, J. Genetic Algorithms for Adaptive Motion Planning of an Autonomous Mobile Robot. In Proceedings of the IEEE International Symposium on Computational Intelligence in Robotics & Automation, Monterey, CA, USA, 10–11 July 1997.
44. Liu, J.; Anavatti, S.; Garratt, M.; Abbass, H.A. Modified continuous Ant Colony Optimisation for multiple Unmanned Ground Vehicle path planning. *Expert Syst. Appl.* **2022**, *196*, 116605. [CrossRef]
45. Raja, R.; Dutta, A.; Venkatesh, K.S. New potential field method for rough terrain path planning using genetic algorithm for a 6-wheel rover. *Robot. Auton. Syst.* **2015**, *72*, 295–306. [CrossRef]
46. Yu, J.; Yang, M.; Zhao, Z.; Wang, X.; Bai, Y.; Wu, J.; Xu, J. Path planning of unmanned surface vessel in an unknown environment based on improved D*Lite algorithm. *Ocean. Eng.* **2022**, *266*, 112873. [CrossRef]

applied sciences

MDPI

Article

A Reinforcement Learning Approach to Dynamic Trajectory Optimization with Consideration of Imbalanced Sub-Goals in Self-Driving Vehicles

Yu-Jin Kim [1], Woo-Jin Ahn [2], Sun-Ho Jang [2,3], Myo-Taeg Lim [2,*] and Dong-Sung Pae [4,*]

[1] Korea Institute of Science and Technology, Korea University, Seoul 02456, Republic of Korea; sally1004k@kist.re.kr
[2] School of Electrical Engineering, Korea University, Seoul 02841, Republic of Korea; wjahn@korea.ac.kr (W.-J.A.); jang1229@kiro.re.kr (S.-H.J.)
[3] Korea Institute of Robotics Technology Convergence, Andong 36728, Republic of Korea
[4] Department of Software, Sangmyung University, Cheonan 31066, Republic of Korea
* Correspondence: mlim@korea.ac.kr (M.-T.L.); paeds915@smu.ac.kr (D.-S.P.)

Abstract: Goal-conditioned Reinforcement Learning (RL) holds promise for addressing intricate control challenges by enabling agents to learn and execute desired skills through separate decision modules. However, the irregular occurrence of required skills poses a significant challenge to effective learning. In this paper, we demonstrate the detrimental effects of this imbalanced skill (sub-goal) distribution and propose a novel training approach, Classified Experience Replay (CER), designed to mitigate this challenge. We demonstrate that adapting our method to conventional RL methods significantly enhances the performance of the RL agent. Considering the challenges inherent in tasks such as driving, characterized by biased occurrences of required sub-goals, our study demonstrates the improvement in trained outcomes facilitated by the proposed method. In addition, we introduce a specialized framework tailored for self-driving tasks on highways, integrating model predictive control into our RL trajectory optimization training paradigm. Our approach, utilizing CER with the suggested framework, yields remarkable advancements in trajectory optimization for RL agents operating in highway environments.

Keywords: reinforcement learning; experience replay; self-driving; trajectory optimization

Citation: Kim, Y.-J.; Ahn, W.-J.; Jang, S.-H.; Lim, M.-T.; Pae, D.-S. A Reinforcement Learning Approach to Dynamic Trajectory Optimization with Consideration of Imbalanced Sub-Goals in Self-Driving Vehicles. *Appl. Sci.* **2024**, *14*, 5213. https://doi.org/10.3390/app14125213

Academic Editors: Fabrizio Giulietti, Nadjim Horri and William Holderbaum

Received: 14 May 2024
Revised: 8 June 2024
Accepted: 11 June 2024
Published: 15 June 2024

1. Introduction

Reinforcement Learning (RL) has made remarkable strides across various domains, particularly in addressing sequential decision-making problems. Initial successes stemmed from simulated game environments [1–3], primarily in discrete action spaces, marking milestones in RL's development. Recent advancements have extended RL applications to continuous action spaces, encompassing control tasks like manipulator robot control [4–6], mobile robot navigation [7], and legged robot locomotion [8,9].

However, when applied to the domain of autonomous driving, RL faces formidable challenges owing to the intricacies of traffic environments. Autonomous driving involves navigating at high speeds while adhering to traffic regulations and responsively reacting to nearby vehicles. This complexity has led to the adoption of goal-conditioned RL in path planning approaches, aiming to separate decision making and trajectory generation [10–12].

A universal value function [13] serves as a standard framework for goal-conditioned RL, particularly useful in environments with multiple sub-tasks [14,15]. However, the irregular occurrence and dominance of certain sub-goals hamper optimal policy learning in such settings, adversely impacting training performance. This imbalance in sub-goal distribution is not unique to RL but is also pervasive in data-centric Artificial Intelligence (AI) systems [16,17] and real-world applications. The highway environment exemplifies this imbalance, where the sub-goal of maintaining the current lane outweighs other sub-goals, such as changing to an adjacent lane. This highlights the necessity of addressing this imbalance to achieve high-performance agents.

Experience replay [18], applied in the deep Q-network algorithm [19], marked a significant advancement in off-policy RL. The experience replay method stores transition samples in a buffer, enabling the training of RL networks using uniformly randomly sampled minibatches from this buffer. However, the uniform sampling method poses limitations in effectively learning optimal policies, especially in environments characterized by imbalanced sub-goal distributions across multiple goals. Addressing this challenge has been the focus of research efforts aiming to enhance training efficiency within experience replay. Prioritizing temporal difference error [20] and reinterpreting unsuccessful experiences using methods like Hindsight Experience Replay (HER) [21] have shown effectiveness in sparse reward environments, yet they struggle to fully resolve severe imbalances inherent in practical applications.

To tackle this fundamental issue, we propose Classified Experience Replay (CER), specifically designed for multi-goal tasks. CER optimizes sample selection by adjusting the probability of choosing transitions associated with each sub-goal. Unlike traditional uniform sampling, CER's mechanism strategically alters the selection probabilities, mitigating the impact of imbalanced sub-goal distributions during training. Our empirical evaluations show the detrimental effects of imbalanced sub-goal distributions on the training policy. Through experimentation, we demonstrate that CER significantly improves sample efficiency and enhances the trained agent's performance in handling the challenges posed by imbalanced sub-goal distributions in practical scenarios.

Another challenge in highway driving tasks lies in designing an effective reward function. Unlike discrete success or fail task completion in game environments, the continuous trajectory generation in driving tasks demands meticulous consideration at each step due to multifaceted constraints. Designing an appropriate reward function for RL in such constraint-rich environments proves challenging, often leading to overlapping reward components and unexpected outcomes [22–25]. Achieving a comprehensive reward function encompassing all objectives and regulations necessitates extensive iterative reward shaping, a process prone to trial and error. Prior studies have attempted fine-grained rewards, aiming to obtain an optimal policy of action for continuous control. Meanwhile, Model Predictive Control (MPC) [26,27] optimizes vehicle control under constraints using objective functions. However, despite yielding optimal solutions, the computational overhead for real-time optimization in high-speed driving scenarios remains a challenge. Complex state space models and longer horizon calculations escalate computational demands significantly.

This paper proposes a framework for achieving stable driving maneuver training with reinforcement learning. To further promote efficient learning, we devise a two-stage hierarchical structure for driving tasks. This structure separates decision-making and trajectory generation modules, enabling online sub-goal generation conditioned for optimal trajectory planning. Leveraging both CER and the hierarchical structure, the RL agent exhibits enhanced training stability and adaptivity to evolving sub-goals. Moreover, we develop an MPC-synchronized reward structure that eliminates the need for extensive reward shaping. This hybrid structure allows concurrent computation of rewards alongside MPC control signals, facilitating training by utilizing optimized references with reduced reliance on manual crafting. Importantly, both the reward function and MPC are bypassed during the evaluation phases, ensuring real-time performance without their computational overhead. To evaluate our approach, agents are trained and tested in a complex gym highway environment [28], showcasing the efficacy of the proposed RL-MPC hybrid approach in navigating challenging highway scenarios efficiently. By demonstrating the adaptability and efficiency of our framework in this robust testing ground, we pave the way for the successful deployment of RL-based autonomous driving systems in dynamic highway settings.

Our contributions in this paper are as follows:

1. We address imbalanced goal distributions in driving environments through a novel off-policy training method called CER, enabling the RL agent to dynamically adapt to changing sub-goals during driving scenarios.
2. We propose a two-stage hierarchical structure for driving tasks, separating decision making and trajectory generation. This approach fosters efficient learning by facilitating online sub-goal generation tailored for optimal trajectory planning.
3. We introduce an MPC-synchronized reward structure that eliminates the need for extensive reward shaping. This hybrid framework allows concurrent reward computation alongside control signals, utilizing optimized references and minimizing manual tuning.

The remainder of the paper is organized as follows: Section 2 presents an overview of related works on the proposed method. Section 3 demonstrates the CER and the framework for a highway driving task, while Section 4 presents the experimental environment and the training results. Subsequently, we demonstrate the driving performance of the proposed method. Our framework is then evaluated qualitatively and quantitatively compared with other methods. Finally, Section 5 closes the paper with the conclusions.

2. Related Works

Many prior studies have employed RL for driving tasks, adopting diverse approaches such as end-to-end architectures [29,30] and hierarchical methods [10–12]. End-to-end approaches often oversimplify the complexity of driving tasks, encompassing multiple decision, planning, and control layers in dynamic environments. For instance, ref. [29] utilized Deep Deterministic Policy Gradients (DDPGs) [31] to generate control inputs like steering and throttle from image inputs on a two-lane track. However, their assumption of an environment with only the ego vehicle limited realism, as other vehicles were neglected. A similar constraint applied to the use of Dueling Deep Q Networks (DQNs) in [30], where the training and testing tracks were identical, hindering generalization. Unfortunately, typical end-to-end approaches intertwine actions with decisions, making maneuvers challenging to interpret and necessitating relearning from scratch when environmental conditions change. In contrast, hierarchical approaches, as demonstrated by [10], segregate the decision-making process from the trajectory generation process, mitigating the drawbacks of the end-to-end paradigm. Such approaches employ multiple agents to address various behaviors in driving situations.

In recent years, transformer [32] has emerged as a promising tool for training sequential actions, with several studies adopting them for self-driving tasks [33,34]. However, their large network size and computational demands pose significant challenges, particularly in real-time control applications such as autonomous vehicles.

Unlike end-to-end architectures, hierarchical approaches face fewer constraints on task complexity. Here, the driving task is divided into decision-making and trajectory generation, with RL used as the decision-making module to generate high-level commands, such as lane keeping or lane changing. The motion planner is designed to capture vehicle dynamics using a two-dimensional grid map planning method [10]. Alternatively, in [35], RL is used for both decision making and motion planning with separate RL agents, initialized using the inverse RL method [36]. Although this approach avoids rule-based methods, designing and shaping rewards for each agent demands considerable time and effort.

In another study, ref. [12] employed a conditional variational autoencoder for high-level decision making and an RL agent for control input generation based on the decision. While RL excels at handling unseen scenarios, it falls short in reflecting changes in traffic regulations and ensuring decision diversity. Moreover, limitations arise if speed limits are not included in the observation during training, requiring the agent to be trained anew. To address these concerns, some approaches leverage RL for decision-making and rule-based models for trajectory generation, as exemplified by [11]. This method utilizes a rule-based decision making module and generates control inputs from an RL agent, specifying decisions more comprehensively in both lateral and longitudinal directions.

However, the efficiency of designing and training multiple specialized agents with this approach is questionable.

To tackle these issues, our research incorporates a straightforward rule-based decision-making module to adapt to changes in traffic conditions and individual preferences. The control input is generated by the RL motion planner, ensuring the adaptive management of situations by utilizing concatenated state inputs from sub-goals and observations.

3. Classified Experience Replay and Driving Framework

Our goal is to enhance decision-making performance in sequential multi-goal problems within an environment characterized by imbalanced goal distribution and to design a motion planning and control framework for highway driving tasks. In this paper, we focus on the off-policy RL method, which employs experience replay. In this section, we introduce the motivation and method of our approach, as well as how we organized the overall framework to apply it to multi-goal tasks.

3.1. Classified Experience Replay

In environments where the sub-goal $g \in \mathcal{G}$ is manually switched, the quantity of transitions stored in the replay buffer varies depending on the distribution of sub-goals. While a uniform distribution is typically used for sampling the minibatch, this approach may lead to an uneven utilization of each sub-goal g and its corresponding transitions, thereby adversely affecting training performance. This effect is exacerbated in cases of more imbalanced sub-goal distributions. To mitigate this issue, we propose classified experience replay. Unlike conventional methods that use a single buffer, we employ multiple buffers, each having the same dimensions as the sub-goal space $\mathcal{G} \in \mathbb{R}^n$. Every transition is stored in the buffer corresponding to its associated g. During policy updates, transitions are equally sampled from all buffers, thereby mitigating the disproportionate sub-goal experience during training and preventing the agent from becoming overly reinforced towards specific goals. Figure 1 illustrates the process of storing and utilizing transitions in CER.

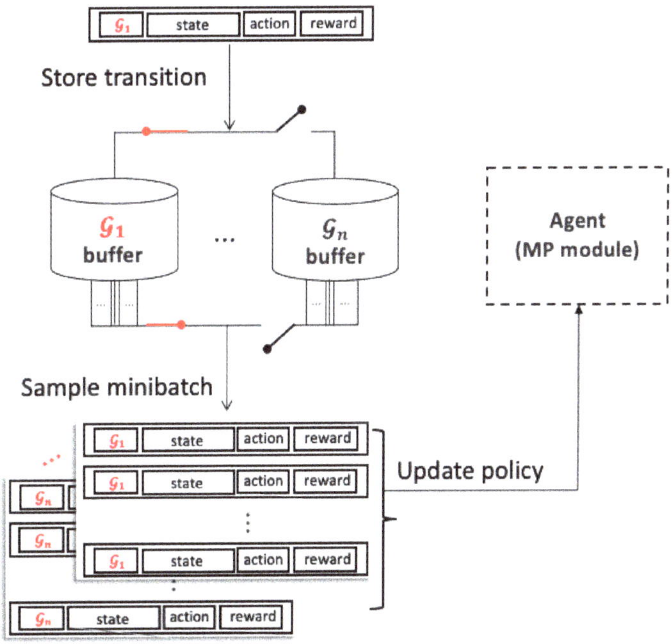

Figure 1. Visualization of classified experience replay. Transitions are stored in the classified buffer and uniformly sampled to train the agent, ensuring a balanced selection of transition samples.

To evaluate the impact of an imbalanced sub-goal distribution on the agent's training, we devised a straightforward cliff walk scenario featuring two cliffs (see Figure 2). Here, the agent aims to navigate from the start state $\mathbb{S}$ to the goal state $\mathbb{G}$ under different sub-goals: "Straight", "Up", and "Down". The optimal trajectory to the goal state varies depending on these sub-goals, which are manually designated before each episode. The agent receives rewards for reaching the goal, with specific rewards varying based on the sub-goal. Figure 2 depicts the reward settings in the environment for each sub-goal, with colors representing the one-step reward received by the agent for stepping on each cell.

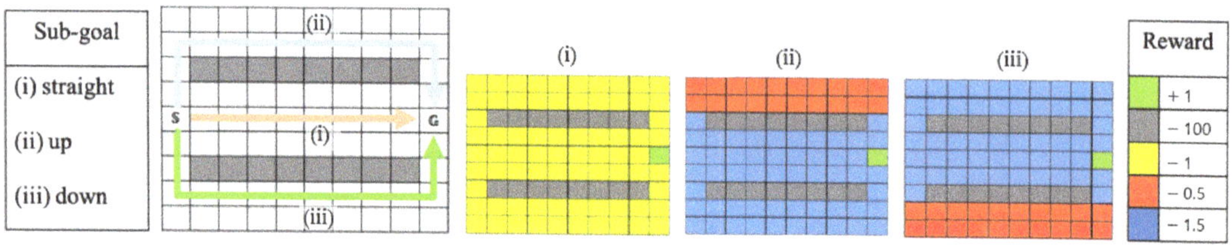

Figure 2. The twin cliff walk environment and reward configuration are depicted. The colored arrows in the left figure illustrate the optimal paths for achieving each sub-goal. The reward maps for each sub-goal are depicted as (**i–iii**).

To demonstrate the effect of sub-goal distribution on training performance, we trained and tested the agent in 10 cases with varying probabilities for g_{down}. The probability distribution is shown in Figure 3a. Initially, the probabilities of the sub-goals were uniform, with $p(g_{straight})$:$p(g_{up})$:$p(g_{down})$ = 1:1:1. Subsequently, we gradually increased the probabilities of $g_{straight}$ and g_{up} until reaching a ratio of $p(g_{straight})$:$p(g_{up})$:$p(g_{down})$ = 10:10:1. We compared the performances of agents trained using DQN and DQN with CER. The total buffer sizes used for DQN + CER were equivalent to the single buffer size used in DQN. A test was considered successful only if the agent found an optimal path for any g, denoted by O for optimal path, SO for sub-optimal path, and F for failure to reach the goal state. Figure 3b presents the test results, with the optimality test indicating whether the agent found an optimal path to the final destination under a given sub-goal. The agent trained with DQN frequently failed to reach the goal state as the probabilities became biased. In contrast, the agent trained with DQN with CER consistently found the optimal path under any sub-goal, irrespective of the non-uniformity in probabilities.

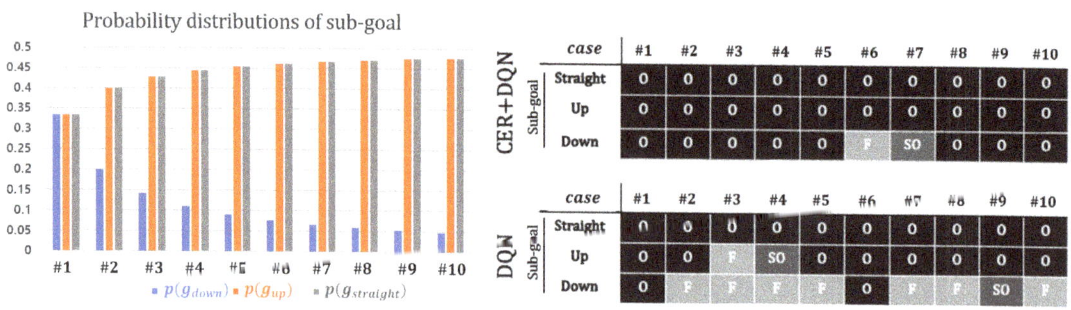

(**a**) Probability of sub-goal distribution per episode

(**b**) CER + DQN vs. DQN trained result comparison

Figure 3. (**a**) shows the probability distribution of each sub-goal from case #1 to case #10. In case #1, the distribution of three sub-goals is uniform, but as the cases progress, the distribution becomes biased towards two sub-goals other than g_{down}. (**b**) demonstrates the success or failure of the trained agent. The symbol in each cell indicates whether the agent found a path from the start cell to the goal cell. O denotes an optimal result, SO denotes a sub-optimal result, and F denotes failure in finding a path.

3.2. Two-Stage Hierarchical Framework for Driving Task

In dynamic environments like highways, the task of determining where to drive and executing detailed maneuvers is intricately linked. This inherent complexity in decision making and control poses significant challenges. To address this, we adopted a two-stage hierarchical framework with separate modules for decision-making and trajectory generation. The overall structure of our framework, depicted in Figure 4, comprises the decision-making and motion planning modules. For trajectory generation, we employ an RL agent conditioned on the sub-goals determined by the decision-making module. Additionally, we utilize MPC to generate reference trajectories during the training of the motion planner. In this section, we demonstrate the proposed framework in detail.

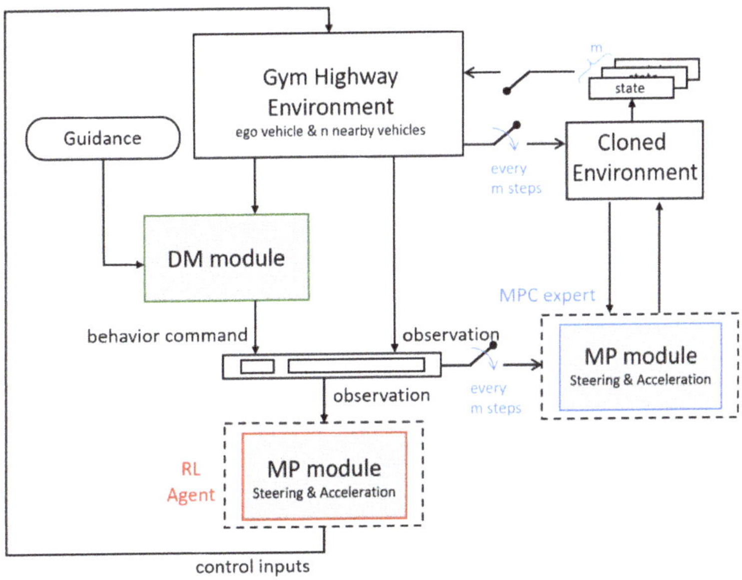

Figure 4. Overview of the hierarchical framework, comprising the Decision-Making (DM) module and the Motion Planning (MP) module. The behavior command determined by the DM module acts as a sub-goal for the RL agent. Additionally, the MPC module regularly (represented as m steps) computes trajectories, providing expert trajectory data for reward calculation.

3.2.1. Decision Making

In driving scenarios, decision-making can vary depending on the specific objectives. In our research, we employed a Decision-Making (DM) module that determines the vehicle's actions, represented as its "turn lamp". Our approach incorporates two primary considerations: adhering to the center of the guidance lane and avoiding unnecessarily slow driving. While these considerations adequately address highway driving situations, certain nuanced decisions remain, such as determining when to accelerate or decelerate, the optimal timing for lane changes, and adjustments to the steering angle for passenger comfort. These nuanced decisions are difficult to address with a complex state machine and lack the necessary generality and robustness. Therefore, our DM module focuses solely on lane-changing decisions, leaving other control decisions to be handled by the RL motion planner.

The DM module receives input in the form of a predetermined guidance lane and the current state of the vehicle. Using odometry information, the DM module calculates the Time-To-Collision (TTC) value between the ego vehicle and its surroundings at each time step. Longitudinal TTC serves as a safety margin for lane changes, permitting lane-change commands only when the measured TTC values exceed predetermined thresholds for both the front and rear directions. These thresholds remain fixed throughout training. The ego vehicle strives to adhere closely to the guidance target lane, but if a preceding vehicle is driving slowly, the agent must decide whether to overtake or maintain its current lane. Guidance lanes are randomly generated in each episode, with a mandatory lane change occurring every 800 m by designating a different target lane. Figure 5 illustrates the operation of the DM module, depicting the ego vehicle as a yellow box and surrounding vehicles as blue boxes. Additionally, specific designations are used: yellow lines represent frontal TTC measurements (Figure 5b), orange lines represent rear time-to-collision measurements (Figure 5b), and red lines represent designated global guidance (Figure 5c,d). The DM module determines lane-changing actions based on the current TTC values and the guidance path. In Figure 5d, for example, the ego vehicle transitions to the right lane to overtake the vehicle ahead.

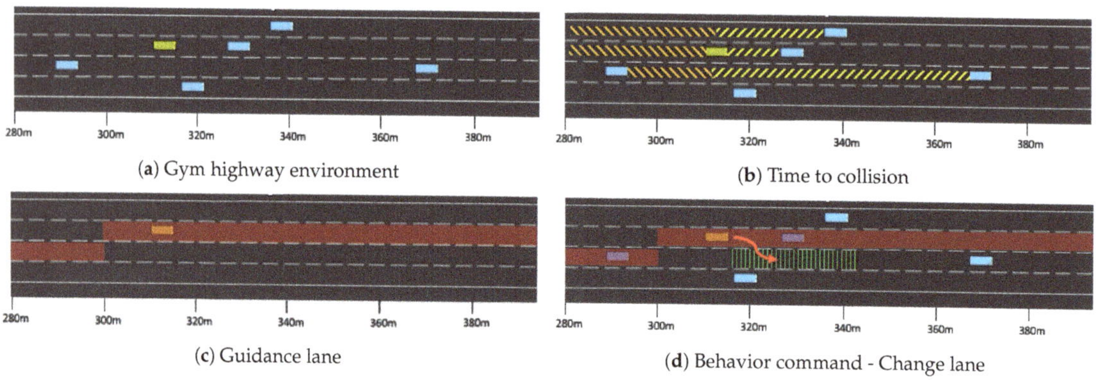

Figure 5. Decision-making process visualization in simulation. (**a**) The simulation environment showing multiple lanes and nearby vehicles. (**b**) Front and rear TTC values are depicted as orange and yellow dashed lines. (**c**) The guidance lane is indicated by red lines. (**d**) The green stripe represents the decision made by the ego vehicle. TTC values at (**b**) and the guidance lane at (**c**) determine the sub-goal of the current state of the ego vehicle. As depicted in (**d**), the DM module can decide to change lanes even when the vehicle is on the guidance lane if the TTC value is too small.

3.2.2. Motion Planning

The decision output from the DM module is transmitted to the RL agent by concatenating it with the observation, subsequently generating the control input of the vehicle. Given the inherent uncertainty in driving scenarios, it is safer to constrain movements within an expected range through decision making rather than directly generating a control signal. The corresponding MDP formulation is presented below.

(1) State Space and Action Space: The state space comprises absolute and relative observations. Absolute observation encompasses the ego vehicle's odometric information and the odometric information of four surrounding vehicles. Odometric information, denoted as $\mathbf{O}_k \in \mathbb{R}^5$, includes two-dimensional position and velocity information, as well as the heading angle: $\mathbf{O}_k = [s_x, s_y, v_x, v_y, \psi]_k^T$. Here, k represents the index of the vehicle, with 0 assigned to the ego vehicle and 1, 2, 3, and 4 representing nearby vehicles in clockwise order from the ego vehicle's center of mass. Relative information includes the time-to-collision value ttc for nearby states and the derived target lane l_{target} from high-level decisions and the current ego vehicle's position. The state accumulates four consecutive decision time steps to ensure the Markov property. The action space $\mathbf{a} \in \mathbb{R}^2$ comprises the steering angle β and acceleration α, which collectively serve as the control input for the system dynamics.

(2) Reward Function: The core component of the reward function involves comparing the ego vehicle's odometric states with those generated by synchronized MPC. To facilitate MPC for vehicle control, we employed a linearized kinematic bicycle model as described in [37]. In this model, we introduced the longitudinal distance d_f between the ego vehicle and the vehicle in front as an additional state variable. This incorporation enables us to

impose a constraint on maintaining a safe front gap. MPC shares the same action space and state space with an RL agent. The objective of the driving task is to follow a reference velocity and desired lateral position while adhering to inequality conditions that satisfy the vehicle's dynamics.

Figure 4 illustrates the synchronization process of MPC, where decisions from the DM module serve as the reference for the MPC trajectory, and the MPC trajectory, in turn, serves as the reference for the RL reward function. This approach enables the trained agent to achieve trajectory optimization performance comparable to that of an MPC controller with reduced computation time. The MPC reference trajectory is computed at every m step from the ego vehicle's current state to ensure reference stability. These m steps of the reference trajectory are calculated within a cloned environment and then transferred to the original environment for reward calculation. In addition to the odometric reward, the change of action is also factored in to prevent oscillation or abrupt changes in the control input. The running reward is calculated as follows:

$$r_{running} = \frac{1}{2}(\mathbf{O}_0 - \mathbf{O}_0^{ref})^T \mathbf{H}(\mathbf{O}_0 - \mathbf{O}_0^{ref}) + \frac{1}{2}\mathbf{a}^T \mathbf{R}\mathbf{a}, \tag{1}$$

where $\mathbf{H}$ and $\mathbf{R}$ are diagonal matrices used to weigh each element. Another element of the reward function involves checking the termination conditions, which occurs in the event of collision or deviation from the traffic road of the ego vehicle, and is defined as follows:

$$r_{terminal} = \begin{cases} -1, & \text{if collision occurs or the vehicle deviates from the track} \\ 0, & \text{otherwise} \end{cases} \tag{2}$$

Building upon the framework detailed in Section 3.2, we trained our motion planning agent with our proposed off-policy training method, CER. Notably, CER is compatible with a range of off-policy algorithms.

4. Experimental Results

In this section, we evaluate how the proposed framework enhances the performance of trajectory optimization in dynamic highway settings. Our agent is trained in the gym highway environment [28], featuring multiple lanes and nearby vehicles, requiring the generation of appropriate control inputs based on odometry information for both its own position and that of nearby vehicles. The objective in this environment is to maintain a target speed of 30 m/s (108 km/h) while staying within the target lane and avoiding collisions. The base RL algorithms are implemented using the open-source RL framework, stable-baselines3 [38] with PyTorch, and the open-source toolbox do-mpc [39] is used for the MPC calculation.

To assess the effectiveness of our training method and driving framework, we implement CER in TD3 [40], which is considered the state-of-the-art deterministic policy gradient RL method. As the baseline method, we employ TD3 with the UVFA [13] structure for the goal-conditioned policy.

In an ablation study, we explore the impact of the behavior policy in the training process by integrating expert demonstrations with the stored transitions in the buffer. Trajectories from the MPC controller are utilized for reward calculation, with these trajectories also serving as expert demonstrations integrated into the training process, denoted as CER[M]. While adhering to the network hyperparameters outlined in [40], additional buffers are employed for our configuration. To ensure training stability, the reward scale is clipped within the range of [−1, +1]. Each episode consists of 1000 steps, with termination triggered if the agent deviates off track or collides with another vehicle.

This section is structured as follows: In Section 4.1, we assess whether CER enhances training performance and compare the performances of TD3 in the highway environment with and without CER. Section 4.2 provides a detailed analysis of trajectories and sub-goal decisions, while Section 4.3 presents an evaluation of action diversity in the trained agent.

We demonstrate the adaptability of our hierarchical module to changes in the decision-making module in this section. Lastly, in Section 4.4, we conduct an ablation study to examine the effect of the behavior policy on the CER training agent.

4.1. Training Performance Evaluation

We conducted training using TD3 both with and without CER on driving tasks to evaluate whether CER enhances training performance. Figure 6a illustrates the average episode length curve per 1000 episodes over the complete training steps, with a maximum episode length of 1000 steps. Additionally, Figure 6b depicts the reward curves for the baseline method and the proposed method. As shown in Figure 6b, the use of CER contributes to enhanced training speed and performance.

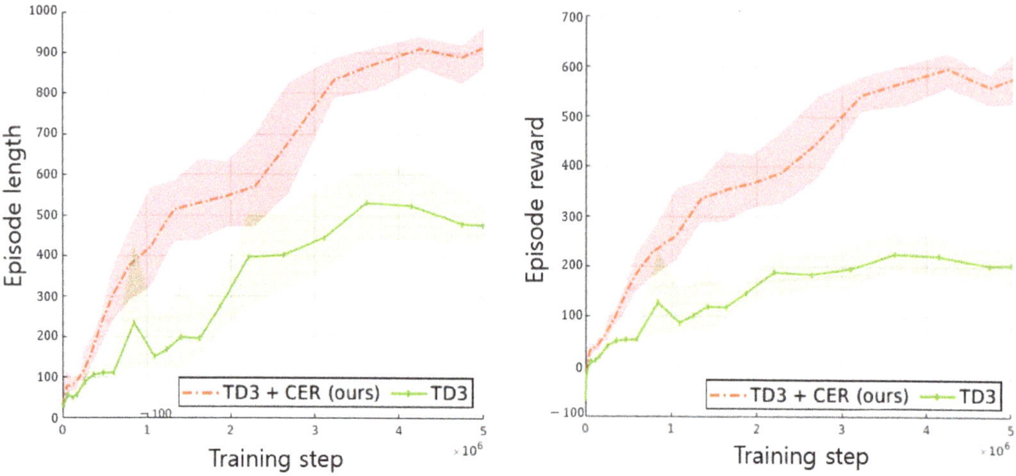

(**a**) Episode length (**b**) Episode reward

Figure 6. Reward curves of baseline method (TD3) and proposed method (TD3 + CER). The colored areas on each graph represents the boundaries of each measure, with averages calculated per 10 episodes, while lines represent averages per 1000 episodes.

4.2. Driving Performance Evaluation

In this section, we demonstrate the superior task competency of our method compared with the baseline method by showcasing several scenarios.

4.2.1. Case 1: Lane Keeping

In Figure 7, the maneuver of the proposed method and the baseline method is shown within the same simulation environment, under identical maneuver conditions of nearby vehicles. Despite this uniformity, noticeable discrepancies in the paths generated by our method and the baseline method are observed. To assess the lane-keeping performance of each path, Figure 8 zooms into a segment spanning from 1100 to 1500 m. The Center of the Lane (CoL) represents the position where the vehicle should be located, determined by the sub-goal output from the DM module referenced in Section 3.2.1. The positional deviation is computed using the mean square error between CoL and the vehicle's lateral position. Based on the maneuver data depicted in Figure 8a, TD3 + CER demonstrates an average deviation of 0.06 m and a maximum deviation of 0.07 m, accompanied by stable steering angles as illustrated in Figure 8c. In comparison, TD3 exhibits an average deviation of 0.49 m and a maximum deviation of 0.7 m under similar conditions. Figure 8d highlights the fluctuation of steering angles. Our method achieves minimal lane deviation through smooth steering control inputs.

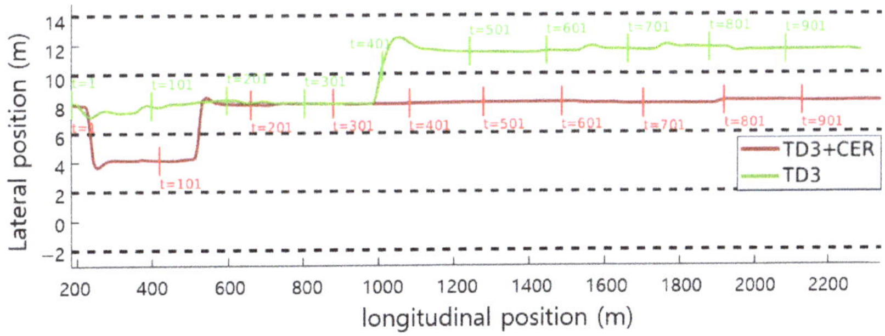

Figure 7. Scenario 1, comparison of trajectories between TD3 + CER agent and TD3 agent.

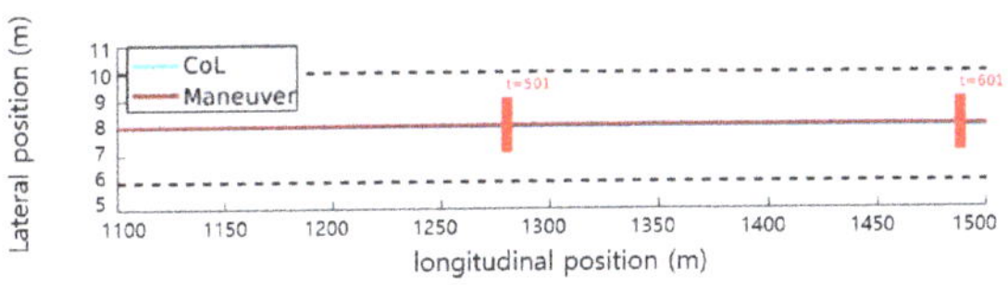

(**a**) TD3 + CER, lane-keeping maneuver.

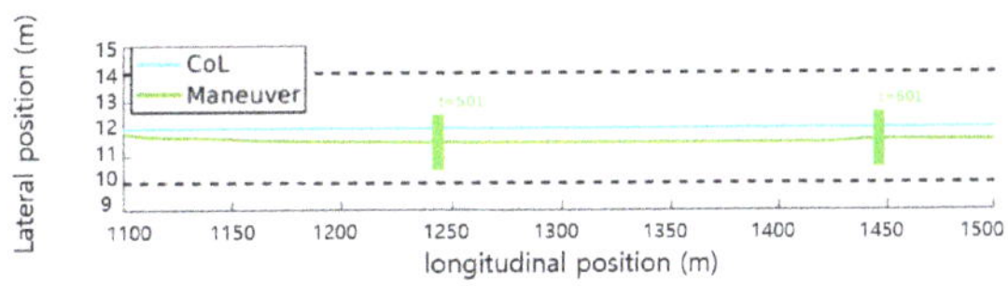

(**b**) TD3, lane-keeping maneuver.

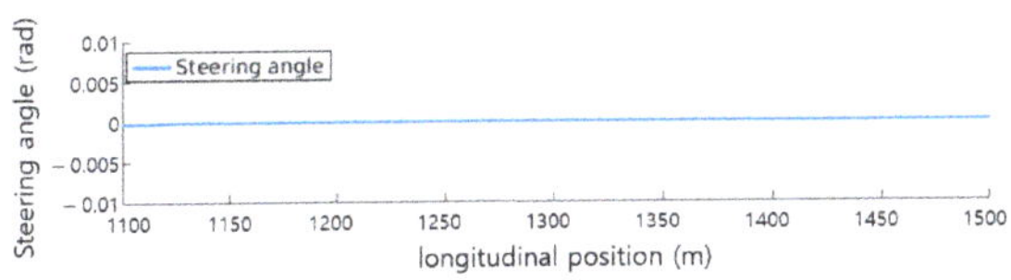

(**c**) TD3 + CER, steering angle.

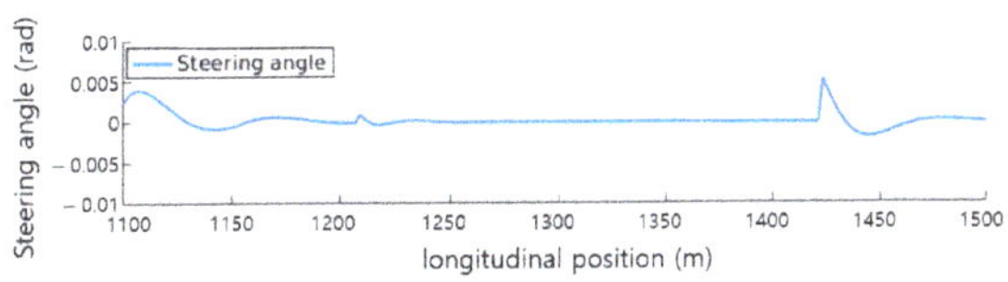

(**d**) TD3, steering angle.

Figure 8. Case study of lane-keeping performance: Section from 1100 m to 1500 m in Scenario 1. (**a**,**b**) depict the lane-keeping maneuvers of our method and the baseline method, respectively. (**c**,**d**) show the steering angles corresponding to the maneuvers in (**a**,**b**). "CoL" denotes the center of the lane, representing the optimal position for the vehicle. The red rectangle in (**a**) and the green rectangle in (**b**) denote the location of the vehicle at each labeled time step. In this lane-keeping scenario, the suggested method demonstrates superior performance, exhibiting stable lane-keeping maneuvers and steering angles, as evidenced by the stable steering angle changes.

4.2.2. Case 2: Lane Changing

In Figure 9, rapid sequential lane changes are observed. Between 1100 and 1300 m, two-lane changes take place consecutively. The detailed trajectory is depicted in Figure 10, where the proposed method adeptly follows the target lane, while the baseline method experiences delay and failure during the lane change. Additionally, between 1400 and 2200 m, the baseline method fails to maintain the lane, indicating deficiencies in steering control. Furthermore, within the desired speed range of 20 to 30 m/s, TD3 also struggles with velocity control.

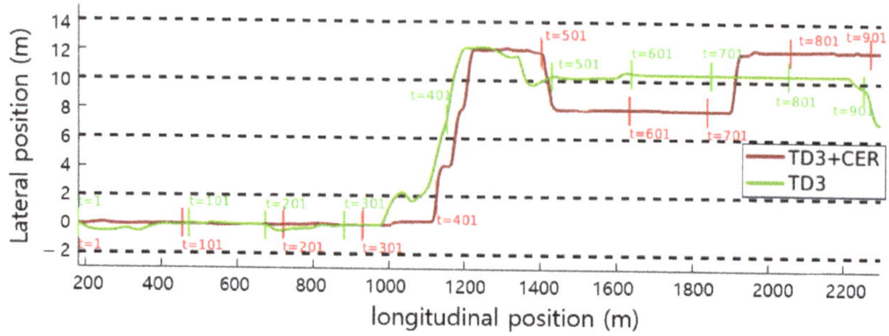

Figure 9. Scenario 2, comparison of trajectories between TD3 + CER agent and TD3 agent.

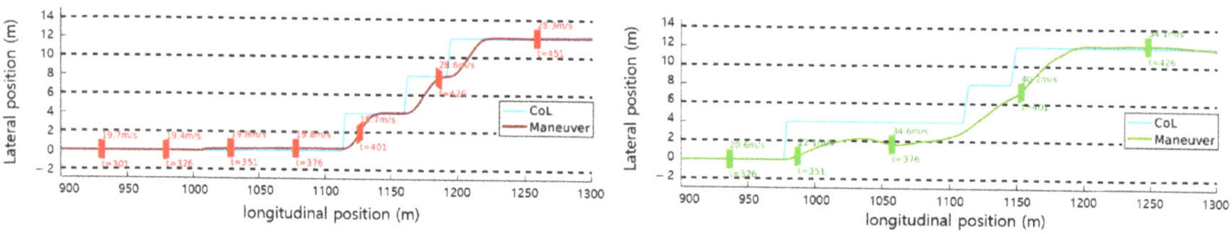

(**a**) TD3 + CER, lane-changing maneuver.

(**b**) TD3, lane-changing maneuver

Figure 10. Case study of lane-changing performance: Section from 1100 to 1300 m in Scenario 2. The lower label of the vehicle (rectangle box) indicates the time step, while the upper label indicates the velocity of the vehicle at each time step. The maneuver of the suggested method in (**a**) follows CoL while maintaining the desired velocity range. In contrast, the maneuver without CER, depicted in (**b**), fails to follow CoL and exceeds the desired velocity range.

4.2.3. Case 3: Overtaking

In Figure 11, both mandatory and discretionary lane changes (overtaking) occur. The gray areas denote the predetermined guidance lane, indicating mandatory lane changes for the ego vehicle. To assess each agent's performance in following sub-goals, Figure 12 shows the maneuvers of each agent with sub-goals at each time step. Both agents must opt to change lanes to the right at 1000 m as a mandatory lane change. While our method successfully follows the target lane, the TD3 agent deviates from the target lane after 1000 m. A closer examination of the overtaking action is provided in the section from 1100 to 1470 m in Figure 13, where the maneuver is depicted in both global Figure 13a,b coordinates. In relative coordinates, the origin is set as the obstacle vehicle. Here, the agent in our method executes an overtaking maneuver while maintaining a safe distance from the obstacle vehicle.

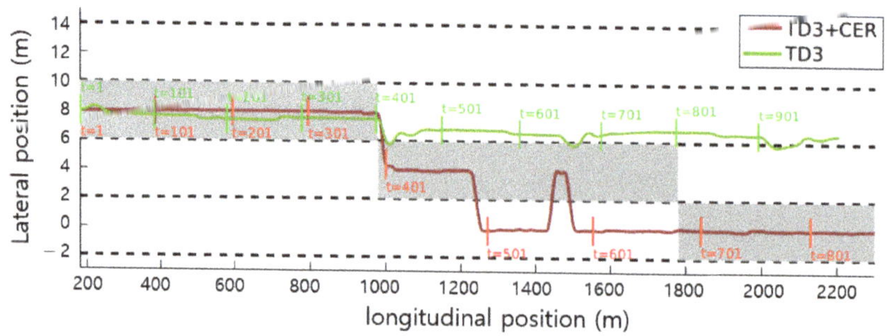

Figure 11. Scenario 3, comparison of drive paths for TD3 + CER agent and TD3 agent.

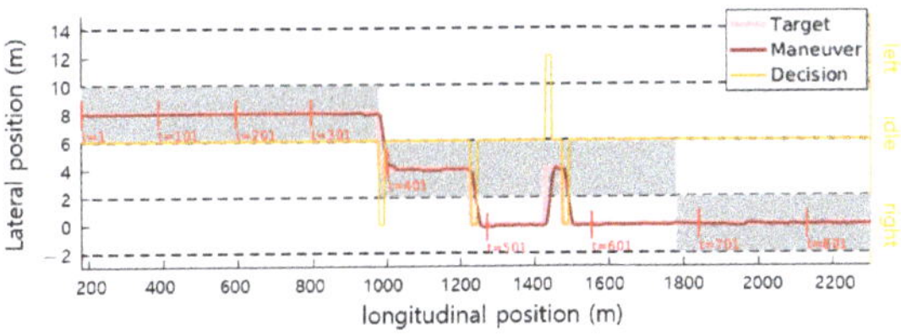

(**a**) TD3 + CER, sub-goal and following maneuver made.

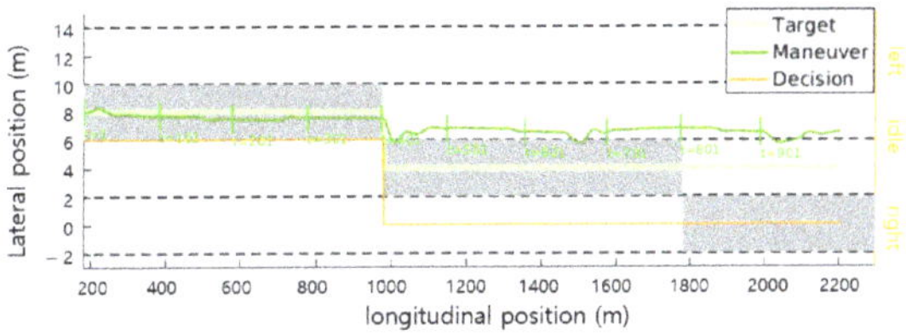

(**b**) TD3, sub-goal and following maneuver made.

Figure 12. Sub-goals and resulting paths of each agent are depicted in the figure. The yellow line represents the sub-goal decisions, labeled as "left", "right", and "idle". As the guidance lane changes at 1000 m, mandatory lane changes must occur for both agents in (**a**,**b**). While a lane change occurs in (**a**), it does not take place in (**b**).

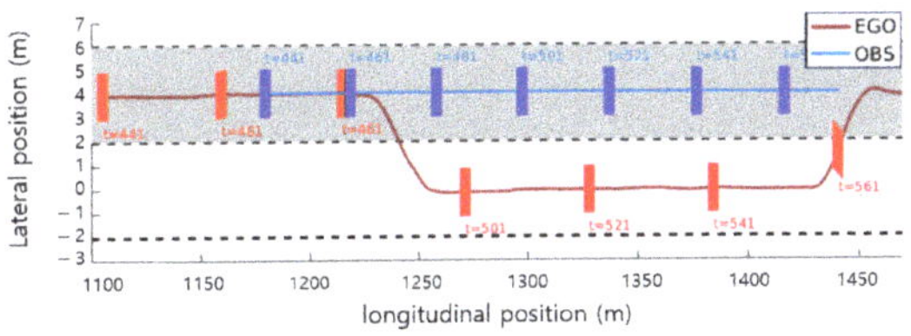

(**a**) TD3 + CER, overtaking maneuver at global coordinate.

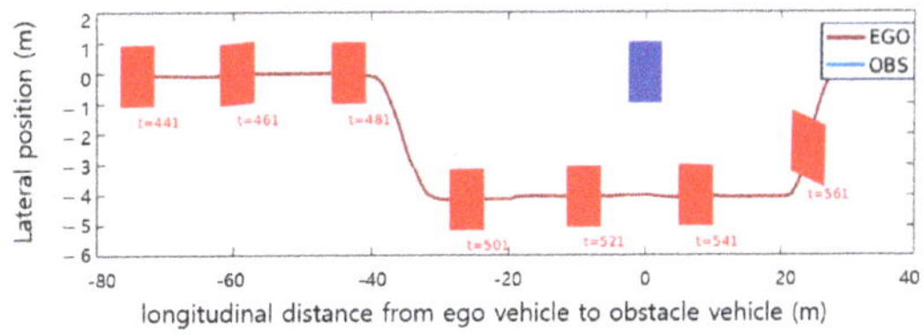

(**b**) TD3 + CER, overtaking maneuver at relative coordinate.

Figure 13. Case study of overtaking performance: Section from 1100 to 1470 m in Scenario 3. The stability of the lane-changing maneuver is shown in (**a**). Examining the relative maneuver of the ego vehicle to the obstacle vehicle during overtaking, illustrated in (**b**), the ego vehicle maintains a safe gap, ensuring a safe overtaking maneuver.

4.3. Action Diversity Evaluation

To assess the adaptability of our trained agent, we evaluated the performance of our best-performing policy from TD3 + CER using three different decision-making modules. Each decision-making module generated distinct sub-goals within the same scenario, allowing us to examine how effectively the motion planner responded to these sub-goals.

Figure 14 illustrates the vehicle trajectory influenced by three distinct decision-making modules, representing different driving styles: original (blue), aggressive (red), and tolerant (yellow). The variation in time-to-collision thresholds resulted in different behaviors within identical situations. Table 1 displays the thresholds for each DM module, denoted by TTC_{front} and TTC_{rear}. Figure 14a presents the trajectories of each vehicle with line markers and highlights the predetermined guidance lanes, represented by discrete gray areas every 800 m. The original decision-making module reflects the approach used during training. The "tolerant" module employed higher thresholds for lane changes, while the "aggressive" module exhibited a propensity for frequent lane changes. These DM modules and the associated maneuver analyses were evaluated under identical conditions on a four-lane highway with 50 nearby vehicles. Table 2 presents an analysis of the trajectories generated by each DM module using an identical motion planner, including the number of lane changes ($N_{\text{l.c.}}$) and the average velocity ($v_{\text{avg.}}$).

Figure 14c depicts the velocities of each vehicle within the preferred speed range of 20 to 30 m/s. The "aggressive" approach involved frequent lane changes to maintain speed, particularly when encountering slower vehicles ahead. Conversely, the "tolerant" approach strictly adhered to the guidance, sometimes even decelerating. The dashed gray line in Figure 14c represents the desired speed limit of 30 m/s. In instances where no vehicle was in front, such as at 100 m in Figure 14c, the vehicles accelerated close to the desired 30 m/s. Our best-performing policy demonstrated adaptive behavior, generating diverse actions within the same scenarios based on the provided sub-goals.

Table 1. DM parameters for different driving styles.

DM Parameters	Original	Aggressive	Tolerant
TTC_{front}	7 s	10 s	3 s
TTC_{rear}	4 s	8 s	2 s

Table 2. Analysis results for different driving styles.

Analysis	Original	Aggressive	Tolerant
$N_{\text{l.c.}}$	9	6	8
$v_{\text{avg.}}$	25.2 m/s	25.1 m/s	22.4 m/s

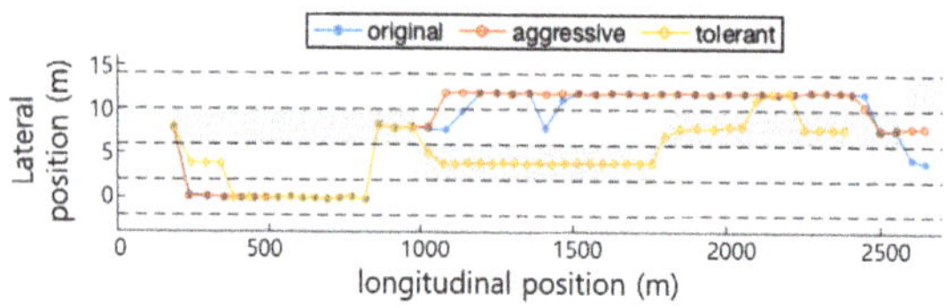

(**a**) Comparison of trajectories over the entire section.

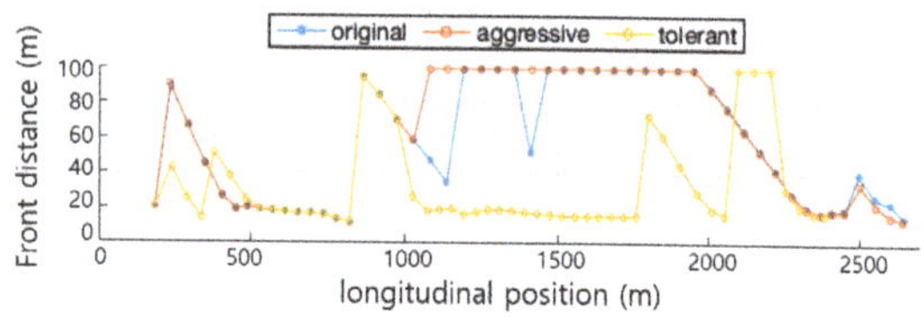

(**b**) Distance from the vehicle in front.

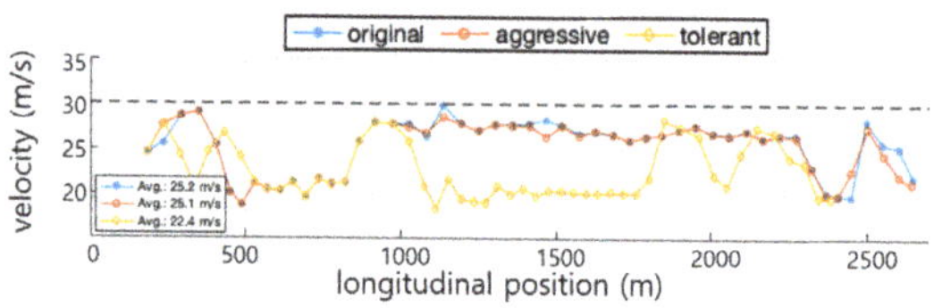

(**c**) Change in velocity over the entire section.

Figure 14. Action diversity evaluation with the TD3 + CER agent. Table 1 displays varying hyperparameters of the decision-making modules: "original", "aggressive", and "tolerant". (**a**) demonstrates different maneuvers made according to the varying decision-making modules. The distance and velocity plots in (**b**,**c**) reveal more evident characteristics of each decision-making module, along with the corresponding trajectories generated from each module. This result indicates that the trained MP module agent does not overfit to specific DM modules ("original") but adheres to sub-goals.

4.4. Ablation Study

In our MDP formulation, we integrated trajectories from the MPC controller as odometry references in the calculation of the running reward. These trajectories were repurposed as expert demonstrations during the training of our motion planner, with the expectation of accelerating convergence in the training process. This variant, referred to as CER[M], employed approximately 10% of each buffer for MPC demonstrations to mitigate the risk of policy failures due to extrapolation errors [41]. Consequently, TD3 + CER[M] was trained with two behavior policies: the agent itself and the MPC policy.

Figure 15 illustrates the reward graph of the proposed methods (TD3 + CER and TD3 + CER[M]) alongside the baseline method (TD3). Notably, TD3 + CER[M] achieved convergence much faster compared with TD3 + CER, indicating that CER can significantly enhance the performance of the trained agent. Moreover, the judicious use of expert demonstrations can expedite convergence speed during training.

In Figure 16, the maneuver and velocity changes of each algorithm's best-performing policy are visualized. Each algorithm generates distinct trajectories under the same scenario. Notably, TD3 + CER[M], trained with MPC demonstrations, exhibits similar maneuver and velocity changes to MPC during the evaluation period.

Table 3 provides a detailed comparison of each trajectory, focusing on trajectory similarity to MPC and deviation from the target lane. The accumulated Euclidean distances between the MPC trajectory and the target line throughout the scenario are presented, reflecting the trajectory similarity criterion. Additionally, the deviation from the target lane, indicative of decision-making module responses at each time step, is analyzed. Among the RL-based approaches, TD3 + CER[M] demonstrates the closest trajectory similarity to MPC and the smallest deviation from the target lane. While exhibiting trajectory similarities with the MPC motion planner, our proposed method achieves a significant reduction in computational time during the scenario, approximately by a factor of one-tenth.

Table 3. Quantitative analysis in Scenario 4. Considering the sum of Euclidean distance to the target lane as an indicator of how well the agent adheres to sub-goals, TD3 + CER outperforms the baseline method. Furthermore, TD3 + CER[M] generates trajectories most similar to MPC, as evidenced by the criteria of Euclidean distance to the MPC trajectory. Notably, the computation time of the RL agent is significantly less than that of MPC, suggesting the potential of substituting MPC with an RL agent trained using our proposed method.

Evaluation	Methods			
	MPC	TD3 + CER[M]	TD3 + CER	TD3
Euclidean Distance to MPC Trajectory	0.00 m	417.20 m	1171.41 m	1530.63 m
Euclidean Distance to Target Line	1.26 m	8.58 m	5.82 m	47.89 m
Average Velocity	27.53 m/s	25.61 m/s	22.63 m/s	20.51 m/s
Number of Lane Changes	11	9	3	0
Computational Time for 1000 Steps	6.99 s	0.76 s	0.78 s	0.72 s

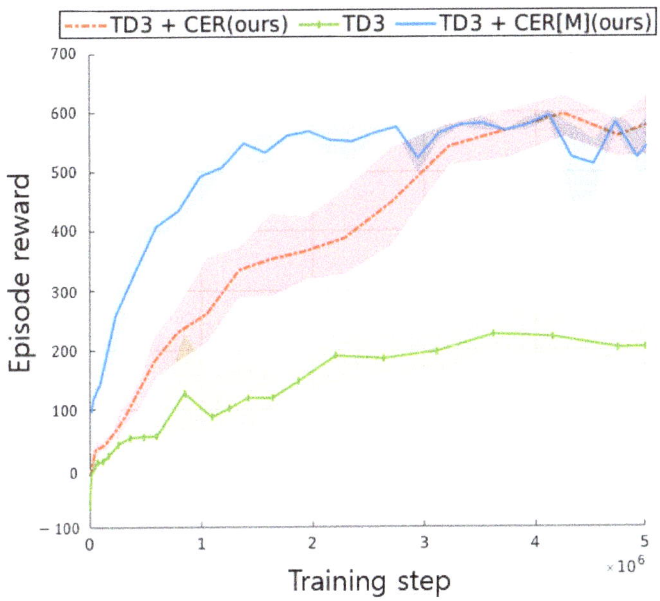

Figure 15. Reward curves of baseline method (TD3) and proposed method with (TD3 + CER) and without MPC expert demonstration (TD3 + CER[M]).

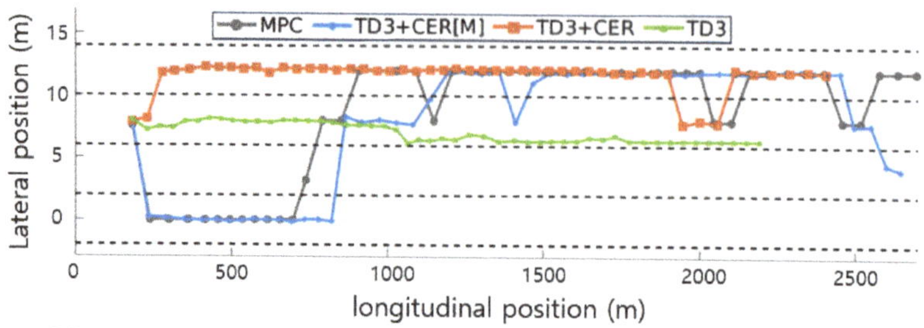

(a) Scenario 4, maneuver of baseline, proposed and expert method

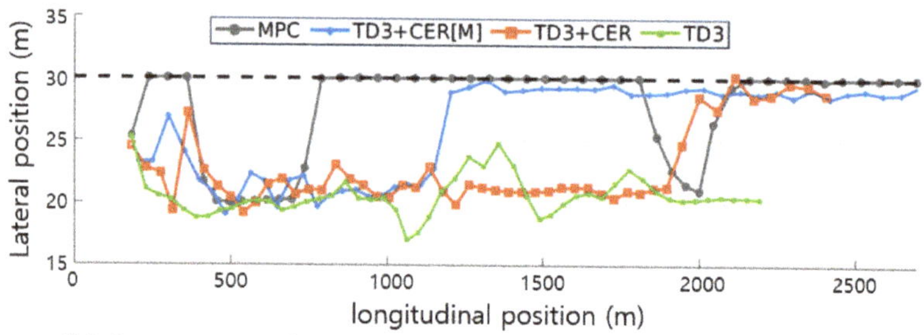

(b) Scenario 4, velocity of baseline, proposed and expert method

Figure 16. Maneuver and velocity analysis with the addition of the MPC expert demonstration method. When the agent employs MPC demonstration, it trains to imitate the MPC's policy more straightforwardly.

We assessed the trained policy of each method across two-, three-, and four-lane highway environments, conducting 100 repetitions for each test. Success for each episode was defined as the agent completing 1000 time steps without encountering forced termination conditions (such as collision or deviation from the track). The success rate, representing the number of successful cases out of the total cases, is presented in Table 4. Additionally, we analyzed the average quality of maneuvers based on deviation and velocity, as outlined in Table 5. Our method consistently exhibited the smallest deviations from the lane center across all environments, while also maintaining velocity within the desired bounds. Collision with other vehicles was the sole cause of failure. In the two-lane environment, the absence of additional lanes made it challenging to overtake the slow-moving vehicles ahead, necessitating more delicate acceleration control.

Table 4. Quantitative analysis: Success rate measured as the number of successful cases out of the total cases. Out of 100 random road environment conditions, each with a corresponding number of lanes, TD3 + CER[M] exhibits the highest success rate.

Domain	Methods		
	TD3	TD3 + CER[M]	TD3 + CER
2-lane	53/100	95/100	94/100
3-lane	78/100	98/100	98/100
4-lane	72/100	96/100	92/100

Table 5. Quantitative analysis of driving performance, including lane-following accuracy and maintenance of desired velocity. Given that the target speed range is 30 m/s, TD3 + CER[M] shows maintaining the most approximate value to the target speed. However, given the deviations from the target lanes, TD3 + CER exhibits the smallest values. Notably, suggested methods are superior to baseline methods in every evaluation criterion.

Domain	Evaluation	Method		
		TD3	TD3 + CER[M]	TD3 + CER
2-lane	Avg. Deviation	0.31 m	0.12 m	0.09 m
	Max. Deviation	0.49 m	0.18 m	0.20 m
	Min. Deviation	0.09 m	0.03 m	0.03 m
	Avg. Velocity	19.36 m/s	24.61 m/s	23.33 m/s
3-lane	Avg. Deviation	0.32 m	0.11 m	0.08 m
	Max. Deviation	0.99 m	0.21 m	0.28 m
	Min. Deviation	0.11 m	0.04 m	0.02 m
	Avg. Velocity	20.87 m/s	26.87 m/s	25.28 m/s
4-lane	Avg. Deviation	0.35 m	0.10 m	0.09 m
	Max. Deviation	0.90 m	0.20 m	0.23 m
	Min. Deviation	0.11 m	0.06 m	0.02 m
	Avg. Velocity	21.53 m/s	27.97 m/s	25.49 m/s

5. Conclusions

This paper has presented an off-policy RL training method to address environments characterized by imbalanced sub-goal distributions. Our work demonstrates how incorporating CER into off-policy algorithms can significantly improve performance. The key contributions and findings of this research are summarized as follows:

1. Imbalanced Sub-Goal Distribution and CER Technique: In Section 3, we utilized the twin cliff walk example to demonstrate how disproportional goal distributions in multi-goal environments can adversely impact training efficiency and optimality. Our adaptation of the CER technique to off-policy algorithms showed improved performance compared with methods that do not incorporate CER, demonstrating the validity of employing our method for multi-goal tasks. We adapt this approach to highway driving scenarios in the subsequent section.
2. Hierarchical Framework for Highway Driving Task: We proposed a hierarchical framework suitable for dynamic highway environments and trained a motion planner using our suggested training method. By employing separate modules, our agent proficiently generates control inputs across various scenarios without necessitating re-training or transfer learning. This flexibility allows for the easy modification of the decision-making module, empowering the agent to accommodate changes in traffic regulations or individual driving preferences.
   Simplifying the reward function by synchronizing the online MPC trajectory led to two notable advantages: a more straightforward design of the reward function and maneuvering capabilities of the RL-trained motion planner comparable to an optimized MPC path planner, with computational demands reduced to less than one-tenth of the original.
3. Discussions and Future Work: The incorporation of CER into the training of the motion planner demonstrated substantial improvements in both sample efficiency and performance. Our proposed method fosters the generation of uniform data in the training process, departing from uniform sampling to yield optimal solutions. Our method consistently outperformed other approaches without CER integration, underscoring its efficacy in addressing tasks, even in continuous control problems where traditional RL methods encounter difficulties in training.

However, there are limitations to the proposed framework. As replay buffers are defined independently for individual sub-goals, the method assumes that the sub-goals are known in advance, necessitating some degree of prior knowledge of the task. This dependency on predefined sub-goals could limit the method's applicability in scenarios where such prior knowledge is not readily available or where some sub-goals may be added or removed. Despite necessitating multiple buffers, the overall buffer capacity remains the same as that of a general single-experience replay buffer. Consequently, our method is adaptable to various off-policy algorithms, offering enhanced performance in multi-goal environments.

Ensuring safety in driving tasks is paramount. Incorporating additional techniques to constrain undesirable control actions, such as oscillations, exceeding speed limits, and executing safety maneuvers like emergency stops, can further enhance the robustness and reliability of the driving task. This can be achieved by adding masking layers or using constraint RL methods.

While our method demonstrates significant advancements, particularly in handling imbalanced sub-goal distributions and improving training efficiency, our future work will involve adding more safety-aware constraints for continuous control in driving tasks. These enhancements aim to improve the robustness and applicability of our framework in diverse and complex traffic scenarios.

Author Contributions: Conceptualization, Y.-J.K., D.-S.P. and M.-T.L.; methodology, Y.-J.K., W.-J.A. and M.-T.L.; software, Y.-J.K., W.-J.A. and S.-H.J.; validation, W.-J.A. and D.-S.P.; formal analysis, Y.-J.K. and D.-S.P.; investigation, W.-J.A., S.-H.J. and M.-T.L.; resources, Y.-J.K.; data curation, Y.-J.K. and W.-J.A.; writing—original draft preparation, Y.-J.K. and D.-S.P.; writing—review and editing, D.-S.P. and M.-T.L.; visualization, Y.-J.K. and W.-J.A.; supervision, D.-S.P. and M.-T.L.; project administration, M.-T.L.; funding acquisition, all authors. All authors have read and agreed to the published version of the manuscript.

Funding: This research was supported by the Basic Science Research Program through the National Research Foundation of Korea (NRF) (grant no. NRF-2022R1F1A1073543).

Institutional Review Board Statement: Not applicable.

Informed Consent Statement: Not applicable.

Data Availability Statement: The original contributions presented in the study are included in the article, further inquiries can be directed to the corresponding author.

Conflicts of Interest: The authors declare no conflicts of interest.

References

1. Mnih, V.; Kavukcuoglu, K.; Silver, D.; Graves, A.; Antonoglou, I.; Wierstra, D.; Riedmiller, M. Playing atari with deep reinforcement learning. *arXiv* **2013**, arXiv:1312.5602.
2. Silver, D.; Huang, A.; Maddison, C.; Guez, A.; Sifre, L.; Van Den Driessche, G.; Schrittwieser, J.; Antonoglou, I.; Panneershelvam, V.; Lanctot, M. et al. Mastering the game of Go with deep neural networks and tree search. *Nature* **2016**, *529*, 484–489. [CrossRef] [PubMed]
3. Berner, C.; Brockman, G.; Chan, B.; Cheung, V.; Dębiak, P.; Dennison, C.; Farhi, D.; Fischer, Q.; Hashme, S.; Hesse, C.; et al. Dota 2 with large scale deep reinforcement learning. *arXiv* **2019**, arXiv:1912.06680.
4. Ahn, M.; Brohan, A.; Brown, N.; Chebotar, Y.; Cortes, O.; David, B.; Finn, C.; Fu, C.; Gopalakrishnan, K.; Hausman, K.; et al. Do as i can, not as i say: Grounding language in robotic affordances. *arXiv* **2022**, arXiv:2204.01691.
5. Pertsch, K.; Lee, Y.; Lim, J. Accelerating reinforcement learning with learned skill priors. *Conf. Robot. Learn.* **2021**, *155*, 188–204.
6. Ehsani, K.; Han, W.; Herrasti, A.; VanderBilt, E.; Weihs, L.; Kolve, E.; Kembhavi, A.; Mottaghi, R. Manipulathor: A framework for visual object manipulation. In Proceedings of the IEEE/CVF Conference On Computer Vision And Pattern Recognition, Nashville, TN, USA, 20–25 June 2021; pp. 4497–4506.
7. Jiang, Y.; Yang, F.; Zhang, S.; Stone, P. Integrating task-motion planning with reinforcement learning for robust decision making in mobile robots. *arXiv* **2018**, arXiv:1811.08955.
8. Hwangbo, J.; Lee, J.; Dosovitskiy, A.; Bellicoso, D.; Tsounis, V.; Koltun, V.; Hutter, M. Learning agile and dynamic motor skills for legged robots. *Sci. Robot.* **2019**, *4*, eaau5872. [CrossRef]

9. Hwangbo, J.; Sa, I.; Siegwart, R.; Hutter, M. Control of a quadrotor with reinforcement learning. *IEEE Robot. Autom. Lett.* **2017**, *2*, 2096–2103. [CrossRef]
10. Rezaee, K.; Yadmellat, P.; Nosrati, M.; Abolfathi, E.; Elmahgiubi, M.; Luo, J. Multi-lane cruising using hierarchical planning and reinforcement learning. In Proceedings of the 2019 IEEE Intelligent Transportation Systems Conference (ITSC), Auckland, New Zealand, 27–30 October 2019; pp. 1800–1806.
11. Gangopadhyay, B.; Soora, H.; Dasgupta, P. Hierarchical program-triggered reinforcement learning agents for automated driving. *IEEE Trans. Intell. Transp. Syst.* **2021**, *23*, 10902–10911. [CrossRef]
12. Li, J.; Tang, C.; Tomizuka, M.; Zhan, W. Hierarchical planning through goal-conditioned offline reinforcement learning. *IEEE Robot. Autom. Lett.* **2022**, *7*, 10216–10223. [CrossRef]
13. Schaul, T.; Horgan, D.; Gregor, K.; Silver, D. Universal value function approximators. In Proceedings of the 32nd International Conference on Machine Learning, Lille, France, 7–9 July 2015; pp. 1312–1320.
14. Plappert, M.; Andrychowicz, M.; Ray, A.; McGrew, B.; Baker, B.; Powell, G.; Schneider, J.; Tobin, J.; Chociej, M.; Welinder, P.; et al. Multi-goal reinforcement learning: Challenging robotics environments and request for research. *arXiv* **2018**, arXiv:1802.09464.
15. Nasiriany, S.; Pong, V.; Lin, S.; Levine, S. Planning with goal-conditioned policies. *Adv. Neural Inf. Process. Syst.* **2019**, *32*, pp. 14776–14787.
16. Zha, D.; Bhat, Z.; Lai, K.; Yang, F.; Jiang, Z.; Zhong, S.; Hu, X. Data-centric artificial intelligence: A survey. *arXiv* **2023**, arXiv:2303.10158.
17. Cui, J.; Zong, L.; Xie, J.; Tang, M. A novel multi-module integrated intrusion detection system for high-dimensional imbalanced data. *Appl. Intell.* **2023**, *53*, 272–288. [CrossRef]
18. Lin, L. Self-improving reactive agents based on reinforcement learning, planning and teaching. *Mach. Learn.* **1992**, *8*, 293–321. [CrossRef]
19. Mnih, V.; Kavukcuoglu, K.; Silver, D.; Rusu, A.; Veness, J.; Bellemare, M.; Graves, A.; Riedmiller, M.; Fidjel, A.; Ostrovski, G.; et al. Human-level control through deep reinforcement learning. *Nature* **2015**, *518*, 529–533. [CrossRef]
20. Schaul, T.; Quan, J.; Antonoglou, I.; Silver, D. Prioritized experience replay. *arXiv* **2015**, arXiv:1511.05952.
21. Andrychowicz, M.; Wolski, F.; Ray, A.; Schneider, J.; Fong, R.; Welinder, P.; McGrew, B.; Tobin, J.; Abbeel, P.; Zaremba, W. Hindsight experience replay. *Adv. Neural Inf. Process. Syst.* **2017**, *30*, 5049–5059
22. Dayal, A.; Cenkeramaddi, L.; Jha, A. Reward criteria impact on the performance of reinforcement learning agent for autonomous navigation. *Appl. Soft Comput.* **2022**, *126*, 109241. [CrossRef]
23. Laud, A.; DeJong, G. The influence of reward on the speed of reinforcement learning: An analysis of shaping. In Proceedings of the 20th International Conference On Machine Learning (ICML-03), Washington, DC, USA, 21–24 August 2003; pp. 440–447.
24. Ng, A.; Harada, D.; Russell, S. Policy invariance under reward transformations: Theory and application to reward shaping. *Icml* **1999**, *99*, 278–287.
25. Gupta, A.; Pacchiano, A.; Zhai, Y.; Kakade, S.; Levine, S. Unpacking reward shaping: Understanding the benefits of reward engineering on sample complexity. *Adv. Neural Inf. Process. Syst.* **2022**, *35*, 15281–15295.
26. Brüdigam, T.; Olbrich, M.; Wollherr, D.; Leibold, M. Stochastic model predictive control with a safety guarantee for automated driving. *IEEE Trans. Intell. Veh.* **2021**, *8*, 22–36. [CrossRef]
27. Williams, G.; Drews, P.; Goldfain, B.; Rehg, J.; Theodorou, E. Aggressive driving with model predictive path integral control. In Proceedings of the 2016 IEEE International Conference On Robotics And Automation (ICRA), Stockholm, Sweden, 16–21 May 2016; pp. 1433–1440.
28. Leurent, E. An Environment for Autonomous Driving Decision-Making. GitHub Repository. 2018. Available online: https://github.com/eleurent/highway-env (accessed on 10 June 2024).
29. Kendall, A.; Hawke, J.; Janz, D.; Mazur, P.; Reda, D.; Allen, J.; Lam, V.; Bewley, A.; Shah, A. Learning to drive in a day. In Proceedings of the 2019 International Conference On Robotics And Automation (ICRA), Montreal, QC, Canada, 20–24 May 2019; pp. 8248–8254.
30. Peng, B.; Sun, Q.; Li, S.; Kum, D.; Yin, Y.; Wei, J.; Gu, T. End-to-End autonomous driving through dueling double deep Q-network. *Automot. Innov.* **2021**, *4*, 328–337. [CrossRef]
31. Lillicrap, T.; Hunt, J.; Pritzel, A.; Heess, N.; Erez, T.; Tassa, Y.; Silver, D.; Wierstra, D. Continuous control with deep reinforcement learning. In Proceedings of the 4th International Conference On Learning Representations, ICLR 2016, San Juan, Puerto Rico, 2–4 May 2016.
32. Vaswani, A.; Shazeer, N.; Parmar, N.; Uszkoreit, J.; Jones, L.; Gomez, A.; Kaiser, Ł; Polosukhin, I. Attention is all you need. *Adv. Neural Inf. Process. Syst.* **2017**, *30*, 5999–6009.
33. Mao, Z.; Liu, Y.; Qu, X. Integrating big data analytics in autonomous driving: An unsupervised hierarchical reinforcement learning approach. *Transp. Res. Part C Emerg. Technol.* **2024**, *162*, 104606. [CrossRef]
34. Sharma, O.; Sahoo, N.; Puhan, N. Transformer based composite network for autonomous driving trajectory prediction on multi-lane highways. *Appl. Intell.* **2024**, *54*, 1–35. [CrossRef]
35. Wang, J.; Wang, Y.; Zhang, D.; Yang, Y.; Xiong, R. Learning hierarchical behavior and motion planning for autonomous driving. In Proceedings of the 2020 IEEE/RSJ International Conference On Intelligent Robots And Systems (IROS), Las Vegas, NV, USA, 25–29 October 2020; pp. 2235–2242.

36. Abbeel, P.; Ng, A. Apprenticeship learning via inverse reinforcement learning. In Proceedings of the Twenty-First International Conference On Machine Learning, Banff, AB, Canada, 4–8 July 2004; p. 1.
37. Rajamani, R. *Vehicle Dynamics and Control*; Springer Science & Business Media: Berlin/Heidelberg, Germany, 2011.
38. Raffin, A.; Hill, A.; Gleave, A.; Kanervisto, A.; Ernestus, M.; Dormann, N. Stable-Baselines3: Reliable Reinforcement Learning Implementations. *J. Mach. Learn. Res.* **2021**, *22*, 1–8. Available online: http://jmlr.org/papers/v22/20-1364.html (accessed on 13 June 2024).
39. Fiedler, F.; Karg, B.; Lüken, L.; Brandner, D.; Heinlein, M.; Brabender, F.; Lucia, S. do-mpc: Towards FAIR nonlinear and robust model predictive control. *Control. Eng. Pract.* **2023**, *140*, 105676. [CrossRef]
40. Fujimoto, S.; Hoof, H.; Meger, D. Addressing function approximation error in actor-critic methods. In Proceedings of the 35th International Conference on Machine Learning, Stockholm Sweden, 10–15 July 2018; pp. 1587–1596.
41. Fujimoto, S.; Meger, D.; Precup, D. Off-policy deep reinforcement learning without exploration. In Proceedings of the International Conference On Machine Learning, Long Beach, CA, USA, 9–15 June 2019; pp. 2052–2062.

applied sciences

MDPI

Article

Effect of Onboard Training for Improvement of Navigation Skill under the Simulated Navigation Environment for Maritime Autonomous Surface Ship Operation Training

Hyoseon Hwang [1], Taemin Hwang [1] and Ik-Hyun Youn [2,*]

1 Department of Maritime Transportation System, Mokpo National Maritime University, Mokpo 58628, Korea
2 Division of Navigation & Information Systems, Mokpo National Maritime University, Mokpo 58628, Korea
* Correspondence: iyoun@mmu.ac.kr

Abstract: As the technology of the maritime autonomous surface ship (MASS) systems has geared toward autonomy, the importance of human operations in the shore control center (SCC) has gained in significance. Accordingly, the effects of the training method, including the traditional and new remote operator training methods have to be investigated in terms of MASS navigation safety. Therefore, this study conducted a comparative analysis to prove the effect of onboard training. The findings include the execution of a simulated navigation experiment, the extraction of rudder steering-related features, selection of significant features, and comparative analysis with network graph visualization. The separate results obtained from the "untrained" group and "trained" group were exhibited as the purpose of research for the effect of onboard training on navigation skills. Then, the authors interpreted the difference in each group allusively in accordance with features considering actual navigation and compared groups using descriptive statistics. Consequently, this study emphasized the importance of proving the effect of training before the new training technologies are used to train MASS remote operators in the future.

Keywords: maritime autonomous surface ship; remote operator; effect of training; simulated navigation experiments; comparative analysis

Citation: Hwang, H.; Hwang, T.; Youn, I.-H. Effect of Onboard Training for Improvement of Navigation Skill under the Simulated Navigation Environment for Maritime Autonomous Surface Ship Operation Training. *Appl. Sci.* **2022**, *12*, 9300. https://doi.org/10.3390/app12189300

Academic Editors: Fabrizio Giulietti, Nadjim Horri and William Holderbaum

Received: 24 August 2022
Accepted: 15 September 2022
Published: 16 September 2022

1. Introduction

With the technological development of the maritime autonomous surface ship (MASS) systems now gearing toward autonomy, it is expected that MASS operations will be carried out via interaction with the shore control center (SCC) under human operation via shore remote operators [1–5]. Before the commercialization of MASS, it is necessary to develop a new training method for remote operation [6–9]. Because human intervention cannot be entirely removed from the operation of autonomous ships, remote operators are identified to play a crucial role in the safe operation of MASS (as the autonomous navigation technology of MASS) [10,11]. As human factors account for the majority of maritime accidents, operators must be trained properly [12–15].

Navigation skills have been identified as the most important human factor in the training of traditional navigators [16]. At the moment, the training opportunities given to navigators are onboard training and SHS simulation training [17,18]. From onboard training, trainees can gain comprehensive ship operation knowledge and experience as an officer, such as ship management, cargo handling, risk management, navigation planning and execution, attitude, and leadership as an officer of the watch, including training for familiarity with the use of navigation equipment and a sense of life on board [15]. Although onboard training indeed includes both theoretical and practical ship handling training, these priceless opportunities are often difficult to provide regularly [19,20]. As a result of the research on maritime education and training, more simulation methods will be used for navigation training [21,22]. In future training, simulation training will account for a greater proportion of navigator training than it does now, and the training must include navigation skill enhancement [23,24].

Existing navigation skills will have to be reinterpreted in line with the new education and training requirements for MASS [25,26]. In a similar manner, other research has emphasized the need for newly added training, such as the mandatory requirements of the future version of STCW [17,27]. Despite the growing debate over the new training, the expected effect of training on navigation skills is yet to be adequately elucidated. Thus, the verification of enhanced navigation skills ought to be conducted, especially before the new method of training is applied. As a result, this study conducted a simulation experiment on two different groups, namely, the "untrained group" and the "trained group," of onboard training as preparation for obtaining the newly added type of training. The proposed methods include related materials from the process of experiment execution to the interpretation of test results.

As a result, this study primarily aims to investigate the impact of onboard training on navigation skills improvement through a comparative analysis of various aspects of navigation evaluation perspectives.

2. Materials and Methods

The proposed methods use the data from the practical experiments. The most effective features in group separation were then applied to the network graph to determine the overall difference between the two groups using the navigation features extracted and selected based on the difference in onboard training. Figure 1 depicts the workflow of our proposed methods.

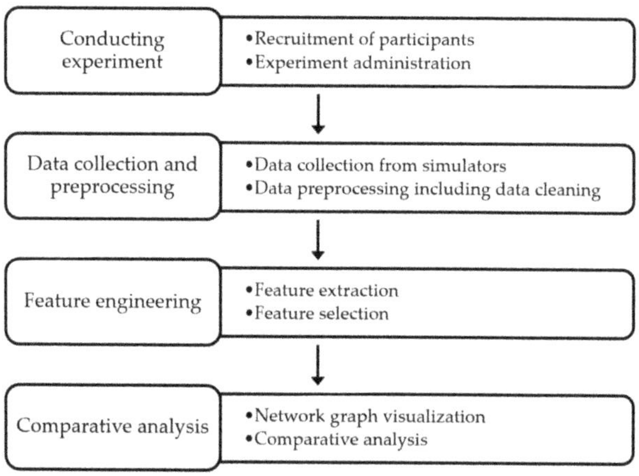

Figure 1. Workflow of the proposed methods.

This research conducted the simulation experiment designed and generated under contemplation. Then, the data collected from the experiment were preprocessed and analyzed for feature engineering. Afterward, selected features were scrutinized to compare data from the separated groups upon the onboard experience.

2.1. *Design of Simulation Experiment*

2.1.1. Participants

This research has recruited the students into two separate groups: "group A" and "group B". Group A consisted of students who had never experienced being on board the ship as cadets. Thus, their knowledge was limited to the theoretical education they acquired in class. On the contrary, group B consisted of students with onboard experience as cadets for one year on average. The specific information on participant recruitment is arranged in Table 1.

Table 1. Participants' information.

Characteristics	Group A	Group B
Number of participants	20	20
Onboard training experience	None	One year

2.1.2. Protocol

The authors created the basic navigation scenario solely to derive the different results of navigation skills, specifically the maneuvering skills of the groups, based on the simulation experiment design. Since groups differ significantly in terms of navigation experience, the experiment scenario used in this research was composed of only "route leg" and "waypoint" without territorial components. As a result, the participants' instructions were as simple as "navigate the ship on the route". The only permitted controller was the steering wheel. The experiment scenario is shown in Figure 2.

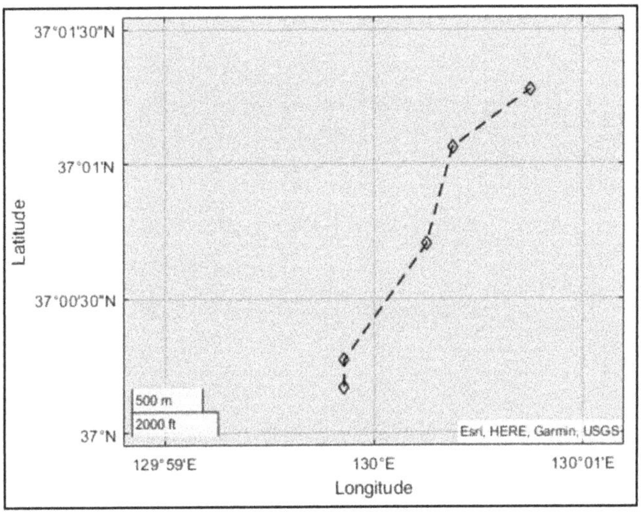

Figure 2. Experiment scenario.

The experiment used the full-mission-ship-handling simulator, and the target ship was the Mokpo National Maritime University training ship. The target ship is 133 m in length and 19.4 m in width and in 9196 GT; this ship is commonly boarded under the education curriculum.

2.2. *Data Collection and Preprocessing*

Data was collected from the simulator. The raw data consisted of 13 elements, as arranged in Table 2.

The raw data contained a temporal element, "Time" of elapsed time; the spatial element of "East" and "North" in the UTM datum; motion elements of "Heaving", "Yawing", "Pitching", and "Rolling" from 6 degrees of freedom; ship speed with RPM; and rudder degrees. Each value was represented numerically. Because the raw data was well-organized numerically, the data preprocessing focused primarily on unit conversion.

Table 2. Elements of raw data.

Name	Sample	Unit
Time	4×10^{-1}	Second (s)
UTM (East)	-6.8094×10^{-4}	Meter (m)
UTM (North)	1.1170×10^{5}	Meter (m)
Heaving	0.5142×10^{-6}	Meter (m)
Yawing	2.2744×10^{-4}	Degree (°)
Pitching	-1.1577×10^{-7}	Degree (°)
Rolling	4.1719×10^{-4}	Degree (°)
Speed	6.7110	Meter per second (m/s)
Drift	-1.0535×10^{-3}	Meter per second (m/s)
Rate of turn	2.0434×10^{-3}	Degree per second (°/s)
Propeller (PORT)	1.0287×10^{2}	RPM
Propeller (STBD)	1.0287×10^{2}	RPM
Rudder angle	−7.2000	Degree (°)

2.3. Feature Extraction

Steering features were commonly related to maneuvering skills: the use of a rudder. The rudder is the primary device that allows the navigator to control the ship directly [28–30]. Thus, rudder movement-based features that can represent how the participant steered the ship are meaningful. Table 3 displays the extracted features.

Table 3. List of steering features.

Name	Description	Unit
Max ROT	The maximum value of "rate of turn"	Degree per minute (°/min)
Mean ROT	The average value of "rate of turn"	Degree per minute (°/min)
STD ROT	Standard deviation value of "rate of turn"	Degree per minute (°/min)
Mean rudder	The average value of rudder angle	Degree (°)
First max rudder	Maximum rudder angle used for first altering	Degree (°)
Midship rudder	The temporal ratio of rudder angle at midship	Ratio (%)
Hard rudder	The temporal ratio of rudder angle over 30 degrees	Ratio (%)
Idle time	Idle time to use of rudder	Second (s)
First max time	Elapsed time to use "first max rudder"	Second (s)
Reverse time	Elapsed time to end first altering	Second (s)

2.4. Feature Selection

Features selection was conducted based on the basic attribute features in the figures. A significant difference was identified from features in the shape of the box, whisker, and interquartile ranges based on aggregation and dispersion.

2.5. Comparative Analysis

Based on the findings, the comparative analysis focused on the difference in the onboard experience. This research used the network graph algorithm and conducted the interpretation to visualize the difference between the groups and explain the effect of onboard training on various aspects of navigation performance. The network graph algorithm is composed of "nodes", "edges", and "weight". When the weight of the edges is high, the connectivity between the nodes becomes strong. Furthermore, if the nodes are strongly connected, the network is visualized as an aggregate shape as shown in Figure 3a, whereas the nodes are dispersed as a dispersion shape as shown in Figure 3b.

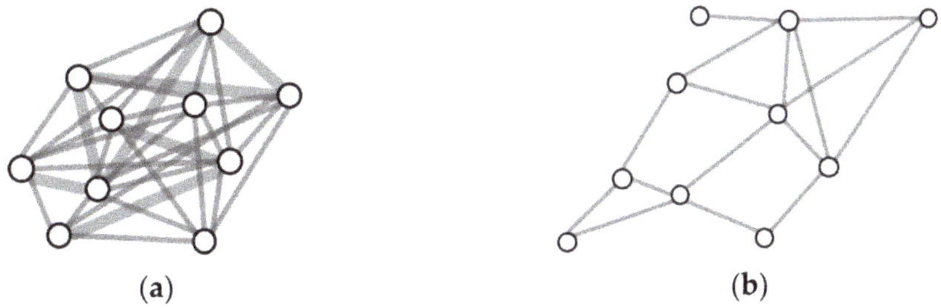

Figure 3. Concept of different shapes of networks: (**a**) aggregate shape and (**b**) dispersion shape.

This study defined each participant as a node, feature similarity as an edge, and absolute difference in feature value as a weight; this is known as connectivity. Then, edges with weak connections based on the average of weights were excluded from taking the verification of navigation skills into account.

3. Results

Steering features in Table 3 indicate fundamental skills, including the familiarization of rudder control. The authors then concentrated on examining the difference between features because rudder control is the most targeted attribute in the verification of enhanced navigation skills.

3.1. Result of Feature Extraction

The extracted feature demonstrated a significant difference as proof of onboarding training. Figure 4 depicts the differences between groups using normalized values of features, while Figures 5 and 6 depicts the aggregated navigation performance results of each group, which is useful for comparing the overall trajectories.

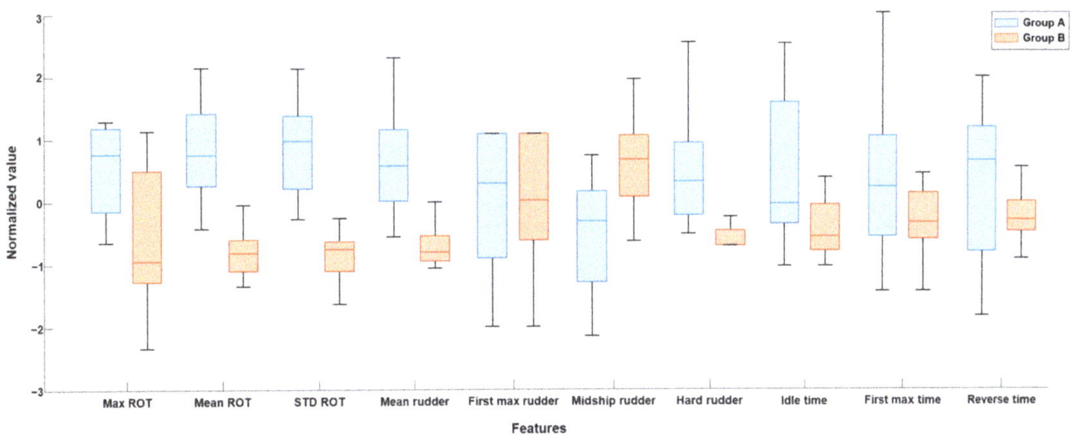

Figure 4. Maneuvering features and domains.

As proof of onboard training, the extracted feature was noted to have a remarkable difference. Figure 4 visualizes the difference of groups in each feature, and Figures 5 and 6 show the accumulated navigation performance results of each group, which is helpful in comparing overall trajectories.

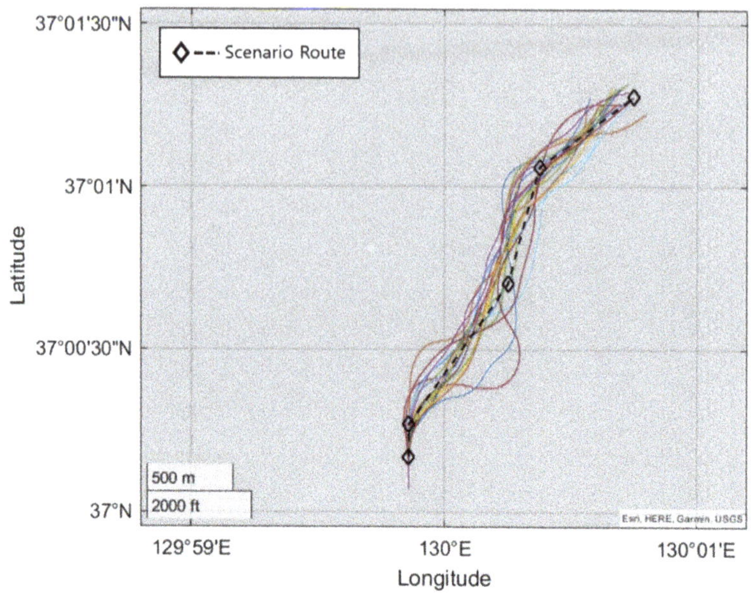

Figure 5. Navigation performance result: group A.

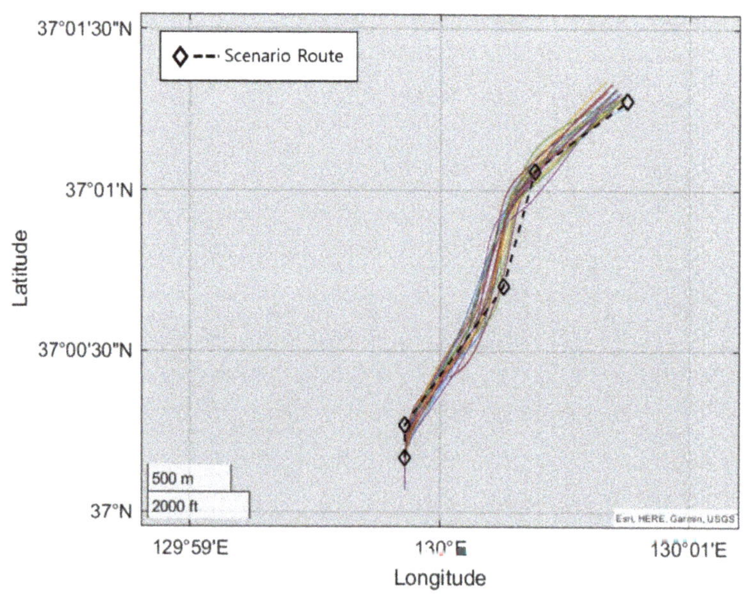

Figure 6. Navigation performance result: group B.

3.2. Result of Feature Selection

Figure 4 shows the basic boxplot attributes "box and whisker" and "interquartile range including skewness and dispersion. "First max rudder", "First max time", and "Reverse time" are the three features that have a non-significant difference in the shape of the box, whisker, and a form in which the interquartile ranges completely overlap in either direction. As the conformation displays, no particular discrepancy between the group was found. Another group of features that partially overlapped in box and whisker range were "Max

ROT", "Midship rudder", and "Idle time", yet their discrepancy was weak even when their skewness and dispersion differed. On the contrary, the "Mean ROT", "STD ROT", "Mean rudder", and "Hard rudder" showed a convincing difference between groups. Boxes were completely independent, and the interquartile ranges were disparate, with group A having large dispersion and group B having high skewness. As a result, features of "Mean ROT", "STD ROT", "Mean rudder", and "Hard rudder" were selected.

From the selected features, the authors found that "Mean ROT" and "STD ROT" are dependent; thus, the "STD ROT" was additionally excluded as Figure 7 shows final three selected features in the red frame.

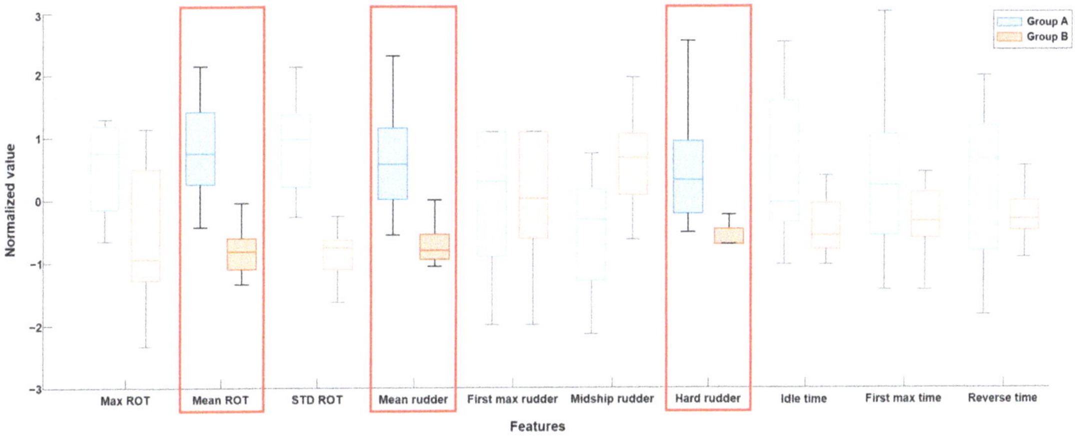

Figure 7. Selected features.

3.3. Comparative Analysis Result

Following the feature selection, this research calculated an absolute difference to represent the weight of edges. As a result, the closest connection had the lowest value, which is appropriate for visualization as connectivity. Then the maximum value was deducted from all values, and the values were transformed into absolute values again to reverse the meaning of the value, as a larger value indicates a stronger connection. As a result, in terms of weight, the reversed value came to mean what this research intended: larger values are stronger in connectivity, and smaller values are weaker. The network graphs of selected features are shown in Figures 8–10.

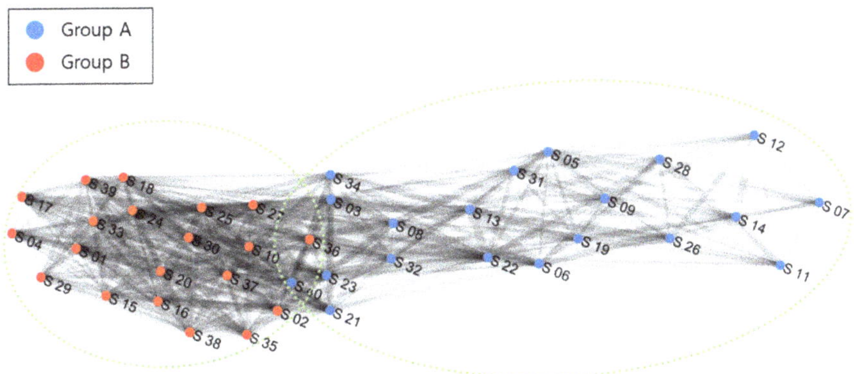

Figure 8. Network graph for the feature "Mean ROT".

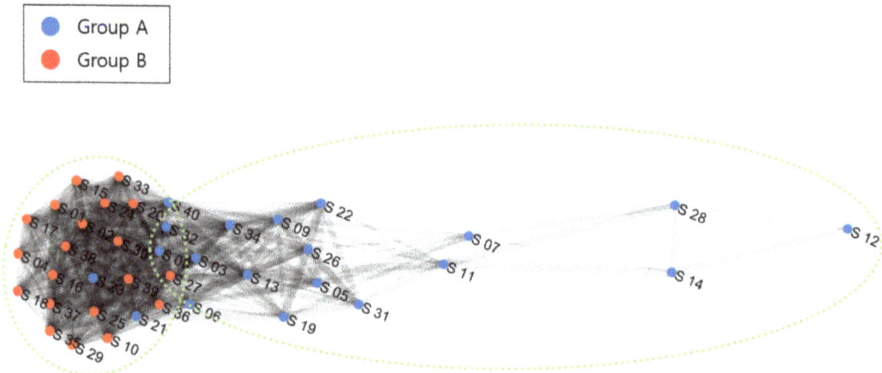

Figure 9. Network graph for the feature "Mean rudder".

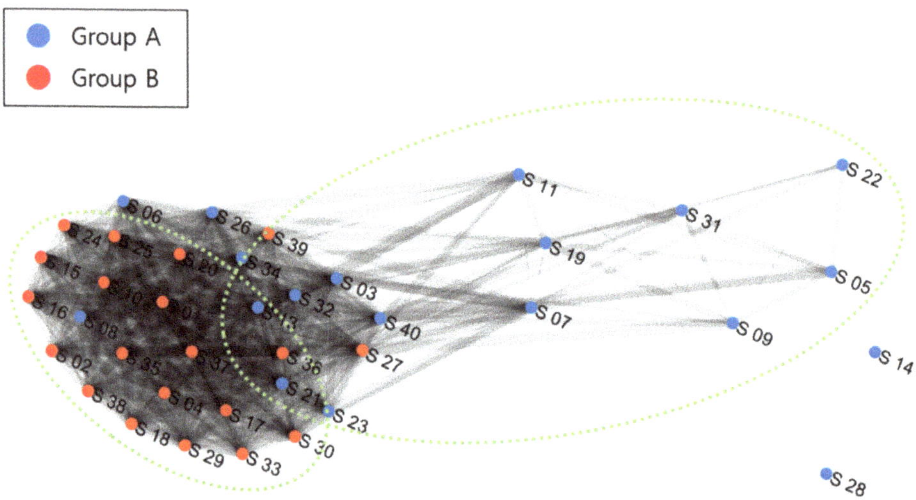

Figure 10. Network graph for the feature "Hard rudder".

4. Discussion

The authors chose features with notable differences based on the familiarization gained from onboard training. After some thought, it was discovered that the criteria for examining the features could vary along the set of factors in scenarios. As a result, when selecting features, we considered the designated factors of an experiment scenario. The network graph algorithm was then used in visualization to explain the significant differences. The obvious differences were thereafter discovered and explained.

4.1. Selected Features

The three preferred features are the most important aspect of rudder use. When the average rate of turn, also known as "Mean ROT", becomes excessive, the ship turns quickly, and maintaining course then becomes difficult. Similarly, the average use of rudder, that is, "Mean rudder", meant the participants swiped the rudder widely, reducing the ship's inertia of motion. Similarly, exaggerated control for the given scenario is a temporal ratio of rudder angle greater than 30 degrees. In terms of navigation skills, "Mean ROT" denotes an understanding of a ship's movements, "Mean rudder" denotes an understanding of the interaction between the rudder and the ship's movement, and "Hard rudder" denotes the proclivity of participants in handling the ship. Those navigation skills are the fructification of the training, that is, the onboard training.

4.2. Comparative Analysis of Groups

4.2.1. Network Graph Interpretation

Although this research did not consider node phases, the examination focused on the shapes: "aggregation" and "dispersion" in network graph interpretation. The aggregation in Figure 3a depicts the network's strong bond. Those strong bonds were found in group B of Figures 8–10. Group B was described in the circle because its nodes were not precluded when edges in weak connection were removed. The network graph's strong connection indicates that the similarity was high, and the values of selected features were consistent within a certain range. Similarly, the dispersion shape was discovered in group A. The circle was crushed and scattered as a result of the preclusion process's severed edges, weakening the network's bond. In other words, there was a little resemblance.

4.2.2. Comparative Analysis

The difference of groups in the perspective of the network graph is well presented in Figures 8–10 with the interpretation in the previous section. Here, the authors specifically examined the selected features by using descriptive statistics in Table 4.

Table 4. Descriptive statistics for comparative analysis.

Descriptive Statistics	Group A	Group B
Mean of "Mean ROT"	29.5155	17.0955
Median of "Mean ROT"	29.1357	16.9824
S.D of "Mean ROT"	5.8821	2.8930
Mean of "Mean rudder"	13.0642	5.7998
Median of "Mean rudder"	12.3258	5.3612
S.D of "Mean rudder"	4.7802	1.4568
Mean of "Hard rudder"	0.1476	0.0143
Median of "Hard rudder"	0.1164	0
S.D of "Hard rudder"	0.1280	0.0252

In Table 4, the mean of "Mean ROT" depicts the entire trend of the specific feature, so the value shows how different the groups were in the comparison. The mean of group B was 0.58 times smaller than group A for the "Mean ROT", indicating that group B had a better understanding of a ship's movement. Similarly, knowledge of the interaction between the rudder and the ship's motion was far superior to group A, and participants' proclivity to handle the ship was reduced.

5. Conclusions

This research was motivated by the need to demonstrate the effectiveness of training, particularly the current training methods. When the era comes for remote operators to be trained and navigate the actual MASS, the training effect will be important for navigation safety. The study objectively identified the navigation skills that could be improved through onboard training in this study. A comparative analysis was conducted to demonstrate the value of onboard training. The participants in the experiment were divided into two groups based on whether or not they were trained onboard. The results of the analysis efficiently explain the difference between the groups in the specific features related to the use of the rudder. To emphasize the meaning of features in actual navigation, this study interpreted the difference in terms of the navigation skills. Although this research used a simple navigation scenario for the experiment and that participant recruitment was not systemic, the effect of onboard training was appropriately revealed. In the future, the training simulator designed explicitly for MASS remote control will be developed.

The authors expect that the remote operator training simulator will have better effects on improving operator navigation skills. Accordingly, the selected navigation performance features; "Mean ROT", "Mean rudder" and "Hard rudder" derived from this research are expected to be applied for the development of simulation performance-based navigation proficiency evaluation methods during the SRCO training period.

Author Contributions: Conceptualization, H.H., T.H. and I.-H.Y. Data curation, I.-H.Y. Formal analysis, H.H. and T.H. Funding acquisition, I.-H.Y. Methodology, H.H. and T.H. Project administration, I.-H.Y. Supervision, I.-H.Y. Visualization, H.H. and T.H. Writing—original draft, H.H. and T.H. Writing—review and editing, I.-H.Y. All authors have read and agreed to the published version of the manuscript.

Funding: This research and APC were funded by the Ministry of Oceans and Fisheries, Korea, through grant number 202106314.

Institutional Review Board Statement: The internal institutional review board determined to exempt the ethical review and approval for this study due to the ship handling simulator used in the experiment being a simulator used for education and training generally on the maritime institute. Also, the experiment was conducted in an environment without any physical restrictions on the subject's body.

Informed Consent Statement: Not applicable.

Data Availability Statement: Not applicable.

Acknowledgments: This research was a part of a project titled "Development of Smart Port-Autonomous Ships Linkage Technology," supported by the Ministry of Oceans and Fisheries, Korea.

Conflicts of Interest: The authors declare no conflict of interest.

References

1. Kim, T.; Schröder-Hinrichs, J. Research developments and debates regarding maritime autonomous surface ship: Status, challenges and perspectives. *New Marit. Bus.* **2021**, *10*, 175–197.
2. Ramos, M.A.; Utne, I.B.; Mosleh, A. Collision avoidance on maritime autonomous surface ships: Operators' tasks and human failure events. *Saf. Sci.* **2019**, *116*, 33–44. [CrossRef]
3. Fan, C.; Montewka, J.; Zhang, D. A risk comparison framework for autonomous ships navigation. *Reliab. Eng. Syst. Saf.* **2022**, *226*, 108709. [CrossRef]
4. Fan, C.; Wróbel, K.; Montewka, J.; Gil, M.; Wan, C.; Zhang, D. A framework to identify factors influencing navigational risk for Maritime Autonomous Surface Ships. *Ocean Eng.* **2020**, *202*, 107188. [CrossRef]
5. Veitch, E.A.; Kaland, T.; Alsos, O.A. Design for resilient human-system interaction in autonomy: The case of a shore control centre for unmanned ships. *Proc. Des. Soc.* **2021**, *1*, 1023–1032. [CrossRef]
6. Munim, Z.H.; Haralambides, H. Advances in maritime autonomous surface ships (MASS) in merchant shipping. *Marit. Econ. Logist.* **2022**, *24*, 181–188. [CrossRef]
7. Bogusławski, K.; Gil, M.; Nasur, J.; Wróbel, K. Implications of autonomous shipping for maritime education and training: The cadet's perspective. *Marit. Econ. Logist.* **2022**, *24*, 327–343. [CrossRef]
8. Kim, T.; Mallam, S. A Delphi-AHP study on STCW leadership competence in the age of autonomous maritime operations. *WMU J. Marit. Aff.* **2020**, *19*, 163–181. [CrossRef]
9. TORESANO, C.C. The next step on the maritime education and training in the era of autonomous shipping: A literature review In Proceedings of the 9th International Conference on Maritime Transport, Barcelona, Spain, 27–29 June 2022.
10. Ahvenjärvi, S. The human element and autonomous ships. *TransNav Int. J. Mar. Navig. Saf. Sea Transp.* **2016**, *10*, 517–521. [CrossRef]
11. Karvonen, H.; Martio, J. Human factors issues in maritime autonomous surface ship systems development. In *SINTEF Proceedings:3, Proceedings of the 1st International Conference on Maritime Autonomous Surface Ships, Busan, Korea, 8–9 November 2018*; SINTEF Academic Press: Trondheim, Norway, 2019.
12. Hanzu-Pazara, R.; Barsan, E.; Arsenie, P.; Chiotoroiu, L.; Raicu, G. Reducing of maritime accidents caused by human factors using simulators in training process. *J. Marit. Res.* **2008**, *5*, 3–18.
13. Fan, S.; Blanco-Davis, E.; Yang, Z.; Zhang, J.; Yan, X. Incorporation of human factors into maritime accident analysis using a data-driven Bayesian network. *Reliab. Eng. Syst. Saf.* **2020**, *203*, 107070. [CrossRef]
14. Shi, X.; Zhuang, H.; Xu, D. Structured survey of human factor-related maritime accident research. *Ocean Eng.* **2021**, *237*, 109561. [CrossRef]

15. Demirel, E.; Mehta, R. Developing an effective maritime education and training system-TUDEV experiment. In *International Maritime Lawyers Association Conference*; Citeseer: Accra, Ghana, 2009; pp. 1–11.
16. Özdemir, Ü.; GÜNEROĞLU, A. Strategic approach model for investigating the cause of maritime accidents. *PROMET-Traffic Transp.* **2015**, *27*, 113–123. [CrossRef]
17. Yoshida, M.; Shimizu, E.; Sugomori, M.; Umeda, A. Regulatory requirements on the competence of remote operator in maritime autonomous surface ship: Situation awareness, ship sense and goal-based gap analysis. *Appl. Sci.* **2020**, *10*, 8751. [CrossRef]
18. Mallam, S.C.; Nazir, S.; Renganayagalu, S.K. Rethinking maritime education, training, and operations in the digital era: Applications for emerging immersive technologies. *J. Mar. Sci. Eng.* **2019**, *7*, 428. [CrossRef]
19. Albayrak, T.; Ziarati, R. Training: Onboard and simulation based familiarisation and skill enhancement to improve the performance of seagoing crew. In Proceedings of the International Conference on Human Performance at Sea (HPAS), Glasgow, UK, 16–18 June 2010; p. 586.
20. DEMİREL, E.; BAYER, D. A Study on the assessment of sea training as an integral part of maritime education and training. *Online J. Qual. High. Educ.* **2016**, *3*, 12.
21. Bhaskaran, B. Importance of simulators in maritime training. *Int. J. Res. Anal. Rev.* **2018**, *5*, 6–8.
22. Kim, T.; Sharma, A.; Bustgaard, M.; Gyldensten, W.C.; Nymoen, O.K.; Tusher, H.M.; Nazir, S. The continuum of simulator-based maritime training and education. *WMU J. Marit. Aff.* **2021**, *20*, 135–150. [CrossRef]
23. Nazir, S.; Øvergård, K.I.; Yang, Z. Towards effective training for process and maritime industries. *Procedia Manuf.* **2015**, *3*, 1519–1526. [CrossRef]
24. Deling, W.; Dongkui, W.; Changhai, H.; Changyue, W. Marine Autonomous Surface Ship–a Great Challenge to Maritime Education and Training. *Am. J. Water Sci. Eng.* **2020**, *6*, 10–16. [CrossRef]
25. Mallam, S.C.; Nazir, S.; Sharma, A. The human element in future Maritime Operations–perceived impact of autonomous shipping. *Ergonomics* **2020**, *63*, 334–345. [CrossRef] [PubMed]
26. Liu, C.; Chu, X.; Wu, W.; Li, S.; He, Z.; Zheng, M.; Zhou, H.; Li, Z. Human–machine cooperation research for navigation of maritime autonomous surface ships: A review and consideration. *Ocean Eng.* **2022**, *246*, 110555. [CrossRef]
27. Pietrzykowski, J.; Pietrzykowski, Z.; Hajduk, J. Operations of maritime autonomous surface ships. *TransNav Int. J. Mar. Navig. Saf. Sea Transp.* **2019**, *13*, 725–733. [CrossRef]
28. Sandaruwan, D. A Ship Simulation System for Maritime Education. *Int. J. Adv. ICT Emerg. Reg.* **2010**, *3*, 34–47.
29. Hasanvand, A.; Hajivand, A. Investigating the effect of rudder profile on 6DOF ship turning performance. *Appl. Ocean Res.* **2019**, *92*, 101918. [CrossRef]
30. Liu, J.; Hekkenberg, R. Sixty years of research on ship rudders: Effects of design choices on rudder performance. *Ships Offshore Struct.* **2017**, *12*, 495–512. [CrossRef]

applied sciences

MDPI

Article

A Study on Longitudinal Motion Scenario Design for Verification of Advanced Driver Assistance Systems and Autonomous Driving Systems

Kunhee Cho [1], Changwoo Park [1] and Hyeongcheol Lee [2,*]

[1] Department of Electrical Engineering, Hanyang University, Seoul 04763, Republic of Korea
[2] Department of Electrical and Biomedical Engineering, Hanyang University, Seoul 04763, Republic of Korea
* Correspondence: hclee@hanyang.ac.kr; Tel.: +82-2-2220-1685

Abstract: This paper proposes a test scenario design method that reflects the longitudinal characteristics of reality for effective verification of advanced driver assistance systems (ADAS) and autonomous driving systems (ADS). Since the target systems interact with the external environment differently from the existing vehicle control system, realistic and various verification scenarios are required for verification. The proposed method consists of a vehicle model for simulating the vehicle behavior and a driver model to actively respond to the driving environment. In particular, the driver model used a model predictive control (MPC) algorithm to reflect the characteristic of human drivers. The longitudinal driving characteristics of human drivers were derived through a large-scale driving database analysis and considered in the driver model. The proposed method was compared with an existing car-following model using computer simulations. It was confirmed that its longitudinal driving behavior is similar to that of human drivers and that various scenarios can be designed by changing the model parameters.

Keywords: advanced driver assistance systems (ADAS); autonomous driving systems (ADS); model predictive control (MPC); scenario design; driver model; naturalistic driving database

Citation: Cho, K.; Park, C.; Lee, H. A Study on Longitudinal Motion Scenario Design for Verification of Advanced Driver Assistance Systems and Autonomous Driving Systems. *Appl. Sci.* **2023**, *13*, 716. https://doi.org/10.3390/app13020716

Academic Editor: Nadjim Horri

Received: 29 October 2022
Revised: 11 December 2022
Accepted: 12 December 2022
Published: 4 January 2023

1. Introduction

A great deal of research has been carried out for decades to develop advanced driver assistance systems (ADAS) and autonomous driving systems (ADS) for drivers' convenience and safety [1–5]. ADAS and ADS are closely related to safety, thus verification is required through various environments and scenarios. Recently, studies conducted using virtual vehicle and traffic simulation programs, such as IPG Carmaker and Carla, to improve the efficiency of ADAS and ADS design and verification are being actively conducted [6–8]. In addition, research on a vehicle in the loop (VIL) models that utilize the dynamic characteristics of real vehicles and interconnect with virtual environments is increasing [9–12]. Especially, ADAS and ADS operate organically with surrounding and external environments, hence the scenarios for verification in a simulation environment must consider various factors.

As scenario design methods for verification of ADAS and ADS, standards assessment-based, test matrix-based, malicious scenario-based, and traffic simulation-based methods are widely known [13–21]. First, there is a standards assessment-based method widely known as ISO and Euro NCAP [13,14]. The scenarios presented by these two organizations focus on the verification of autonomous emergency braking systems (AEBS), adaptive cruise control systems (ACCS), and lane support systems among ADAS. Although different models have different purposes in terms of vehicle safety assessment or establishing ADAS standards, they are similar in that they present minimum standards that ADAS must satisfy. As the scenarios provided by the two organizations are limited, these scenarios need to be expanded for drivers' safety in various driving situations. A test matrix-based method can be used for designing various scenarios, which defines factors to be evaluated based on a large-scale database [15,20]. Many projects have suggested test scenarios using actual driving/accident data [15–18]. However, ADAS and ADS designed with known scenarios inevitably have lower reliability for scenarios that have not occurred ever. A malicious scenario-based method is one that defines scenarios near the limit of the mathematical models of the ego and surrounding vehicles [19–21]. This method can be verified in a short time because the verification is performed in the unstable regions of the ego and surrounding vehicles. However, when the system is very complex or designing a model of the system is not possible, it is difficult to design malicious scenarios. Last but not least, a traffic simulation-based method can be used to validate ADAS and ADS in a comprehensive environment. A traffic simulation is widely known as a method for macroscopic analysis of traffic volumes for road or signal system design [12,22,23]. This method is based on car-following models to simulate the traffic flow, and its application has recently been increasing for the verification of ADAS and ADS. Table 1 represents the list of widely known car-following models [24–29]. Each car-following model commonly uses the time constant τ to express the driver's delay. In Table 1, v_f, a_f, R, $\dot{R}$, R_{min}, and V_d represent the speed and acceleration of the ego vehicle, relative distance, speed and minimum relative distance with a leading vehicle, and the maximum speed of the ego vehicle, respectively. Other parameters (C, C_v, C_s, C_t, A, B, $\hat{B}$) are configured to express various relative distances and speeds with a leading vehicle. Traffic simulation software is designed by improving the car-following models described in Table 1 [12,22,23]. Among the car-following models, only the Gipps model considers both acceleration and deceleration, which are the basic parameters governing the motion of the vehicle [28]. Therefore, it is not appropriate to use the remaining car-following models in the verification of ADAS and ADS, where the behavior of surrounding vehicles is important. As studies of ADAS and ADS became active in the 2000s, a large-scale naturalistic database was built to design the behavior of surrounding vehicles [18,30–37]. The 100-car naturalistic driving study (NDS) is a database of 3 million km of driving data from 241 drivers led by Virginia Tech [30]. Safety pilot model deployment (SPMD) is a database using roadside equipment and vehicle-mounted radar and vision to demonstrate the technology of connected vehicles in various environments and analyze driver acceptance of vehicle safety systems [35]. The High D dataset has collected driving data from more than 110,500 vehicles on highways in Germany and is being used in various ADAS and ADS studies [21,38]. Recently, using these large-scale driving databases, research on drivers' driving characteristics and their application to car-following models is being actively conducted [29,39–42]. In [39], Bifulco defined the leading vehicle from the driving database and derived parameters of the Gipps model that follows it using artificial neural networks (ANNs). In [42], Przybyla deduced drivers' characteristics through analysis of the driving database and applied them to the Newell model to analyze driving risk [42]. In [40], based on the NDS, Zheng derived the longitudinal acceleration and yaw angle of the following vehicle using reinforcement learning and a fuzzy algorithm. These methods are suitable for analyzing the traffic system by simulating the driving of the vehicle in a simulation environment like one facing an actual human driving. However, since it is difficult to design models for implementing malicious scenarios and control them, the traffic simulation-based method has limitations in ADAS and ADS verification.

Therefore, to overcome the limitations of the analyzed ADAS and ADS verification scenario design methods studied thus far, we propose a longitudinal motion scenario design method. In this method, the longitudinal driving characteristics of human drivers were reflected in the surrounding vehicles of the ego vehicle for more realistic scenario design in a simulation environment. First, a simplified vehicle dynamic model expresses the dynamic behavior of the vehicle, and a driver model imitates human driving by using model predictive control (MPC). With the vehicle and the driver model, the test vehicle can adapt to the surrounding environment and drive realistically, and various scenarios can be actively produced. Especially, to improve the consistency of the driver model, we determined the acceleration/deceleration standards for typical human drivers derived from a large-scale driving database analysis. Lastly, the proposed method was verified using computer simulations to confirm human driver imitations and their significance. The computer simulations assumed a driving situation on a straight-line road with a high road frictional coefficient. As the driving characteristics of human drivers vary depending on the driving situation, it was limited to the longitudinal acceleration/deceleration situation. The acceleration/deceleration and jerk, which are the longitudinal driving characteristics of human drivers, are derived in this situation. The characteristics of the sailing/braking start point are derived in this situation as well. To verify this and confirm consistency, the velocity, acceleration/deceleration, and jerk of the proposed method were compared with the acceleration/deceleration section of the randomly selected driver data. This study confirmed the possibility of simulating various drivers by changing the weights and constraints setting of the driver model. Through this, it is possible to create various longitudinal driving scenarios that overcome the shortcoming of the existing scenario design methods. In addition, if the constraints of the proposed driver model are set close to the limit situation, such as close distance to the surrounding vehicle or sudden stop, malicious scenarios can be easily designed with simple parameter changes.

This paper is organized as follows. Section 2 introduces a simplified vehicle model in more detail. Section 3 describes the error dynamics between the ego vehicle and surrounding vehicles and the design of an MPC-based driver model. Section 4 explains human driving characteristics derived from a large-scale database analysis. Section 5 shows the computer simulation results for the validation of the proposed method, and a conclusion is conducted in Section 6.

Table 1. Deterministic classical car-following models.

Researcher (Year)	Model
Pipe [24] (1953)	$a_F(t+\tau) = C\dot{R}(t)$
Gazis [25] (1961)	$a_F(t+\tau) = C\dot{R}(t)\frac{v_F(t)^m}{R(t)^l}$
Newell [26] (1961)	$v_F(t+\tau) = V_d \cdot \left\{1 - \exp\left(-C \cdot \frac{R(t)-R_{min}}{V_d}\right)\right\}$
Tyler [27] (1964)	$a_F(t+\tau) = C_V \cdot \dot{R}(t) + C_S \cdot (R(t) - C_t \cdot v_F(t))$
Gipps [28] (1981)	$v_F(t+\tau) = \min\begin{cases} v_F(t) + 2.5 \cdot A \cdot \tau \cdot \left(1 - \frac{v_F(t)}{v_d}\right) \cdot \sqrt{0.025 + \frac{v_F(t)}{V_d}} \\ B \cdot \tau + \sqrt{B^2 \cdot \tau^2 - B \cdot \left\{2 \cdot (R(t) - R_{min}) - v_F(t) \cdot \tau - \frac{v_L(t)^2}{\hat{B}}\right\}} \end{cases}$

2. Simplified Vehicle Model

A simplified vehicle model for the longitudinal motion scenario design can simulate vehicle motion by inputting the required acceleration ($a_{x,des}$), which is designed by the driver model proposed in Section 3. First, the simplified vehicle model assumes that the vehicle is traveling on flat ground, thus the relationship between the acceleration and velocity is shown in Equation (1) [43].

$$\dot{v}_x = a_x \tag{1}$$

$$\begin{gathered} ma_x = F_x - F_r - F_a \\ w.r.t \begin{cases} F_r = N\left(C_{r,0} + C_{r,1}v_x + C_{r,2}v_x^2\right) \\ F_a = 0.5C_d\rho A v_x^2 \end{cases} \end{gathered} \tag{2}$$

$$a_x = \begin{cases} a_{x,\,U} & if\; a_{x,\,des} > a_{x,U} \\ a_{x,\,L} & else\,if\; a_{x,des} < a_L\, x \\ a_{x,\,des} & else \end{cases} \tag{3}$$

To consider the dynamic characteristics of the vehicle, a longitudinal dynamic vehicle model is described in Equation (2). The vehicle acceleration is deduced by the longitudinal force (F_x), rolling resistance (F_r), and aerodynamic resistance (F_a) of the vehicle. The rolling resistance (F_r) is composed of the normal force of the vehicle (N), the constant coefficient ($C_{r,0}$), and velocity coefficients ($C_{r,1}$, $C_{r,2}$). The aerodynamic resistance (F_a) is composed of the aerodynamic drag coefficient (C_d), the density of the air (ρ), and the area of the vehicle (A). The rolling resistance and the aerodynamic resistance are dependent on the vehicle velocity. The longitudinal force is constrained by the performance of the vehicle, which is related to the vehicle's powertrain and braking system. For this reason, the longitudinal acceleration of the vehicle was constrained by an upper bound ($a_{x,U}$) and lower bound ($a_{x,L}$). Figure 1a describes the upper constraint of the longitudinal force that varies with the transmission gear ratio. For simplification, a simplified acceleration constraint, as shown in Figure 1b, is designed to consider dynamic characteristics of the vehicle in the longitudinal motion scenario design. It constrains the desired acceleration ($a_{x,des}$) from the driver model to describe realistic vehicle motions. In other words, the actual vehicle is composed of complex dynamic systems, such as powertrain and hydraulic systems. However, in the case of this simplified vehicle model, it is used to simulate vehicle movement by considering acceleration and its related constraints. Acceleration and deceleration situations can occur depending on the sign of the longitudinal acceleration (a_x), and the simplified vehicle model moves by causing a change in vehicle speed.

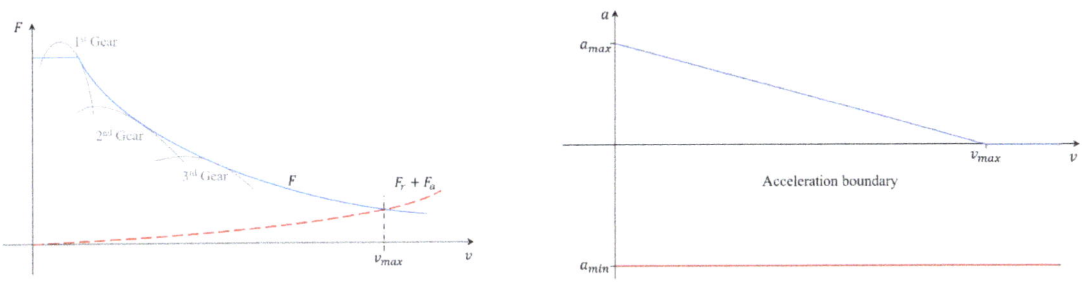

(a) (b)

Figure 1. Constraints of longitudinal force and acceleration: (**a**) the constraint of longitudinal force varies with the transmission gear ratio; (**b**) the simplified acceleration constraint for the vehicle model.

3. Longitudinal Driver Model

3.1. Error Dynamics for the Car-Following Model

To design a control input of the simplified vehicle model, Figure 2 a state-space equation is derived using the relative distance/velocity to the leading vehicle [43–45].

$$\begin{bmatrix} e_x \\ e_v \end{bmatrix} = \begin{bmatrix} x_{L,r} \\ v_L \end{bmatrix} - \begin{bmatrix} x_{EGO}, f \\ v_{EGO} \end{bmatrix} - \begin{bmatrix} c_{TG} v_{EGO} \\ 0 \end{bmatrix} \tag{4}$$

$$\begin{gathered} \frac{d}{dt}\begin{bmatrix} e_x \\ e_v \end{bmatrix} = \begin{bmatrix} 0 & 1 \\ 0 & 0 \end{bmatrix}\begin{bmatrix} e_x \\ e_v \end{bmatrix} + \begin{bmatrix} -c_{TG} \\ -1 \end{bmatrix} a_x + \begin{bmatrix} 0 \\ 1 \end{bmatrix} a_L \\ \rightarrow \dot{x} = Ax + Bu \\ s.t \begin{cases} a_{min} & \leq a_x \leq & a_{max} \\ x_{min} - c_{tg} v_{EGO} & \leq e_x \leq & \infty \\ v_L - v_{EGP,\, max} & \leq e_v \leq & v_L \end{cases} \end{gathered} \tag{5}$$

In Equation (4), the desired relative distance ($c_{TG} v_{EGO}$) to the leading vehicle is determined using the constant time gap (CTG) policy to satisfy the string stability of a row of cars [44,45]. The distance error (e_x) is designed using the desired relative distance and the difference between the rear of the leading vehicle ($x_{L,r}$) and the front of the ego vehicle ($x_{EGO,f}$). The velocity error (e_v) is set as the difference between the velocity of the leading vehicle (v_L) and the velocity of the ego vehicle (v_{EGO}). The state equation constructed using the distance/velocity error is composed of the system matrices A, B, and the ego vehicle's acceleration as input u. In order to simulate the movement of the ego vehicle for designing scenarios, a virtual target for deriving the distance/velocity error is required. For this reason, it is assumed that there is a virtual stopped vehicle (v_L, $a_L \approx 0$) in front of the vehicle model at a future stopping position or a virtual vehicle with constant velocity ($a_L \approx 0$) to derive the distance/velocity error for simplification. The control input is constrained by the a_{max}, a_{min} that are defined by the characteristics of the vehicle and human drivers. In addition, the constraints of the distance/velocity error are defined to satisfy the CTG policy.

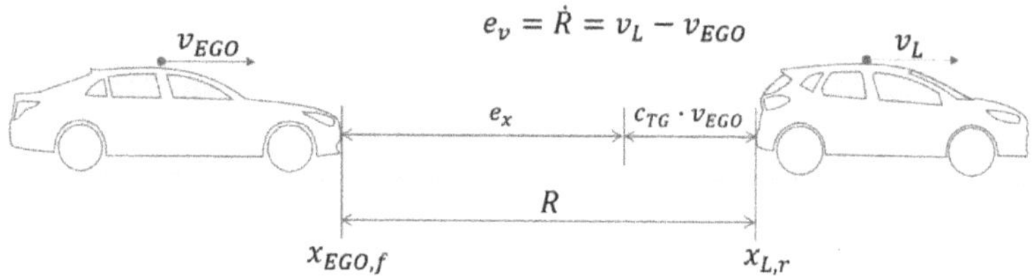

Figure 2. The scheme of error dynamics for the car-following model.

3.2. *MPC-Based Driver Model Design*

To design a driver model that considers human driving characteristics and various constraints using the simplified vehicle model, it is required that a controller should reflect future information and satisfy constraints. A model predictive control (MPC) algorithm is a well-known optimal controller design method to consider future states and constraints. The MPC algorithm is based on the discrete-time state-space model, and it calculates sub-optimal control inputs to minimize the cost function with various potential optimization methods [46]. The driver model of the proposed longitudinal motion scenario design method is designed by the MPC algorithm that minimizes the cost function ($J(u,\ x_0)$) by satisfying the constraints in Equation (6).

$$\begin{gathered} J(u, x_0) = x_N^T P x_N + \sum_{k=0}^{N-1} (x_k^T Q x_k + u_k^T R u_k) \\ s.t.\ x_{k+1} = A_d x_k + B_d u_k,\ k = 0, \ldots,\ N-1 \\ x_k \in \mathcal{X},\ u_k \in \mathcal{U},\ k = 0,\ \ldots, N-1 \\ x_0 = x(0),\ P = P^T > 0,\ Q = Q^T > 0,\ R = R^T > 0 \end{gathered} \tag{6}$$

The continuous time error dynamics model shown in Equation (4) is discretized by the Euler method, denoted by the discrete-time state-space model A_d, B_d to construct the MPC problem. $\mathcal{X}, \mathcal{U}$, and N are the constraints on the state, input, and prediction horizon, respectively. In this paper, a quadratic cost function is used, where P, Q, and R are symmetric and positive definite. The control input u^* that minimizes the cost function $J(u, x_0)$ and satisfies the constraints can be calculated using a quadradic programming (QP) solver, such as "quadprog" in MATLAB, which is used in this paper. The driver model is set to 0.2 s for the sampling time and 60 m for the prediction horizon, which means the prediction time is about 12 s to consider the time that a vehicle traveling at 30 m/s can stop with a maximum deceleration of -2.5 m/s^2. Since this paper focuses on deriving a human driving characteristics analysis and longitudinal motion scenario design, the process for solving the MPC problem is omitted, and the existing MPC-related literature is referred to for the solution process [46].

The driver model in this study is used to simulate longitudinal driving situations, such as acceleration/deceleration. It is designed using the fixed weights and constraints in Equation (6). The driving tendency of human drivers can vary, mainly depending on the driving situation [47,48]. Accordingly, the weights and the constraints of the driver model for imitating a specific driver can be changed. Therefore, it is limited to expressing the acceleration/deceleration situation, in which the longitudinal driving characteristics of human drivers are well expressed, and it is globally optimal in this situation through the optimal control problem in Equation (6).

4. Naturalistic Driving Data-Based Longitudinal Driver Model Parameter Design

4.1. Driving Data Pre-Processing

The safety pilot model deployment (SPMD) database, an open naturalistic driving database, is used to derive the driving characteristics of human drivers [35]. SPMD consists of data acquisition systems (DAS), basic safety messages (BSM), roadside equipment (RSE), and contextual data. Among the vast amount of data in this database, only a part of DAS, which is closely related to this paper, is selected for human driving characteristics analysis. DAS consists of DAS1 and DAS2. Both DAS1 and DAS2 provide data for ego vehicles and GPS measurements for each vehicle. However, for measurement of the vehicle in front, DAS1 used a vision sensor and DAS2 used a radar sensor. Between them, only the radar sensor-based DAS2, which has relatively high measurement accuracy for longitudinal driving, is used. In DAS2, 14,346 driving datapoints of 64 vehicles are measured at a 10 Hz cycle. It is confirmed that there are omissions or noise in some of the data. Therefore, the data are preferentially classified through a classification process shown in Table 2 in consideration of its reliability and efficiency. For more detail, the data classification process is described in Appendix A.

Table 2. Data classification process to ensure reliability.

Step	Exclusion Conditions	Remain Time (s)
(1) Original	-	8,338,419.4
(2) Trip time	Trip data less than 300 s	7,589,392.0
(3) Radar	Radar data nonexistence	7,523,291.5
(4) 0 to 0	Zero speed nonexistence	7,277,452.8
(5) Data	Continuous data nonexistence over 10 s	6,884,491.7

A smoothing process is performed to remove noise from the classified data. For this purpose, the locally estimated scatterplot smoothing (LOESS) algorithm, which is a widely known local regression method, is used [49]. LOESS is a regression method that constructs a quadratic polynomial using a weight function w for regression smoothing and a robust weight function G based on median values (see Equations (7)–(9)). In Equation (9), the coefficients $a, b,$ and c of the quadratic polynomial are derived in the direction of minimizing the product of the weights and the square error of the actual data and the quadratic polynomial.

$$w(x_k) = \left(1 - \left|\frac{x_i - x_k}{d_i}\right|^3\right)^3 \; for\, k = 1, \ldots, n \tag{7}$$

$$G(x_k) = \begin{cases} 1 - 6MID & |6MID| < 1 \\ 0 & |6MID| \geq 1 \end{cases} \quad s.t\; 6MID = \frac{|y_i - \hat{y}_i|}{6\; median\;(|y_i - \hat{y}_i|)} \tag{8}$$

$$RSS_x(a,b,c) = \sum_k w(x_k) G(x_k) \left(y_k - a - bx_k - cx_k^2\right)^2 \tag{9}$$

Figure 3 is an example of smoothing results obtained by applying LOESS to the classified data. As shown in Figure 3, the smoothing process was carried out only to the extent that the characteristics of the data were not loose. In the case of the acceleration data, there is an offset within the data, thus there is an error even when the smoothing process is performed. For example, although the speed increases in the 65–70 s section in Figure 3a, the acceleration in the same section in Figure 3b is shown as 0, indicating that there exists an offset of the acceleration data. It is judged that this offset is caused by problems such as the measurement equipment setting and the mounting error when gathering the SPMD data set. However, in the case of the speed data, the smoothing processing showed a better result than the acceleration data. For this reason, acceleration for the data analysis is used by differentiating the smoothed speed of the data.

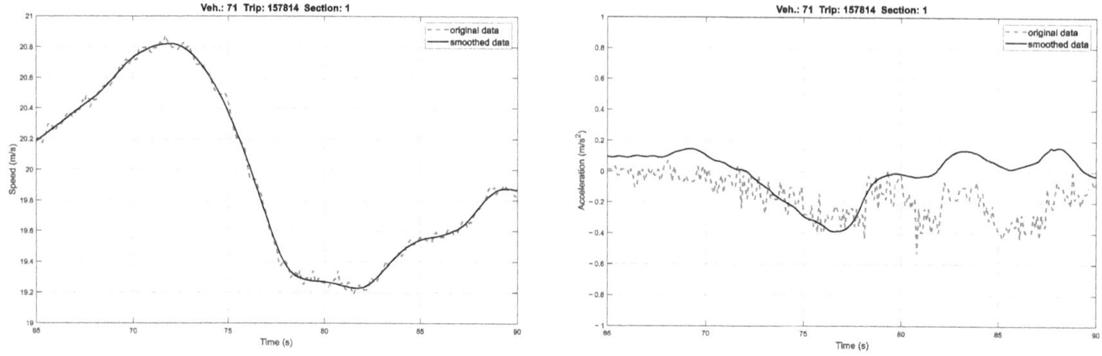

(**a**) (**b**)

Figure 3. Example of pre-processed driving data: (**a**) speed; (**b**) acceleration.

4.2. Analysis of Free Acceleration/Deceleration Point

Using the pre-processed data in Section 4.1, we analyzed the human drivers' free acceleration/deceleration driving data in a situation where the vehicle in front is far away and not obstructed. Figure 4 shows the maximum acceleration/deceleration distributions of vehicles driven by human drivers derived from the randomly chosen 300 free acceleration/deceleration datasets. The speed, acceleration, and jerk data used for the analysis are presented in Figure A2. Moreover, the distributions of maximum acceleration/deceleration for each maximum section during free acceleration/deceleration are summarized in Tables A1 and A2.

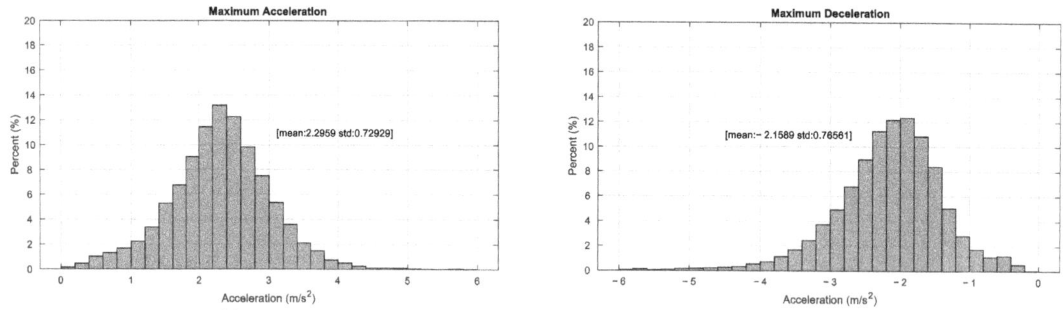

(**a**) (**b**)

Figure 4. Distribution of maximum acceleration/deceleration (300 random vehicles): (**a**) case of free acceleration; (**b**) case of free deceleration.

With the analysis of the free acceleration/deceleration data, human drivers generally have similar patterns during free acceleration/deceleration (see Figure A2), and the mean values of maximum acceleration/deceleration are 2.30 m/s^2 and -2.16 m/s^2, respectively. Longitudinal driving characteristics of human drivers can be derived mainly from the mean and median values of acceleration/deceleration, and the mean was used in this study [47,48]. Figure 5 shows the driving of different drivers arbitrarily selected to see the human drivers' free acceleration/deceleration patterns in detail, and the maximum speed and maximum acceleration/deceleration for each driver are expressed in Table 3.

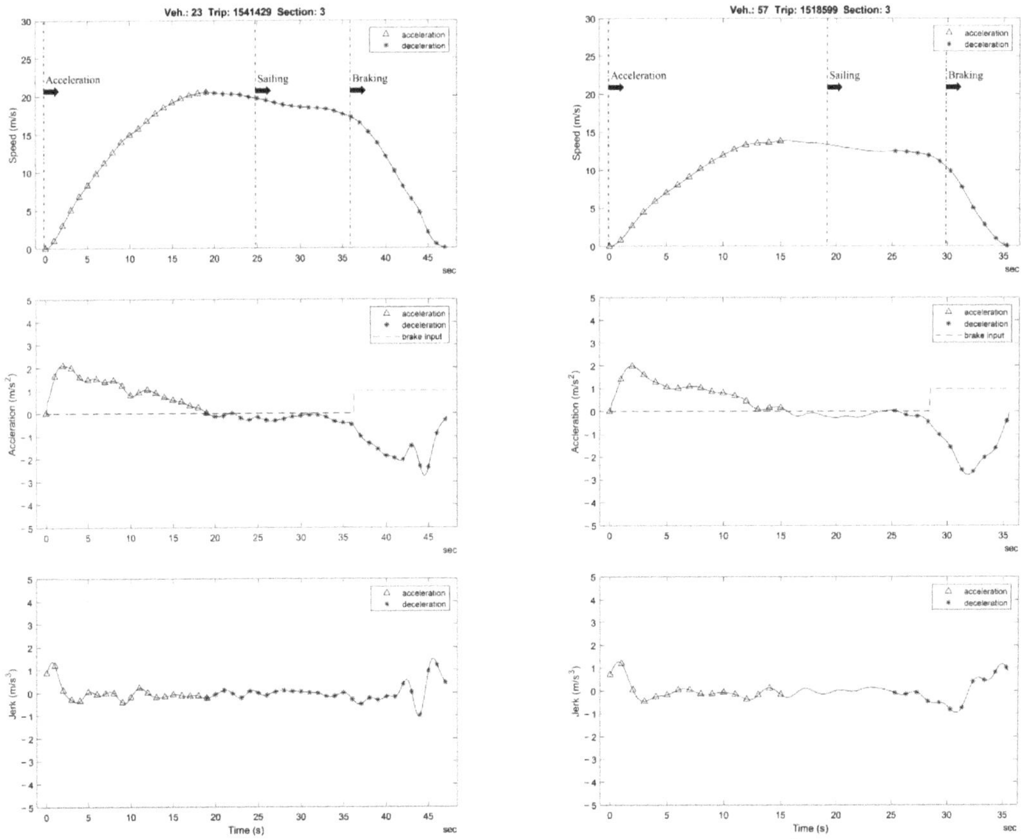

(**a**) (**b**)

Figure 5. Examples of free acceleration/deceleration driving cases: (**a**) case of SPMD driver #23; (**b**) case of SPMD driver #57.

Table 3. Summary of two drivers' trips.

	Trip Time (s)	Max. Speed (m/s)	Max. Acceleration (m/s^2)	Max. Deceleration (m/s^2)
Driver #23	47.2	20.5	2.10	−2.75
Driver #57	35.5	13.9	1.98	−1.92

In Figure 5, the maximum acceleration/deceleration of drivers #23 and #57, which are arbitrarily chosen, demonstrate similar results to the mean values shown in Figure 4. That is, the acceleration/deceleration characteristics of typical human drivers can be defined through the criteria in Figure 4 and Tables A1 and A2, and various acceleration/deceleration scenarios can be configured according to the analyzed distributions.

4.3. Analysis of Sailing/Braking Start Time

Next, we analyzed the sailing and braking start times of human drivers. In Figure 5, shown in the previous section, human drivers generally coast down before stopping, which is related to the timing of pressing the brake pedal. In other words, typical human drivers have a similar driving pattern, which sequences as acceleration, sailing, and braking. Such driving can be viewed as one of the driving characteristics derived from human drivers.

To analyze the driving characteristics at the beginning of the time of braking, the required deceleration for braking is calculated and analyzed from data. The required deceleration means the deceleration required for a complete stop when the relative distance is R and the relative speed is $\dot{R}$. Figure 6 and Equation (10) express the required deceleration calculation.

$$a_{req} = sign\left(\dot{R}\right)\frac{\dot{R}^2}{2R} \tag{10}$$

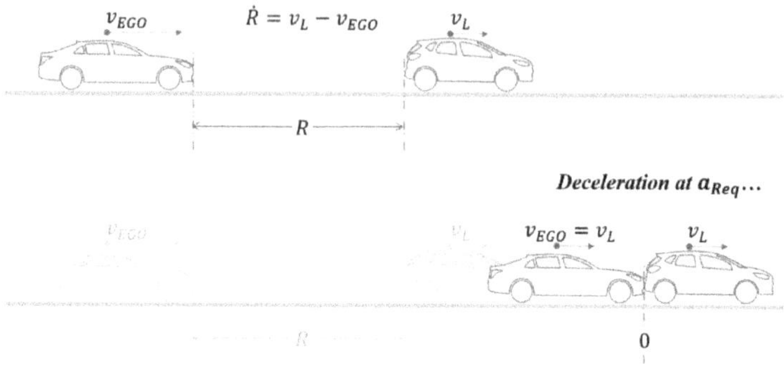

Figure 6. Scheme of required deceleration.

The beginning point of braking is defined as the point at which the first brake pedal signal occurs in the deceleration section before stopping. For the case where there is no leading vehicle, it is assumed that the virtual vehicle is stopped at the stop position of the ego vehicle. The analysis of the required deceleration calculated from the data at the beginning point of braking is specified in Figure 7a. The required deceleration at the start point of braking is -1.37 m/s^2 on average with a 0.55 m/s^2 standard deviation. In the same way, the required deceleration at the sailing start point before braking is analyzed. The sailing start point can be expressed as the time when the acceleration pedal is released to brake, and this can be regarded as the start point of braking. As described in Equations (2) and (3), the rolling/aerodynamic resistances affect the motion of the vehicle. Therefore, the tendency to decelerate during sailing is different depending on the size or type of the vehicle. As this study develops the longitudinal motion scenario by imitating the overall driving pattern of a human driver, the sailing behaviors of various vehicles can be expressed with the variation of the required deceleration at the sailing start point. The analysis result of the required deceleration at the start of sailing is specified in Figure 7b, and the average value is -0.89 m/s^2 with a standard deviation of 0.44 m/s^2. Moreover, the distributions of required deceleration at braking/sailing start point are summarized in Tables A3 and A4.

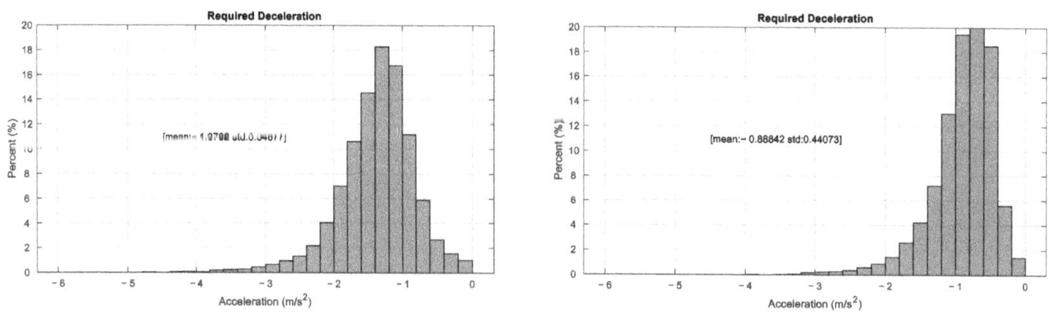

(**a**) (**b**)

Figure 7. Distribution of required deceleration (300 random vehicles): (**a**) case of braking start point; (**b**) case of sailing start point.

To analyze the reason why the required deceleration is necessary to imitate the driving of human drivers, the distributions of braking/sailing start points are represented with respect to relative distance/velocity (see Figure 8). Human drivers show a similar tendency to trend toward the average value of the required deceleration specified in Tables A3 and A4. In addition, the time-to-collision (TTC), which is widely used in vehicle risk assessment standards, is approximated near the origin and compared with the specified average required deceleration. As the TTC has a linear relationship with both the relative distance and the relative speed, it cannot adequately represent the driving patterns of human drivers. In other words, to express the characteristics of the sailing and braking start points, the driver model should be an acceleration/deceleration-based model.

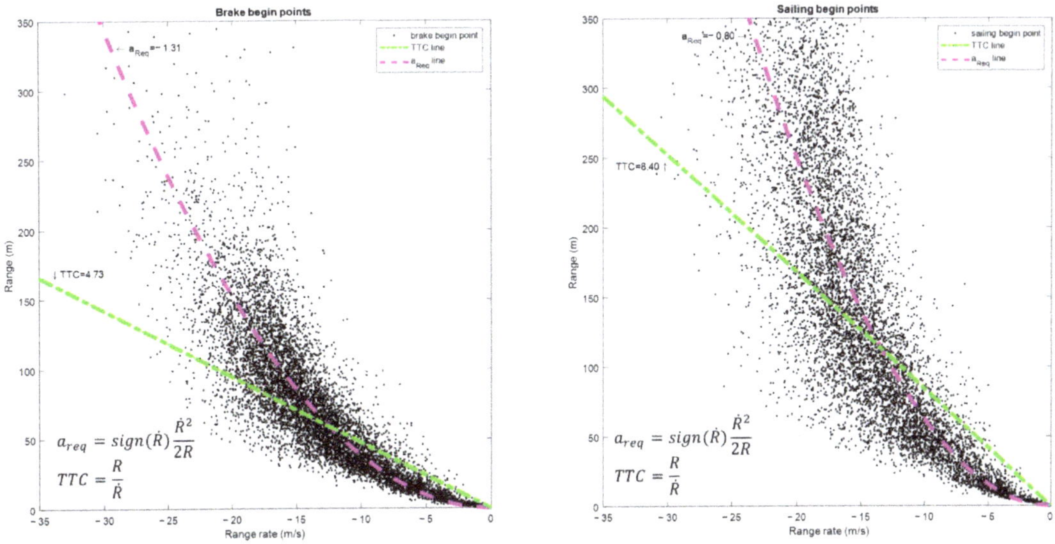

(**a**) (**b**)

Figure 8. Distributions of braking/sailing start point w.r.t. relative distance/velocity: (**a**) the case of the braking start point; (**b**) the case of the sailing start point.

4.4. Driver Model Parameter Adaptation from the Analysis of Driving Data

In Sections 4.1–4.3, parameters expressing the driving characteristics of human drivers are derived. These parameters are the maximum acceleration (a_{max}), the maximum deceleration or minimum acceleration (a_{min}), and the deceleration at the sailing start point (a_{sail}). The three parameters are analyzed in terms of the mean and standard deviation in the previous sections. A typical human driver can be regarded as having the mean value of each parameter, and the more atypical the driver is, the further their behavior is from the mean. Therefore, the driving situations of various drivers can be simulated using the parameters derived from the analysis results. This means that atypical scenarios and various human driving patterns can be expressed according to changing parameters, such as the constraint for acceleration/deceleration and the weights of the driver model in the proposed longitudinal motion scenario design method. In other words, it is possible to create various longitudinal scenarios by defining the mode transition conditions of the driver model, as shown in Figure 9, and changing the constraint settings presented in Equation (6) for the required acceleration/deceleration corresponding to each mode.

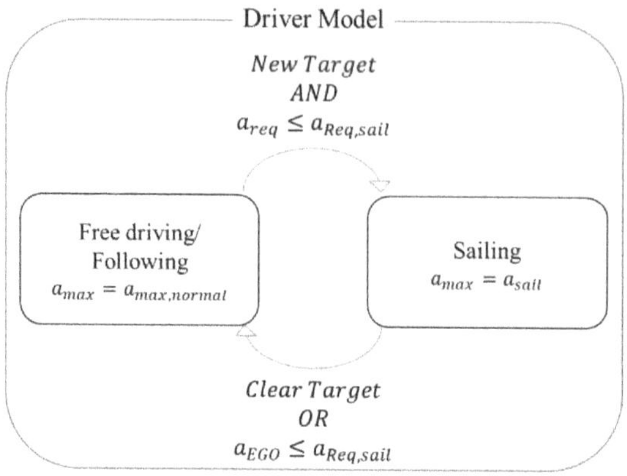

Figure 9. Scheme of the driver model for the proposed longitudinal motion scenario design.

There is a limit to expressing the driving patterns of various drivers only by setting the constraints. Thus, as shown in Equation (11), they can be expressed by changing the weights of the MPC used in the driver model design.

$$Q = \begin{bmatrix} q_1 & 0 \\ 0 & q_2 \end{bmatrix}, \quad R = r_1 \tag{11}$$

q_1 and q_2 are weights for the distance error and velocity error, respectively, and r_1 is a weight for acceleration, which is a control input. In linear quadratic-based optimal control, suitable control performance is derived with the ratio of each weight. Therefore, in this study, the driving pattern is simulated by changing only q_1 and r_1, while fixing the weight of the velocity error q_2 as 1 [46]. When the driver's velocity to be imitated in the simulation is $v_{velo,data}$, the weights q_1 and r_1 of the driver model can be derived using the following equation. Here, T is the length of the time interval to be imitated, and v_{model} is the speed of the driver model.

$$q_1^*, r_1^* = \underset{q1,\, r1}{\operatorname{argmin}} \left(\sqrt{\frac{1}{T} \sum_{k=1}^{T} (v_{velo,data}(k) - v_{model}(k))^2} \right) \tag{12}$$

5. Computer Simulation Results

To verify the proposed longitudinal motion scenario design method, we set appropriate parameters in the driver model to check whether the driving is like that of a human driver. The Gipps model introduced in Table 1 is used for comparison, and the parameters common to both the Gipps model and the MPC-based driver model are set to the same values [28]. The computer simulation was conducted assuming the driving situation on a dry, asphalt, straight road.

The simulations are carried out with Intel® CoreTM i9-11900CPU@2.50 GHz, 32 GB RAM. The execution time was derived with 22 ms as the maximum, 0.64 ms as the minimum, and 0.92 ms as the mean per step, which is less than the step time specified as 200 ms (0.2 s) for the overall simulation results. For this reason, real-time performance can be guaranteed during simulation. If the execution is performed with explicit MPC, it is expected to derive less than the preciously derived execution time. Generally, the driving tendency of human drivers can vary depending on the driving situation [47,48]. For this reason, the simulation results were expressed as fixed parameters and constraints for the randomly selected acceleration/deceleration section of the human driving data.

The simulation results are compared with the results imitating the driving data of drivers #23, #57, #61, and #13 of SPMD, and are presented in Figures 10 and 11. The weights of the proposed MPC-based driver model were derived using Equation (12) to show the longitudinal driving pattern most similar to the actual driver's speed profile. The parameters set in the MPC-based driver model are specified in Table 4. It can be confirmed that the accelerations shown in Table 4 are included in the acceleration range of human drivers derived from Section 4. In these results, the proposed model has mostly lower RMSE than the Gipps model's, which means that the proposed model shows a more similar driving pattern than the Gipps model. With the simulation results from Figures 10 and 11a, it is shown that the proposed model can simulate both the acceleration and the deceleration of the target driver, as well as the timing and deceleration of the driving speed. However, the Gipps model shows a sudden acceleration/deceleration change that is physically impossible due to the absence of constraints. In other words, the Gipps model shows a large value of jerks when the vehicle starts acceleration and braking. Moreover, the sailing shown by the human driver is not considered in the Gipps model, resulting in differences between the Gipps model predictions and the actual human driver behavior. If the Gipps model is used in the verification of ADAS and ADS, the ADAS and ADS can act abnormally because the Gipps model acts in a manner that is physically impossible in the real world. For this reason, the proposed method can be reasonable in the verification of ADAS and ADS.

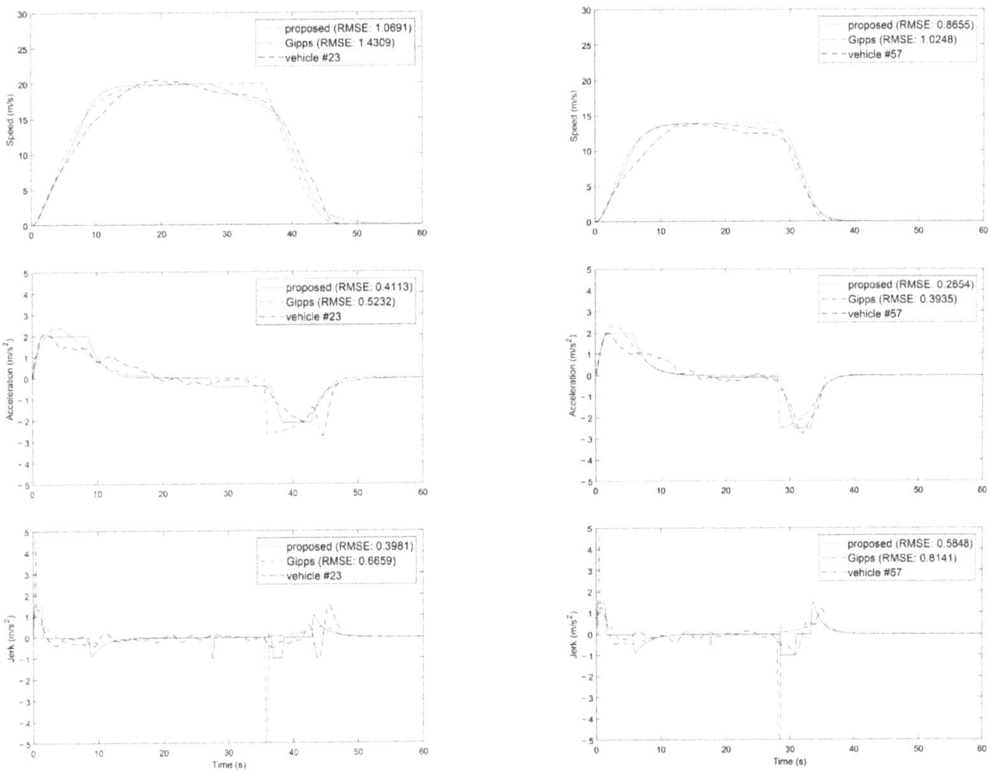

(a) (b)

Figure 10. Computer simulation results to compare the Gipps model (compared) and the proposed driver model (proposed): (**a**) the case of SPMD driver #23; (**b**) the case of SPMD driver #57.

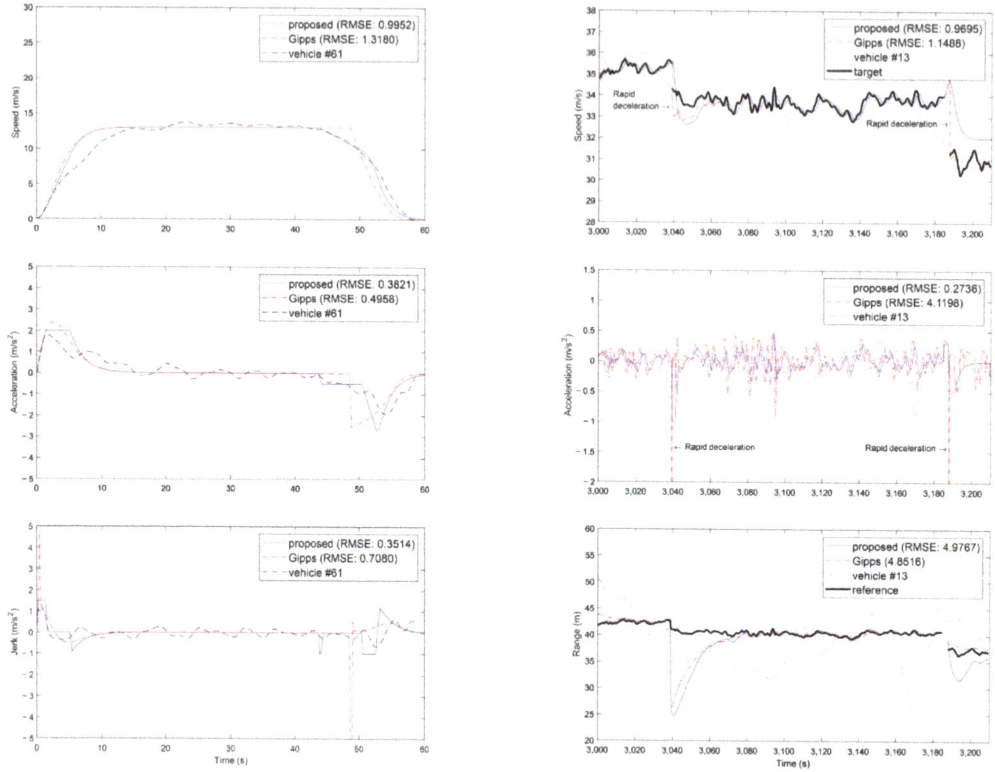

(a) (b)

Figure 11. Computer simulation results to compare the Gipps model (compared) and the proposed driver model (proposed): (**a**) the case of SPMD driver #61; (**b**) the case of SPMD driver #13 (cut-in/out situation).

Table 4. Driver model parameters for simulation.

	a_{max}(m/s^2)	a_{min}(m/s^2)	$a_{req,\ sail}$(m/s^2)	a_{sail}(m/s^2)	q_1(-)	r_1(-)
Driver #23	2	−2.1	−0.5	−0.4	3	45
Driver #57	2	−2.5	−0.5	−0.1	5	45
Driver #61	2	−2.7	−0.8	−0.5	2	40
Driver #13	2	−1	−0.5	−0.1	5	40

The proposed driver model is verified with the presence of multiple lane changes in the SPMD data, which is driver #13's driving, to confirm the driving reliability in the traffic environment (see Figure 11b). The leading vehicle with a relatively low speed changes into a lane ahead of the ego vehicle at 3040 s and 3190 s. In the case of the proposed model, the change in acceleration is similar to that of driver #13's driving, and stable driving is achieved for the leading vehicle. However, the Gipps model shows rapid acceleration changes due to the mode transition when a lane change occurs. This shows the same result as the problem shown in Figures 10 and 11a. The RMSE of the proposed method is similar to the Gipps'. This is because the proposed method is assumed with the constant velocity motion of the leading vehicle (v_L, $a_L \approx 0$) for simplification. Moreover, driver #13's driving is not shown to have the CTG policy to the leading vehicle as the proposed model. However, the proposed model has a similar acceleration tendency as the human driver, despite the presence of the velocity/acceleration of the leading vehicle. If the leading vehicle has a sudden velocity change, the acceleration of the leading vehicle should be considered in the driver model, as mentioned in Section 3.1, for more realistic driving motion.

With the computer simulation results, the proposed longitudinal motion scenario design method not only simulates sailing and the longitudinal driving characteristics of human drivers, but it also satisfies the constraints. In addition, with variations of the constraints and weights of the driver model, it is confirmed that various human drivers can be imitated. Moreover, the proposed model can induce an atypical scenario if the constraints and weights are suitably chosen. For these reasons, this method can simulate the longitudinal movement of surrounding vehicles of the ego vehicle more diversely and realistically in a simulation environment. With these results, we confirmed that the possibility of designing various longitudinal scenarios with human driving characteristics. This method is focused on typical longitudinal driving situations. However, it is judged that malicious driving situations or edge cases can be dealt with if the constraint is set to approximate the vehicle's limit situation or collision risk situation, or if the target acceleration/deceleration is approximated to the limit situation, by diversifying the weights of the driver model. Therefore, the proposed method can provide more realistic scenarios by overcoming the shortcomings in the existing scenario design methods. However, as these results were compared with the randomly selected drivers in the limited driving section, additional verification is required for the robustness of the proposed method. Due to this, further improvement and verification will be conducted in future work.

6. Conclusions

In this paper, a longitudinal motion scenario design method was proposed to overcome the shortcomings of the existing scenarios for ADAS and ADS verification. A simplified vehicle model was designed, including the dynamic characteristics of the vehicle that provide physical constraints, and the longitudinal driver model to control it was designed based on the MPC. The longitudinal driver model parameters were derived from a large-scale driving database to overcome the limitations of the conventional traffic models. To obtain reliable model parameters, the human drivers' acceleration/deceleration tendency and constraints were derived through the analysis of pre-processed data. The proposed longitudinal motion scenario design method showed an organic relationship with the test vehicle and other vehicles in the scenario, and at the same time can define atypical conditions more clearly and intuitively, which is meaningful in conducting more free and realistic tests. In addition, the proposed method can simulate various driving situations by simply changing parameters, without having to configure a complex vehicle dynamics model or driver model.

The proposed method was validated in the computer simulation environment. The MPC, which is used in the driver model design, may have difficulty satisfying real-time performance due to the high amount of computation if the driving situation is more complex. For this reason, it can be supplemented by using explicit MPC or other constraint control methods, such as the control barrier function method [46]. Moreover, in this study, only the longitudinal aspect of the vehicle was considered to derive the driving characteristics of human drivers. However, human drivers also have lateral characteristics in the real world, such as two-dimensional trajectories in a lane change situation. Therefore, future work should develop an adaptive scenario design method to consider both longitudinal and lateral driving scenarios.

Author Contributions: Formal analysis, K.C. and C.P.; project administration, K.C., C.P. and H.L.; software, K.C. and C.P.; validation, K.C. and C.P.; writing—original draft preparation, K.C. and C.P.; writing—review and editing, K.C. and H.L.; supervision, H.L. All authors have read and agreed to the published version of the manuscript.

Funding: This research received no external funding.

Institutional Review Board Statement: Not applicable.

Informed Consent Statement: Not applicable.

Data Availability Statement: Not applicable.

Acknowledgments: This work was supported by the Autonomous Driving Development Innovation Program (20014476, Development of Mixed Reality-based Autonomous Driving Parts and System Evaluation Technology) funded by the Ministry of Trade, Industry & Energy (MOTIE, Korea).

Conflicts of Interest: The authors declare no conflict of interest.

Appendix A

This is a description of the data classification process of SPMD for human driver driving characteristic analysis. For data classification, it is based on speed and radar data. First, as the driving characteristics of human drivers are not well expressed when the length of the data is short, data of less than 300 s were excluded. Moreover, due to the problem of the dataset itself, the case where there was no radar data measuring surrounding vehicles was excluded. Next, the data was classified by excluding the situation where the speed, which is judged as stop, was 0 and the data did not continue for more than 10 s. For example, section #3 of Figure A1 is excluded from the data for driving characteristics analysis because the data does not continue for more than 10 s.

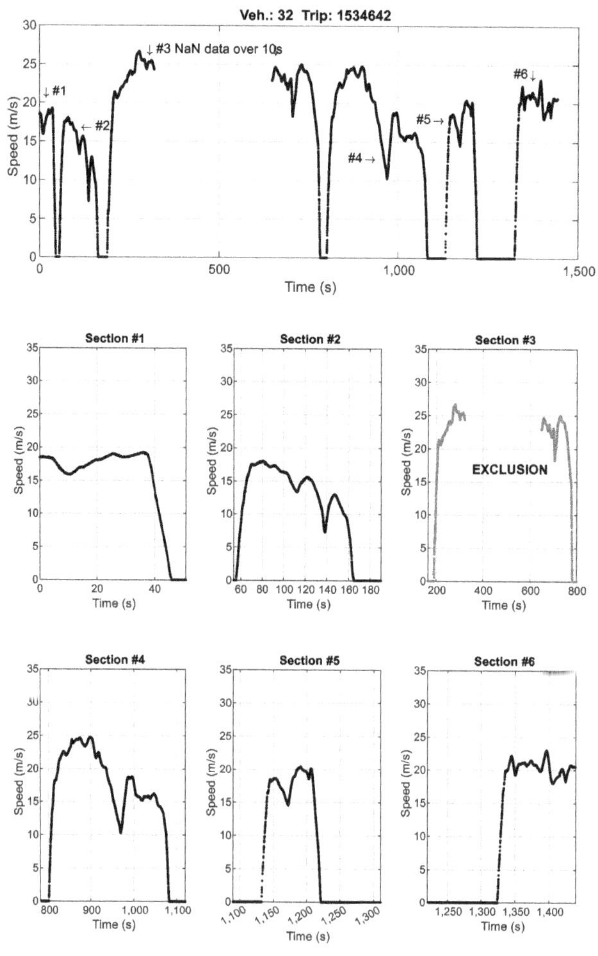

Figure A1. Example of the data exclusion process for the case of SPME driver #32.

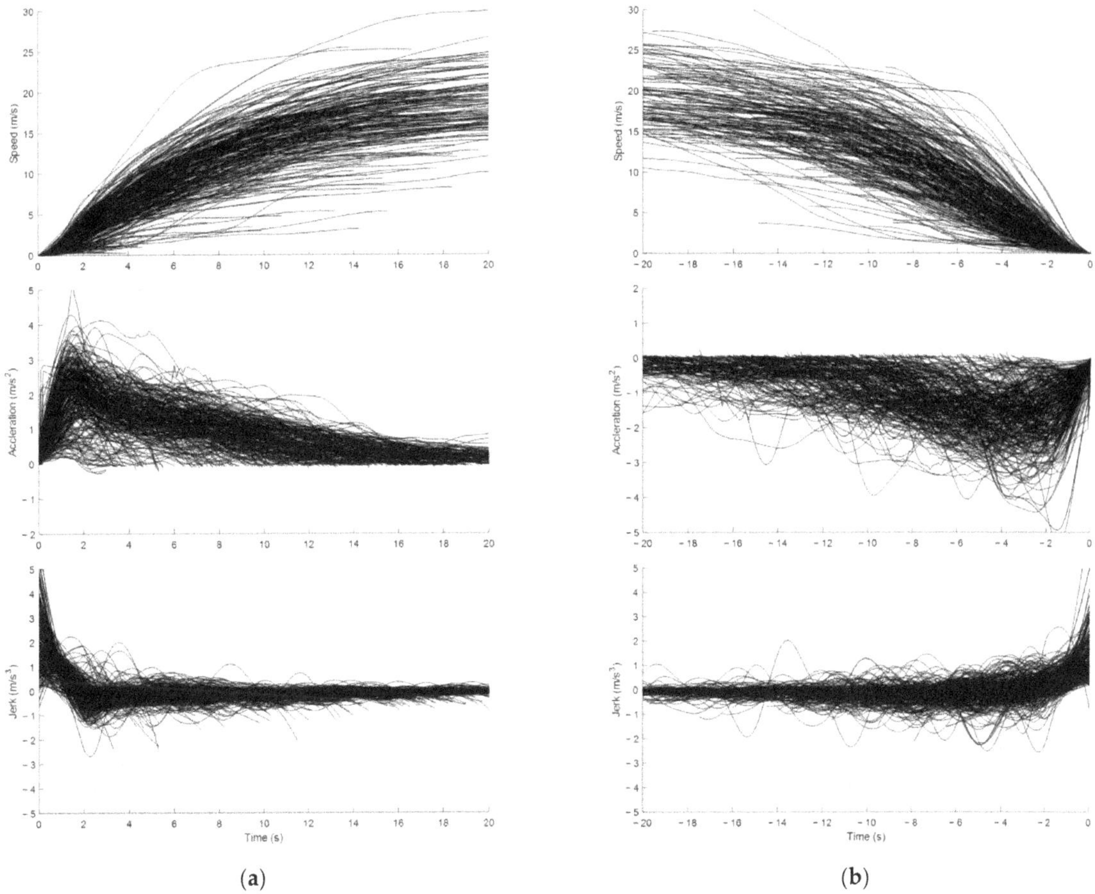

Figure A2. 300 randomly sampled data of free acceleration/deceleration situation from SPMD: (**a**) case of free acceleration; (**b**) case of free deceleration.

Figure A2 shows 300 randomly sampled datasets from SPMD for the free acceleration/deceleration characteristics of human drivers in Section 4.2. Tables A1–A4 are the distributions of maximum acceleration, deceleration, required deceleration at the braking start point, and required deceleration at the sailing start point, respectively.

Table A1. Distribution of maximum acceleration (300 random vehicles, free acceleration).

Range (m/s)	Bottom	75%	Med.	25%	Top
10	0.13	1.11	1.57	2.06	3.47
10–20	1.03	2.04	2.38	2.76	3.83
20–30	1.23	2.32	2.67	3.09	4.24
30–40	1.63	2.39	2.73	3.18	4.076
ALL	0.567	1.86	2.30	2.73	4.02

Table A2. Distribution of maximum deceleration (300 random vehicles, free deceleration).

Range (m/s)	Bottom	75%	Med.	25%	Top
10	−4.85	−1.90	−1.49	−1.08	−0.15
10–20	−3.73	−2.57	−2.16	−1.80	−0.95
20–30	−4.36	−3.03	−2.52	−2.13	−0.98
30–40	−3.99	−3.08	−2.62	−2.21	−1.38
ALL	−3.85	−2.56	−2.10	−1.68	−0.36

Table A3. The distribution of the required acceleration at the braking start point.

Range (m/s)	Bottom	75%	Med.	25%	Top
10	−2.16	−1.27	−0.98	−0.67	−0.03
10–20	−2.56	−1.67	−1.36	−1.11	−0.33
20–30	−2.80	−1.88	−1.54	−1.27	−0.38
30–40	−2.47	−1.69	−1.20	−0.86	−0.46
ALL	−2.57	−1.65	−1.31	−1.04	−0.12

Table A4. The distribution of the required acceleration at the sailing start point.

Range (m/s)	Bottom	75%	Med.	25%	Top
10	−1.67	−0.96	−0.70	−0.47	−0.03
10–20	−1.88	−1.13	−0.84	−0.63	−0.21
20–30	−0.83	−1.12	−0.84	−0.65	−0.20
30–40	−1.24	−0.82	−0.64	−0.50	−0.17
ALL	−1.81	−1.08	−0.80	−0.60	−0.03

References

1. Pendleton, S.D.; Andersen, H.; Du, X.; Shen, X.; Meghjani, M.; Eng, Y.H.; Rus, D.; Ang, M.H. Perception, Planning, Control, and Coordination for Autonomous Vehicles. *Machines* **2017**, *5*, 6. [CrossRef]
2. Fényes, D.; Németh, B.; Gáspár, P. A Novel Data-Driven Modeling and Control Design Method for Autonomous Vehicles. *Energies* **2021**, *14*, 517. [CrossRef]
3. Nie, Z.; Farzaneh, H. Adaptive Cruise Control for Eco-Driving Based on Model Predictive Control Algorithm. *Appl. Sci.* **2020**, *10*, 5271. [CrossRef]
4. Wang, P.; Gao, S.; Li, L.; Sun, B.; Cheng, S. Obstacle Avoidance Path Planning Design for Autonomous Driving Vehicles Based on an Improved Artificial Potential Field Algorithm. *Energies* **2019**, *12*, 2342. [CrossRef]
5. Kuutti, S.; Bowden, R.; Jin, Y.; Barber, P.; Fallah, S. A Survey of Deep Learning Applications to Autonomous Vehicle Control. *IEEE Trans. Intell. Transp. Syst.* **2021**, *22*, 712–733. [CrossRef]
6. Dosovitskiy, A.; Ros, G.; Codevilla, F.; Lopez, A.; Koltun, V. CARLA: An Open Urban Driving Simulator. In Proceedings of the 1st Annual Conference on Robot Learning; PMLR, Mountain View, CA, USA, 13–15 November 2017; pp. 1–16.
7. Zapridou, E.; Bartocci, E.; Katsaros, P. Runtime Verification of Autonomous Driving Systems in CARLA. In Proceedings of the Runtime Verification, Los Angeles, CA, USA, 6–9 October 2020; Deshmukh, J., Ničković, D., Eds.; Springer International Publishing: Cham, Switzerland, 2020; pp. 172–183.
8. Arcidiacono, A. ADAS Virtual Validation: ACC and AEB Case Study with IPG CarMaker. Master's Thesis, Politecnico di Torino, Turin, Italy, 2018.
9. Zhou, J.; Schmied, R.; Sandalek, A.; Kokal, H.; del Re, L. A Framework for Virtual Testing of ADAS. *SAE Int. J. Passeng. Cars Electron. Electr. Syst.* **2016**, *9*, 66–74. [CrossRef]
10. Park, C.; Chung, S.; Lee, H. Vehicle-in-the-Loop in Global Coordinates for Advanced Driver Assistance System. *Appl. Sci.* **2020**, *10*, 2645. [CrossRef]
11. Siegl, S.; Ratz, S.; Düser, T.; Hettel, R. Vehicle-in-the-Loop at Testbeds for ADAS/AD Validation. *ATZ Electron. Worldw* **2021**, *16*, 62–67. [CrossRef]
12. Tettamanti, T.; Szalai, M.; Vass, S.; Tihanyi, V. Vehicle-in-the-Loop Test Environment for Autonomous Driving with Microscopic Traffic Simulation. In Proceedings of the 2018 IEEE International Conference on Vehicular Electronics and Safety (ICVES), Madrid, Spain, 12–14 September 2018; pp. 1–6.
13. Van Ratingen, M.R. The Euro Ncap Safety Rating. In *Karosseriebautage Hamburg 2017*; Springer: Berlin, Germany, 2017; pp. 11–20.
14. ISO. *Intelligent Transport Systems–Adaptive Cruise Control Systems–Performance Requirements and Test Procedures*; ISO: Geneva, Switzerland, 2018.

15. Hulshof, W.; Knight, I.; Edwards, A.; Avery, M.; Grover, C. Autonomous Emergency Braking Test Results. In Proceedings of the 23rd International Technical Conference on the Enhanced Safety of Vehicles (ESV), Seoul, Republic of Korea, 27–30 May 2013; National Highway Traffic Safety Administration: Washington, DC, USA, 2013; pp. 1–13.
16. Zhao, D. Accelerated Evaluation of Automated Vehicles. Ph.D. Thesis, University of Michigan, Ann Arbor, MI, USA, 2016.
17. Peng, H.; Leblanc, D. Evaluation of the Performance and Safety of Automated Vehicles. White Paper on NSF Transportation CPS Work. 2012. Available online: https://cps-vo.org/sites/default/files/webform/Evaluation_of_the_Performance_and_Safety_of_Automated_Vehicles_0.pdf (accessed on 16 December 2022).
18. Lee, K. *Longitudinal Driver Model and Collision Warning and Avoidance Algorithms Based on Human Driving Databases*; University of Michigan: Ann Arbor, MI, USA, 2004.
19. Klischat, M.; Althoff, M. Generating Critical Test Scenarios for Automated Vehicles with Evolutionary Algorithms. In Proceedings of the 2019 IEEE Intelligent Vehicles Symposium (IV), Paris, France, 9–12 June 2019; pp. 2352–2358.
20. Koopman, P.; Wagner, M. Challenges in Autonomous Vehicle Testing and Validation. *SAE Int. J. Transp. Safety* **2016**, *4*, 15–24. [CrossRef]
21. Riedmaier, S.; Ponn, T.; Ludwig, D.; Schick, B.; Diermeyer, F. Survey on Scenario-Based Safety Assessment of Automated Vehicles. *IEEE Access* **2020**, *8*, 87456–87477. [CrossRef]
22. Fellendorf, M.; Vortisch, P. Microscopic Traffic Flow Simulator VISSIM. In *Fundamentals of Traffic Simulation*; Springer: Berlin, Germany, 2010; pp. 63–93.
23. Kaths, J.; Krause, S. Integrated Simulation of Microscopic Traffic Flow and Vehicle Dynamics. In Proceedings of the IPG Apply & Innovate 2016, Karlsruhe, Germany, 20–21 September 2016.
24. Pipes, L.A. An Operational Analysis of Traffic Dynamics. *J. Appl. Phys.* **1953**, *24*, 274–281. [CrossRef]
25. Gazis, D.C.; Herman, R.; Rothery, R.W. Nonlinear Follow-the-Leader Models of Traffic Flow. *Oper. Res.* **1961**, *9*, 545–567. [CrossRef]
26. Newell, G.F. Nonlinear Effects in the Dynamics of Car Following. *Oper. Res.* **1961**, *9*, 209–229. [CrossRef]
27. Tyler, J. The Characteristics of Model-Following Systems as Synthesized by Optimal Control. *IEEE Trans. Autom. Control.* **1964**, *9*, 485–498. [CrossRef]
28. Gipps, P.G. A Behavioural Car-Following Model for Computer Simulation. *Transp. Res. Part B Methodol.* **1981**, *15*, 105–111. [CrossRef]
29. Ahmed, H.U.; Huang, Y.; Lu, P. A Review of Car-Following Models and Modeling Tools for Human and Autonomous-Ready Driving Behaviors in Micro-Simulation. *Smart Cities* **2021**, *4*, 314–335. [CrossRef]
30. Dingus, T.A.; Klauer, S.G.; Neale, V.L.; Vicki, L.; Petersen, A.; Lee, S.E.; Sudweeks, J.; Perez, M.A.; Hankey, J.; Ramsey, D.; et al. *The 100-Car Naturalistic Driving Study, Phase II Results of the 100-Car Field Experiment*; National Highway Traffic Safety Administration: Washington, DC, USA, 2006.
31. Ervin, R.; Sayer, J.; LeBlanc, D.; Bogard, S.; Mefford, M.; Hagan, M.; Bareket, Z.; Winkler, C. *Automotive Collision Avoidance System Field Operational Test Report: Methodology and Results*; Transportation Research Board: Washington, DC, USA, 2005.
32. LeBlanc, D.J. *Road Departure Crash Warning System Field Operational Test: Methodology and Results. Volume 1: Technical Report*; Transportation Research Institute, University of Michigan: Ann Arbor, MI, USA, 2006.
33. Victor, T.; Bärgman, J.; Hjälmdahl, M.; Kircher, K.; Svanberg, E.; Hurtig, S.; Gellerman, H.; Moeschlin, F. *Sweden-Michigan Naturalistic Field Operational Test (Semifot) Phase 1: Final Report*; Chalmers: Goteborg, Sweden, 2010.
34. Sayer, J.; LeBlanc, D.; Bogard, S.; Funkhouser, D.; Bao, S.; Buonarosa, M.L.; Blankespoor, A. *Integrated Vehicle-Based Safety Systems Field Operational Test: Final Program Report*; Transportation Research Institute, University of Michigan: Ann Arbor, MI, USA, 2011.
35. Bezzina, D.; Sayer, J. Safety Pilot Model Deployment: Test Conductor Team Report. *Report No. DOT HS* **2014**, *812*, 18.
36. Traffic Analysis Tools: Next Generation Simulation-FHWA Operations. Available online: https://ops.fhwa.dot.gov/trafficanalysistools/ngsim.htm (accessed on 20 October 2022).
37. Open Dataset. Available online: https://waymo.com/open/ (accessed on 20 October 2022).
38. Krajewski, R.; Bock, J.; Kloeker, L.; Eckstein, L. The HighD Dataset: A Drone Dataset of Naturalistic Vehicle Trajectories on German Highways for Validation of Highly Automated Driving Systems. In Proceedings of the 2018 21st International Conference on Intelligent Transportation Systems (ITSC), Maui, HI, USA, 4–7 November 2018; pp. 2118–2125.
39. Bifulco, G.N.; Simonelli, F.; Di Pace, R. Experiments toward an Human-like Adaptive Cruise Control. In Proceedings of the 2008 IEEE Intelligent Vehicles Symposium, Eindhoven, The Netherlands, 4–6 June 2008; pp. 919–924.
40. Zheng, J.; Suzuki, K.; Fujita, M. Car-Following Behavior with Instantaneous Driver–Vehicle Reaction Delay: A Neural-Network-Based Methodology. *Transp. Res. Part C Emerg. Technol.* **2013**, *36*, 339–351. [CrossRef]
41. He, Z.; Zheng, L.; Guan, W. A Simple Nonparametric Car-Following Model Driven by Field Data. *Transp. Res. Part B Methodol.* **2015**, *80*, 185–201. [CrossRef]
42. Przybyla, J.; Taylor, J.; Jupe, J.; Zhou, X. Estimating Risk Effects of Driving Distraction: A Dynamic Errorable Car-Following Model. *Transp. Res. Part C Emerg. Technol.* **2015**, *50*, 117–129. [CrossRef]
43. Rajamani, R. *Vehicle Dynamics and Control*; Mechanical Engineering Series; Springer: Boston, MA, USA, 2012; ISBN 978-1-4614-1432-2.
44. Kiencke, U.; Nielsen, L. *Automotive Control Systems: For Engine, Driveline, and Vehicle*; IOP Publishing Ltd.: Bristol, UK, 2000.

45. Yi, K.; Song, B. Automated Driving Vehicles. In *Vehicle Dynamics: Fundamentals and Ultimate Trends*; Lenzo, B., Ed.; CISM International Centre for Mechanical Sciences; Springer International Publishing: Cham, Switzerland, 2022; pp. 289–387. ISBN 978-3-030-75884-4.
46. Borrelli, F.; Bemporad, A.; Morari, M. *Predictive Control for Linear and Hybrid Systems*; Cambridge University Press: Cambridge, UK, 2017.
47. Bosetti, P.; Da Lio, M.; Saroldi, A. On the Human Control of Vehicles: An Experimental Study of Acceleration. *Eur. Transp. Res. Rev.* **2014**, *6*, 157–170. [CrossRef]
48. Xia, L.; Li, P.; Su, Z.; Chen, T.; Deng, Z.; Sun, D. Longitudinal Driving Behavior before, during, and after a Left-Turn Movement at Signalized Intersections: A Naturalistic Driving Study in China. *Sustainability* **2022**, *14*, 11630. [CrossRef]
49. Loader, C. *Local Regression and Likelihood*; Springer Science & Business Media: Berlin, Germany, 2006.

MDPI AG
Grosspeteranlage 5
4052 Basel
Switzerland
Tel.: +41 61 683 77 34

Applied Sciences Editorial Office
E-mail: applsci@mdpi.com
www.mdpi.com/journal/applsci

www.ingramcontent.com/pod-product-compliance
Lightning Source LLC
Chambersburg PA
CBHW061542220926
51850CB00005B/231
* 9 7 8 3 7 2 5 8 2 1 8 3 9 *